Theory and Applications of
Cellular Automata

Advanced series on complex systems — volume 1

Theory and Applications of Cellular Automata

(including selected papers 1983 – 1986)

Stephen Wolfram

World Scientific

Published by

World Scientific Publishing Co. Pte. Ltd.
P. O. Box 128, Farrer Road, Singapore 9128

The editor and publisher are grateful to the authors and the following publishers for their assistance and permission to reproduce the reprinted papers included in this volume: Academic Press Inc. (*Adv. Applied Math., Ann. Phys.*); American Association for the Advancement of Science (*Science*); American Physical Society (*Phys. Rev. Lett., Rev. Mod. Phys.*); Elsevier Science Publishing Co. (*Math. Biosciences*); Institute of Physics, Sweden (*Phys. Scripta*); KTK Scientific Publishers, Tokyo (*Proceedings of the First International Symposium for Science on Form*); North-Holland Publishing Co. (*Physica D*); Plenum Publishing Co. (*J. Stat. Phys.*); Springer-Verlag (*Commun. Math. Phys., Z. Phys.*); The Institute of Physics (*J. Phys. A*).

Library of Congress Cataloging-in-Publication data is available.

THEORY AND APPLICATIONS OF CELLULAR AUTOMATA

ISBN 9971-50-123-6
9971-50-124-4 pbk

Printed in Singapore by Kyodo-Shing Loong Printing Industries Pte Ltd.

Preface

The understanding of complexity and its origins is a fundamental challenge for modern science. Research in physics, biology and other fields has been immensely successful in finding the basic components of most of the systems of everyday experience. What must now be done is to discover how large numbers of these components, often each quite simple, can act together to produce the complex behaviour that is seen. The basic laws which govern the dynamics of fluids have for example long been known. But how these laws lead to the intricate patterns of turbulent flow remains unknown. Science has traditionally concentrated on analysing systems by breaking them down into simple constituent parts. A new form of science is now developing which addresses the problem of how those parts act together to produce the complexity of the whole.

Fundamental to the approach is the investigation of models which are as simple as possible in construction, yet capture the essential mathematical features necessary to reproduce the complexity that is seen. Cellular automata provide probably the best examples of such models. Their basic construction is very simple. Yet their overall behaviour is found to be highly complex, and can reproduce the complex phenomena observed in many physical and other systems.

Traditional mathematical models and methods are not well suited to studies of complex systems. Exact mathematical formulae or analysis can be carried out only in rather simple cases. In other cases, the consequences of models must be found by direct simulation. Digital computers are the basic tools for such simulations.

The central role played by digital computers suggests that models should be constructed so as to be as easy to implement by digital computer as possible. Cellular automata are some of the models which are most directly suitable for implementation both on conventional digital computers, and on the forthcoming generation of massively parallel computers.

The computational aspects of cellular automata are important not only for practical implementation, but also for fundamental analysis. Cellular automata can themselves be viewed as computational systems, which process the data encoded in their configurations. This view leads to the application of the mathematical theory of computation to cellular automata, and to the physical and other systems that can be modelled with them. Such applications provide some of the most promising approaches to the problem of finding general principles which govern the behaviour of complex systems.

Aspects of cellular automata have long been studied (see the Bibliography). But it is only in the last few years, particularly as a result of the availability of interactive graphical computer simulation, that the properties of cellular automata most relevant for the general study of complex systems have been extensively investigated.

This book includes recent papers that describe some of the phenomenology, theory, and applications of cellular automata. These papers establish some of the basic results and directions for cellular automaton approaches to complex systems. But much remains to be done.

Stephen Wolfram
Cambridge, Mass.
May 3, 1986

Contents

Preface *v*

Outline *1*

1. Basic theory and phenomenology

1.1: S. Wolfram, "Statistical mechanics of cellular automata", Rev. Mod. Phys. 55 (1983) 601. *7*

1.2: O. Martin, A. Odlyzko and S. Wolfram, "Algebraic properties of cellular automata", Commun. Math. Phys. 93 (1984) 219. *51*

1.3: S. Wolfram, "Universality and complexity in cellular automata", Physica 10D (1984) 1. *91*

1.4: N. Packard and S. Wolfram, "Two-dimensional cellular automata", J. Stat. Phys. 38 (1985) 901. *126*

1.5: S. Wolfram, "Twenty problems in the theory of cellular automata", Phys. Scripta T9 (1985) 170. *172*

2. Computation theoretical approaches

2.1: S. Wolfram, "Computation theory of cellular automata", Commun. Math. Phys. 96 (1984) 15. *189*

2.2: N. Margolus, "Physics-like models of computation", Physica 10D (1984) 81. *232*

2.3: S. Wolfram, "Random sequence generation by cellular automata", Adv. Applied Math. 7 (1986) 123. *247*

2.4: S. Wolfram, "Undecidability and intractability in theoretical physics", Phys. Rev. Lett. 54 (1985) 735. *294*

2.5: S. Wolfram, "Origins of randomness in physical systems", Phys. Rev. Lett. 55 (1985) 449. *298*

3. Some applications

3.1: N. Packard, "Lattice models for solidification and aggregation", to appear in Proc. First International Symposium for Science on Form, (Tsukuba, Japan, 1985). *305*

3.2: B. Madore and W. Freedman, "Computer simulations of the Belousov-Zhabotinsky reaction", Science 222 (1983) 615. *311*

3.3: A. Winfree, E. Winfree and H. Seifert, "Organizing centers in a cellular excitable medium", Physica 17D (1985) 109. *313*

3.4: D. Young, "A local activator-inhibitor model of vertebrate skin patterns", Math. Biosciences 72 (1984) 51. *320*

3.5: Y. Oono and M. Kohmoto, "Discrete model of chemical turbulence", Phys. Rev. Lett. 55 (1985) 2927. *328*

3.6: J. Park, K. Steiglitz and W. Thurston, "Soliton-like behaviour in automata", Physica 19D (1986) 423. *333*

3.7: Y. Pomeau, "Invariant in cellular automata", J. Phys. A17 (1984) L415. *343*

3.8: M. Creutz, "Deterministic Ising dynamics", Ann. Phys. 167 (1986) 62. *347*

3.9: U. Frisch, B. Hasslacher and Y. Pomeau, "Lattice gas automata for the Navier-Stokes equation", Phys. Rev. Lett. 56 (1986) 1505. *358*

3.10: J. Salem and S. Wolfram, "Thermodynamics and hydrodynamics of cellular automata". *362*

3.11: K. Kaneko, "Attractors, basin structures and information processing in cellular automata".   *367*

3.12: S. Wolfram, "Approaches to complexity engineering", to be published in Physica D.   *400*

4. Probabilistic cellular automata

4.1: W. Kinzel, "Phase transitions of cellular automata", Z. Phys. B58 (1985) 229.   *419*

4.2: P. Grassberger, F. Krause and T. von der Twer, "A new type of kinetic critical phenomenon", J. Phys. A17 (1984) L105.   *435*

4.3: K. Kaneko and Y. Akutsu, "Phase transitions in two-dimensional stochastic cellular automata", J. Phys. A19 (1986) L69.   *440*

4.4: E. Domany and W. Kinzel, "Equivalence of cellular automata to Ising models and directed percolation", Phys. Rev. Lett. 53 (1984) 311.   *447*

4.5: G. Grinstein, C. Jayaprakash and Y. He, "Statistical mechanics of probabilistic cellular automata", Phys. Rev. Lett. 55 (1985) 2527.   *451*

4.6: C. Bennett and G. Grinstein, "Role of irreversibility in stabilizing complex and nonergodic behaviour in locally interacting discrete systems", Phys. Rev. Lett. 55 (1985) 657.   *455*

An annotated bibliography of cellular automata

Introduction   *460*

1.   Some general references   *460*

2.   Pure mathematical approaches   *462*

3.   Phenomenological and constructional approaches   *466*

4.   Computation theoretical approaches   *468*

5.   Probabilistic cellular automata   *470*

6.   Physical applications of cellular automata   *472*

7.   Biological applications of cellular automata   *475*

8.   Practical computation with cellular automata   *476*

9.   Further applications of cellular automata   *479*

10.   Some systems related to cellular automata   *479*

Author index   *481*

Appendix: Properties of the $k=2$, $r=1$ cellular automata

Introduction   *485*

Tables:

1.   Rule forms and equivalences   *487*

2.   Patterns from disordered states   *493*

3.   Blocked patterns from disordered states   *498*

4.   Difference patterns   *502*

5.   Patterns from single site seeds   *506*

6.   Statistical properties   *513*

7.   Blocking transformation equivalences   *516*

8.   Factorizations of rules   *520*

9.   Lengths of newly-excluded blocks   *521*

10. Regular language complexities   *523*

11. Measure theoretical complexities   *527*

12. Iterated rule expression sizes   *529*

13. Finite lattice state transition diagrams   *531*

14. Global properties for finite lattices   *537*

15. Structures in rule 110   *547*

16. Patterns generated by second-order rules   *550*

Index   *559*

Outline

Definitions

In a simple case, a cellular automaton consists of a line of cells or sites, each with value 0 or 1. These values are updated in a sequence of discrete time steps, according to a definite, fixed, rule. Denoting the value of a site at position i by a_i, a simple rule gives its new value as

$$a'_i = \phi(a_{i-1}, a_i, a_{i+1}) \ . \tag{1}$$

Here ϕ is a Boolean function which specifies the rule.

Despite the simplicity of their construction, many cellular automata produce behaviour of considerable complexity. The Appendix gives extensive examples and tables of the properties of cellular automata with the form (1). Such cellular automata already capture the essential features of many complex phenomena. But detailed, realistic, models often involve slightly more complicated cellular automata.

In general, the sites in a cellular automaton may have any finite number k of possible values. The rules for updating these sites may depend on values up to any finite distance r away. In addition, the cellular automaton sites may be arranged not on a line, but on a regular lattice in any number of dimensions. Many of the generic features of behaviour are however largely unaffected by such additional complications.

Cellular automata have a number of basic defining characteristics.

Discrete in space. They consist of a discrete grid of spatial cells or sites.

Discrete in time. The value of each cell is updated in a sequence of discrete time steps.

Discrete states. Each cell has a finite number of possible values.

Homogeneous. All cells are identical, and are arranged in a regular array.

Synchronous updating. All cell values are updated in synchrony, each depending on the previous values of neighbouring cells.

Deterministic rule. Each cell value is updated according to a fixed, deterministic, rule.

Spatially local rule. The rule at each site depends only on the values of a local neighbourhood of sites around it.

Temporally local rule. The rule for the new value of a site depends only on values for a fixed number of preceding steps (usually just one step).

Each of these characteristics represents a simplifying feature in the basic construction of cellular automata. They make many kinds of analysis and simulation much easier. But they yield few, if any, restrictions on overall cellular automaton behaviour. While each cell is discrete, collections of large numbers of cells may for example show effectively continuous behaviour.

Basic theory and phenomenology

Section 1 of this book discusses the general forms of behaviour found in cellular automata, and describes various approaches to their analysis.

Computer simulation is a crucial tool in these studies. The overall properties of a cellular automaton are usually not readily apparent from its basic rules. But given these rules, its behaviour can always be determined by explicit simulation on a digital computer. And from such mathematical experiments, empirical and theoretical results can be abstracted.

Paper 1.1 introduces cellular automata, and describes some of their basic phenomenology and theory. It discusses some general approaches to the application of cellular automata, and includes references to earlier and related work. The Bibliography in this book gives more extensive references.

Paper 1.2 discusses the particular class of additive cellular automata, for which a rather complete algebraic analysis can be given. This paper can be viewed as giving "exact solutions" for the behaviour of these comparatively simple cellular automata.

Paper 1.3 gives more general observations and results on cellular automata. It identifies four basic classes of generic behaviour. This classification is at first an empirical one, derived from computer experiments. But paper 1.3 describes various approaches to providing a mathematical basis for it. Many are based on dynamical systems theory, which characterizes the statistical properties of information transmission and storage. Cellular automata in fact yield some of the simplest and most readily analysable examples of many phenomena, such as chaos, commonly studied with dynamical systems theory.

Papers 1.1, 1.2 and 1.3 are primarily concerned with one-dimensional cellular automata. Paper 1.4 discusses the basic theory and phenomenology of two-dimensional cellular automata. Many details are different, but overall features are similar to those found in one-dimensional cases.

Paper 1.5 summarizes some of the theoretical results on the generic behaviour of cellular automata obtained so far, and lists some outstanding problems.

Computation theoretical approaches

Cellular automata can be viewed not only as mathematical systems, or as models for physical and other systems, but also as essentially computational systems. Their evolution can be thought of as processing the data given as their initial conditions. As a result, it is possible to use ideas and methods from the mathematical theory of computation to analyse their behaviour. The papers in section 2 discuss various approaches that have been taken.

Paper 2.1 uses formal language theory to obtain more complete characterizations of sets of configurations generated by cellular automaton evolution. These characterizations extend those derived from dynamical systems theory.

Paper 2.2 describes a simple two-dimensional cellular automaton which can be shown to be capable of universal computation. As a result, its evolution should in principle be able to mimic an arbitrary computational procedure, and thus to show the most complicated conceivable behaviour.

Paper 2.3 gives a rather detailed analysis of a particular simple cellular automaton which generically produces behaviour so complicated as to seem apparently random. The evolution of this cellular automaton can be considered as a computation which "encrypts" its initial conditions.

Viewing cellular automata both as models of physical systems, and as computational systems, suggests some fundamental connections between the principles of physics and computation. Papers 2.4 and 2.5 discuss some of these connections. Paper 2.4 introduces the notion of "computational irreducibility", and suggests that the behaviour of many physical systems, like many cellular automata, may be sufficiently complex that it can be reproduced only through an irreducible amount of computation, essentially equivalent to direct simulation. Paper 2.5 gives a computation theoretical perspective on the origins of randomness in physical processes.

Some applications

Cellular automata potentially provide direct models for many systems which contain large numbers of similar components with local interactions. Typically these models are simpler in construction than those traditionally used. But there is increasing evidence that the cellular automaton models can indeed capture the essential features necessary to reproduce complex behaviour that is observed. Section 3 includes papers which discuss a variety of applications of cellular automaton models.

Cellular automata are related to many models such as partial differential equations. Numerical approximations to partial differential equations typically involve a discrete grid of sites, updated in discrete time steps according to definite rules. Such finite difference systems differ from cellular automata in allowing sites to have continuous values. In a cellular automaton, the sites would have just a few discrete values. Nevertheless, it is common to find that collections of sites yield average behaviour that can accurately mimic continuous variables. The cellular automaton evolution leads to effective randomization of individual site values, but maintains smooth macroscopic average behaviour.

Here is a list of a few common classes of models, and their most significant differences from cellular automata:

Partial differential equations: space, time and values are continuous.

Finite difference equations/lattice dynamical systems: site values are continuous.

Dynamic spin systems: rules are often probabilistic, and updates may be asynchronous.

Directed percolation: rules are intrinsically probabilistic.

Markov random fields: sites carry probabilities not definite values.

Particle models: particles can usually have continuously variable positions and velocities.

Iterated mappings of real numbers: non-local operations are performed on digit sequences.

Boolean delay equations: only a few symbols or sites are considered.

Feedback shift registers: rules can be non-local, and boundary conditions can be different.

Systolic arrays: sites can store extensive information, and connections are often unidirectional.

Turing machines: operations are carried out only on a few sites at each time step.

L systems: rules can change the number of sites.

Production systems: operations are carried out asynchronously, and rules can change number of sites.

Random Boolean networks/neural network models: sites can be connected in an arbitrary network, and updates are usually asynchronous.

Paper 3.1 considers models based on cellular automata for pattern formation in growth processes, particularly of dendritic structures. The models it considers are simple are construction, yet seem to incorporate the essential physical effects necessary to reproduce global properties of behaviour.

Papers 3.2 through 3.5 consider cellular automaton models for reaction-diffusion processes. Simple cellular automaton models again seem capable of reproducing complex patterns seen in experiments on chemical and biological systems.

Paper 3.6 discusses how soliton behaviour, found in various nonlinear partial differential equations, can have an analogue in systems related to one-dimensional cellular automata.

Paper 3.7 considers reversible cellular automata, and discusses conserved quantities which might represent macroscopic properties.

Paper 3.8 describes a cellular automaton model for the dynamic Ising spin model.

Papers 3.9 and 3.10 discuss the construction of cellular automaton models which on a micrsoscopic scale give simple idealizations of molecular dynamics, and on a macroscopic scale reproduce standard hydrodynamic behaviour.

Papers 3.11 and 3.12 discuss some background to the application of cellular automata for practical computational tasks. In particular, paper 3.12 considers some of the concepts involved in "engineering" cellular automata to have particular forms of behaviour.

Cellular automata are very general models. The papers of section 3 give some specific examples of their applications. Many other diverse applications can be imagined.

Probabilistic cellular automata

Standard cellular automata have definite rules for updating the values of sites. Often, however, their microscopic behaviour can seem apparently random, and can mimic the irregularity commonly associated for example with thermodynamic heat baths. Section 4 includes some papers which discuss cellular automata in which intrinsic probabilistic elements have been inserted. Such cellular automata can potentially be analysed in direct analogy with various statistical mechanical systems.

Paper 4.1 gives a detailed discussion of the statistical properties of simple probabilistic cellular automata, and the phase transitions which can occur in them. Papers 4.2 and 4.3 discuss some of the structures that can occur in various one- and two-dimensional probabilistic cellular automata.

Papers 4.4 and 4.5 consider mathematical equivalences between probabilistic cellular automata and spin system models.

Paper 4.6 discusses some of the peculiar statistical mechanics which can occur in probabilistic cellular automata. It describes a particular cellular automaton rule which can support definite, perhaps complex, behaviour, even when probabilistic elements are present.

1. Basic theory and phenomenology

1.1: S. Wolfram, "Statistical mechanics of cellular automata", Rev. Mod. Phys. 55 (1983) 601.

1.2: O. Martin, A. Odlyzko and S. Wolfram, "Algebraic properties of cellular automata", Commun. Math. Phys. 93 (1984) 219.

1.3: S. Wolfram, "Universality and complexity in cellular automata", Physica 10D (1984) 1.

1.4: N. Packard and S. Wolfram, "Two-dimensional cellular automata", J. Stat. Phys. 38 (1985) 901.

1.5: S. Wolfram, "Twenty problems in the theory of cellular automata", Phys. Scripta T9 (1985) 170.

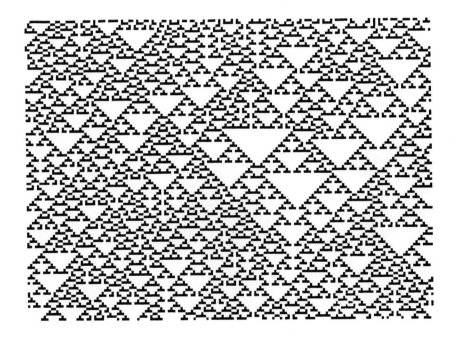

Chaotic pattern generated by the evolution of a simple one-dimensional cellular automaton. The value of each cell is determined by the value of the cells immediately above it, and to its left and right, according to a definite cellular automaton rule ($k=2$, $r=1$, number 22). Starting from a randomly-chosen initial state, a pattern that shows a mixture of order and chaos is produced.

Statistical mechanics of cellular automata

Stephen Wolfram

The Institute for Advanced Study, Princeton, New Jersey 08540

Cellular automata are used as simple mathematical models to investigate self-organization in statistical mechanics. A detailed analysis is given of "elementary" cellular automata consisting of a sequence of sites with values 0 or 1 on a line, with each site evolving deterministically in discrete time steps according to definite rules involving the values of its nearest neighbors. With simple initial configurations, the cellular automata either tend to homogeneous states, or generate self-similar patterns with fractal dimensions $\simeq 1.59$ or $\simeq 1.69$. With "random" initial configurations, the irreversible character of the cellular automaton evolution leads to several self-organization phenomena. Statistical properties of the structures generated are found to lie in two universality classes, independent of the details of the initial state or the cellular automaton rules. More complicated cellular automata are briefly considered, and connections with dynamical systems theory and the formal theory of computation are discussed.

CONTENTS

I. Introduction   601
II. Introduction to Cellular Automata   602
III. Local Properties of Elementary Cellular Automata   607
IV. Global Properties of Elementary Cellular Automata   621
V. Extensions   630
VI. Discussion   639
Acknowledgments   641
References   642

I. INTRODUCTION

The second law of thermodynamics implies that isolated microscopically reversible physical systems tend with time to states of maximal entropy and maximal "disorder." However, "dissipative" systems involving microscopic irreversibility, or those open to interactions with their environment, may evolve from "disordered" to more "ordered" states. The states attained often exhibit a complicated structure. Examples are outlines of snowflakes, patterns of flow in turbulent fluids, and biological systems. The purpose of this paper is to begin the investigation of cellular automata (introduced in Sec. II) as a class of mathematical models for such behavior. Cellular automata are sufficiently simple to allow detailed mathematical analysis, yet sufficiently complex to exhibit a wide variety of complicated phenomena. Cellular automata are also of sufficient generality to provide simple models for a very wide variety of physical, chemical, biological, and other systems. The ultimate goal is to abstract from a study of cellular automata general features of "self-organizing" behavior and perhaps to devise universal laws analogous to the laws of thermodynamics. This paper concentrates on the mathematical features of the simplest cellular automata, leaving for future study more complicated cellular automata and details of applications to specific systems. The paper is largely intended as an original contribution, rather than a review. It is presented in this journal in the hope that it may thereby reach a wider audience than would otherwise be possible. An outline of some of its results is given in Wolfram (1982a).

Investigations of simple "self-organization" phenomena in physical and chemical systems (Turing, 1952; Haken, 1975, 1978, 1979, 1981; Nicolis and Prigogine, 1977; Lan-

dauer, 1979; Prigogine, 1980; Nicolis *et al.*, 1981) have often been based on the Boltzmann transport differential equations (e.g., Lifshitz and Pitaevskii, 1981) (or its analogs) for the time development of macroscopic quantities. The equations are obtained by averaging over an ensemble of microscopic states and assuming that successive collisions between molecules are statistically uncorrelated. For closed systems (with reversible or at least unitary microscopic interactions) the equations lead to Boltzmann's H theorem, which implies monotonic evolution towards the macroscopic state of maximum entropy. The equations also imply that weakly dissipative systems (such as fluids with small temperature gradients imposed) should tend to the unique condition of minimum entropy production. However, in strongly dissipative systems, several final states may be possible, corresponding to the various solutions of the polynomial equations obtained from the large time limit of the Boltzmann equations. Details or "fluctuations" in the initial state determine which of several possible final states are attained, just as in a system with multiple coexisting phases. Continuous changes in parameters such as external concentrations or temperature gradients may lead to discontinuous changes in the final states when the number of real roots in the polynomial equations changes, as described by catastrophe theory (Thom, 1975). In this way, "structures" with discrete boundaries may be formed from continuous models. However, such approaches become impractical for systems with very many degrees of freedom, and therefore cannot address the formation of genuinely complex structures.

More general investigations of self-organization and "chaos" in dynamical systems have typically used simple mathematical models. One approach (e.g., Ott, 1981) considers dissipative nonlinear differential equations (typically derived as idealizations of Navier-Stokes hydrodynamic equations). The time evolution given particular initial conditions is represented by a trajectory in the space of variables described by the differential equations. In the simplest cases (such as those typical for chemical concentrations described by the Boltzmann transport equations), all trajectories tend at large times to a small number of isolated limit points, or approach simple periodic limit cycle orbits. In other cases, the trajectories may instead concentrate on complicated and apparently chaotic surfaces ("strange attractors"). Nearly linear systems typically exhibit simple limit points or cycles. When nonlinearity is increased by variation of external parameters, the number of limit points or cycles may increase without bound, eventually building up a strange attractor (typically exhibiting a statistically self-similar structure in phase space). A simpler approach (e.g., Ott, 1981) involves discrete time steps, and considers the evolution of numbers on an interval of the real line under iterated mappings. As the nonlinearity is increased, greater numbers of limit points and cycles appear, followed by essentially chaotic behavior. Quantitative features of this approach to chaos are found to be universal to wide classes of mappings. Notice that for both differential equations and iterated mappings, initial conditions are specified by real numbers with a potentially infinite number of significant digits. Complicated or seemingly chaotic behavior is a reflection of sensitive dependence on high-order digits in the decimal expansions of the numbers.

Models based on cellular automata provide an alternative approach, involving discrete coordinates and variables as well as discrete time steps. They exhibit complicated behavior analogous to that found with differential equations or iterated mappings, but by virtue of their simpler construction are potentially amenable to a more detailed and complete analysis.

Section II of this paper defines and introduces cellular automata and describes the qualitative behavior of elementary cellular automata. Several phenomena characteristic of self-organization are found. Section III gives a quantitative statistical analysis of the states generated in the time evolution of cellular automata, revealing several quantitative universal features. Section IV describes the global analysis of cellular automata and discusses the results in the context of dynamical systems theory and the formal theory of computation. Section V considers briefly extensions to more complicated cellular automata. Finally, Sec. VI gives some tentative conclusions.

II. INTRODUCTION TO CELLULAR AUTOMATA

Cellular automata are mathematical idealizations of physical systems in which space and time are discrete, and physical quantities take on a finite set of discrete values. A cellular automaton consists of a regular uniform lattice (or "array"), usually infinite in extent, with a discrete variable at each site ("cell"). The state of a cellular automaton is completely specified by the values of the variables at each site. A cellular automaton evolves in discrete time steps, with the value of the variable at one site being affected by the values of variables at sites in its "neighborhood" on the previous time step. The neighborhood of a site is typically taken to be the site itself and all immediately adjacent sites. The variables at each site are updated simultaneously ("synchronously"), based on the values of the variables in their neighborhood at the preceding time step, and according to a definite set of "local rules."

Cellular automata were originally introduced by von Neumann and Ulam (under the name of "cellular spaces") as a possible idealization of biological systems (von Neumann, 1963, 1966), with the particular purpose of modelling biological self-reproduction. They have been applied and reintroduced for a wide variety of purposes, and referred to by a variety of names, including "tessellation automata," "homogeneous structures," "cellular structures," "tessellation structures," and "iterative arrays."

Physical systems containing many discrete elements with local interactions are often conveniently modelled as cellular automata. Any physical system satisfying differential equations may be approximated as a cellular automaton by introducing finite differences and discrete variables.[1] Nontrivial cellular automata are obtained whenever the dependence on the values at each site is nonlinear, as when the system exhibits some form of "growth inhibition." A very wide variety of examples may be considered; only a few are sketched here. In the most direct cases, the cellular automaton lattice is in position space. At a microscopic level, the sites may represent points in a crystal lattice, with values given by some quantized observable (such as spin component) or corresponding to the types of atoms or units. The dynamical Ising model (with kinetic energy terms included) and other lattice spin systems are simple cellular automata, made nondeterministic by "noise" in the local rules at finite temperature. At a more macroscopic level, each site in a cellular automaton may represent a region containing many molecules (with a scale size perhaps given by an appropriate correlation length), and its value may label one of several discrete possible phases or compositions. In this way, cellular automata may be used as discrete models for nonlinear chemical systems involving a network of reactions coupled with spatial diffusion (Greenberg *et al.*, 1978). They have also been used in a (controversial) model for the evolution of spiral galaxies (Gerola and Seiden, 1978; Schewe, 1981). Similarly, they may provide models for kinetic aspects of phase transitions (e.g., Harvey *et al.*, 1982). For example, it is possible that growth of dendritic crystals (Langer, 1980) may be described by aggregation of discrete "packets" with a local growth inhibition effect associated with local releases of latent heat, and thereby treated as a cellular automaton [Witten and Sander (1981) discuss a probabilistic model of this kind, but there are indications that the probabilistic elements are inessential]. The spatial structure of turbulent fluids may perhaps be modelled using cellular automata by approximating the velocity field as a lattice of cells, each containing one or no eddies, with interactions between neighboring cells. Physical systems may also potentially be described by cellular automata in wave-vector or momentum space, with site values representing excitations in the corresponding modes.

---

[1] The discussion here concentrates on systems first order in time; a more general case is mentioned briefly in Sec. IV.

Many biological systems have been modelled by cellular automata (Lindenmayer, 1968; Herman, 1969; Ulam, 1974; Kitagawa, 1974; Baer and Martinez, 1974; Rosen, 1981) (cf. Barricelli, 1972). The development of structure and patterns in the growth of organisms often appears to be governed by very simple local rules (Thompson, 1961; Stevens, 1974) and is therefore potentially well described by a cellular automaton model. The discrete values at each site typically label types of living cells, approximated as growing on a regular spatial lattice. Short-range or contact interactions may lead to expression of different genetic characteristics, and determine the cell type. Simple nonlinear rules may lead to the formation of complex patterns, as evident in many plants and animals. Examples include leaf and branch arrangements (e.g., Stevens, 1974) and forms of radiolarian skeletons (e.g., Thompson, 1961). Simple behavior and functioning of organisms may be modelled by cellular automata with site values representing states of living cells or groups of cells [Burks (1973) and Flanigan (1965) discuss an example in heart fibrillation]. The precise mathematical formulation of such models allows the behavior possible in organisms or systems with particular construction or complexity to be investigated and characterized (e.g., von Neumann, 1966). Cellular automata may also describe populations of non-mobile organisms (such as plants), with site values corresponding to the presence or absence of individuals (perhaps of various types) at each lattice point, with local ecological interactions.

Cellular automata have also been used to study problems in number theory and their applications to tapestry design (Miller, 1970, 1980; ApSimon, 1970a, 1970b; Sutton, 1981). In a typical case, successive differences in a sequence of numbers (such as primes) reduced with a small modulus are taken, and the geometry of zero regions is investigated.

As will be discussed in Sec. IV, cellular automata may be considered as parallel processing computers (cf. Manning, 1977; Preston *et al.*, 1979). As such, they have been used, for example, as highly parallel multipliers (Atrubin, 1965; Cole, 1969), sorters (Nishio, 1981), and prime number sieves (Fischer, 1965). Particularly in two dimensions, cellular automata have been used extensively for image processing and visual pattern recognition (Deutsch, 1972; Sternberg, 1980; Rosenfeld, 1979). The computational capabilities of cellular automata have been studied extensively (Codd, 1968; Burks, 1970; Banks, 1971; Aladyev, 1974, 1976; Kosaraju, 1974; Toffoli, 1977b), and it has been shown that some cellular automata could be used as general purpose computers, and may therefore be used as general paradigms for parallel computation. Their locality and simplicity might ultimately permit their implementation at a molecular level.

The notorious solitaire computer game "Life" (Conway, 1970; Gardner, 1971, 1972; Wainwright, 1971−1973; Wainwright, 1974; Buckingham, 1978; Berlekamp *et al.*, 1982; R. W. Gosper, private communications) (qualitatively similar in some respects to the game of "Go") is an example of a two-dimensional cellular automaton, to be

discussed briefly in Sec. V.

Until Sec. V, we shall consider exclusively one-dimensional cellular automata with two possible values of the variables at each site ("base 2") and in which the neighborhood of a given site is simply the site itself and the sites immediately adjacent to it on the left and right. We shall call such cellular automata elementary. Figure 1 specifies one particular set of local rules for an elementary cellular automaton. On the top row, all $2^3 = 8$ possible values of the three variables in the neighborhood are given, and below each one is given the value achieved by the central site on the next time step according to a particular local rule. Figure 2 shows the evolution of a particular state of the cellular automaton through one time step according to the local rule given in Fig. 1.

The local rules for a one-dimensional neighborhood-three cellular automaton are described by an eight-digit binary number, as in the example of Fig. 1. (In specifying cellular automata, we use this binary number interchangeably with its decimal equivalent.) Since any eight-digit binary number specifies a cellular automaton, there are $2^8 = 256$ possible distinct cellular automaton rules in one dimension with a three-site neighborhood. Two inessential restrictions will usually be imposed on these rules. First, a cellular automaton rule will be considered "illegal" unless a "null" or "quiescent" initial state consisting solely of 0 remains unchanged. This forbids rules whose binary specification ends with a 1 (and removes symmetry in the treatment of 0 and 1 sites). Second, the rules must be reflection symmetric, so that 100 and 001 (and 110 and 011) yield identical values. These restrictions[2] leave 32 possible "legal" cellular automaton rules of the form $\alpha_1 \alpha_2 \alpha_3 \alpha_4 \alpha_2 \alpha_5 \alpha_4 0$.

The local rules for a cellular automaton may be considered as a Boolean function of the sites within the neighborhood. Let $s_n(m)$ be the value of site m at time step n. As a first example consider the "modulo-two" rule 90 (also used as the example for Fig. 1). According to this rule, the value of a particular site is simply the sum modulo two of the values of its two neighboring sites on the previous time step. The Boolean equivalent of this rule is therefore

$$s_{n+1}(m) = s_n(m-1) \oplus s_n(m+1) \qquad (2.1)$$

or schematically $s_+ = s^- \oplus s^+$, where $\oplus$ denotes addition modulo two ("exclusive disjunction" or "inequality"). Similarly, rule 18 is equivalent to $s_+ = s \vee (s^- \oplus s^+)$ [where s denotes $s_n(m)$], rule 22 to $s_+ = s \vee (s^- \wedge s^+)$, rule 54 to $s_+ = s \oplus (s^- \vee s^+)$, rule 150 to $s_+ = s^- \oplus s \oplus s^+$, and so on. Designations s^- and s^+ always enter symmetrically in legal cellular automaton rules by virtue of re-

---

[2] The quiescence condition is required in many applications to forbid "instantaneous propagation" of value-one sites. The reflection symmetry condition guarantees isotropy as well as homogeneity in cellular automaton evolution.

$1\,1\,1$	$1\,1\,0$	$1\,0\,1$	$1\,0\,0$	$0\,1\,1$	$0\,1\,0$	$0\,0\,1$	$0\,0\,0$
0	1	0	1	1	0	1	0

FIG. 1. Example of a set of local rules for the time evolution of a one-dimensional elementary cellular automaton. The variables at each site may take values 0 or 1. The eight possible states of three adjacent sites are given on the upper line. The lower line then specifies a rule for the time evolution of the cellular automaton by giving the value to be taken by the central site of the three on the next time step. The time evolution of the complete cellular automaton is obtained by simultaneous application of these rules at each site for each time step. The rule given is the modulo-two rule: the value of a site at a particular time step is simply the sum modulo two of the values of its two neighbors at the previous time step. Any possible sequence of eight binary digits specifies a cellular automaton.

0 1 0 1 1 0 1 1 0 1 0 1 0 1 1 1 0 0 0 1 0
0 0 1 1 0 1 1 0 0 0 0 0 1 0 1 1 0 1 0

FIG. 2. Evolution of a configuration in one-dimensional cellular automaton for one time step according to the modulo-two rule given in Fig. 1. The values of the two end sites after the time step depend on the values of sites not shown here.

flection symmetry. The Boolean function representation of cellular automaton rules is convenient for practical implementation on standard serial processing digital computers.[3]

Some cellular automaton rules exhibit the important simplifying feature of "additive superposition" or "additivity." Evolution according to such rules satisfies the superposition principle

$$s_0 = t_0 \oplus u_0 \quad \leftrightarrow \quad s_n = t_n \oplus u_n , \tag{2.2}$$

which implies that the configurations obtained by evolution from any initial configuration are given by appropriate combinations of those found in Fig. 3 for evolution from a single nonzero site. Notice that such additivity does not imply linearity in the real number sense of Sec. I, since the addition is over a finite field. Cellular automata satisfy additive superposition only if their rule is of the form $\alpha_1\alpha_2 0 \alpha_3\alpha_2\alpha_1\alpha_3 0$ with $\alpha_3 = \alpha_1 \oplus \alpha_2$. Only rules 0, 90, 150, and 204 are of this form. Rules 0 and 204 are trivial; 0 erases any initial configuration, and 204 maintains any initial configuration unchanged (performing the identity transformation at each time step). Rule 90 is the modulo-two rule discussed above, and takes a particular site to be the sum modulo two of the values of its two neighbors at the previous time step, as in Eq. (2.1). Rule 150 is similar. It takes a particular site to be the sum modulo two of the values of its two neighbors and its own

---

[3] The values of a sequence of (typically 32) sites are represented by bits in a single computer word. Copies of this word shifted one bit to the left and one bit to the right are obtained. Then the cellular automaton rule may be applied in parallel to all bits in the words using single machine instructions for each word-wise Boolean operation. An analogous procedure is convenient in simulation of two-dimensional cellular automata on computer systems with memory-mapped displays, for which application of identical Boolean operations to each display pixel is usually implemented in hardware or firmware.

value at the previous time step $(s_+ = s^- \oplus s \oplus s^+)$.

The additive superposition principle of Eq. (2.2) combines values at different sites by addition modulo two (exclusive disjunction). Combining values instead by conjunction (Boolean multiplication) yields a superposition principle for rules 0, 4, 50, and 254. Combining values by (inclusive) disjunction (Boolean addition) yields a corresponding principle for rules 0, 204, 250, and 254. It is found that no other legal cellular automaton rules satisfy superposition principles with any combining function.

The Boolean representation of cellular automaton rules reveals that some rules are "peripheral" in the sense that the value of a particular site depends on the values of its two neighbors at the previous time step, but not on its own previous value. Rules 0, 90, 160, and 250 are of the form $\alpha_1\alpha_2\alpha_1\alpha_2 0 \alpha_2 0 \alpha_2 0$ and exhibit this property.

Having discussed features of possible local rules we now outline their consequences for the evolution of elementary cellular automata. Sections III and IV present more detailed quantitative analysis.

Figure 3 shows the evolution of all 32 possible legal cellular automata from an initial configuration containing a single site with value 1 (analogous to the growth of a "crystal" from a microscopic "seed"). The evolution is shown until a particular configuration appears for the second time (a "cycle" is detected), or for at most 20 time steps. Several classes of behavior are evident. In one class, the initial 1 is immediately erased (as in rules 0 and 160), or is maintained unchanged (as in rules 4 and 36). Rules of this class are distinguished by the presence of the local rules $100 \rightarrow 0$ and $001 \rightarrow 0$, which prevent any propagation of the initial 1. A second class of rules (exemplified by 50 or 122) copies the 1 to generate a uniform structure which expands by one site in each direction on each time step. These two classes of rules will be termed "simple." A third class of rules, termed "complex," and exemplified by rules 18, 22, and 90, yields nontrivial patterns.

As a consequence of their locality, cellular automaton rules define no intrinsic length scale other than the size of a single site (or of a neighborhood of three sites) and no intrinsic time scale other than the duration of a single time step. The initial state consisting of a single site with value 1 used in Fig. 3 also exhibits no intrinsic scale. The cellular automaton configurations obtained in Fig. 3 should therefore also exhibit no intrinsic scale, at least in the infinite time limit. Simple rules yield a uniform final state, which is manifestly scale invariant. The scale invariance of the configurations generated by complex rules

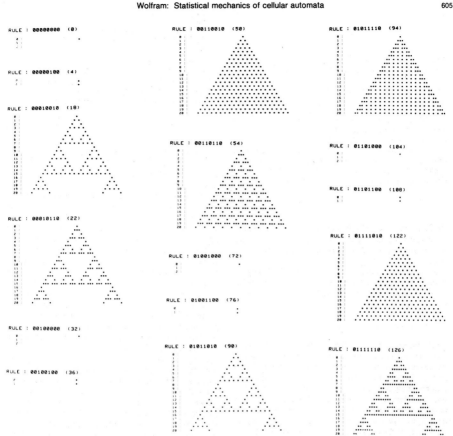

FIG. 3. Evolution of one-dimensional elementary cellular automata according to the 32 possible legal sets of rules, starting from a state containing a single site with value 1. Sites with value 1 are represented by stars, and those with value 0 by blanks. The configurations of the cellular automata at successive time steps are shown on successive lines. The time evolution is shown up to the point where the system is detected to cycle (visiting a particular configuration for the second time), or for at most 20 time steps. The process is analogous to the growth of a crystal from a microscope seed. A considerable variety of behavior is evident. The cellular automata which do not tend to a uniform state yield asymptotically self-similar fractal configurations.

is nontrivial. In the infinite time limit, the configurations are "self-similar" in that views of the configuration with different "magnifications" (but with the same "resolution") are indistinguishable. The configurations thus exhibit the same structure on all scales.

Consider as an example the modulo-two rule 90 (also used as the example for Fig. 1 and in the discussion above). This rule takes each site to be the sum modulo two of its two nearest neighbors on the previous time step. Starting from an initial state containing a single site with value 1, the configuration it yields on successive time steps is thus simply the lines of Pascal's triangle modulo

two, as illustrated in Fig. 4 (cf. Wolfram, 1982b). The values of the sites are hence the values of binomial coefficients [or equivalently, coefficients of x^i in the expansion of $(1+x)^n$] modulo two. In the large time limit, the pattern of sites with value 1 may be obtained by the recursive geometrical construction (cf. Sierpinski, 1916; Abelson and diSessa, 1981, Sec. 2.4) shown in Fig. 5. This geometrical construction manifests the self-similarity (Mandelbrot, 1977, 1982; Geffen *et al.*, 1981) or "scale invariance" of the resulting curve. Figure 3 shows that evolution of other complex cellular automata from a single nonzero site yields essentially identical self-similar pat-

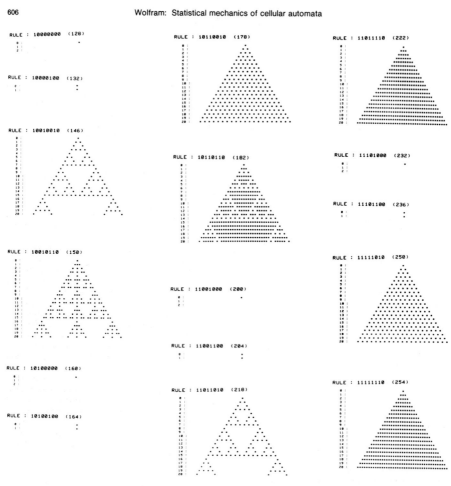

FIG. 3. (*Continued.*)

FIG. 4. An algebraic construction for the configurations of a cellular automaton starting from a state containing a single site with value 1 and evolving according to the modulo-two rule 90. The rule is illustrated in Fig. 1, and takes the value of a particular site to be the sum modulo two of the values of its two neighboring sites at the previous time step. The value of a site at a given time step is then just the value modulo two of the corresponding binomial coefficient in Pascal's triangle.

terns. An exception is rule 150, for which the value of each site is determined by the sum modulo two of its own value and the values of its two neighbors on the previous time step. The sequence of binary digits obtained by evolution from a single-site initial state for n time steps with this rule is thus simply the coefficients of x^i in the expansion of $(x^2+x+1)^n$ modulo two. A geometrical construction for the pattern obtained is given in Fig. 6.

Figure 7 shows examples of time evolution for some cellular automata with illegal local rules (defined above) which were omitted from Fig. 3. When the quiescence condition is violated, successive time steps involve alternation of 0 and 1 at infinity. When reflection symmetry is violated, the configurations tend to undergo uniform

FIG. 5. Sequence of steps in a geometrical construction for the large time behavior of a cellular automaton evolving according to the modulo-two rule 90. The final pattern is the limit of the sequence shown here. It is a self-similar figure with fractal dimension $\log_2 3$.

FIG. 6. Sequence of steps in a geometrical construction for the large time behavior of a cellular automaton evolving according to the modulo-two rule 150. The final pattern is the limit of the sequence shown here. It is a self-similar figure with fractal dimension $\log_2 2\varphi \simeq 1.69$ [where $\varphi = (1 + \sqrt{5})/2$ is the golden ratio]. An analogous construction for rule 90 was given in Fig. 5.

shifting. The self-similar patterns seen in Fig. 3 are also found in cases such as rule 225, but are sheared by the overall shifting. It appears that consideration of illegal as well as "legal" cellular automaton rules introduces no qualitatively new features.

Figure 3 showed the growth of patterns by cellular automaton evolution from a very simple initial state containing a single nonzero site (seed). Figure 8 now illustrates time evolution from a disordered or "random" initial state according to each of the 32 legal cellular automaton rules. A specific "typical" initial configuration was taken, with the value of each site chosen independently, with equal probabilities for values 0 and 1.[4] Just as in Fig. 3, several classes of behavior are evident. The simple rules exhibit trivial behavior, either yielding a uniform final state or essentially preserving the form of the initial state. Complex rules once again yield nontrivial behavior. Figure 8 illustrates the remarkable fact that time evolution according to these rules destroys the independence of the initial sites, and generates correlations between values at separated sites. This phenomenon is the essence of self-organization in cellular automata. An initially random state evolves to a state containing long-range correlations and structure. The bases of the "triangles" visible in Fig. 8 are fluctuations in which a sequence of many adjacent cells have the same value. The length of these correlated sequences is reduced by one site per time step, yielding the distinctive triangular structure. Figure 8 suggests that triangles of all sizes are generated. Section III confirms this impression through a quantitative analysis and discusses universal features of the structures obtained.

The behavior of the cellular automata shown in Fig. 8 may be characterized in analogy with the behavior of dynamical systems (e.g., Ott, 1981): simple rules exhibit simple limit points or limit cycles, while complex rules exhibit phenomena analogous to strange attractors.

The cellular automata shown in Fig. 8 were all assumed to satisfy periodic boundary conditions. Instead of treating a genuinely infinite line of sites, the first and last sites are identified, as if they lay on a circle of finite radius. Cellular automata can also be rendered finite by imposing null boundary conditions, under which sites beyond each end are modified to maintain value zero, rather than evolving according to the local rules. Figure 9 compares results obtained with these two boundary conditions in a simple case; no important qualitative differences are apparent.

Finite one-dimensional cellular automata are similar to a class of feedback shift registers (e.g., Golomb, 1967; Berlekamp, 1968).[5] A feedback shift register consists of a sequence of sites ("tubes") carrying values $a(i)$. At each time step, the site values evolve by a shift $a(i) = a(i-1)$ and feedback $a(0) = \mathbf{F}[a(j_1), a(j_2), \dots]$ where, j_i give the positions of "taps" on the shift register. An elementary cellular automaton of length N corresponds to a feedback shift register of length N with site values 0 and 1 and taps at positions $N-2$, $N-1$, and N. The Boolean function F defines the cellular automaton rule. [The additive rules 90 and 150 correspond to linear feedback shift registers in which $\mathbf{F}$ is addition modulo two (exclusive disjunction).] At each shift register time step, the value of one site is updated according to the cellular automaton rule. After N time steps, all N sites have been updated, and one cellular automaton time step is complete. All interior sites are treated exactly as in a cellular automaton, but the two end sites evolve differently (their values depend on the two preceding time steps).

III. LOCAL PROPERTIES OF ELEMENTARY CELLULAR AUTOMATA

We shall examine now the statistical analysis of configurations generated by time evolution of "elementary" cellular automata, as illustrated in Figs. 3 and 8. This section considers statistical properties of individual such configurations; Sec. IV discusses the ensemble of all possible configurations. The primary purpose is to obtain a quantitative characterization of the "self-organization" pictorially evident in Fig. 8.

---

[4] Here and elsewhere a standard linear congruential pseudorandom number generator with recurrence relation $x_{n+1} = (1103515245x_n + 12345) \bmod 2^{31}$ was used. Results were also obtained using other pseudorandom number generation procedures and using random numbers derived from real-time properties of a time-shared computer system.

---

[5] This similarity may be used as the basis for a simple hardware implementation of one-dimensional cellular automata (Pearson et al., 1981; Hoogland et al., 1982; Toffoli, 1983).

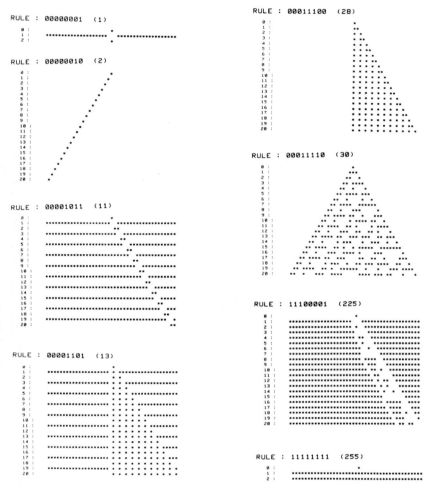

FIG. 7. Evolution of a selection of one-dimensional elementary cellular automata obeying illegal rules. Rules are considered illegal if they violate reflection symmetry, which requires identical rules for 100 and 001 and for 110 and 011, or if they violate the quiescence condition which requires that an initial state containing only 0 sites should remain unchanged. For example, rule 2 violates reflection symmetry, and thus yields a uniformly shifting pattern, while rule 1 violates the quiescence condition and yields a pattern which "flashes" from all 0 to all 1 in successive time steps.

A configuration may be considered disordered (or essentially random) if values at different sites are statistically uncorrelated (and thus behave as "independent random variables"). Such configurations represent a discrete form of "white noise." Deviations of statistical measures for cellular automaton configurations from their values for corresponding disordered configurations indicate order, and signal the presence of correlations between values

at different sites. An (infinite) disordered configuration is specified by a single parameter, the independent probability p for each site to have value 1. The description of an ordered configuration requires more parameters.

Figure 10 shows a set of examples of disordered configurations with probabilities $p=0.25$, 0.5, and 0.75. Such disordered configurations were used as the initial configurations for the cellular automaton evolution shown in Fig.

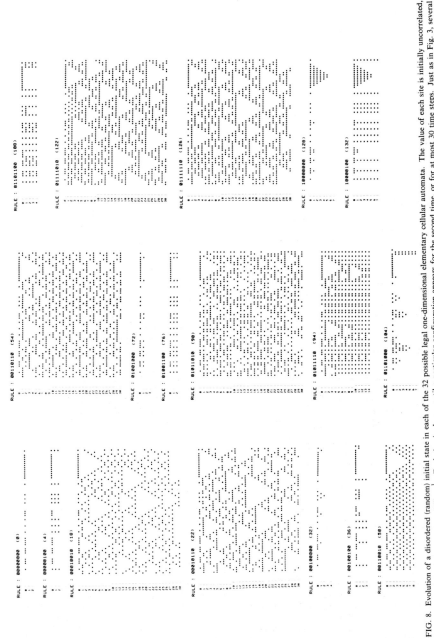

FIG. 8. Evolution of a disordered (random) initial state in each of the 32 possible legal one-dimensional elementary cellular automata. The value of each site is initially uncorrelated, and is taken to be 0 or 1 with probability $\frac{1}{2}$. Evolution is shown until a particular configuration appears for the second time, or for at most 30 time steps. Just as in Fig. 3, several classes of behavior are evident. In one class, time evolution generates long-range correlations and fluctuations, yielding distinctive "triangular" structures, and exhibiting a simple form of self-organization. All the cellular automata shown are taken to satisfy periodic boundary conditions, so that their sites are effectively arranged on a circle.

FIG. 8. *(Continued.)*

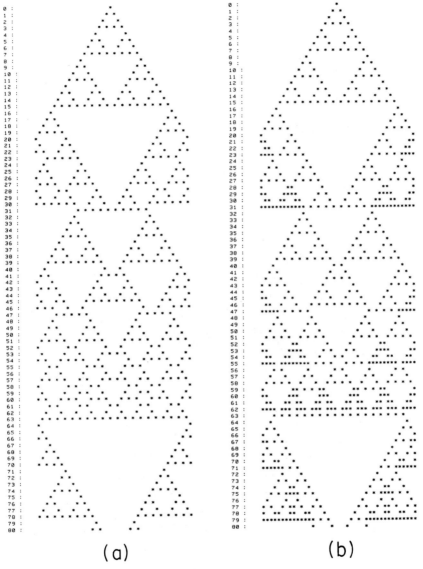

(a)   (b)

FIG. 9. Time evolution of a simple initial state according to the modulo-two rule 90, on a line of sites satisfying (a) periodic boundary conditions (so that first and last sites are identified, and the sites are effectively arranged on a circle), and (b) null boundary conditions (so that sites not shown are assumed always to have value 0). Changes in boundary conditions apparently have no significant qualitative effect.

8. Qualitative comparison of the configurations obtained by this evolution with the disordered configurations of Fig. 10 strongly suggests that cellular· automata indeed generate more ordered configurations, and exhibit a simple form of self-organization.

The simplest statistical quantity with which to characterize a cellular automaton configuration is the average fraction (density) of sites with value 1, denoted by ρ. For a disordered configuration, ρ is given simply by the independent probability p for each site to have value 1.

We consider first the density ρ_1 obtained from a disordered configuration by cellular automaton evolution for one time step. When $p = \rho = \frac{1}{2}$ (as in Fig. 8), a disordered configuration contains all eight possible three-site neighborhoods (illustrated in Fig. 1) with equal probability. Applying a cellular automaton rule (specified, say, by a binary sequence $\mathbf{R}$, as in Fig. 1) to this initial state for one time step ($\tau = 1$) yields a configuration in which the fraction of sites with value 1 is given simply by the fraction of the eight possible neighborhoods which yield 1 according to the cellular automaton rule. This fraction is given by

$$\rho_1 = \#_1(\mathbf{R})/(\#_0(\mathbf{R}) + \#_1(\mathbf{R})) = \#_1(\mathbf{R})/8 , \quad (3.1)$$

where $\#_d(S)$ denotes the number of occurrences of the digit d in the binary representation of S. Hence, for example, $\#_1(10110110) = \#_1(182) = 5$ and $\#_0(10110110) = \#_0(182) = 3$. With cellular automaton rule 182, therefore, the density ρ after the first time step shown in Fig. 8 is $\frac{5}{8}$ if an infinite number of sites is included. The result (3.1) may be generalized to initial states with $p \neq \frac{1}{2}$ by using the probabilities $p(\sigma) = p^{\#_1(\sigma)}(1-p)^{\#_0(\sigma)}$, for each of the eight possible three-site neighborhoods σ (such as 110) shown in Fig. 1, and adding the probabilities for those σ which yield 1 on application of the cellular automaton rule.

The function $\#_1(n)$ will appear several times in the analysis given below. A graph of it for small n is given in Fig. 11, and is seen to be highly irregular. For any n, $\#_1(n) + \#_0(n)$ is the total number of digits ($\lceil \log_2 n \rceil$) in the binary representation of n, so that $\#_1(n) \leq \log_2 n$. Furthermore, $\#_1(2^k n) = \#_1(n)$ and for $n < 2^k$, $\#_1(n + 2^k) = \#_1(n) + 1$. Finally, one finds that

$$\#_1(n) = n - \sum_{i=1} \lfloor n/2^i \rfloor .$$

References to further results are given in McIlroy (1974) and Stolarsky (1977).

We now consider the behavior of the density ρ_τ obtained after τ time steps in the limit of large τ. When $\tau > 1$, correlations induced by cellular automaton evolution invalidate the approach used in Eq. (3.1), although a similar approach may nevertheless be used in deriving statistical approximations, as discussed below.

Figure 8 suggests that with some simple rules (such as 0, 32, or 72), any initial configuration evolves ultimately to the null state $\rho = 0$, although the length of transient varies. For rule 0, it is clear that $\rho = 0$ for all $\tau > 0$. Similarly, for rule 72, $\rho = 0$ for $\tau > 1$. For rule 32, infinite

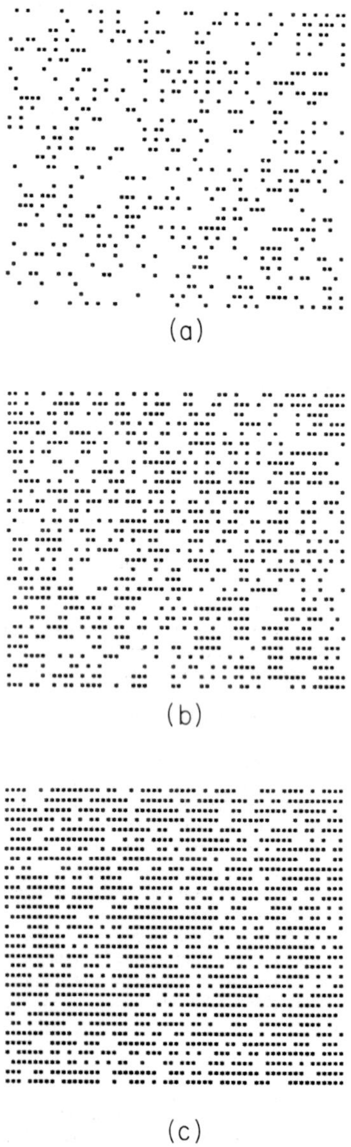

(a)

(b)

(c)

FIG. 10. Examples of sets of disordered configurations in which each site is chosen to have value 1 with independent probability (a) 0.25, (b) 0.5, and (c) 0.75. Successive lines are independent. The configurations are to be compared with those generated by cellular automaton evolution as shown in Fig. 8.

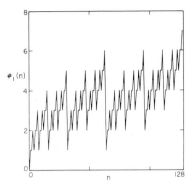

FIG. 11. The number of occurrences $\#_1(n)$ of the binary digit 1 in the binary representation of the integer n [$\#_1(1)=1$, $\#_1(2)=1$, $\#_1(3)=2$, $\#_1(4)=1$, and so on]. The function is defined only for integer n: values obtained for successive integer n have nevertheless been joined by straight lines.

transients may occur, but the probability that a nonzero value survives at a particular site for τ time steps assuming an initial disordered state with $\rho=\frac{1}{2}$ is $2^{-3(2\tau+1)}$. Rule 254 yields $\rho_\infty=1$, with a probability $(1-\rho_0)^{2\tau+1}$ for a transient of length $\geq \tau$. Rule 204 is the "identity rule," which propagates any initial configuration unchanged and yields $\rho_\infty=\rho_0$. The "disjunctive superposition" principle for rule 250 discussed in Sec. II implies $\rho_\infty=1$. For rule 50, the "conjunctive superposition" principle yields $\rho_\infty=\frac{1}{2}$.

Other simple rules serve as "filters" for specific initial sequences, yielding final densities proportional to the initial density of the sequences to be selected. For rule 4, the final density is equal to the initial density of 101 sequences, so that $\rho_\infty=\rho_0^2(1-\rho_0)$. For rule 36, ρ_∞ is determined by the density of initial 00100 and ...1010101... sequences and is approximately $\frac{1}{16}$ for $\rho_0=\frac{1}{2}$.

Exact results for the behavior of ρ_τ with the modulo-two rule 90 may be derived using the additive superposition property discussed in Sec. II.

Consider first the number of sites $N_\tau^{(1)}$ with value 1 obtained by evolution according to rule 90 from an initial state containing a single site with value 1, as illustrated in Fig. 3. Geometrical considerations based on Fig. 5 yield the result[6]

$$N_\tau^{(1)} = 2^{\#_1(\tau)}, \qquad (3.2)$$

where the function $\#_1(\tau)$ gives the number of occurrences of the digit 1 in the binary representation of the integer τ, as defined above, and is illustrated in Fig. 11. Equation (3.2) may be derived as follows. Consider the

figure generated by $\lceil\log_2\tau\rceil$ (the number of digits in the binary representation of τ) steps in the construction of Fig. 5. The configuration obtained after τ time steps of cellular automaton evolution corresponds to a slice through this figure, with a 1 at each point crossed by a line of the figure, and 0 elsewhere. By construction, the slice must lie in the lower half of the figure. Successive digits in the binary representation of τ determine whether the slice crosses the upper (0) or lower (1) halves of successively smaller triangles. The number of lines of the figure crossed is multiplied by a factor each time the lower half is chosen. The total number of sites with value 1 encountered is then given by a product of the factors of two associated with each 1 digit in the binary representation of τ. Inspection of Fig. 5 also yields a formula for the positions of all sites with value 1. With the original site at position 0, the positions of sites with value 1 after τ time steps are given by $\pm(2^{j_1}\pm(2^{j_2}\pm\cdots))$, where all possible combinations of signs are to be taken, and the j_i correspond to the positions at which the digit 1 appears in the binary representation of τ, defined so that $\tau = 2^{j_1} + 2^{j_2} + \cdots$ and $j_1 > j_2 > \cdots$.

Equation (3.2) shows that the density averaged over the region of nonzero sites ("light cone") in the rule 90 evolution of Fig. 3 is given by $\rho_\tau=N_\tau^{(1)}/(2\tau+1)$ and does not tend to a definite limit for large τ. Nevertheless, the time-average density

$$\bar{\rho}_T=(1/T)\sum_{\tau=0}^{\tau=T}\rho_\tau$$

tends to zero (as expected from the geometrical construction of Fig. 5) like $T^{\log_2 3-2}\sim T^{-0.42}$.[7] Results for initial states containing a finite number of sites with value 1 may be obtained by additive superposition. If the initial configuration is one which would be reached by evolution from a single site after, say, τ_0 time steps, then the resulting density is given by Eq. (3.2) with the replacement $\tau\to\tau-\tau_0$. Only a very small fraction of initial configurations may be treated in this way, since evolution from a single site generates only one of the 2^k possible configurations in which the maximum separation between nonzero sites is k. For small or highly regular initial configurations, results analogous to (3.2) may nevertheless be derived. Statistical results for evolution from disordered initial states may also be derived. Equation (3.2) implies that after exactly $\tau=2^j$ time steps, an initial state containing a single nonzero site evolves to a configuration with only two nonzero sites. At this point, the value of a particular site at position n is simply the sum modulo two of the initial values of sites at positions $n-\tau$ and $n+\tau$. If we start from a disordered initial configuration, the density at such time steps is thus given by $\rho_{\tau=2^j}=2\rho_0(1-\rho_0)$.

---

[6] This result has also been derived by somewhat lengthy algebraic means in Glaisher (1899), Fine (1947), Roberts (1957), Kimball *et al.* (1958), and Honsberger (1976).

[7] This form is strictly correct only for $T=2^k$. For $T=2^k(1+\delta)$, there is a correction factor $\simeq(1+\delta^{\log_2 3})/(1+\delta)^{\log_2 3}$, which lies between 0.86 and 1, with a broad minimum around $\delta=0.3$.

In general, the value of a site at time step τ is a sum modulo two of the initial values of $N_\tau^{(1)} = 2^{\#_1(\tau)}$ sites, which each have value 1 with probability ρ_0. If each of a set of k sites has value 1 with probability p, then the probability that the sum of the values at the sites will be odd (equal to 1 modulo two) is

$$\sum_{i \text{ odd}} \binom{k}{i} p^i (1-p)^{k-i} = \frac{1}{2}[1-(1-2p)^k] .$$

Thus the density of sites with value 1 obtained by evolution for τ time steps from an initial state with density ρ_0 according to cellular automaton rule 90 is given by

$$\rho_\tau = \frac{1}{2}[1-(1-2\rho_0)^{2^{\#_1(\tau)}}] . \tag{3.3}$$

This result is shown as a function of τ for the case $\rho_0 = 0.2$ in Fig. 12. For large τ, $\#_1(\tau) = O(\log_2 \tau)$, except at a set of points of measure zero, and Eq. (3.3) implies that $\rho_\tau \to \frac{1}{2}$ as $\tau \to \infty$ for almost all τ (so long as $\rho_0 \neq 0$).

Cellular automaton rule 150 shares with rule 90 the property of additive superposition. Inspection of the results for rule 150 given in Fig. 3 indicates that the value of a particular site depends on the values of at least three initial sites (this minimum again being achieved when $\tau = 2^k$), so that $|\rho_\tau - \frac{1}{2}| \leq |1-2\rho_0|^3$. Between the exceptional time steps $\tau = 2^k$, the ρ_τ for rule 150 tends to be

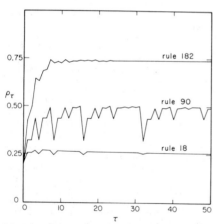

FIG. 12. Average density ρ_τ of sites with value 1 obtained by time evolution according to various cellular automaton rules starting from a disordered initial state with $\rho_0 = 0.2$. The additivity of the modulo-two rule 90 may be used to derive the exact result (3.2) for ρ_τ. The irregularities appear for time steps at which the value of each site depends on the values of only a few initial sites. For the nonadditive complex rules exemplified by 18 and 182, the values of sites at time step τ depend on the values of $O(\tau)$ initial sites, and ρ_τ tends smoothly to a definite limit. This limit is independent of the density of the initial disordered state.

much flatter than that for rule 90 (illustrated in Fig. 12). An exact result may be obtained, but is more complicated than in the case of rule 90. The geometrical construction of Fig. 6 shows that for rule 150, $N_\tau^{(1)}$ is a product of factors $X(j)$ associated with each sequence of j ones (delimited by zeroes) in the binary representation of τ. The expression $X(j)$ is given by the recurrence relation $X(j) = (2j \pm 1) X(j-1)$ where the upper (lower) sign is taken for j odd (even), and $X(1) = 3$ [so that $X(2) = 5$, $X(3) = 11$ and so on]. [$N_\tau^{(1)}$ thus measures "sequence correlations" in τ.] The density is then given in analogy with Eq. (3.3) by $\rho_\tau = \frac{1}{2}[1-(1-2\rho_0)^{N_\tau^{(1)}}]$.

Some aspects of the large-time behavior of nonadditive complex cellular automata may be found using a correspondence between nonadditive and additive rules (Grassberger, 1982). Special classes of configurations in nonadditive cellular automata effectively evolve according to additive rules. For example, with the nonadditive complex rule 18, a configuration in which, say, all even-numbered sites have value zero evolves after one time step to a configuration with all odd-numbered sites zero, and with the values of even-numbered sites given by the sums modulo two of their odd-numbered neighbors on the previous time step, just as for the additive rule 90. An arbitrary initial configuration may always be decomposed into a sequence of (perhaps small) "domains," in each of which either all even-numbered sites or all odd-numbered sites have value zero. These domains are then separated by "domain walls" or "kinks." The kinks move in the cellular automaton evolution and may annihilate in pairs. The motion of the kinks is determined by the initial configuration; with a disordered initial configuration, the kinks initially follow approximately a random walk, so that their mean displacement increases with time according to $\langle x^2 \rangle = t$ (Grassberger, 1982), and the paths of the kinks are fractal curves. This implies that the average kink density decreases through annihilation as if by diffusion processes according to the formula $\langle \rho_{\text{kink}} \rangle \sim (4\pi t)^{-1/2}$ (Grassberger, 1982). Thus after a sufficiently long time all kinks (at least from any finite initial configuration) must annihilate, leaving a configuration whose alternate sites evolve according to the additive cellular automaton rule 90. Each point on the "front" formed by the kink paths yields a pattern analogous to Fig. 5. The superposition of such patterns, each diluted by the insertion of alternate zero sites, yields configurations with an average density $\frac{1}{4}$ (Grassberger, 1982). The large number of sites on the "front" suppresses the fluctuations found for complete evolution according to additive rule 90. Starting with a disordered configuration of any nonzero density, evolution according to cellular automaton rule 18 therefore yields an asymptotic density $\frac{1}{4}$. The existence of a universal ρ_∞, independent of initial density ρ_0, is characteristic of complex cellular automaton rules.

Straightforward transformations on the case of rule 18 above then yield asymptotic densities $\rho_\infty = \frac{1}{4}$ for the complex nonadditive rules 146, 122, and 126, and an asymptotic density $\frac{3}{4}$ for rule 182, again all independent of the initial density ρ_0 (Grassberger, 1982). No simple domain structure appears with rule 22, and the approach fails. Simulations yield a numerical estimate $\rho_\infty = 0.35 \pm 0.02$ for evolution from disordered configurations with any nonzero ρ_0.

Figure 12 shows the behavior of ρ_τ for the complex nonadditive cellular automata 18 and 182 with $\rho_0 = 0.2$, and suggests that the final constant values $\rho_\infty = 0.25$ and $\rho_\infty = 0.75$ are approached roughly exponentially with time.

One may compare exact results for limiting densities of cellular automata with approximations obtained from a statistical approach (akin to "mean-field theory"). As discussed above, cellular automaton evolution generates correlations between values at different sites. Nevertheless, as a simple approximation, one may ignore these correlations, and parametrize all configurations by their average density ρ, or, equivalently, by the probabilities p and $q = 1 - p$, assumed independent, for each site to have value 1 and 0, respectively. With this approximation, the time evolution of the density is given by a master equation

$$\frac{\delta\rho}{\delta\tau} = \Gamma(0 \rightarrow 1) - \Gamma(1 \rightarrow 0) ,$$

$$\Gamma(0 \rightarrow 1) = \mathbf{P} \cdot (00110011 \wedge \mathbf{R}) ,$$

$$\Gamma(1 \rightarrow 0) = \mathbf{P} \cdot (11001100 \wedge \sim \mathbf{R}) , \qquad (3.4)$$

$$\mathbf{P} = \{p^3, p^2q, p^2q, pq^2, p^2q, pq^2, pq^2, q^3\} .$$

The term $\Gamma(0 \rightarrow 1)$ represents the average fraction of sites whose values change from 0 to 1 in each time step, and $\Gamma(1 \rightarrow 0)$ the fraction changing from 1 to 0. $\mathbf{R}$ is the binary specification of a cellular automaton rule, and the binary number with which it is "masked" (digitwise conjunction) selects local rules for three-site neighborhoods with appropriate values at the center site. $\mathbf{P}$ is the vector of probabilities for the possible three-site neighborhoods, assuming each site independently to have value 1 with probability $p = \rho$, and to have value 0 with probability $q = 1 - p = 1 - \rho$. The dot indicates that each element of this vector is to be multiplied by the corresponding digit of the binary sequence, and the results are to be added together. The equilibrium density ρ_∞ is achieved when

$$\frac{\delta\rho}{\delta\tau} = 0 .$$

This condition yields a polynomial equation for p and thus ρ_∞ for each of the legal cellular automaton rules. For rule 90, the equation is $pq^2 - p^3 = p - 2p^2 = p(1 - 2p) = 0$, which has solutions $p = 0$ (null state for all time) and $p = \frac{1}{2}$. Rule 18 yields the equation $pq^2 - 2p^2q - p^3 = p(1 - 4p + 2p^2) = 0$, which has the solutions $p = 0$ and $p = 1 - 1/\sqrt{2} \simeq 0.293$, together with the irrelevant solution $p = 1 + 1/\sqrt{2} > 1$. Rule 182 yields $2pq^2 - p^2q = p(2 - 3p)(1 - p) = 0$, giving $p = 0, 1, \frac{2}{3}$. For rules 90 and 18, these approximate results are close to the exact results 0.5 and 0.25. For rule 182, there is a significant discrepancy from the exact value 0.75. Nevertheless, for all complex cellular automaton rules, it appears

that the master equation (3.4) yields equilibrium densities within 10–20% of the exact values. The discrepancies are a reflection of the violation of the Markovian approximation required to derive Eq. (3.4) and thus of the presence of correlations induced by cellular automaton evolution.

In the discussion above, a definite value for the density ρ_τ at each time step was found by averaging over all sites of an infinite cellular automaton. If instead the density is estimated by averaging over blocks containing a finite number of sites b, a distribution of density values is obtained. In a disordered state, the central limit theorem ensures that for large b, these density estimates follow a Gaussian distribution with standard deviation $\simeq 1/\sqrt{b}$. Evolution according to any of the complex cellular automaton rules appears accurately to maintain this Gaussian distribution, while shifting its mean as illustrated in Fig. 12. Density in cellular automaton configurations thus obeys the "law of large numbers." Instead of taking many blocks of sites at a single time step, one might estimate the density at "equilibrium" by averaging results for a single block over many time steps. For nonadditive complex cellular automaton rules, it appears that these two procedures yield the same limiting results. However, the large fluctuations in average density visible in Fig. 12 at particular time steps for additive rules (90 and 150) would be lost in a time average.

Cellular automaton evolution is supposed to generate correlations between values at different sites. The very simplest measure of these correlations is the two-point correlation function $C^{(2)}(r) = \langle S(m)S(m+r)\rangle - \langle S(m)\rangle \times \langle S(m+r)\rangle$, where the average is taken over all possible positions m in the cellular automaton at a fixed time, and $S(k)$ takes on values -1 and $+1$ when the site at position k has values 0 and 1, respectively. A disordered configuration involves no correlations between values at different sites and thus gives $C^{(2)}(r) = 0$ for $r > 0$ $[C^{(2)}(0) = 1 - (2p - 1)^2]$. With the single-site initial state of Fig. 3, evolution of complex cellular automata yields configurations with definite periodicities. These periodicities give rise to peaks in $C^{(2)}(r)$. At time step τ, the largest peaks occur when $r = 2^k$ and the digit corresponding to 2^k appears in the binary decomposition of τ; smaller peaks occur when $r = 2^{k_1} \pm 2^{k_2}$, and so on. For the additive cellular automaton rules 90 and 150, a convolution of this result with the correlation function for any initial state gives the form of $C^{(2)}(r)$ after evolution for τ time steps. With these rules, the correlation function obtained by evolution from a disordered initial configuration thus always remains zero. For nonadditive rules, nonzero short-range correlations may nevertheless be generated from disordered initial configurations. The form of $C^{(2)}(r)$ for rule 18 at large times is shown in Fig. 13, and is seen to fall roughly exponentially with a correlation length ~ 2. The existence of a nonzero correlation length in this case is our first indication of the generation of order by cellular automaton evolution.

Figures 3 and 8 show that the evolution of complex cellular automata generates complicated patterns with a distinctive structure. The average density and the two-point correlation function are too coarse as statistical measures to be sensitive to this structure. Individual configurations appear to contain long sequences of correlated sites, punctuated by disordered regions. The two-dimensional picture formed by the succession of configurations in time is characteristically peppered with triangle structures. These triangles are formed when a long sequence of sites which suddenly all attain the same value, as if by a fluctuation, is progressively reduced in length by "ambient noise." Let $T_{(i)}(n)$ denote the density of triangles (in position and time) with base length n and filled with sites of value i. It is convenient to begin by considering the behavior of this density and then to discuss its consequences for the properties of individual configurations, whose long sequences typically correspond to sections through the triangles.

Consider first evolution from a simple initial state containing a single site with value 1. Figure 3 shows that in this case, all complex cellular automata (except rule 150) generate a qualitatively similar pattern, containing many congruent triangles whose bases have lengths 2^k. A geometrical construction for the limiting pattern obtained at large times was given in Fig. 5. At each successive stage in the construction, the linear dimensions (base lengths) of the triangles added are halved, and their number is multiplied by a factor 3. In the limit, therefore, $T(n/2) \sim 3T(n)$, (with $n = 2^k$), and hence

$$T(n) \sim n^{-\log_2 3} \sim n^{-1.59} \qquad (3.5)$$

[requiring exactly one triangle of size $\tau/2$ at time step τ

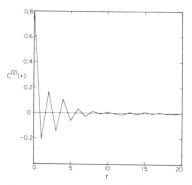

FIG. 13. Two-point correlation function $C^{(2)}(r)$ for configurations generated at large times by evolution according to cellular automaton rule 18 from any disordered initial configuration. $C^{(2)}(r)$ is defined as $\langle S(m)S(m+r)\rangle - \langle S(m)\rangle\langle S(m+r)\rangle$, where the average is taken over all sites m of the cellular automaton, and $S(k) = \pm 1$ when site k has values 1 and 0, respectively. No correlations are present in a disordered configuration, so that $C^{(2)}(r) = 0$ for $r > 0$. Evolution according to certain complex cellular automaton rules, such as 18, yield nonzero but exponentially damped correlations.

fixes the normalization as $T(n) = (2n/\tau)^{-\log_2 3}$]. The result (3.5) demonstrates that the patterns obtained from complex cellular automata in Fig. 3 not only contain structure on all scales (in the form of triangles of all sizes), but also exhibit a scale invariance or self similarity which implies the same structure on all scales (cf. Mandelbrot, 1982; Willson, 1982). The power law form of the triangle density (3.5) is independent of the absolute scale of n.

Self-similar figures on, for example, a plane may in general be characterized as follows. Find the minimum number $N(a)$ of squares with side a necessary to cover all parts of the figure (all sites with nonzero values in the cellular automaton case). The figure is self-similar or scale invariant if rescaling a changes $N(a)$ by a constant factor independent of the absolute size of a. In this case, $N(a) \sim a^{-D}$, where D is defined to be the Hausdorff-Besicovitch or fractal dimension (Mandelbrot, 1977, 1982) of the figure. A figure filling the plane would give $D = 2$, while a line would give $D = 1$. Intermediate values of D indicate clustering or intermittency. According to this definition, the cellular automaton pattern of Fig. 5 has fractal dimension $D = \log_2 3 \simeq 1.59$.

Figure 6 gives the construction analogous to Fig. 5 for the pattern generated by rule 150 in Fig. 3. In this case, the triangle density satisfies the two-term recurrence relation $T(n = 2^k) = 2T(2^{k+1}) + 4T(2^{k+2})$ with, say, $T(1) = 0$ and $T(2) = 2$. For large k, this yields (in analogy with the Fibonacci series)[8]

$$T(n) \sim n^{-\log_2(2\varphi)} = n^{-\log_2(1+\sqrt{5})} \sim n^{1.69}, \qquad (3.6)$$

where $\varphi = (1+\sqrt{5})/2 \simeq 1.618$ is the "golden ratio" which solves the equation $x^2 = x + 1$. The limiting fractal dimension of the pattern in Fig. 6 generated by cellular automaton rule 150 is thus $\log_2(2\varphi) = 1 + \log_2(\varphi) \simeq 1.69$.

The self similarity of the patterns generated by time evolution with complex cellular automaton rules in Fig. 3 is shared by almost all the configurations appearing at particular time steps and corresponding to lines through the patterns. If the fractal dimension of the two-dimensional patterns is D, then the fractal dimension of almost all the individual configurations is $D-1$. The configurations obtained at, for example, time steps τ of the form 2^k are members of an exceptional set of measure zero, for which no fractal dimension is defined. Almost all configurations generated from a single initial site by complex cellular automaton rules are thus self-similar, and (except for rule 150) are characterized by a fractal di-

---

[8] For small k, the triangle density in this case does not behave as a pure power of 2^k. Whereas the solution to any one-term recurrence relation, of the type found for cellular automaton rule 90, is a pure power, the solution to a p-term recurrence relation is in general a sum of p powers, with each exponent given by a root of the characteristic polynomial equation. In the high-order limit, the solutions are dominated by the term with the highest exponent (corresponding to the largest root of the equation). Complex roots yield oscillatory behavior [as in $f(k) = -f(k-1) + f(k-2); f(0) = 0, f(1) = 1$].

mension $D = \log_2 3 - 1 = \log_2(\frac{3}{2}) \simeq 0.59$. The second form may be deduced directly from the geometrical construction of Fig. 5. For rule 150, the configurations have fractal dimension $D = \log_2 \varphi$.

Figure 14 shows patterns generated by evolution with a selection of complex cellular automaton rules from initial states containing a few sites with value 1, extending over a region of size n_0. Comparison with Fig. 3 demonstrates that in most cases the patterns obtained even after many time steps differ from those generated with a single initial site. A few exceptional initial configurations (such as the one used for the first rule 90 example in Fig. 14) coincide with configurations reached by evolution from a single initial site and therefore yield a similar pattern, appropriately shifted in time. In the general case, Fig. 14 suggests that the form of the initial state determines the number of triangles with size $n \lesssim n_0$, but does not affect the density of triangles with $n \gg n_0$. As a simple example consider the modulo-two rule 90, whose additive superposition property implies that the final pattern obtained from an arbitrary initial state is simply a superposition of the patterns which would be generated from each of the nonzero initial sites in isolation. These latter patterns were shown in Fig. 5, and involve the generation of a triangle of size 2^k at time step 2^k. The superposition of such patterns yields at time step 2^k a triangle of size at least $2^k - 2n_0$. This conclusion apparently holds also for nonadditive complex cellular automata, so that, in general, for $n \gg n_0$, the density of triangles follows the form (3.5), as for a single site initial state. The patterns thus exhibit self-similarity for features large compared to the intrinsic scale defined by the "size" of the initial state. One therefore concludes that patterns which "grow" from any simple initial state according to any of the "complex" cellular automaton rules (except 150) share the universal feature of self similarity, characterized by a fractal dimension $\log_2 3$. On this basis, one may then conjecture that given suitable geometry (perhaps in more than one dimension, and possibly with more than three sites in a neighborhood), many of the wide variety of systems found to exhibit self-similar structure (Mandelbrot, 1977, 1982) attain this structure through local processes which follow cellular automaton rules.

Having considered the case of simple initial configurations, we now turn to the case of evolution from disordered initial configurations, illustrated in Fig. 8. Figure 15 shows the first 300 time steps in the evolution of cellular automaton 126, starting from a disordered initial state with density $\rho = 0.5$. Triangles of all sizes appear to be generated (the largest appearing in the figure has $n = 27$). Figure 16 shows the density of triangles $T(n)$ obtained at large times by evolution according to rule 126 and all of the other complex cellular automaton rules. The figure reveals the remarkable fact that for large n, all nonadditive rules yield the same $T(n)$, distinct from that for the additive rules (90 and 150). All the results are well fit by the form

$$T(n) \sim \lambda^{-n}. \qquad (3.7)$$

For nonadditive rules $\lambda \sim \frac{4}{3}$, while for the additive rules $\lambda \sim 2$. The same results are obtained at large times regardless of the density of the initial state. Thus the spectrum of triangles generated by complex cellular automaton evolution is universal, independent both of the details of the initial state, and of the precise cellular automaton rule used.

The behavior (3.8) of the triangle density with disordered initial states is to be contrasted with that of (3.5) for simple initial states. The precise form of an initial state of finite extent n_0 affects the pattern generated only at length scales $\lesssim n_0$: at larger length scales the pattern takes on a universal self-similar character. A disordered initial state of infinite extent affects the pattern generated at all length scales and for all times. Triangles of all sizes are nevertheless obtained, so that structure is generated on all scales, as suggested by Fig. 15. However, the pattern is not self-similar, but depends on the absolute scale defined by the spacing between sites.

Disordered configurations are defined to involve no statistical correlations between values at different sites. They thus correspond to a discrete form of white noise and yield a flat spatial frequency spectrum. One may also consider "pseudodisordered" configurations in which the value of each individual site is chosen randomly, but according to a distribution which yields statistical correlations between different sites, and a nontrivial spatial Fourier spectrum. For example, a Brownian configuration (with spatial frequency spectrum $1/k^2$) is obtained by assigning a value to each site in succession, with a certain probability for the value to differ from one site to the next (as in a random walk). The patterns generated by cellular automaton from such initial configurations may differ from those obtained with disordered (white noise) initial configurations. Complex nonadditive cellular automata evolving from a Brownian initial state yield patterns whose triangle density $T(n)$ decreases less rapidly at large n than for disordered initial configurations: the "long-range order" of the initial state leads to the generation of longer-range fluctuations. In the extreme limit of a homogeneous initial state (such as $\ldots 11111 \ldots$ or $\ldots 10101 \ldots$), cellular automaton evolution preserves the homogeneity, and no finite structures are generated.

The appearance of triangles over a series of time steps in the evolution of complex cellular automata from disordered initial states reflects the generation of long sequences of correlated sites in individual cellular automaton configurations. This effect is measured by the "sequence density" $Q_{(i)}(n)$, defined as the density of sequences of exactly n adjacent sites with the same value i (bordered by sites with a different value). Thus, for example, $Q_{(0)}(4)$ gives the density of 100001 sequences. $Q_{(0)}(n)$ clearly satisfies the sum rule

$$\sum_{n=1}^{\infty} n Q_{(0)}(n) = 1 - \rho.$$

In a disordered configuration with density $p = 1 - q$, $Q_{(0)}(n) \sim p^2 q^n$ for large n. Any sequence longer than two sites in a complex cellular automaton must yield a trian-

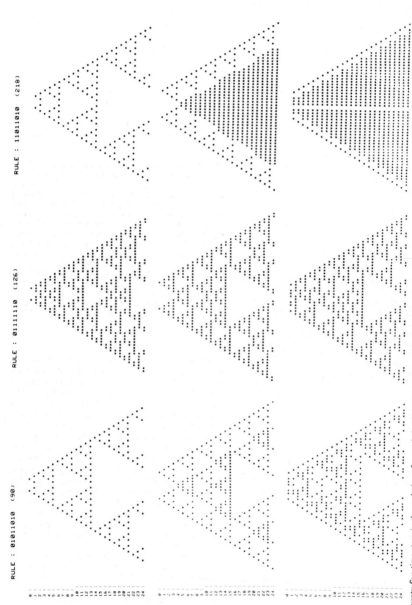

FIG. 14. Twenty-five time steps in the evolution of several simple initial configurations according to cellular automaton rules 90, 126, and 218. Configurations generated by rule 90 obey additive superposition (under addition modulo two). The first initial state taken is exceptional for rules 90 and 218, since it occurs in evolution from a single initial site, as shown in Fig. 3, so that the final pattern is a shifted form of that found in Fig. 3. For other initial states, the patterns obtained deviate substantially from those of Fig. 3. However, features with sizes much larger than the extent of the initial state remain unchanged. For complex cellular automaton rules such as 90 and 126, such features share the self-similarity found in Fig. 3.

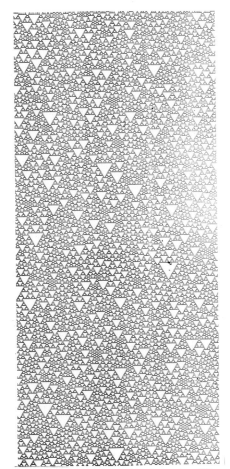

FIG. 15. Configurations obtained by evolution for 300 time steps from an initial disordered configuration with $\rho = 0.5$ according to cellular automaton rule 126. The fluctuations visible in the form of triangles and apparent at small scales in Fig. 8 are seen here to occur on all scales. The largest triangle in this sample has a base length of 27 sites.

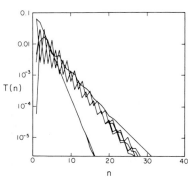

FIG. 16. Density $T(n)$ of triangle structures generated in the evolution of all the possible complex cellular automata from disordered initial configurations with density $\rho_0 = 0.5$. Triangles are evident in Figs. 8 and 16. They are formed when a sequence of sites suddenly attain the same value, but the length of the sequence is progressively reduced on subsequent time steps, until the apex of the triangle is reached. The appearance of triangles is a simple indication of self-organization. The triangle density $T(n)$ is defined only at integer values of n, but these points have been joined in the figure. For large n, the triangle densities for all complex cellular automata are seen to tend towards one of two limiting forms. The group tending to the upper curve are the nonadditive complex cellular automata 18, 22, 122, 126, 146, and 182. The additive rules 90 and 150 follow the lower curve. In both cases, $T(n)$ falls off exponentially with n, in contrast to the power law form found for the self-similar patterns of Figs. 3, 5, and 14.

gle, leading to the sum rule

$$Q(n) \simeq \sum_{i=n}^{\infty} [2T(i)/i] .$$

Thus the $Q(n)$ obtained at large times by evolution from a disordered initial state should follow the same exponential form (3.8) as $T(n)$.

Figure 17 shows the sequence density $Q_{(0)}(n)$ obtained at various time steps in the evolution of rule 126 from a

disordered initial state, as illustrated in Fig. 15. At each time step, the $Q_{(0)}(n)$ for a disordered configuration (illustrated in Fig. 10) with the same average density has been subtracted. The resulting difference vanishes by definition at $\tau = 0$, but Fig. 17 shows that for $\tau \geq 1$, the cellular automaton evolution yields a nonzero difference. After a few time steps, the cellular automaton tends to an equilibrium state containing an excess of long sequences of sites with value 0, and a deficit of short ones. This final equilibrium $Q_{(0)}(n)$ does not depend on the density of the initial disordered configuration. Starting from any disordered initial state (random noise), repeated application of the local cellular automaton rules thus generates ordered configurations whose statistical properties, as measured by sequence densities, differ from those of corresponding disordered configurations. The impression of self-organization in individual configurations given by Fig. 8 is thus quantitatively confirmed.

As suggested by the sum rule, the $Q_{(0)}(n)$ for complex cellular automata with disordered initial states follow the exponential behavior (3.7) found for the $T(n)$. Again, the parameter λ has a universal value $\sim \frac{4}{3}$ for all nonadditive cellular automaton rules and ~ 2 for additive ones. If all configurations of the cellular automata were disordered, then the sequence density would behave at large n as

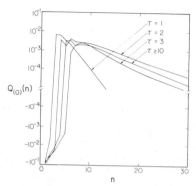

FIG. 17. Density $Q_{(0)}(n)$ of sequences of exactly n successive sites with value 0 (delimited by sites with value 1) in configurations generated by τ steps in time evolution according to cellular automaton rule 126, starting from an initial disordered state with density $\rho=0.5$. [The function $Q_{(0)}(n)$ is defined only for integer n: points are joined for ease of identification.] At each time step, the density of sequences in a disordered configuration with the same average total density has been subtracted. This difference vanishes for $\tau=0$ by definition. The nonzero value shown in the figure for $\tau \geq 1$ is a manifestation of self-organization in the cellular automaton, suggested qualitatively by comparison of Figs. 8 and 10. For large τ, an equilibrium state is reached, which exhibits an excess of long sequences and a deficit of short ones.

$(1-\rho)^n$ and depend on total average density ρ for the configurations. The form (3.5) yields sequence correlations with the same exponential behavior, but with a fixed λ, universal to all the nonadditive complex cellular automaton rules, and irrespective of the final densities to which they lead. (The universal form may be viewed as corresponding to an "effective density" $\simeq 0.25$.)

Cellular automata are usually defined to evolve according to definite deterministic local rules. In modelling physical or biological systems it is, however, sometimes convenient to consider cellular automata whose local rules involve probabilistic elements or noise (cf. Griffeath, 1970; Schulman and Seiden, 1978; Gach *et al.*, 1978). The simplest procedure is to prescribe that at each time step the value obtained by application of the deterministic rule at each site is to be reversed with a probability κ (and with each site treated independently). (If an energy is associated with the reversal of a site, κ gives the Boltzmann factor corresponding to a finite temperature heat bath.) Figure 18 shows the effects of introducing such noise in the evolution of cellular automaton rule 126. The structures generated are progressively destroyed as κ increases. Investigation of densities and correlation functions indicates that the transition to disorder is a continuous one, and no phenomenon analogous to a "phase transition" is found.

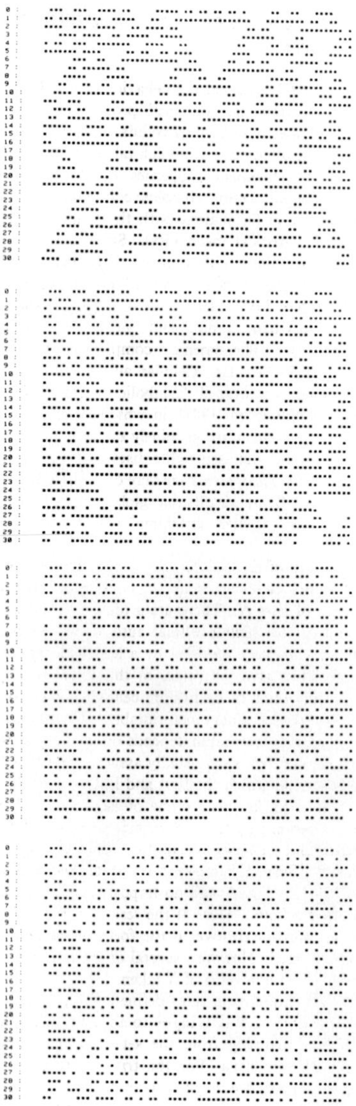

FIG. 18. Configurations generated from a disordered initial state (with $\rho_0=0.5$) by the evolution of the complex nonadditive cellular automaton 126, in the presence of noise which causes values obtained at each site to be reversed with probability κ at every time step. (a) is for $\kappa=0$ (no "noise"), (b) for $\kappa=0.1$, (c) for $\kappa=0.2$, and (d) for $\kappa=0.5$. As κ increases, the structure generated is progressively destroyed. No discontinuity in behavior as a function of κ is found.

IV. GLOBAL PROPERTIES OF ELEMENTARY CELLULAR AUTOMATA

Section III analyzed the behavior of cellular automata by considering the statistical properties of the set of values of sites in individual cellular automaton configurations. The alternative approach taken in this section considers the statistical properties of the set (ensemble) comprising all possible complete configurations of a cellular automaton (in analogy with the Γ-space approach to classical statistical mechanics). Such an approach provides connections with dynamical systems theory (Ott, 1981) and the formal theory of computation (Minsky, 1967; Arbib, 1969; Manna, 1974; Hopcroft and Ullman, 1979; Beckman, 1980), and yields a view of self-organization phenomena complementary to that developed in Sec. III. Cellular automaton rules may be considered as a form of "symbolic dynamics" (e.g., Alekseev and Yakobson, 1981), in which the degrees of freedom in the system are genuinely discrete, rather than being continuous but assigned to discrete "bins."

As in Sec. III, we examine here only elementary cellular automata. Some results on global properties of more complicated cellular automata will be mentioned in Sec. V.

For most of this section, it will be convenient to consider "finite" cellular automata, containing only a finite number of sites N. There are a total of 2^N possible configurations for such a cellular automaton. Each configuration is uniquely specified by a length N binary integer whose digits give the values of the corresponding sites.[9] (A configuration of an infinite cellular automaton would correspond to a binary real number.) The evolution of a finite cellular automaton depends on the boundary conditions applied. We shall usually assume periodic boundary conditions, in which the first and last sites are identified, as if the sites lay on a circle of circumference N. One could alternatively take an infinite sequence of sites, but assume that all those outside the region of length N have value 0. Results obtained with these two choices were compared in Fig. 9, and no important qualitative differences were found. Most of the results derived in this section are also insensitive to the form of boundary conditions assumed. However, several of the later ones depend sensitively on the value of N taken.

Cellular automaton rules define a transformation from one sequence of binary digits to another. The rules thus provide a mapping from the set of binary numbers of length N onto itself. For the trivial case of rule 0, all binary numbers are mapped to zero. Figure 19 shows the mappings corresponding to evolution for one and five

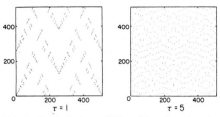

FIG. 19. Mapping in the set of 512 possible configurations of a length nine finite cellular automaton corresponding to evolution for τ time steps according to the modulo-two rule 90. Each possible configuration is represented by the decimal equivalent of the binary number whose digits give the values at each of its sites. The horizontal axis gives the number specifying the initial configuration; the vertical axes that for the final configuration. Each initial configuration is mapped to a unique final configuration.

time steps according to cellular automaton rule 90 with $N=9$. The mapping corresponding to one time step is seen to maintain some nearby sets of configurations. After five time steps, however, the evolution is seen to map configurations roughly uniformly, so that the final configurations obtained from nearby initial configurations are essentially uncorrelated.

A convenient measure of distance in the space of cellular automaton configurations is the "Hamming distance" $H(s_1, s_2)$ [familiar from the theory of error-correcting codes (Peterson and Weldon, 1972)], defined as the number of digits (bits) which differ between the binary sequences s_1 and s_2. [Thus in Boolean form, $H(s_1, s_2) = \#_1(s_1 \oplus s_2)$.] Particular configurations correspond to points in the space of all possible configurations. Under cellular automaton evolution, each initial configuration traces out a trajectory in time. If cellular automaton evolution is "stochastic," then the trajectories of nearby points (configurations) must diverge (exponentially) with time. Consider first the case of two initial configurations (say, S_1 and S_2) which differ by a change in the value at one site (and are thus separated by unit Hamming distance). After τ time steps of cellular automaton evolution, this initial difference may affect the values of at most 2τ sites (so that $H \le 2\tau$). However, for simple cellular automaton rules, the difference remains localized to a few sites, and the total Hamming distance tends rapidly to a small constant value. The behavior of complex cellular automaton rules differs radically between additive rules (such as 90 and 150) and nonadditive ones. For additive rules, the difference obtained after τ time steps is given simply by the evolution of the initial difference (in this case a single nonzero site) for τ time steps. The Hamming distance at time step τ is thus given by the number of nonzero sites in the configuration obtained by evolution from a single site, and for rule 90 has the form $H_\tau = 2^{\#_1(\tau)}$, as illustrated in Fig. 20(a). The average Hamming distance, smoothed over many time steps,

---

[9] An alternative specification would take each configuration to correspond to one of the 2^N vertices of an N-dimensional hypercube, labeled by coordinates corresponding to the values of the N sites. Points corresponding to configurations differing by values at a single site are then separated by a unit distance in N-dimensional space.

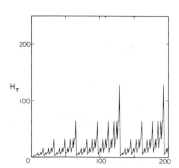

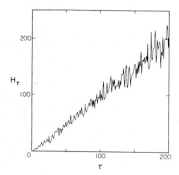

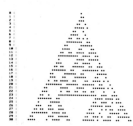

FIG. 20. Divergence in behavior of disordered configurations initially differing by a change in the value of a single site under cellular automaton evolution. The Hamming distance H between two configurations is defined as the number of bits (site values) which differ between the configurations. (a) shows the evolution of the Hamming distance between two configurations of the additive cellular automaton 90 (modulo-two rule); (b) shows the corresponding Hamming distance for the nonadditive cellular automaton 126; and (c) gives the actual difference (modulo two) between the configurations of cellular automaton 126 for the first few time steps. For nonadditive rules [case (b)], $H_\tau \sim \tau$, while for additive rules [case (a)], after time averaging, $H_\tau \sim \tau^{0.59}$.

behaves as $H_\tau = \tau^{\log_2 3 - 1} \simeq \tau^{0.59}$. For nonadditive rules, the difference between configurations obtained through cellular automaton evolution no longer depends only on the difference between the initial configurations. Figure 20(c) shows the difference between configurations obtained by evolution according to the nonadditive cellular automaton rule 126. The lack of symmetry in the pattern is a reflection of the dependence on the values of multiple initial sites. Figure 20(b) shows the Hamming distance corresponding to this difference. Apart from small fluctuations, it is seen to increase linearly with τ, tending at large τ to the form $H_\tau \simeq \tau$. This Hamming distance is the same as would be obtained by comparing sequences of 2τ sites in two disordered configurations with density 0.5. Thus a change in the value of a small number of initial sites is amplified by the evolution of a nonadditive cellular automaton, and leads to configurations with a linearly increasing number of essentially uncorrelated sites. (Changes in single sites may sometimes be eradicated after a single time step; this exceptional behavior occurs for cellular automaton rule 18, but is always absent if more than one adjacent site is reversed.) A bundle of initial trajectories therefore diverges with time into an exponentially increasing volume.

One may specify a statistical ensemble of states for a finite cellular automaton by giving the probability for each of the 2^N possible configurations. In a collection of many disordered states with density $\rho = \frac{1}{2}$, each possible cellular automaton configuration is asymptotically populated with equal probability. Such a collection of states will be termed an "equiprobable ensemble," and may be considered "completely disorganized." Cellular automaton evolution modifies the probabilities for states in an ensemble, thereby generating "organization." Figure 21 shows the probabilities for the 1024 possible configurations of a finite cellular automaton with $N = 10$ obtained after evolution for ten time steps according to rule 126 from an initial equiprobable ensemble. Figure 22 shows the evolution of these probabilities over ten time steps for several complex cellular automata. At each time step, dots are placed in positions corresponding to configurations occurring with nonzero probabilities. At $\tau = 0$, all configurations are taken to be equally probable. Cellular automaton evolution modifies the probabilities for different configurations, reducing the probabilities for some to zero, and leading to "gaps" in Fig. 22. In the initial ensemble, all configurations were assigned equal *a priori* probabilities. After evolution (or "processing") for a few time steps, an equilibrium ensemble is attained in which different configurations carry different probabilities, according to a definite distribution. Properties of the more probable configurations dominate statistical averages over the ensemble, giving rise to the distinctive average local features of equilibrium configurations described in Sec. III.

In the limit $N \rightarrow \infty$, a cellular automaton configuration may be specified by real number in the interval 0 to 1 whose binary decomposition consists of a sequence of digits corresponding to the values of the cellular automa-

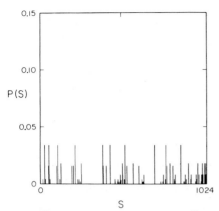

FIG. 21. Probabilities for each of the 1024 possible configurations in a finite (circular) cellular automaton with length $N = 10$ obtained by evolution according to rule 126 for ten time steps from an initial ensemble containing each possible configuration with equal probability. On the horizontal axis, each configuration S is labeled by a ten-digit binary integer (marked in decimal form) whose digits give the values of the corresponding sites. The null configuration (with value zero at all sites) is labeled by the integer 0, and occurs with the largest probability $\simeq 0.13$. The inequality of the probabilities for initially equiprobable configurations is a reflection for self-organization.

ton sites. Then the equilibrium ensemble of cellular automaton configurations analogous to those of Fig. 22 corresponds to a set of points on the real line. The unequal probabilities for appearance of 0 and 1 digits, together with higher-order correlations, implies that the points form a Cantor set (Farmer, 1982a, 1982b). The fractal dimensionality of the Cantor set is given by the negative of the entropy discussed below, associated with the ensemble of cellular automaton configurations (and hence real-number binary digit sequences) (Farmer, 1982a, 1982b). For rule 126 the fractal dimension of the Cantor set is then 0.5.

An important feature of the elementary cellular automata considered here and in Sec. III is their "local irreversibility." Cellular automaton rules may transform several different initial configurations into the same final configuration. A particular configuration thus has unique descendants, but does not necessarily have unique ancestors (predecessors). Hence the trajectories traced out by the time evolution of several cellular automaton configurations may coalesce, but may never split. A trivial example is provided by cellular automaton rule 0, under which all possible initial configurations evolve after one time step to the unique null configuration. In a reversible system, each state has a unique descendant and a unique ancestor, so that trajectories representing time evolution of different states may never intersect or meet. Thus in a reversible system, the total number of possible configura-

tions must remain constant with time (Liouville's theorem). However, in an irreversible system, the number of possible configurations may decrease with time. This effect is responsible for the "thinning" phenomenon visible in Fig. 22. The trajectories corresponding to the evolution of cellular automaton configurations are found to become concentrated in limited regions, and do not asymptotically fill the available volume densely and uniformly. This behavior makes self-organization possible, by allowing some configurations to occur with larger probabilities than others even in the large-time equilibrium limit.

One consequence of local irreversibility evident from Fig. 22 is that some cellular automaton configurations may appear as initial conditions but may never be reached as descendants of other configurations through cellular automaton time evolution.[10] Such configurations carry zero weight in the ensemble obtained by cellular automaton evolution. In the trivial case of cellular automaton rule 0, only the null state with all sites zero may be reached by time evolution; all other configurations are unreachable. Rule 4 generates only those configurations in which no two adjacent sites have the same value. The fraction of the 2^N possible configurations which satisfy this criterion tends to zero as N tends to infinity, so that in this limit, a vanishingly small fraction of the configurations are reached. Cellular automaton rule 204 is an identity transformation, and is unique among cellular automaton rules in allowing all configurations to be reached. (The rule is trivially reversible.) Assuming periodic boundary conditions, one finds that with N odd, the complex additive rule 90 generates only configurations in which an even number of sites have value one, and thus allows exactly half of the 2^N possible configurations to be reached. For even N, $\frac{1}{4}$ of the possible configurations may be reached. A finite fraction of all the configurations are thus reached in the limit $N \to \infty$. For the complex nonadditive rule 126, inspection of Fig. 8 shows that only configurations in which nonzero sites appear in pairs may be reached. Figure 23 shows the fraction of unreachable configuration for this cellular automaton rule as a function of N. The fraction tends steadily to one as $N \to \infty$. A complete characterization of the unreachable configurations for this case is given in Martin *et al.* (1983); these configurations are enumerated there, and their fraction is shown to behave as $1 - \lambda^N$ for large N, where $\lambda \simeq 0.88$ is determined as the root of a cubic equation. Similar behavior is found for other nonadditive rules.

Irreversible behavior in cellular automata may be analyzed by considering the behavior of their "entropy" S or "information content" $-S$. Entropy is defined as usual as the logarithm (here taken to base two) of the average

---

[10] The existence of unreachable or "garden-of-Eden" configurations in cellular automata is discussed in Moore (1962) and Aggarwal (1973), where criteria (equivalent to irreversibility) for their occurrence are given.

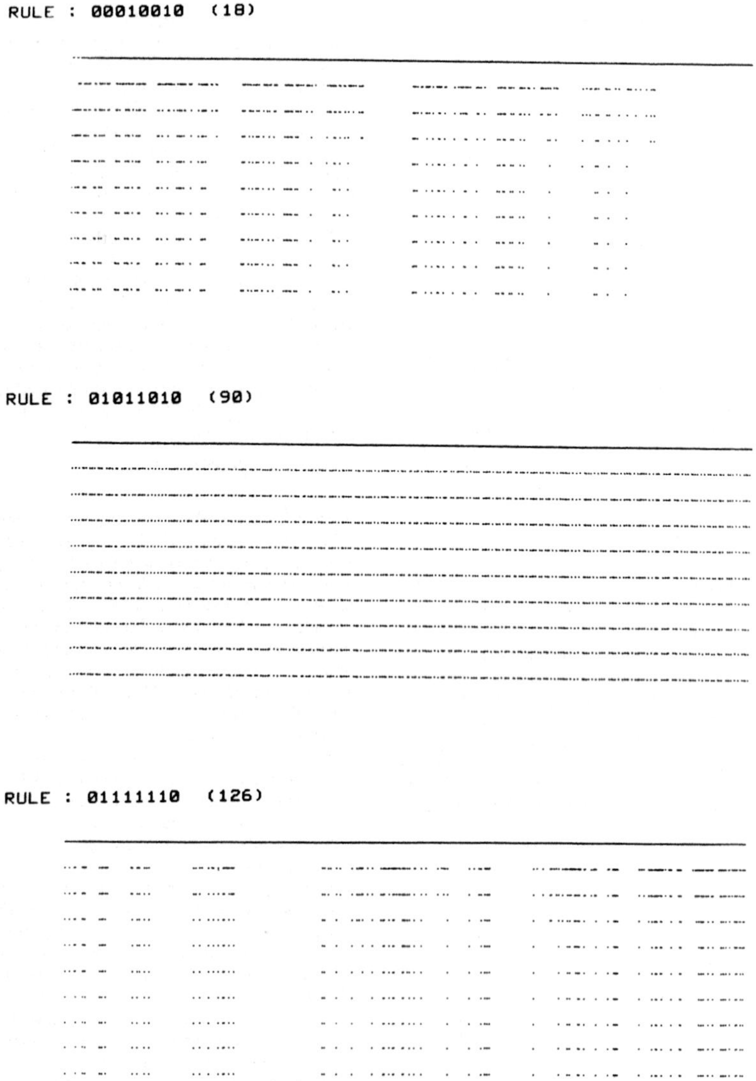

624   Wolfram: Statistical mechanics of cellular automata

RULE : 00010010   (18)

RULE : 01011010   (90)

RULE : 01111110   (126)

FIG. 22. Time evolution of the probabilities for each of the 1024 possible configurations of several length 10 cellular automata starting from an initial ensemble containing all 1024 configurations with equal probabilities. The configurations are specified by binary
integers whose digits form the sequence of values at the sites of the cellular automaton. The history of a particular configuration is
given on successive lines in a vertical column: a dot appears at a particular time step if the configuration occurs with nonzero probability at that time step. In the initial ensemble, all configurations occur with equal nonzero probabilities, and dots appear in all positions. Cellular automaton evolution modifies the probabilities for the configurations, making some occur with zero probability, yielding gaps in which no dots appear. The probabilities obtained by evolution for ten time steps according to cellular automaton rule 126
were given in Fig. 21: dots appear in the tenth line of the rule 126 part of this figure at the positions corresponding to configurations
with nonzero probabilities.

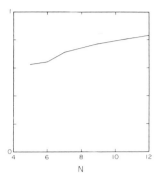

FIG. 23. Fraction of the 2^N possible configurations of a length N cellular automaton (with periodic boundary conditions) not reached by evolution from an arbitrary initial configuration according to cellular automaton rule 126. The existence of unreachable configurations is a consequence of the irreversibility of cellular automaton evolution. The fraction of such configurations is seen to increase steadily towards one as N increases.

number of possible states of a system, or

$$S = \sum_i p_i \log_2 p_i \qquad (4.1)$$

where p_i is the probability for state i. The entropy may equivalently be considered as the average number of binary bits necessary to specify one state in an ensemble of possible states. The total entropy of a system is the sum of the entropies of statistically independent subsystems. Entropy is typically maximized when a system is completely disorganized, and the maximum number of subsystems act independently. The entropy of a cellular automaton takes on its maximal value of one bit per site

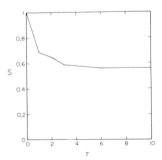

FIG. 24. Time evolution of average entropy per site for an ensemble of finite cellular automata with $N = 10$ evolving according to rule 126 from an initial equiprobable ensemble. The entropy gives the logarithm of the average number of possible configurations. Its decrease with time is a reflection of the local irreversibility of the cellular automaton.

for an equiprobable ensemble. For reversible systems, time evolution almost always leads to an increase in entropy. However, for irreversible systems, such as cellular automata, the entropy may decrease with time. Figure 24 shows the time dependence of the entropy for a finite cellular automaton with $N = 10$, evolving according to rule 126, starting from an initial equiprobable ensemble. The entropy is seen to decrease with time, eventually reaching a constant equilibrium value. The decrease is a direct signal of irreversibility.

The entropy for a finite cellular automaton given in Fig. 24 is obtained directly from Eq. (4.1) by evaluating the probabilities for each of the finite set of 2^N possible configurations. For infinite cellular automata, enumeration of all configurations is no longer possible. However, so long as values of sufficiently separated sites are statistically independent, the average entropy per site may nevertheless be estimated by a limiting procedure. Define a "block entropy" [or "Renyi entropy" (Renyi, 1970; Farmer, 1982a, 1982b)]

$$S_b = (1/b) \sum_i p_i^{(b)} \log p_i^{(b)} \ ,$$

where $p_i^{(b)}$ denotes the probability for a sequence i of b values in an infinite cellular automaton configuration. The limit $S_{b \to \infty}$ gives the average total entropy per site. This limit is approached rapidly for almost all cellular automaton configurations, reflecting the exponential decrease of correlations with distance discussed in Sec. III. [Similar results are obtained in estimating the entropy of printed English from single letter, digram, trigram and so on frequencies (Shannon, 1951). Typical results (for example, for the text of this paper) are $S_1 \simeq 4.70$, $S_2 \simeq 4.15$, $S_3 \simeq 3.57$, and $S_\infty \sim 2.3$.]

Irreversibility is not a necessary feature of cellular automata. In the case of the elementary cellular automata considered here, the irreversibility results from the assumption that a configuration S_n at a particular time step n depends only on its immediate predecessor so that evolution may be represented schematically by $S_n = F[S_{n-1}]$. Except in the trivial case of the identity transformation (rule 204), F is not invertible. The cellular automata are discrete analogs of systems governed by partial differential equations of first order in time (such as the diffusion equation), and exhibit the same local irreversibility. One may construct reversible one-dimensional cellular automata (Fredkin, 1982; Margolus, 1982)[11] by allowing a particular configuration to depend on the previous two configurations, in analogy with reversible second-order differential equations such as the wave equation. The evolution of these cellular automata may be represented schematically by $S_n = F[S_{n-1}] \oplus S_{n-2}$. The

---

[11] Reversible cellular automata may be constructed in two (or more) dimensions by allowing arbitrary evolution along a line, but generating a sequence of copies ("history") in the orthogonal direction of the configurations on the line at each time step (Toffoli, 1977a, 1980).

invertibility of modulo-two addition allows S_{n-2} to be obtained uniquely from S_n and S_{n-1}, so that all pairs of successive configurations have unique descendants and unique ancestors. For infinite reversible cellular automata, the entropy (4.1) (evaluated for the appropriate successive pairs of configurations) almost always increases with time. Finite reversible cellular automata may exhibit globally irreversible behavior when dissipative boundary conditions are imposed. Such boundary conditions are obtained if sites beyond the boundary take on random values at each time step. If all sites beyond the boundary have a fixed or predictable value as a function of time, the system remains effectively reversible. With simple initial configurations, reversible cellular automata generate self-similar patterns analogous to those found for irreversible ones.[12] A striking difference is that reversible rules yield diamond-shaped structures symmetrical in time, rather than the asymmetrical triangle structures found with irreversible rules.

Since a finite cellular automaton has a total of only 2^N possible configurations, the sequence of configurations reached by evolution from any initial configuration must become periodic after at most 2^N time steps (the "Poincare recurrence time"). After an initial transient, the cellular automaton must enter a cycle in which a set of configurations is generated repeatedly, as illustrated in Fig.

25. Figure 8 suggests that simple cellular automata yield short cycles containing only a few configurations, while complex cellular automata may yield much longer cycles. Simple rules such as 0 or 72 evolve after a fixed small number of time steps from any configuration to the stationary null configuration, corresponding to a trivial length-one cycle. Other simple cellular automaton rules, such as 36, 76, or 104 evolve after $\lesssim N$ time steps to nontrivial stationary configurations (with cycle length one). Rules such as 94 or 108 yield (after a transient of $\lesssim N$ steps) a state consisting of a set of small independent regions, each of which independently follows a short cycle (usually of length one or two and at most of length 2^b, where b is the number of sites in the region). In general, simple cellular automata evolve to cycles whose length remains constant as N increases. On the other hand, complex cellular automata may yield cycles whose length increases without bound as N increases. Figure 26 shows the distribution in the number of time steps before evolution from each possible initial configuration according to the complex rule 126 leads to repetition of a configuration. Only a small fraction of the 2^N possible configurations is seen to be reached in evolution from a particular initial configuration. For example, in the case $N=8$, a maximum of eight distinct configurations (out of 256) are generated by evolution from any specific initial state.

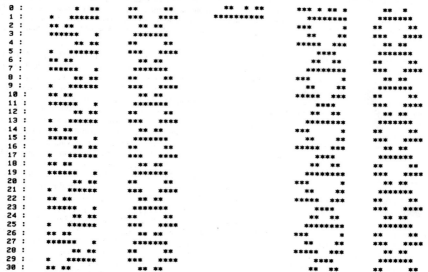

FIG. 25. Evolution of typical initial configurations in a finite cellular automaton with $N = 8$ (and periodic boundary conditions) according to rule 126. Evolution from a particular initial state could generate up to $2^8 = 256$ distinct configurations before entering a cycle and returning to a configuration already visited. Much shorter cycles, however, are seen to occur in practice.

[12] For example, evolution from a pair of successive configurations containing zero and one nonzero sites according to the reversible analog of rule 150 yields a self-similar pattern with fractal dimension $\log_2[4/(\sqrt{17}-3)] \simeq 1.84$.

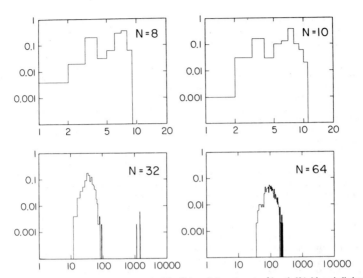

FIG. 26. Distribution in the number of time steps required for finite cellular automata of length N (with periodic boundary conditions) evolving according to rule 126 to reach a particular configuration for the second time, signaling the presence of a cycle. The cycle times found are much smaller than the value 2^N obtained if evolution from a particular initial configuration eventually visited all 2^N possible configurations. The results for $N=8$ and $N=10$ include all 256 and 1024 possible initial configurations; those for $N=32$ and $N=64$ are obtained by uniform Monte Carlo sampling from the space of possible initial configurations. In all cases, the number of configurations visited in transients before entering a cycle is very much smaller than the number of configurations in the cycle.

After a transient of at most two time steps, the cellular automaton enters a cycle, which repeats after at most six further time steps. Apart from the trivial one-cycle corresponding to the null configuration, six distinct cycles (containing nonintersecting sets of configurations) occur. Four have length six, and two have length two. A total of 29 distinct "final" configurations appear in these cycles. The number of configurations reached by evolution from a particular initial state increases with N as shown in Fig. 26. For $N=10$, the maximum is 38 states, while for $N=32$, it is at least 1547. Similar behavior is found for most other complex nonadditive rules.

Analytical results for transient and cycle lengths may be given for finite cellular automata (with periodic boundary conditions) evolving according to the additive rules 90 and 150 (Martin *et al.*, 1983). A complete and general derivation may be obtained using algebraic methods and is given in Martin *et al.* (1983). The additive superposition principle implies that the evolution of any initial configuration is a superposition of evolution from single nonzero sites (in each of the N cyclically equivalent possible positions). The period of any cycle must therefore divide the period Π_N obtained by evolution from a single nonzero site. Similarly, the length of any transient must divide the length Υ_N obtained with a single nonzero initial site. It is found that Π_N is identical for rules 90 and 150, but Υ_N in general differs. The first few values of Π_N for rules 90 and 150 (for $N=3$ through $N=30$) are 1,

1, 3, 2, 7, 1, 7, 6, 31, 4, 63, 14, 15, 1, 15, 14, 511, 12, 63, 62, 2047, 8, 1023, 126, 511, 28, 16383, and 30. Consider rule 90; derivations for rule 150 are similar. Whenever N is of the form 2^α, the cellular automaton ultimately evolves from any initial configuration to the null configuration, so that $\Pi_N=1$ in this case. When N is odd, it is found that the first configuration in the cycle always consists of two nonzero sites, separated by a single zero site. The nonzero sites may be taken at positions ± 1 modulo N. Equation (3.2) implies that configurations obtained by evolution for 2^j time steps again contain exactly two nonzero sites, at positions $\pm 2^j$ modulo N. A cycle occurs when $2^j \equiv \pm 1 \bmod N$. Π_N then divides $\overset{\bullet}{\Pi}_N$ given by $2^{sord_N(2)}-1$ where $sord_N(k)$ is defined as the minimum j for which $2^j=\pm 1 \bmod N$, and $sord_N(k)=ord_N(k)/2$ or $sord_N(k)=ord_N(k)$. The multiplicative order function $ord_N(k)$ (e.g., MacWilliams and Sloane, 1977) is defined as the minimum j for which $2^j=1 \bmod N$. It is found in fact that $\Pi_N=\overset{\bullet}{\Pi}_N$ for most N; the first exception occurs for $N=37$, in which case $\Pi_{37}=\overset{\bullet}{\Pi}_{37}/3$. For $N=k^\alpha-1$, $ord_N(k)=\alpha$, so that when $N=2^\alpha-1$, $\overset{\bullet}{\Pi}_N=N$. Similarly, when $N=k^\alpha+1$, $k^\alpha \equiv -1 \bmod N$ so that $k^{2\alpha} \equiv +1 \bmod N$ and $ord_N(k)=2\alpha$, yielding $\overset{\bullet}{\Pi}_N=N-2$ for $N=2^\alpha+1$. In general, $N=p_1^{\alpha_1}p_2^{\alpha_2}\cdots$, where the p_i are primes not equal to k, $ord_N(k)=\mathrm{lcm}[ord_{p_1^{\alpha_1}}(k),\,ord_{p_2^{\alpha_2}}(k),\ldots]$. $ord_N(k)$ divides the Euler totient function $\varphi(N)$, defined as the number of integers less than N which are relatively prime to N (e.g., Apostol, 1976; Hardy and Wright, 1979, Sec. 5.5). [$\varphi(N)$ is even for all $N > 1$.] $\varphi(N)$ satisfies the Euler-Fermat relation $k^{\varphi(N)} \equiv 1 \bmod N$. It is clear that $\pi(n) \leq \varphi(n) \leq n-1$, where $\pi(n)$ denotes the number of primes less than n, and the upper bound is saturated when n is prime. If $ord_N(k)$ is even, then $ord_N(k) \leq \varphi(N)$, while for $ord_N(k)$ odd, $ord_N(k) \leq \varphi(N)/2$. Thus $\Pi_N \leq 2^{(N-1)/2} - 1$, where the bound is saturated for some prime N. Such a Π_N is the maximum possible cycle length for configurations with reflection symmetry, but is approximately the square root of the maximum possible length $2^N - 1$ for an arbitrary system with N binary sites.[13] When N is even, $\Pi_N = 2\Pi_{N/2}$. Notice that Π_N is an irregular function of N: its value depends not only on the magnitude of N, but also on its number theoretical properties.

When Π_N is prime, all possible cycles must have a period of one or exactly Π_N. When Π_N is composite, any of its divisors may occur as a cycle period. Thus, for example, with $N = 10$, $\Pi_N = 6$, and in evolution from the $2^{10} - 1$ possible non-null initial configurations, forty distinct cycles of length 6 appear, and five of length 3. In general it appears that for large N, an overwhelming fraction of cycles have the maximal length Π_N.

As mentioned above, for the additive rules 90 and 150, the length of the transients before a cycle is entered in evolution from an arbitrary initial configuration must divide Υ_N, the length of transient with a single nonzero initial site. For rule 90, $\Upsilon_N = 1$ for N odd, and $\Upsilon_N = D_2(N)/2$ otherwise, where $D_2(n)$ is the largest 2^j which divides n. For rule 150, $\Upsilon_N = 0$ if N is not a multiple of three, $\Upsilon_N = 1$ if N is odd, and $\Upsilon_N = D_2(N)$ otherwise. Since, as discussed above, evolution from all 2^N possible initial configurations according to rule 90 visits 2^{N-1} configurations for odd N, the result $\Upsilon_N = 1$ implies that in this case, exactly half of the 2^N possible configurations appear on cycles.

Configurations in cellular automata may be divided into essentially three classes according to the circumstances under which they may be generated. One class discussed above consists of configurations which can appear only as initial states, but can never be generated in the course of cellular automaton evolution. A second class contains configurations which cannot arise except within the first, say τ, time steps. For $\tau = 2$, such configurations have "parents" but no "grandparents." The third class of configurations is those which appear in cycles, and may be visited repeatedly. Such configurations may be generated at any time step (for example, by choosing an initial configuration at the appropriate point in the cycle, and then allowing the necessary number of cycle steps to occur. The second class of configurations appears as transients leading to cycles. The cycles may be considered as attractors eventually attained in evolution from any initial

configuration. The 2^N possible configurations of a finite cellular automaton may be represented as nodes in a graph, joined by arcs representing transitions corresponding to cellular automaton evolution. Cycles in the graph correspond to cycles in cellular automaton evolution. As shown in Martin *et al.* (1983), the transient configurations for the additive rules 90 and 150 appear on balanced quaternary trees, rooted on the cycles. The leaves of the trees correspond to unreachable configurations. The height of the trees is given by Υ_N. The balanced structure of the trees implies that the number of configurations which may appear after τ time steps decreases as $4^{-\tau}$; $4^{-\Upsilon_N}$ configurations appear on cycles and may therefore be generated at arbitrarily large times.

The algebraic techniques of Martin *et al.* (1983) apply only to additive rules. For nonadditive cellular automaton rules, the periods of arbitrary cycles do not necessarily divide the periods Π_N of cycles generated by evolution from configurations with one nonzero site. Empirical investigations nevertheless reveal many regularities.

Cyclic behavior is inevitable for finite cellular automata which allow only a finite number of possible states. Infinite cellular automata exhibit finite cycles only under exceptional circumstances. For a wide class of initial states, simple cellular automaton rules can yield nontrivial cyclic behavior. Cycles occur in complex cellular automata only with exceptional initial conditions. Any initial configuration with a finite number of nonzero sites either evolves ultimately to the null state, or yields a pattern whose size increases progressively with time. Most infinite initial configurations do not lead to cyclic behavior. However, if the values of the initial sites form an infinite periodic sequence (c.f. Miller, 1970, 1980), with period k, then the evolution of the infinite cellular automaton will be identical to that of a finite cellular automaton with $k = N$, and cycles with length $\ll 2^k$ will be found.

The transformation of a finite cellular automaton configuration according to cellular automaton rules defines a mapping in the set of 2^N binary integers representing the cellular automaton configurations. An example of such a mapping was given in Fig. 19. Repeated applications of the mapping yield successive time steps in the evolution of the cellular automaton. One may compare the results with those obtained for a system which evolves by iteration of a random mapping among the 2^N integers (cf. Kauffman, 1969). Random mappings of K elements are obtained by choosing one of the K possible images independently for each integer and with equal probabilities. The mapping is permitted to take an element to itself. In this way, all K^K possible mappings are generated with equal probability. The probability of a particular element's having no preimage (predecessor) under a random mapping between K elements is $(K-1)^K / K^K = (1 - 1/K)^K$. In the limit $K \to \infty$ this implies that a fraction $1/e \simeq 0.37$ of the possible states are not reached in evolution by iteration of a random mapping. For complex nonadditive cellular automata, it appears that as $N \to \infty$, almost all configurations become unreachable, indicating that cellular automaton evolution

---

[13] The result is therefore to be contrasted with the behavior of linear feedback shift registers, analogous to cellular automata except for end effects, in which cycles (de Bruijn sequences) of period $2^N - 1$ may occur (e.g., Golomb, 1967; Berlekamp, 1968).

is "more irreversible" than iteration of a random mapping would imply. A system evolving according to a random mapping exhibits cycles analogous to those found in actual cellular automata. The probability of a length r cycle's occurring by iteration of a mapping between K elements is found to be

$$\sum_{i=r}^{K} \frac{(K-1)!}{(K-i)!K^i}$$

(Harris, 1960; Knuth, 1981, Sec. 3.1, Ex. 6, 11-16; Levy, 1982). Cycles of the maximum length K occur with finite probability. In the large K limit, the average cycle length becomes $\simeq \sqrt{\pi K/8} \simeq 0.63\sqrt{K}$, while the standard deviation of the cycle length distribution is $\simeq \sqrt{(2/3 - \pi/8)K} \simeq 0.52\sqrt{K}$. The length of transients follows exactly the same distribution. The number of distinct cycles $\sim \sqrt{\pi/2} \log K$. If we take $K = 256$ for comparison with an $N = 8$ cellular automaton, this implies an average cycle length $\simeq 10$, an average transient length $\simeq 10$, $\simeq 94$ unreachable configurations, and $\simeq 7$ distinct cycles. Cellular automaton rule 126 yields in this case an average cycle length $\simeq 3.2$, an average transient length $\simeq 2.5$, 190 unreachable configurations, and 7 distinct cycles. Any agreement with results for random mappings appears to be largely fortuitous: even for large N cellular automata do not behave like random mappings.

This section has thus far considered cellular automata which evolve according to definite deterministic local rules. However, as discussed in Sec. III, one may introduce probabilistic elements or noise into cellular automata rules—for example, by reversing the value of a site at each time step with probability κ. Section III showed that the local properties of cellular automata change continuously as κ is increased from zero. Global properties may,

however, change discontinuously when a nonzero κ is introduced. An example of such behavior is shown in Fig. 27, which gives the fraction of configurations visited as a function of time for a cellular automaton evolving according to rule 126 with various values of κ, starting from a single typical initial configuration. When $\kappa = 0$, only six distinct configurations are generated before the cellular automaton enters a cycle. When $\kappa \neq 0$, the cellular automaton ultimately visits every possible configuration (cf. Gach *et al.*, 1978). For $\kappa \simeq 0.5$, one may approximate each configuration as being chosen from the 2^N possible configurations with equal probabilities: in this case, the average number of configurations visited after τ time steps is found to be $1 - ([1 - 2^{-N}]^{2^N})^{\tau/2^N} \simeq 1 - e^{-\tau/2^N}$.

Cellular automata may be viewed as simple idealizations of physical systems. They may also be interpreted as "computers" (von Neumann, 1966; Baer and Martinez, 1974; Burks, 1970; Aladyev, 1974, 1976; Toffoli, 1977b) and analyzed using methods from the formal theory of computation (Minsky, 1967; Arbib, 1969; Manna, 1974; Hopcroft and Ullman, 1979; Beckman, 1980). With this interpretation, the initial configuration of a cellular automaton represents a "program" and "initial data," processed by cellular automaton time evolution to give a configuration corresponding to the "output" or "result" of the "computation." The cellular automaton rules represent the basic mechanism of the computer; different programs may be "run" (or different "functions evaluated") by giving different initial or "input" configurations. This process is analogous to the "evolution" of the sequence of symbols on the tape of a Turing machine (Turing, 1936). However, instead of considering a single "head" which modifies one square of the tape at each time step, the cellular automaton evolution simultaneously affects all sites at each time step. As discussed in Sec. V, there exist "universal" cellular automata analogous to universal Turing machines, for which changes in the initial configuration alone allow any computable (or "recursive") function to be evaluated. A universal Turing machine may simulate any other Turing machine using an "interpreter program" which describes the machine to be simulated. Each "instruction" of the simulated machine is simulated by running the appropriate part of the interpreter program on the universal machine. Universal cellular automata may similarly simulate any other cellular automata. The interpreter consists of an encoding of the configurations for the cellular automaton to be simulated on the universal automaton. A crucial point is that so long as the encoding defined by the interpreter is sufficiently simple, the statistical characteristics of the evolution of configurations in the universal cellular automaton will be shared by the cellular automaton being simulated. This fact potentially forms the basis for universality in the statistical properties of complicated cellular automata.

The simplest encodings which allow one cellular automaton to represent or simulate others are pure substitution or "linear" ones, under which the value of a single site is represented by a definite sequence of site values.

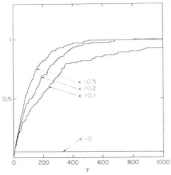

FIG. 27. Fraction of configurations visited after τ time steps in a finite cellular automaton (with $N = 7$) evolving from a single typical initial state according to rule 126 in the presence of noise which randomly reverses the values of sites at each time step with probability κ. When $\kappa = 0$, the cellular automaton enters a cycle after visiting only six distinct configurations. When $\kappa \neq 0$, the cellular automaton eventually visits all 128 possible configurations.

(Such encodings are analogous to the correspondences between complex cellular automaton rules mentioned in Sec. III.) For example, a cellular automaton A evolving according to rule 22 may be used to simulate another cellular automaton B evolving according to rule 146. For every 0 in the initial configuration of B, a sequence 00 is taken in the initial configuration of A, and for every 1 in B, 01 is taken in A. Then after 2τ time steps, the configuration of A under this encoding is identical to that obtained by evolution of B for τ time steps. If cellular automaton B instead evolved according to rule 182, 01 (or 10) in A would correspond to 0 in B, and 00 to 1. The simplicity of the interpreter necessary to represent rules 146 and 182 under rule 22 is presumably responsible for the similarities in their statistical behavior found in Sec. III. Figure 28 gives a network which describes the simulation capabilities of the complex elementary cellular automaton rules using length two linear encodings and with the simulated rule running at half the speed of the simulator. Many of these complex rules may also simulate simple rules under such an encoding. Simulations possible with longer linear encodings appear to be described by indirection through the network. Not all complex cellular automaton rules are thus related by linear encodings of any length.

As discussed in Sec. V, the elementary cellular automata considered here and in Secs. II and III are not of sufficient complexity to be capable of universal computation. However, some of the more complicated cellular automata described in Sec. V are "universal," and may therefore in principle represent any other cellular automata. The necessary encoding must be of finite length, but may be very long. The shorter or simpler the encoding, the closer will be the statistical properties of the simulating and simulated cellular automata.

V. EXTENSIONS

The results of Secs. II—IV have for the most part been restricted to elementary cellular automata consisting of a sequence of sites in one dimension with each site taking on two possible values, and evolving at each time step according to the values of its two nearest neighbors. This

FIG. 28. Network describing simulation capabilities of complex elementary cellular automata with length two pure substitution or linear encodings. Cellular automata evolving according to the destination rule are simulated by giving an encoded initial configuration in a cellular automaton evolving according to the source rule. Representability of one cellular automaton by another under a simple encoding implies similar statistical properties for the two cellular automata, and forms potentially the basis for universality in statistical properties of cellular automata.

section gives a brief discussion of the behavior of more complicated cellular automata. Fuller development will be given in future publications.

We consider first cellular automata in which the number of possible values k at each site is increased from two, but whose sites are still taken to lie on a line in one dimension. The evolution of each site at each time step is for now assumed to depend on its own value and on the values of its two nearest neighbors. In this case, the total number of possible sets of local rules is $k^{(k^3)}$. Imposition of the reflection symmetry and quiescence "legality conditions" discussed in Sec. II introduces $\frac{1}{2}k^2(k-1)+1$ constraints, yielding $k^{[k^2(1+k)-1]/2}$ "legal" sets of rules. For $k=2$, this implies $2^5=32$ legal rules, as considered in Sec. II. The number of possible legal rules increases rapidly with k. For $k=3$, there are $3^{17}=129\,140\,163\simeq1.3\times10^8$ rules, for $k=4$, $\simeq3\times10^{24}$, and for $k=10$, 10^{549}.

As a very simple example of cellular automata with $k>2$, consider the family of "modulo-k" rules in which at each time step, the value of a site is taken to be the sum modulo k of the values of its two neighbors on the previous time step. This is a generalization of the modulo-two rule (90) discussed on several occasions in Secs. II—IV. Figure 29 shows the evolution of initial states containing a single site with value one according to several modulo-k rules. In all cases, the pattern of nonzero sites is seen to tend to a self-similar fractal figure in the large time limit. The pattern in general depends on the value of the nonzero initial site, but in all cases yields an asymptotically self-similar figure. When k is prime, independent of the value of the initial nonzero site, a very regular pattern is generated, in which the density $T(n)$ of "triangle structures" is found to satisfy a one-term recurrence relation yielding a fractal dimension

$$D_k = \log_k \sum_{i=1}^{k} i = 1 + \log_k \left[\frac{k+1}{2} \right],$$

so that $D_3=1+\log_32\simeq1.631$, $D_5\simeq1.683$, and so on. When k is a composite number, the pattern generated depends on the value s of the initial nonzero site. If the greatest common divisor (s,k) of k and s is greater than one (so that s and k share nontrivial prime factors), then the pattern is identical to that obtained by evolution from an initial site with value one according to a modulo-$k/(s,k)$ rule. In general, the density of triangles satisfies a multiple-term recurrence relation. In all cases, the fractal dimension for large k behaves as $D\sim2-1/\log_2k$ [assuming $(s,k)\ll k$]. When $k\to\infty$, the values of sites become ordinary integers, all with nonzero values by virtue of the nonvanishing values of binomial coefficients, yielding a figure of dimension two.

All modulo-k rules obey the additive superposition principle discussed for the modulo two in Secs. II and III. The number of sites with value r after evolution for τ steps from an initial state containing a single site with value one is found [on analogy to Eq. (3.2)] to be $N_\tau^{(r)}=2^{\#_r^{[k]}(\tau)}$, where the function $\#_r^{[k]}(\tau)$ gives the number of occurrences of the digit r in the base-k decomposi-

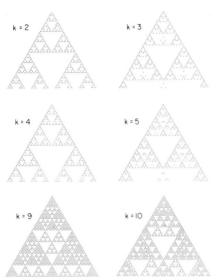

FIG. 29. Patterns generated by evolution of one-dimensional cellular automata with k states per site according to a modulo-k rule, starting from an initial configuration containing a single nonzero site with value one. At each time step, the value of a site is the sum of the values of its two nearest neighbors at the previous time step. Configurations obtained at successive time steps are shown on successive lines. Sites with value zero are indicated as blanks; *, +, and − represent, respectively, values one, two, and three, and in the lower two patterns, ρ represents any nonzero value. In the large time limit, all the patterns tend to a self-similar form, with definite fractal dimensions.

tion of the integer τ and generalizes the function $\#_r(\tau)$ introduced in Sec. III.

Figure 30 shows typical examples of the behavior of some cellular automata with $k=3$. Considerable diversity is evident. However, with simple initial states, self-similar patterns are obtained at asymptotically large times, just as in the $k=2$ case of Sec. III. (Notice that the length and time scales before self-similarity are typically longer than those found for $k=2$: in the limit $k\to\infty$ where each site takes on an arbitrary integer value, self-similarity may not be apparent at any finite time.) Evolution of disordered initial states also again appears to generate nontrivial structure, though several novel phenomena are present. First, alternation of value-one and value-two sites on successive time steps can lead to "half-speed propagation" as in rule

000000000000001002001010020 .

Second, rules such as

000000000000001011002010010

lead to a set of finite regions containing only sites with values zero and one, separated by "impermeable mem-

branes" of value-two sites. The evolution within each region is independent, with the membranes enforcing boundary conditions, and leading to cycles after a finite number of time steps. Third, even for legal rules such as

000000121022002210021020100

and

211000122121012200112021200 ,

illustrated in Fig. 30, there exist patterns which display a uniform shifting motion. For example, with rule

211000122121012200112021200

an isolated 12 shifts to the right by one site every time step, while an isolated 21 shifts to the left; when 21 and 12 meet, they cross without interference. Uniform shifting motion is impossible with legal rules when $k=2$, since sequences of zero and one sites cannot define suitable directions (evolution of 1101 and 1011 always yield a pattern spreading in both directions).

An important feature of some cellular automata with more than two states per site is the possibility for the formation of a membrane which "protects" sites within it from the effects of noise outside. In this way, there may exist seeds from which very regular patterns may grow, shielded by membranes from external noise typical in a disordered configuration. Examples of such behavior are to be found in Fig. 30. Only when two protective membranes meet is the structure they enclose potentially destroyed. The size of the region affected by a particular seed may grow linearly with time. Even if seeds occur with very low probability, any sufficiently long disordered configuration will contain at least one, and the large time behavior of the cellular automaton will be radically affected by its presence.

In addition to increasing the number of states per site, the cellular automata discussed above may be generalized by increasing the number of sites whose values affect the evolution of a particular site at each time step. For example, one may take the neighborhood of each site to contain the site itself, its nearest neighbors, and its next-nearest neighbors. With two states per site, the number of possible sets of legal local rules for such cellular automata is $2^{26}\simeq7\times10^7$ (for $k=3$, this number increases to $3^{174}\simeq10^{83}$). Figure 31 shows patterns generated by these cellular automata for two typical sets of local rules. With simple initial states, self-similar patterns are obtained at large times. With disordered initial states, less structure is apparent than in the three-site neighborhood cellular automata discussed above. The patterns obtained with such cellular automata are again qualitatively similar to those shown in Sec. II.

The cellular automata discussed so far have all involved a line of sites in one dimension. One may also consider cellular automata in which the sites lie on a regular square or (hyper)cubic lattice in two or more space dimensions. As usual, the value of each site is determined by the values of a neighborhood of sites at the previous time step. In the simplest case, the neighborhood includes a site

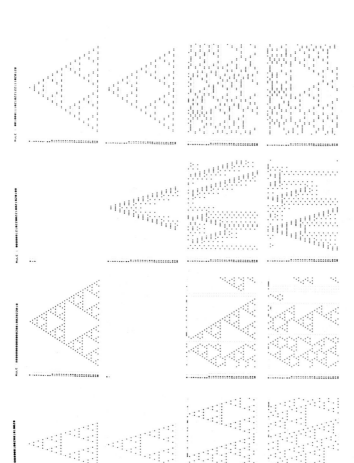

FIG. 30. Examples of the evolution of several typical cellular automata with three states per site. Sites with value zero are shown as blanks, while values one and two are indicated by *
and ', respectively. The value of a site at each time step is determined in analogy with Fig. 1 by the digit in the ternary specification of the rule corresponding to the values of the site
and its two nearest neighbors at the previous time step. The evolution is shown until a configuration is reached for the second time (signaling a cycle) or for at most thirty time steps.
The initial configurations in the lower two rows are typical of disordered configurations in which each site is statistically independent and takes on its three possible values with equal
probabilities.

Wolfram: Statistical mechanics of cellular automata

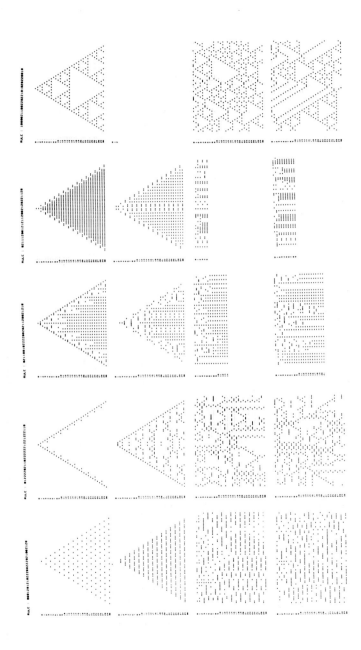

FIG. 30. (*Continued.*)

FIG. 30. (*Continued.*)

Wolfram: Statistical mechanics of cellular automata   635

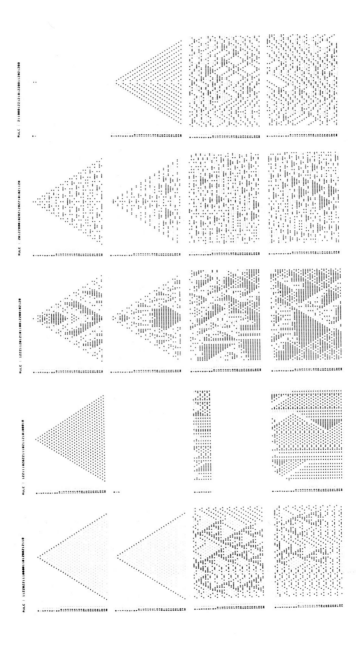

FIG. 30. (*Continued*.)

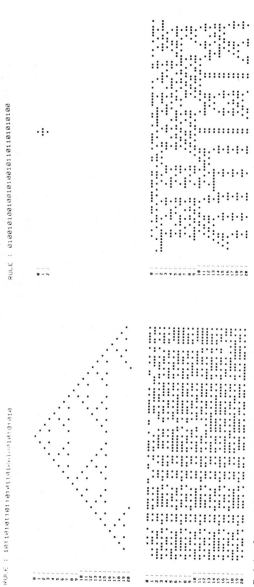

FIG. 31. Evolution of two typical one-dimensional cellular automata with two states per site in which the value of a site at a particular time step is determined by the preceding values of a neighborhood of five sites containing the site, its nearest neighbors, and its next-nearest neighbors. The initial configurations in the lower row are typical of disordered configurations, in which each site has value one with probability $\frac{1}{2}$.

and its nearest neighbors. However, in $d > 1$ dimensions two possible identifications of nearest neighbors can be made. First, sites may be considered neighbors if one of their coordinates differ by one unit, and all others are equal, so that the sites are "orthogonally" adjacent. In this case, a "type-I" cellular automaton neighborhood containing $2d + 1$ sites is obtained. Second, sites may be considered neighbors if none of their coordinates differ by more than one unit, so that the sites are "orthogonally" or "diagonally" adjacent. This case yields a "type-II" cellular automaton neighborhood containing 3^d sites. When $d = 1$, type-I and -II neighborhoods are identical and each contains three sites. For $d = 2$, the type-I neighborhood contains five sites, while the type-II neighborhood contains nine sites.[14] Cellular automaton rules may be considered legal if they satisfy the quiescence condition and are invariant under the rotation and reflection symmetries of the lattice. For $d = 2$, the number of possible legal type-I rules with k states per site is found to be $k^{(k^5 + k^3 + 2k^2 - 4)/4}$, yielding $2^{11} = 2048$ rules for $k = 2$ and $3^{71} \simeq 8 \times 10^{33}$ for $k = 3$. The number of type-II rules with $k = 2$ in two dimensions is found to be $2^{59} \simeq 6 \times 10^{17}$ (or 2^{71} if reflection symmetries are not imposed).

Figure 32 shows the evolution of an initial configuration containing a single nonzero site according to two-dimensional (type-I) modulo-two rules. In case (a) the value of a site is taken to be the sum modulo two of the values of its four neighbors on the previous time step, in analogy with one-dimensional cellular automaton rule 90. In case (b), the previous value of the site itself included in the sum (and the complement is taken), in analogy with rule 150. The sequence of patterns obtained at successive time steps may be "stacked" to form pyramidal structures in three-dimensional space. These structures become self-similar at large times: in case (a) they exhibit a fractal dimension $\log_2 5 \simeq 2.32$, and in case (b) a dimension $1 + \log_2(1 + \sqrt{3}) \simeq 2.45$. The patterns found on vertical slices containing the original nonzero site through the pyramids (along one of the two lattice directions) are the same as those generated by the one-dimensional modulo-two rules discussed in Secs. II and III. The patterns obtained at each time step in Fig. 31 are almost always self-similar in the large time limit. For case (a), the number of sites with value one generated after τ time steps in Fig. 31 is found to be $4^{\#_1(\tau)}$, where $\#_1(\tau)$ gives the number of occurrences of the digit one in the binary decomposition of the integer τ, as discussed in Sec. III (cf. Butler and Ntafos, 1977). The type-I modulo-two rules may be generalized to d-dimensional cellular automata. In case (a) the patterns obtained by evolution from a single nonzero initial site have fractal dimension $\log_2(2d + 1)$ and give $(2d)^{\#_1(\tau)}$ nonzero sites at time step τ. In case (b), the asymptotic fractal dimension is found to be $\log_2[d(\sqrt{1 + 4/d} + 1)]$. Once again, simple initial states always yield self-similar structures in the large time limit.

A particular type-II two-dimensional cellular automaton whose evolution has been studied extensively is the game of "Life" (Conway, 1970; Gardner, 1971, 1972; Wainwright, 1971–1973; Wainwright, 1974; Buckingham, 1978; Berlekamp *et al.*, 1982, Chap. 25; R. W. Gosper, private communications). The local rules take a site to "die" (attain value zero) unless two or three of its neighbors are "alive" (have value one). If two neighbors are alive, the value of the site is left unchanged; if three are alive, the site always takes on the value one. Many configurations exhibiting particular properties have been found. The simplest isolated configurations invariant under time evolution are the "square" (or "block") consisting of four adjacent live sites, and the "hexagon" (or "beehive") containing six live sites. "Oscillator" configurations which cycle through a sequence of states are also known. The simplest is the "blinker" consisting of a line of three live sites, which cycles with a period of two time steps. Oscillators with periods 3, 5, and 7 are also known; other periods may be obtained by composition. So long as they are separated by four or more unfilled sites, many of these structures may exist without interference in the configurations of a cellular automaton, and their effects are localized. There also exist configurations which "move" uniformly across the lattice, executing a cycle of a few internal states. The simplest example is the "glider" which contains five live sites and undergoes a cycle of length two. The number of filled sites in all the configurations mentioned so far is bounded as a function of time. However, "glider gun" configurations have been found which generate infinite streams of gliders, yielding a continually increasing number of live sites. The simplest known glider gun configuration evolves from a configuration containing 26 live cells. Monte Carlo simulation suggests that a disordered state of N^2 cells usually evolves to a steady state within about N^2 time steps (and typically an order of magnitude quicker); very few of the 2^{N^2} possible configurations are visited. Complicated structures such as glider guns are very rarely produced. Rough empirical investigation suggests that the density of structures containing L live sites generated from a disordered initial state (cf. Buckingham, 1981) decreases like e^{-L_-}/L, where L_- is the size of the minimal distinct configuration which evolves to the required structure in one time step. Just as for the one-dimensional cellular automata discussed in Sec. IV, the irreversibility of "Life" leads to configurations which cannot be reached by evolution from any other configurations, and can appear only as initial states. However, the simplest known "unreachable" configuration contains around 300 sites (Wainwright, 1971–1973; Hardouin-Duparc, 1974; Berlekamp *et al.*, 1982, Chap. 25).

The game of "Life" is an example of a special class of "totalistic" cellular automata, in which the value of a site depends only on the sum of the values of its neighbors at the previous time step, and not on their individual values. Such cellular automata may arise as models of systems involving additive local quantities, such as chemical concentrations. In one dimension with $k = 2$ (and three sites

---

[14] In the case $d = 2$, neighborhoods of types I and II are known as von Neumann and Moore neighborhoods, respectively.

FIG. 32. Evolution of an initial state containing a single nonzero site in a two-dimensional cellular automaton satisfying type-I modulo-two rules. In case (a) the value of each site is taken to be the sum modulo two of the values of its four (orthogonally adjacent) neighbors at the previous time step, while in case (b) the previous value of the site itself is included in the sum, and the complement is taken. Case (a) is the two-dimensional analog of a one-dimensional cellular automaton evolving according to local rule 90, and case (b) of one evolving according to rule 150. The pyramidal structure obtained in each case by stacking the patterns generated at successive time steps is self-similar in the large time limit.

in each neighborhood) all cellular automaton rules are totalistic. In general, the number of totalistic (legal) sets of rules for cellular automata with v neighbors for each site is $k^{(k-1)(vk+1)}$. In one dimension with $k=3$, $\simeq 5 \times 10^6$ of the $\simeq 10^8$ possible rules are therefore totalistic. Only 243 of the totalistic rules are also periperal in the sense defined in Sec. II. With $k=2$ in two dimensions, 2^9 of the 2^{11} possible rules in a type-I neighborhood are totalistic (and 32 are also peripheral), and 2^{17} of the 2^{59} in a type-II neighborhood.

A potentially important feature of cellular automata is the capability for "self-reproduction" through which the evolution of a configuration yields several separated identical copies of the configuration. Figure 33 illustrates a very simple form of self-reproduction with the elementary one-dimensional modulo-two rule (see Waksman, 1969; Amoroso and Cooper, 1971; Fredkin, 1981). With a single nonzero site in the initial state, a configuration containing exactly two nonzero sites is obtained after 2^j time steps[15] as indicated by Eq. (3.2). The additive superposition property of the modulo-two rule implies that results for more complicated initial states are obtained by superposition of those for single-site initial states. Thus after $\tau = 2^j$ time steps, for sufficiently large j, the cellular automaton generates two exact copies of any initial sequence of site values. After a further 2^{j-1} time steps, four copies are obtained. However, after another 2^{j-1} time steps, the innermost pair of these copies meet again, and annihilate, leaving only two copies when $\tau = 2^{j+1}$. Purely geometrical "overcrowding" thus prevents exponential multiplication of copies by self-reproduction in this case. An exactly analogous phenomenon occurs with the two-dimensional modulo-two rule illustrated in Fig. 32, and its higher-dimensional analogs. In general, the number of sites in a d-dimensional cellular automaton configuration grows with time at most as fast as $(2\tau)^d$, which is asymptotically slower than the number $> (2d)^{\alpha\tau}$ required for an exponentially increasing number of copies to be generated. Exponential self-reproduction can thus occur only if the copies generated are not precisely identical, but exhibit variability, and for example execute a random walk motion in response to external noise or contain a "counter" which causes later generations to "live" longer before reproducing.

Section IV mentioned the view of cellular automata as computers. An important class of computers is those with the property of "computational universality," for which changes in input alone allow any "computable function" to be evaluated, without any change in internal construction. Universal computers can simulate the operation of any other computer if their input is suitably encoded. Many Turing machines have been shown to be computationally universal. The simplest has seven internal states, and allows four possible "symbols" in each square of its tape. One method for demonstrating computational universality of cellular automata shows correspondence with a universal Turing machine. The head of the Turing machine is typically represented by a phononlike structure which propagates along the cellular automaton. It may be shown (Smith, 1971) that an eighteen-state one-dimensional cellular automaton with a three-site neighborhood can simulate the seven-state four-symbol Turing machine in this way, and is therefore computationally universal. Simpler computationally universal cellular automata must be found by other methods. The most straightforward method is to show correspondence with a standard digital computer or electronic circuit by identifying cellular automaton structures which act like "wires," carrying signals without dissipation and crossing without interference, and structures representing NAND gates at intersections between wires. "Memories" which maintain the same state for all time are also required. In the Life-game cellular automaton discussed above, streams of gliders generated by glider guns may be used as wires, with bits in the signal represented by the presence or absence of gliders. At the points where "glider streams" meet, other structures determine whether the corresponding wires cross or interact through a "NAND gate." The Life-game cellular automaton is thus computationally universal. "Circuits" such as binary adders (Buckingham, 1978) may be constructed from Life configurations. It appears that such circuits run at a speed slower than the digital computers to which they correspond only by a constant multiplicative factor. The "Life game" is a type-II two-dimensional cellular automaton with two states per site. A computationally universal type-I two-dimensional cellular automaton has been constructed with three states per site (Banks, 1971); only two states are required if the initial configuration is permitted to contain an infinite "background" of nonzero sites (Toffoli, 1977a). In one dimension, with a neighborhood of three sites, there are some preliminary indications that a universal cellular automaton may be constructed with five states per site. The details and implications of this cellular automaton will be described in a future publication.

VI. DISCUSSION

This paper represents a first step in the investigation of cellular automata as mathematical models for self-organizing statistical systems. The bulk of the paper consisted in a detailed analysis of elementary cellular automata involving a sequence of sites on a line, with a binary variable at each site evolving in discrete time steps according to the values of its nearest neighbors. Despite the simplicity of their construction, these systems were found to exhibit very complicated behavior.

The 32 possible (legal) elementary cellular automata were found to fall into two broad classes. The first class consisted of simple cellular automata whose time evolu-

---

[15] An analogous result holds for all modulo-k rules with k prime by virtue of the relation $\binom{k^j}{i} \bmod k = 0$, $0 < i < k^j$ valid for all primes k. The relation is a special case of the general result (Knuth, 1973, Sec. 1.2.6, Ex. 10)

$$\begin{bmatrix} j \\ i \end{bmatrix} = \begin{bmatrix} \lfloor j/k \rfloor \\ \lfloor i/k \rfloor \end{bmatrix} \begin{bmatrix} j \bmod k \\ i \bmod k \end{bmatrix} \bmod k \; .$$

640

Wolfram: Statistical mechanics of cellular automata

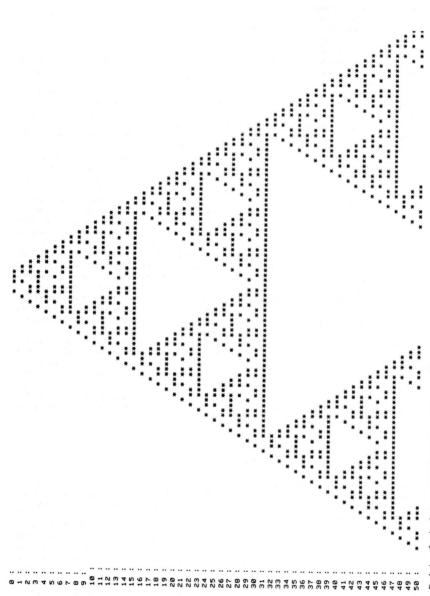

FIG. 33. Evolution of a simple pattern according to the modulo-two cellular automaton rule (number 90), exhibiting a simple self-reproduction phenomenon. The additive superposition property of the cellular automaton leads to the generation of two exact copies of the initial 1011 pattern at time steps 8, 16, 32, ... "Geometrical overcrowding" prevents exponential increase in the number of copies produced.

tion led eventually to simple, usually homogeneous, final states. The second class contained complex cellular automata capable of generating quite complicated structures even from simple initial states. Figure 3 showed the patterns of growth obtained with the very simplest initial state in which only one site had a nonzero value. The complex rules were found to yield self-similar fractal patterns. For all but one of the rules, the patterns exhibited the same fractal dimension $\log_2 3 \simeq 1.59$ (the remaining rule gave a fractal dimension $\log_2 2\varphi \simeq 1.69$). With more complicated initial states, the patterns obtained after evolution for many time steps remained self-similar—at least on scales larger than the region of nonzero initial sites. The generation of self-similar patterns was thus found to be a generic feature of complex cellular automata evolving from simple initial states. This result may provide some explanation for the widespread occurrence of self-similarity in natural systems.

Section III discussed the evolution of cellular automata from general initial states, in which a finite fraction of the infinite number of initial sites carried value one. Regardless of the initial density of nonzero sites, definite densities were found in the large time limit. Markovian master equation approximations to the density development were found inadequate because of the importance of "feedback" in the cellular automaton evolution. Even with disordered or random initial states, in which the values of different sites are statistically uncorrelated, the evolution of complex cellular automata was found to lead to the formation of definite structures, as suggested in Figs. 8 and 15. One characteristic of this self-organization was the generation of long sequences of correlated sites. The spectrum of these sequences was found to reach an equilibrium form after only a few time steps, extending to arbitrarily large scales, but with an exponential damping. The exponents were again found to be universal for all initial states and almost all complex cellular automata (with the exception of two special additive cellular automata).

Any initial cellular automaton state was found to lead at large times to configurations with the same statistical structures. However, in complex cellular automata, the trajectories of almost all specific nearby initial configurations (differing by changes in the values at a few sites) were found to diverge exponentially with time in the phase space of possible configurations. After a few time steps, the mapping from initial to final configurations becomes apparently random (although there are quantitative deviations from a uniform random mapping). Cellular automaton rules may map several initial configurations into the same final configuration, and thus lead to microscopically irreversible time evolution in which trajectories of different states may merge. In the limit of an infinite number of sites, a negligible fraction of all the possible cellular automaton configurations are reached by evolution from any of the possible initial states after a few time steps. Starting even from an ensemble in which each possible configuration appears with equal probability, the cellular automaton evolution concentrates the probabilities for particular configurations, thereby reducing entropy. This phenomenon allows for the possibility of self-organization by enhancing the probabilities of organized configurations and suppressing disorganized configurations.

Many of the qualitative features found for elementary cellular automata appear to survive in more complicated cellular automata (considered briefly in Sec. V), although several novel phenomena may appear. For example, in one-dimensional cellular automata with three or more possible values at each site, protective membranes may be generated which shield finite regions from the effects of external noise, and allow very regular patterns to grow from small seeds.

Cellular automata may be viewed as computers, with initial configurations considered as input programs and data processed by cellular automaton time evolution. Sufficiently complicated cellular automata are known to be universal computers, capable of computing any computable function given appropriate input. Such cellular automata may be considered as capable of the most complicated behavior conceivable and are presumably capable of simulating any physical system given a suitable input encoding and a sufficiently long running time. In addition, they may be used to simulate the evolution of any other cellular automaton. If the necessary encoding is sufficiently simple, the statistical properties of the simulated cellular automaton should follow those of the universal cellular automaton. Although not capable of universal simulation, simpler cellular automata may often simulate each other. This capability may well form a basis for the universality found in the statistical properties of various cellular automata.

Cellular automata have been developed in this paper as general mathematical models. One may anticipate their application as simple models for a wide variety of natural processes. Their nontrivial features are typically evident only when some form of growth inhibition is present. Examples are found in aggregation processes in which aggregation at a particular point prevents further aggregation at the same point on the next time step.

ACKNOWLEDGMENTS

I am grateful for suggestions and assistance from J. Ambjorn, N. Margolus, O. Martin, A. Odlyzko, and T. Shaw, and for discussions with J. Avron, C. Bennett, G. Chaitin, J. D. Farmer, R. Feynman, E. Fredkin, M. Gell-Mann, R. W. Gosper, A. Hoogland, T. Toffoli, and W. Zurek. I thank S. Kauffman, R. Landauer, P. Leyland, B. Mandelbrot, and A. Norman for suggesting references. The symbolic manipulation computer language SMP (Wolfram et al., 1981) was used in some of the calculations. Some of this work was done before I resigned from Caltech; computer calculations performed at Caltech were supported in part by the U.S. Department of Energy under Contract Number DE-AC-03-81-ER40050.

REFERENCES

Abelson, H. and A. A. diSessa, 1981, *Turtle Geometry: The Computer as a Medium for Exploring Mathematics* (MIT Press, Cambridge).

Aggarwal, S., 1973, "Local and global Garden of Eden theorems," University of Michigan technical report No. 147.

Aladyev, V., 1974, "Survey of research in the theory of homogeneous structures and their applications," Math. Biosci. 22, 121.

Aladyev, V., 1976, "The Behavioural Properties of Homogeneous Structures," Math. Biosci. 29, 99.

Alekseev, V. M., and M. V. Yakobson, 1981, "Symbolic dynamics and hyperbolic dynamic systems," Phys. Rep. 75, 287.

Amoroso, S. and G. Cooper, 1971, "Tessellation structures of reproduction of arbitrary patterns," J. Comput Syst. Sci. 5, 455.

Apostol, T. M., 1976, *Introduction to Analytic Number Theory* (Springer, Berlin).

ApSimon, H. G., 1970a, "Periodic forests whose largest clearings are of size 3," Philos. Trans. R. Soc. London, Ser. A 266, 113.

ApSimon, H. G., 1970b, "Periodic forests whose largest clearings are of size $n \geq 4$," Proc. R. Soc. London, Ser. A 319, 399.

Arbib, M. A., 1969, *Theories of Abstract Automata* (Prentice-Hall, Englewood Cliffs).

Atrubin, A. J., 1965, "A one-dimensional real-time iterative multiplier," IEEE Trans. Comput. EC-14, 394.

Baer, R. M., and H. M. Martinez, 1974, "Automata and biology," Ann. Rev. Biophys. 3, 255.

Banks, E. R., 1971, "Information processing and transmission in cellular automata," MIT Project MAC report No. TR-81.

Barricelli, N. A., 1972, "Numerical testing of evolution theories," J. Statist. Comput. Simul. 1, 97.

Berlekamp, E. R., 1968, *Algebraic Coding Theory* (McGraw-Hill, New York).

Berlekamp, E. R., J. H. Conway, and R. K. Guy, 1982, *Winning Ways for Your Mathematical Plays* (Academic, New York), Vol. 2, Chap. 25.

Buckingham, D. J., 1978, "Some facts of life," Byte 3, 54.

Burks, A. W., 1970, *Essays on Cellular Automata* (University of Illinois, Urbana).

Burks, A. W., 1973, "Cellular Automata and Natural Systems," Proceedings of the 5th Congress of the Deutsche Gessellschaft für Kybernetik, Nuremberg.

Butler, J. T., and S. C. Ntafos, 1977, "The vector string descriptor as a tool in the analysis of cellular automata systems," Math. Biosci. 35, 55.

Codd, E. F., 1968, *Cellular Automata* (Academic, New York).

Cole, S. N., 1969, "Real-time computation by n-dimensional iterative arrays of finite-state machines," IEEE Trans. Comput. C-18, 349.

Conway, J. H., 1970, unpublished.

Deutsch, E. S., 1972, "Thinning algorithms on rectangular, hexagonal and triangular arrays," Commun. ACM 15, 827.

Farmer, J. D., 1982a, "Dimension, fractal measures, and chaotic dynamics," in *Evolution of Order and Chaos in Physics, Chemistry and Biology*, edited by H. Haken (Springer, Berlin).

Farmer, J. D., 1982b, "Information dimension and the probabilistic structure of chaos," Z. Naturforsch. 37a, 1304.

Fine, N. J., 1947, "Binomial coefficients modulo a prime," Am. Math. Mon. 54, 589.

Fischer, P. C., 1965, "Generation of primes by a one-dimensional real-time iterative array," J. ACM 12, 388.

Flanigan, L. K., 1965, "An experimental study of electrical conduction in the mammalian atrioventricular node," Ph.D. thesis (University of Michigan).

Fredkin, E., 1981, unpublished, and PERQ computer demonstration (Three Rivers Computer Corp.).

Gach, P., G. L. Kurdyumov, and L. A. Levin, 1978, "One-dimensional uniform arrays that wash out finite islands," Probl. Peredachi. Info., 14, 92.

Gardner, M., 1971, "Mathematical Games," Sci. Amer. 224, February, 112; March, 106; April, 114.

Gardner, M., 1972, "Mathematical Games," Sci. Amer. 226, January, 104.

Geffen, Y., A. Aharony, B. B. Mandelbrot, and S. Kirkpatrick, "Solvable fractal family, and its possible relation to the backbone at percolation," Phys. Rev. Lett. 47, 1771.

Gerola, H. and P. Seiden, 1978, "Stochastic star formation and spiral structure of galaxies," Astrophys. J. 223, 129.

Glaisher, J. W. L., 1899, "On the residue of a binomial-theorem coefficient with respect to a prime modulus," Q. J. Math. 30, 150.

Golomb, S. W., 1967, *Shift Register Sequences* (Holden-Day, San Francisco).

Grassberger, P., 1982, "A new mechanism for deterministic diffusion," Wuppertal preprint WU B 82-18.

Greenberg, J. M., B. D. Hassard, and S. P. Hastings, 1978, "Pattern formation and periodic structures in systems modelled by reaction-diffusion equations," Bull. Am. Math. Soc. 84, 1296.

Griffeath, D., 1970, *Additive and Cancellative Interacting Particle Systems* (Springer, Berlin).

Haken, H., 1975, "Cooperative phenomena in systems far from thermal equilibrium and in nonphysical systems," Rev. Mod. Phys. 47, 67.

Haken, H., 1978, *Synergetics*, 2nd ed. (Springer, Berlin).

Haken, H., 1979, *Pattern Formation by Dynamic Systems and Pattern Recognition* (Springer, Berlin).

Haken, H., 1981, *Chaos and Order in Nature* (Springer, Berlin).

Hardouin-Duparc, J., 1974, "Paradis terrestre dans l'automate cellulaire de conway," R.A.I.R.O. 8 R-3, 63.

Hardy, G. H. and E. M. Wright, 1979, *An Introduction to the Theory of Numbers*, 5th ed. (Oxford University Press, Oxford).

Harris, B., 1960, "Probability distributions related to random mappings," Ann. Math. Stat. 31, 1045.

Harvey, J. A., E. W. Kolb, and S. Wolfram, 1982, unpublished.

Herman, G. T., 1969, "Computing ability of a developmental model for filamentous organisms," J. Theor. Biol. 25, 421.

Honsberger, R., 1976, "Three surprises from combinatorics and number theory," in *Mathematical Gems II*, Dolciani Math. Expositions (Mathematical Association of America, Oberlin), p.1.

Hoogland, A., *et al.*, 1982, "A special-purpose processor for the Monte Carlo simulation of Ising spin systems," Delft preprint.

Hopcroft, J. E., and J. D. Ullman, 1979, *Introduction to Automata Theory, Languages and Computation* (Addison-Wesley, Reading).

Kauffman, S. A., 1969, "Metabolic stability and epigenesis in randomly constructed genetic nets," J. Theor. Biol. 22, 437.

Kimball, S. H., *et al.*, 1958, "Odd binomial coefficients," Am. Math. Mon. 65, 368.

Kitagawa, T., 1974, "Cell space approaches in biomathematics," Math. Biosci. 19, 27.

Knuth, D. E., 1973, *Fundamental Algorithms* (Addison-Wesley, Reading).

Knuth, D. E., 1981, *Seminumerical Algorithms*, 2nd ed. (Addison-Wesley, Reading).

Kosaraju, S. R., 1974, "On some open problems in the theory of cellular automata," IEEE Trans. Comput. C-23, 561.

Landauer, R., 1979, "The role of fluctuations in multistable systems and in the transition to multistability," Ann. N.Y. Acad. Sci. **316**, 433.

Langer, J. S., 1980, "Instabilities and pattern formation in crystal growth," Rev. Mod. Phys. **52**, 1.

Levy, Y. E., 1982, "Some remarks about computer studies of dynamical systems," Phys. Lett. A **88**, 1.

Lifshitz, E. M., and L. P. Pitaevskii, 1981, *Physical Kinetics* (Pergamon, New York).

Lindenmayer, A., 1968, "Mathematical models for cellular interactions in development," J. Theoret. Biol. **18**, 280.

MacWilliams, F. J., and N. J. A. Sloane, *Theory of Error-Correcting Codes* (North-Holland, Amsterdam).

Mandelbrot, B., 1977, *Fractals: Form, Chance and Dimension* (Freeman, San Francisco).

Mandelbrot, B., 1982, *The Fractal Geometry of Nature* (Freeman, San Francisco).

Manna, Z., 1974, *Mathematical Theory of Computation* (McGraw-Hill, New York).

Manning, F. B., 1977, "An approach to highly integrated, computer-maintained cellular arrays," IEEE Trans. Comput. C-26, 536.

Margolus, N., 1982, private communication.

Martin, O., A. Odlyzko, and S. Wolfram, 1983, "Algebraic properties of cellular automata," Bell Laboratories report (January, 1983).

McIlroy, M. D., 1974, "The numbers of 1's in binary integers: bounds and extremal properties," SIAM J. Comput. **3**, 255.

Miller, J. C. P., 1970, "Periodic forests of stunted trees," Philos. Trans. R. Soc. London, Ser. A **266**, 63.

Miller, J. C. P., 1980, "Periodic forests of stunted trees," Philos. Trans. R. Soc. London Ser. A **293**, 48.

Minsky, M. L., 1967, *Computation: Finite and Infinite Machines* (Prentice-Hall, Englewood Cliffs).

Moore, E. F., 1962, "Machine Models of Self-Reproduction," Proceedings of a Symposium on Applied Mathematics **14**, 17, reprinted in *Essays on Cellular Automata*, edited by A. W. Burks (University of Illinois, Urbana, 1970), p. 187.

Nicolis, G. and Prigogine, I., 1977, *Self-Organization in Nonequilibrium Systems* (Wiley, New York).

Nicolis, G., G. Dewel, and J. W. Turner, editors, 1981, *Order and Fluctuations in Equilibrium and Nonequilibrium Statistical Mechanics*, Proceedings of the XVIIth International Solvay Conference on Physics (Wiley, New York).

Nishio, H., 1981, "Real time sorting of binary numbers by 1-dimensional cellular automata," Kyoto University report.

Ott, E., 1981, "Strange attractors and chaotic motions of dynamical systems," Rev. Mod. Phys. **53**, 655.

Pearson, R., J. Richardson, and D. Toussaint, 1981, "A special purpose machine for Monte-Carlo simulation," Santa Barbara preprint NSF-ITP-81-139.

Peterson, W. W., and E. J. Weldon, *Error-Correcting Codes*, 2nd ed. (MIT Press, Cambridge).

Preston, K., M. J. B. Duff, S. Levialdi, Ph. E. Norgren, and J.-I. Toriwaki, 1979, "Basics of Cellular Logic with Some Applications in Medical Image Processing," Proc. IEEE **67**, 826.

Prigogine, I., 1980, *From Being to Becoming* (Freeman, San Francisco).

Renyi, A., 1970, *Probability Theory* (North-Holland, Amsterdam).

Roberts, J. B., 1957, "On Binomial Coefficient Residues," Can. J. Math. **9**, 363.

Rosen, R., 1981, "Pattern Generation in Networks," Prog. Theor. Biol. **6**, 161.

Rosenfeld, A., 1979, *Picture Languages* (Academic, New York).

Schewe, P. F., editor, 1981, "Galaxies, the Game of Life, and Percolation," in *Physics News*, Amer. Inst. Phys. Pub. R-302, 61.

Schulman, L. S., and P. E. Seiden, 1978, "Statistical mechanics of a dynamical system based on Conway's game of life," J. Stat. Phys. **19**, 293.

Shannon, C. E., 1951, "Prediction and entropy of printed English," Bell Syst. Tech. J., **30**, 50.

Sierpinski, W., 1916, "Sur une courbe dont tout point est un point de ramification," Pr. Mat.-Fiz. **27**, 77; *Oeuvres Choisis*, (Państwowe Wydawnictwo Naukowe, Warsaw) Vol. II, p. 99.

Smith, A. R., 1971, "Simple computation-universal cellular spaces," J. ACM **18**, 339.

Sternberg, S. R., 1980, "Language and architecture for parallel image processing," in *Pattern Recognition in Practice*, edited by E. S. Gelesma and L. N. Kanal (North-Holland, Amsterdam), p. 35.

Stevens, P. S., 1974, *Patterns in Nature* (Little, Brown, Boston).

Stolarsky, K. B., 1977, "Power and exponential sums of digital sums related to binomial coefficient parity," SIAM J. Appl. Math. **32**, 717.

Sutton, C., 1981, "Forests and numbers and thinking backwards," New Sci. **90**, 209.

Thom, R., 1975, *Structural Stability and Morphogenesis* (Benjamin, New York).

Thompson, D'A. W., 1961, *On Growth and Form*, abridged ed. edited by J. T. Bonner (Cambridge University, Cambridge, England).

Toffoli, T., 1977a, "Computation and construction universality of reversible cellular automata," J. Comput. Sys. Sci. **15**, 213.

Toffoli, T., 1977b, "Cellular automata mechanics," Ph.D. thesis, Logic of Computers Group, University of Michigan.

Toffoli, T., 1980, "Reversible computing," MIT report MIT/LCS/TM-151.

Toffoli, T., 1983, "Squareland: a hardware cellular automaton simulator," MIT LCS preprint, in preparation.

Turing, A. M., 1936, "On computable numbers, with an application to the Entscheidungsproblem," Proc. London Math. Soc. Ser. 2, **42**, 230; **43**, 544E, reprinted in *The Undecidable*, edited by M. David (1965; Hewlett, New York), p. 115.

Turing, A. M., 1952, "The chemical basis of morphogenesis," Philos. Trans. R. Soc. London, Ser. B **237**, 37.

Ulam, S., 1974, "Some ideas and prospects in biomathematics," Ann. Rev. Bio., **255**.

von Neumann, J., 1963, "The general and logical theory of automata," in J. von Neumann, *Collected Works*, edited by A. H. Taub, **5**, 288.

von Neumann, J., 1966, *Theory of Self-Reproducing Automata*, edited by A. W. Burks (University of Illinois, Urbana).

Wainwright, R. T., 1971−73, Lifeline, 1−11.

Wainwright, R. T., 1974, "Life is Universal!," Proceedings of the Winter Simulation Conference, Washington, D.C., ACM, p. 448.

Waksman, A., 1969, "A model of replication," J. ACM **16**, 178.

Willson, S., 1982, "Cellular automata can generate fractals," Iowa State University, Department of Mathematics, preprint.

Witten, T. A. and L. M. Sander, 1981, "Diffusion-limited aggregation, a kinetic critical phenomenon," Phys. Rev. Lett. **47**, 1400.

Wolfram, S., *et al.*, 1981, "SMP Handbook," Caltech.

Wolfram, S., 1982a, "Cellular automata as simple self-organizing systems," Caltech preprint CALT-68-938 (submitted to Nature).

Wolfram, S., 1982b, "Geometry of binomial coefficients," to be published in Am. Math. Monthly.

Commun. Math. Phys. 93, 219–258 (1984)

Communications in
**Mathematical
Physics**
© Springer-Verlag 1984

Algebraic Properties of Cellular Automata

Olivier Martin[1,*], Andrew M. Odlyzko[2], and Stephen Wolfram[2,3,**]

1 California Institute of Technology, Pasadena, CA 91125, USA
2 Bell Laboratories, Murray Hill, NJ 07974, USA
3 The Institute for Advanced Study, Princeton, NJ 08540, USA

Abstract. Cellular automata are discrete dynamical systems, of simple construction but complex and varied behaviour. Algebraic techniques are used to give an extensive analysis of the global properties of a class of finite cellular automata. The complete structure of state transition diagrams is derived in terms of algebraic and number theoretical quantities. The systems are usually irreversible, and are found to evolve through transients to attractors consisting of cycles sometimes containing a large number of configurations.

1. Introduction

In the simplest case, a cellular automaton consists of a line of sites with each site carrying a value 0 or 1. The site values evolve synchronously in discrete time steps according to the values of their nearest neighbours. For example, the rule for evolution could take the value of a site at a particular time step to be the sum modulo two of the values of its two nearest neighbours on the previous time step. Figure 1 shows the pattern of nonzero sites generated by evolution with this rule from an initial state containing a single nonzero site. The pattern is found to be self-similar, and is characterized by a fractal dimension $\log_2 3$. Even with an initial state consisting of a random sequence of 0 and 1 sites (say each with probability $\frac{1}{2}$), the evolution of such a cellular automaton leads to correlations between separated sites and the appearance of structure. This behaviour contradicts the second law of thermodynamics for systems with reversible dynamics, and is made possible by the irreversible nature of the cellular automaton evolution. Starting from a maximum entropy ensemble in which all possible configurations appear with equal probability, the evolution increases the probabilities of some configurations at the expense of others. The configurations into which this concentration occurs then dominate ensemble averages and the system is "organized" into having the

---

* Address from September 1983: Physics Department, Columbia University, New York, NY 10027, USA
** Address from January 1983

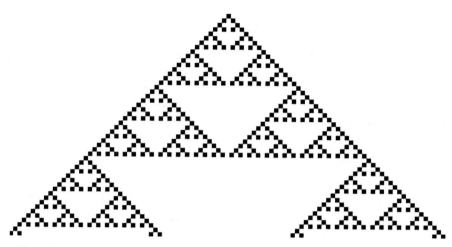

Fig. 1. Example of evolution of a one-dimensional cellular automaton with two possible values at each site. Configurations at successive time steps are shown as successive lines. Sites with value one are black; those with value zero are left white. The cellular automaton rule illustrated here takes the value of a site at a particular time step to be the sum modulo two of the values of its two nearest neighbours on the previous time step. This rule is represented by the polynomial $\mathbb{T}(x) = x + x^{-1}$, and is discussed in detail in Sect. 3

properties of these configurations. A finite cellular automaton with N sites (arranged for example around a circle so as to give periodic boundary conditions) has 2^N possible distinct configurations. The global evolution of such a cellular automaton may be described by a state transition graph. Figure 2 gives the state transition graph corresponding to the cellular automaton described above, for the cases $N = 11$ and $N = 12$. Configurations corresponding to nodes on the periphery of the graph are seen to be depopulated by transitions; all initial configurations ultimately evolve to configurations on one of the cycles in the graph. Any finite cellular automaton ultimately enters a cycle in which a sequence of configurations are visited repeatedly. This behaviour is illustrated in Fig. 3.

Cellular automata may be used as simple models for a wide variety of physical, biological and computational systems. Analysis of general features of their behaviour may therefore yield general results on the behaviour of many complex systems, and may perhaps ultimately suggest generalizations of the laws of thermodynamics appropriate for systems with irreversible dynamics. Several aspects of cellular automata were recently discussed in [1], where extensive references were given. This paper details and extends the discussion of global properties of cellular automata given in [1]. These global properties may be described in terms of properties of the state transition graphs corresponding to the cellular automata.

This paper concentrates on a class of cellular automata which exhibit the simplifying feature of "additivity". The configurations of such cellular automata satisfy an "additive superposition" principle, which allows a natural representation of the configurations by characteristic polynomials. The time evolution of

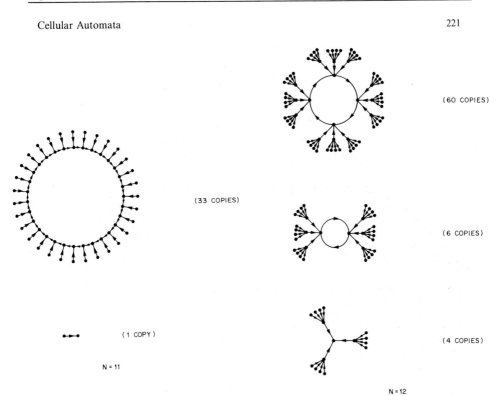

Fig. 2. Global state transition diagrams for finite cellular automata with size N and periodic boundary conditions evolving according to the rule $\mathbb{T}(x) = x + x^{-1}$, as used in Fig. 1, and discussed extensively in Sect. 3. Each node in the graphs represents one of the 2^N possible configurations of the N sites. The directed edges of the graphs indicate transitions between these configurations associated with single time steps of cellular automaton evolution. Each cycle in the graph represents an "attractor" for the configurations corresponding to the nodes in trees rooted on it

the configurations is represented by iterated multiplication of their characteristic polynomials by fixed polynomials. Global properties of cellular automata are then determined by algebraic properties of these polynomials, by methods analogous to those used in the analysis of linear feedback shift registers [2, 3]. Despite their amenability to algebraic analysis, additive cellular automata exhibit many of the complex features of general cellular automata.

Having introduced notation in Sect. 2, Sect. 3 develops algebraic techniques for the analysis of cellular automata in the context of the simple cellular automaton illustrated in Fig. 1. Some necessary mathematical results are reviewed in the appendices. Section 4 then derives general results for all additive cellular automata. The results allow more than two possible values per site, but are most complete when the number of possible values is prime. They also allow influence on the evolution of a site from sites more distant than its nearest neighbours. The results are extended in Sect. 4D to allow cellular automata in which the sites are arranged in a square or cubic lattice in two, three or more dimensions, rather than

222   O. Martin, A. M. Odlyzko, and S. Wolfram

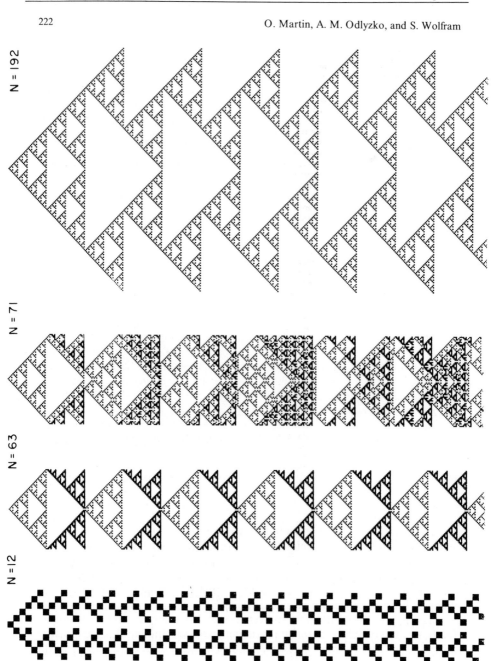

Fig. 3. Evolution of cellular automata with N sites arranged in a circle (periodic boundary conditions) according to the rule $\mathbb{T}(x) = x + x^{-1}$ (as used in Fig. 1 and discussed in Sect. 3). Finite cellular automata such as these ultimately enter cycles in which a sequence of configurations are visited repeatedly. This behaviour is evident here for $N = 12$, 63, and 192. For $N = 71$, the cycle has length $2^{35} - 1$

just on a line. Section 4E then discusses generalizations in which the cellular automaton time evolution rule involves several preceding time steps. Section 4F considers alternative boundary conditions. In all cases, a characterization of the global structure of the state transition diagram is found in terms of algebraic properties of the polynomials representing the cellular automaton time evolution rule.

Section 5 discusses non-additive cellular automata, for which the algebraic techniques of Sects. 3 and 4 are inapplicable. Combinatorial methods are nevertheless used to derive some results for a particular example.

Section 6 gives a discussion of the results obtained, comparing them with those for other systems.

2. Formalism

We consider first the formalism for one-dimensional cellular automata in which the evolution of a particular site depends on its own value and those of its nearest neighbours. Section 4 generalizes the formalism to several dimensions and more neighbours.

We take the cellular automaton to consist of N sites arranged around a circle (so as to give periodic boundary conditions). The values of the sites at time step t are denoted $a_0^{(t)}, ..., a_{N-1}^{(t)}$. The possible site values are taken to be elements of a finite commutative ring $\mathbb{R}_k$ with k elements. Much of the discussion below concerns the case $\mathbb{R}_k = \mathbb{Z}_k$, in which site values are conveniently represented as integers modulo k. In the example considered in Sect. 3, $\mathbb{R}_k = \mathbb{Z}_2$, and each site takes on a value 0 or 1.

The complete configuration of a cellular automaton is specified by the values of its N sites, and may be represented by a characteristic polynomial (generating function) (cf. [2, 3])

$$A^{(t)}(x) = \sum_{i=0}^{N-1} a_i^{(t)} x^i, \qquad (2.1)$$

where the value of site i is the coefficient of x^i, and all coefficients are elements of the ring $\mathbb{R}_k$. We shall often refer to configurations by their corresponding characteristic polynomials.

It is often convenient to consider generalized polynomials containing both positive and negative powers of x: such objects will be termed "dipolynomials". In general, $H(x)$ is a dipolynomial if there exists some integer m such that $x^m H(x)$ is an ordinary polynomial in x. As discussed in Appendix A, dipolynomials possess divisibility and congruence properties analogous to those of ordinary polynomials.

Multiplication of a characteristic polynomial $A(x)$ by $x^{\pm j}$ yields a dipolynomial which represents a configuration in which the value of each site has been transferred (shifted) to a site j places to its right (left). Periodic boundary conditions in the cellular automaton are implemented by reducing the characteristic dipolynomial modulo the fixed polynomial $x^N - 1$ at all stages, according to

$$\sum_i a_i x^i \bmod (x^N - 1) = \sum_{i=0}^{N-1} \left(\sum_j a_{i+jN} \right) x^i. \qquad (2.2)$$

Note that any dipolynomial is congruent modulo $(x^N - 1)$ to a unique ordinary polynomial of degree less than N.

In general, the value $a_i^{(t)}$ of a site in a cellular automaton is taken to be an arbitrary function of the values $a_{i-1}^{(t-1)}$, $a_i^{(t-1)}$, and $a_{i+1}^{(t-1)}$ at the previous time step. Until Sect. 5, we shall consider a special class of "additive" cellular automata which evolve with time according to simple linear combination rules of the form (taking the site index i modulo N)

$$a_i^{(t)} = \alpha_{-1} a_{i-1}^{(t-1)} + \alpha_0 a_i^{(t-1)} + \alpha_{+1} a_{i+1}^{(t-1)}, \tag{2.3}$$

where the α_j are fixed elements of $\mathbb{R}_k$, and all arithmetic is performed in $\mathbb{R}_k$. This time evolution may be represented by multiplication of the characteristic polynomial by a fixed dipolynomial in x,

$$\mathbb{T}(x) = \alpha_{-1} x + \alpha_0 + \alpha_{+1} x^{-1}, \tag{2.4}$$

according to

$$A^{(t)}(x) \equiv \mathbb{T}(x) A^{(t-1)}(x) \quad \mod(x^N - 1), \tag{2.5}$$

where arithmetic is again performed in $\mathbb{R}_k$. Additive cellular automata obey an additive superposition principle which implies that the configuration obtained by evolution for t time steps from an initial configuration $A^{(0)}(x) + B^{(0)}(x)$ is identical to $A^{(t)}(x) + B^{(t)}(x)$, where $A^{(t)}(x)$ and $B^{(t)}(x)$ are the results of separate evolution of $A^{(0)}(x)$ and $B^{(0)}(x)$, and all addition is performed in $\mathbb{R}_k$. Since any initial configuration can be represented as a sum of "basis" configurations $\Delta(x) = x^j$ containing single nonzero sites with unit values, the additive superposition principle determines the evolution of all configurations in terms of the evolution of $\Delta(x)$. By virtue of the cyclic symmetry between the sites it suffices to consider the case $j = 0$.

3. A Simple Example

A. Introduction

This section introduces algebraic techniques for the analysis of additive cellular automata in the context of a specific simple example. Section 4 applies the techniques to more general cases. The mathematical background is outlined in the appendices.

The cellular automaton considered in this section consists of N sites arranged around a circle, where each site has value 0 or 1. The sites evolve so that at each time step the value of a site is the sum modulo two of the values of its two nearest neighbours at the previous time step:

$$a_i^{(t)} = a_{i-1}^{(t-1)} + a_{i+1}^{(t-1)} \quad \mod 2. \tag{3.1}$$

This rule yields in many respects the simplest non-trivial cellular automaton. It corresponds to rule 90 of [1], and has been considered in several contexts elsewhere (e.g. [4]).

The time evolution (3.1) is represented by multiplication of the characteristic polynomial for a configuration by the dipolynomial

$$\mathbb{T}(x) = x + x^{-1} \tag{3.2}$$

according to Eq. (2.5). At each time step, characteristic polynomials are reduced modulo $x^N - 1$ (which is equal to $x^N + 1$ since all coefficients are here, and throughout this section, taken modulo two). This procedure implements periodic boundary conditions as in Eq. (2.2) and removes any inverse powers of x.

Equation (3.2) implies that an initial configuration containing a single nonzero site evolves after t time steps to a configuration with characteristic dipolynomial

$$\mathbb{T}(x)^t 1 = (x + x^{-1})^t = \sum_{i=0}^{t} \binom{t}{i} x^{2i-t}. \tag{3.3}$$

For $t < N/2$ (before "wraparound" occurs), the region of nonzero sites grows linearly with time, and the values of sites are given simply by binomial coefficients modulo two, as discussed in [1] and illustrated in Fig. 1. (The positions of nonzero sites are equivalently given by $\pm 2^{j_1} + 2^{j_2} \pm ...$, where the j_i give the positions of nonzero digits in the binary decomposition of the integer t.) The additive superposition property implies that patterns generated from initial configurations containing more than one nonzero site may be obtained by addition modulo two (exclusive disjunction) of the patterns (3.3) generated from single nonzero sites.

B. Irreversibility

Every configuration in a cellular automaton has a unique successor in time. A configuration may however have several distinct predecessors, as illustrated in the state transition diagram of Fig. 2. The presence of multiple predecessors implies that the time evolution mapping is not invertible but is instead "contractive". The cellular automaton thus exhibits irreversible behaviour in which information on initial states is lost through time evolution. The existence of configurations with multiple predecessors implies that some configurations have no predecessors[1]. These configurations occur only as initial states, and may never be generated in the time evolution of the cellular automaton. They appear on the periphery of the state transition diagram of Fig. 2. Their presence is an inevitable consequence of irreversibility and of the finite number of states.

Lemma 3.1. *Configurations containing an odd number of sites with value 1 can never be generated in the evolution of the cellular automaton defined in Sect. 3A, and can occur only as initial states.*

Consider any configuration specified by characteristic polynomial $A^{(0)}(x)$. The successor of this configuration is $A^{(1)}(x) = \mathbb{T}(x) A^{(0)}(x) = (x + x^{-1}) A^{(0)}(x)$, taken, as always, modulo $x^N - 1$. Thus

$$A^{(1)}(x) = (x^2 + 1) B(x) + R(x) (x^N - 1)$$

for some dipolynomials $R(x)$ and $B(x)$. Since $x^2 + 1 = x^N - 1 = 0$ for $x = 1$, $A^{(1)}(1) = 0$. Hence $A^{(1)}(x)$ contains an even number of terms, and corresponds to a configuration with an even number of nonzero sites. Only such configurations can therefore be reached from some initial configuration $A^{(0)}(x)$.

An extension of this lemma yields the basic theorem on the number of unreachable configurations:

---

1   Such configurations have been termed "Gardens of Eden" [5]

Theorem 3.1. *The fraction of the* 2^N *possible configurations of a size N cellular automaton defined in Sect. 3A which can occur only as initial states, and cannot be reached by evolution, is* $1/2$ *for N odd and* $3/4$ *for N even.*

A configuration $A^{(1)}(x)$ is reachable after one time step of cellular automaton evolution if and only if for some dipolynomial $A^{(0)}(x)$,

$$A^{(1)}(x) \equiv \mathbb{T}(x)A^{(0)}(x) \equiv (x+x^{-1})A^{(0)}(x) \quad \mod(x^N-1), \tag{3.4}$$

so that

$$A^{(1)}(x) = (x^2+1)B(x) + R(x)(x^N-1) \tag{3.5}$$

for some dipolynomials $R(x)$ and $B(x)$. To proceed, we use the factorization of (x^N-1) given in Eq. (A.7), and consider the cases N even and N odd separately.

(a) N even. Since by Eq. (A.4), $(x^2+1)=(x+1)^2=(x-1)^2$ (taken, as always, modulo 2), and by Eq. (A.7),

$$(x-1)^2 \,|\, (x^{N/2}-1)^2 = (x^N-1)$$

for even N, Eq. (3.5) shows that

$$(x-1)^2 \,|\, A^{(1)}(x)$$

in this case. But since $(x-1)^2$ contains a constant term, $A^{(1)}(x)/(x-1)^2$ is thus an ordinary polynomial if $A^{(1)}(x)$ is chosen as such. Hence all reachable configuratons represented by a polynomial $A^{(1)}(x)$ are of the form

$$A^{(1)}(x) = (x-1)^2 C(x),$$

for some polynomial $C(x)$. The predecessor of any such configuration is $xC(x)$, so any configuration of this form may in fact be reached. Since $\deg A(x) < N$, $\deg C(x) < N-2$. There are thus exactly 2^{N-2} reachable configurations, or $1/4$ of all the 2^N possible configurations.

(b) N odd. Using Lemma 3.1 the proof for this case is reduced to showing that all configurations containing an even number of nonzero sites have predecessors. A configuration $A^{(1)}(x)$ with an even number of nonzero sites can always be written in the form $(x+1)D(x)$. But

$$A^{(1)}(x) = (x+1)D(x) \equiv (x+x^{-1})(x^2+x^4+ \ldots +x^{N-1})D(x) \quad \mod(x^N-1)$$
$$\equiv \mathbb{T}(x)(x^2+x^4+ \ldots +x^{N-1})D(x) \quad \mod(x^N-1),$$

giving an explicit predecessor for $A^{(1)}(x)$.

The additive superposition principle for the cellular automaton considered in this section yields immediately the result:

Lemma 3.2. *Two configurations* $A^{(0)}(x)$ *and* $B^{(0)}(x)$ *yield the same configuration* $C(x) \equiv \mathbb{T}(x)A^{(0)}(x) \equiv \mathbb{T}(x)B^{(0)}$ *after one time step in the evolution of the cellular automaton defined in Sect. 3A if and only if* $A^{(0)}(x) = B^{(0)}(x) + Q(x)$, *where* $\mathbb{T}(x)Q(x) \equiv 0$.

Theorem 3.2. *Configurations in the cellular automaton defined in Sect. 3A which have at least one predecessor have exactly two predecessors for N odd and exactly four for N even.*

This theorem is proved using Lemma 3.2 by enumeration of configurations $Q(x)$ which evolve to the null configuration after one time step. For N odd, only the configurations 0 and $1 + x + \dots + x^{N-1} = \dfrac{x^N - 1}{x - 1}$ (corresponding to site values $11111 \dots$) have this property. For N even, $Q(x)$ has the form

$$(1 + x^2 + \dots + x^{N-2})S_i(x) = \frac{x^N - 1}{x^2 - 1} S_i(x),$$

where the $S_i(x)$ are the four polynomials of degree less than two. Explicitly, the possible forms for $Q(x)$ are 0, $1 + x^2 + \dots + x^{N-2}$, $x + x^3 + \dots + x^{N-1}$, and $1 + x + x^2 + \dots + x^{N-1}$.

C. Topology of the State Transition Diagram

This subsection derives topological properties of the state transition diagrams illustrated in Fig. 2. The results determine the amount and rate of "information loss" or "self organization" associated with the irreversible cellular automaton evolution.

The state transition network for a cellular automaton is a graph, each of whose nodes represents one of the possible cellular automaton configurations. Directed arcs join the nodes to represent the transitions between cellular automaton configurations at each time step. Since each cellular automaton configuration has a unique successor, exactly one arc must leave each node, so that all nodes have out-degree one. As discussed in the previous subsection, cellular automaton configurations may have several or no predecessors, so that the in-degrees of nodes in the state transition graph may differ. Theorems 3.1 and 3.2 show that for N odd, $1/2$ of all nodes have zero in-degree and the rest have in-degree two, while for N even, $3/4$ have zero in-degree and $1/4$ in-degree four.

As mentioned in Sect. 1, after a possible "transient", a cellular automaton evolving from any initial configuration must ultimately enter a loop, in which a sequence of configurations are visited repeatedly. Such a loop is represented by a cycle in the state transition graph. At every node in this cycle a tree is rooted; the transients consist of transitions leading towards the cycle at the root of the tree.

Lemma 3.3. *The trees rooted at all nodes on all cycles of the state transition graph for the cellular automaton defined in Sect. 3A are identical.*

This result is proved by showing that trees rooted on all cycles are identical to the tree rooted on the null configuration. Let $A(x)$ be a configuration which evolves to the null configuration after exactly t time steps, so that $\mathbb{T}(x)^t A(x) \equiv 0 \bmod (x^N - 1)$. Let $R(x)$ be a configuration on a cycle, and let $R^{(-t)}(x)$ be another configuration on the same cycle, such that $\mathbb{T}(x)^t R^{(-t)}(x) \equiv R(x) \bmod (x^N - 1)$. Then define

$$\Psi_{R(x)}[A(x)] = A(x) + R^{(-t)}(x).$$

We first show that as $A(x)$ ranges over all configurations in the tree rooted on the null configuration, $\Psi_{R(x)}[A(x)]$ ranges over all configurations in the tree rooted at $R(x)$. Since

$$\mathbb{T}(x)^t \Psi_{R(x)}[A(x)] = \mathbb{T}(x)^t A(x) + \mathbb{T}(x)^t R^{(-t)}(x) \equiv R(x) \qquad \bmod (x^N - 1),$$

it is clear that all configurations $\Psi_{R(x)}[A(x)]$ evolve after t time steps [where the value of t depends on $A(x)$] to $R(x)$. To show that these configurations lie in the tree rooted at $R(x)$, one must show that their evolution reaches no other cycle configurations for any $s < t$. Assume this supposition to be false, so that there exists some $m \neq 0$ for which

$$R^{(-m)}(x) \equiv \mathbb{T}(x)^s \Psi_{R(x)}[A(x)] = \mathbb{T}(x)^s A(x) + R^{(s-t)}(x) \quad \mod(x^N - 1).$$

Since $\mathbb{T}(x)^t A(x) \equiv 0 \mod(x^N - 1)$, this would imply $R^{(t-s-m)}(x) = R^{(0)}(x) = R(x)$, or $R^{(-m)}(x) = R^{(s-t)}(x)$. But $R^{(-m)}(x) - R^{(s-t)}(x) \equiv \mathbb{T}(x)^s A(x)$, and by construction $\mathbb{T}(x)^s A(x) \neq 0$ for any $s < t$, yielding a contradiction. Thus $\Psi_{R(x)}$ maps configurations at height t in the tree rooted on the null configuration to configurations at height t in the tree rooted at $R(x)$, and the mapping Ψ is one-to-one. An analogous argument shows that Ψ is onto. Finally one may show that Ψ preserves the time evolution structure of the trees, so that if $\mathbb{T}(x)A^{(0)}(x) = A^{(1)}(x)$, then

$$\mathbb{T}(x)\Psi_{R(x)}[A^{(0)}(x)] = \Psi_{R(x)}[A^{(1)}(x)],$$

which follows immediately from the definition of Ψ. Hence Ψ is an isomorphism, so that trees rooted at cycle configurations are all isomorphic to that rooted at the null configuration.

Notice that this proof makes no reference to the specific form (3.2) chosen for $\mathbb{T}(x)$ in this section; Lemma 3.3 thus holds for any additive cellular automaton.

Theorem 3.3. *For N odd, a tree consisting of a single arc is rooted at each node on each cycle in the state transition graph for the cellular automaton defined in Sect. 3A.*

By virtue of Lemma 3.3, it suffices to show that the tree rooted on the null configuration consists of a single node corresponding to the configuration $111 \dots 111$. This configuration has no predecessors by virtue of Lemma 3.1.

Corollary. *For N odd, the fraction of the 2^N possible configurations which may occur in the evolution of the cellular automaton defined in Sect. 3A is $1/2$ after one or more time steps.*

The "distance" between two nodes in a tree is defined as the number of arcs which are visited in traversing the tree from one node to the other (e.g. [6]). The "height" of a (rooted) tree is defined as the maximum number of arcs traversed in a descent from any leaf or terminal (node with zero in-degree) to the root of the tree (formally node with zero out-degree). A tree is "balanced" if all its leaves are at the same distance from its root. A tree is termed "quaternary" ("binary") if each of its non-terminal nodes has in-degree four (two).

Let $D_2(N)$ be the maximum 2^j which divides N (so that for example $D_2(12) = 4$).

Theorem 3.4. *For N even, a balanced tree with height $D_2(N)/2$ is rooted at each node on each cycle in the state transition graph for the cellular automaton defined in Sect. 3A; the trees are quaternary, except that their roots have in-degree three.*

Theorem 3.2 shows immediately that the tree is quaternary. In the proof of Theorem 3.1, we showed that a configuration $Q_1(x)$ can be reached from some configuration $Q_0(x)$ if and only if $(1+x^2)|Q_1(x)$; Theorem 3.2 then shows that if $Q_1(x)$ is reachable, it is reachable from exactly four distinct configurations $Q_0(x)$. We now extend this result to show that a configuration $Q_m(x)$ can be reached from some configuration $Q_0(x)$ by evolution for m time steps, with $m \leq D_2(N)/2$, if and only if $(1+x^2)^m|Q_m(x)$. To see this, note that if

$$Q_m(x) \equiv \mathbb{T}(x)^m Q_0(x) \quad \mod(x^N - 1), \tag{3.6}$$

then

$$(x^N - 1)|Q_m(x) + (x^2 + 1)^m x^{N-m} Q_0(x), \tag{3.7}$$

and so, since by Eq. (A.7), $(x^2 + 1)^m|(x^N - 1)$ for $m \leq D_2(N)/2$, it follows that

$$(x^2 + 1)^m | Q_m(x) \tag{3.8}$$

for $m \leq D_2(N)/2$. On the other hand, if $(x^2 + 1)^m|Q_m(x)$, say $Q_m(x) = (x^2 + 1)^m Q_0(x)$, then $Q_m(x) \equiv \mathbb{T}(x)^m x^m Q_0(x)$, which shows that $Q_m(x)$ is reachable in m steps. The balance of the trees is demonstrated by showing that for $m < D_2(N)/2$, if $(x^2 + 1)^m|Q_m(x)$, then $Q_m(x)$ can be reached from exactly 4^m initial configurations $Q_0(x)$. This may be proved by induction on m. If

$$(1+x^2)^m | Q_m(x) \quad (1 \leq m < D_2(N)/2), $$

then all of the four states $Q_{m-1}(x)$ from which $Q_m(x)$ may be reached in one step satisfy $(x^2 + 1)^{m-1}|Q_{m-1}(x)$. Consider now the configurations $Q(x)$ which satisfy

$$(x^2 + 1)^{D_2(N)/2} | Q(x). \tag{3.9}$$

If we write $Q(x) = (x+1)^{D_2(N)} R(x)$, then as in Theorem 3.2, the four predecessors of $Q(x)$ are exactly

$$Q_{-1}(x) = (x+1)^{D_2(N)-2} R^*(x) + \left(\frac{x^{N/2} - 1}{x - 1}\right)^2 S_i(x), \tag{3.10}$$

where $xR(x) \equiv R^*(x) \mod(x^N - 1)$. $S_i(x)$ ranges over the four polynomials of degree less than two, as in Theorem 3.2. Exactly one of these polynomials satisfies Eq. (3.9), whereas the other three satisfy only

$$(x+1)^{D_2(N)-2} | Q_{-1}(x). $$

Any state satisfying Eq. (3.9) thus belongs to a cycle, since it can be reached after an arbitrary number of steps. Conversely, since any cycle configuration must be reachable after $D_2(N)/2$ time steps, any and all configurations $Q_{-1}(x)$ satisfying Eq. (3.9) are indeed on cycles. But, as shown above, the three $Q_{-1}(x)$ which do not satisfy Eq. (3.9) are roots of balanced quaternary trees of height $D_2(N)/2-1$. The proof of the theorem is thus completed.

Corollary. *For N even, a fraction 4^{-t} of the 2^N possible configurations appear after t steps in the evolution of the cellular automaton defined in Sect. 3A for $t \leq D_2(N)/2$. A fraction $2^{-D_2(N)}$ of the configurations occur in cycles, and are therefore generated at arbitrarily large times.*

Corollary. *All configurations $A(x)$ on cycles in the cellular automaton of Sect. 3A are divisible by $(1+x)^{D_2(N)}$.*

This result follows immediately from the proof of Theorems 3.3 and 3.4.

Entropy may be used to characterize the irreversibility of cellular automaton evolution (cf. [1]). One may define a set (or topological) entropy for an ensemble of configurations i occurring with probabilities p_i according to

$$s = \frac{1}{N} \log_2 \sum_i \theta(p_i), \qquad (3.11)$$

where $\theta(p) = 1$ for $p > 0$, and 0 otherwise. One may also define a measure entropy

$$s_\mu = -\frac{1}{N} \sum_i p_i \log_2 p_i. \qquad (3.12)$$

For a maximal entropy ensemble in which all 2^N possible cellular automaton configurations occur with equal probabilities,

$$s = s_\mu = 1.$$

These entropies decrease in irreversible cellular automaton evolution, as the probabilities for different configurations become unequal. However, the balance property of the state transition trees implies that configurations either do not appear, or occur with equal nonzero probabilities. Thus the set and measure entropies remain equal in the evolution of the cellular automaton of Sect. 3A. Starting from a maximal entropy ensemble, both nevertheless decrease with time t according to

$$s(t) = s_\mu(t) = 1 - 2t/N, \qquad 0 \le t \le D_2(N)/2,$$

$$s(t) = s_\mu(t) = 1 - D_2(N)/N, \qquad t \ge D_2(N)/2.$$

D. Maximal Cycle Lengths

Lemma 3.4. *The lengths of all cycles in a cellular automaton of size N as defined in Sect. 3A divide the length Π_N of the cycle obtained with an initial configuration containing a single site with value one.*

This follows from additivity, since any configuration can be considered as a superposition of configurations with single nonzero initial sites.

Lemma 3.5. *For the cellular automaton defined in Sect. 3A, with N of the form 2^j, $\Pi_N = 1$.*

In this case, any initial configuration evolves ultimately to a fixed point consisting of the null configuration, since

$$(x + x^{-1})^{2^j} 1 \equiv (x^{2^j} + x^{-2^j}) \equiv (x^N + x^{-N}) \equiv 0 \quad \bmod(x^N - 1).$$

Lemma 3.6. *For the cellular automaton defined in Sect. 3A, with N even but not of the form 2^j, $\Pi_N = 2\,\Pi_{N/2}$.*

A configuration $A(x)$ appears in a cycle of length π if and only if

$$\mathbb{T}(x)^\pi A(x) \equiv A(x) \quad \bmod(x^N - 1),$$

and therefore

$$(x^N - 1) \mid [(x^2 + 1)^\pi + x^\pi] A(x).$$

After t time steps, the configuration obtained by evolution from an initial state containing a single nonzero site is $(x + x^{-1})^t$; by Theorems 3.3 and 3.4 and the additive superposition principle, the configuration

$$A(x) \equiv (x + x^{-1})^{D_2(N)/2}$$

is therefore on the maximal length cycle. Thus the maximal period Π_N is given by the minimum π for which

$$(x^N - 1) \mid [(x^2 + 1)^\pi + x^\pi] (x + 1)^{D_2(N)},$$

and so

$$\left(\frac{x^n - 1}{x + 1} \right)^{D_2(N)} \mid [(x^2 + 1)^{\Pi_N} + x^{\Pi_N}], \tag{3.13}$$

with $N = D_2(N)n$, n odd. Similarly,

$$(x^{N/2} - 1) \mid [(x^2 + 1)^{\Pi_{N/2}} + x^{\Pi_{N/2}}] (x + 1)^{D_2(N/2)},$$

$$\left(\frac{x^n - 1}{x + 1} \right)^{D_2(N)/2} \mid [(x^2 + 1)^{\Pi_{N/2}} + x^{\Pi_{N/2}}]. \tag{3.14}$$

Squaring this yields

$$\left(\frac{x^n - 1}{x + 1} \right)^{D_2(N)} \mid [(x^2 + 1)^{2\Pi_{N/2}} + x^{2\Pi_{N/2}}],$$

from which it follows that

$$\Pi_N \mid 2\Pi_{N/2}. \tag{3.15}$$

Since $x^N - 1$ divides $[(x^2 + 1)^{\Pi_N} + x^{\Pi_N}] (x + 1)^{D_2(N)}$, so does its square root, $x^{N/2} - 1$, and therefore

$$\Pi_{N/2} \mid \Pi_N. \tag{3.16}$$

Combining Eqs. (3.15) and (3.16) implies that either $\Pi_N = 2\Pi_{N/2}$ or $\Pi_N = \Pi_{N/2}$. To exclude the latter possibility, we use derivatives. Using Eq. (A.6), and the fact that the derivative of $x^2 + 1$ vanishes over $GF(2)$, one obtains from (3.13),

$$\left(\frac{x^n - 1}{x + 1} \right) \mid \Pi_N x^{\Pi_N - 1}.$$

If Π_N were odd, the right member would be non-trivial, and the divisibility condition could not hold. Thus Π_N must be even. But then the right member of (3.13) is a perfect square, so that

$$\left(\frac{x^{N/2} - 1}{(x + 1)^{D_2(N)/2}} \right)^2 \mid [(x^2 + 1)^{\Pi_{N/2}} + x^{\Pi_{N/2}}]^2.$$

Thus $\Pi_{N/2} \mid \Pi_N/2$, and the proof is complete.

Theorem 3.5. *For the cellular automaton defined in Sect. 3A, with N odd,* $\Pi_N | \Pi_N^* = 2^{\text{sord}_N(2)} - 1$ *where* $\text{sord}_N(2)$ *is the multiplicative "sub-order" function of 2 modulo N, defined as the least integer j such that* $2^j = \pm 1 \bmod N$. *(Properties of the suborder functions are discussed in Appendix B.)*

By Lemma 3.1, an initial configuration containing a single nonzero site cannot be reached in cellular automaton evolution. The configuration $(x + x^{-1}) \bmod (x^N - 1)$ obtained from this after one time step can be reached, and in fact appears again after $2^{\text{sord}_N(2)} - 1$ time steps, since

$$\mathbb{T}(x)^{2^{\text{sord}_N(2)}} 1 \equiv (x + x^{-1})^{2^{\text{sord}_N(2)}} \equiv (x^{2^{\text{sord}_N(2)}} + x^{-2^{\text{sord}_N(2)}})$$

$$\equiv (x^{\pm 1} + x^{\mp 1}) \equiv (x + x^{-1}) \qquad \bmod(x^N - 1).$$

The maximal cycle lengths Π_N for the cellular automaton considered in this section are given in the first column of Table 1. The values are plotted as a function of N in Fig. 4. Table 1 together with Table 4 show that $\Pi_N = \Pi_N^*$ for almost all odd N. The first exception appears for $N = 37$, where $\Pi_N = \Pi_N^*/3$; subsequent exceptions are $\Pi_{95} = \Pi_{95}^*/3$, $\Pi_{101} = \Pi_{101}^*/3$, $\Pi_{141} = \Pi_{141}^*/3$, $\Pi_{197} = \Pi_{197}^*/3$, $\Pi_{199} = \Pi_{199}^*/7$, $\Pi_{203} = \Pi_{203}^*/105$ and so on.

As discussed in Appendix B, $\text{sord}_N(2) \leq (N-1)/2$. This bound can be attained only when N is prime. It implies that the maximal period is $2^{(N-1)/2} - 1$. Notice that this period is the maximum that could be attained with any reflection symmetric initial configuration (such as the single nonzero site configuration to be considered by virtue of Lemma 3.4).

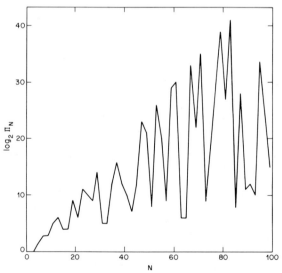

Fig. 4. The maximal length Π_N of cycles generated in the evolution of a cellular automaton with size N and $\mathbb{T}(x) = x + x^{-1}$, as a function of N. Only values for integer N are plotted. The irregular behaviour of Π_N as a function of N is a consequence of the dependence of Π_N on number theoretical properties of N

Table 1. Maximal cycle lengths Π_N for one-dimensional nearest-neighbour additive cellular automata with size N and k possible values at each site. Results for all possible nontrivial symmetrical rules with $k \leq 4$ are given. For $k=2$, the fixed time evolution polynomials are $\mathbb{T}(x) = x + x^{-1}$ and $x + 1 + x^{-1}$ (corresponding to rules 90 and 150 of [1], respectively). For $k=3$, the polynomials are $x + x^{-1}$, $x + 1 + x^{-1}$, and $x + 2 + x^{-1}$, while for $k=4$, they are $x + x^{-1}$, $x + 1 + x^{-1}$, $x + 2 + x^{-1}$, and $x + 3 + x^{-1}$

N	$k=2$		$k=3$			$k=4$			
3	1	1	6	1	3	2	2	1	1
4	1	2	2	2	2	1	4	1	4
5	3	3	8	8	4	6	6	3	6
6	2	1	6	6	3	2	2	2	2
7	7	7	26	26	13	14	14	7	14
8	1	4	4	8	8	1	8	1	8
9	7	7	18	1	9	14	14	7	14
10	6	6	8	8	8	6	12	6	12
11	31	31	242	121	121	62	62	31	62
12	4	2	6	6	6	4	4	4	4
13	63	21	26	13	13	126	42	63	42
14	14	14	26	26	13	14	28	14	28
15	15	15	24	24	12	30	30	15	30
16	1	8	16	80	80	1	16	1	16
17	15	15	1,640	6,560	820	30	30	15	30
18	14	14	18	18	9	14	28	14	28
19	511	511	19,682	19,682	9,841	1,022	1,022	511	1,022
20	12	12	16	40	40	12	24	12	24
21	63	63	78	78	39	126	126	63	126
22	62	62	242	242	242	62	124	62	124
23	2,047	2,047	177,146	88,573	88,573	4,094	4,094	2,047	4,094
24	8	4	12	24	24	8	8	8	8
25	1,023	1,023	59,048	59,048	29,524	2,046	2,046	1,023	2,046
26	126	42	26	26	26	126	84	126	84
27	511	511	54	1	27	1,022	1,022	511	1,022
28	28	28	26	26	26	28	56	28	56
29	16,383	16,383	4,782,968	4,782,968	2,391,484	32,766	32,766	16,383	32,766
30	30	30	24	24	24	30	60	30	60
31	31	31	1,103,762	14,348,906	551,881	62	62	31	62
32	1	16	160	6,560	6,560	1	32	1	32
33	31	31	726	363	363	62	62	31	62
34	30	30	1,640	6,560	6,560	30	60	30	60
35	4,095	4,095	265,720	265,720	132,860	8,190	8,190	4,095	8,190
36	28	28	18	18	18	28	56	28	56
37	87,381	29,127	19,682	19,682	9,841	174,762	58,254	87,381	58,254
38	1,022	1,022	19,682	19,682	9,841	1,022	2,044	1,022	2,044
39	4,095	4,095	78	39	39	8,190	8,190	4,095	8,190
40	24	24	80	40	40	24	48	24	48

E. Cycle Length Distribution

Lemma 3.4 established that all cycle lengths must divide Π_N and Theorems 3.3 and 3.4 gave the total number of states in cycles. This section considers the number of distinct cycles and their lengths.

Lemma 3.7. *For the cellular automaton defined in Sect. 3A, with N a multiple of 3, there are four distinct fixed points (cyles of length one); otherwise, only the null configuration is a fixed point.*

For $N = 3n$, the only stationary configurations are 000000 ... (null configuration), 0110110 ..., 1011011 ..., and 1101101 ....

Table 2 gives the lengths and multiplicities of cycles in the cellular automaton defined in Sect. 3A, for various values of N. One result suggested by the table is that the multiplicity of cycles for a particular N increases with the length of the cycle, so that for large N, an overwhelming fraction of all configurations in cycles are on cycles with the maximal length.

When Π_N is prime, the only possible cycle lengths are Π_N and 1. Then, using Lemma 3.7, the number of cycles of length Π_N is $(2^{(N-1)}-4)/\Pi_N$ for $N = 3n$, and is $(2^{(N-1)}-1)/\Pi_N$ otherwise.

When Π_N is not prime, cycles may exist with lengths corresponding to various divisors of Π_N. It has not been possible to express the lengths and multiplicities of cycles in this case in terms of simple functions. We nevertheless give a computationally efficient algorithm for determining them.

Theorems 3.3 and 3.4 show that any configuration $A(x)$ on a cycle may be written in the form

$$A(x) = (1+x)^{D_2(N)}B(x),$$

where $B(x)$ is some polynomial. The cycle on which $A(x)$ occurs then has a length given by the minimum π for which

$$\mathbb{T}(x)^{\pi}B(x) \equiv (x+x^{-1})^{\pi}B(x) \equiv B(x) \quad \mod\left(\frac{x^n-1}{x+1}\right)^{D_2(N)}, \qquad (3.17)$$

where $N = D_2(N)n$ with n odd, and $(x^n-1)^{D_2(N)} = x^N - 1$. Using the factorization [given in Eq. (A.8)]

$$x^n - 1 = (x-1) \prod_{\substack{d|n \\ d\neq 1}} \prod_{i=1}^{\frac{\phi(d)}{\mathrm{ord}_d(2)}} C_{d,i}(x), \qquad (3.18)$$

where the $C_{d,i}(x)$ are the irreducible cyclotomic polynomials over $\mathbb{Z}_2$ of degree $\mathrm{ord}_d(2)$, Eq. (3.17) can be rewritten as

$$(x+x^{-1})^{\pi}B(x) \equiv B(x) \quad \mod C_{d,i}(x)^{D_2(N)} \qquad (3.19)$$

for all $d|n$, $d \neq 1$, and for all i such that $1 \leq i \leq \phi(d)/\mathrm{ord}_d(2)$. Let $\pi_{d,i}[B(x)]$ denote the smallest π for which (3.19) holds with given d, i. Then the length of the cycle on which $A(x)$ occurs is exactly the least common multiple of all the $\pi_{d,i}[B(x)]$. If $C_{d,i}(x)^{D_2(N)}|B(x)$, then clearly Eq. (3.19) holds for $\pi = 1$, and $\pi_{d,i}[B(x)] = 1$. If $C_{d,i}(x)^{r_{d,i}[B(x)]}||B(x)$ (and $0 \leq r_{d,i}[B(x)] < D_2(N)$), then Eq. (3.19) is equivalent to

$$(x+x^{-1})^{\pi} \equiv 1 \quad \mod C_{d,i}(x)^{D_2(N)-r_{d,i}[B(x)]}. \qquad (3.20)$$

The values of $\pi_{d,i}$ for configurations with $r_{d,i}[B(x)] = s$ are therefore equal, and will be denoted $\pi_{d,i,s}$ ($0 \leq s \leq D_2(N)$). Since $C_{d,i}(x)|(x^d-1)/(x+1)$ ($d \neq 1$), the value of $\pi_{d,i,1}$ divides the minimum π for which $(x+x^{-1})^{\pi} \equiv 1 \mod(x^d-1)/(x+1)$. This

Cellular Automata   235

Table 2. Multiplicities and lengths of cycles in the cellular automaton of Sect. 3A with size N. The notation $g_i \times \pi_i$ indicates the occurence of g_i distinct cycles each of length π_i. The last column of the table gives the total number of distinct cycles or "attractors" in the system

N		
3	4×1	4
4	1×1	1
5	1×1; 5×3	6
6	4×1; 6×2	10
7	1×1; 9×7	10
8	1×1	1
9	4×1; 36×7	40
10	1×1; 5×3; 40×6	46
11	1×1; 33×31	34
12	4×1; 6×2; 60×4	70
13	1×1; 65×63	66
14	1×1; 9×7; 288×14	298
15	4×1; 20×3; $1,088 \times 15$	1,112
16	1×1	1
17	1×1; 51×5; $4,352 \times 15$	4,404
18	4×1; 6×2; 36×7; $4,662 \times 14$	4,708
19	1×1; 513×511	514
20	1×1; 5×3; 40×6; $5,440 \times 12$	5,486
21	4×1; 36×7; $16,640 \times 63$	16,680
22	1×1; 33×31; $16,896 \times 62$	16,930
23	1×1; $2,049 \times 2,047$	2,050
24	4×1; 6×2; 60×4; $8,160 \times 8$	8,230
25	1×1; 5×3; $16,400 \times 1,023$	16,406
26	1×1; 65×63; $133,120 \times 126$	133,186
27	4×1; 36×7; $131,328 \times 511$	131,368
28	1×1; 9×7; 288×14; $599,040 \times 28$	599,338
29	1×1; $16,385 \times 16,383$	16,386
30	4×1; 6×2; 20×3; 670×6; $1,088 \times 15$; $8,947,168 \times 30$	8,948,956
31	1×1; $34,636,833 \times 31$	34,636,834
32	1×1	1
33	4×1; $138,547,332 \times 31$	138,547,336
34	1×1; 51×5; $6,528 \times 10$; $4,352 \times 15$; $143,161,216 \times 30$	143,172,148
35	1×1; 5×3; 9×7; 45×21; $4,195,328 \times 4,095$	4,195,388
36	4×1; 6×2; 60×4; 36×7; $4,662 \times 14$; $153,389,340 \times 28$	153,394,108
37	1×1; $786,435 \times 87,381$	786,436
38	1×1; 513×511; $67,239,936 \times 1,022$	672,340,450
39	4×1; 260×63; $49,164 \times 1,365$; $67,108,860 \times 4,095$	67,158,288
40	1×1; 5×3; 40×6; $5,440 \times 12$; $178,954,240 \times 24$	178,959,726

equation is the same as the one for the maximal cycle length of a size d cellular automaton: the derivation of Theorem 3.5 then shows that

$$\pi_{d,i,1} \mid 2^{\mathrm{sord}_d(2)} - 1 . \tag{3.21}$$

It can also be shown that $\pi_{d,i,2s} = \pi_{d,i,s}$ or $\pi_{d,i,2s} = 2\pi_{d,i,s}$.

As an example of the procedure described above, consider the case $N=30$. Here,

$$x^{30}+1=(x^{15}+1)^2=C_{1,1}(x)^2 C_{3,1}(x)^2 C_{5,1}(x)^2 C_{15,1}(x)^2 C_{15,2}(x)^2, \quad (3.22)$$

where

$$C_{1,1}(x)=x+1,$$
$$C_{3,1}(x)=x^2+x+1,$$
$$C_{5,1}(x)=x^4+x^3+x^2+x+1,$$
$$C_{15,1}(x)=x^4+x+1,$$
$$C_{15,2}(x)=x^4+x^3+1.$$

Then

$$\pi_{d,i,2}=1,$$
$$\pi_{3,1,1}=1, \quad \pi_{3,1,0}=2,$$
$$\pi_{5,1,1}=3, \quad \pi_{5,1,0}=6, \qquad\qquad (3.23)$$
$$\pi_{15,1,1}=\pi_{15,2,1}=15,$$
$$\pi_{15,1,0}=\pi_{15,2,0}=30.$$

Thus the cycles which occur in the case $N=30$ have lengths 1, 2, 3, 6, 15, and 30.

To determine the number of distinct cycles of a given length, one must find the number of polynomials $B(x)$ with each possible set of values $r_{d,i}[B(x)]$. This number is given by

$$\prod_{\substack{d|n \\ d \neq 1}} \prod_i V(r_{d,i}, d, D_2(N)),$$

where $V(D_2(N), d, D_2(N))=1$ and

$$V(r, d, D_2(N)) = 2^{\mathrm{ord}_d(2)(D_2(N)-r)} - 2^{\mathrm{ord}_d(2)(D_2(N)-r-1)}$$

for $0 \leq r < D_2(N)$. The cycle lengths of these polynomials are determined as above by the least common multiple of the $\pi_{d,i,r_{d,i}}$.

In the example $N=30$ discussed above, one finds that configurations on cycles of length 3 have $(r_{3,1}, r_{5,1}, r_{15,1}, r_{15,2})=(1,1,2,2)$ or $(2,1,2,2)$, implying that 60 such configurations exist, in 20 distinct cycles.

4. Generalizations

A. Enumeration of Additive Cellular Automata

We consider first one-dimensional additive cellular automata, whose configurations may be represented by univariate characteristic polynomials. We assume that the time evolution of each site depends only on its own value and the value of its two nearest neighbours, so that the time evolution dipolynomial $\mathbb{T}(x)$ is at most of degree two. Cyclic boundary conditions on N sites are implemented by reducing the characteristic polynomial at each time step modulo x^N-1 as in Eq. (2.2). There are taken to be k possible values for each site. With no further constraints imposed, there are k^3 possible $\mathbb{T}(x)$, and thus k^3 distinct cellular automaton rules. If the coefficients of x and x^{-1} in $\mathbb{T}(x)$ both vanish, then the characteristic polynomial is at most multiplied by an overall factor at each time step, and the behaviour of the cellular automaton is trivial. Requiring nonzero coefficients for x and x^{-1} in $\mathbb{T}(x)$ reduces the number of possible rules to $k^3 - 2k^2 + k$. If the cellular automaton evolution is assumed reflection symmetric, then $\mathbb{T}(x) = \mathbb{T}(x^{-1})$, and only $k^2 - k$ rules are possible. Further characterisation of possible rules depends on the nature of k.

(a) k Prime. In this case, integer values $0, 1, ..., k-1$ at each site may be combined by addition and multiplication modulo k to form a field (in which each nonzero element has a unique multiplicative inverse) $\mathbb{Z}_k$. For a symmetrical rule, $\mathbb{T}(x)$ may always be written in the form

$$\mathbb{T}(x) = x + s + x^{-1} \tag{4.1}$$

up to an overall multiplicative factor. For $k=2$, the rule $\mathbb{T}(x) = x + x^{-1}$ was considered above; the additional rule $\mathbb{T}(x) = x + 1 + x^{-1}$ is also possible (and corresponds to rule 150 of [1]).

(b) k Composite.

Lemma 4.1. *For $k = p_1^{\alpha_1} p_2^{\alpha_2} ...$, with p_i prime, the value $a^{[k]}$ of a site obtained by evolution of an additive cellular automaton from some initial configuration is given uniquely in terms of the values $a^{[p_i^{\alpha_i}]}$ attained by that site in the evolution of the set of cellular automata obtained by reducing $\mathbb{T}(x)$ and all site values modulo $p_i^{\alpha_i}$.*

This result follows from the Chinese remainder theorem for integers (e.g. [8, Chap. 8]), which states that if k_1 and k_2 are relatively prime, then the values n_1 and n_2 determine a unique value of n modulo $k_1 k_2$ such that $n \equiv n_i \bmod k_i$ for $i = 1, 2$.

Lemma 4.1 shows that results for any composite k may be obtained from those for k a prime or a prime power.

When k is composite, the ring $\mathbb{Z}_k$ of integers modulo k no longer forms a field, so that not all commutative rings $\mathbb{R}_k$ are fields. Nevertheless, for k a prime power, there exists a Galois field $GF(k)$ of order k, unique up to isomorphism (e.g. [9; Chap. 4]). For example, the field $GF(4)$ may be taken to act on elements $0, 1, \kappa, \kappa^2$ with multiplication taken modulo the irreducible polynomial $\kappa^2 + \kappa + 1$. Time evolution for a cellular automaton with site values in this Galois field can be reduced to that given by $x + \sigma + x^{-1}$, where σ is any element of the field. The behaviour of this subset of cellular automata with k composite is directly analogous to those over $\mathbb{Z}_p$ for prime p.

It has been assumed above that the value of a site at a particular time step is determined solely by the values of its nearest neighbours on the previous time step. One generalization allows dependence on sites out to a distance $r > 1$, so that the evolution of the cellular automaton corresponds to multiplication by a fixed dipolynomial $\mathbb{T}(x)$ of degree $2r$. Most of the theorems to be derived below hold for any r.

B. Cellular Automata over $\mathbb{Z}_p$ (p Prime)

Lemma 4.2. *The lengths of all cycles in any additive cellular automaton over $\mathbb{Z}_p$ of size N divide the length Π_N of the cycle obtained for an initial configuration containing a single site with value 1.*

O. Martin, A. M. Odlyzko, and S. Wolfram

This lemma is a straightforward generalization of Lemma 3.4, and follows directly from the additivity assumed for the cellular automaton rules.

Lemma 4.3. *For N a multiple of p, $\Pi_N | p\Pi_{N/p}$ for an additive cellular automaton over* $\mathbb{Z}_p$.

Remark. For N a multiple of p, but not a power of p, it can be shown that $\Pi_N = p\Pi_{N/p}$ for an additive cellular automaton over $\mathbb{Z}_p$ with $\mathbb{T}(x) = x + x^{-1}$. In addition, $\Pi_{p^j} = 1$ in this case.

Theorem 4.1. *For any N not a multiple of p, $\Pi_N | \Pi_N^* = p^{\mathrm{ord}_N(p)} - 1$, and $\Pi_N | \Pi_N^* = p^{\mathrm{sord}_N(p)} - 1$ if $\mathbb{T}(x)$ is symmetric, for any additive cellular automaton over* $\mathbb{Z}_p$.

The period Π_N divides Π_N^* if

$$[\mathbb{T}(x)]^{\Pi_N^* + 1} \equiv \mathbb{T}(x) \quad \mod (x^N - 1). \tag{4.2}$$

Taking

$$\mathbb{T}(x) = \sum_i \alpha_i x^{\gamma_i},$$

Eq. (A.3) yields

$$[\mathbb{T}(x)]^{p^{\mathrm{ord}_N(p)}} \equiv \sum_i \alpha_i x^{\gamma_i p^{\mathrm{ord}_N(p)}} \equiv \sum_i \alpha_i x^{\gamma_i} = \mathbb{T}(x) \quad \mod (x^N - 1),$$

since $\alpha^{p^\lambda} \equiv \alpha \bmod p$ and $p^{\mathrm{ord}_N(p)} \equiv 1 \bmod N$, and the first part of the theorem follows. Since $x^{p^{\mathrm{sord}_N(p)}} \equiv x^{\pm 1} \bmod p$, Eq. (4.2) holds for

$$\Pi_N^* = p^{\mathrm{sord}_N(p)} - 1$$

if $\mathbb{T}(x)$ is symmetric, so that $\mathbb{T}(x) = \mathbb{T}(x^{-1})$.

This result generalizes Theorem 3.5 for the particular $k = 2$ cellular automaton considered in Sect. 3.

Table 1 gives the values of Π_N for all non-trivial additive symmetrical cellular automata over $\mathbb{Z}_2$ and $\mathbb{Z}_3$. Just as in the example of Sect. 3 (given as the first column of Table 1), one finds that for many values of N not divisible by p

$$\Pi_N = p^{\mathrm{sord}_N(p)} - 1. \tag{4.3}$$

When $p = 2$, all exceptions to (4.3) when $\mathbb{T}(x) = x + x^{-1}$ are also exceptions for $\mathbb{T}(x) = x + 1 + x^{-1}$ [19]. We outline a proof for the simplest case, when N is relatively prime to 6 (as well as 2). Let $\Pi_N(x + x^{-1})$ be the maximal period obtained with $\mathbb{T}(x) = x + x^{-1}$, equal to the minimum integer π for which

$$(x + 1)^{2\pi} \equiv x^\pi \quad \mod \left(\frac{x^N - 1}{x + 1} \right). \tag{4.4}$$

We now show that $\Pi_N(x + x^{-1})$ is a multiple of the maximum period $\Pi_N(x + 1 + x^{-1})$ obtained with $\mathbb{T}(x) = x + 1 + x^{-1}$. Since the mapping $x \to x^3$ is a homomorphism in the field of polynomials with coefficients in $GF(2)$, one has

$$(x^3+1)^{2\pi} \equiv x^{3\pi} \quad \mathrm{mod}\left(\frac{x^N-1}{x+1}\right)$$

for any π such that $\Pi_N(x+x^{-1})|\pi$. Dividing by Eq. (4.4), and using the fact that N is odd to take square roots, yields

$$\left(\frac{x^3+1}{x+1}\right)^\pi \equiv x^\pi \quad \mathrm{mod}\left(\frac{x^N-1}{x+1}\right) \tag{4.5}$$

for any π such that $\Pi_N(x+x^{-1})|\pi$. But since $x+1+x^{-1}=x^{-1}\left(\frac{x^3+1}{x+1}\right)$, Eq. (4.5) is the analogue of Eq. (4.4) for $\mathbb{T}(x)=x+1+x^{-1}$, and the result follows.

More exceptions to Eq. (4.3) are found with $p=3$ than with $p=2$.

Lemma 4.4. *A configuration $A(x)$ is reachable in the evolution of a size N additive cellular automaton over $\mathbb{Z}_p$, as described by $\mathbb{T}(x)$ if and only if $A(x)$ is divisible by $\Lambda_1(x)=(x^N-1, \mathbb{T}(x))$.*

Appendix A.A gives conventions for the greatest common divisor $(A(x), B(x))$.

If $A^{(1)}(x)$ can be reached, then

$$A^{(1)}(x)=\mathbb{T}(x)A^{(0)}(x) \quad \mathrm{mod}(x^N-1)$$

for some $A^{(0)}(x)$, so that

$$(x^N-1)\,|\,A^{(1)}(x)-\mathbb{T}(x)A^{(0)}(x).$$

But $\Lambda_1(x)|x^N-1$ and $\Lambda_1(x)|\mathbb{T}(x)$, and hence if $A^{(1)}(x)$ is reachable,

$$\Lambda_1(x)\,|\,A^{(1)}(x). \tag{4.6}$$

We now show by an explicit construction that all $A^{(1)}(x)$ satisfying (4.6) in fact have predecessors $A^{(0)}(x)$. Using Eq. (A.10), one may write

$$\Lambda_1(x)=r(x)\mathbb{T}(x)+\xi(x)\,(x^N \to 1)$$

for some dipolynomials $r(x)$ and $\xi(x)$, so that

$$\Lambda_1(x)\equiv r(x)\mathbb{T}(x) \quad \mathrm{mod}(x^N-1).$$

Then taking $A^{(1)}(x)=\Lambda_1(x)B(x)$, the configuration given by the polynomial obtained by reducing the dipolynomial $r(x)B(x)$ satisfies

$$\mathbb{T}(x)r(x)B(x)\equiv\Lambda_1(x)B(x)\equiv A^{(1)}(x) \quad \mathrm{mod}(x^N-1)$$

and thus provides an explicit predecessor for $A^{(1)}(x)$.

Corollary. *$A(x)$ is reachable in j steps if and only if $\Lambda_j(x)=(x^N-1, \mathbb{T}^j(x))$ divides $A(x)$.*

This is a straightforward extension of the above lemma.

Theorem 4.2. *The fraction of possible configurations which may be reached by evolution of an additive cellular automaton over $\mathbb{Z}_p$ of size N is $p^{-\deg\Lambda_1(x)}$, where $\Lambda_1(x)=(x^N-1, \mathbb{T}(x))$.*

By Lemma 4.4, only configurations divisible by $\Lambda_1(x)$ may be reached. The number of such configurations is $p^{N-\deg \Lambda_1(x)}$, while the total number of possible configurations is p^N.

Let $D_p(N)$ be the maximum p^j which divides N and let v_i denote the multiplicity of the ith irreducible factor of $\Lambda_1(x)$ in $\mathbb{T}^*(x)$, where $\mathbb{T}^*(x) = x^r \mathbb{T}(x)$ is a polynomial with a nonzero constant term. We further define $\chi = \min_i v_i$, so that $0 \leq \chi \leq D_p(N)$.

Theorem 4.3. *The state transition diagram for an additive cellular automaton of size N over $\mathbb{Z}_p$ consists of a set of cycles at all nodes of which are rooted identical $p^{\deg \Lambda_1(x)}$-ary trees. A fraction $p^{-D_p(N) \deg \Lambda_1(x)}$ of the possible configurations appear on cycles. For $\chi > 0$, the height of the trees is $\lceil D_p(N)/\chi \rceil$. The trees are balanced if and only if (a) $v_i \geq D_p(N)$ for all i, or (b) $v_i = v_j$ for all i and j, and $v_i | D_p(N)$.*

To determine the in-degrees of nodes in the trees, consider a configuration $A(x)$ with predecessors represented by the polynomials $B_1(x)$ and $B_2(x)$, so that

$$A(x) \equiv \mathbb{T}(x) B_i(x) \quad \mod (x^N - 1).$$

Then since

$$\mathbb{T}(x)(B_1(x) - B_2(x)) \equiv 0 \quad \mod (x^N - 1),$$

and $\Lambda_1(x) | x^N - 1$, it follows that

$$B_1(x) - B_2(x) \equiv 0 \quad \mod \left(\frac{x^N - 1}{\Lambda_1(x)} \right).$$

Since $C(x) = (x^N - 1)/\Lambda_1(x)$ has a non-zero constant term, $(B_1(x) - B_2(x))/C(x)$ is an ordinary polynomial. The number of solutions to this congruence and thus the number of predecessors $B_i(x)$ of $A(x)$ is $p^{\deg \Lambda_1(x)}$.

The proof of Lemma 3.3 demonstrates the identity of the trees. The properties of the trees are established by considering the tree rooted on the null configuration. A configuration $A(x)$ evolves to the null configuration after j steps if $\mathbb{T}(x)^j A(x) \equiv 0 \mod (x^N - 1)$, so that

$$\frac{x^N - 1}{\Lambda_j(x)} \ \bigg| \ A(x). \tag{4.7}$$

Hence all configurations on the tree are divisible by $(x^N - 1)/\Lambda_\infty(x)$, where $\Lambda_\infty(x) = \lim_{j \to \infty} \Lambda_j(x)$. All configurations in the tree evolve to the null configuration after at most $\lceil D_p(N)/\chi \rceil$ steps, which is thus an upper bound on the height of the trees. But since the configuration $(x^N - 1)/\Lambda_\infty(x)$ evolves to the null configuration after exactly $\lceil D_p(N)/\chi \rceil$ steps, this quantity gives the height of the trees. The tree of configurations which evolve to the null configuration (and hence all other trees in the state transition diagram) is balanced if and only if all unreachable (terminal) configurations evolve to the null configuration after the same number of steps. First suppose that neither condition (a) nor (b) is true. One possibility is that some irreducible factor $\sigma(x)$ of $\Lambda_1(x)$ satisfies $\sigma^v(x) \| \Lambda_1(x)$ with $v < D_p(N)$ but v does not divide $D_p(N)$. The configuration $(x^N - 1)/\sigma^{D_p(N)}(x)$ reaches 0 *in* $\lceil D_p(N)/v \rceil$ steps whereas $(x^N - 1)/\sigma^{D_p(N)+1-v}(x)$ reaches 0 in one step fewer, yet both are unreachable, so that the tree cannot be balanced. The only other possibility is that there exist two irreducible factors $\sigma_1(x)$ and $\sigma_2(x)$ of multiplicities v_1 and v_2, respectively, with v_1 and v_2 dividing $D_p(N)$ but $v_1 \neq v_2$. Then $(x^N - 1)/\sigma_1^{D_p(N)}(x)$ reaches 0 in $D_p(N)/v_1$ steps, whereas $(x^N - 1)/\sigma_2^{D_p(N)}(x)$ reaches 0 in $D_p(N)/v_2$ steps. Neither of these configurations is reachable, so again the trees cannot be balanced. This establishes that in all cases either condition (a) or (b) must hold. The sufficiency of condition (a) is evident. If the condition (b) is true, then

$$\Lambda_1(x) = [\textstyle\prod \sigma(x)]^v, \qquad \Lambda_\infty(x) = [\textstyle\prod \sigma(x)]^{D_p(N)},$$

and $\Lambda_j(x) = \Lambda_1^j(x)$. Equation (4.7) shows that any configuration $A(x)$ which evolves to the null configuration after j steps is of the form

$$A(x) = \frac{x^N - 1}{\Lambda_1^j(x)} R(x),$$

where $R(x)$ is some polynomial. The proof is completed by showing that all such configurations $A(x)$ with $j < D_p(N)/v$ are indeed reachable. To construct an explicit predecessor for $A(x)$, define the dipolynomial $S(x)$ by $\mathbb{T}(x) = \Lambda_1(x)S(x)$, so that $(S(x), x^N - 1) = 1$. Then there exist dipolynomials $r(x)$ and $\xi(x)$ such that

$$r(x)S(x) + \xi(x)(x^N - 1) = 1.$$

The configuration given by the dipolynomial

$$B(x) = \frac{x^N - 1}{\Lambda_1^{j+1}(x)} r(x)R(x)$$

then provides a predecessor for $A(x)$.

Notice that whenever the balance condition fails, the set and measure entropies of Eqs. (3.11) and (3.12) obtained by evolution from an initial maximal entropy ensemble become unequal.

The results of Theorems 4.2 and 4.3 show that if $\deg \Lambda_1(x) = 0$, then the evolution of an additive cellular automaton if effectively reversible, since every configuration has a unique predecessor.

In general,

$$\deg \Lambda(x) \leq \deg \mathbb{T}^*(x),$$

so that for the one-dimensional additive cellular automata considered so far, the maximum decrease in entropy starting from an initial equiprobable ensemble is $D_p(N)$.

Note that for a cellular automaton over $\mathbb{Z}_p$ $(p > 2)$ of length N with $\mathbb{T}(x) = x + x^{-1}$, $\deg \Lambda(x) = 2$ if $4|N$ and $\deg \Lambda(x) = 0$ otherwise. Such cellular automata are thus effectively reversible for $p > 2$ whenever N is not a multiple of 4.

Remark. A configuration $A(x)$ lies on a cycle in the state transition diagram of an additive cellular automaton if and only if $\Lambda_\infty(x)|A(x)$.

This may be shown by the methods used in the proof of Theorem 4.3.

C. Cellular Automata over $\mathbb{Z}_k$ (k Composite)

Theorem 4.4. *For an additive cellular automaton over* $\mathbb{Z}_k$,

$$\Pi_N(\mathbb{Z}_k; \mathbb{T}_k(x)) = \mathrm{lcm}(\Pi_N(\mathbb{Z}_{p_1^{\alpha_1}}; \mathbb{T}_{p_1^{\alpha_1}}(x)), \Pi_N(\mathbb{Z}_{p_2^{\alpha_2}}; \mathbb{T}_{p_2^{\alpha_2}}(x)), \ldots),$$

where $k = p_1^{\alpha_1} p_2^{\alpha_2} \ldots$, *and in* $\mathbb{T}_j(x)$ *all coeffficients are reduced modulo j.*

This result follows immediately from Lemma 4.1.

Theorem 4.5. $\Pi_N(\mathbb{Z}_{p^{\alpha+1}}; \mathbb{T}_{p^{\alpha+1}}(x))$ *is equal to either* (a) $p\Pi_N(\mathbb{Z}_{p^\alpha}; \mathbb{T}_{p^\alpha}(x))$ *or* (b) $\Pi_N(\mathbb{Z}_{p^\alpha}; \mathbb{T}_{p^\alpha}(x))$ *for an additive cellular automaton.*

First, it is clear that

$$\Pi_N(\mathbb{Z}_{p^\alpha}; \mathbb{T}_{p^\alpha}(x)) \mid \Pi_N(\mathbb{Z}_{p^{\alpha+1}}; \mathbb{T}_{p^{\alpha+1}}(x)).$$

To complete the proof, one must show that in addition

$$\Pi_N(\mathbb{Z}_{p^{\alpha+1}}; \mathbb{T}_{p^{\alpha+1}}(x)) \mid p\Pi_N(\mathbb{Z}_{p^\alpha}; \mathbb{T}_{p^\alpha}(x)).$$

$\Pi_N(\mathbb{Z}_{p^\alpha}; \mathbb{T}_{p^\alpha}(x))$ is the smallest positive integer π for which a positive integer m and dipolynomials $U(x)$ and $V(x)$ satisfying

$$\mathbb{T}(x)^{m+\pi} = \mathbb{T}(x)^m + (x^N - 1)U(x) + p^\alpha V(x) \tag{4.8}$$

exist, where all dipolynomial coefficients (including those in $\mathbb{T}(x)$) are taken as ordinary integers in $\mathbb{Z}$, and irrelevant powers of x on both sides of the equation have been dropped. Raising both sides of Eq. (4.8) to the power p, one obtains

$$\mathbb{T}(x)^{mp+\pi p} = (x^N - 1)W(x) + (\mathbb{T}(x)^m + p^\alpha V(x))^p$$
$$= (x^N - 1)W(x) + \mathbb{T}(x)^{mp} + p^{\alpha+1}Q(x).$$

Reducing modulo $p^{\alpha+1}$ yields the required result.

For $p = 2$ and $\alpha = 1$, it can be shown that case (a) of Theorem 4.5 always obtains if $\mathbb{T}(x) = x + x^{-1}$, but case (b) can occur when $\mathbb{T}(x) = x + 1 + x^{-1}$.

Theorem 4.6. *With* $k = k_1 k_2 \ldots$ *(all* k_i *relatively prime), the number of configurations which can be reached by evolution of an additive cellular automaton over* $\mathbb{Z}_k$ *is equal to the product of the numbers reached by evolution of cellular automata with the same* $\mathbb{T}(x)$ *over each of the* $\mathbb{Z}_{k_i}$. *The state transition diagram for the cellular automaton over* $\mathbb{Z}_k$ *consists of a set of identical trees rooted on cycles. The in-degrees of non-terminal nodes in the trees are the product of those for each of the* $\mathbb{Z}_{k_i}$ *cases. The height of the trees is the maximum of the heights of trees for the* $\mathbb{Z}_{k_i}$ *cases, and the trees are balanced only if all these heights are equal.*

These results again follow directly from Lemma 4.1.

Theorem 4.6 gives a characterisation of the state transition diagram for additive cellular automata over $\mathbb{Z}_k$ when k is a product of distinct primes. No general results are available for the case of prime power k. However, for example, with $\mathbb{T}(x) = x + x^{-1}$, one may obtain the fraction of reachable states by direct combinatorial methods. With $k = 2^\alpha$ one finds in this case that the fraction is $1/2$ for N odd, $1/4$ for $N \equiv 2 \bmod 4$, and $2^{-2\alpha}$ for $4|N$. With $k = p^\alpha$ ($p \neq 2$) the systems are reversible (all configurations reachable) unless $4|N$, in which case a fraction $p^{-2\alpha}$ may be reached.

D. Multidimensional Cellular Automata

The cellular automata considered above consist of a sequence of sites on a line. One generalization takes the sites instead to be arranged on a square lattice in two dimensions. The evolution of a site may depend either on the values of its four orthogonal neighbours (type I neighbourhood) or on the values of all eight neighbours including those diagonally adjacent (type II neighbourhood) (e.g. [1]). Configurations of two-dimensional cellular automata may be represented by bivariate characteristic polynomials $A(x_1, x_2)$. Time evolution for additive cellular automaton rules is obtained by multiplication of these characteristic polynomials by a fixed bivariate dipolynomial $\mathbb{T}(x_1, x_2)$. For a type I neighbourhood, $\mathbb{T}(x_1, x_2)$ contains no $x_1 x_2$ cross-terms; such terms may be present for a type II neighbourhood. Periodic boundary conditions with periods N_1 and N_2 may be implemented by reduction modulo $x_1^{N_1} - 1$ and modulo $x_2^{N_2} - 1$ at each time step. Cellular automata may be generalized to an arbitrary d-dimensional cubic or hypercubic lattice. A type I neighbourhood in d dimensions contains $2d + 1$ sites, while a type II neighbourhood contains 3^d sites. As before, we consider cellular automata with k possible values for each site.

Theorem 4.7. *For an additive cellular automaton over $\mathbb{Z}_k$ on a d-dimensional cubic lattice, with a type I or type II neighbourhood, and with periodicities $N_1, N_2, ..., N_d$,*
$$\text{lcm}(\Pi_{N_1}(\mathbb{Z}_k; \mathbb{T}(x_1, 1, ..., 1)), ..., \Pi_{N_d}(\mathbb{Z}_k; \mathbb{T}(1, ..., 1, x_d))) | \Pi_{N_1, ..., N_d}(\mathbb{Z}_k; \mathbb{T}(x_1, ..., x_d)).$$

The result may be proved by showing that

$$\Pi_{N_i}(\mathbb{Z}_i; \mathbb{T}(1, ..., 1, x_i, 1, ..., 1)) | \Pi_{N_1, ..., N_d}(\mathbb{Z}_k, \mathbb{T}(x_1, ..., x_d)) \qquad (4.9)$$

for all i (such that $1 \le i \le d$). The right member of Eq. (4.9) is given by the smallest integer π for which there exists a positive integer m such that

$$[\mathbb{T}(x_1, ..., x_d)]^{\pi+m} = [\mathbb{T}(x_1, ..., x_d)]^m + \sum_{j=1}^{d} (x_j^{N_j} - 1)U_j(x_1, ..., x_d) \qquad (4.10)$$

for some dipolynomials U_j. Taking $x_j = 1$ with $j \ne i$ in Eq. (4.10), all terms in the sum vanish except for the one associated with x_i, and the resulting value of π corresponds to the left member of Eq. (4.9).

Theorem 4.8. *For an additive cellular automaton over $\mathbb{Z}_p$ on a d-dimensional cubic lattice (type I or type II neighbourhood) with periodicities $N_1, N_2, ..., N_d$ none of which are multiples of p,*

$$\Pi_{N_1, ..., N_d}(\mathbb{Z}_p; \mathbb{T}(x_1, ..., x_d)) | \Pi^*_{N_1, ..., N_d} = p^{\text{ord}_{N_1, ..., N_d}(p)} - 1.$$

If $\mathbb{T}(x_1, ..., x_d)$ is symmetrical, so that

$$\mathbb{T}(x_1, ..., x_i, ..., x_d) = \mathbb{T}(x_1, ..., x_i^{-1}, ..., x_d)$$

for all i, then

$$\Pi^*_{N_1, ..., N_d} = p^{\text{sord}_{N_1, ..., N_d}(p)} - 1.$$

The $\text{ord}_{n_1, \ldots, n_d}(p)$ *and* $\text{sord}_{n_1, \ldots, n_d}(p)$ *are multidimensional generalizations of the multiplicative order and suborder functions, described in Appendix B.*

This theorem is proved by straightforward extension of the one-dimensional Theorem 4.1.

Using the result (B.13), one finds for symmetrical rules

$$\Pi^*_{N_1, \ldots, N_d} = p^{\text{lcm}(\text{sord}_{N_1}(p), \ldots, \text{sord}_{N_d}(p))} - 1 .$$

The maximal cycle length is thus bounded by

$$\Pi_{N_1, \ldots, N_d} \leq p^{\text{lcm}((N_1-1)/2, \ldots, (N_d-1)/2)} - 1 \leq p^{(N_1-1)\ldots(N_d-1)/2^d} - 1 ,$$

with the upper limits achieved only if all the N_i are prime. (For example,

$$\Pi_{83,59} = 2^{1189} \simeq 10^{358}$$

saturates the upper bound.)

Algebraic determination of the structure of state transition diagrams is more complicated for multi-dimensional cellular automata than for the one dimensional cellular automata considered above[2]. The generalization of Lemma 4.4 states that a configuration $A(x_1, \ldots, x_d)$ is reachable only if $A(z_1, \ldots, z_d)$ vanishes whenever the z_i are simultaneously roots of $\mathbb{T}(x_1, \ldots, x_d)$, $x^{N_1} - 1, \ldots, x^{N_d} - 1$. The root sets z_i form an algebraic variety over $\mathbb{Z}_k$ (cf. [9]).

E. Higher Order Cellular Automata

The rules for cellular automaton evolution considered above took configurations to be determined solely from their immediate predecessors. One may in general consider higher order cellular automaton rules, which allow dependence on say s preceding configurations. The time evolution for additive one-dimensional higher-order cellular automata (with N sites and periodic boundary conditions) may be represented by the order s recurrence relation

$$A^{(t)}(x) = \sum_{j=1}^{s} \mathbb{T}_j(x) A^{(t-j)}(x) \quad \text{mod}(x^N - 1). \tag{4.11}$$

This may be solved in analogy with order s difference equations to yield

$$A^{(t)}(x) = \sum_{j=1}^{s} c_j(x) [U_j(x)]^t ,$$

where the $U_j(x)$ are solutions to the equation

$$[U(x)]^s = \sum_{j=1}^{s} [U(x)]^{s-j} \mathbb{T}_j(x) ,$$

and the $c_j(x)$ are analogous to "constants of integration" and are determined by the initial configurations $A^{(0)}(x), \ldots, A^{(s-1)}(x)$. The state of an order s cellular

---

2   In the specific case $\mathbb{T}(x_1, x_2) = x_1 + x_1^{-1} + x_2 + x_2^{-1}$, one finds that the in-degrees I_{N_1, N_2} of trees in the state transition diagrams for a few $N_1 \times N_2$ cellular automata are: $I_{2,2} = 16$, $I_{2,3} = 4$, $I_{2,4} = 16$, $I_{2,5} = 4$, $I_{2,6} = 16$, $I_{3,3} = 32$, $I_{3,4} = 4$, $I_{3,5} = 2$, $I_{4,4} = 256$

automaton depends on the values of its N sites over a sequence of s time steps; there are thus a total of k^{Ns} possible states. The transition diagram for these states can in principle be derived by algebraic methods starting from Eq. (4.11). In practice, however, the $U_j(x)$ are usually not polynomials, but elements of a more general function field, leading to a somewhat involved analysis not performed here.

For first-order additive cellular automata, any configuration may be obtained by superposition of the configuration 1 (or its translates x^j). For higher-order cellular automata, several "basis" configurations must be included. For example, when $s=2$, $\{0,1\}$, $\{1,0\}$, and $\{x^j,1\}$ are all basis configurations, where in $\{A_1(x), A_2(x)\}$, $A_1(x)$, and $A_2(x)$ represent configurations at successive time steps.

As discussed in Sect. 4B, some first-order cellular automata over $\mathbb{Z}_p$ ($p>2$) are effectively reversible for particular values of N, so that all states are on cycles. The class of second-order cellular automata with $\mathbb{T}_2(x) = -1$ is reversible for all N and k, and for any $\mathbb{T}_1(x)$ [10]. In the simple case $\mathbb{T}_1(x) = x + x^{-1}$, one finds $U_1(x) = x$, $U_2(x) = x^{-1}$. It then appears that

$$\Pi_N = kN/2 \quad (k \text{ even}, N \text{ even})$$
$$= kN \quad (\text{otherwise}).$$

(The proof is straightforward when $k=2$.) In the case $\mathbb{T}_1(x) = x + 1 + x^{-1}$, the $U_j(x)$ are no longer polynomials. For the case $k=2$, the results for Π_N with N between 3 and 30 are: 6, 6, 15, 12, 9, 12, 42, 30, 93, 24, 63, 18, 510, 24, 255, 84, 513, 60, 1170, 186, 6141, 48, 3075, 126, 3066, 36, 9831, 1020.

F. Other Boundary Conditions

The cellular automata discussed above were taken to consist of N indistinguishable sites with periodic boundary conditions, as if arranged around a circle. This section considers briefly cellular automata with other boundary conditions. The discussion is restricted to the case of symmetric time evolution rules $\mathbb{T}(x) = \mathbb{T}(x^{-1})$.

The periodic boundary conditions considered above are not the only possible choice which preserve the translation invariance of cellular automata (or the indistinguishability of their sites)[3]. One-dimensional cellular automata may in general be viewed as $\mathbb{R}_k$ bundles over $\mathbb{Z}_N$. Periodic boundary conditions correspond to trivial bundles. Non-trivial bundles are associated with "twisted" boundary conditions. Explicit realizations of such boundary conditions require a twist to be introduced at a particular site. The evolution of particular configurations then depends on the position of the twist, but the structure of the state transition diagram does not.

A twist of value R at position $i=\sigma$ causes sites with $i \geq \sigma$ to appear multiplied by R in the time evolution of sites with $i < \sigma$, and correspondingly, for sites with $i < \sigma$ to appear multiplied by R^{-1} in the evolution of sites with $i \geq \sigma$. In the presence of a twist taken at position $\sigma=0$, the time evolution formula (2.5) becomes

$$A^{(t)}(x) = \mathbb{T}(x)A^{(t-1)}(x) \quad \text{mod}(x^N - R). \tag{4.12}$$

---

3   We are grateful to L. Yaffe for emphasizing this point

Multiple twists are irrelevant; only the product of their values R_j is significant for the structure of the state transition diagram. If $\mathbb{R}_k = \mathbb{Z}_p$ with p prime, then $\mathbb{R}_k$ (with the zero element removed) forms a multiplicative group, and twists with any value R not equal to 0 or 1 yield equivalent results. When $\mathbb{R}_k = \mathbb{Z}_k$ with k composite, several equivalence classes of R values may exist.

Using Eq. (4.12) one may obtain general results for twisted boundary conditions analogous to those derived above for the case of periodic boundary conditions (corresponding to $R = 1$). When $\mathbb{R}_k = \mathbb{Z}_p$ (p prime), one finds for example,

$$\Pi_N^{[R \neq 1]} \mid \Pi_{N(p-1)}^{[R=1]}.$$

An alternative class of boundary conditions introduces fixed values at particular cellular automaton sites. One may consider cellular automata consisting of N sites with values $a_1, ..., a_N$ arranged as if along a line, bounded by sites with fixed values a_0 and a_{N+1}. Maximal periods obtained with such boundary conditions will be denoted $\Pi_N^{(a_0, a_{N+1})}$. The case $a_0 = a_{N+1} = 0$ is simplest. In this case, configurations

$$A(x) = \sum_{i=1}^{N} a_i x^i$$

of the length N system with fixed boundary conditions may be embedded in configurations

$$\tilde{A}(x) = \sum_{i=1}^{N} a_i x^i + \sum_{i=1}^{N} (k - a_{N+1-i}) x^{N+1+i} \tag{4.13}$$

of a length $\tilde{N} = 2N + 2$ system with periodic boundary conditions. The condition $a_0 = a_{N+1} = 0$ is preserved by time evolution, so that one must have

$$\Pi_N^{(0,0)} \mid \Pi_{2N+2}.$$

The periods are equal if the configurations obtained by evolution from a single nonzero initial site have the symmetry of Eq. (4.13). (The simplest cellular automaton defined in Sect. 3A satisfies this condition.)

Fixed boundary conditions $a_0 = r$, $a_{N+1} = 0$, may be treated by constructing configurations $\tilde{A}(x)$ of the form (4.13), with periodic boundary conditions, but now with time evolution

$$\tilde{A}^{(t)}(x) \equiv [\mathbb{T}(x) \tilde{A}^{(t-1)}(x) + r(1 - \alpha_0)] \mod (x^{\tilde{N}} - 1),$$

where $\mathbb{T}(x)$ is taken of the form $x + \alpha_0 + x^{-1}$. Iteration generates a geometric series in $\mathbb{T}(x)$, which may be summed to yield a rational function of x. For $k = 2$, $r = 1$, one may then show that with $\mathbb{T}(x) = x + 1 + x^{-1}$, $\Pi_N^{(0,1)} = \Pi_{2N+2}$, while with $\mathbb{T}(x) = x + x^{-1}$ (the case of Sect. 3A), $\Pi_N^{(0,1)} \mid \Pi_{2(2N+2)}$.

5. Non-Additive Cellular Automata

Equation (2.3) defines the time evolution for a special class of "additive" cellular automata, in which the value of a site is given by a linear combination (in $\mathbb{R}_k$) of the values of its neighbours on the previous time step. In this section we discuss "non-additive" cellular automata, which evolve according to

$$a_i^{(t)} = \mathbb{F}[a_{i-1}^{(t-1)}, a_i^{(t-1)}, a_{i+1}^{(t-1)}], \qquad (5.1)$$

where $\mathbb{F}[a_{-1}, a_0, a_{+1}]$ is an arbitrary function over $\mathbb{R}_k$, not reducible to linear form. The absence of additivity in general prevents use of the algebraic techniques developed for additive cellular automata in Sects. 3 and 4. The difficulties in the analysis of non-additive cellular automata are analogous to those encountered in the analysis of non-linear feedback shift registers (cf. [11]). In fact, the possibility of universal computation with sufficiently complex non-additive cellular automata demonstrates that a complete analysis of these systems is fundamentally impossible. Some results are nevertheless available (cf. [12]). This section illustrates some methods which may be applied to the analysis of non-additive cellular automata, and some of the results which may be obtained.

As in [1], most of the discussion in this section will be for the case $k=2$. In this case, there are 32 possible functions $\mathbb{F}$ satisfying the symmetry condition

$$\mathbb{F}[a_{-1}, a_0, a_{+1}] = \mathbb{F}[a_{+1}, a_0, a_{-1}]$$

and the quiescence condition

$$\mathbb{F}[0, 0, 0] = 0.$$

Reference [1] showed the existence of two classes of these "legal" cellular automata. The "simple" class evolved to fixed points or short cycles after a small number of time steps. The "complex" class (which included the additive rules discussed above) exhibited more complicated behaviour.

We consider as an example the complex non-additive $k=2$ rule defined by

$$\mathbb{F}[1, 0, 0] = \mathbb{F}[0, 0, 1] = 1,$$
$$\mathbb{F}[a_{-1}, a_0, a_{+1}] = 0 \quad \text{otherwise}, \qquad (5.2)$$

and referred to as rule 18 in [1]. This function yields a time evolution rule equivalent to

$$a_i^{(t)} \equiv (1 + a_i^{(t-1)})(a_{i-1}^{(t-1)} + a_{i+1}^{(t-1)}) \quad \mod 2. \qquad (5.3)$$

The rule does not in general satisfy any superposition principle. However, for the special class of configurations with $a_{2j} = 0$ or $a_{2j+1} = 0$, Eq. (5.3) implies that the evolution of even (odd) sites on even (odd) time steps is given simply by the rule defined in Sect. 3A. Any configuration may be considered as a sequence of "domains" in which all even (or odd) sites have value zero, separated by "domain walls" or "kinks" [13]. In the course of time the kinks annihilate in pairs. If sites are nonzero only in some finite region, then at sufficiently large times in an infinite cellular automaton, all kinks (except perhaps one) will have annihilated, and an effectively additive system will result. However, out of all 2^N possible initial configurations for a cellular automaton with N sites and periodic boundary conditions, only a small fraction are found to evolve to this form before a cycle is reached: in most cases, "kinks" are frozen into cycles, and contribute to global behaviour in an essential fashion.

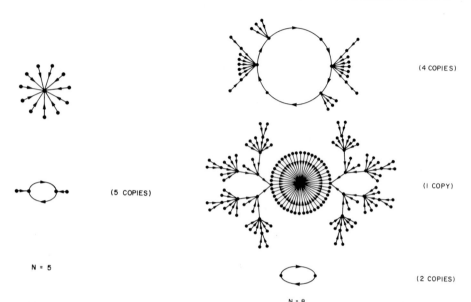

Fig. 5. Global state transition diagrams for a typical finite non-additive cellular automaton discussed in Sect. 5

Typical examples of the state transition diagrams with the rule (5.3) are shown in Fig. 5. They are seen to be much less regular than those for additive rules illustrated in Fig. 2. In particular, not all transient trees are identical, and few of the trees are balanced. Just as for the additive rules discussed in Sects. 3 and 4, only a fraction of the 2^N possible configurations may be reached by evolution according to Eq. (5.3); the rest are unreachable and appear as nodes with zero in-degree on the periphery of the state transition diagram of Fig. 5. An explicit characterization of these unreachable configurations may be found by lengthy but straightforward analysis.

Lemma 5.1. *A configuration is unreachable by cellular automaton time evolution according to Eq. (5.3) if and only if one of the following conditions holds:*
   (a) *The sequence of site values* 111 *appears.*
   (b) *No sequence* 11 *appears, but the total number of* 1 *sites is odd.*
   (c) *A sequence* $11a_1a_2 \ldots a_n11$ *appears, with an odd number of the* a_i *having value* 1. *The two* 11 *sequences may by cyclically identified.*

The number of reachable configurations may now be found by enumerating the configurations defined by Lemma 5.1. This problem is analogous to the enumeration of legal sentences in a formal language. As a simple example of the techniques required (e.g. [14]), consider the enumeration of strings of N symbols 0 or 1 in which no sequence 111 appears (no periodicity is assumed). Let the number of such strings be α. In addition, let β_N be the number of length N strings containing no 111 sequences in their first $N-1$ positions, but terminating with the sequence 111. Then

$$\beta_0 = \beta_1 = \beta_2 = 0, \quad \beta_3 = 1, \quad \alpha_0 = 1, \quad \alpha_1 = 2, \tag{5.4a}$$

and

$$2\alpha_N = \alpha_{N+1} + \beta_{N+1} \quad (N \geq 0), \qquad (5.4b)$$

$$\alpha_N = \beta_{N+1} + \beta_{N+2} + \beta_{N+3} \quad (N \geq 0). \qquad (5.4c)$$

The recurrence relations (5.4) may be solved by a generating function technique. With

$$A(z) = \sum_{n=0}^{\infty} \alpha_n z^n, \qquad B(z) = \sum_{n=0}^{\infty} \beta_n z^n, \qquad (5.5a)$$

Eq. (5.4) may be written as

$$2A(z) = z^{-1}(A(z)-1) + z^{-1}B(z),$$

$$A(z) = z^{-3}B(z) + z^{-2}B(z) + z^{-1}B(z).$$

Solving these equations yields the result

$$A(z) = \frac{1+z+z^2}{1-z-z^2-z^3}. \qquad (5.5b)$$

Results for specific N are obtained as the coefficients of z^N in a series expansion of $A(z)$. Taking

$$A(z) = \frac{A_N(z)}{A_D(z)},$$

Eq. (5.5a) may be inverted to yield

$$\alpha_N = \sum_i \left(\frac{-A_N(z_i)}{z_i A_D'(z_i)} \right) (1/z_i)^N, \qquad (5.5c)$$

where the z_i are the roots of $A_D(z)$ (all assumed distinct), and prime denotes differentiation. This yields finally

$$\alpha_N \simeq 1.14(1.84)^N + 0.283(0.737)^N \cos(2.176N + 2.078). \qquad (5.6)$$

The behaviour of the coefficients for large N is dominated by the first term, associated with the smallest root of $A_D(N)$. The first ten values of α_N are 1, 2, 4, 7, 13, 24, 44, 81, 149, 274, 504.

A lengthy calculation shows that the number of possible strings of length N which do not satisfy the conditions in Lemma 5.1, and may therefore be reached by evolution of the cellular automaton defined by Eq. (5.3), is given as the coefficient of z^N in the expansion of the generating function

$$P(z) = \frac{z - 3z^2 + 6z^3 - 8z^4 + 4z^5 - z^7}{1 - 4z + 6z^2 - 5z^3 + 2z^4 + z^5 - z^6 + z^7}$$

$$= \frac{3-4z+z^2}{1-2z+z^2-z^3} - \frac{2-z}{2(1-z+z^2)} + \frac{2-z}{2(-1+z+z^2)} - 1. \qquad (5.7)$$

Inverting according to Eq. (5.5c), the number of reachable configurations of length N is given by

$$\varrho_N = \kappa^N - (\phi^N + (-\phi)^{-N}) - \cos(N\pi/3) + 2\mu^N \cos(N\theta), \qquad (5.8)$$

O. Martin, A. M. Odlyzko, and S. Wolfram

Table 3. Fraction of configurations appearing in cycles for the non-additive cellular automaton of Eq. (5.2)

N	ϱ_N^∞
4	0.3125
5	0.3438
6	0.1094
7	0.0078
8	0.1133
9	0.1426
10	0.0791
11	0.0435
12	0.0466
13	0.0350
14	0.0163
15	0.00308
16	0.00850
17	0.00857

where $\kappa \simeq 1.7548$ is the real root of $z^3 - z^2 + 2z - 1 = 0$, $\phi = (1 + \sqrt{5})/2 = 1.6182$, and $\mu \simeq 0.754$, $\theta \simeq 1.408$. The first ten values of ϱ_N are 1, 1, 4, 7, 11, 19, 36, 67, 121, 216. For large N, $\varrho_N \sim \kappa^N$. Equation (5.8) shows that corrections decrease rapidly and smoothly with N. This behaviour is to be contrasted with the irregular behaviour as a function of N found for additive cellular automata in Theorems 3.1 and 4.2.

Equation (5.8) shows that the fraction of all 2^N possible configurations which are reachable after one time step in the evolution of the cellular automaton of Eq. (5.2) is approximately $(\kappa/2)^N \simeq 0.92^N$. Thus, starting from an initial maximal entropy ensemble with $s = 1$, evolution for one time step according to Eq. (5.2) yields a set entropy

$$s(t = 1) \simeq \log_2 \kappa \simeq 0.88. \tag{5.9}$$

The irregularity of the transient trees illustrated in Fig. 5 implies a measure entropy $s_\mu < s$.

The result (5.9) becomes exact in the limit $N \to \infty$. A direct derivation in this limit is given in [17, 18], where it is also shown that the set of infinite configurations generated forms a regular formal language. The set continues to contract with time, so that the set entropy decreases below the value given by Eq. (5.9) [18].

Techniques similar to those used in the derivation of Eq. (5.5) may in principle be used to deduce the number of configurations reached after any given number of steps in the evolution of the cellular automaton (5.2). The fraction of configurations which appear in cycles is an irregular function of N; some results for small N are given in Table 3.

6. Discussion

The analysis of additive cellular automata in Sects. 3 and 4 yielded results on the global behaviour of additive cellular automata more complete than those available for most other dynamical systems. The extensive analysis was made possible by the discrete nature of cellular automata, and by the additivity property which led to the algebraic approach developed in Sect. 3. Similar algebraic techniques should be applicable to some other discrete dynamical systems.

The analysis of global properties of cellular automata made in this paper complements the analysis of local properties of ref. [1].

One feature of the results on additive cellular automata found in Sects. 3 and 4, is the dependence of global quantities not only on the magnitude of the size parameter N, but also on its number theoretical properties. This behaviour is shared by many dynamical systems, both discrete and continuous. It leads to the irregular variation of quantities such as cycle lengths with N, illustrated in Table 1 and Fig. 3. In physical realizations of cellular automata with large size N, an average is presumably performed over a range of N values, and irregular dependence on N is effectively smoothed out. A similar irregular dependence is found on the number k of possible values for each site: simple results are found only when k is prime.

Despite such detailed dependence on N, results such as Theorem 4.1–4.3 show that global properties of additive cellular automata exhibit a considerable universality, and independence of detailed aspects of their construction. This property is again shared by many other dynamical systems. It potentially allows for generic results, valid both in the simple cases which may easily be analysed, and in the presumably complicated cases which occur in real physical systems.

The discrete nature of cellular automata makes possible an explicit analysis of their global behaviour in terms of transitions in the discrete phase space of their configurations. The results of Sect. 4 provide a rather complete characterization of the structure of the state transition diagrams for additive cellular automata. The state transition diagrams consists of trees corresponding to irreversible "transients", leading to "attractors" in the form of distinct finite cycles. The irreversibility of the cellular automata is explicitly manifest in the convergence of several distinct configurations to single configurations through motion towards the roots of the trees. This irreversibility leads to a decrease in the entropy of an initially equiprobable ensemble of cellular automaton configurations; the results of Sect. 4 show that in most cases the entropy decreases by a fixed amount at each time step, reflecting the balanced nature of the trees. Theorem 4.3 gives an algebraic characterization of the magnitude of the irreversibility, in terms of the in-degrees of nodes in the trees. The length of the transients during which the entropy decreases is given by the height of the trees in Theorem 4.3, and is found always to be less than N. After these transients, any initial configurations evolve to configurations on attractors or cycles. Theorem 4.3 gives the total number of configurations on cycles in terms of N and algebraic properties of the cellular automaton time evolution polynomial. At one extreme, all configurations may be on cycles, while at the other extreme, all initial configurations may evolve to a single limit point consisting simply of the null configuration.

Theorem 4.1 gives a rather general result on the lengths of cycles in additive cellular automata. The maximum possible cycle length is found to be of order the square root of the total number of possible configurations. Rather long cycles are therefore possible. No simple results on the total number of distinct cycles or attractors were found; however, empirical results suggest that most cycles have a length equal to the maximal length for a particular cellular automaton.

The global properties of additive cellular automata may be compared with those of other mathematical systems. One closely related class of systems are linear feedback shift registers. Most results in this case concentrate on analogues of the cellular automaton discussed in Sect. 3, but with the values at a particular time step in general depending on those of a few far-distant sites. The boundary conditions assumed for feedback shift registers are typically more complicated than the periodic ones assumed for cellular automata in Sect. 3 and most of Sect. 4. The lack of symmetry in these boundary conditions allows for maximal length shift register sequences, in which all $2^N - 1$ possible configurations occur on a single cycle [2, 3].

A second mathematical system potentially analogous to cellular automata is a random mapping [15]. While the average cycle length for random mappings is comparable to the maximal cycle length for cellular automata, the probability for a node in the state transition diagram of a random mapping to have in-degree d is $\sim 1/d!$ and is much more sharply peaked at low values than for a cellular automaton, leading to many differences in global properties.

Non-additive cellular automata are not amenable to the algebraic techniques used in Sects. 3 and 4 for the additive case. Section 5 nevertheless discussed some properties of non-additive cellular automata concentrating on a simple one-dimensional example with two possible values at each site. Figure 5 indicates that the state transition diagrams for such non-additive cellular automata are less regular than those for additive cellular automata. Combinatorial methods were nevertheless used to derive the fraction of configurations with no predecessors in these diagrams, giving the irreversibility and thus entropy decrease associated with one time step in the cellular automaton evolution. Unlike the case of additive cellular automata, the result was found to be a smooth function of N.

Appendix A: Notations and Elementary Results on Finite Fields

Detailed discussion of the material in this appendix may be found in [8].

A. Basic Notations

$a \bmod b$ denotes a reduced modulo b, or the remainder of a after division by b.

(a, b) or $\gcd(a, b)$ denotes the greatest common divisor of a and b. When a and b are polynomials, the result is taken to be a polynomial with unit leading coefficient (monic).

$a|b$ represents the statement that a divides b (with no remainder).

$a^n \| b$ indicates that a^n is the highest power of a which divides b.

Exponentiation is assumed right associative, so that a^{bc} denotes $a^{(bc)}$ not $(a^b)^c$.

p usually denotes a prime integer.

$\mathbb{R}_k$ denotes an arbitrary commutative ring of k elements.

$\mathbb{Z}_k$ denotes the ring of integers modulo k.

$\deg P(x)$ denotes the highest power of x which appears in $P(x)$.

B. Finite Fields

There exists a finite field unique up to isomorphism with any size p^α (p prime), denoted $\mathrm{GF}(p^\alpha)$. p is termed the characteristic of the field.

The ring $\mathbb{Z}_k$ of integers modulo k forms a field only when k is prime, since only in this case do unique inverses under multiplication modulo k exist for all nonzero elements. (For example, in $\mathbb{Z}_4$, 2 has no inverse.) $\mathrm{GF}(p)$ is therefore isomorphic to $\mathbb{Z}_p$.

The field $\mathrm{GF}(p^\alpha)$ is conveniently represented by the set of polynomials of degree less than α with coefficients in $\mathbb{Z}_p$, with all polynomial operations performed modulo a fixed irreducible polynomial of degree α over $\mathrm{GF}(p)$. For example, $\mathrm{GF}(4)$ may be represented by elements 0, 1, κ, $\kappa+1$ with operations performed modulo 2 and modulo $\kappa^2+\kappa+1$. In this case for example $\kappa \times \kappa \equiv \kappa+1$. Notice that, as mentioned in Sect. A.C below, polynomials over a field form a unique factorization domain.

Any field of size q yields a group of size $q-1$ under multiplication if the zero element is removed. Thus for any element of $\mathrm{GF}(q)$,

$$x^q = x, \tag{A.1}$$

and $x^{q-1} = 1$ for $x \neq 0$. Notice that if $x \in \mathrm{GF}(p^\alpha)$ and $x^{p^\beta} = x$, then $x \in \mathrm{GF}(p^\beta)$.

C. Polynomials over Finite Fields

Polynomials in any number of variables with coefficients in $\mathrm{GF}(q)$ form a unique factorization domain. For such polynomials, therefore $A(x)B(x) \equiv A(x)C(x) \bmod P(x)$ implies $B(x) \equiv C(x) \bmod P(x)$ if $A(x), P(x)) = 1$.

For any polynomials $A(x)$ and $B(x)$ with coefficients in $\mathrm{GF}(q)$, there exist polynomials $\alpha(x)$ and $\beta(x)$ such that

$$C(x) = (A(x), B(x)) = \alpha(x)A(x) + \beta(x)B(x). \tag{A.2}$$

There are exactly q^n univariate polynomials over $\mathrm{GF}(q)$ with degree less than n. With a polynomial $Q(x)$ of degree m, the number of polynomials $P(x)$ with degree not exceeding n for which $Q(x)|P(x)$ is q^{n-m} for $m \leq n$.

For any prime p, and for elements a_i of $\mathrm{GF}(p^\beta)$,

$$(\sum a_i x^i)^{p^\alpha} = \sum (a_i x^i)^{p^\alpha}. \tag{A.3}$$

Thus for example,

$$(x^{2^\alpha} + 1) \equiv (x+1)^{2^\alpha} \qquad \bmod 2, \tag{A.4}$$

a result used extensively in Sect. 3.

If $P(x)|Q(x)$, then every root of $P(x)$ must be a root of $Q(x)$. If $\lambda \geq 2$ and

$$[P(x)]^\lambda \mid Q(x), \tag{A.5}$$

then

$$P(x) \mid Q'(x), \tag{A.6}$$

where $Q'(x)$ is the formal derivative of $Q(x)$, obtained by differentiation of each term in the polynomial. [Note that integration is not defined for polynomials over $\mathrm{GF}(q)$.]

O. Martin, A. M. Odlyzko, and S. Wolfram

The number of roots (not necessarily distinct) of a polynomial over $GF(q)$ is equal to the degree of the polynomial. The roots may lie in an extension of $GF(q)$. Over the field $GF(p)$,

$$x^N - 1 = (x^n - 1)^{D_p(N)}, \tag{A.7}$$

where $N = D_p(N)n$, with $D_p(N)$ defined in Sects. 3 and 4 as the maximum power of p which divides N. The polynomial $x^n - 1$ with n not a multiple of p then factorizes over $GF(p)$ according to

$$x^n - 1 = (x - 1) \prod_{\substack{d|n \\ d \neq 1}} \prod_{i=1}^{\frac{\phi(d)}{\mathrm{ord}_d(p)}} C_{d,i}(x), \tag{A.8}$$

where the $C_{d,i}(x)$ are irreducible cyclotomic polynomials of degree $\mathrm{ord}_d(p)$. Note that the multiplicity of any irreducible factor of $x^N - 1$ is exactly $D_p(N)$, and that

$$C_{d,i}(x) \mid x^d - 1. \tag{A.9}$$

D. Dipolynomials over Finite Fields

A dipolynomial $A(x)$ is taken to divide a dipolynomial $B(x)$ if there exists a dipolynomial $C(x)$ such that $B(x) = A(x)C(x)$. Hence if $A(x)$ and $B(x)$ are polynomials, with $A(0) \neq 0$, and if $A(x) \mid B(x)$ are dipolynomials, then $A(x) \mid B(x)$ are polynomials.

Congruence in the ring of dipolynomials is defined as follows: $A(x) \equiv B(x) \bmod C(x)$ for dipolynomials $A(x)$, $B(x)$, and $C(x)$ if $C(x) \mid A(x) - B(x)$.

The greatest common divisor of two nonzero dipolynomials $A_1(x)$ and $A_2(x)$ is defined as the ordinary polynomial $(A_1^*(x), A_2^*(x))$, where $A_i^*(x) = x^{m_i} A_i(x)$ and m_i is chosen to make $A_i^*(x)$ a polynomial with nonzero constant term. Note that by analogy with Eq. (A.2), for any dipolynomials $A_1(x)$ and $A_2(x)$, there exist dipolynomials $\alpha_1(x)$ and $\alpha_2(x)$ such that

$$(A_1(x), A_2(x)) = \alpha_1(x)A_1(x) + \alpha_2(x)A_2(x). \tag{A.10}$$

Appendix B: Properties and Values of some Number Theoretical Functions

A. Euler Totient Function $\phi(N)$

$\phi(N)$ is defined as the number of integers less than N which are relatively prime to N [7]. $\phi(N)$ is a multiplicative function, so that

$$\phi(mn) = \phi(m)\phi(n), \quad (m, n) = 1. \tag{B.1}$$

For p prime,

$$\phi(p^\alpha) = p^{\alpha - 1}(p - 1). \tag{B.2}$$

Hence

$$\phi(n) = \prod_{p^\alpha \| n} p^{\alpha - 1}(p - 1), \tag{B.3}$$

providing a formula by which $\phi(N)$ may be computed. Some values of $\phi(N)$ are given in Table 4.

$\phi(N)$ is bounded (for $N > 1$) by

$$cN/\log\log N \leqq \phi(N) \leqq N-1, \tag{B.4}$$

where c is some positive constant, and the upper bound is achieved if and only if N is prime. For large N, $\phi(N)/N$ tends on average to a constant value.

$\phi(n)$ satisfies the Euler-Fermat theorem

$$k^{\phi(n)} = 1 \quad \bmod n \quad (k, n) = 1. \tag{B.5}$$

B. *Multiplicative Order Function* $\mathrm{ord}_N(k)$

The multiplicative order function $\mathrm{ord}_N(k)$ is defined as the minimum positive integer j for which [8]

$$k^j = 1 \quad \bmod N. \tag{B.6}$$

This condition can only be satisfied if $(k, N) = 1$.

By the Euler-Fermat theorem (B.5),

$$\mathrm{ord}_N(k) \mid \phi(N). \tag{B.7}$$

In addition, $\mathrm{ord}_{mn}(k) = \mathrm{lcm}(\mathrm{ord}_n(k), \mathrm{ord}_m(k))$, $(n, k) = (m, k) = (n, m) = 1$.

Some special cases are

$$\mathrm{ord}_{k^\alpha - 1}(k) = \alpha,$$

$$\mathrm{ord}_{k^\alpha + 1}(k) = 2\alpha.$$

A rigorous bound on $\mathrm{ord}_N(k)$ is

$$\log_k(N) \leqq \mathrm{ord}_N(k) \leqq N-1, \tag{B.8}$$

where the upper bound is attained only if N is prime. It can be shown that on average, for large N, $\mathrm{ord}_N(k) \gtrsim \sqrt{N}$; the actual average is presumably closer to N. Nevertheless, for large N, $\mathrm{ord}_N(k)/N$ tends to zero on average.

Some values of the multiplicative order function are given in Table 4.

The multidimensional generalization $\mathrm{ord}_{N_1,\ldots,N_d}(k)$ of the multiplicative order function is defined as the minimum positive integer j for which $k^j = 1$ simultaneously modulo $N_1, N_2, \ldots,$ and N_d. It is clear that

$$\mathrm{ord}_{N_1,\ldots,N_d}(k) = \mathrm{lcm}(\mathrm{ord}_{N_1}(k), \ldots, \mathrm{ord}_{N_d}(k)) = \mathrm{ord}_{\mathrm{lcm}(N_1,\ldots,N_d)}(k),$$
$$(k, N_1) = \ldots = (k, N_d) = 1. \tag{B.9}$$

C. *Multiplicative Suborder Function* $\mathrm{sord}_N(k)$

The multiplicative suborder function is defined as the minimum j for which

$$k^j = \pm 1 \bmod N, \tag{B.10}$$

again assuming $(k, N) = 1$. Comparison with (B.6) yields

$$\mathrm{sord}_N(k) = \mathrm{ord}_N(k), \tag{B.11a}$$

or

$$\mathrm{sord}_N(k) = \tfrac{1}{2}\mathrm{ord}_N(k). \tag{B.11b}$$

The second case becomes comparatively rare for large N; the fraction of integers less than X for which it is realised may be shown to be asymptotic to $c/[\log X]^\lambda$

Table 4. Values of the multiplicative order $\mathrm{ord}_N(k)$ and suborder $\mathrm{sord}_N(k)$ functions defined in Eqs. (B.6) and (B.10), respectively, together with values of the Euler totient function $\phi(N)$. Each column gives values of the pair $\mathrm{ord}_N(k)$, $\mathrm{sord}_N(k)$

N	k=2		k=3		k=4		k=5		φ(N)
1									1
2			1	1			1	1	1
3	2	1			1	1	2	1	2
4			2	1			1	1	2
5	4	2	4	2	2	1			4
6							2	1	2
7	3	3	6	3	3	3	6	3	6
8			2	2			2	2	4
9	6	3			3	3	6	3	6
10			4	2					4
11	10	5	5	5	5	5	5	5	10
12							2	2	4
13	12	6	3	3	6	3	4	2	12
14			6	3			6	3	6
15	4	4			2	2			8
16			4	4			4	4	8
17	8	4	16	8	4	2	16	8	16
18							6	3	6
19	18	9	18	9	9	9	9	9	18
20			4	4					8
21	6	6			3	3	6	3	12
22			5	5			5	5	10
23	11	11	11	11	11	11	22	11	22
24							2	2	8
25	20	10	20	10	10	5			20
26			3	3			4	2	12
27	18	9			9	9	18	9	18
28			6	3			6	6	12
29	28	14	28	14	14	7	14	7	28
30									8
31	5	5	30	15	5	5	3	3	30
32			8	8			8	8	16
33	10	5			5	5	10	10	20
34			16	8			16	8	16
35	12	12	12	12	6	6			24
36							6	6	12
37	36	18	18	9	18	9	36	18	36
38			18	9			9	9	18
39	12	12			6	6	4	4	24
40			4	4					16

[16], where c and λ are constants determined by k.

In general,

$$\log_k(N) \leqq \mathrm{sord}_N(k) \leqq (N-1)/2, \qquad (B.12)$$

the upper limit again being achieved only if N is prime. For large N, $\mathrm{sord}_N(k)/N \to 0$ on average.

The multidimensional generalization $\mathrm{sord}_{N_1,\ldots,N_d}(k)$ of the multiplicative suborder function is defined as the minimum positive integer j for which $k^j = \pm 1$ simultaneously modulo $N_1, \ldots, N_d$, with $+1$ and -1 perhaps taken variously for the different N_i. The analogue of Eq. (B.9) for this function is

$$\mathrm{sord}_{N_1,\ldots,N_d}(k) = \mathrm{lcm}(\mathrm{sord}_{N_1}(k), \ldots, \mathrm{sord}_{N_d}(k)), \tag{B.13a}$$

and

$$\mathrm{lcm}(\mathrm{sord}_{N_1}(k), \ldots, \mathrm{sord}_{N_d}(k)) = \mathrm{sord}_{\mathrm{lcm}(N_1,\ldots,N_d)}(k), \tag{B.13b}$$

or

$$\mathrm{lcm}(\mathrm{sord}_{N_1}(k), \ldots, \mathrm{sord}_{N_d}(k)) = \tfrac{1}{2}\mathrm{sord}_{\mathrm{lcm}(N_1,\ldots,N_d)}(k). \tag{B.13c}$$

Acknowledgement. We are grateful to O. E. Lanford for several suggestions.

References

1. Wolfram, S.: Statistical mechanics of cellular automata. Rev. Mod. Phys. **55**, 601 (1983)
2. Golomb, S.W.: Shift register sequences. San Francisco: Holden-Day 1967
3. Selmer, E.S.: Linear recurrence relations over finite fields. Dept. of Math., Univ. of Bergen, Norway (1966)
4. Miller, J.C.P.: Periodic forests of stunted trees. Philos. Trans. R. Soc. Lond. A**266**, 63 (1970); A**293**, 48 (1980)
   ApSimon, H.G.: Periodic forests whose largest clearings are of size 3. Philos. Trans. R. Soc. Lond. A**266**, 113 (1970)
   ApSimon, H.G.: Periodic forests whose largest clearings are of size $n \geq 4$. Proc. R. Soc. Lond. A**319**, 399 (1970)
   Sutton, C.: Forests and numbers and thinking backwards. New Sci. **90**, 209 (1981)
5. Moore, E.F.: Machine models of self-reproduction. Proc. Symp. Appl. Math. **14**, 17 (1962) reprinted in: Essays on cellular automata, A. W. Burks. Univ. of Illinois Press (1966)
   Aggarwal, S.: Local and global Garden of Eden theorems. Michigan University technical rept. 147 (1973)
6. Knuth, D.: Fundamental algorithms, Reading, MA: Addison-Wesley 1968
7. Hardy, G.H., Wright, E.M.: An introduction to the theory of numbers. Oxford: Oxford University Press 1968
8. Mac Williams, F.J., Sloane, N.J.A.: The theory of error-correcting codes. Amsterdam: North-Holland 1977
9. Griffiths, P., Harris, J.: Principles of algebraic geometry. New York: Wiley 1978
10. Fredkin, E., Margolus, N.: Private communications
11. Ronse, C.: Non-linear shift registers: A survey. MBLE Research Lab. report, Brussels (May 1980)
12. Harao, M., Noguchi, S.: On some dynamical properties of finite cellular automaton. IEEE Trans. Comp. C-**27**, 42 (1978)
13. Grassberger, P.: A new mechanism for deterministic diffusion. Phys. Rev. A (to be published)
14. Guibas, L.J., Odlyzko, A.M.: String overlaps, pattern matching, and nontransitive games. J. Comb. Theory (A) **30**, 83 (1981)
15. Knuth, D.: Seminumerical algorithms. 2nd ed. Reading, MA: Addison-Wesley 1981
    Gelfand, A.E.: On the cyclic behavior of random transformations on a finite set. Tech. rept. 305, Dept. of Statistics, Stanford Univ. (August 1981)
16. Odlyzko, A.M.: Unpublished

258   O. Martin, A. M. Odlyzko, and S. Wolfram

17. Lind, D.A.: Applications of ergodic theory and sofic systems to cellular automata. Physica D **10** (to be published)
18. Wolfram, S.: Computation theory of cellular automata. Institute for Advanced Study preprint (January 1984)
19. Lenstra, H.W., Jr.: Private communication

Communicated by O.E. Lanford

Received February 11, 1983; in revised form September 7, 1983

Physica 10D (1984) 1–35
North-Holland, Amsterdam

UNIVERSALITY AND COMPLEXITY IN CELLULAR AUTOMATA

Stephen WOLFRAM*

The Institute for Advanced Study, Princeton NJ 08540, USA

Cellular automata are discrete dynamical systems with simple construction but complex self-organizing behaviour. Evidence is presented that all one-dimensional cellular automata fall into four distinct universality classes. Characterizations of the structures generated in these classes are discussed. Three classes exhibit behaviour analogous to limit points, limit cycles and chaotic attractors. The fourth class is probably capable of universal computation, so that properties of its infinite time behaviour are undecidable.

1. Introduction

Cellular automata are mathematical models for complex natural systems containing large numbers of simple identical components with local interactions. They consist of a lattice of sites, each with a finite set of possible values. The value of the sites evolve synchronously in discrete time steps according to identical rules. The value of a particular site is determined by the previous values of a neighbourhood of sites around it.

The behaviour of a simple set of cellular automata were discussed in ref. 1, where extensive references were given. It was shown that despite their simple construction, some cellular automata are capable of complex behaviour. This paper discusses the nature of this complex behaviour, its characterization, and classification. Based on investigation of a large sample of cellular automata, it suggests that many (perhaps all) cellular automata fall into four basic behaviour classes. Cellular automata within each class exhibit qualitatively similar behaviour. The small number of classes implies considerable university in the qualitative

* Work supported in part by the Office of Naval Research under contract number N00014-80-C0657.

behaviour of cellular automata. This universality implies that many details of the construction of a cellular automaton are irrelevant in determining its qualitative behaviour. Thus complex physical and biological systems may lie in the same universality classes as the idealized mathematical models provided by cellular automata. Knowledge of cellular automaton behaviour may then yield rather general results on the behaviour of complex natural systems.

Cellular automata may be considered as discrete dynamical systems. In almost all cases, cellular automaton evolution is irreversible. Trajectories in the configuration space for cellular automata therefore merge with time, and after many time steps, trajectories starting from almost all initial states become concentrated onto "attractors". These attractors typically contain only a very small fraction of possible states. Evolution to attractors from arbitrary initial states allows for "self-organizing" behaviour, in which structure may evolve at large times from structureless initial states. The nature of the attractors determines the form and extent of such structures.

The four classes mentioned above characterize the attractors in cellular automaton evolution. The attractors in classes 1, 2 and 3 are roughly analogous respectively to the limit points, limit cycles and chaotic ("strange") attractors found in continuous dynamical systems. Cellular automata of the fourth class behave in a more complicated manner, and are conjectured to be capable of universal computation, so that their evolution may implement any finite algorithm.

The different classes of cellular automaton behaviour allow different levels of prediction of the outcome of cellular automaton evolution from particular initial states. In the first class, the outcome of the evolution is determined (with probability 1), independent of the initial state. In the second class, the value of a particular site at large times is determined by the initial values of sites in a limited region. In the third class, a particular site value depends on the values of an ever-increasing number of initial sites. Random initial values then lead to chaotic behaviour. Nevertheless, given the necessary set of initial values, it is conjectured that the value of a site in a class 3 cellular automaton may be determined by a simple algorithm. On the other hand, in class 4 cellular automata, a particular site value may depend on many initial site values, and may apparently be determined only by an algorithm equivalent in complexity to explicit simulation of the cellular automaton evolution. For these cellular automata, no effective prediction is possible; their behaviour may be determined only by explicit simulation.

This paper describes some preliminary steps towards a general theory of cellular automaton behaviour. Section 2 below introduces notation and formalism for cellular automata. Section 3 discusses general qualitative features of cellular automaton evolution illustrating the four behaviour classes mentioned above. Section 4 introduces entropies and dimensions which characterize global features of cellular automaton evolution. Successive sections consider each of the four classes of cellular automata in turn. The last section discusses some tentative conclusions.

This paper covers a broad area, and includes many conjectures and tentative results. It is not intended as a rigorous mathematical treatment.

2. Notation and formalism

$a_i^{(t)}$ is taken to denote the value of site i in a one-dimensional cellular automaton at time step t. Each site value is specified as an integer in the range 0 through $k-1$. The site values evolve by iteration of the mapping

$$a_i^{(t)} = \mathbf{F}[a_{i-r}^{(t-1)}, a_{i-r+1}^{(t-1)}, \ldots, a_i^{(t-1)}, \ldots, a_{i+r}^{(t-1)}]. \quad (2.1)$$

$\mathbf{F}$ is an arbitrary function which specifies the cellular automaton rule.

The parameter r in eq. (2.1) determines the "range" of the rule: the value of a given site depends on the last values of a neighbourhood of at most $2r + 1$ sites. The region affected by a given site grows by at most r sites in each direction at every time step; propagating features generated in cellular automaton evolution may therefore travel at most r sites per time step. After t time steps, a region of at most $1 + 2rt$ sites may therefore be affected by a given initial site value.

The "elementary" cellular automata considered in ref. 1 have $k = 2$ and $r = 1$, corresponding to nearest-neighbour interactions.

An alternative form of eq. (2.1) is

$$a_i^{(t)} = \mathbf{f}\left[\sum_{j=-r}^{j=r} \alpha_j a_{i+j}^{(t-1)}\right], \quad (2.2)$$

where the α_j are integer constants, and the function $\mathbf{f}$ takes a single integer argument. Rules specified according to (2.1) may be reproduced directly by taking $\alpha_j = k^{r-j}$.

The special class of additive cellular automaton rules considered in ref. 2 correspond to the case in which $\mathbf{f}$ is a linear function of its argument modulo k. Such rules satisfy a special additive superposition principle. This allows the evolution of any initial configuration to be determined by superposition of results obtained with a few basis configurations, and makes possible the algebraic analysis of ref. 2.

"Totalistic" rules defined in ref. 1, and used in several examples below, are obtained by taking

$$\alpha_j = 1 \qquad (2.3)$$

in eq. (2.2). Such rules give equal weight to all sites in a neighbourhood, and imply that the value of a site depends only on the total of all preceding neighbourhood site values. The results of section 3 suggest that totalistic rules exhibit behaviour characteristic of all cellular automata.

Cellular automaton rules may be combined by composition. The set of cellular automaton rules is closed under composition, although composition increases the number of sites in the neighbourhood. Composition of a rule with itself yields patterns corresponding to alternate time steps in time evolution according to the rule. Compositions of distinct results do not in general commute. However, if a composition $F_1 F_2$ of rules generates a sequence of configurations with period π, then the rule $F_2 F_1$ must also allow a sequence of configurations with period π. As discussed below, this implies that the rules $F_1 F_2$ and $F_2 F_1$ must yield behaviour of the same class.

The configuration $a_i = 0$ may be considered as a special "null" configuration ("ground state"). The requirement that this configuration remain invariant under time evolution implies

$$F[0, 0, \ldots, 0] = 0 \qquad (2.4a)$$

and

$$f[0] = 0 . \qquad (2.4b)$$

All rules satisfy this requirement if iterated at most k times, at least up to a relabelling of the k possible values.

It is convenient to consider symmetric rules, for which

$$F[a_{i-r}, \ldots, a_{i+r}] = F[a_{i+r}, \ldots, a_{i-r}] . \qquad (2.5)$$

Once a cellular automaton with symmetric rules has evolved to a symmetric state (in which $a_{n+i} = a_{n-i}$ for some n and all i), it may subsequently generate only symmetric states (as-

suming symmetric boundary conditions), since the operation of space reflection commutes with time evolution in this case.

Rules satisfying the conditions (2.4) and (2.5) will be termed "legal".

The cellular automaton rules (2.1) and (2.2) may be considered as discrete analogues of partial differential equations of order at most $2r + 1$ in space, and first order in time. Cellular automata of higher order in time may be constructed by allowing a particular site value to depend on values of a neighbourhood of sites on a number s of previous time steps. Consideration of "effective" site values $\sum_{n=0}^{s-1} m^n a_i^{(t-n)}$ always allows equivalent first-order rules with $k = m^s - 1$ to be constructed.

The form of the function F in the time evolution rule (2.1) may be specified by a "rule number" [1]

$$R_F = \sum_{\{a_{i-r}, a_{i+r}\}} F[a_{i-r}, \ldots, a_{i+r}] k^{\sum_{i=-r}^{r} k^{r-i} a_{i+j}} . \qquad (2.6)$$

The function f in eq. (2.2) may similarly be specified by a numerical "code"

$$C_f = \sum_{n=0}^{(2r+1)(k-1)} k^n f[n] . \qquad (2.7)$$

The condition (2.4) implies that both R_F and C_f are multiples of k.

In general, there are a total of $k^{k^{(2r+1)}}$ possible cellular automaton rules of the form (2.1) or (2.2). Of these, $k^{k^{r+1}(kr+1)/2-1}$ are legal. The rapid growth of the number of possible rules with r implies that an exponentially small fraction of rules may be obtained by composition of rules with smaller r.

A few cellular automaton rules are "reducible" in the sense that the evolution of sites with particular values, or on a particular grid of positions and times, are independent of other site values. Such cellular automata will usually be excluded from the classification described below.

Very little information on the behaviour of a cellular automaton can be deduced directly from simple properties of its rule. A few simple results are nevertheless clear.

First, necessary (but not sufficient) conditions for a rule to yield unbounded growth are

$$F[a_{i-r}, a_{i-r+1}, \ldots, a_{i-1}, 0, 0, \ldots, 0] \neq 0,$$

$$F[0, \ldots, 0, 0, a_{i+1}, \ldots, a_{i+r}] \neq 0, \qquad (2.8)$$

for some set of a_i. If these conditions are not fulfilled then regions containing nonzero sites surrounded by zero sites can never grow, and the cellular automaton must exhibit behaviour of class 1 or 2. For totalistic rules, the condition (2.8) becomes

$$f[n] \neq 0 \qquad (2.9)$$

for some $n < r$.

Second, totalistic rules for which

$$f[n_1] \geq f[n_2] \qquad (2.10)$$

for all $n_1 > n_2$ exhibit no "growth inhibition" and must therefore similarly be of class 1 or 2.

One may consider cellular automata both finite and infinite in extent.

When finite cellular automata are discussed below, they are taken to consist of N sites arranged around a circle (periodic boundary conditions). Such cellular automata have a finite number k^N of possible states. Their evolution may be represented by finite state transition diagrams (cf. [2]), in which nodes representing each possible configuration are joined by directed arcs, with a single arc leading from a particular node to its successor after evolution for one time step. After a sufficiently long time (less than k^N), any finite cellular automaton must enter a cycle, in which a sequence of configurations is visited repeatedly. These cycles represent attractors for the cellular automaton evolution, and correspond to cycles in the state transition graph. At nodes in the cycles may be rooted trees representing transients. The transients are irreversible in the sense that nodes in the tree have a single successor, but may have several predecessors. In the course of time evolution, all

states corresponding to nodes in the trees ultimately evolve through the configurations represented by the roots of the trees to the cycles on which the roots lie. Configurations corresponding to nodes on the periphery of the state transition diagram (terminals or leaves of the transient trees) are never reached in the evolution: they may occur only as initial states. The fraction of configurations which may be reached after one time step in cellular automaton evolution, and which are therefore not on the periphery of the state transition diagram, gives a simple measure of irreversibility.

The configurations of infinite cellular automata are specified by (doubly) infinite sequences of site values. Such sequences are naturally identified as elements of a Cantor set (e.g. [3]). (They differ from real numbers through the inequivalence of configurations such as .111111... and 1.0000...). Cellular automaton rules define mappings from this Cantor set to itself. The mappings are invariant under shifts by virtue of the identical treatment of each site in eqs. (2.1) and (2.2). With natural measures of distance in the Cantor set, the mappings are also continuous. The typical irreversibility of cellular automaton evolution is manifest in the fact that the mapping is usually not injective, as discussed in section 4.

Eqs. (2.1) and (2.2) may be generalized to several dimensions. For $r = 1$, there are at least two possible symmetric forms of neighbourhood, containing $2d + 1$ (type I) and 3^d (type II) sites respectively; for larger r other "unit cells" are possible.

3. Qualitative characterization of cellular automaton behaviour

This section discusses some qualitative features of cellular automaton evolution, and gives empirical evidence for the existence of four basic classes of behaviour in cellular automata. Section 4 introduces some methods for quantitative analysis of cellular automata. Later sections use these methods to suggest fundamental characterizations of the four cellular automaton classes.

Fig. 1 shows the pattern of configurations generated by evolution according to each of the 32 possible legal totalistic rules with $k = 2$ and $r = 2$, starting from a "disordered" initial configuration (in which each site value is independently chosen as 0 or 1 with probability $\frac{1}{2}$). Even with such a structureless initial state, many of the rules are seen to generate patterns with evident structure. While the patterns obtained with different rules all differ in detail, they appear to fall into four qualitative classes:

1) Evolution leads to a homogeneous state (realized for codes 0, 4, 16, 32, 36, 48, 54, 60 and 62).

2) Evolution leads to a set of separated simple stable or periodic structures (codes 8, 24, 40, 56 and 58).

3) Evolution leads to a chaotic pattern (codes 2, 6, 10, 12, 14, 18, 22, 26, 28, 30, 34, 38, 42, 44, 46 and 50).

4) Evolution leads to complex localized structures, sometimes long-lived (codes 20 and 52).

Some patterns (e.g. code 12) assigned to class 3 contain many triangular "clearings" and appear more regular than others (e.g. code 10). The degree of regularity is related to the degree of irreversibility of the rules, as discussed in section 7.

Fig. 2 shows patterns generated from several different initial states according to a few of the cellular automaton rules of fig. 1. Patterns obtained with different initial states are seen to differ in their details, but to exhibit the same characteristic qualitative features. (Expectional initial states giving rise to different behaviour may exist with low or zero probability.) Fig. 3 shows the differences between patterns generated by various cellular automaton rules from initial states differing in the value of a single site.

*This sampling and many other investigations reported in this paper were performed using the C language computer program[4]. Requests for copies of this program should be directed to the author.

Figs. 4, 5 and 6 show examples of various sets of totalistic cellular automata. Fig. 4 shows some $k = 2$, $r = 3$ rules, fig. 5 some $k = 3$, $r = 1$ rules, and fig. 6 some $k = 5$, $r = 1$ rules. The patterns generated are all seen to be qualitatively similar to those of fig. 1, and to lie in the same four classes.

Patterns generated by all possible $k = 2$, $r = 1$ cellular automata were given in ref. 1, and are found to lie in classes 1, 2 and 3. Totalistic $k = 2$, $r = 1$ rules are found to give patterns typical of all $k = 2$, $r = 1$ rules. In general, totalistic rules appear to exhibit no special simplifications, and give rise to behaviour typical of all cellular automaton rules with given k and r.

An extensive sampling of many other cellular automaton rules supports the general conjecture that the four classes introduced above cover all one-dimensional cellular automata*.

Table I gives the fractions of various sets of cellular automata in each of the four classes. With increasing k and r, class 3 becomes overwhelmingly the most common. Classes 1 and 2 are decreasingly common. Class 4 is comparatively rare, but becomes more common for larger k and r.

"Reducible" cellular automata (mentioned in section 2) may generate patterns which contain features from several classes. In a typical case, fixed or propagating "membranes" consisting of sites with a particular value may separate regions containing patterns from classes 3 or 4 formed from sites with other values.

This paper concerns one-dimensional cellular automata. Two-dimensional cellular automata also appear to exhibit a few distinct classes of behaviour. Superficial investigations [5] suggest

Table I
Approximate fractions of legal totalistic cellular automaton rules in each of the four basic classes

Class	$k = 2$ $r = 1$	$k = 2$ $r = 2$	$k = 2$ $r = 3$	$k = 3$ $r = 1$
1	0.50	0.25	0.09	0.12
2	0.25	0.16	0.11	0.19
3	0.25	0.53	0.73	0.60
4	0	0.06	0.06	0.07

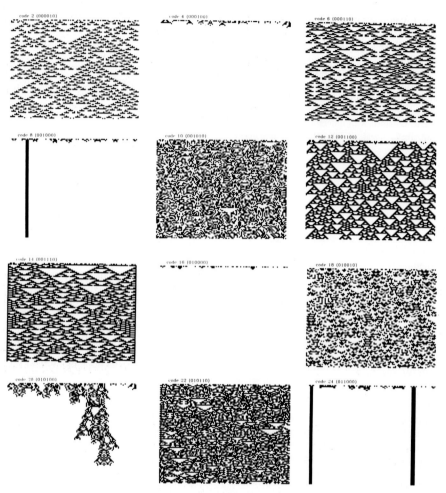

Fig. 1a.

S. Wolfram / Universality and complexity in cellular automata   7

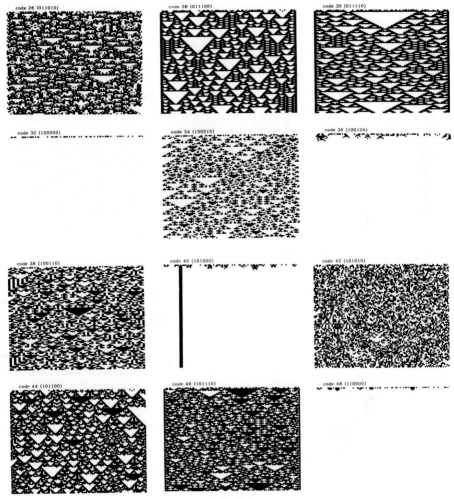

Fig. 1b.

8   *S. Wolfram / Universality and complexity in cellular automata*

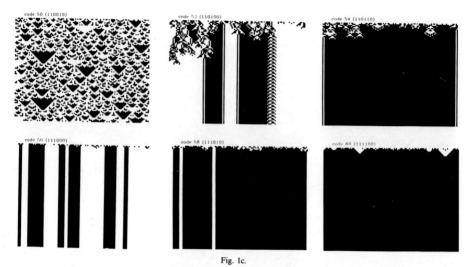

Fig. 1c.

Fig. 1a–c. Evolution of all possible legal one-dimensional totalistic cellular automata with $k = 2$ and $r = 2$. k gives the number of possible values for each site, and r gives the range of the cellular automaton rules. A range $r = 2$ allows the nearest and next-nearest neighbours of a site to affect its value on the next time step. Time evolution for totalistic cellular automata is defined by eqns. (2.2) and (2.7). The initial state is taken disordered, each site having values 0 and 1 with independent equal probabilities. Configurations obtained at successive time steps in the cellular automaton evolution are shown on successive horizontal lines. Black squares represent sites with value 1; white squares sites with value 0. All the cellular automaton rules illustrated are seen to exhibit one of four qualitative classes of behaviour.

that these classes may in fact be identical to the four found in one-dimensional cellular automata.

4. Quantitative characterizations of cellular automaton behaviour

This section describes quantitative statistical measures of order and chaos in patterns generated by cellular automaton evolution. These measures may be used to distinguish the four classes of behaviour identified qualitatively above.

Consider first the statistical properties of configurations generated at a particular time step in cellular automaton evolution. A disordered initial state, in which each site takes on its k possible values with equal independent probabilities, is statistically random. Irreversible cellular

automaton evolution generates deviations from statistical randomness. In a random sequence, all k^X possible subsequences ("blocks") of length X must occur with equal probabilities. Deviations from randomness imply unequal probabilities for different subsequences. With probabilities $p_j^{(x)}$ for the k^X possible sequences of site values in a length X block, one may define a specific "spatial set entropy"

$$s^{(x)}(X) = \frac{1}{X} \log_k \left(\sum_{j=1}^{k^X} \theta(p_j^{(x)}) \right), \tag{4.1}$$

where $\theta(p) = 1$ for $p > 0$ and $\theta(0) = 0$, and a specific "spatial measure entropy"

$$s_\mu^{(x)}(X) = -\frac{1}{X} \sum_{j=1}^{k^X} p_j^{(x)} \log_k p_j^{(x)}. \tag{4.2}$$

S. Wolfram / Universality and complexity in cellular automata   9

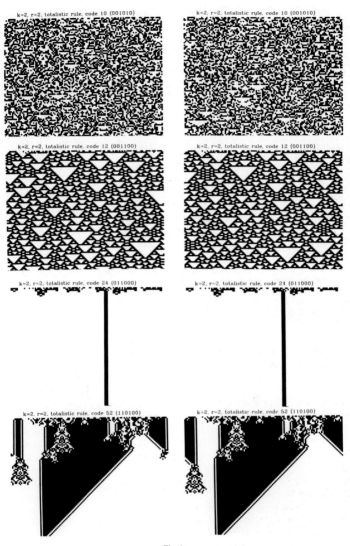

Fig. 2a.

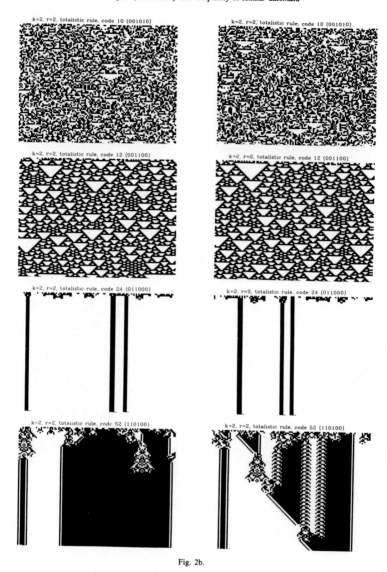

Fig. 2b.

Fig. 2. Evolution of some cellular automata illustrated in fig. 1 from several disordered states. The first two initial states shown differ by a change in the values of two sites, the next by a change in the values of ten sites. The last state is completely different.

S. Wolfram / Universality and complexity in cellular automata   11

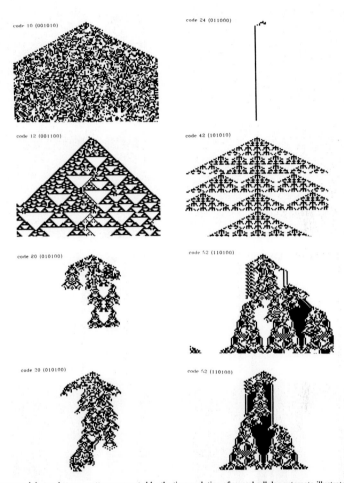

Fig. 3. Differences modulo two between patterns generated by the time evolution of several cellular automata illustrated in fig. 1 with disordered states differing by a change in the value of a single site.

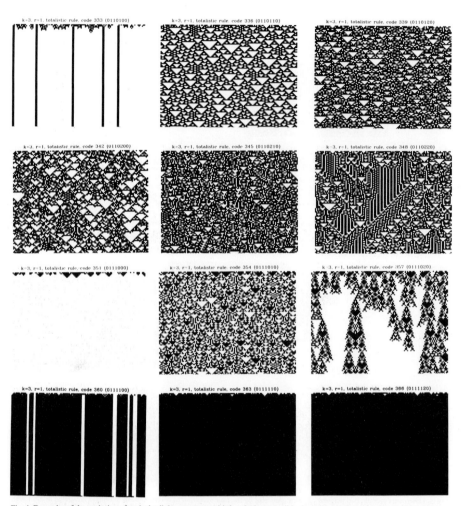

Fig. 4. Examples of the evolution of typical cellular automata with $k = 3$ (three possible site values) and $r = 1$ (only nearest neighbours included in time evolution rules). White squares represent value 0, grey squares value 1, and black squares value 2. The initial state is taken disordered, with each site having values 0, 1 and 2 with equal independent probabilities.

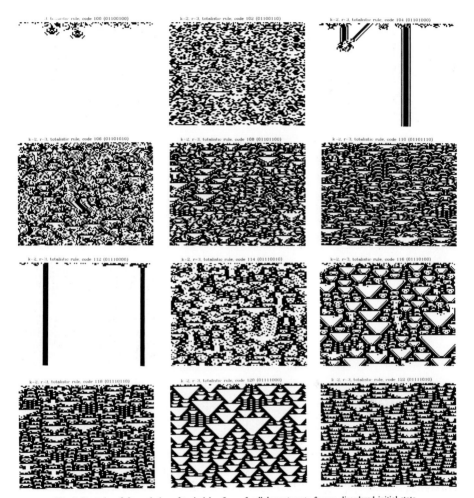

Fig. 5. Examples of the evolution of typical $k = 2$, $r = 3$ cellular automata from a disordered initial state.

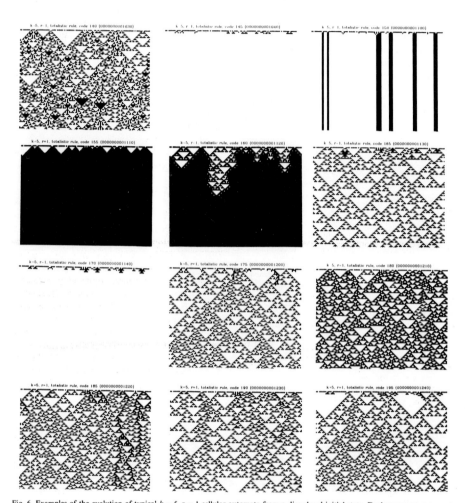

Fig. 6. Examples of the evolution of typical $k = 5$, $r = 1$ cellular automata from a disordered initial state. Darker squares represent sites with larger values.

In both cases, the superscript (x) indicates that "spatial" sequences (obtained at a particular time step) are considered. The "set entropy" (4.1) is determined directly by the total number $N^{(x)}(X)$ of length X blocks generated (with any nonzero probability) in cellular automaton evolution, according to

$$s^{(x)}(X) = \frac{1}{X} \log_k N^{(x)}(X) . \tag{4.3}$$

In the "measure entropy" (4.2) each block is weighted with its probability, so that the result depends explicitly on the probability measure for different cellular automaton configurations, as indicated by the subscript μ. Set entropy is often called "topological entropy"; measure entropy is sometimes referred to as "metric entropy"* (e.g. [6]). For blocks of length 1, the measure entropy $s_\mu^{(x)}(1)$ is related to the densities ρ_i of sites with each of the k possible values i. $s_\mu^{(x)}(2)$ is related to the densities of "digrams" (blocks of length 2), and so on. In general, the measure entropy gives the average "information content" per site computed by allowing for correlations in blocks of sites up to length X. Note that the entropies (4.1) and (4.2) may be considered to have units of (k-ary) bits per unit distance.

In the equation below, $s_{(\mu)}^{(x)}$ stands for either set entropy $s^{(x)}$ or for measure entropy $s_\mu^{(x)}$.

The definitions (4.1) and (4.2) yield immediately

$$s_\mu^{(x)}(X) \le s^{(x)}(X) \le 1 . \tag{4.4}$$

The first inequality is saturated (equality holds) only for "equidistributed" systems, in which all nonzero block probabilities $p_i^{(x)}$ are equal. The second inequality is saturated if all possible length X blocks of site values occur, but perhaps with

*The terms "set" and "measure" entropy, together with "set" and "measure" dimension, are introduced here to rationalize nomenclature.

unequal probabilities. $s_\mu(X) = 1$ only for "X-random" sequences [7], in which all k^X possible sequences of X site values occur with equal probabilities. In addition to (4.4), the definitions (4.1) and (4.2) imply

$$0 \le s_\mu^{(x)}(X) \le s^{(x)}(X) . \tag{4.5}$$

$s_\mu^{(x)}(X) = 0$ if and only if just one length X block occurs with nonzero probability, so that $s^{(x)}(X) = 0$ also. As discussed below, the inequality (4.5) is saturated for class 1 cellular automata.

Both set and measure entropies satisfy the subadditivity condition

$$(X_1 + X_2)s_{(\mu)}^{(x)}(X_1 + X_2) \le X_1 s_{(\mu)}^{(x)}(X_1) + X_2 s_{(\mu)}^{(x)}(X_2) . \tag{4.6}$$

The inequality is saturated if successive blocks of sites are statistically uncorrelated. In general, it implies some decrease in $s_{(\mu)}^{(x)}(X)$ with X (for example, $s_{(\mu)}^{(x)}(2X) \le s_{(\mu)}^{(x)}(X)$). For cellular automata with translation invariant initial probability measures, stronger constraints may be obtained (analogous to those for "stationary" processes in communication theory [8]). First, note that bounds on $s_{(\mu)}^{(x)}(X)$ valid for any set of probabilities $p_i^{(x)}$ also apply to $s^{(x)}(X)$, since $s^{(x)}(X)$ may formally be reproduced from the definition (4.2) for $s_\mu^{(x)}(X)$ by a suitable (extreme) choice of the $p_i^{(x)}$. The probability $p^{(x)}[a_1, \ldots, a_X]$ for the sequence of site values $a_1, \ldots, a_X$ is given in general by

$$p^{(x)}[a_1, \ldots, a_X]$$
$$= p^{(x)}[a_1, \ldots, a_{X-1}] p^{(x)}[a_X | a_1, \ldots, a_{X-1}] , \tag{4.7}$$

where $p^{(x)}[a_X | a_1, \ldots, a_{X-1}]$ denotes the conditional probability for a site value a_X, preceded by site values $a_1, \ldots, a_{X-1}$. Defining a total entropy

$$S_\mu^{(x)}[a_1, \ldots, a_X] =$$
$$-\sum p^{(x)}[a_1, \ldots, a_X] \log_k p^{(x)}[a_1, \ldots, a_X] , \tag{4.8}$$

and corresponding conditional total entropy

$$S^{(x)}_\mu[a_x | a_1, \ldots, a_{x-1}]$$
$$= - \sum p^{(x)}[a_1, \ldots, a_x] \log_k p^{(x)}[a_x | a_1, \ldots, a_{x-1}]$$
$$\leq S^{(x)}_\mu[a_1, \ldots, a_x] , \tag{4.9}$$

one obtains

$$Xs^{(x)}_\mu(X) = S^{(x)}_\mu(X) \leq \frac{X-1}{X} S^{(x)}_\mu(X-1)$$
$$+ \frac{1}{X} S^{(x)}_\mu(X) . \tag{4.10}$$

Hence,

$$s^{(x)}_{(\mu)}(X) \leq s^{(x)}_{(\mu)}(X-1) , \tag{4.11}$$

so that the set and measure entropies for a translationally invariant system decrease monotonically with the block size X. One finds in addition in this case that

$$\Delta^2_X (X s^{(x)}_{(\mu)}(X)) = (X+1)s^{(x)}_{(\mu)}(X+1) - 2Xs^{(x)}_{(\mu)}(X)$$
$$+ (X-1)s^{(x)}_{(\mu)}(X-1) \leq 0 , \tag{4.12}$$

so that $Xs^{(x)}_{(\mu)}(X)$ is a convex function of X.

With the definition $s^{(x)}(0) = 1$, this implies that there exists a critical block size X_c, such that

$$s^{(x)}(X) = 1, \quad \text{for } X < X_c ,$$
$$s^{(x)}(X) < 1, \quad \text{for } X \geq X_c . \tag{4.13}$$

The significance and values of the critical block size X_c will be discussed in section 7 below.

The entropies $s^{(x)}$ and $s^{(x)}_\mu$ may be evaluated either for many blocks in a single cellular automaton configuration, or for blocks in an ensemble of different configurations. For smooth probability measures on the ensemble of possible initial configurations, the results obtained in these two ways are almost always the same. (A probability measure will be considered "smooth" if changes in the values of a few sites in an infinite configuration lead only to infinitesimal changes in the probability for the configuration.) The set entropy $s^{(x)}$ is

typically independent of the probability measure on the ensemble, for any smooth measure. The measure entropy $s^{(x)}_\mu$ in general depends on the probability measure for initial configurations, although for class 3 cellular automata, it is typically the same for at least a large classes of smooth measures. Notice that with smooth measures, the values of $s^{(x)}(X)$ and $s^{(x)}_\mu(X)$ are the same whether the length X blocks used in their computation are taken disjoint or overlapping.

The entropies (4.1) and (4.2) are defined for infinite cellular automata. A corresponding definition may be given for finite cellular automata, with a maximum block length given by the total number of sites N the cellular automaton. The entropies $s^{(x)}(N)$ and $s^{(x)}_\mu(N)$ are related to global properties of the state transition diagram for the finite cellular automaton. The value of $s^{(x)}(N)$ at a particular time is determined by the fraction of possible configurations which may be reached at that time by evolution from any initial configuration. The limiting value of $s^{(x)}(N)$ at large times is determined by the fraction of configuration on cycles in the state transition graph. Starting from an initial ensemble in which all kN configurations occur with equal probabilities, the limiting value of $s^{(x)}_\mu(N)$ is equal to the limiting value of $s^{(x)}(N)$ if all transient trees in the state transition graph for the finite cellular automaton are identical, so that all configurations with non-zero probabilities are generated with the same probability (cf. [2]).

As mentioned in section 2, the configurations of an infinite cellular automaton may be considered as elements of a Cantor set. For an ensemble of disordered configurations (in which each site takes on its k possible values with equal independent probabilities), this Cantor set has fractal dimension 1. Irreversible cellular automaton evolution may lead to an ensemble of configurations corresponding to elements of a Cantor set with dimension less than one. The limiting value of $s^{(x)}(X)$ as $X \to \infty$ gives the fractal or "set" dimension of this set.

Relations between entropy and dimension may be derived in many ways (e.g. [6, 9]). Consider a set of numbers in the interval [0, 1] of the real line. Divide this interval into k^b bins of width k^{-b}, and let the fraction of bins containing numbers in the set be $N(b)$. For large b (small bin width), this number grows as k^{db}. The exponent d is the Kolmogorov dimension (or "capacity" (cf. [8])) of the set. If the set contains all real numbers in the interval [0, 1], then $N(b) = k^b$, and $d = 1$, as expected. If the set contains only a finite number of points, then $N(b)$ must tend to a constant for large b, yielding $d = 0$. The classic Cantor set consists of real numbers in the interval [0, 1], whose ternary decomposition contains only the digits 0 and 2. Dividing the interval into 3^b equal bins, it is clear that 2^b of these bins contain points in the set. The dimension of the set is thus $\log_3 2$. This dimension may also be found by an explicit recursive geometrical construction, using the fact that the set is "self-similar", in the sense that with appropriate magnification, its parts are identical to the whole.

The example above suggests that one may define a "set dimension" d according to

$$d = \lim_{b \to \infty} \frac{1}{b} \log_k N(b), \qquad (4.14)$$

where $N(b)$ is the number of bins which contain elements of the set. The bins are of equal size, and their total number is taken as k^b. Except in particularly pathological examples*, the dimension obtained with this definition is equal to the more usual Hausdorff (or "fractal") dimension (e.g. [11]) obtained by considering the number of patches at arbitrary positions required to cover the set (rather than the number of fixed bins containing elements of the set).

The definition (4.14) may be applied directly to cellular automaton configurations. The k^b "bins" may be taken to consist of cellular automaton configurations in which a block of b sites has a

particular sequence of values. The definition (4.3) of set entropy then shows that the set dimension is given by

$$d^{(x)} = \lim_{X \to \infty} s^{(x)}(X). \qquad (4.15)$$

A disordered cellular automaton configuration, in which all possible sequences of site values occur with nonzero probability (or an ensemble of such configurations), gives $d^{(x)} = 1$, as expected. Similarly, a homogeneous configuration, such as the null configuration, gives $d^{(x)} = 0$.

The set of configurations which appear at large times in the evolution of a cellular automaton constitute the attractors for the cellular automaton. The set dimension of these attractors is given in terms of the entropies for configurations appearing at large times by eq. (4.15).

Accurate direct evaluation of the set entropy $s^{(x)}(X)$ from cellular automaton configurations typically requires sampling of many more than k^X length X blocks. Inadequate samples yield systematic underestimates of $s^{(x)}(X)$. Direct estimates are most accurate when all nonzero probabilities for length X blocks are equal. In this case, a sample of k^h blocks yields an entropy underestimated on average by approximately

$$\log_k(1 - \exp(-k^{b - Xs(X)})). \qquad (4.16)$$

Unequal probabilities increase the magnitude of this error, and typically prevent the generation of satisfactory estimates of $d^{(x)}$ from direct simulations of cellular automaton evolution. (If the probabilities follow a log normal distribution, as in many continuous chaotic dynamical systems [12], then the exponential in eq. (4.16) is apparently replaced by a power [13].)

The dimension (4.15) is given as the limiting exponent with which $N^{(x)}(X)$ increases for large X. In the formula (4.15), this exponent is obtained as the limit of $\log_k[N(X)^{1/X}]$ for large X. If $N^{(x)}(X)$ indeed increases roughly exponentially with X,

---

* Such as the set formed from the end points of the intervals at each stage in the geometrical construction of the classic Cantor set. This set has zero Hausdorff dimension, but Kolmogorov dimension $\log_3 2$ [9].

then the alternative formula

$$d^{(x)} = \lim_{X \to \infty} \frac{X s^{(x)}(X)}{(X-1) s^{(x)}(X-1)}$$

$$= \lim_{X \to \infty} \log_k \left[\frac{N^{(x)}(X)}{N^{(x)}(X-1)} \right] \qquad (4.17)$$

is typically more accurate if entropy values are available only for small X.

The set dimension (4.15) may be used to characterize the set of configurations occurring on the attractor for a cellular automaton, without regard to their probabilities. One may also define a "measure dimension" $d_\mu^{(x)}$ which characterizes the probability measure for the configurations (cf. [12]):

$$d_\mu^{(x)} = \lim_{X \to \infty} s_\mu^{(x)}(X) . \qquad (4.18)$$

It is clear that

$$0 \le d_\mu^{(x)} \le d^{(x)} \le 1 . \qquad (4.19)$$

The measure dimension $d_\mu^{(x)}$ is equal to the "average information per symbol" contained in the sequence of site values in a cellular automaton configuration. If the sequence is completely random (or "∞-random" [7]), then the probabilities $p_i^{(x)}$ for all k^X sequences of length X must be equal for all X, so that $d_\mu^{(x)} = 1$. In this case, there is no redundancy or pattern in the sequence of site values, so that determination of each site value represents acquisition of one (k-ary) bit of information. A cellular automaton configuration with any structure or pattern must give $d_\mu^{(x)} < 1$.

In direct simulations of cellular automaton evolution, the probabilities $p_i^{(x)}$ for each possible length X block are estimated from the frequencies with which the blocks occur. These estimated probabilities are thus subject to Gaussian errors. Although the individual estimated probabilities are unbiased, the measure entropy deduced from them according to eq. (4.2), is systematically biased. Its mean typically yields a systematic underestimate of the true measure entropy, and with fixed sample

size, the underestimate deteriorates rapidly with increasing X, making an accurate estimate of $d_\mu^{(x)}$ impossible. However, since an unbiased estimate may be given for any polynomial function of the $p_i^{(x)}$, unbiased estimated upper and lower bounds for the measure entropy may be obtained from estimates for polynomials in $p_i^{(x)}$ just larger and just smaller than $-p_i^{(x)} \log_k p_i^{(x)}$ for $0 \le p_i^{(x)} \le 1$ [14]. In this way, it may be possible to obtain more accurate estimates of $s_\mu^{(x)}$ for large X, and thus of $d_\mu^{(x)}$.

The "spatial" entropies (4.1) and (4.2) were defined in terms of the sequence of site values in a cellular automaton configuration at a particular time step. One may also define "temporal" entropies which characterize the sequence of values taken on by a particular site though many time steps of cellular automaton evolution, as illustrated in fig. 7. With probabilities $p_i^{(t)}$ for the k^T possible sequences of values for a site at T successive time steps, one may define a specific temporal set entropy in analogy with eq. (4.1) by

$$s^{(t)}(T) = \frac{1}{T} \log_k \left(\sum_{j=1}^{k^T} \theta(p_j^{(t)}) \right), \qquad (4.20)$$

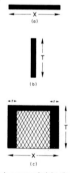

Fig. 7. Space-time regions sampled in the computation of (a) spatial entropies, (b) temporal entropies and (c) patch or mapping entropies. In case (c), the values of sites in the cross-hatched area are completely determined by values in the black "rind".

and a specific temporal measure entropy in analogy with eq. (4.2) by

$$s_\mu^{(t)}(T) = -\frac{1}{T} \sum_{j=1}^{k^T} p_j^{(t)} \log_k p_j^{(t)} . \qquad (4.21)$$

These entropies satisfy relations directly analogous to these given in eqs. (4.3) through (4.6) for spatial entropies. They obey relations analogous to (4.11) and (4.12) only for cellular automata in "equilibrium", statistically independent of time. The temporal entropies (4.20) and (4.21) may be considered to have units of (k-ary) bits per unit time.

Sequences of values in particular cellular automaton configurations typically have little similarity with the "time series" of values attained by a particular site under cellular automaton evolution. The spatial and temporal entropies for a cellular automaton are therefore in general quite different. Notice that the spatial entropy of a cellular automaton configuration may be considered as the temporal entropy of a pure shift mapping applied to the cellular automaton configuration.

Just as dimensions may be assigned to the set of spatial configurations generated in cellular automaton evolution, so also one may assign dimensions to the set of temporal sequences generated by the evolution. The temporal set dimension may be defined in analogy with eq. (4.15) by

$$d^{(t)} = \lim_{T \to \infty} s^{(t)}(T) , \qquad (4.22)$$

and the temporal measure dimension may be defined by

$$d_\mu^{(t)} = \lim_{T \to \infty} s_\mu^{(t)}(T) . \qquad (4.23)$$

If the evolution of a cellular automaton is periodic, so that each site takes on a fixed cycle of values, then

$$d^{(t)} = d_\mu^{(t)} = 0 . \qquad (4.24)$$

As discussed in section 6 below, class 2 cellular automata yield periodic structures at large times, so that the correspondingly temporal entropies vanish.

As a generalization of the spatial and temporal entropies introduced above, one may consider entropies associated with space-time "patches" in the patterns generated by cellular automaton evolution, as illustrated in fig. 7. With probabilities $p_i^{(t, x)}$ for the k^{XT} possible patches of spatial width X and temporal extent T, one may define a set entropy

$$s^{(t, x)}(T; X) = \frac{1}{T} \log_k \left(\sum_{j=1}^{k^{XT}} \theta(p_j^{(t, x)}) \right), \qquad (4.25)$$

and a measure entropy

$$s_\mu^{(t, x)}(T; X) = -\frac{1}{T} \sum_{j=1}^{k^{XT}} p_j^{(t, x)} \log_k p_j^{(t, x)} . \qquad (4.26)$$

Clearly,

$$s_{(\mu)}^{(t)}(T) = s_{(\mu)}^{(t, x)}(T; 1) , \qquad (4.27)$$

$$s_{(\mu)}^{(x)}(X) = \frac{1}{X} s_{(\mu)}^{(t, x)}(1; X) .$$

If no relation existed between configurations at successive time steps then the entropies (4.25) and (4.26) would be bounded simply by

$$s_\mu^{(t, x)}(T; X) \le s^{(t, x)}(T; X) \le X . \qquad (4.28)$$

The cellular automaton rules introduce definite relations between successive configurations and tighten this bound. In fact, the values of all sites in a $T \times X$ space-time patch are determined according to the cellular automaton rules by the values in the "rind" of the patch, as indicated in fig. 7. The rind contains only $X + 2r(T - 1)$ sites (where r is the "range" of the cellular automaton rule, defined in section 2), so that

$$s_\mu^{(t, x)}(T; X) \le s^{(t, x)}(T; X) \le [X + 2r(T - 1)]/T .$$
$$(4.29)$$

For large T (and fixed X), therefore

$$s_\mu^{(t;x)}(T; X) \le s^{(t;x)}(T; X) \le 2r. \qquad (4.30)$$

If both X and T tend to infinity with T/X fixed, eq. (4.30) implies that the "information per site" $s_\mu^{(t;x)}(T; X)/X$ in a $T \times X$ patch must tend to zero. The evolution of cellular automata can therefore never generate random space-time patterns.

With $T \to \infty$, X fixed, the length X horizontal section of the rind makes a negligible contribution to the entropies. The entropy is maximal if the $2r$ vertical columns in the rind are statistically independent, so that

$$s_{(\mu)}^{(t;x)}(\infty; X) \le 2rs_{(\mu)}^{(t)}(\infty) = 2rd_{(\mu)}^{(t)}. \qquad (4.31)$$

In addition,

$$s_{(\mu)}^{(t;x)}(\infty; X) \le s_{(\mu)}^{(t;x)}(\infty; X+1), \qquad (4.32)$$

where the bounds are saturated for large X if the time series associated with different sets of sites are statistically uncorrelated.

The limiting set entropy

$$\mathbf{h} = \lim_{\substack{T \to \infty \\ X \to \infty \\ T/X \to \infty}} s^{(t;x)}(T; X) \qquad (4.33)$$

for temporally-extended patches is a fundamental quantity equivalent to the set (or topological) entropy of the cellular automaton mapping in symbolic dynamics. $\mathbf{h}$ may be considered as a dimension for the mapping. It specifies the asymptotic rate at which the number of possible histories for the cellular automaton increases with time. The limiting measure entropy

$$\mathbf{h}_\mu = \lim_{\substack{T \to \infty \\ X \to \infty \\ T/X \to \infty}} s_\mu^{(t;x)}(T; X) \qquad (4.34)$$

gives the average amount of "new information" contained in each cellular automaton configuration, and not already determined from previous

configurations. Eqs. (4.31) and (4.32) show that

$$d_{(\mu)}^{(t)} \le \mathbf{h}_{(\mu)} \le 2rd_{(\mu)}^{(t)}. \qquad (4.35)$$

In addition,

$$\mathbf{h}_{(\mu)} \le 2rd_{(\mu)}^{(x)}. \qquad (4.36)$$

The basic cellular automaton time evolution rule (2.1) implies that the value a_i of a site i at a particular time step depends on sites a maximum distance r away on the previous time step according to the function $\mathbf{F}[a_{i-r}, \dots, a_{i+r}]$. After T time steps, the values of the site could depend on sites at distances up to rT, so that features in patterns generated by cellular automaton evolution could propagate at "speeds" up to r sites per time step. For many rules, however, the value of a site after many time steps depends on fewer initial site values, and features may propagate only at lower speeds. In general, let $\|\mathbf{F}^T\|$ denote the minimum R for which the value of site i depends only on the initial values of sites $i - R, \dots, i + R$. Then the maximum propagation speed associated with the cellular automaton rule $\mathbf{F}$ may be defined as

$$\lambda_+ = \overline{\lim_{T \to \infty}} \|\mathbf{F}^T\|/T. \qquad (4.37)$$

(The rule is assumed symmetric; for nonsymmetric rules, distinct left and right propagation speeds may be defined.) Clearly,

$$\lambda_+ \le r. \qquad (4.38)$$

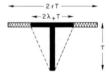

Fig. 8. Pattern of dependence of temporal sequences on spatial sequences, used in the proof of inequalities between spatial and temporal entropies.

When $\lambda_+ = 0$, finite regions of the cellular automaton must ultimately become isolated, so that

$$d_{(\mu)}^{(t)} = \mathbf{h}_{(\mu)}^{(t)} = 0 . \tag{4.39}$$

The construction of fig. 8 shows that for any T,

$$s_{(\mu)}^{(t)}(T) \le 2rs_{(\mu)}^{(x)}(2rT) . \tag{4.40}$$

In the limit $T \to \infty$, the construction implies

$$d_{(\mu)}^{(t)} \le 2\lambda_+ d_{(\mu)}^{(x)} , \tag{4.41}$$

The ratio of temporal to spatial entropy is thus bounded by the maximum propagation speed in the cellular automaton. The relation is consistent with the assignment of units to the spatial and temporal entropies mentioned above.

The corresponding inequalities for mapping entropies are:

$$\begin{aligned} d_{(\mu)}^{(t)} \le \mathbf{h}_{(\mu)} \le 2\lambda_+ d_{(\mu)}^{(x)} , \\ \mathbf{h}_{(\mu)} \le 2rd_{(\mu)}^{(t)} . \end{aligned} \tag{4.42}$$

The quantity λ_+ defined by eq. (4.37) gives the maximum speed with which any feature in a cellular automaton may propagate. With many cellular automaton rules, however, almost all "features" propagate much more slowly. To define an appropriate maximum average propagation speed, consider the effect after many time steps of changes in the initial state. Let $G(|x - x'|; t)$ denote the probability that the value of a site at position x' is changed when the value of a site at position x is changed t time steps before. The form of $G(|x - x'|; t)$ for various cellular automaton rules is suggested by fig. 3. $G(|x - x'|; t)$ may be considered as a Green function for the cellular automaton evolution. For large t, $G(|x - x'|; t)$ typically vanishes outside a "cone" defined by $|x - x'| = \bar{\lambda}_+ t$. $\bar{\lambda}_+$ may then be considered as a maximum average propagation speed. In analogy with eqs. (4.41) and (4.42), one expects

$$d_{(\mu)}^{(t)} \le \mathbf{h}_{(\mu)} \le 2\bar{\lambda}_+ d_{(\mu)}^{(t)} . \tag{4.43}$$

Mapping and temporal entropies thus vanish for cellular automata with zero maximum average propagation speed. Cellular automata in class 2 have this property.

The maximum average propagation speed $\bar{\lambda}_+$ specifies a cone outside which $G(|x - x'|; t)$ almost always vanishes. One may also define a minimum average propagation speed $\bar{\lambda}_-$, such that $G(|x - x'|; t) > 0$ for almost any $|x - x'| < \bar{\lambda}_-$.

The Green function $G(|x - x'|; t)$ gives the probability that a particular site is affected by changes in a previous configuration. The total effect of changes may be measured by the "Hamming distance" $H(t)$ between configurations before and after the changes, defined as the total number of site values which differ between the configurations after t time steps. ($H(t)$ is analogous to Lyapunov exponents for continuous dynamical systems.) Changing the values of initial sites in a small region, $H(t)$ may be given as a space integral of the Green function, and for large t obeys the inequality

$$H(t)/t \le 2\bar{\lambda}_+ , \tag{4.44}$$

to be compared with the result (4.43) obtained above.

The definitions and properties of dimension given above suggests that the behaviour these quantities determines the degree of "chaotic" behaviour associated with cellular automaton evolution. "Spatial chaos" occurs when $d_{(\mu)}^{(x)} > 0$, and "temporal chaos" when $d_{(\mu)}^{(t)} > 0$. Temporal chaos requires a nonzero maximum average propagation speed for features in cellular automaton patterns, and implies that small changes in initial conditions lead to effects ever-increasing with time.

5. Class 1 cellular automata

Class 1 cellular automata evolve after a finite number of time steps from almost all initial states to a unique homogeneous state, in which all sites have the same value. Such cellular automata may be considered to evolve to simple "limit points" in phase space; their evolution completely destroys any information on the initial state. The spatial and temporal dimensions for such attractors are zero.

Rules for class 1 cellular automata typically take the function **F** of eq. (2.1) to have the same value for almost all of its $k^{(2r+1)}$ possible sets of arguments.

Some exceptional configurations in finite class 1 cellular automata may not evolve to a homogeneous state, but may in fact enter non-trivial cycles. The fraction of such exceptional configurations appears to decrease very rapidly with the size N, suggesting that for infinite class 1 cellular automata the set of exceptional configurations is always of measure zero in the set of all possible configurations. For (legal) class 1 cellular automata whose usual final state has $a_i = n$, $n \neq 0$ (such as code 60 in fig. 1), the null configuration is exceptional for any size N, and yields $a_i = 0$.

6. Class 2 cellular automata

Class 2 cellular automata serve as "filters" which generate separated simple structures from particular (typically short) initial site value sequences*. The density of appropriate sequences in a particular initial state therefore determines the statistical properties of the final state into which it evolves. (There is therefore no unique large-time (invariant) probability measure on the set of possible configurations.) Changes of site values in the initial state almost always affect final site values only within a finite range, typically of order r. The maximum average propagation speed $\bar{\lambda}_+$ defined in section 4 thus vanishes for class 2 cellular automata. The temporal and mapping (but not spatial) dimensions for such automata therefore also vanish.

*They are thus of direct significance for digital image processing.

Although $\bar{\lambda} = 0$ for all class 2 cellular automata, λ is often nonzero. Thus exceptional initial state may exist, from which, for example, unbounded growth may occur. Such initial states apparently occur with probability zero for ensembles of (spatially infinite) cellular automata with smooth probability measures.

The simple structures generated by class 2 cellular automata are either stable, or are periodic, typically with small periods. The class 2 rules with codes 8, 24, 40 and 56 illustrated in fig. 1 all apparently exhibit only stable persistent structures. Examples of class 2 cellular automata which yield periodic, rather than stable, persistent structures include the $k = 2$, $r = 1$ cellular automaton with rule number 108 [1], and the $k = 3$, $r = 1$ totalistic cellular automaton with code 198. The periods of persistent structures generated in the evolution of class 2 cellular automata are usually less than $k!$. However, examples have been found with larger periods. One is the $k = 2$, $r = 3$ totalistic cellular automata with code 228, in which a persistent structure with period 3 is generated.

The finiteness of the periods obtained at large times in class 2 cellular automata implies that such systems have $d_{(\mu)}^{(t)} = \mathbf{h}_{(\mu)} = 0$, as deduced above from the vanishing of $\bar{\lambda}_+$. The evolution of class 2 cellular automata to zero (temporal) dimension attractors is analogous to the evolution of some continuous dynamical systems to limit cycles.

The set of persistent structures generated by a given class 2 cellular automaton is typically quite simple. For some rules, there are only a finite number of persistent structures. For example, for the code 8 and code 40 rules of fig. 1, only the sequence 111 (surrounded by 0 sites) appears to be persistent. For code 24, 111 and 1111 are both persistent. Other rules yield an infinite sequence of peristent structures, typically constructed by a simple process. For example, with code 56 in fig. 1, any sequence of two or more consecutive 1 sites is persistent.

In general, it appears that the set of persistent structures generated by any class 2 cellular automaton corresponds to the set of words generated by a regular grammar. A regular grammar [15–18] (or "sofic system" [19]) specifies a regular language, whose legal works may be recognized by a finite automaton, represented by a finite state transition graph. A sequence of symbols (site values) specifies a particular traversal of the state transition graph. The traversal begins at a special "start" node; the symbol sequence represents a legal word only if the traversal does not end at an absorbing "stop" node. Each successive symbol in the sequence causes the automaton to make a transition from one state (node) to one of k others, as specified by the state transition graph. At each step, the next state of the automaton depends only on its current state, and the current symbol read, but not on its previous history.

The set of configurations (symbol sequences) generated from all possible initial configurations by one time step of cellular automaton evolution may always be specified by a regular grammar. To determine whether a particular configuration $a^{(1)}$ may be generated after one time step of cellular automaton evolution, one may attempt to construct an explicit predecessor $a^{(0)}$ for it. Assume that a predecessor configuration has been found which reproduces all site values up to position i. Definite values $a_j^{(0)}$ for all $j \leq i - r$ are then determined. Several of the total of k^{2r} sequences of values $a_{i-r+1}^{(0)}, \ldots, a_{i+r+1}^{(0)}$ may be possible. Each sequence may be specified by an integer $q = \Sigma_{j=0}^{2r} k^j a_{i-r+j+1}^{(0)}$. An integer ψ_i between 0 and $2^{k^{2r}}$ may then be defined, with the qth binary bit in ψ_i equal to one if sequence q is allowed, and 0 otherwise. Each possible value of ψ may be considered to correspond to a state in a finite automaton. $\psi = 0$ corresponds to a "stop" state, which is reached if and only if $a^{(1)}$ has no predecessors. Possible values for $a_{i+r+1}^{(0)}$ are then found from ψ_i and the value of $a_{i+1}^{(0)}$. These possible values then determine the value of ψ_{i+1}. A finite state transition graph, determined by the cellular automaton rules, gives the possible transitions $\psi_i \rightarrow \psi_{i+1}$. Configurations reached after one time step of cellular automaton evolution may thus be recognized by a finite automaton with at most $2^{k^{2r}}$ states.

The set of such configurations is thus specified by a regular grammar.

In general, if the value of a given site after t steps of cellular automaton evolution depends on m initial site values, then the set of configurations generated by this evolution may be recognized by a finite automaton with at most 2^{k^m} states. The value of m may increase as $2rt$, potentially requiring an infinite number of states in the recognizing automaton, and preventing the specification of the set of possible configurations by a regular grammar. However, as discussed above, the value of m for a class 2 cellular automaton apparently remains finite as $t \rightarrow \infty$. Thus the set of configurations which may persist in such a cellular automaton may be recognized by a finite automaton, and are therefore specified by a regular grammar. The complexity of this grammar (measured by the minimum number of states required in the state transition graph for the recognizing automaton) may be used to characterize the complexity of the large time behaviour of the cellular automaton.

Finite class 2 cellular automata usually evolve to short period cycles containing the same persistent structures as are found in the infinite case. The fraction of exceptional initial states yielding other structures decreases rapidly to zero as N increases.

7. Class 3 cellular automata

Evolution of infinite class 3 cellular automata from almost all possible initial states leads to aperiodic ("chaotic") patterns. After sufficiently many time steps, the statistical properties of these patterns are typically the same for almost all initial states. In particular, the density of nonzero sites typically tends to a fixed nonzero value (often close to $1/k$). In infinite cellular automata, "equilibrium" values of statistical quantities are approached roughly exponentially with time, and are typically attained to high accuracy after a very few time steps. For a few rules (such as the $k = 2$, $r = 1$ rule with rule number 18 [20]), however,

24   *S. Wolfram / Universality and complexity in cellular automata*

Fig. 9. Evolution of some class 3 totalistic cellular automata with $k = 2$ and $r = 2$ (as illustrated in fig. 1) from initial states containing one or a few nonzero sites. Some cases yield asymptotically self-similar patterns, while others are seen to give irregular patterns.

"defects" consisting of small groups of sites may exist, and may execute approximate random walks, until annihilating, usually in pairs. Such processes lead to transients which decrease with time only as $t^{-1/2}$.

Fig. 1 showed examples of the patterns generated by evolution of some typical class 3 cellular automata from disordered initial states. The patterns range from highly irregular (as for code 10), to rather regular (as for code 12). The most obvious regularity is the appearance of large triangular "clearings" in which all sites have the same value. These clearings occur when a "fluctuation" in which a sequence of consequence of consecutive sites have the same value, is progressively destroyed by the effects of other sites. The rate at which "information" from other sites may "flow" into the fluctuation, and thus the slope of the boundaries of the clearing, may range from $1/k$ to r sites per time step. The qualitative regularity of patterns generated by some class 3 rules arises from the high density of long sequences of correlated site values, and thus of triangular clearings. In general, however, it appears that the density of clearings decreases with their size n roughly as σ^{-n}. Different cellular automata appear to yield a continuous range of σ values. Those with larger σ yield more regular patterns, while those with smaller σ yield more irregular patterns. No sharp distinction appears to exist between class 3 cellular automata yielding regular and irregular patterns.

The first column in fig. 9 shows patterns obtained by evolution with typical class 3 cellular automaton rules from initial states containing a single nonzero site. Unbounded growth, leading to an asymptotically infinite number of nonzero sites, is evident in all cases. Some rules are seen to give highly regular patterns, others lead to irregular patterns.

The regular patterns obtained with rules such as code 2 are asymptotically self-similar fractal curves (cf. [11]). Their form is identical when viewed at different magnifications, down to length scales of order r sites. The total number of nonzero sites in such patterns after t time steps approaches t^d,

where d gives the fractal dimension of the pattern. Many class 3 $k = 2$ rules generate a similar pattern, illustrated by codes 2 and 34 in fig. 9, with $d = \log_2 3 \approx 1.59$. Some rules yield self-similar patterns with other fractal dimensions (for example, code 38 yields $d \approx 1.75$), but all self-similar patterns have $d < 2$, and lead to an asymptotic density of sites which tends to zero as t^{d-2}.

Rule such as code 10 are seen to generate irregular patterns by evolution even from a single site initial state. The density of nonzero sites in such patterns is found to tend asymptotically to a nonzero value; in some, but not all, cases the value is the same as would be obtained by evolution from a disordered initial state. The patterns appear to exhibit no large-scale structure.

Cellular automata contain no intrinsic scale beyond the size of neighbourhood which appears in their rules. A configuration containing a single nonzero site is also scale invariant, and any pattern obtained by evolution from it with cellular automaton rules must be scale invariant. The regular patterns in fig. 9 achieve this scale invariance by their self-similarity. The irregular patterns presumably exhibit correlations only over a finite range, and are therefore effectively uniform and scale invariant at large distances.

The second and third columns in fig. 11 shows the evolution of several typical class 3 cellular automata from initial states with nonzero sites in a small region. In some cases (such as code 12), the regular fractal patterns obtained with single nonzero sites are stable under addition of further nonzero initial sites. In other cases (such as code 2) they are seen to be unstable. The numbers of rules yielding stable and unstable fractal patterns are found to be roughly comparable.

Many but not all rules which evolve to regular fractal patterns from simple initial states generate more regular patterns in evolution from disordered initial states. Similarly, many but not all rules which produce stable fractal patterns yield more regular patterns from disordered initial states. For example, code 42 in figs. 1 and 9 generates stable fractal patterns from simple initial state, but

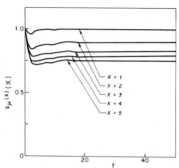

Fig. 10. Evolution of spatial measure entropies $s_\mu^{(x)}(X)$ as a function of time for evolution of the class 3 cellular automaton with code 12 illustrated in fig. 1 from a disordered initial state. The irreversibility of cellular automaton evolution results in a decrease of the entropies with time. Rapid relaxation to an "equilibrium" state is nevertheless seen.

leads to an irregular patterns under evolution from a disordered state. (Although not necessary for such behaviour, this rule possesses the additivity property mentioned in section 2.)

The methods of section 4 may be used to analyse the general behaviour of class 3 cellular automata evolving from typical initial states, in which all

sites have nonzero values with nonzero probability. Class 3 cellular automata apparently always exhibit a nonzero minimum average propagation speed $\bar{\lambda}_-$. Small changes in initial states thus almost always lead to increasingly large changes in later states. This suggests that both spatial and temporal dimensions $d_{(\mu)}^{(x)}$ and $d_{(\mu)}^{(t)}$ should be nonzero for all class 3 cellular automata. These dimensions are determined according to eqs. (4.15), (4.18), (4.22) and (4.23) by the limiting values of spatial and temporal entropies.

A disordered or statistically random initial state, in which each site takes on its k possible values with equal independent probabilities, has maximal spatial entropy $s_{(\mu)}^{(x)}(X) = 1$ for all block lengths X. Fig. 10 shows the behaviour of $s_\mu^{(x)}(X)$ as a function of time for several block lengths X in the evolution of a typical class 3 cellular automaton from a disordered (maximal entropy) initial state. The entropies are seen to decrease for a few time steps, and then to reach "equilibrium" values. The "equilibrium" values of $s_\mu^{(x)}(X)$ for class 3 cellular automata are typically independent of the probability measure on the ensemble of possible initial states, at least for "smooth" measures. The decrease in

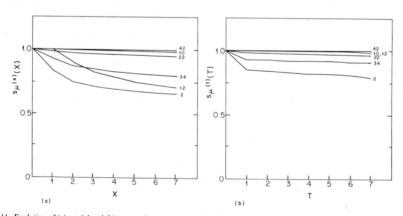

Fig. 11. Evolution of (a) spatial and (b) temporal measure entropies $s_\mu^{(x)}(X)$ and $s_\mu^{(t)}(T)$ obtained at equilibrium by evolution of several class 3 cellular automata illustrated in fig. 1, as a function of the spatial and temporal block lengths X and T. The entropies are evaluated for the region indicated in figs. 7(a) and 7(b). The limit of $s_\mu^{(x)}(X)$ as $X \to \infty$ is the spatial measure dimension of the attractor for the system; the limit of $s_\mu^{(t)}(T)$ as $T \to \infty$ is the temporal measure dimension.

entropy with time manifests the irreversible nature of the cellular automaton evolution. The decrease is found to be much greater for class 3 cellular automata which generate regular patterns (with many triangular clearings) than for those which yield irregular patterns. The more regular patterns require a higher degree of self-organization, with correspondingly greater irreversibility, and larger entropy decrease.

As discussed in section 4, the dependence of $s_{(\mu)}^{(x)}(X)$ on X measures spatial correlations in cellular automaton configurations. $s_{(\mu)}^{(x)}(X)$ therefore tends to a constant if X is larger than the range of any correlations between site values. In the presence of correlations, $s_{(\mu)}^{(x)}(X)$ always decreases with X. Available data from simulations provide reliable accurate estimates for $s_{(\mu)}^{(x)}(X)$ only for $0 \le X \le 8$. Fig. 11 shows the behaviour of the equilibrium value of $s_{\mu}^{(x)}(X)$ as a function of X over this range for several typical class 3 cellular automata. For rules which yield irregular patterns the equilibrium value of $s_{\mu}^{(x)}(X)$ typically remains $\gtrsim 0.9$ for $X \lesssim 8$. $s_{\mu}^{(x)}(X)$ at equilibrium typically decreases much more rapidly for class 3 cellular automata which generate more regular patterns. At least for small X, $s_{\mu}^{(x)}(X)$ for such cellular automata typically decreases roughly as $X^{-\eta}$ with $\eta \approx 0.1$.

The values of the spatial set entropy $s^{(x)}(X)$ provide upper bounds on the spatial measure entropy $s_{\mu}^{(x)}(X)$. The distribution of nonzero probabilities $p_i^{(x)}$ for possible length X blocks is typically quite broad, yielding an $s_{\mu}^{(x)}(X)$ significantly smaller than $s^{(x)}(X)$. Nevertheless, the general behaviour of $s_{\mu}^{(x)}(X)$ with X usually roughly follows $s^{(x)}(X)$, but with a slight X delay.

As discussed in section 4, the set entropy $s^{(x)}(X)$ attains its maximum value of 1 if and only if all k^X sequences of length X appear (with nonzero probability) in evolution from some initial state. Notice that if $s^{(x)}(X) = 1$ after one time step, then $s^{(x)}(X) = 1$ at any time. In general, $s^{(x)}(X)$ takes on value 1 for blocks up to some critical length X_c (perhaps infinite), as defined in eq. (4.13).

Since a block of length X is completely determined by a sequence of length $X + 2r$ in the previous configuration, any predecessors for the block may in principle be found by an exhaustive search of all k^{X+2r} possible length $X + 2r$ sequences. The procedure for progressive construction of predecessors outlined in section 6 provides a more efficient procedure [21]. The critical block length X_c is determined by the minimum number of nodes in the finite automaton state transition graph visited on any path from the "start" to "stop" node. The state transition graph is determined by the set of transition rules $\Psi_i \to \Psi_{i+1}$. Starting with length 1 blocks, these transition rules may be found by considering construction of all possible progressively longer blocks, but ignoring blocks associated with values Ψ_i for which the transition rules have already been found. If X_c is finite, the "stop" node $\Psi = 0$ is reached in the construction of length X_c blocks. Alternatively, the state transition graph may be found to consist of closed cycles, not including $\Psi = 0$. In this case, X_c is determined to be infinite. Since the state transition graph contains at most $2^{k^{2r}}$ nodes, the value of X_c may be found after at most this many tests. The procedure thus provides a finite algorithm for determining whether all possible arbitrarily long sequences of site values may be generated by evolution with a particular cellular automaton rule.

Table II gives the critical block lengths X_c for the cellular automata illustrated in fig. 1. Class 3 cellular automata with smaller X_c tend to generate more regular patterns. Those with larger X_c presumably give systematically larger entropies and their evolution is correspondingly less irreversible.

For additive cellular automata (such as code 42 in fig. 1 and table II), all possible blocks of any length X may be reached, and have exactly k^{2r} predecessors of length $X + 2r$. In this case, therefore, evolution from a disordered initial state gives $s^{(x)}(X) = 1$ for all X (hence $X_c = \infty$). The equality of the number of predecessors for each block implies in addition in this case that $s_{\mu}^{(x)}(X) = 1$, at least for evolution from disordered initial states. Hence for additive cellular automata

$$d^{(x)} = d_{\mu}^{(x)} = 1 . \tag{7.1}$$

Table II

Values of critical block length X_c for legal totalistic $k = 2$, $r = 2$ cellular automata as illustrated in fig. 1. For $X < X_c$, all k^X possible blocks of X site values appear with nonzero probability in configurations generated after any number of time steps in evolution from disordered initial states, while for $X \geq X_c$, some blocks are absent, so that the spatial set entropy $s^{(x)}(X) < 1$

Code	X_c	Code	X_c
2	5	32	3
4	12	34	5
6	7	36	12
8	12	38	7
10	36	40	12
12	5	42	∞
14	5	44	5
16	5	46	5
18	5	48	5
20	36	50	5
22	12	52	22
24	7	54	12
26	12	56	7
28	5	58	12
30	3	60	5

The configurations generated by additive cellular automata are thus maximally chaotic.

In general cellular automata evolving according to eq. (2.1) yield $s^{(x)}(X) = 1$ for all X, so that $d^{(x)} = 1$, if $\mathbf{F}$ is an injective (one-to-one) function of either its first or last argument (or can be obtained by composition of functions with such a property). This may be proved by induction. Assume that all the blocks of length X are reachable, with predecessors of lengths $X + 2r$. Then form a block of length $X + 1$ by adding a site at one end. To obtain all possible length $X + 1$ blocks, the value a' of this additional site must range over k possibilities. Any predecessors for length $X + 1$ blocks must be obtained by adding a $(X + 2r + 1)$-th site (with value a) at one end. For all length $X + 1$ blocks to be reachable, all values of a' must be generated when a runs over its k possible values, and the result follows. Notice that not all length $X + 1$ blocks need have the same (nonzero) number of predecessors, so that the measure entropy $s_\mu^{(x)}(X)$ may be less than the set entropy $s^{(x)}(X)$.

While injectivity of the rule function $\mathbf{F}$ for a cellular automaton in its first or last arguments is sufficient to give $d^{(x)} = 1$, it is apparently not necessary. A necessary condition is not known.

In section 6 it was shown that the set of configurations obtained by cellular automaton evolution for a finite number of time steps from any initial state could be specified by a regular grammar. In general the complexity of the grammar may increase rapidly with the number of time steps, potentially leading at infinite time to a set not specifiable by a regular grammar. Such behaviour may generically be expected in class 3 cellular, for which the average minimum propagation speed $\bar{\lambda} > 0$.

As discussed in section 4, one may consider the statistics of temporal as well as spatial sequences of site values. The temporal aperiodicity of the patterns generated by evolution of class 3 cellular automata from almost all initial states suggests that these systems should have nonvanishing temporal entropies $s_{(\mu)}^{(t)}(T)$ and nonvanishing temporal dimensions $d_{(\mu)}^{(t)}$. Once again, the temporal entropies for blocks starting at progressively later times quickly relax to equilibrium values. Notice that the dimension $d_{(\mu)}^{(t)}$ obtained from the large T limit of the $s_{(\mu)}^{(t)}(T)$ is always independent of the starting times for the blocks. This is to be contrasted with the spatial dimensions $d_{(\mu)}^{(x)}$, which depend on the time at which they are evaluated. Just as for spatial entropies, it found that the equilibrium temporal entropies are essentially independent of probability measure for initial configurations.

The temporal entropies $s_{(\mu)}^{(t)}(T)$ decrease slowly with T. In fact, it appears that in all cases

$$s_{(\mu)}^{(t)}(Z) \geq s_{(\mu)}^{(x)}(Z) \,. \tag{7.2}$$

The ratio $s_{(\mu)}^{(t)}(Z)/s_{(\mu)}^{(x)}(Z)$ is, however, typically much smaller than its maximum value (4.38) equal to the maximum propagation speed λ_+. Notice that the value of λ_+ determines the slopes of the edges of triangular clearings in the patterns generated by cellular automaton evolution.

At least for the class 3 cellular automata in fig. 1 which generate irregular patterns, the equi-

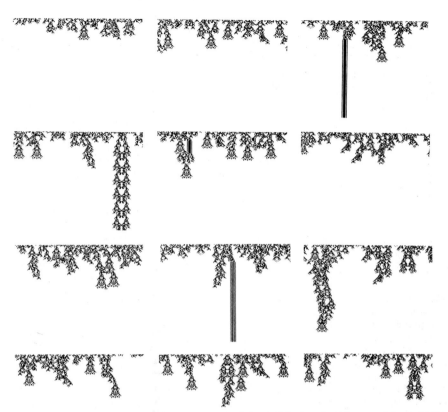

Fig. 12. Examples of the evolution of a class 4 cellular automaton (totalistic code 20 $k = 2$, $r = 2$ rule) from several disordered initial states. Persistent structures are seen to be generated in a few cases. The evolution is truncated after 120 time steps.

librium set entropy $s^{(t)}(T) = 1$ for all $T \lesssim 8$ for which data are available. Note that the result $s^{(t)}(T) = 1$ holds for all T for any additive cellular automaton rule. One may speculate that class 3 cellular automata which generate apparently irregular patterns form a special subclass, characterized by temporal dimension $d^{(t)} = 1$.

For class 3 cellular automata which generate more regular patterns, $s_{(t)}(T)$ appears to decrease, albeit slowly, with T. Just as for spatial sequences, one may consider whether the temporal sequences

which appear form a set described by a regular grammar. For the particular case of the $k = 2$, $r = 1$ cellular automaton with rule number 18, there is some evidence [21] that all possible temporal sequences which contain no 11 subsequences may appear, so that $N^{(t)}(T) = F_T$ where F_T is the Tth Fibonacci number $(F_T = F_{T-1} + F_{T-2}, F_0 = F_1 = 1)$. This implies that $N_{(t)}(T) \sim \phi^T$ ($\phi = (\sqrt{5} + 1)/2 \simeq 1.618$) for large T, suggesting a temporal set dimension $d^{(t)} = \log_2 \phi \approx 0.694$. In general, however, the set of possible temporal sequences is not expected to be described by a regular grammar.

The nonvanishing value of the average minimum propagation speed $\bar{\lambda}_-$ for class 3 cellular automata, suggests that in all cases the value of a particular site depends on an ever-increasing number of initial site values. However, the complexity of the dependence is not known. The value of a site after t time steps can always be specified by a table with an entry for each of $k^{2\lambda t+1}$ relevant initial sequences. Nevertheless, it is possible that a finite state automaton, specified by a finite state transition graph, could determine the value of sites at any time

The behaviour of finite class 3 cellular automata with additive rules was analysed in some detail in ref. 2. It was shown there that the maximal cycle length for additive cellular automata grows on average exponentially with the size N of the cellular automaton. Most cycles were found to have maximal length, and the number of distinct cycles was found also to grow on average exponentially with N. The lengths of transients leading to cycles were found to grow at most linearly with N. The fraction of states on cycles was found on average to tend a finite limit.

For most class 3 cellular automata, the average cycle length grows quite slowly with N, although in some cases, the absolute maximum cycle length appears to grow rapidly. The lengths of transients are typically short for cellular automata which generate more regular patterns, but often become very long as N increases for cellular automata which generate more irregular patterns. The fractions of states on cycles are typically much larger for finite class 3 cellular automata which generate irregular patterns than for those which generate more regular patterns. This is presumably a reflection of the lower irreversibility and larger

*Each site in this cellular automaton can take on one of two possible values; the time evolution rule involves nine site (type II) neighbourhoods. If the values of less than 2 or more than 3 of the eight neighbours of a particular site are nonzero then the site takes on value 0 at the next time step; if 2 neighbouring sites are nonzero the site takes the same value as on the previous time steps; if exactly 3 neighbouring sites are nonzero, the site takes on value 1.

attractor dimension found for the former case in the infinite size limit.

8. Class 4 cellular automata

Fig. 12 shows the evolution of the class 4 cellular automaton with $k = 2$, $r = 2$ and code number 20, from several disordered initial configurations. In most cases, all sites are seen to "die" (attain value zero) after a finite time. However, in a few cases, stable or periodic structures which persist for an infinite time are formed. In addition, in some cases, propagating structures are formed. Fig. 13 shows the persistent structures generated by this cellular automaton from all initial configurations whose nonzero sites lie in a region of length 20 (reflected versions of the last three structures are also found). Table III gives some characteristics of these structures. An important feature, shared by other class 4 cellular automata, is the presence of propagating structures. By arranging for suitable reflections of these propagating structures, final states with any cycle lengths may be obtained.

The behaviour of the cellular automata illustrated in fig. 13, and the structures shown in fig. 14 are strongly reminiscent of the two-dimensional (essentially totalistic) cellular automaton known as the "Game of Life"* (for references see [1]). The Game of Life has been shown to have the important property of computational universality. Cellular automata may be viewed as computers, in which data represented by initial configurations is processed by time evolution. Computational universality (e.g. [15–18]) implies that suitable initial configurations can specify arbitrary algorithmic procedures. The system can thus serve as a general purpose computer, capable of evaluating any (computable) function. Given a suitable encoding, the system may therefore in principle simulate any other system, and in this sense may be considered capable of arbitrarily complicated behaviour.

The proof of computational universality for the Game of Life [22] uses the existence of cellular

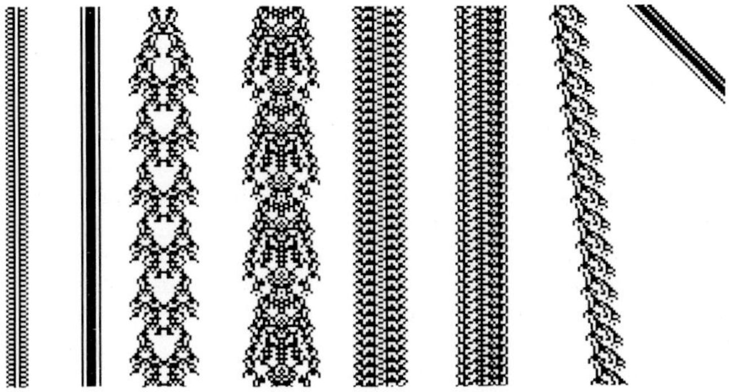

S. Wolfram / Universality and complexity in cellular automata   31

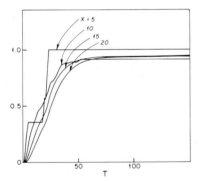

Fig. 13. Persistent structures found in the evolution of the class 4 cellular automaton illustrated in fig. 12 from initial states with nonzero sites in a region of 20 or less sites. Reflected versions of the last three structures are also found. Some properties of the structures are given in table III. These structures are almost sufficient to provide components necessary to demonstrate a universal computation capability for this cellular automaton.

Fig. 14. Fraction of configurations in the class 4 cellular automaton of figs. 12 and 13 which evolve to the null configuration after T time steps, from initial states with nonzero sites in a region of length less than X (translates of configurations are not included). The asymptotic "halting probability" is around 0.93; 7% of initial configurations generate the persistent structures of fig. 13 and never evolve to the null configuration.

automaton structures which emulate components (such as "wires" and "NAND gates") of a standard digital computer. The structures shown in fig. 14 represent a significant fraction of those necessary. A major missing element is a configuration

(dubbed the "glider gun" in the Game of Life) which acts like a clock, and generates an infinite sequence of propagating structures. Such a configuration would involve a finite number of initial nonzero sites, but would lead to unbounded growth, and an asymptotically infinite number of nonzero sites. There are however indications that the required initial configuration is quite large, and is very difficult to find.

These analogies lead to the speculation that class 4 cellular automata are characterized by the capability for universal computation. $k = 2, r = 1$ cellular automata are too simple to support universal computation; the existence of class 4 cellular automata with $k = 2$, $r = 2$ (cf. figs. 12 and 13) and $k = 3$, $r = 1$ suggests that with suitable time evolution rules even such apparently simple systems may be capable of universal computation.

There are important limitations on predictions which may be made for the behaviour of systems capable of universal computation. The behaviour of such systems may in general be determined in detail essentially only by explicit simulation of their time evolution. It may in general be predicted using other systems only by procedures ultimately equivalent to explicit simulation. No finite algo-

S. Wolfram / Universality and complexity in cellular automata

Table III

Persistent structures arising from initial configurations with length less than 20 sites in the class 4 totalistic cellular automaton with $k = 2$, $r = 2$ and code number 20, illustrated in figs. 12, 13 and 14. $\phi(X)$ gives the fraction of initial configurations with nonzero sites in a region less than X sites in length which generate a particular structure. When an initial configuration yields multiple structures, each is included in this fraction.

Period	Minimal predecessor	$\phi(10)$	$\phi(20)$
2	10010111 (151)	0.027	0.024
9R	10111011 (187)	0.012	0.0061
1	10111101 (189)	0.014	0.0075
22	11000011 (195)	0.018	0.017
9L	11011101 (221)	0.012	0.0061
1R	1001111011 (635)	0.0020	0.00066
1L	1101111001 (889)	0.0020	0.00066
38	11110100100101111 (125231)	0	2.9×10^{-5}
4	10010001011011110111 (595703)	0	7.6×10^{-6}
4	10010101001010110111 (610999)	0	7.6×10^{-6}
4	10011000011111101111 (624623)	0	7.6×10^{-6}

rithm or procedure may be devised capable of predicting detailed behaviour in a computationally universal system. Hence, for example, no general finite algorithm can predict whether a particular initial configuration in a computationally universal cellular automaton will evolve to the null configuration after a finite time, or will generate persistent structures, so that sites with nonzero values will exist at arbitrarily large times. (This is analogous to the insolubility of the halting problem for universal Turing machines (e.g. [15–18]).) Thus if the cellular automaton of figs. 12 and 13 is indeed computationally universal, no finite algorithm could predict whether a particular initial state would ultimately "die", or whether it would ultimately give rise to one of the persistent structures of fig. 13. The result could not be determined by explicit simulation, since an arbitrarily large time might elapse before one of the required states was reached. Another universal computer could also in general determine the result effectively only by simulation, with the same obstruction.

If class 4 cellular automata are indeed capable of universal computation, then their evolution involves an element of unpredictability presumably not present in other classes of cellular automata.

Not only does the value of a particular site after many time steps potentially depend on the values of an increasing number of initial site values; in addition, the value cannot in general be determined by any "short-cut" procedure much simpler than explicit simulation of the evolution. The behaviour of a class 4 cellular automaton is thus essentially unpredictable, even given complete initial information: the behaviour of the system may essentially be found only by explicitly running it.

Only infinite cellular automata may be capable of universal computation; finite cellular automata involve only a finite number of internal states, and may therefore evaluate only a subset of all computable functions (the "space-bounded" ones).

The computational universality of a system implies that certain classes of general predictions for its behaviour cannot be made with finite algorithms. Specific predictions may nevertheless often be made, just as specific cases of generally noncomputable function may often be evaluated. Hence, for example, the behaviour of all configurations with nonzero sites in a region of length 20 or less evolving according to the cellular automaton rules illustrated in figs. 12 and 13 has been completely determined. Fig. 14 shows the fraction of initial configurations which evolve to the null state within T time steps, as a function of T, for various sizes X of the region of nonzero sites. For large X and large T, it appears that the fraction of configurations which generate no persistent structures (essentially the "halting probability") is approximately 0.93. It is noteworthy that the curves in fig. 14 as a function of T appear to approach a fixed form at large X. One may speculate that some aspects of the form of such curves may be universal to all systems capable of universal computation.

The sets of persistent structures generated by class 4 cellular automata typically exhibit no simple patterns, and do not appear to be specified, for example, by regular grammars. Specification of persistent structures by a finite procedure is necessarily impossible if class 4 cellular automata are indeed capable of universal computation. Strong support of the conjecture that class 4 cellular automata are capable of universal computation would be provided by a demonstration of the equivalence of systematic enumeration of all persistent structures in particular class 4 cellular automata to the systematic enumeration of solutions to generally insoluble Diophantine equations or word problems.

Although one may determine by explicit construction that specific cellular automata are capable of universal computation, it is impossible to determine in general whether a particular cellular automaton is capable of universal computation. This is a consequence of the fact that the structures necessary to implement universal computation may be arbitrarily complicated. Thus, for example, the smallest propagating structure might involve an arbitrarily long sequence of site values.

For class 1, 2 and 3 cellular automata, fluctuations in statistical quantities are typically found to become progressively smaller as larger numbers of sites are considered. Such systems

*This feature allows practical simulation of such cellular automata to be made more efficient by storing information on the evolution of the specific sequences of sites which occur with larger probabilities (cf. [23]).

therefore exhibit definite properties in the "infinite volume" limit. For class 4 cellular automata, it seems likely that fluctuations do not decrease as larger number of sites are considered, and no simple smooth infinite volume limit exists. Important qualitative effects can arise from special sequences appearing with arbitrarily low probabilities in the initial state. Consider for example the class 4 cellular automaton illustrated in figs. 12 and 13. The evolution of the finite sequences in this cellular automaton shown in fig. 12 (and many thousands of other finite sequences tested) suggests that the average density of nonzero sites in configurations of this cellular automaton should tend to a constant at large times. However, in a sufficiently long finite initial sequence, there should exist a subsequence from which a "glider gun" structure evolves. This structure would generate an increasing number of nonzero sites at large times, and its presence would completely change the average large time density. As a more extreme example, it seems likely that a sufficiently long (but finite) initial sequence should evolve to behave as a self-reproducing "organism", capable of eventually taking over its environment, and leading to completely different large time behaviour. Very special, and highly improbable, initial sequences may thus presumably result in large changes in large time properties for class 4 cellular automata. These sequences must appear in a truly infinite (typical) initial configuration. Although their density is perhaps arbitrarily low, the sequences may evolve to structures which come to dominate the statistical properties of the system. The possibility of such phenomena suggest that no smooth infinite volume exists for class 4 cellular automata.

Some statistical results may be obtained from large finite class 4 cellular automata, although the results are expected to be irrelevant in the truly infinite volume limit. The evolution of most class 4 cellular automata appears to be highly irreversible*. This irreversibility is reflected in the small set of persistent structures usually generated as end-products of the evolution. Changes in small regions of the initial state may affect many sites at

large times. There are however very large
fluctuations in the propagation speed, and no
meaningful averages may be obtained. It should be
noted that groups of class 4 cellular automata with
different rules often yield qualitatively similar behaviour, and similar sets of persistent structures,
suggesting further classification.

The frequency with which a particular structure
is generated after an infinite time by the evolution
of a universal computer from random (disordered)
input gives the "algorithmic probability" p_A [24]
for that structure. This algorithmic probability has
been shown to be invariant (up to constant multiplicative factors) for a wide class of universal
computers. In general, one may define an "evolutionary probability" $p_E(t)$ which gives the probability for a structure to evolve after t time steps
from a random initial state. Complex structures
formed by cellular automata will typically have
evolutionary probabilities which are initially small,
but later grow. As a simple example, the probability for the sequence which yields a period 9
propagating structure in the cellular automaton of
figs. 12 and 13 begins small, but later increases to
a sufficiently large value that such structures are
almost always generated from disordered states of
2000 or more sites. In a much more complicated
example, one may imagine that the probability for
a self-reproducing structure begins small, but later
increases to a substantial value. Structures whose
evolutionary probability becomes significant only
after a time $>T$ may be considered to have
"logical depth" [25] T.

9. Discussion

Cellular automata are simple in construction,
but are capable of very complex behaviour. This
paper has suggested that a considerable universality exists in this complex behaviour. Evidence
has been presented that all one-dimensional cellular automata fall into four basic classes. In the first
class, evolution from almost all initial states leads
ultimately to a unique homogeneous state. The

second class evolves to simple separated structures.
Evolution of the third class of cellular automata
leads to chaotic patterns, with varying degrees of
structure. The behaviours of these three classes of
cellular automata are analogous to the limit points,
limit cycles and chaotic ("strange") attractors
found in continuous dynamical systems. The
fourth class of cellular automata exhibits still more
complicated behaviour, and its members are conjectured to be capable of universal computation.

Even starting from disordered or random initial
configurations, cellular automata evolve to generate characteristic patterns. Such self-organizing
behaviour occurs by virtue of the irreversibility of
cellular automaton evolution. Starting from almost any initial state, the evolution leads to attractors containing a small subset of all possible states.
At least for the first three classes of cellular automata, the states in these attractors form a Cantor
set, with characteristic fractal and other dimensions. For the first and second classes, the states in
the attractor may be specified as sentences with a
regular grammar. For the fourth class, the attractors may be arbitrarily complicated, and no simple
statistical characterizations appear possible.

The four classes of cellular automata may be
distinguished by the level of predictability of their
"final" large time behaviour given their initial
state. For the first class, all initial states yield the
same final state, and complete prediction is trivial.
In the second class, each region of the final state
depends only on a finite region of the initial state;
knowledge of a small region in the initial state thus
suffices to predict the form of a region in the final
state. In the evolution of the third class of cellular
automata, the effects of changes in the initial state
almost always propagate forever at a finite speed.
A particular region thus depends on a region of the
initial state of ever-increasing size. Hence any
prediction of the "final" state requires complete
knowledge of the initial state. Finally, in the fourth
class of cellular automata, regions of the final state
again depend on arbitrarily large regions of the
initial state. However, if cellular automata in the
class are indeed capable of universal computation,

then this dependence may be arbitrarily complex, and the behaviour of the system can be found by no procedure significantly simpler than direct simulation. No meaningful prediction is therefore possible for such systems.

Acknowledgements

I am grateful to many people for discussions, including C. Bennett, J. Crutchfield, D. Friedan, P. Gacz, E. Jen, D. Lind, O. Martin, A. Odlyzko, N. Packard, S^2. Shenker, W. Thurston, T. Toffoli and S. Willson. I am particularly grateful to J. Milnor for extensive discussions and suggestions.

References

[1] S. Wolfram, 'Statistical mechanics of cellular automata", Rev. Mod. Phys. 55 (1983) 601.

[2] O. Martin, A.M. Odlyzko and S. Wolfram, "Algebraic properties of cellular automata", Bell Laboratories report (January 1983); Comm. Math. Phys., to be published.

[3] D. Lind, "Applications of ergodic theory and sofic systems to cellular automata", University of Washington preprint (April 1983); Physica 10D (1984) 36 (these proceedings).

[4] S. Wolfram, "CA: an interactive cellular automaton simulator for the Sun Workstation and VAX", presented and demonstrated at the Interdisciplinary Workshop on Cellular Automata, Los Alamos (March 1983).

[5] T. Toffoli, N. Margolus and G. Vishniac, private demonstrations.

[6] P. Billingsley, Ergodic Theory and Information (Wiley, New York, 1965).

[7] D. Knuth, Seminumerical Algorithms, 2nd. ed. (Addison-Wesley, New York, 1981), section 3.5.

[8] R.G. Gallager, Information Theory and Reliable Communications (Wiley, New York, 1968).

[9] J.D. Farmer, "Dimension, fractal measures and the probabilistic structure of chaos", in: Evolution of Order and Chaos in Physics, Chemistry and Biology, H. Haken, ed. (Springer, Berlin, 1982).

[10] J.D. Farmer, private communication.

[11] B. Mandelbrot, The Fractal Geometry of nature (Freeman, San Francisco, 1982).

[12] J.D. Farmer, "Information dimension and the probabilistic structure of chaos", Z. Naturforsch. 37a (1982) 1304.

[13] P. Grassberger, to be published.

[14] P. Diaconis, private communication; C. Stein, unpublished notes.

[15] F.S. Beckman, "Mathematical Foundations of Programming (Addison-Wesley, New York, 1980).

[16] J.E. Hopcroft and J.D. Ullman, Introduction to Automata Theory, Languages, and Computation (Addison-Wesley, New York, 1979).

[17] Z. Manna, Mathematical Theory of Computation (McGraw-Hill, New York, 1974).

[18] M. Minsky, Computation: Finite and Infinite Machines (Prentice-Hall, London, 1967).

[19] B. Weiss, "Subshifts of finite type and sofic systems", Monat. Math. 17 (1973) 462. E.M. Coven and M.E. Paul, "Sofic systems", Israel J. Math. 20 (1975) 165.

[20] P. Grassberger, "A new mechanism for deterministic diffusion", Wuppertal preprint WU B 82–18 (1982).

[21] J. Milnor, unpublished notes.

[22] R.W. Gosper, unpublished; R. Wainwright, "Life is universal!", Proc. Winter Simul. Conf., Washington D.C., ACM (1974). E.R. Berlekamp, J.H. Conway and R.K. Guy, Winning Ways, for Your Mathematical Plays, vol. 2 (Academic Press, New York, 1982), chap. 25.

[23] R.W. Gosper, "Exploiting regularities in large cellular spaces", Physica 10D (1984) 75 (these proceedings).

[24] G. Chaitin, "Algorithmic information theory", IBM J. Res. & Dev., 21 (1977) 350; "Toward a mathematical theory of life", in: The Maximum Entropy Formalism, R.D. Levine and M. Tribus, ed. (MIT press, Cambridge, MA, 1979).

[25] C. Bennett, "On the logical "depth" of sequences and their reducibilities to random sequences", IBM report (April 1982) (to be published in Info. & Control).

Journal of Statistical Physics, Vol. 38, Nos. 5/6, 1985

Two-Dimensional Cellular Automata

Norman H. Packard[1] and Stephen Wolfram[1]

Received October 10, 1984

A largely phenomenological study of two-dimensional cellular automata is reported. Qualitative classes of behavior similar to those in one-dimensional cellular automata are found. Growth from simple seeds in two-dimensional cellular automata can produce patterns with complicated boundaries, characterized by a variety of growth dimensions. Evolution from disordered states can give domains with boundaries that execute effectively continuous motions. Some global properties of cellular automata can be described by entropies and Lyapunov exponents. Others are undecidable.

KEY WORDS: Discrete models; dynamical systems; pattern formation; computation theory.

1. INTRODUCTION

Cellular automata are mathematical models for systems in which many simple components act together to produce complicated patterns of behavior. One-dimensional cellular automata have now been investigated in several ways (Ref. 1 and references therein). This paper presents an exploratory study of two-dimensional cellular automata.[2] The extension to two dimensions is significant for comparisons with many experimental results on pattern formation in physical systems. Immediate applications include dendritic crystal growth,[6] reaction-diffusion systems, and turbulent flow patterns. (The Navier–Stokes equations for fluid flow appear to admit turbulent solutions only in two or more dimensions.)

A cellular automaton consists of a regular lattice of sites. Each site takes on k possible values, and is updated in discrete time steps according

---

[1] The Institute for Advanced Study, Princeton, New Jersey 08540.
[2] Some aspects of two-dimensional cellular automata were discussed in Refs. 2 and 3, and mentioned in Ref. 4. Additive two-dimensional cellular automata were considered in Ref. 5.

to a rule ϕ that depends on the value of sites in some neighborhood around it. The value a_i of a site at position i in a one-dimensional cellular automata with a rule that depends only on nearest neighbors thus evolves according to

$$a_i^{(t+1)} = \phi[a_{i-1}^{(t)}, a_i^{(t)}, a_{i+1}^{(t)}] \qquad (1.1)$$

There are several possible lattices and neighborhood structures for two-dimensional cellular automata. This paper considers primarily square lattices, with the two neighborhood structures illustrated in Fig. 1. A five-neighbor square cellular automaton then evolves in analogy with Eq. (1.1) according to

$$a_{i,j}^{(t+1)} = \phi[a_{i,j}^{(t)}, a_{i,j+1}^{(t)}, a_{i+1,j}^{(t)}, a_{i,j-1}^{(t)}, a_{i-1,j}^{(t)}] \qquad (1.2)$$

Here we often consider the special class of totalistic rules, in which the value of a site depends only on the sum of the values in the neighborhood:

$$a_{i,j}^{(t+1)} = f[a_{i,j}^{(t)} + a_{i,j+1}^{(t)} + a_{i+1,j}^{(t)} + a_{i,j-1}^{(t)} + a_{i-1,j}^{(t)}] \qquad (1.3)$$

These rules are conveniently specified by a code[7]

$$C = \sum_n f(n)k^n \qquad (1.4)$$

 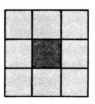

Fig. 1.   Neighborhood structures considered for two-dimensional cellular automata. In the cellular automaton evolution, the value of the center cell is updated according to a rule that depends on the values of the shaded cells. Cellular automata with neighborhood (a) are termed "five-neighbor square;" those with neighborhood (b) are termed "nine-neighbor square." (These neighborhoods are sometimes referred to as the von Neumann and Moore neighborhoods, respectively.) Totalistic cellular automaton rules take the value of the center site to depend only on the sum of the values of the sites in the neighborhood. With outer totalistic rules, sites are updated according to their previous values, and the sum of the values of the other sites in the neighborhood. Triangular and hexagonal lattices are also possible, but are not used in the examples given here. Notice that five-neighbor square, triangular, and hexagonal cellular automaton rules may all be considered as special cases of general nine-neighbor square rules.

**Table 1.   Numbers of Possible Rules of Various Kinds for
Cellular Automata with Two States Per Site,
and Neighborhoods of the Form Shown in Fig. 1**[a]

Rule type	5-neighbor square	9-neighbor square	Hexagonal
General	$2^{32} \simeq 4 \times 10^9$	$2^{512} \simeq 10^{154}$	$2^{128} \simeq 3 \times 10^{38}$
Rotationally symmetric	$2^{12} = 4096$	$2^{140} \simeq 10^{42}$	$2^{64} \simeq 2 \times 10^{19}$
Reflection symmetric	$2^{24} \simeq 2 \times 10^7$	$2^{288} \simeq 5 \times 10^{86}$	$2^{80} \simeq 10^{24}$,
			$2^{74} \simeq 2 \times 10^{22}$
Completely symmetric	$2^{12} = 4096$	$2^{102} \simeq 5 \times 10^{30}$	$2^{28} \simeq 3 \times 10^8$
Outer totalistic	$2^{10} = 1024$	$2^{18} \simeq 3 \times 10^5$	$2^{14} = 16384$
Totalistic	$2^5 = 32$	$2^9 = 512$	$2^7 = 128$

[a] The two entries for reflectional symmetries of the hexagonal lattice refer to reflections across
a cell and across a boundary, respectively. The number of quiescent rules (defined to leave
the null configuration invariant) is always half the total number of rules of a given kind.

We also consider outer totalistic rules, in which the value of a site depends
separately on the sum of the values of sites in a neighborhood, and on the
value of the site itself:

$$a_{i,j}^{(t+1)} = \tilde{f}(a_{i,j}^{(t)}, a_{i,j+1}^{(t)} + a_{i+1,j}^{(t)} + a_{i,j-1}^{(t)} + a_{i-1,j}^{(t)}) \qquad (1.5)$$

Such rules are specified by a code

$$\tilde{C} = \sum_n \tilde{f}[a, n] k^{kn+a} \qquad (1.6)$$

$K = 2$ This paper considers two-dimensional cellular automata with values 0 or 1
at each site, corresponding to $k = 2$. Table I gives the number of possible
rules of various kinds for such cellular automata. A notorious example of
an outer totalistic nine-neighbor square cellular automaton is the "Game of
Life",[8] with a rule specified by code $\tilde{C} = 224$.

   Despite the simplicity of their construction, cellular automata are
found to be capable of very complicated behavior. Direct mathematical
analysis is in general of little utility in elucidating their properties. One
must   at   first   resort   to   empirical   means.   This   paper   gives   a
phenomenological study of typical two-dimensional cellular automata. Its
approach is largely experimental in character: cellular automaton rules are
selected and their evolution from various initial states is traced by direct
simulation.[3] The emphasis is on generic properties. Typical initial states are

[3] Several computer systems were used. The first was the spcial-purpose pipelined TTL
  machine built by the M.I.T. Information Mechanics group.[9] This machine updates all sites
  on a 256×256 square cellular automaton lattice 60 times per second. It is controlled by a

chosen. Except for some restricted kinds of rules, Table I shows that the number of possible cellular automaton rules is far too great for each to be investigated explicitly. For the most part one must resort to random sampling, with the expectation that the rules so selected are typical. The phenomena identified by this experimental approach may then be investigated in detail using analytical approximations, and by conventional mathematical means. Generic properties are significant because they are independent of precise details of cellular automaton construction, and may be expected to be universal to a wide class of systems, including those that occur in nature.

Empirical studies strongly suggest that the qualitative properties of one-dimensional cellular automata are largely independent of such features of their construction as the number of possible values for each site, and the size of the neighborhood. Four qualitative classes of behavior have been identified in one-dimensional cellular automata.[7] Starting from typical initial configurations, class-1 cellular automata evolve to homogeneous final states. Class-2 cellular automata yield separated periodic structures. Class-3 cellular automata exhibit chaotic behavior, and yield aperiodic patterns. Small changes in initial states usually lead to linearly increasing regions of change. Class-4 cellular automata exhibit complicated localized and propagating structures. Cellular automata may be considered as information-processing systems, their evolution performing some computation on the sequence of site values given as the initial state. It is conjectured that class-4 cellular automata are generically capable of universal computation, so that they can implement arbitrary information-processing procedures.

Dynamical systems theory methods may be used to investigate the global properties of cellular automata. One considers the set of configurations generated after some time from any possible initial configuration. Most cellular automaton mappings are irreversible (and not surjective), so that the set of configurations generated contracts with time. Class-1 cellular automata evolve from almost all initial states to a unique final state, analogous to a fixed point. Class-2 cellular automata evolve to collections of periodic structures, analogous to limit cycles. The contraction

---

microcomputer, with software written in FORTH. It allows for five- and nine-neighbor rules, with up to four effective values for each site. The second system was a software program running on the Ridge 32 computer. The kernel is written in assembly language; the top-level interface in the C programming language. A 128×128 cellular automaton lattice is typically updated about seven times per second. Variants of the program, with kernels written in C and FORTRAN, were used on Sun Workstations, VAX, and Cray 1 computers. One-dimensional cellular automaton simulations were carried out with our CA cellular automaton simulation package, written in C, usually running on a Sun Workstation.

of the set of configurations generated by a cellular automaton is reflected in
a decrease in its entropy or dimension. Starting from all possible initial
configurations (corresponding to a set defined to have dimension one),
class-3 cellular automata yield sets of configurations with smaller, but
positive, dimensions. These sets are directly analogous to the chaotic (or
"strange") attractors found in some continuous dynamical systems (e.g.,
Ref. 10).

Entropy or dimension gives only a coarse characterization of sets of
cellular automaton configurations. Formal language theory (e.g. Ref. 11)
provides a more complete and detailed characterization.[12] Configurations
may be considered as words in a formal language; sets of configurations are
specified by the grammatical rules of the language. The set of configurations generated after any finite number of time steps in the evolution
of a one-dimensional cellular automaton can be shown to form a regular
language: the possible configurations thus correspond to possible paths
through a finite graph. For most class-3 and -4 cellular automata, the complexity of this graph grows rapidly with time, so that the limit set is
presumably not a regular language (cf. Ref. 13).

This paper reports evidence that certain global properties of two-
dimensional cellular automata are very similar to those of one-dimensional
cellular automata. Many of the local phenomena found in two-dimensional
cellular automata also have analogs in one dimension. However, there are
a variety of phenomena that depend on the geometry of the two-dimensional lattice. Many of these phenomena involve complicated boundaries
and interfaces, which have no direct analog in one dimension.

Section 2 discusses the evolution of two-dimensional cellular automata
from simple "seeds," consisting of a few nonzero initial sites. Just as in one
dimension, some cellular automata give regular and self-similar patterns;
others yield complicated and apparently random patterns. A new feature in
two dimensions is the generation of patterns with dendritic boundaries,
much as observed in many natural systems. Most two-dimensional patterns
generated by cellular automaton growth have a polytopic boundary that
reflects the structure of the neighborhood in the cellular automaton rule (cf.
Ref. 14). Some rules, however, yield slowly growing patterns that tend to a
circular shape independent of the underlying cellular automaton lattice.

Section 3 considers evolution from typical disordered initial states.
Some cellular automata evolve to stationary structures analogous to
crystalline forms. The boundaries between domains of different phases may
behave as if they carry a surface tension: positive surface tensions lead to
large smooth-walled domains; negative surface tensions give rise to
labyrinthine structures with highly convoluted walls. Other cellular
automata yield chaotic, class-3, behavior. Small changes in their initial con-

Fig. 2.   Examples of classes of patterns generated by evolution of two-dimensional cellular automata from a single-site seed. Each part corresponds to a different cellular automaton rule. All the rules shown are both rotation and reflection symmetric. For each rule, a sequence of frames shows the two-dimensional configurations generated by the cellular automaton evolution after the indicated number of time steps. Black squares represent sites with value 1; white squares sites with value 0. On the left is a space-time section showing the time evolution of the center horizontal line of sites in the two-dimensional lattice. Successive lines correspond to successive time steps. The cellular automaton rules shown are five-neighbor square outer totalistic, with codes (a) 1022, (b) 510, (c) 374, (d) 614 (sum modulo 2 rule), (e) 174, (f) 494.

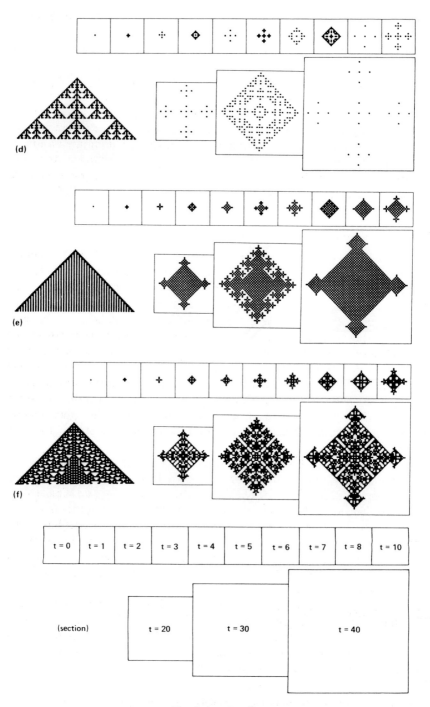

Fig. 2 (*continued*)

figurations lead to linearly increasing regions of change, usually circular or at least rounded.

Section 4 discusses some quantitative characterizations of the global properties of two-dimensional cellular automata. Many definitions are carried through directly from one dimension, but some results are rather different. In particular, the sets of configurations that can be generated after a finite number of time steps of cellular automaton evolution are no longer described by regular languages, and may in fact be nonrecursive. As a consequence, several global properties that are decidable for one-dimensional cellular automata become undecidable in two dimensions (cf. Ref. 15).

2. EVOLUTION FROM SIMPLE SEEDS

This section discusses patterns formed by the evolution of cellular automata from simple seeds. The seeds consist of single nonzero sites, or small regions containing a few nonzero sites, in a background of zero sites. The growth of cellular automata from such initial conditions should provide models for a variety of physical and other phenomena. One example is crystal growth.[6] The cellular automaton lattice corresponds to the crystal lattice, with nonzero sites representing the presence of atoms or regions of the crystal. Different cellular automaton rules are found to yield both faceted (regular) and dendritic (snowflake-like) crystal structures. In other systems the seed may correspond to a small initial disturbance, which grows with time to produce a complicated structure. Such a phenomenon presumably occurs when fluid turbulence develops downstream from an obstruction or orifice.[4]

Figure 2 shows some typical examples of patterns generated by the evolution of two-dimensional cellular automata from initial states containing a single nonzero site. In each case, the sequence of two-dimensional patterns formed is shown as a succession of "frames." A space-time "section" is also shown, giving the evolution of the center horizontal line in the two-dimensional lattice with time. Fig. 3 shows a view of the complete three-dimensional structures generated. Fig. 4 gives some examples of space-time sections generated by typical one-dimensional cellular automata.

With some cellular automaton rules, simple seeds always die out, leaving the null configuration, in which all sites have value zero. With other rules, all or part of the initial seed may remain invariant with time, yielding a fixed pattern, independent of time. With many cellular automaton rules, however, a growing pattern is produced.

---

[4] A cellular automaton approximation to the Euler equations is given in Ref. 16.

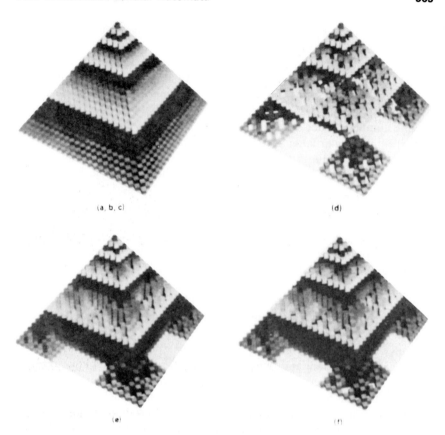

(a, b, c)   (d)

(e)   (f)

Fig. 3. View of three-dimensional structures formed from the configurations generated in the first 24 time steps of the evolution of the two-dimensional cellular automata shown in Fig. 2. Rules (a), (b), and (c) all give rise to configurations with regular, faceted, boundaries. Rules (d), (e), and (f) yield dendritic patterns. In this and other three-dimensional views, the shading ranges periodically from light to dark when the number of time steps increases by a factor of two. The three-dimensional graphics here and in Figs. 10 and 14 is courtesy of M. Prueitt at Los Alamos National Laboratory.

Rule (a) in Figs. 2 and 3 is an example of the simple case in which the growing pattern is uniform. At each time step, a regular pattern with a fixed density of nonzero sites is produced. The boundary of the pattern consists of flat (linear) "facets," and traces out a pyramid in space-time, whose edges lie along the directions of maximal growth. Sections through this pyramid are analogous to the space-time pattern generated by the one-dimensional cellular automaton of Fig. 4(a).

Cellular automaton rule (b) in Figs. 2 and 3 yields a pattern whose boundary again has a simple faceted form, but whose interior is not

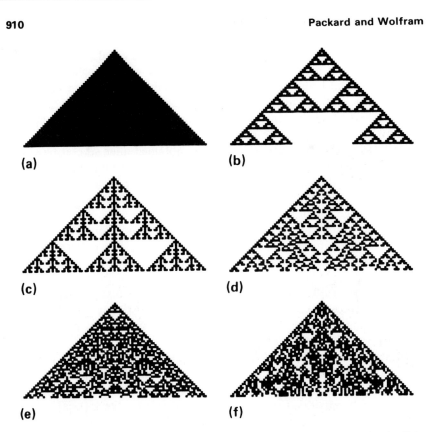

Fig. 4. Examples of classes of patterns generated by evolution of one-dimensional cellular automata from a single-site seed. Successive time steps are shown on successive lines. Nonzero sites are shown black. The cellular automaton rules shown are totalistic nearest-neighbor ($r = 1$), with k possible values at each site: (a) $k = 2$, code 14, (b) $k = 2$, code 6, (c) $k = 2$, code 10, (d) $k = 3$, code 21, (e) $k = 3$, code 102, (f) $k = 3$, code 138. Irregular patterns are also generated by some $k = 2$, $r = 2$ rules (such as that with totalistic code 10), and by asymmetric $k = 2$, $r = 1$ rules (such as that with rule number 30).

uniform. Space-time sections through the pattern exhibit an asymptotically self-similar or fractal form: pieces of the pattern, when magnified, are indistinguishable from the whole. Figure 4(b) shows a one-dimensional cellular automaton that yields sections of the same form. The density of nonzero sites in these sections tends asymptotically to zero. The pattern of nonzero sites in the sections may be characterized by a Hausdorff or fractal dimension that is found by a simple geometrical construction to have value $\log_2 3 \simeq 1.59$.

Self-similar patterns are generated in cellular automata that are invariant under scale or blocking transformations.[17,18] Particular blocks of sites in a cellular automaton often evolve according to a fixed effective cellular automaton rule. The overall behavior of the cellular automaton is then left invariant by a replacement of each block with a single site and of the original cellular automaton rule by the effective rule. In some cases, the effective rule may be identical to the original rule. Then the patterns generated must be invariant under the blocking transformation, and are therefore self-similar. (All the rules so far found to have this property are additive.) In many cases, the effective rule obtained after several blocking transformations with particular blocks may be invariant under further blocking transformations. Then if the initial state contains only the appropriate blocks, the patterns generated must be self-similar, at least on sufficiently large length scales.

Cellular automaton (c) gives patterns that are not homogeneous, but appear to have a fixed nonzero asymptotic density. The patterns have a complex, and in some respects random, appearance. It is remarkable that simple rules, even starting from the simple initial conditions shown, can generate patterns of such complexity. It seems likely that the iteration of the cellular automaton rule is essentially the simplest procedure by which these patterns may be specified. The cellular automaton rule is thus "computationally irreducible" (cf. Ref. 19).

Cellular automata (a), (b), and (c) in Figs 2 and 3 all yield patterns whose boundaries have a simple faceted form. Cellular automata (d), (e), and (f) give instead patterns with corrugated, dendritic, boundaries. Such complicated boundaries can have no analog in one-dimensional cellular automata: they are a first example of a qualitative phenomenon in cellular automata that requires two or more dimensions.

Cellular automaton (d) follows the simple additive rule that takes the value of each site to be the sum modulo two of the previous values of all sites in its five-site neighborhood. The space-time pattern generated by this rule has a fractal form. The fractal dimension of this pattern, and its analogs on d-dimensional lattices, is given by[4]: $\log_2\{d[(1+4/d)^{1/2}+1]\}$, or approximately 2.45 for $d=2$. The average density of nonzero sites in the pattern tends to zero with time.

Rules (e) and (f) give patterns with nonzero asymptotic densities. The boundaries of the patterns obtained at most time steps are corrugated, and have fractal forms analogous to Koch curves. The patterns grow by producing "branches" along the four lattice directions. Each of these branches then in turn produces side branches, which themselves produce side branches, and so on. This recursive process yields a highly corrugated boundary. However, as the process continues, the side branches grow into each other, forming an essentially solid region. In fact, after each 2^j time steps the boundary takes on an essentially regular form. It is only between such times that a dendritic boundary is present.

Cellular automaton (e) is an example of a "solidification" rule,[6] in which any site, once it attains value one, never reverts to value zero. Such rules are of significance in studies of processes such as crystal growth. Notice that although the interior of the pattern takes on a fixed form with time, the possibility of a simple one-dimensional cellular automaton model for the boundary alone is precluded by nonlocal effects associated with interactions between different side branches.

The boundaries of the patterns generated by cellular automata (a), (b), and (c) expand with time, but maintain the same faceted form. So after a rescaling in linear dimensions by a factor of t, the boundaries take on a fixed form: the pattern obtained is a fixed point of the product of the cellular automaton mapping and the rescaling transformation (cf. Refs. 20 and 21). The boundaries of Figs. 2(d, e, f) and 3(d, e, f) continually change with time; a fixed limiting form after rescaling can be obtained only by considering a particular sequence of time steps, such as those of the form 2^j. The result depends critically on the sequence considered: some sequences yield dendritic limiting forms, while other yield faceted forms. The complete space-time patterns illustrated in Figs 3(d, e, f) again approach a fixed limiting form after rescaling only when particular sequences of times are considered. It appears, however, that the forms obtained with different sequences have the same overall properties: they are asymptotically self-similar and have definite fractal dimensions.

The limiting structure of patterns generated by the growth of cellular automata from simple seeds can be characterized by various "growth dimensions." Two general types may be defined. The first, denoted generically D, depend on the overall space-time pattern. The second, denoted $\bar{D}$, depend only on the boundary of the pattern. The boundary may be defined as the set of sites that can be reached by some path on the lattice that begins at infinity and does not cross any nonzero sites. The boundary can thus be found by a simple recursive procedure (cf. Ref. 22). For rules that depend on more than nearest-neighboring sites, paths that pass within the range of the rule of any nonzero site are also excluded, and so no paths can enter any "pores" in the surface of the pattern.

Growth dimensions in general describe the logarithmic asymptotic scaling of the total sizes of patterns with their linear dimensions. For example, the spatial growth dimension D_x is defined in terms of the total number of sites n (interior and boundary) contained in patterns generated by a cellular automaton as a function of time t by the limit of $\log n/\log t$ as $t \to \infty$. Figure 5 shows the behavior of $\log n$ as a function of $\log t$ for the cellular automata of Figs 2 and 3. For those with faceted boundaries, $D_x = \log n/\log t = 2$ for all sufficiently large t: the total size of the patterns scales as the square of the parameter t that determines their linear dimensions.

When the boundaries can be dendritic, however, $\log n$ varies irregularly with $\log t$. In case (d), for example, $\log n$ depends on the number of non-zero digits in the binary decomposition of the integer t (cf. Ref. 4): $\log n/\log t$ is thus maximal when $t = 2^j - 1$, and is minimal when $t = 2^j$. One may define upper and lower spatial growth dimensions D_x^+ and D_x^- in terms of the upper and lower limits (lim *sup* and lim *inf*) of $\log n/\log t$ as $t \to \infty$. For case (d), $D_x^+ = 2$, while $D_x^- = 0$. For cases (e) and (f), $\log n/\log t$ oscillates with time, achieving its maximal value at $t = 2^j - 1$, and its minimal value at or near $t = 3/2 \times 2^j$. However, in these cases numerical results suggest that the upper and lower growth dimensions are in fact equal, and in both cases have a values $\simeq 2$.

An alternative definition of the spatial growth dimension includes only nonzero sites in computing the total sizes of patterns generated by cellular automaton evolution. With this definition, the spatial growth dimension has no definite limit even for cellular automata such as that of case (b) which give patterns with faceted boundaries.

The spatial growth dimensions $\bar{D}_x$ for the boundaries of patterns generated by cellular automata are obtained from the limits of $\log \bar{n}/\log t$ at large t, where $\bar{n}$ gives the number of sites in the boundary at time t (cf. Ref. 23). Figure 5 shows the behavior of $\log \bar{n}$ with $\log t$ for the cellular automata of Figs. 2 and 3. For the faceted boundary cases (a), (b), and (c), $\bar{D}_x = 1$. In cases (d), (e), and (f), where dendritic boundaries occur, $\log \bar{n}$ varies irregularly with $\log t$. $\log \bar{n}/\log t$ is minimal when $t = 2^j$ and the boundary is faceted, and is maximal when the boundary is maximally dendritic, typically at $t = 2^j - 1$. No unique limit for $\bar{D}_x$ exists. In case (d), $\bar{D}_x^+ = 1.62 \pm 0.02$, while $\bar{D}_x^- = 0$. In case (e), $\bar{D}_x^+ = 1.65 \pm 0.02$ and $\bar{D}_x^- = 1$, while in case (f), $\bar{D}_x^+ = 1.53 \pm 0.02$ and $\bar{D}_x^- = 1$.

The limiting forms obtained after rescaling for the spatial patterns generated by the dendritic cellular automata (d), (e), and (f) depend on the sequences of time steps used in the limiting procedure, so that there are no unique values for their spatial growth dimensions. On the other hand, the overall forms of the complete space-time patterns generated by these cellular automata do have definite limits, so that the growth dimensions that characterize them have definite values. The total growth dimensions D and $\bar{D}$[5] may be defined as $\lim_{T \to \infty} \log N/\log T$ and $\lim_{T \to \infty} \log \bar{N}/\log T$, where N is the total number of sites contained in the space-time pattern generated up to time step T, and $\bar{N}$ is the number of sites in its boundary. [Notice that $N = \sum_{t=0}^{T} n(t)$.] Figure 5 shows the behavior of $\log N$ and $\log \bar{N}$ as a function of $\log T$ for the cellular automata of Figs. 2 and 3. Unique values of D and $\bar{D}$ are indeed found in all cases. Rules that give pat-

---

[5] This quantity is referred to as the "growth rate dimension" in Ref. 20.

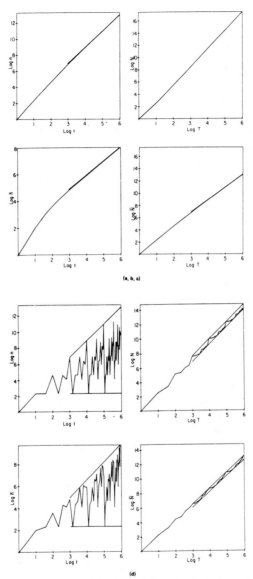

Fig. 5. Sizes of structures generated by the two-dimensional cellular automata of Fig. 2 growing from single nonzero initial sites as a function of time. (Although the sizes are defined only at integer times, their successive values are shown joined by straight lines.) $\bar{n}$ gives the number of sites on the boundaries of patterns obtained at time t. n gives the total number of sites contained within these boundaries. $\bar{N}$ is the number of sites in the boundary (surface) of the complete three-dimensional space-time structures illustrated in Fig. 3 up to time T, and N is the number of sites in their interior. The large-t limits of log n/log t and so on give various growth dimensions for the structures. In cases (a), (b), and (c), structures with faceted boundaries are produced, and the growth dimensions have unique values. In cases (d), (e), and (f) the structures have dendritic boundaries, and the slopes of the bounding lines shown give upper (lim *sup*) and lower (lim *inf*) limits for the growth dimensions. In many of the cases shown, the numerical values of these upper and lower limits appear to coincide.

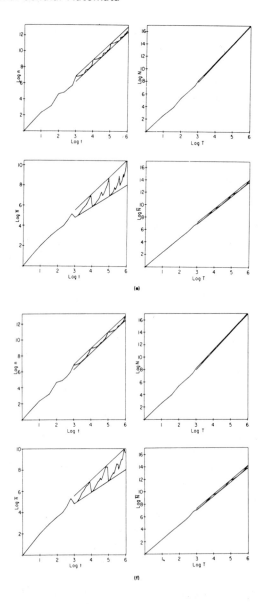

Fig. 5 (*continued*)

terns with faceted boundaries have $D = 3$, $\bar{D} = 2$. The additive rule of case (d) gives $D = 2.36 \pm 0.02$, $\bar{D} = 2.19 \pm 0.02$. Cases (e) and (f) both give $D = 3$, $\bar{D} = 2.27 \pm 0.02$.

Growth dimensions may be defined in general by considering the intersection of the complete space-time pattern, or its boundary, with various families of hyperplanes. With fixed-time hyperplanes one obtains the spatial growth dimensions D_x and $\bar{D}_x$. Temporal growth dimensions $D_t^{(x)}$ and $\bar{D}_t^{(x)}$ are obtained by considering sections through the space-time pattern in spatial direction **x**. (The section typically includes the site of the original seed.) The total growth dimension may evidently be obtained as an appropriate average over temporal growth dimensions in different directions. (The average must be taken over pattern sizes n, and so requires exponentiation of the growth dimensions.) The values of the temporal growth dimensions for the patterns of Figs. 2 and 3 depend on their internal structure. Cases (a), (c), (e), and (f) have $D_t = 2$; case (b) has $D_t = \log_2 3 \simeq 1.59$, and case (d) has $D_t = \log_2(1 + \sqrt{5}) \simeq 1.69$. The temporal growth dimensions $\bar{D}_t^{(x)}$ for the boundaries of the patterns are equal to one for the faceted boundary cases. These dimensions vary with direction in cases with dendritic boundaries. They are equal to one in directions of maximal growth, but are larger in other directions.

In general the values of growth dimensions associated with particular hyperplanes are bounded by the topological dimensions of those hyperplanes. Empirical studies indicate that among all (symmetric) two-dimensional cellular automata, patterns with the form of case (c), characterized by $D = 3$, $\bar{D} = 2$, $D_t = 2$ are the most commonly generated. Fractal boundaries are comparatively common, but their growth dimensions $\bar{D}$ are usually quite close to the minimal value of two. Fractal sections with $D_t < 2$ are also comparatively common for five-neighbor rules, but become less common for nine-neighbor rules.

The rules for the two-dimensional cellular automata shown in Figs. 2 and 3 are completely invariant under all the rotation and reflection symmetry transformations on their neighborhoods. Figure 6 shows patterns generated by cellular automaton rules with lower symmetries. These patterns are often complicated both in their boundaries and internal structure. Even though the patterns grow from completely symmetric initial states consisting of single nonzero sites, they exhibit definite directionalities and vorticities as a consequence of asymmetries in the rules. Asymmetric patterns may be obtained with symmetrical rules from asymmetric initial states containing several nonzero sites. For example, some rules should support periodic structures that propagate in particular directions with time. Other rules should yield spiral patterns with definite vorticities. Structures of these kinds are expected to be simpler in many $k > 2$ rules than for

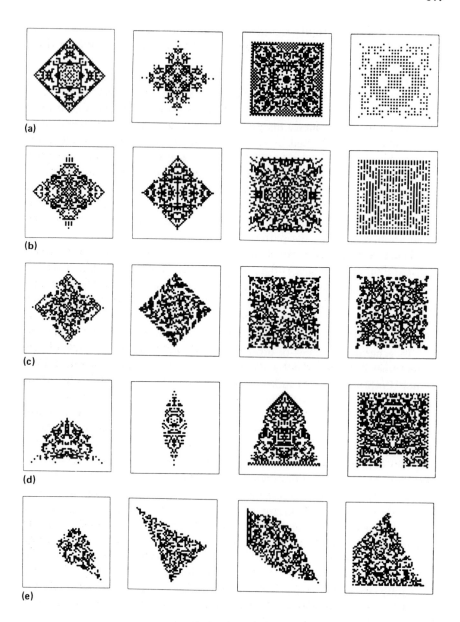

Fig. 6. Examples of patterns generated by growth from single-site seeds for 24 time steps according to general nine-neighbor square rules, with symmetries: (a) all, (b) horizontal and vertical reflection, (c) rotation, (d) vertical reflection, (e) none.

$k = 2$ rules (cf. Ref. 24) just as in one-dimensional cellular automata. Notice that spiral patterns in two-dimensional cellular automata have total growth dimensions $D = \bar{D} = 2$.

Figure 7 shows the evolution of various two-dimensional cellular automata from initial states containing both single nonzero sites, and small regions with a few nonzero sites. In most cases, the overall patterns generated after a sufficiently long time are seen to be largely independent of the particular form of the initial state. In cases such as (c) and (e), features in the initial seed lead to specific dislocations in the final patterns. Nevertheless, deformations in the boundaries of the patterns usually occur only on length scales of order the size of the seed, and presumably become negligible in the infinite time limit. As a consequence, the growth dimensions for the resulting patterns are usually independent of the form of the initial seed (cf. Ref. 20 for additive rules).

There are nevertheless some cellular automaton rules for which slightly different seeds can lead to very different patterns. This phenomenon occurs when a cellular automaton whose configurations contain only certain blocks of site values satisfies an effective rule with special properties such as scale invariance. If the initial seed contains only these blocks, then the pattern generated follows the effective rule. However, if other blocks are present, a pattern of a different form may be generated. An example of this behavior for a one-dimensional cellular automaton is shown in Fig. 8. Patterns produced with one type of seed have temporal growth dimension $\log_2 3 \simeq 1.59$, while those with another type of seed have dimension 2.

Cellular automaton rules embody a finite maximum information propagation speed. This implies the existence of a "bounding surface" expanding at this finite speed. All nonzero sites generated by cellular automaton evolution from a localized seed must lie within this bounding surface. (The cellular automata considered here leave a background of zero sites invariant; such a background must be mapped to itself after at most k time steps with any cellular automaton rule.) Thus the pattern generated after t time steps by any cellular automaton is always bounded by the polytope (planar-faced surface) corresponding to the "unit cell" formed from the set of vectors specifying the displacements of sites in the neighborhood, magnified by a factor t in linear dimensions (cf. Ref. 14). Thus patterns generated by five-neighbor cellular automaton rules always lie within an expanding diamond-shaped region, while those with nine-neighbor rules may fill out a square region.

The actual minimal bounding surface for a particular cellular automaton rule often lies far inside the surface obtained by magnifying the unit cell. A sequence of better approximations to the bounding surface may be found as follows. First consider a set of sites representing the

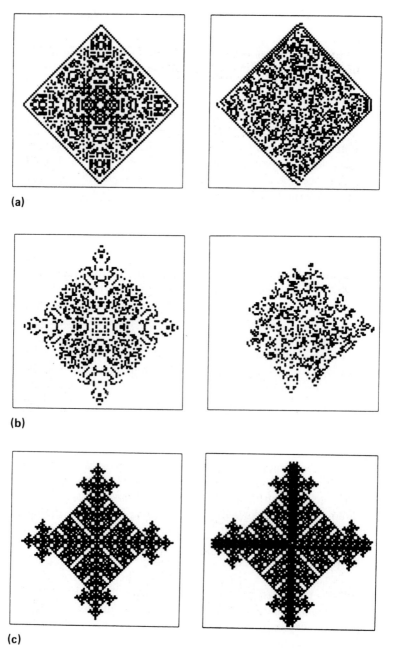

Fig. 7. Examples of patterns generated by evolution of two-dimensional cellular automata from minimal seeds and small disordered regions. In most cases, growth is initiated by a seed consisting of a single nonzero site; for some of the rules shown, a square of four nonzero sites is required. The cellular automaton rules shown are nine-neighbor square outer totalistic, with codes (a) 143954, (b) 50224, (c) five-neighbor 750, (d) 15822, (e) 699054, (f) 191044, (g) 11202, (h) 93737, (i) 85507.

(d)

(e)

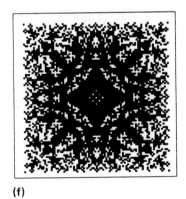

(f)

Fig. 7 (*continued*)

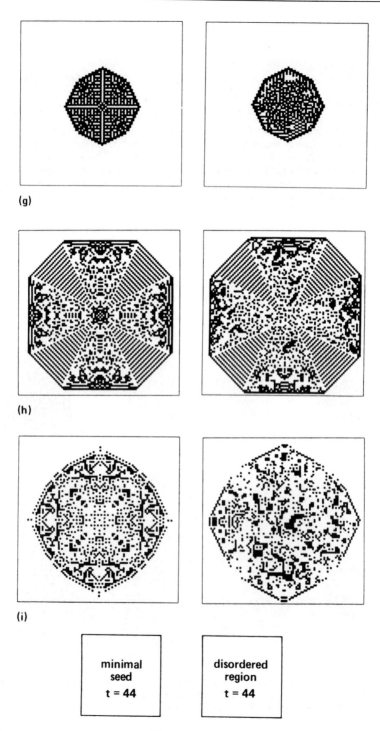

Fig. 7 (*continued*)

Fig. 8. Example of a one-dimensional cellular automaton in which space-time patterns with different temporal growth dimensions are obtained with different initial seeds. The cellular automaton has $k = 2$, $r = 1$, and rule number 218. With an initial state containing only the blocks 00 and 10, it behaves like the additive rule 90, and yields a self-similar space-time pattern with fractal dimension $\log_2 3$. But when the initial state contains 10 and 11 blocks, it behaves like rule 128, and yields a uniform space-time pattern.

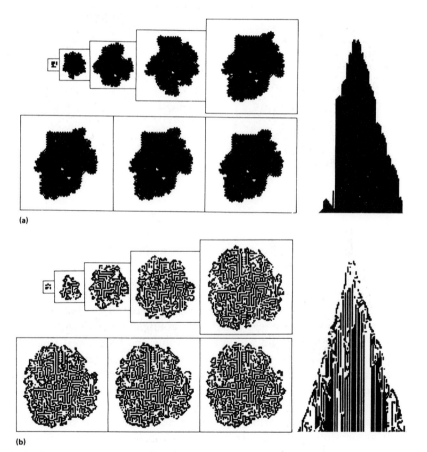

(a)

(b)

Fig. 9. Examples of two-dimensional cellular automata that exhibit slow diffusive growth from small disordered regions. The cellular automaton rules shown are nine-neighbor square outer totalistic, with codes (a) 256746, (b) 736, (c) 291552.

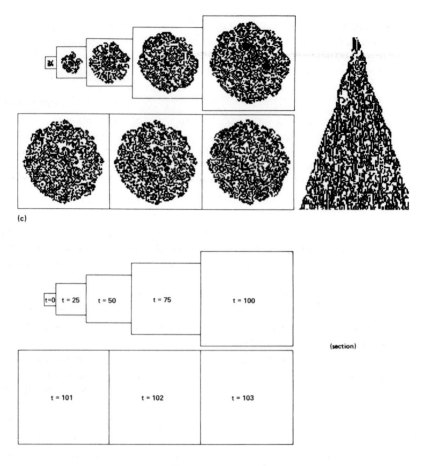

(c)

(section)

Fig. 9 (*continued*)

neighborhood for a cellular automaton rule. If the center site has value one at a particular time step, there could exist configurations for which all of the sites in the neighborhood would attain value one on the next time step. However, there may be some sites whose values cannot change from zero to one in a single time step with any configuration. Growth does not occur along directions corresponding to such sites. The polytope formed from sites in the neighborhood, excluding such sites, may be magnified by a factor t to yield a first approximation to the actual bounding surface for a cellular automaton rule. A better approximation is given by the polytope obtained after two time steps of cellular automaton evolution, magnified by a factor $t/2$.

The actual bounding surfaces for five-neighbor two-dimensional cellular automaton rules usually have their maximal diamond-shaped form. However, many nine-neighbor rules have a diamond-shaped form, rather than their maximal square form. Some nine-neighbor rules, such as those of Figs. 7(g) and 7(h) have octagonal bounding surfaces, while still others, such as those of Fig. 7(i) have dodecagonal bounding surfaces. The cellular automata rules with lower symmetries illustrated in Fig. 6 in many cases exhibit more complicated boundaries, with lower symmetries.

Patterns that maintain regular boundaries with time typically fill out their bounding surface at all times. Dendritic patterns, however, usually expand with the bounding surface only along a few axes. In other directions, they meet the bounding surface only at specific times, typically of the form 2^j. At other times, they lie within the bounding surface.

Dendritic boundaries seem to be associated with cellular automaton rules that exhibit "growth inhibition" (cf. Ref. 14). Growth inhibition occurs if there exist some a_i for which $\phi(a_1,..., 0,..., a_n) = 1$, but $\phi(a_1,..., 1,..., a_n) = 0$, or vice versa. Such behavior appears to be common in physical and other systems.

Figures 9 and 10 show examples of two-dimensional cellular automata that exhibit the comparatively rare phenomenon of slow, diffusive, growth from simple seeds. Figure 11 gives a one-dimensional cellular automaton with essentially analogous behavior.

The phenomenon is most easily discussed in the one-dimensional case. The pattern shown in Fig. 11 is such that it expands by one site at a particular time step only if the site on the boundary has value one. If the boundary site has one of its other three possible nonzero values, then on average, no expansion occurs. The cellular automaton rule is such that the boundary sites have values one through four with roughly equal frequencies. Thus the pattern expands on average at a speed of about 1/4 sites per time step (on each side).

The origin of diffusive growth is similar in the two-dimensional case. Growth occurs there only when some particular several-site structure appears on the boundary. For example, in the cellular automaton of Fig. 9(a), a linear interface propagates at maximal velocity. Deformations of the interface slow its propagation, and a maximally corrugated interface with a "battlement" form does not propagate at all. Since many boundary structures occur with roughly equal probabilities, the average growth rate is small. In the cases investigated, the growth rate is asymptotically constant, so that the growth dimensions have definite values. A remarkable feature is that the boundaries of the patterns produced do not follow the polytopic form suggested by the underlying lattice construction of the cellular automaton. Instead, in many cases, asymptotically circular patterns appear to be produced.

Fig. 10.   View of three-dimensional structure formed from the configurations generated in the first 24 time steps of evolution according to the two-dimensional cellular automaton rule of Fig. 9(a).

Fig. 11.   Example of a one-dimensional cellular automaton that exhibits slow growth. The rule shown is totalistic $k = 5$, $r = 1$, with code 985707700. All nonzero sites are shown black. The initial state contains a single site with value 3. Growth occurs when a site with value 1 appears on the boundary.

3. EVOLUTION FROM DISORDERED INITIAL STATES

In this section, we discuss the evolution of cellular automata from disordered initial states, in which each site is randomly chosen to have value zero or one (usually with probability 1/2). Such disordered configurations are typical members of the set of all possible configurations. Patterns generated from them are thus typical of those obtained with any initial state. The presence of structure in these patterns is an indication of self-organization in the cellular automaton.

As mentioned in Section 1, four qualitative classes of behavior have been identified in the evolution of one-dimensional cellular automata from disordered initial states. Examples of these classes are shown in Fig. 12. Figure 13 shows the evolution of some typical two-dimensional cellular automata from disordered initial states. The same four qualitative classes of behavior may again be identified here. In fact, the space-time sections for two-dimensional cellular automata have a striking qualitative similarity to sections obtained from one-dimensional cellular automata, perhaps with some probabilistic noise added.

Just as in one dimension, some two-dimensional cellular automata evolve from almost all initial states to a unique homogeneous state, such as the null configuration. The final state for such class 1 cellular automata is usually reached after just a few time steps, but in some rare cases, there may be a long transient.

Figures 13(a) and 14(a) give an example of a two-dimensional cellular automaton with class-2 behavior. The disordered initial state evolves to a collection of separated simple structures, each stable or oscillatory with a small period. Each of these structures is a remnant of a particular feature in the initial state. The cellular automaton rule acts as a "filter" which preserves only certain features of the initial state. There is usually a simple pattern to the set of features preserved, and to the set of persistent structures produced. It should in fact be possible to devise cellular automaton rules that recognize particular sets of features, and to use such class-2 cellular automata for practical image processing tasks (cf. Ref. 25).

The patterns generated by evolution from several different disordered configurations according to a particular cellular automaton rule are almost always qualitatively similar. Yet in many cases the cellular automaton evolution is unstable, in that small changes in the initial state lead to increasing changes in the patterns generated with time. Figures 12 and 13 include difference patterns that illustrate the effect of changing the value of a single site in the initial state. For class-2 cellular automata, such a change affects only a finite region, and the difference pattern remains bounded with time. Information propagates only a finite distance in class-2 cellular automata, so that a particular region of the final state is determined from a bounded region in the initial state. For class-3 cellular automata, on the other hand, information generically propagates at a nonzero speed forever, and a small change in the initial state affects an ever-increasing region. The difference patterns for class-3 cellular automata thus grow without bound, usually at a constant rate.

The locally periodic patterns generated after many time steps by class-2 cellular automata such as in Fig 13(a) consist of many separated structures located at essentially arbitrary positions. Figure 13(b) shows another form of class-2 cellular automaton. There are four basic "phases." Two phases have vertical stripes, with either on even or odd sites. The other two phases have horizontal stripes. Regions that take on forms corresponding to one of these phases are invariant under the cellular automaton rule. Starting from a typical disordered state, each region in the cellular automaton lattice evolves toward a particular phase. At large times, the cellular automaton thus "crystallizes" into a patchwork of "domains." The domains consist of regions in particular phases. They are separated by domain walls. In the example of Fig. 13(b), these domain walls become essentially stationary after a finite time.

A change in a single initial site produces a difference pattern that ultimately spreads only along the domain walls. The spread continues only so long as each successive region on the domain wall contains only particular arrangements of site values. The spread stops if a "pinning defect," corresponding to other arrangements of site values, is encountered. The arrangement of site values on the domain walls may in a first approximation be considered random. The difference pattern will thus spread forever only if the arrangements of site values necessary to support its propagation occur with a probability above the percolation threshold (e.g., Ref. 26), so that they form an infinite connected cluster with probability one.

Phases in cellular automata may in general be described by "order parameters" that specify the spatially periodic patterns of sites corresponding to each phase. The size of domains generated by evolution from disordered initial states depends on the length of time before the domains become "frozen": slower relaxation leads to larger domains, as in annealing. A final state reached after any finite time can contain only finite size domains, and therefore cannot be a pure phase. States generated by two-dimensional cellular automata may contain "point" and "line" defects. Point defects are localized regions within domains. An example is the "L-shaped" region of zero sites in domains of the value one phase for the cellular automaton illustrated in Fig. 13(e). Line defects correspond to walls separating domains.

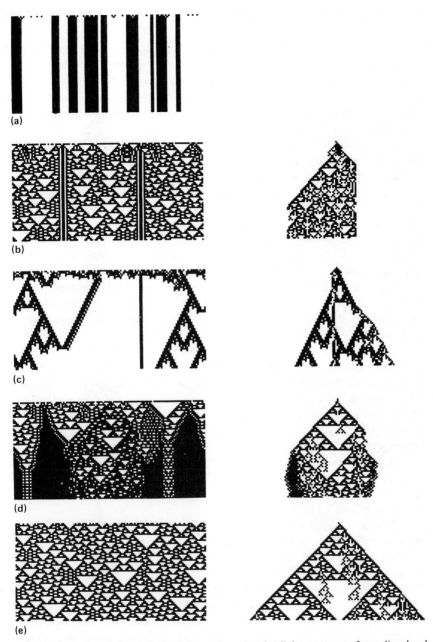

Fig. 12. Examples of the evolution of one-dimensional cellular automata from disordered initial states. The difference patterns on the right show site values that change when a single initial site value is changed. All nonzero sites are shown black. The cellular automaton rules shown are totalistic nearest neighbor ($r = 1$), with k possible values at each site: (a) $k = 2$, code 12, (b) $k = 5$, code 7530, (c) $k = 3$, code 681, (d) $k = 5$, code 3250, (e) $k = 2$, code 6, (f) $k = 3$, code 348, (g) $k = 3$, code 138, (h) $k = 3$, code 318, (i) $k = 3$, code 792.

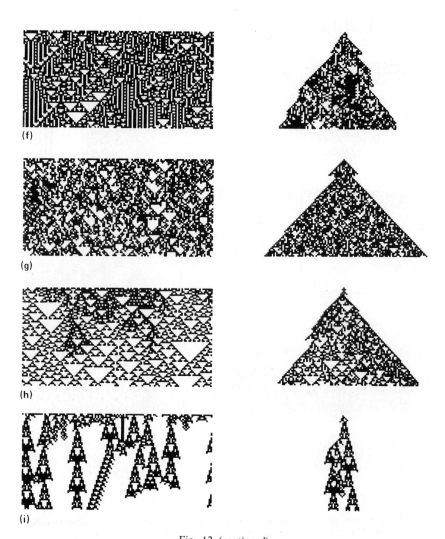

Fig. 12 (*continued*)

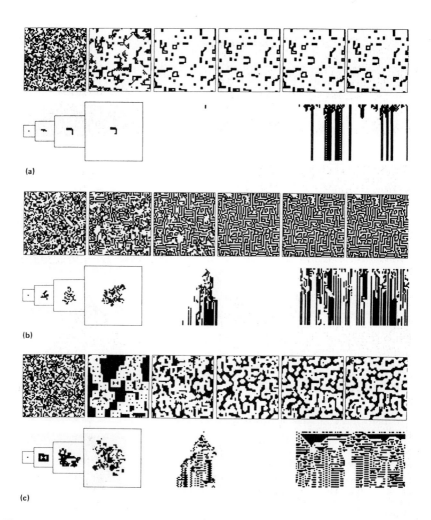

Fig. 13. Examples of the evolution of two-dimensional cellular automata from disordered initial states. The cellular automaton rules shown are totalistic five-neighbor square with codes: (a) 24, (d) 510, (e) 52; and outer totalistic nine-neighbor with codes: (b) 736, (c) 196623, (f) 152822, (g) 143954, (h) 3276, (i) 224 (the "Game of Life").

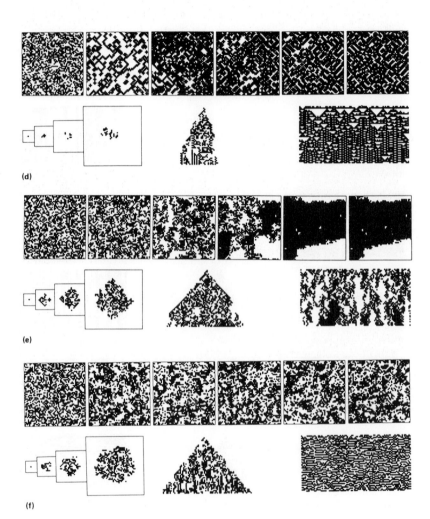

(d)

(e)

(f)

Fig. 13 (*continued*)

932   **Packard and Wolfram**

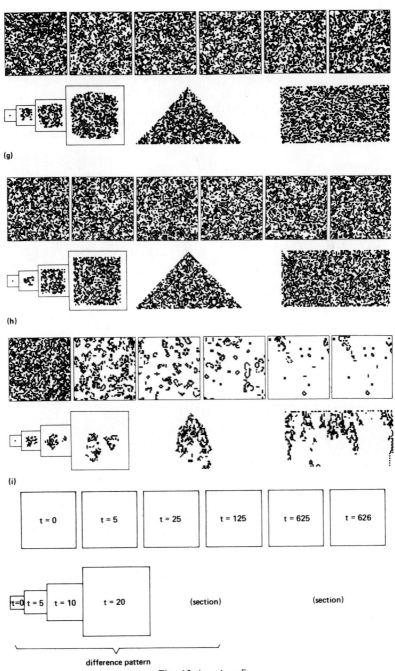

Fig. 13 (*continued*)

Two-Dimensional Cellular Automata

Fig. 14. View of three-dimensional structures formed by configurations generated in the first 24 times of evolution from disordered initial states (in a finite region) according to the cellular automaton rules of Figs. 13(a) and 13(i).

In the cellular automaton of Fig. 13(b), the domains become stationary after a few time steps. In the case of Fig. 13(e), however, the domains can continue to move forever, essentially by a diffusion process. Figure 12(d) shows a one-dimensional cellular automaton with domain walls that exhibit analogous behavior (cf. Refs. 27 and 17). In both cases, some domains become progressively larger with time, while others eventually disappear completely. The domain walls in Fig. 13(e) behave as if they carry a positive surface tension (cf. Ref. 28); the diffusion process responsible for their movement is biased to reduce the local curvature of the interface. A linear interface is stable under the cellular automaton rule of Fig. 13(e). In addition, the heights of any protrusions or intrusions cannot increase with time. In general, they decay, often quite slowly, until they are of height at most one. Deformations of height one, analogous to surface waves, do not decay further, and are governed by a one-dimensional cellular automaton rule (with $k = 2$, $r = 1$, and rule number 150). At large times, therefore, a domain must either shrink to zero size, or must have walls with continually decreasing curvatures.

Figure 13(c) shows a two-dimensional cellular automaton with structures analogous to domain walls that carry a negative surface tension. More and more convoluted patterns are obtained with time. The resulting labyrinthine state is strongly reminiscent of behavior observed with ferrofluids or magnetic bubbles.[29]

Figures 13(f), 13(g), and 13(h) are examples of two-dimensional cellular automata that exhibit class-3 behavior. Chaotic aperiodic patterns are obtained at all times. Moreover, the difference patterns resulting from changes in single initial site values expand at a fixed rate forever. A remarkable feature is that in almost all cases (Fig. 13(h) is an exception), the expansion occurs at the same speed in all directions, resulting in an asymptotically circular difference pattern. For some rules, the expansion occurs at maximal speed; but often the speed is about 0.8 times the maximum. When the difference patterns are not exactly circular, they tend to have rounded corners. And even with asymmetrical rules, circular difference patterns are often obtained. A rough analog of this behavior is found in asymmetric one-dimensional cellular automata which generate symmetrical difference patterns. Such behavior is found to become increasingly common as k and r increase, or as the number of independent parameters in the rule ϕ increases.

An argument based on the central limit theorem suggests an explanation for the appearance of circular difference patterns in two-dimensional class-3 cellular automata. Consider the set of sites corresponding to the neighborhood for a cellular automaton rule. For each site, compute the probability that the value of that site changes after one time step of cellular automaton evolution when the value of the center site is changed, averaged over all possible arrangements of site values in the neighborhood. An approximation to the probability distribution of differences is then obtained as a multiple convolution of this kernel. (This approximation is effectively a linear one, analogous to Huygens' principle in optics.) The number of convolutions performed increases with time. If the number of neighborhood arrangements is sufficiently large, the kernel tends to be quite smooth. Convolutions of the kernel thus tend to a Gaussian form, independent of direction.

Some asymmetric class-3 cellular automata yield difference patterns that expand, say, in the horizontal direction, but contract in the vertical direction. At large times, such cellular automata produce patterns consisting of many independent horizontal lines, each behaving essentially as a one-dimensional class-3 cellular automaton.

Class-3 behavior is considerably the commonest among two-dimensional cellular automata, just as it is for one-dimensional cellular automata with large k and r. It appears that as the number of parameters or degrees of freedom in a cellular automaton rule increases, there is a higher probability for some degree of freedom to show chaotic behavior, leading to overall chaotic behavior.

Figure 12(i) shows an example of a class-4 one-dimensional cellular automaton. A characteristic feature of class-4 cellular automata is the existence of a complicated set of persistent structures, some of which propagate through space with time. Class-4 rules appear to occur with a frequency of a few per cent among all one-dimensional cellular automaton rules. Often one suspects that some degrees of freedom in a cellular automaton exhibit class-4 behavior, but they are masked by overall chaotic class-3 behavior.

Class-4 cellular automata appear to be much less common in two dimensions than in one dimension. Figures 13(i) and 13(b) show the evolution of a two-dimensional cellular automaton known as the "Game of Life".[8] Many persistent structures, some propagating, have been identified in this cellular automaton. It has in addition been shown that these structures can be combined to perform arbitrary information processing, so that the cellular automaton supports universal computation.[8] Starting from a disordered initial state, the density of propagating structures ("gliders") produced is about one per 2000 site region.

Except for a few simple variants on the Game of Life, no other definite class-4 two-dimensional cellular automata were found in a random sample of several thousand outer totalistic rules.[6] Some of the rules that appeared

---

[6] A few examples of class-4 behavior were however found among general rules. Requests for copies of the relevant rule tables should be directed to the authors.

to be of class 2 were found to have long transients, characteristic of class-4 behavior, but no propagating structures were seen. Other rules seemed to exhibit some class-4 features, but they were overwhelmed by dominant class-3 behavior.

4. GLOBAL PROPERTIES

Section 3 discussed the typical behavior of cellular automata evolving from particular initial states. This section considers the global properties of cellular automata, determined by evolution from all possible initial states. Studies of the global properties of one-dimensional cellular automata have been made using methods both from dynamical systems theory[7] and from computation theory.[12] Here these studies are generalized to the case of two-dimensional cellular automata. For those based on dynamical systems theory the generalization is quite straightforward; but in the computation theory approach substantial additional complications occur. Whereas the sets of configurations generated after any finite number of steps in the evolution of a one-dimensional cellular automata always correspond to regular formal languages,[12] the corresponding sets in two-dimensional cellular automata may be nonrecursive.[15]

Most cellular automaton rules are irreversible, so that several different initial states may evolve to the same final state. As a consequence, even starting from all possible initial states, only a subset of possible states may be generated with time. The properties of this set then determine the overall behavior of the cellular automaton, and the self-organization that occurs in it.

Entropy and dimension provide quantitative characterizations of the "sizes" of sets generated by cellular automaton evolution (e.g., Ref. 7). The spatial set entropy for a set of two-dimensional cellular automaton configurations is defined by considering a $X \times Y$ patch of sites. If the set contains all possible configurations, then all k^{XY} possible different arrangements of sites values must occur in the patch. In general $N(X, Y) \leqslant k^{XY}$ different arrangements will occur. Then the set entropy (or dimension) is defined as

$$s = \lim_{X, Y \to \infty} \frac{1}{XY} \log_k N(X, Y) \tag{4.1}$$

If the cellular automaton mapping is surjective, so that all possible configurations occur, then this entropy is equal to one. In general it decreases with time in the evolution of the cellular automaton.

Spatial set entropy characterizes the set of configurations that can possibly be generated in the evolution of a cellular automaton, regardless of their probabilities of occurrence. One may also define a spatial measure entropy in terms of the probabilities p_i for possible $X \times Y$ patches as

$$s_\mu = \lim_{X,Y \to \infty} \frac{-1}{XY} \sum_{i=1}^{k^{XY}} p_i \log_k p_i \qquad (4.2)$$

The limiting value of s_μ at large times is typically nonzero for all but class-1 cellular automaton rules. Notice that in cases where domains with positive "surface tension" are formed, s_μ tends only very slowly to zero with time.

To find the spatial set entropy after, say, one time step in the evolution of a cellular automaton one must identify what configurations can be generated. In a one-dimensional cellular automaton, one can specify the set of configurations that can be generated in terms of rules that determine which sequences of site values can appear. These rules correspond to a regular formal grammar, and give the state transition graph for a finite state machine. The set of configurations that can be generated in a two-dimensional cellular automaton is more difficult to specify. In many circumstances in fact the occurrence of a particular patches of site values requires a global consistency that cannot be verified in general by any finite computation. As a consequence, many propositions concerning sets of configurations generated after even a finite number of steps in the evolution of two-(and higher-)dimensional cellular automata can be formally undecidable.

In a one-dimensional cellular automaton with a range-r rule, a particular sequence of X site values can be generated (reached) after one time step only if there exists some length $X + 2r$ sequence of initial site values that evolves to it. The locality of the cellular automaton rule ensures that in determining whether a length $X + 1$ sequence obtained by appending one new site can also be generated, it suffices to test only those length $X + 2r + 1$ predecessor configurations that differ in their last $2r + 1$ site values. In determining whether sequences of progressively greater lengths can be generated it suffices at each stage to record with which length $2r$ overlaps in the predecessor configuration a particular new site value can be appended. Since there are only k^{2r} possible sequences of site values in the overlaps, only a finite amount of information must be recorded, and a finite procedure can be given for determining whether any given sequence can be generated (cf. Ref. 12). Hence in particular there is a finite procedure to determine whether any given cellular automaton rule is surjective, so that all possible configurations can be reached in its evolution.[30]

In two-dimensional cellular automata there is no such simple iterative procedure for determining whether progressively larger patches of site values can be generated. An $X \times Y$ patch of site values is generated after one step in the evolution of a two-dimensional cellular automaton with a range-r rule if there exists some $(X + 2r)(Y + 2r)$ patch of initial site values that evolves to it. Progressively larger patches can be generated if appropriate progressively large predecessor patches exist. The number of sites in the overlap between such progressively larger predecessor patches is not fixed, as in one-dimensional cellular automata, but instead grows essentially like the perimeter of the patch, $2r(X + Y + 2r)$. With this procedure, there is thus no upper bound on the amount of information that must be recorded to determine whether progressively larger patches can be generated. To find whether a patch of any particular size $X \times Y$ can be generated, it suffices to test all $k^{(X + 2r)(Y + 2r)}$ candidate predecessor patches. (As mentioned below, this is in fact an NP-complete problem, and therefore presumably cannot be solved in general in a time polynomial in the patch size.) However, questions concerning complete configurations can be answered only be considering arbitrarily large patches, and may require arbitrarily complex computations. As a consequence, there are global questions about configurations generated by two-dimensional cellular automata after a finite number of time steps that can posed, but cannot in general be answered by any finite computational process, and are therefore formally undecidable.[15]

Some examples of such undecidable questions about two-dimensional cellular automata are (i) whether a particular complete (but finitely specified) configuration can be generated after one time step from any initial configuration; (ii) whether a particular cellular automaton rule is surjective, so that all possible configurations can be generated; (iii) whether the set of complete configurations generated after say one time step has a nonempty intersection with some recursive formal language such as a regular language, whose words can be recognized by a finite computation; (iv) whether there exist configurations that have a particular period in time (and are thus invariant under some number of iterations of the cellular automaton rule).

It seems that global questions about the finite time behaviour of one-dimensional cellular automata are always decidable. Questions about their ultimate infinite time behavior may nevertheless be undecidable. To show this, one considers one-dimensional cellular automata whose evolution emulates that of a universal Turing machine. The successive arrangements of symbols on the Turing machine tape correspond to successive configurations of site values generated in the evolution of the cellular automaton. Undecidable questions such as halting for the Turing machines are then shown to be undecidable for the corresponding one-dimensional cellular automaton.[13]

In two-dimensional cellular automata, questions about global properties on infinite spatial scales can be undecidable even at finite times. This is proved[15] by considering the line-by-line construction of configurations. The rules used to obtain each successive line from the last can correspond to the rules for a universal Turing machine. The construction of the configuration can then be continued to infinity and completed only if this Turing machine does not halt with the input given, which is in general undecidable. Sets of configurations generated at finite times in the evolution of two-dimensional cellular automata can thus be nonrecursive.

Many global questions about two-dimensional cellular automata are closely analogous to geometrical questions associated with tilings of the plane. Consider for example the problem of finding configurations that remain invariant under a particular cellular automaton rule. All the neighborhoods in such configurations must be such that the values of their center sites are left unchanged by the cellular automaton rule. Each such neighborhood may be considered as a "tile." Complete invariant configurations are constructed from an array of tiles, with each adjacent pair of tiles subject to a consistency condition that the overlapping sites in the neighborhoods to which they correspond should agree. In a one-dimensional cellular automaton, the set of possible arrangements of tiles or configurations that satisfy the conditions can be enumerated immediately, and form a finite complement regular language (subshift of finite type).[12] In a two-dimensional cellular automaton, the problem of finding invariant configurations is equivalent to tiling the plane with a set of "dominoes" corresponding to the possible allowed neighborhoods, and subject to constraints that can be cast in the form of requiring adjacent pairs of edges to have complementary colours.[31] The problem of determining wether a particular set of dominoes can in fact be used to tile the plane is however known to be undecidable[32,33] (cf. Ref. 34). The problem of finding whether there exist invariant configurations under a particular two-dimensional cellular automaton rule is likewise undecidable.

If any infinite sequence can be constructed from some set of dominoes in one dimension, then it is clear that a spatially periodic sequence can be found. Hence if there are to be any configurations with a particular temporal period in a one-dimensional cellular automaton, then there must be spatially periodic configurations with this temporal period. (The maximum necessary spatial period for configurations with temporal period p is k^{2rp+1} [12]: the existence of such spatially periodic configurations can be viewed as a consequence of the pumping lemma (e.g., Ref. 11) for regular languages.) In two dimensions, however, there are sets of dominoes for which a tiling of the plane is possible, but the tiling cannot be spatially periodic.[32,33,35] In the examples known, it appears that the basic arrangement of tiles is always self-similar, so that it is almost periodic. In the simplest known examples, six square dominoes[33] or just two irregularly shaped dominoes[35] are required for this phenomenon to occur. (The simplest known example in three dimensions involves seven polyhedral "dominoes".[36])

The problem of whether a set of dominoes can tile a finite, say, $X \times X$ region of the plane is clearly decidable, but is *NP* complete.[37] The analogous problem of determining whether a particular patch can occur in an invariant configuration for a two-dimensional cellular automaton, or can in fact be generated by one time step of evolution from any initial state, is thus also *NP* complete. These problems can presumably be solved only by computations whose complication increases faster than a polynomial in X, and are essentially equivalent to explicit testing of all $O(k^{X^2})$ possible cases.

In addition to considering configurations of site values generated at a particular step in the evolution of a cellular automaton, one may also discuss sequences of site values obtained with time. In general one may consider the number of possible arrangements $N(\mathbf{v}_1,..., \mathbf{v}_p)$ of site values in a space-time volume consisting of a parallelepiped with generator vectors $\mathbf{v}_i$. The set entropy may than be defined as the exponential rate of increase of N as the lengths of certain generators are taken to infinity (cf. Refs. 38 and 39):

$$s = \lim_{\alpha_1 \to \infty} \cdots \lim_{\alpha_p \to \infty} \frac{1}{\alpha_1 \cdots \alpha_{p'}} \log_k N(\alpha_1 \mathbf{v}_1, \alpha_2 \mathbf{v}_2,..., \alpha_p \mathbf{v}_p) \qquad (4.3)$$

where the α_i are scalar parameters, and $p' \leqslant p \leqslant d$. These entropies are in fact functions of the unit p forms obtained as the exterior products of the generator vectors $\mathbf{v}_i$ considered as one-forms in space-time. Certain convergence properties of the limits in Eq. (4.3) can be proved from the fact that the number of arrangements $N(V)$ of site values in a volume V is submultiplicative, so that $N(V_1 \cup V_2) \leqslant N(V_1) N(V_2)$. A measure-theoretical analog of the set entropy (4.3) may be defined in correspondence with Eq. (4.2).

The spatial entropy (4.1) for two-dimensional cellular automata is obtained from the general definition (4.3) by choosing $p = 3$, $p' = 2$ and taking $\mathbf{v}_1$ and $\mathbf{v}_2$ to be orthogonal purely spacelike vectors along the two lattice directions. The generator vector $\mathbf{v}_3$ is taken to be in the positive time direction, but the number of arrangements N is independent of α_3 since a complete configuration at one time step determines all future configurations.

For a d-dimensional cellular automaton, there are critical values of p and p' such that entropies corresponding to higher or lower-dimensional parallelepipeds are zero or infinity. Entropies with exactly those critical values may be nonzero and bounded by quantities that depend on the cellular automaton neighborhood size.

Entropies are essentially determined by the correlations between values of sites at different space-time points. These correlations depend on the propagation of information in the cellular automaton. The difference patterns discussed in Section 3 provide measures of such information propagation. They can be considered analogs of Green's functions (cf. Ref. 4) which describe the change produced at some space-time point x' in a cellular automaton as a consequence changes at another point x. The set-theoretical Green's function is defined to be nonzero whenever a change at x in any configuration could lead to a change at x'. In the measure-theoretical Green's function the possible configurations are weighted with their probabilities. The maximum rate of information propagation is determined by the slope of the space-time ("light") cone within which the Green's function is nonzero. The slope corresponding to propagation in a particular spatial direction in say a two-dimensional cellular automaton gives the Lyapunov exponent in that direction for the cellular automaton evolution.[7,40] In most cases it appears that the space-time structure corresponding to the set of sites on which the Green's function is nonzero tends to a fixed form after rescaling at large times, so that the structure has a unique growth dimension, and the Lyapunov exponents have definite values. Exceptions may occur in rules where difference patterns spread along domain boundaries, typically producing asymptotically self-similar structures analogous to percolation clusters (e.g., Ref. 26).

The Green's functions describe not only how a change at some time affects site values at later times, but also how the value of a particular site is affected by the previous values of other sites. The backward light cone of a site contains all the sites whose values can affect it. (Notice that the backward light cone for a bijective rule in general has little relation with the forward light cone for the inverse rule.[41]) The values of all sites in a volume V are thus determined by the values of sites on a surface S that "absorbs" (covers) all the backward light cones of points in V. The number of possible configurations in V is then bounded from above by the number of possible configurations of the set of sites within one cellular automaton neighborhood of the surface S. The entropy associated with the volume V is then not greater than the entropy associated with the volume around S. By choosing various "absorbing surfaces" S, whose sizes are determined by the rates of information propagation in different directions, one can derive various inequalities between entropies.

Many entropies can be defined for cellular automata using Eq. (4.3). One significant class is those that are invariant under continuous invertible transformations on the space of cellular automaton configurations. Such entropies can be used to identify topologically inequivalent cellular automaton rules. For one-dimensional cellular automata, an invariant entropy may be defined by taking $p = 2$, $p' = 1$ in Eq. (4.3), and choosing $\mathbf{v}_1$ in the positive time direction, and $\mathbf{v}_2$ in the space direction. The entropy may be generalized by taking the $\mathbf{v}_i$ to be an arbitrary pair of orthogonal spacetime vectors (with $\mathbf{v}_1$ having a positive time component).[38] The most direct generalization of these invariant entropies to two-dimensional cellular automata would have $p = 3$, $p' = 1$, and take the $\mathbf{v}_i$ to be an orthogonal triple of space-time vectors with $\mathbf{v}_1$ having a positive time component. If $\mathbf{v}_1$ were chosen purely timelike, then this entropy would have no dependence on spatial direction, and would correspond to the standard invariant entropy defined for the cellular automaton mapping. In general however, there is no upper bound on its value, and it is apparently infinite for most cellular automata that have positive Lyapunov exponents in more than one spatial direction. A finite entropy can nevertheless be constructed by choosing $p' = 2$. This entropy depends on the spatial (or in general space-time) vector $\mathbf{v}_1 \times \mathbf{v}_2$. To obtain an invariant entropy, one must perform some average over this vector (accounting for the fact that the entropy is a homogeneous function of degree one in the length of the vector). One possibility is to form the integral of the quantity (4.3) over those values of the vector for which the quantity is less than some constant (say, 1).

5. DISCUSSION

This paper has presented an exploratory study of two-dimensional cellular automata. Much remains to be done, but a few conclusions can be already be given.

A first approach to the study of cellular automaton behavior is statistical: one considers the average properties of evolution from typical initial configurations. Statistical studies of one-dimensional cellular automata have suggested that four basic qualitative classes of behavior can be identified. This paper has given analogs of these classes in two-dimensional cellular automata. One expects that the qualitative classification will also apply in three- and higher-dimensional systems.

Entropies and Lyapunov exponents are statistical quantitites that measure the information content and rate of information transmission in cellular automata. Their definitions for one-dimensional cellular automata are closest to those used in smooth dynamical systems. But rather direct generalizations can nevertheless be found for two- and higher-dimensional cellular automata.

Beyond statistical properties, one may consider geometrical aspects of patterns generated by cellular automaton evolution. Even though the basic construction of a cellular automaton is discrete, its "macroscopic" behavior at large times and on large spatial scales may be a close approximation to that of a continuous system. In particular domains of correlated sites may be formed, with boundaries that at a large scale seem to show continuous motions and deformations. While some such phenomena do occur in one dimension, they are most significant in two and higher dimensions. Often their motion appears to be determined by attributes such as curvature, that have no analog in one dimension.

The structures generated by two- and higher-dimensional cellular automata evolving from simple seeds show many geometrical phenomena. The most significant is probably the formation of dendritic patterns, characterized by noninteger growth dimensions.

Statistical measurements provide one method for comparing cellular automaton models with experimental data. Geometrical properties provide another. The geometry of patterns formed by cellular automata may be compared directly with the geometry of patterns generated by natural systems.

Topology is another aspect of cellular automaton patterns. When domains or regions containing many correlated sites exist, one may approximate them as continuous structures, and consider their topology. For example, domains produced by cellular automaton evolution may exhibit topological defects that are stable under the cellular automaton rule. In two-dimensional cellular automata, only point and line defects occur. But in three dimensions, knotted line defects (e.g., Ref. 42) and other complicated topological forms are possible. The topology of the structures supported by a cellular automaton rule may be compared directly with the topology of structures that arise in natural systems (cf. Ref. 43).

Geometry and topology provide essentially local descriptions of the behavior of cellular automata. Computation theory potentially provides a more global characterization. One may classify the behavior and properties of cellular automata in terms of the nature of the computations required to reproduce them. Even in one dimension, there are cellular automata that can perform arbitrary computations, so that at least some of their properties can be reproduced only by direct simulation or observation, and their limiting behavior is formally undecidable. The range of properties for which undecidability can occur in much larger in two dimensions than in one dimension. In particular, properties that involve a limit of infinite spatial size, even at finite times, can be undecidable. As higher-dimensional cellular automata are considered, the degree of undecidability that can be encountered in studies of particular properties increases.

ACKNOWLEDGMENTS

The research reported here made essential use of several computer systems other than our own. We are grateful to those who made the systems available, and helped us in using them. We thank the M.I.T. Information Mechanics Group (E. Fredkin, N. Margolus, T. Toffoli, and G. Vichniac) for the use of their special-purpose two-dimensional cellular automaton simulation system, and for their hospitality and assistance. We thank R. Shaw for writing the kernel of our software simulation system for two-dimensional cellular automata in Ridge assembly language. We thank the Theoretical Division and the Center for Nonlinear Studies at Los Alamos National Laboratory for hospitality during the final stages of this work. We thank M. Prueitt at Los Alamos for making the three-dimensional illustrations, and D. Umberger for help with some Cray-1 programming. We are grateful to those mentioned and to C. Bennett, J. Crutchfield, H. Hartman, L. Hurd, J. Milnor, S. Willson, and others for discussions.

This work was supported in part by the U.S. Office of Naval Research under Contract No. N00014-80-C-0657.

REFERENCES

1. S. Wolfram, Cellular automata as models for complexity, *Nature* **311**:419 (1984).
2. T. Toffoli, Cellular automata mechanics, Ph.D. thesis and Technical Report 208, The Logic of Computers Group, University of Michigan (1977).
3. G. Vichniac, Simulating physics with cellular automata, *Physica* **10D**:96 (1984).
4. S. Wolfram, Statistical mechanics of cellular automata, *Rev. Mod. Phys.* **55**:601 (1983).
5. O. Martin, A. Odlyzko, and S. Wolfram, Algebraic properties of cellular automata, *Commun. Math. Phys.* **93**:219 (1984).
6. N. Packard, Cellular automaton models for dendritic crystal growth, Institute for Advanced Study, preprint (1985).
7. S. Wolfram, Universality and complexity in cellular automata, *Physica* **10D**:1 (1984).
8. E. R. Berlekamp, J. H. Conway, and R. K. Guy, *Winning Ways for Your Mathematical Plays*, Vol. 2 (Academic Press, New York, 1982), Chap. 25; M. Gardner, *Wheels, Life and Other Mathematical Amusements* (Freeman, San Francisco, 1983).
9. T. Toffoli, CAM: A high-performance cellular-automaton machine, *Physica* **10D**:195 (1984).
10. J. Guckenheimer and P. Holmes, *Nonlinear Oscillations, Dynamical Systems, and Bifurcations of Vector Fields* (Springer, New York, 1983).
11. J. E. Hopcroft and J. D. Ullman, *Introduction to Automata Theory, Languages, and Computation* (Addison-Wesley, Reading, Massachusetts, 1979).
12. S. Wolfram, Computation theory of cellular automata, *Commun. Math. Phys.* **96**:15 (1984).
13. L. Hurd, Formal language characterizations of cellular automaton limit sets, to be published.
14. S. Willson, On convergence of configurations, *Discrete Math.* **23**:279 (1978).
15. T. Yaku, The constructability of a configuration in a cellular automaton, *J. Comput. Syst.*

Sci. **7**:481 (1973); U. Golze, Differences between 1- and 2-dimensional cell spaces, in *Automata, Languages, Development*, A. Lindenmayer and G. Rozenberg, eds. (North-Holland, Amsterdam, 1976).

16. J. Hardy, O. de Pazzis, and Y. Pomeau, Molecular dynamics of a classical lattice gas: transport properties and time correlation functions, *Phys. Rev.* **A13**:1949 (1976).

17. S. Wolfram, Cellular automata, *Los Alamos Science* (Fall 1983); Some recent results and questions about cellular automata, Institute for Advanced Study preprint (September 1983).

18. S. Wolfram, Twenty problems in the theory of cellular automata, *Phys. Scripta* (to be published).

19. S. Wolfram, Computer software in science and mathematics, *Sci. Am.* **251**(3):188 (1984).

20. S. Willson, Comparing limit sets for certain increasing cellular automata, Mathematics Department, Iowa State University preprint (June 1984).

21. S. Willson, Growth rates and fractional dimensions in cellular automata, *Physica* **10D**:69 (1984).

22. N. Packard, Notes on a Go-playing program, unpublished (1984).

23. Y. Sawada, M. Matsushita, M. Yamazaki, and H. Kondo, Morphological phase transition measured by "surface kinetic dimension" of growing random patterns, *Phys. Scripta* (to be published).

24. J. M. Greenberg, B. D. Hassard, and S. P. Hastings, Pattern formation and periodic structures in systems modelled by reaction-diffusion equations, *Bull. Am. Math. Soc.* **84**:1296 (1975); B. Madore and W. Freedman, Computer simulations of the Belousov-Zhabotinsky reaction, *Science* **222**:615 (1983).

25. K. Preston *et al.*, Basics of cellular logic with some applications in medical image processing, *Proc. IEEE* **67**:826 (1979).

26. J. W. Essam, Percolation theory, *Rep. Prog. Phys.* **43**:833 (1980).

27. P. Grassberger, Chaos and diffusion in deterministic cellular automata, *Physica* **10D**:52 (1984).

28. G. Vichniac, Cellular automaton dynamics for interface motion and ordering, M.I.T. report, to appear.

29. R. Rosensweig, Fluid dynamics and science of magnetic liquids, *Adv. Electronics Electron Phys.* **48**:103 (1979); Magnetic fluids, *Sci. Am.* **247**(4):136 (1982).

30. G. A. Hedlund, Endomorphisms and automorphisms of the shift dynamical system, *Math. Syst. Theory* **3**:320 (1969); G. A. Hedlund, Transformations commuting with the shift, in *Topological Dynamics*, J. Auslander and W. H. Gottschalk, eds. (Benjamin, New York, 1968); S. Amoroso and Y. N. Patt, Decision procedures for surjectivity and injectivity of parallel maps for tessellation structures, *J. Comp. Syst. Sci.* **6**:448 (1972); M. Nasu, Local maps inducing surjective global maps of one-dimensional tessellation automata, *Math. Syst. Theory* **11**:327 (1978).

31. H. Wang, Proving theorems by pattern recognition—II, *Bell. Syst. Tech. J.* **40**:1 (1961).

32. R. Berger, The undecidability of the domino problem, *Mem. Am. Math. Soc.*, No. 66 (1966).

33. R. Robinson, Undecidability and nonperiodicity for tilings of the plane, *Inventiones Math.* **12**:177 (1971).

34. D. Ruelle, *Thermodynamic Formalism* (Addison-Wesley, Reading, Massachusetts, 1978), p. 68.

35. R. Penrose, Pentaplexity: a class of nonperiodic tilings of the plane, *Math. Intelligencer* **2**:32 (1979); M. Gardner, Extraordinary nonperiodic tiling that enriches the theory of tiles, *Sci. Am.* **236**(1):110 (1977); N. de Bruijn, Algebraic theory of Penrose's nonperiodic tilings of the plane, *Nederl. Ǎkad. Wetensch. Indag. Math.* **43**:39 (1981).

36. P. Kramer, Non-periodic central space filling with icosahedral symmetry using copies of seven elementary cells, *Acta Crystallogr.* **A38**:257 (1982).

37. M. Garey and D. Johnson, *Computers and Intractability: A Guide to the Theory of NP-completeness* (Freeman, San Francisco, 1979), p. 257.

38. J. Milnor, Entropy of cellular automaton-maps, Institute for Advanced Study preprint (May 1984).

39. J. Milnor, Directional entropies in higher dimensions, rough notes (September 1984).

40. N. Packard, Complexity of growing patterns in cellular automata, Institute for Advanced Study preprint (October 1983).

41. J. Milnor, Notes on surjective cellular automaton-maps, Institute for Advanced Study preprint (June 1984).

42. N. D. Mermin, The topological theory of defects in ordered media, *Rev. Mod. Phys.* **51**:591 (1979).

43. A. Winfree and E. Winfree, Organizing centers in a cellular excitable medium, Purdue University preprint (July 1984) and *Physica D* (to be published).

Physica Scripta. Vol. T9, 170–183, 1985

Twenty Problems in the Theory of Cellular Automata*

Stephen Wolfram

The Institute for Advanced Study, Princeton NJ 08540, U.S.A.

Received August 7, 1984; accepted August 13, 1984

Abstract

Cellular automata are simple mathematical systems that exhibit very complicated behaviour. They can be considered as discrete dynamical systems or as computational systems. Progress has recently been made in studying several aspects of them. Twenty central problems that remain unsolved are discussed.

Many of the complicated systems in nature have been found to have quite simple components. Their complex overall behaviour seems to arise from the cooperative effect of a very large number of parts that each follow rather simple rules. Cellular automata are a class of mathematical models that seem to capture the essential features of this phenomenon. From their study one may hope to abstract some general laws that could extend the laws of thermodyamics to encompass complex and self-organizing systems.

There has been recent progress in analysing some aspects of cellular automata. But many important problems remain. This paper discusses some of the ones that have so far been identified. The problems are intended to be broad in scope, and are probably not easy to solve. To solve any one of them completely will probably require a multitude of subsidiary questions to be asked and answered. But when they are solved, substantial progress towards a theory of cellular automata and perhaps of complex systems in general should have been made.

The emphasis of the paper is on what is not known: for expositions of what is already known about cellular automata, see [1–4]. The paper concentrates on theoretical aspects of cellular automata. There is little discussion of models for actual natural systems. But many of the theoretical issues discussed should have direct consequences for such models.

Cellular automata consist of a homogeneous lattice of sites, with each site taking on one of k possible values. The sites are updated according to a definite rule that involves a neighbourhood of sites around each one. So in a one-dimensional cellular automaton the value $a_i^{(t)}$ of a site at position i evolves according to

$$a_i^{(t+1)} = \phi[a_{i-r}^{(t)}, a_{i-r+1}^{(t)}, \ldots, a_{i+r}^{(t)}].$$

The local rule ϕ has a range of r sites. Its form determines the behaviour of the cellular automaton. Some examples of patterns generated by cellular automata are shown in Figs. 1 and 2. Figure 1 shows examples of the four basic classes of behaviour seen in the evolution of cellular automata from disordered initial states. Figure 2 shows patterns generated by evolution from initial configurations containing a single nonzero site.

Cellular automata may be considered as discrete dynamical

* Work supported in part by the U.S. Office of Naval Research under contract number N00014-80-C-0657.

systems. Their global properties are studied by considering evolution from the set of all possible initial configurations (e.g., [5]). Since most cellular automata are irreversible, the set of configurations that is generated typically contracts with time. Its limiting form at large times determines the asymptotic behaviour of the cellular automaton, and is dominated by the attractors for the evolution. Some of the properties of cellular automata may be characterized in terms of quantities such as entropies and Lyapunov exponents that are used in studies of continuous dynamical systems (e.g., [6]).

An alternative view of cellular automata is as information-processing systems [7]. Cellular automaton evolution may be considered to carry out a computation on data represented by the initial sequence of site values. The nature of the evolution may then be characterized using methods from the theory of computation (e.g., [8]). So for example the sets of configurations generated in the evolution may be described as formal languages: a one-dimensional cellular automaton gives a regular formal language after any finite number of time steps [7]. One suspects that in many cases the computations corresponding to cellular automaton evolution are sufficiently complicated as to be irreducible (cf. [9]). In that case, there can be essentially no short-cut to determining the outcome of the cellular automaton evolution by explicit simulation or observation of each step. This implies that certain limiting properties of the cellular automaton are undecidable, since to find them would require an infinite computation.

The problems discussed here address both dynamical systems theory and computation theory aspects of cellular automata. But probably the most valuable insights will come from the interplay between these two aspects.

Problem 1

What overall classification of cellular automaton behaviour can be given?

Experimental mathematics provides a first approach to this problem. One performs explicit simulations of cellular automata, and tries to find empirical rules for their behaviour. These may then suggest results that can be investigated by more conventional mathematical methods.

An extensive experimental study [5] suggests that the patterns generated in the evolution of cellular automata from disordered initial states can be grouped into four general classes, illustrated in Fig. 1:

(1) Evolves to homogeneous state.
(2) Evolves to simple separated periodic structures.
(3) Yields chaotic aperiodic patterns.
(4) Yields complex pattern of localized structures.

The classification is at first qualitative. But there are several

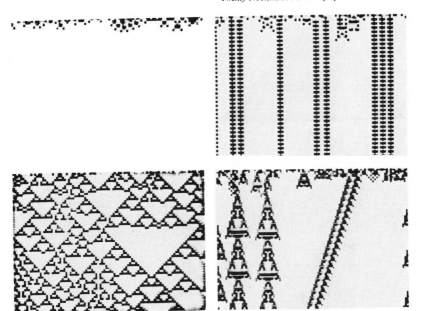

Fig. 1. Examples of the four qualitative classes of behaviour seen in the evolution of one-dimensional cellular automata from disordered initial states. Successive time steps are shown on successive lines. Complex and varied behaviour is evident. The sites in the cellular automata illustrated have three possible values ($k = 3$); value 0 is shown blank, 1 is grey, and 2 is black. The value of each site at each time step is given by rules that depend on the sum of its own and its nearest neighbours' old values ($r = 1$ totalistic). The cases shown have rules specified by code numbers [5] 1302, 1005, 444 and 792, respectively.

ways to make it more quantitative, and to formulate precise definitions for the four classes. For some cellular automaton rules, one expects that all definitions will agree. But there are likely to be borderline cases where definitions will disagree.

Continuous dynamical systems provide analogues for the classes of behaviour seen in cellular automata. Class 1 cellular automata show limit points, while class 2 cellular automata may be considered to evolve to limit cycles. Class 3 cellular automata exhibit chaotic behaviour analogous to that found with strange attractors. Class 4 cellular automata effectively have very long transients, and no direct analogue for them has been identified among continuous dynamical systems.

Dynamical systems theory gives a first approach to the quantitative characterization of cellular automaton behaviour. Various kinds of entropy may be defined for cellular automata. Each counts the number of possible sequences of site values corresponding to some spacetime region. For example, the spatial entropy gives the dimension of the set of configurations that can be generated at some time step in the evolution of the cellular automaton, starting from all possible initial states. There are in general $N(X) \leqslant k^X$ (k is the number of possible values for each site) possible sequences of values for a block of X sites in this set of configurations. The spatial topological entropy $d^{(x)}$ is given by $\lim_{X \to \infty} (1/X) \log_k N(X)$. One may also define a spatial measure entropy $d_\mu^{(x)}$ formed from the probabilities of possible sequences. Temporal entropies $d^{(t)}$ may then be defined to count the number of sequences that occur

in the time series of values taken on by each site. Topological entropies reflect the possible configurations of a system; measure entropies reflect those that are probable, and are insensitive to phenomena that occur with zero probability. A tentative definition of the four classes of cellular automaton behaviour may be given in terms of measure entropies. Class 1 has zero spatial and temporal measure entropy. Class 2 has zero temporal measure entropy, since it almost always yields periodic structures, but has positive spatial measure entropy. Class 3 has positive spatial and temporal measure entropies.

Another property of cellular automata is their stability under small perturbations in initial conditions. Figure 3 shows differences in patterns generated by cellular automata induced by changes in a single initial site value. Such differences almost always die out in class 1 cellular automata. In class 2 cellular automata, they may persist, but remain localized. In class 3 cellular automata, however, they typically expand at an asymptotically constant rate. The rate of this expansion gives the Lyapunov exponent for the evolution [5, 10], and measures the speed of propagation of information about the initial configuration in the cellular automaton. Class 4 cellular automata give rise to a pattern of differences that typically expands irregularly with time.

The four classes of cellular automaton behaviour identified here can be defined to be complete. But there are some cellular automata whose behaviour · should probably be considered intermediate between the classes. In particular, there are many

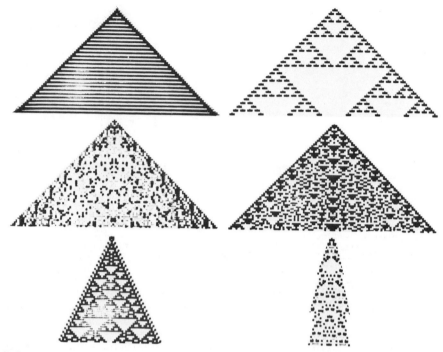

Fig. 2. Examples of patterns generated by the evolution of various cellular automata starting from single site seeds. In the second case shown, a fractal pattern is generated. The subsequent cases shown illustrate the remarkable phenomenon that complicated and in some cases apparently random patterns can be generated by cellular automaton rules even from simple initial states. The cellular automata shown have $k = 3$, $r = 1$ totalistic rules with code numbers 1443, 312, 1554, 1617, 1410 and 600, respectively.

where there is a clear superposition of two classes of behaviour. So for example sites with values 0 and 1 can exhibit class 2 behaviour, while sites with values 0 and 2 show class 3 behaviour. The result is a sequence of chaotic regions separated by rigid "walls".

Even at a qualitative level, it is possible that definite subclasses of the four classes of cellular automaton behaviour may be identified. Some class 3 cellular automata in one dimension seem to give patterns with large triangular clearings and low but presumably nonzero entropies; others give highly irregular patterns with no long-range structure. No clear statistical difference between these kinds of class 3 cellular automata has yet been found. But it is possible that one exists. Among class 4 cellular automata there seem to be some definite subclasses in which persistent or almost persistent structures of rather particular kinds occur.

Problem 2

What are the exact relations between entropies and Lyapunov exponents for cellular automata?

Using the finite information density of cellular automaton configurations, and the finite rate of information propagation

in cellular automata, a number of inequalities may be derived between entropies and Lyapunov exponents (λ). An example is $d^{(t)}/d^{(x)} \leqslant 2\lambda$ [5]. Preliminary numerical evidence suggests that for some cellular automata these inequalities may in fact be equalities. This would imply an important connection between the static properties of cellular automata, as embodied in entropies, and their dynamic properties, as measured by Lyapunov exponents. One is hampered in these studies by the lack of an efficient method for computing entropies. The best approach so far uses a conditional entropy method [11].

Lyapunov exponents can be considered to measure the rate of divergence of trajectories in the space of configurations. In continuous dynamical systems, a geometry is defined for this space, and one can identify Lyapunov exponents for various directions.

Problem 3

What is the analogue of geometry for the configuration space of a cellular automaton?

Several simple observations may be made. First, if the cellular automaton lattice is more than one-dimensional, one may consider Lyapunov exponents in different directions on this

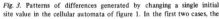

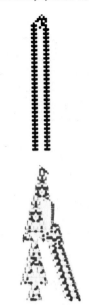

Fig. 3. Patterns of differences generated by changing a single initial site value in the cellular automata of figure 1. In the first two cases, the difference (shown modulo three) is seen to remain localized. In the second two cases, it grows progressively with time.

lattice. A remarkable empirical observation is that for most cellular automata these exponents are approximately equal in all directions, even those not along the axes of the lattice, and even for cellular automata with asymmetric rules [12]. Second, in a one-dimensional cellular automaton one may consider Lyapunov exponents for subsets of configurations, or for particular components of configurations. For example, for a cellular automaton in which a class 1 component involving sites with values 0 and 2 is superimposed on class 3 behaviour involving sites with values 0 and 1, the Lyapunov exponent is positive in the "value 1" direction, and negative in the "value 2" direction. In general it seems that the cellular automaton evolution induces a form of geometry on the configuration space [13]. But the details are unclear; one does not know, for example, the analogue of the tangent space considered in continuous dynamical systems.

Problem 4

What statistical quantities characterize cellular automaton behaviour?

There are several direct statistical measurements that can be made on cellular automaton configurations. Very simple examples are densities of sites or blocks of sites with particular values. Such densities are closely related to block entropies; their limit for large block sizes is the spatial entropy of the cellular automaton configurations, equal to the dimension of the Cantor set formed by the configurations (e.g. [5]). Another direct statistical measurement that can be made is of correlation

functions, which describe the interdependence of the values of separated sites [2]. For class 1 and 2 cellular automata, one expects that the correlation functions vanish beyond some critical distance. For class 3 cellular automata there are indications that the correlations functions typically fall off exponentially with distance. For class 4 cellular automata, the large distance part of the correlation function is dominated by propagating persistent structures, and many decrease slowly.

Power spectra or Fourier transforms provide other statistical measures of cellular automaton configurations. (Entirely discrete Walsh-Hadamard transforms [14] may be slightly more suitable.) Their form is not yet known. But many processes in cellular automata occur on a variety of spatial or temporal scales, so one expects definite structure in their transforms.

Beyond entropies and Lyapunov exponents, dynamical systems theory suggests that zeta functions may give a characterization of the global behaviour of cellular automata. Zeta functions measure the density of periodic sequences in cellular automaton configurations, and may possibly be related to Fourier transforms. The fact that the set of configurations generated from all possible initial states at a particular time step in the evolution of a cellular automaton forms a regular language (or "sofic system") implies that the corresponding zeta function is rational [15].

Problem 5

What invariants are there in cellular automaton evolution?

The existence of invariants or conservation laws in the evolution of a cellular automaton would imply a partitioning of its state space, much as energy provides a partitioning of the state space for Hamiltonian (energy-conserving) dynamical systems. For some class 1 and 2 cellular automata it is straightforward to identify invariants. In other cases, one can specifically construct cellular automaton rules that exhibit certain conservation laws [16–18]. For example, the cellular automata may evolve as if on several disjoint spatial lattices. Or it may support a set of persistent structures or "particles" that interact in simple ways. But in general, the identification of numerical invariants in cellular automata will probably be as difficult as it is in other non-linear dynamical systems.

It is nevertheless often possible to find partitionings of the state space for a cellular automaton that are left invariant by its evolution. The partitionings may be formed for example from sets of configurations corresponding to particular regular formal languages (cf. [7]). For example, the set of configurations with a particular period under a cellular automaton mapping is invariant, and in one dimension forms a finite-complement regular language (or "subshift of finite type"). Different elements in such partitionings may be considered to carry different values of what is often an infinite set of conserved quantities.

A particular cellular automaton rule usually evolves to give qualitatively similar behaviour from almost all initial states (each site is chosen to have each of the k possible values with equal probabilities). Often there are sets of initial states that occur with probability zero (for example, states in which all sites have the same value) that evolve differently from the rest. Such states may be distinguished by invariant or conserved quantities. But most initial states evolve to configurations with the same statistical properties. This suggests that even if the possible states could be partitioned according to the value of some invariant, they would essentially equivalent. It remains conceivable, however, that there exist cellular automata in which two sets of initial states that occur with nonzero probabilities could lead to two qualitatively different forms of behaviour.

Problem 6

How does thermodynamics apply to cellular automata?

Thermodynamics is supposed to describe the average overall behaviour of physical systems with many components. The microscopic dynamics of these systems is assumed to be reversible, so that the mapping from one state to another with time is invertible. Most cellular automata are irreversible, so that a particular configuration may arise from several distinct predecessors. However, a small subset of cellular automaton rules are bijective or invertible. Complete tables of invertible rules exist for $k = 2$, $r \leqslant 2$ [19, 20] and for $k = 3$, $r = 1$ [20], but in general no efficient procedure for finding such rules is known. Nevertheless, it is possible to construct particular classes of invertible rules [16, 21].

To apply thermodynamics one must also "coarse-grain" the system, grouping together many microscopically-different states to mimic the effect of imprecise measurements. Coarse-graining in cellular automata may be achieved by applying an irreversible transformation, perhaps a cellular automaton rule, to the cellular automaton configurations. A simple example would be to map the value of every other site to zero.

Coarse-grained entropy in reversible cellular automata should follow the second law of thermodynamics, and be on average non-decreasing with time. One may start from a set or ensemble of configurations with non-maximal coarse-grained entropy. The degrees of freedom that do not affect the coarse-grained entropy are undetermined, and are assumed to have maximal (fine-grained) entropy. In reversible class 2 cellular automata, the determined and undetermined degrees of freedom do not mix significantly with time, and the coarse-grained entropy remains essentially constant. But for class 3 and 4 cellular automata, the degrees of freedom mix, and the coarse-grained entropy increases towards its maximum possible value.

As in all applications of thermodynamics, the question arises of what coarse-graining prescriptions and ensembles of initial states are permissible. The initial states could for example be specially chosen so as to be the predecessors of a low coarse-grained entropy ensemble. The coarse-grained entropy would then decrease. Such examples do not seem physically reasonable. But it has never been clear exactly what mathematical criteria should be imposed to exclude them. One possibility is that one could require the coarse-graining procedure and the initial ensemble to be computationally simple (cf. [22]). If the cellular automaton evolution were computationally irreducible, then such a criterion could exclude ensembles obtained by reversing the evolution for many steps.

For the usual case of irreversible cellular automata, coarse-graining is usually of little consequence: the progressive contraction in the number of states generated by the cellular automaton evolution soon far outweighs the reduction associated with coarse-graining.

Problem 7

How is different behaviour distributed in the space of cellular automaton rules?

Random sampling yields some empirical indications of the frequencies of different classes of behaviour among cellular automaton rules of various kinds. For symmetric one-dimensional cellular automata, class 1 and 2 cellular automata appear to become progressively less common as k and r increase; class 3 becomes more common, and class 4 slowly becomes less common. In two-dimensional cellular automata, class 3 is overwhelmingly the most common; class 4 is very rare [12]. It seems that class 3 behaviour in any "direction" in the cellular automaton state space leads to overall class 3 behaviour. And as the number of degrees of freedom in the rules increases, the chance that this happens for one of the directions increases. For very large k and r a direct statistical treatment of the set of cellular automaton rules may well be possible.

There are many common features in the behaviour of cellular automata with apparently very different rules. It is not clear to what extent a direct equivalence exists between rules with qualitatively similar behaviour. In some cases, different rules may be related through invertible cellular automaton mappings. The nature of the equivalence classes of cellular automata generated in this way is presumably determined largely by the structure of the group of invertible cellular automaton mappings.

There are various ways to define distances in the space of cellular automaton rules. There are often cellular automata whose rules differ only slightly, but whose behaviour is very different. Nevertheless, it should be possible to find families of cellular automaton rules with closely related behaviour. For example, one may consider totalistic rules [5] in which the function that gives the new value of a site in terms of the sum of the old values in its neighbourhood is a discrete approximation to a function that involves a continuous parameter [23]. The behaviour of different cellular automaton rules obtained by changing this parameter may be compared with the behaviour found in iterated mappings of an interval of the real line (e.g. [24]) according to the same function. There are indications of a significant correspondence [23]. As the parameter is increased, regular periodic (class 2) cellular automaton behaviour can exhibit period doubling. Then as the parameter is further increased, chaotic (class 3) behaviour can occur. Class 4 seems to appear as an intermediate phenomenon.

Problem 8

What are the scaling properties of cellular automata?

Scaling transformations change the number of sites in a cellular automaton. Under such transformations, one cellular automaton rule may simulate another one. For example, if each site with value 0 is replaced by a pair of sites 00, and each 1 is replaced by 01, a new cellular automaton rule is obtained [2]. In some cases, this rule may have the same k and r as the original rule; in other cases it may not. The inverse transformation, in which 00 is replaced by 0, and 01 by 1, may be considered as a "blocking transformation" analogous to a block spin transformation (e.g. [25]), and yields a cellular automaton with fewer degrees of freedom. However, the transformation may be applied only to those special configurations in which just 00 and 01 site value pairs occur.

One may develop a network that shows the results of blocking transformations on rules of a particular kind, say with $k = 2$ and $r = 1$ [4, 26]. Some rules are found to be invariant under blocking transformations. Examples are the additive rules numbers 90 and 150 with $k = 2$ and $r = 1$. Patterns generated by these rules are thus scale invariant, so that they have the same form when viewed with different magnifications. If the initial configuration consists of a simple seed, say a single nonzero site, then regular scale-invariant patterns are obtained. These fractal patterns [27] have the property that pieces of them, when magnified, are indistinguishable from the whole pattern. (The fractal dimensions of the patterns are related to the parameters of the blocking transformations.) When the initial state is disordered, the patterns generated are instead statistically scale invariant, in the sense that their statistical properties are invariant under blocking transformations. So, for example, the pattern obtained by considering every site in the cellular automaton may have the same statistical properties as the pattern obtained by considering only every other site on every other time step.

Blocking transformations typically apply only to configurations that contain specific blocks in a given cellular automaton. So for example, different simple initial seeds in a cellular automaton may lead to rather different behaviour if they contain blocks that allow for different blocking transformations. Under certain blocking transformations, many of the $k = 2$, $r = 1$ cellular automata simulate the additive rules 90 or 150, which are invariant under blocking transformations. An initial state containing a single nonzero site is often one for which this simulation occurs, so that the pattern to which it leads is

self-similar, just as for rule 90 or rule 150. With more complicated initial states, however, patterns with different forms may be obtained.

Starting from a disordered initial state, in which all possible sequences of site values occur with equal probabilities, the irreversible evolution of many cellular automata leads to states in which only particular sequences actually occur. If these sequences correspond to those for which some blocking transformation applies, then the overall behaviour of the cellular automaton will be given by the result of this blocking transformation. In a typical case, a cellular automaton rule supports a number of "phases". Each phase consists of sequences to which some blocking transformation applies, and under which the cellular automaton beahves just like one with a different rule. So for example [28], in the $k = 2$, $r = 1$ rule number 18, sequences containing only 00 and 01, or only 00 and 10, constitute two phases with behaviour just like the additive rule 90. An arbitrary disordered state consists of a series of small domains, each in one of these phases, separated by "domain walls", consisting of 11 blocks. These domain walls execute approximately random walks with time, and annihilate in pairs, leaving larger and larger domains in a pure phase [28]. In two and higher dimensional cellular automata, the domains may have complicated geometrical structures [12]. The domain walls often behave as if they have a surface tension. When the surface tension is positive, the domains tend to become spherical. When the surface tension is negative, the domains take on a highly-convoluted labyrinthine form.

It seems that one may in general define a quantity analogous to free energy, or essentially pressure, for each possible phase in a cellular automaton. Domains containing phases with higher pressures typically expand linearly with time through domains with lower pressures, sometimes following biased random walks. The walls between domains with equal pressures typically execute unbiased random walks. After a long time, the phases with the highest pressure (or lowest free energy) dominate the behaviour of the cellular automaton, and thus determine the form of the limiting set of configurations. One may speculate that the phases that survive in this limit should be fixed points of the blocking transformation, and thus should exhibit some form of scale invariance. This is evident in some cases, where there are phases that behave say like rule 90. It is not clear how general the phenomenon is. If, however, it were widespread, then the overall large time behaviour of cellular automata would be dominated by fixed points of the blocking transformations, much as critical phenomena in spin systems are dominated by fixed points of the renormalization group or block spin transformation. Then there would be a universality in the properties of the many different cellular automata attracted to a particular fixed point rule. (So far the only fixed points of the blocking transformation that have been found are additive rules, but one suspects that not all fixed point rules need in fact be additive.) The spatial measure entropies for the different cellular automata would for example presumably then be related by simple rational factors.

One rule whose scaling properties remain unclear is the $k = 2$, $r = 1$ rule number 22. This rule simulates rule 90 under the blocking transformation $0000 \rightarrow 0$, $0001 \rightarrow 1$, and its rotated equivalents. But the simulation is not an attractive one: starting from a disordered initial state, domains of these phases do not grow. It may be possible to describe the configurations obtained as domains of phases corresponding to some other blocking transformation. A generalization of blocking transformations may be required. One may consider a blocking transformation as a translation from one formal language to another. In simple cases, such a translation may be achieved with a finite automaton that reads symbols sequentially from the "input" configuration, and writes symbols into the "output" configuration according to the internal state that it reaches. Blocking transformations that consist of simple substitutions correspond to very simple finite automata of this kind. More complicated finite automata may be necessary to describe phases in cellular automata such as rule number 22. In general, the irreversible nature of most cellular automata implies that only a subset of possible configurations are generated with time. As a consequence, only certain neighbourhoods of site values may appear, so that some of the elements of the cellular automaton rule are never used, and a different rule would give identical results.

The description of cellular automaton configurations in terms of domains of different phases is related to a description in terms of "elementary excitations". Just as for a spin system, one may consider decomposing a cellular automaton configuration into a "ground state" part, together with "phonons" or excitations. The excitations may for example correspond to domain walls. Or they could be persistent structures in class 4 cellular automata. But if their interactions are comparatively simple, then they can be used to provide an overall description of the cellular automaton behaviour, and can perhaps allow for example a computation of entropies.

Problem 9

What is the correspondence between cellular automata and continuous systems?

Cellular automata are discrete in several respects. First, they consist of a discrete spatial lattice of sites. Second, they evolve in discrete time steps. And finally, each site has only a finite discrete set of possible values.

The first two forms of discreteness are addressed in the numerical analysis of approximate solutions to, say, differential equations. It is known that so long as a "stable" discretization is used, the exact continuum results are approximated more and more closely as the number of sites and the number of time steps is increased. It is possible to devise cellular automaton rules that provide approximations to partial differential equations in this way. In the simplest cases, however, the approximations are of the Jacobi, rather than the Gauss-Seidel kind, in that the algorithm for calculating new site values uses the old values of all the neighbours, rather than the new values of some of them. This can lead to slow convergence and instabilities in some cases.

The third form of discreteness in cellular automata is not so familiar from numerical analysis. It is an extreme form of round-off, in which each "number" can have only a few possible values (rather than the usual say 2^{16} or 2^{32}). It is not clear what aspects of, say, differential equations are preserved in such an approximation. However, preliminary studies in a few cases suggest that the overall structure of solutions to the equations are remarkably insensitive to such approximations. If the cellular automaton approximates for example a continuous field, then the value of the field at a particular point could correspond roughly to the density of say nonzero sites

around that point: the values of individual field points would be represented in a distributed manner, just as they often are in actual physical systems. Explicit examples of cellular automaton approximations to partial differential equations of physical importance would be valuable.

There are some aspects of nonlinear differential equations that may well have rather direct analogues in cellular automata. For example, the persistent propagating structures found in class 4 cellular automata may well be related to solitons in nonlinear differential equations, at least in their solitary persistence, if not in their interactions. Similarly, the overall topological forms of some of the patterns generated by two and higher dimensional cellular automata [29] may correspond to those generated say by reaction-diffusion equations [30]. Moreover, many highly-nonlinear partial differential equations give solutions that exhibit discrete or cellular structure on some characteristic length scale (e.g., [31]). The interactions between components in the cellular structure cannot readily be described by a direct discretization of the original differential equation, but a cellular automaton model for them can be constructed.

Continuum descriptions may be given of many of the large-scale structures that occur in cellular automata. For example, the motion of domain walls between phases may be described by diffusion-like differential equations. A very direct continuum approximation to a cellular automaton is provided by a mean field theory, in which only the average density of sites, and not their individual values, is considered [2]. Presumably in the limit of large spatial dimensionality, this approximation should become accurate. But in one or two dimensions, it is usually quite inadequate, and gives largely misleading results. Large-scale phenomena in cellular automata occur as collective effects involving many individual sites, and the particular rules that relate the values of these sites are significant.

Problem 10

What is the correspondence between cellular automata and stochastic systems?

Cellular automata satisfy deterministic rules. But their initial states can have a random form. And the patterns they generate can have many of the properties of statistical randomness. As a consequence, the behaviour of cellular automata may have a close correspondence with the behaviour of systems usually described by basic rules that involve noise or probabilities. So for example domain walls in cellular automata execute essentially random walks, even though the evolution of the cellular automaton as a whole is entirely deterministic. Similarly, one can construct a cellular automaton that mimics say an Ising spin system with a fixed total energy (microcanonical ensemble) [32]. Apparently random behaviour occurs as a consequence of randomly-chosen initial conditions, just as in many systems governed by the deterministic laws of classical physics.

Even models that involve explicit randomness are in practice simulated in computer experiments using pseudorandom sequences generated by some definite algorithm. These sequences are not unlike the sequences of site values produced by many cellular automata. In fact, the linear feedback shift registers often used in practice to produce pseudorandom sequences are exactly equivalent to certain additive cellular automata (cf. [33]). Empirical evidence suggests that the properties of many supposedly stochastic models are quite insensitive to the detailed form of the randomness used in their simulation. It should be possible to find entirely deterministic forms for such models, based say on cellular automata. One expects in general that just as with algorithms say for primality testing the fundamental capabilities of stochastic and deterministic models should be equivalent.

Problem 11

How are cellular automata affected by noise and other imperfections?

Many mathematical approaches to the analysis of cellular automata make essential use of their simple deterministic structure. One must find out to what extent results for the overall behaviour of cellular automata are changed when imperfections are introduced into them. The imperfections can be of several kinds. First, the cellular automaton rules can have a probabilistic element (e.g., [17, 34, 35]). Then for example each site may be updated at each time step according to one rule with probability p, and according to another rule with probability $1 - p$. A second class of imperfections modifies the homogeneous cellular automaton lattice. One may for example take different sites to follow different rules. Or one may take the connections that specify the rules on the lattice to be different at different sites. In an ordinary cellular automaton, the values of all the sites are updated simultaneously, using the previous values of the sites in their neighbourhoods. One may consider the effect of deviations from this synchronization, allowing different sites to be updated at different times [36]. Finally, each site is usually taken to have a discrete set of possible values. One could instead allow the sites to have a continuum of values, but take the rules to be continuous functions with sharp thresholds.

Several classes of models can be considered as imperfect cellular automata. Directed percolation is directly analogous to certain cellular automata in the presence of noise [35]. The patterns generated with time by noisy cellular automata also correspond to the equilibrium configurations of spin systems at finite temperature [35]. And if inhomogeneities are introduced into the cellular automata, they give spin glass configurations. When nonlocal connections and asynchronous updates are introduced, models analogous to Boolean or neural networks are obtained (e.g., [37]).

Even an arbitrarily small imperfection in a cellular automaton can have a large effect at arbitrarily large times. However, small imperfections very often do not affect the overall behaviour of a cellular automaton. There is often a critical magnitude of imperfection at which essentially a phase transition occurs, and the behaviour of the cellular automaton changes suddenly. One can presumably find such transitions as a function of noise and other imperfections in many different cellular automata (cf. [34, 35]). Often the transitions should be associated with critical exponents; one expects that several universality classes may be identified. Note that even one-dimensional cellular automata can exhibit phase transitions at nonzero values of imperfection parameters if imperfections are introduced in such a way that for example certain initial states still evolve as they would without the imperfections.

Given a pattern generated by a cellular automaton with imperfections, as might be obtained in a physical experiment, one may consider how the basic cellular automaton rule could be deduced. One could lay down a definite grid, and then accumulate histograms of the new site values obtained with all

neighbourhoods, and thereby deduce the cellular automaton rule (it will not necessarily be unique, since certain neighbourhoods may never appear) [13]. This procedure accounts for imperfections due to noise, but not for imperfections such as deformations of the lattice. It appears that an iterative optimization approach must be used to treat such imperfections.

Problem 12

Is regular language complexity generically non-decreasing with time in one-dimensional cellular automata?

The sets of configurations generated by cellular automaton evolution, starting say from all possible initial states, can be considered as formal languages. Each configuration corresponds to a word in the language, formed from a sequence of symbols representing site values, according to a definite set of grammatical rules. For one-dimensional cellular automata, it can be shown that the set of configurations generated after any finite number of time steps forms a regular formal language [7]. Thus the configurations correspond to the possible paths through a finite directed graph, whose arcs are labelled by the values that occur at each site. There is an algorithm to find the graph with the minimal number of nodes that represents a particular regular language [8, 38], in such a way that each word in the language corresponds to a unique path through the graph (deterministic finite automaton). This minimal graph provides a complete canonical description of the set generated by the cellular automaton evolution. From it properties such as topological entropy may be deduced. The entropy is in fact given by the logarithm of the largest eigenvalue of the adjacency matrix for the graph, which is an algebraic integer.

One characteristic of a regular language is the total size or number of nodes Ξ in its minimal graph. This quantity can be considered as a measure of the complexity of the regular language. The larger it is, the more complicated a subset of the space of possible symbol sequences the language corresponds to. Ξ gives in a sense the size of the shortest description of this subset, at least in terms of regular languages. The value of Ξ is in general bounded above by $2^{k^{2rt}} - 1$. The empirical studies done so far suggest that for class 1 and 2 cellular automata, Ξ in fact becomes constant after a few time steps, or increases at most as a polynomial with t. For most class 3 and 4 cellular automata, however, Ξ appears to increase rapidly with time, though it usually stays far below the upper bound. There are a few cases where Ξ decreases slightly at a particular time step, but in general it seems that Ξ is usually non-decreasing with time. If this is indeed a general result, it gives a quantitative form to the qualitative statement that complexity seems to increase with time. It could be a principle for self-organizing systems analogous in generality but complementary in content to the law of entropy increase in thermodynamic systems.

If the non-decrease of Ξ is indeed a general result, then it should have a simple proof that depends on few of the properties of the system considered. A crucial property of cellular automata may be irreversibility, which leads to a progressive contraction in the set of configurations generated. As a consequence of this contraction, the set generated at each time step must correspond to a different regular language. But there are only a limited number of regular languages with complexities less than any particular value, and so the complexity of the language generated must increase, albeit slowly, with time. To find a complete bound, one must study the structure of the space of possible regular languages. It is clear that the number of regular languages of complexity Ξ is less than the number of labelled directed graphs with Ξ nodes, $2^{k\Xi^2}$. The minimal graph for a regular language must have a trivial automorphism group; but the number of graphs with a given automorphism group does not appear to be known (e.g., [39]). Beyond the total number of regular languages, one may consider the network that represents the containment of regular languages, divided into zones of different Ξ. One suspects that this network is close to a tree, with a number of nodes increasing perhaps exponentially with depth Ξ.

Problem 13

What limit sets can cellular automata produce?

Not all possible sets of configurations can be produced as limit sets of cellular automata. For the number of distinct cellular automaton rules, while infinite, is countable. Yet the number of possible sets of configurations is uncountable.

At each step in the evolution of an irreversible cellular automaton, a new set of configurations is excluded. The limit set consists of those configurations that are never excluded. The set of all excluded configurations is recursively enumerable, since each of its elements is found by a finite computation. Thus the limit sets for cellular automata are always the complements of recursively enumerable (co-r.e.) sets, and are therefore countable in number. Nevertheless, not every co-r.e. set is the limit set for a cellular automaton: one additional condition is that they must be translationally invariant. For example, cellular automaton limit sets must contain either one configuration, or an infinite number of distinct configurations, and cannot consist of some other finite number of configurations [40]. Not every possible real number value of dimension or entropy can be realized by cellular automata; but the set that is realized presumably includes some values that are non-computable.

After any finite number of time steps, the set of configurations generated by a one-dimensional cellular automaton forms a regular formal language. For some cellular automata (essentially those in classes 1 and 2), the limit set is also a regular language. But in other cases, the limit set probably corresponds to a more complicated formal language. Explicit examples are known in which context-free and context-sensitive languages are obtained as limit sets [40]. In addition, cellular automata that are capable of universal computation can generate limit sets that are not recursive [40]. The generic behaviour is however not known: some more examples would be valuable.

When the limit set forms a regular language, the simplest description of it, in terms of a regular grammar or graph, can be found by a finite algorithm. The size Ξ of this description can be used as a measure of the complexity of the set. However, for languages more complicated than regular ones, there is in general no finite algorithm to find the simplest grammar (e.g., [8]). The size of such a minimal grammar is thus formally non-computable. One may test a sequence of grammars, but the languages to which they lead cannot in general be enumerated by a computation of any bounded length.

Minimum grammar size is thus not a useful measure of complexity for complicated cellular automaton limit sets. Some other measure must be found. And in terms of this measure,

one should be able to determine how the complexity of the behaviour of a cellular automaton, as revealed by the structure of its limit set, depends on the complexity of its local rule, or the values of k and r.

One may wonder what features of the local rule for a cellular automaton determine its global properties, and the structure of its limit set. Some simple observations may be made. For example, unless the local rule contains elements that give value 1 with neighbourhoods such as 001, no information can propagate in the cellular automaton, and class 1 or 2 behaviour must occur. But in general one expects that the problem is undecidable: the only way to determine many of the limiting properties of a cellular automaton is probably by explicit simulation of its evolution, for an infinite time.

As a practical matter, one may ask whether cellular automaton rules may be constructed to yield particular limit sets (cf. [41]), so that their evolution serves to filter out the components that appear in these limit sets. It is probably possible to construct cellular automata that yield any of some class of regular languages as limit sets. But one suspects that a construction for more complicated limit sets can be carried out only in very special cases.

Problem 14

What are the connections between the computational and statistical characteristics of cellular automata?

The rate of information transmission is one attribute of cellular automata that potentially affects both computational and statistical properties. On the statistical side, the rate of information transmission gives the Lyapunov exponent for the cellular automaton evolution. Class 1 and 2 cellular automata have zero Lyapunov exponents, so that information almost always remains localized, and the value of a particular site at any time can almost always be determined from the initial values of a bounded neighbourhood of initial sites. As a consequence, the limit sets for one-dimensional such cellular automata correspond to regular languages. The configurations can thus be generated by an essentially Markovian process, in which there are no long-range correlations between different parts.

Class 3 and 4 cellular automata have positive Lyapunov exponents, so that a small initial change expands with time. The value of a particular site after many time steps thus depends in general on an ever-increasing region in the initial state. The limit sets for such cellular automata can thus involve long-range correlations, and need not correspond to regular languages. If class 4 cellular automata are generically capable of universal computation, then their limit sets should be unrestricted, in general non-recursive, formal languages. Some arguments can be given that class 3 cellular automata should yield limit sets that correspond to context-sensitive languages. In general, one suspects that dynamical systems that exhibit chaotic behaviour characterized by positive Lyapunov exponents should yield limit sets that are more complicated than regular languages.

When the limit set for a cellular automaton is a regular language, its spatial entropy can be computed, and is given by the logarithm of an algebraic integer. If the limit set is a context-free language, then it seems that the entropy is always the logarithm of some algebraic number. But for context-sensitive and more complicated languages, the entropy is in general non-computable. It may thus be common to find class 3 and 4 cellular automata for which the entropy of their limit sets is non-computable.

The computational structure of sets generated in the evolution of two and higher dimensional cellular automata can be very complicated even after a finite number of time steps. In particular, while in one-dimensional cellular automata the set of configurations that can be generated at any finite time forms a regular formal language, this set can be non-recursive in two-dimensional cellular automata [12, 42]. The essential origin of this difference is that there is an iterative procedure to find the possible predecessors of arbitrarily long sequences in one-dimensional cellular automata, but no such procedure exists for two-dimensional cellular automata. In fact, even the problem of finding configurations that evolve periodically in time in a two-dimensional cellular automaton appears to be equivalent to the domino tiling problem, which is known to be formally undecidable [43]. Nevertheless, it seems likely that only two-dimensional cellular automata in which information transmission can occur throughout the plane, as revealed by positive Lyapunov exponents in all directions, exhibit such complications, and give non-recursive sets at finite times.

The grammar for a formal language specifies which sequences occur in the language, but not how often they occur. It does not for example distinguish sequences that occur with zero probability from those that occur with positive probability. However, it is the probable, rather than the possible, behaviour of cellular automata that is most significant in determining their statistical properties, such as Lyapunov exponents and measure entropies. There are class 1 and 2 cellular automata in which a set of states of measure zero yields class 3 behaviour: this is irrelevant in the Lyapunov exponent or the measure entropy, but affects the topological entropy, and the structure of the grammar for the limit set. One should construct formal languages that include probabilities for configurations. A suitable approach may be to consider stochastic automata, closely related to standard Markov chains.

Problem 15

How random are the sequences generated by cellular automata?

The spatial sequences obtained after a finite number of steps in the evolution of a one-dimensional cellular automaton starting from all possible initial states are known to form a regular formal language. But no such characterization is known for the temporal sequences generated by cellular automata. At least for cellular automata capable of universal computation, these sequences can be non-recursive. But the generic behaviour is not known, and no non-trivial examples have yet been given.

One question is to what extent the initial state of a cellular automaton can be reconstructed from a knowledge of the time series of values of a few sites. An essentially equivalent question is how wide a patch of sites need to be considered to compute the invariant entropy of the cellular automaton mapping. When the mapping is surjective and expansive (so that roughly information transmission occurs at a posivie rate), only a finite width is required (e.g., [44]). Nevertheless, the transformation necessary to find the initial state from the temporal sequence may be very complicated. In particular, there may be effectively no better method than to try all exponentially many possible initial states. Temporal sequences in cellular automata are thus

candidates for use in pseudorandom number generation and in cryptography [20].

The patterns generated by some cellular automata evolving from initial states consisting of simple seeds have a simple form. They may be asymptotically homogeneous, or may correspond to regular fractals. But many cellular automata yield complicated patterns even starting from an initial state as simple as a single nonzero site. Some examples are shown in Fig. 2. It is remarkable that such complicated and intricate patterns can be generated in such a simple system.

Often the temporal sequences that appear in these patterns have a seemingly random form, and satisfy many statistical tests for randomness. There is empirical evidence that in many cases the sequence of values taken on say by the centre site in the pattern contains all possible subsequences with equal frequences, so that the whole sequence effectively has maximal measure entropy. A simple example of this phenomenon occurs in the $k = 2$, $r = 1$ rule number 30 $(a_i^{(t+1)} = a_{i-1}^{(t)} \oplus \max (a_i^{(t)}, a_{i+1}^{(t)}))$.

Systems that exhibit chaotic behaviour usually start from initial conditions that contain an infinite amount of information, either in the form of an infinite sequence of cellular automaton site values, or the infinite sequence of digits in a real number. Their irregular behaviour with time can then be viewed as a progressive excavation of the initial conditions. The chaotic behaviour seen in Fig. 2 is however of another kind: it occurs as a consequence of the dynamics of the system, even though the initial conditions are simple. It may well be that this kind of chaos is central to physical phenomena such as fluid turbulence.

It is important to investigate the mathematical bases for such behaviour. The closest analogies seem to lie in number theory. The integers generated for example by repeated application of a linear congruence transformation form a pseudorandom sequence (e.g., [45]), often used in practical applications. The linearity of this system makes it amenable to a rather complete number theoretical analysis, which provides formulae for computing the nth integer in the sequence directly from the original seed, with working out all the intermediates. It seems likely that such analyses, and the resulting short cuts, are not possible in most nonlinear cellular automata. The randomness produced in these systems may be more like the randomness say of the digits of π. In some cases it is in fact possible to cast essentially number theoretical problems in terms of questions about patterns generated by cellular automata. One example concerns the sequence of leading binary digits in the fractional parts of successive powers of $3/2$ [46]. There is empirical evidence that all possible blocks of digits occur in this sequence, so that in a sense it has maximal entropy. The sequence corresponds to the time series of values of the central site in the pattern generated by a particular cellular automaton from a simple initial state.

Problem 16

How common are computational universality and undecidability in cellular automata?

If a system is capable of universal computation, then with appropriate initial conditions, its evolution can carry out any finite computational process. A computationally universal system can thus mimic the behaviour of any other system, and so can in a sense exhibit the most complicated possible behaviour.

Several specific cellular automata are known to be capable of universal computation. The two-dimensional nearest-neighbour cellular automaton with two possible values at each site known as the "Game of Life" has been proved computation universal [47]. The proof was carried out by showing that the cellular automaton could support structures that correspond to all the components of an idealized digital electronic computer, and that these components could be connected so to implement any algorithm. Some one-dimensional nearest-neighbour cellular automata with $k = 18$ have been shown to be computationally equivalent to the simplest known universal Turing machines, and are thus capable of universal computation [48].

One speculates that cellular automata identified on statistical grounds as class 4 are in fact generically capable of universal computation. This would imply that there exist one-dimensional computationally universal cellular automata in cases as simple as $k = 2$, $r = 2$ or $k = 3$, $r = 1$. But it remains to prove the computational universality of any particular such rule. Several methods could be used for such a proof. One is to identify a set of persistent structures in the cellular automaton that could act as the components of digital computer, or like combinations of symbols and internal states for a Turing machine. Structures that remain fixed, propagate, and interact in various ways have been found. A structure that can act as a "clock", producing an infinite sequence of "signals", has not yet been found in such cellular automata. Another method of proving universality would be a direct demonstration that this cellular automaton rule could simulate any other cellular automaton rule with an appropriate encoding of initial states. Blocking transformations may provide the necessary encodings: so one must find out whether a particular cellular automaton rule is connected to all others in the simulation networks constructed from blocking transformations.

If class 4 cellular automata are indeed capable of universal computation, then the capability for universal computation is quite common among one-dimensional cellular automata. Class 4 behaviour is however much rarer in two dimensional cellular automata–the "Game of Life" is almost the only known example (cf. [12]).

There may well be cellular automata whose behaviour is usually computationally simple, but which with very special initial states can perform arbitrary computations. It is certainly possible to construct cellular automata in which universal computation occurs only with initial states in which say every other site has value zero (cf. [49]), a condition that occurs in disordered states with probability zero. Such phenomena may be common in class 3 cellular automata.

Any predictions about the behaviour of a cellular automaton must be made by performing some computation. But if the cellular automaton is capable of universal computation, then this computation must in general reduce to a direct simulation of the cellular automaton evolution. So questions about the infinite time limiting behaviour of cellular automata may require infinite computations, and therefore be formally undecidable.

For example, one may consider the question of whether the patterns generated from particular finite initial seeds ever die out in the evolution of the cellular automaton. One may simulate the evolution explicitly to find out whether a pattern dies out after say a thousand time steps; but to determine its ultimate fate in general requires a computation of unbounded length.

The question is therefore formally undecidable.

The set of finite configurations that evolve to the null configuration after a fixed finite time can be specified by a regular formal language (cf. [50]). But there is no such finite specification for the set of finite configurations that evolve after any time to the null configuration. Even the fraction of configurations in this set is in general a non-computable number.

A similar problem is to determine whether a particular finite sequence of site values occurs in any configurations in the limit set for a cellular automaton. Again this problem is in general undecidable [40]. An explicit finite calculation can show that a sequence is forbidden after say three time steps. But a particular sequence may only be forbidden after some arbitrarily large number of time steps. In a one-dimensional cellular automaton, the length $L^{(t)}$ of the shortest sequence newly excluded at a given time step in the evolution is bounded by $L^{(t)} \geqslant L^{(t-1)} - 2r$. In most actual examples $L^{(t)}$ seems to increase monotonically with time, so that the exclusion of a particular finite sequence must occur before some predictable finite time. But in some cases $L^{(t)}$ is not monotonic, and the occurrence of particular sequences may be undecidable.

The capability for universal computation can be used to establish the undecidability of questions about the behaviour of a system. But undecidability can occur even in systems not capable of full universal computation. For example, one may arrange to disable all computations that give results of a certain form. In this way, the system fails to be able to perform arbitrary computations. Nevertheless, there may be undecidable questions about the class of computations that it still can perform. These may well occur in cellular automata. Proofs of undecidability usually use a diagonal argument based essentially on universal computation. To establish undecidability in a system not itself capable of universal computation, one must usually find another system that is capable of universal computation, and show that a reduction of its capabilities does not affect undecidability.

Rice's theorem states that almost all questions about an arbitrary recursively-enumerable set are undecidable (e.g., [8]). However, it may be that natural or simple questions, which can be stated in say a few logical symbols, are usually decidable. So for example the halting of all simple initial seeds in a particular cellular automaton might be easy to determine, and it might only be very large and specially-chosen initial seeds whose halting was difficult to determine. There are certainly examples in which the halting problem appears to be difficult to answer even for simple seeds. One must establish in general not only whether there are any undecidable propositions about the behaviour of a particular cellular automaton, but whether simple propositions about it are in fact undecidable.

Problem 17

What is the nature of the infinite size limit for cellular automata?

Statistical averages in many systems converge to definite values when the infinite size or thermodynamic limit is taken. Several complications can however arise in cellular automata.

Different seeds can lead to very different behaviour in class 4 cellular automata. Some may die out; others may yield periodic patterns; still others may produce propagating structures. Propagating structures usually involve at least five or ten sites, and appear only with seeds of such a size. One expects that when larger seeds are used, new kinds of structures can begin to occur. For example, there may be structures that periodically generate propagating patterns, giving an asymptotically infinite number of nonzero sites. If the cellular automaton is capable of universal computation, then it should support structures with arbitrarily complicated behaviour. So for example there may be self-reproducing structures, which replicate even in the presence of a disordered background. Any such structure present in an initial state would yield offspring that could eventually dominate the behaviour of the system. In a given class 4 cellular automaton, the simplest self-reproducing structure may have a size of say 100 sites. The density at which the structure would occur in a disordered state is then k^{-100}. So in practical simulations, there is an overwhelming probability that no such structure would ever been seen. But if configurations of size much larger than k^{100} were considered such a structure would occur in almost every case. And after a long time, the behaviour of the system would almost always be dominated by the self-reproducing structures. Statistical results obtained with smaller configurations would then be misleading. And as the idealized limit of infinite size is taken, more and more complicated phenomena may occur, and statistical quantities have no simple limits.

Since a finite description in terms of regular formal languages can be given for the set of configurations generated at any finite time in the evolution of a one-dimensional cellular automaton, definite infinite size limits for statistical quantities presumably exist in this case. With time the limits may however become more complicated, and be reached more slowly. One expects that most statistical quantities will continue to show simple behaviour for class 3 cellular automata. But for class 4 cellular automata, in which different structure appears to be manifest on every different scale, the limits may become progressively more complicated, and may not exist at infinite times.

Two-dimensional cellular automata exhibit complicated infinite size limits even after a finite number of time steps. The sets of configurations that they generate can be non-recursive in the infinite size limit [12, 42], and some statistical quantities may have no limits as a consequence.

It is in general undecidable how large the smallest structure with some property such as self-reproduction can be in a particular cellular automaton. In some cases, the cellular automaton rule may be specially constructed to allow such structures. But for simple rules, one is reduced to an essentially experimental search for the structures. In several class 4 one-dimensional clelular automata with $k = 2$, all configurations of less than 21 sites have been tested, and all those up to about 30 sites are probably accessible with special-purpose computer hardware [51]. In the Game of Life, a number of complex structures were found through extensive experimentation. Further examples, particularly in one-dimensional cellular automata, would be valuable. One may imagine that each capability such as self-reproduction has a logical description of some length. Then the size of the smallest configuration that has the capability may be related in some way to this length. Obviously particular cellular automata may have special properties with respect to particular capabilities, but the result may hold as some average over all possible capabilities. If so, the very large number of particles in the universe could be essential for very complex physical and biological phenomena to occur.

For direct simulation and other practical purposes one is often concerned with cellular automata of finite size. When an infinite size limit exists, the local properties deduced from studies of finite cellular automata are likely to correspond directly with the infinite size case. But for global properties the correspondence is less clear. For the rather special case of finite cellular automata with additive rules, algebraic methods provide a complete description of the state transition diagram [33]. There are typically about $k^{N/2}$ cycles, each of length about $k^{N/2}$ steps. The cycles are reached after transients of length less than N. In the limit $N \to \infty$, the system exhibits chaotic behaviour, but the mapping is surjective, so that all configurations are generated. Presumably in this limit there are an infinite number of infinite cycles, perhaps each characterized by a particular form of some invariant algebraic function. In general, some cellular automata that show chaotic behaviour in the infinite size limit exhibit exponentially long cycles at small finite sizes. Others exhibit exponentially long transients. Some show neither. The general connections between the structure of finite state transition diagrams, and the behaviour of cellular automata in the infinite size limit remain to be established.

Problem 18

How common is computational irreducibility in cellular automata?

One way to find out the behaviour of a cellular automaton is to simulate each step in its evolution explicitly. The question is how often there are better ways.

Cellular automaton evolution can be considered as a computation. A procedure can short cut this evolution only if it involves a more sophisticated computation. But there are cellular automata capable of universal computation that can perform arbitrarily sophisticated computations. So at least in these cases no short cut procedure can in general be found. The cellular automaton evolution corresponds to an irreducible computation, whose outcome can be found effectively only by carrying it out explicitly.

A number of complications arise in giving a precise definition of such computational irreducibility. In general one should compare the number of steps in the evolution of a system such a cellular automaton with the number of steps required to reproduce the evolution using another computational system. However, by making the computational system more complicated, it is always possible to reduce the number of steps required by an arbitrary constant factor, or even an arbitrary function. For example, if a computer can apply the square of a cellular automaton mapping at each step, then it can always simulate T steps of cellular automaton evolution in $T/2$ steps.

Nevertheless, no amount of additional complication in the computer can allow it to find in a finite time the outcome of an infinite number of steps in the evolution of a cellular automata that is for example capable of universal computation. As a consequence, there are undecidable propositions about the ultimate behaviour of the cellular automaton. The occurrence of such undecidable propositions may be viewed as a consequence of computational irreducibility. But to give a complete definition of computational irreducibility for finite time processes, one must in some way exclude arbitrary complication in the computer used for predictions.

One approach is to consider finite cellular automata and to

use methods from computational complexity theory. A cellular automaton with N sites can evolve for a time up to k^N before retracing its steps. The computation corresponding to this evolution is performed in a bounded space, and is therefore in the class *PSPACE* (e.g., [8]), but it can take a time exponential in N. However if the computation were reducible, then it could be possible to find the outcome of the evolution in a time polynomial in N, or in other words to reduce the problem to one in the class *P*. It is believed that $PSPACE \neq P$, so that there exist problems that can be solved in polynomial space that cannot be solved in polynomial time. Determining the outcome of the evolution of some cellular automata may be a problem of this kind (cf. [52]).

Conventional computational complexity theory concerns computations in finite systems. It may well be that the definition of computational irreducibility for cellular automata can be sharpened in the infinite size limit.

The evolution of class 1 and 2 cellular automata yielding periodic configurations is computationally reducible. But one suspects that the evolution of most class 3 and 4 cellular automata is computationally irreducible. In fact, it may well be in general that most systems that show apparently complex or chaotic behaviour are computationally irreducible.

Even if the detailed behaviour of a system can effectively be found only by direct simulation, it could be that many of its overall properties can be found by more efficient procedures. It is this possibility that makes investigations of cellular automata worthwhile even when computational irreducibility is present. But what should be done is to find a characterization of those properties whose behaviour can be found by efficient methods, and those for which computational irreducibility makes explicit simulation the only possible approach, and precludes a simple description.

Problem 19

How common are computationally intractable problems about cellular automata?

Questions concerning the finite time behaviour of finite cellular automata can always be answered by finite computations. But as the phenomenon of computational irreducibility suggests, there may be questions for which the computations are necessarily very long. One may consider for example the question of whether a particular sequence of X site values can occur after T time steps in the evolution of a one-dimensional cellular automaton, starting from any initial state. Then one may ask whether there exists any algorithm that can determine the answer in a time given by some polynomial in X and T. The question can certainly be answered by testing all k^{X+2rT} sequences of initial site values that determine the length X sequence, but this procedure requires a time that grows exponentially with X and T. Nevertheless, if an initial sequence could be guessed, then it could be tested in a time polynomial in X and T. As a consequence, the problem is in the class *NP*. Now if $P \neq NP$, then there may be no polynomial time algorithm for the problem, and the best method of solution may essentially be to try all the exponentially many possible cases explicitly, so that the problem rapidly becomes intractable. In the infinite time limit, the analogous problem is in general undecidable.

Just as undecidability in a system can be proved by establishing a capability for universal computation, so, assuming $P \neq NP$,

computational intractability can be proved through *NP*-completeness. A problem is *NP*-complete if specific instances of its correspond to arbitrary problems in the class *NP* [8, 53]. This can be shown by establishing equivalence to a known *NP*-complete problem. Thus for example it has been possible to give a specific example of a cellular automaton in which the problem of determining whether particular sequences can occur after T time steps is equivalent the *NP*-complete problem of finding a set of truth values for variables so that a particular logical expression is satisfied [54]. How widespread *NP*-completeness is in problems concerning cellular automata has yet to be established. But one suspects that it is common in many class 3 and 4 systems.

Problem 20

What higher-level descriptions of information processing in cellular automata can be given?

Cellular automaton evolution can in principle carry out arbitrary information processing. An important problem of theory and practice is to find a way of organizing this information processing. In specific cases one can devise cellular automaton rules that allow particular computations to be carried out (e.g. [55]). Or one can identify within a cellular automaton structures that can interact so as to mimic the components of conventional digital computers. But all these approaches are strongly based on analogues with conventional serial-processing computers. Information processing in cellular automata occurs however in a fundamentally distributed and parallel fashion, and one must invent a new framework to make use of it. Such a framework would likely be valuable in studying the many physical systems in which information processing is also distributed.

One approach is statistical in nature. It consists in devising and describing attractors for the global evolution of cellular automata. All initial configurations in a particular basin of attraction may be thought of as instances of some pattern, so that their evolution towards the same attractor may be considered as a recognition of the pattern. This approach is probably effective when the basins of attraction are local in space, as in image processing (e.g., [56]). But the construction of attractors for more general problems is likely to be very difficult. An attempt in this direction might be made by considering basins of attraction as sets of sequences corresponding to particular formal languages (cf. [50]).

Another approach is to use symbolic representations for various attributes or components of cellular automaton configurations. But the structures used in conventional computer languages are largely inappropriate. The definite organization of computer memory into named areas, stacks, and so one, is not suitable for cellular automata in which processing elements are not distinguished from memory elements. Rather perhaps data could be represented by an object like a graph, on which transformations can be performed in parallel. But the simple organizing principles that are required still remain to be found. It seems likely that a radically new approach is needed [57].

Acknowledgements

I have benefitted from discussions about cellular automata with many people, too numerous to list here. For recent discussions I am particularly grateful, to: C. Bennett, M. Feigenbaum, E. Fredkin, D. Hillis, L. Hurd, L. Kadanoff, D. Lind, O. Martin, J. Milnor, N. Packard, D. Ruelle, R. Shaw, and K. Steiglitz.

References

1. Wolfram, S., Nature **311**, 419 (1984).
2. Wolfram, S., Rev. Mod. Phys. **55**, 601 (1983).
3. Farmer, D., Toffoli, T. and Wolfram, S. (editors), "Cellular automata: proceedings of an interdisciplinary workshop", Physica **10D** numbers 1 and 2 (1984), North-Holland Publishing Co. (1984).
4. Wolfram, S., "Cellular automata", Los Alamos Science (fall 1983).
5. Wolfram, S., Physica **10D**, 1 (1984).
6. Guckenheimer, J. and Holmes, P., Nonlinear Oscillations, Dynamical Systems, and Bifurcations of Vector Fields, Springer-Verlag (1983).
7. Wolfram, S., Comm. Math. Phys. **96**, 15 (1984).
8. Hopcroft, J. E. and Ullman, J. D., Introduction to Automata Theory, Languages, and Computation, Addison-Wesley (1979).
9. Wolfram, S., "Computer software in science and mathematics", Scientific American (September 1984).
10. Packard, N., "Complexity of growing patterns in cellular automata", Institute for Advanced Study preprint (October 1983).
11. Milnor, J., "Entropy of cellular automaton-maps". Institute for Advanced Study preprint (May 1984).
12. Packard, N. and Wolfram, S., "Two-dimensional cellular automata", to be published in J. Stat. Phys.
13. Packard, N., Private communication.
14. Ahmed, N. and Rao, K. R., Orthogonal Transforms for Digital Signal Processing, Springer-Verlag (1975).
15. Franks, J. and Fried, D., Private communications.
16. Margolus, N., Physica **10D**, 81 (1984).
17. Vichniac, G., Physica **10D**, 96 (1984).
18. Pomeau, Y., J. Phys. **A17**, L415 (1984).
19. Hedlund, G., Private communication.
20. Milnor, J. and Wolfram, S., In preparation.
21. Milnor, J., "Notes on surjective cellular automaton-maps", Institute for Advanced Study preprint (June 1984).
22. Chaitin, G., "Towards a mathematical definition of life", in The Maximum Entropy Formalism, R. D. Levine and M. Tribus (eds.), MIT press (1979).
23. Crutchfield, J. and Packard, N., Private communication.
24. Collet, P. and Eckmann, J. -P., Iterated Maps on the Interval as Dynamical Systems, Birkhauser (1980).
25. Amit, D., Field Theory, the Renormalization Group, and Critical Phenomena, McGraw-Hill (1978).
26. Milnor, J., Unpublished.
27. Mandelbrot, B., The Fractal Geometry of Nature, Freeman (1982).
28. Grassberger, P., Physica **10D**, 52 (1984).
29. Greenberg, J. M., Hassard, B. D. and Hastings, S. P., Bull. Amer. Math. Soc. **84**, 1296 (1975); Madore, B. and Freedman, W., Science **222**, 615 (1983).
30. Winfree, A. and Winfree, E., "Organizing centers in a cellular excitable medium", to be published.
31. Packard, N., "Cellular automaton models for dendritic growth", Institute for Advanced Study preprint, in preparation.
32. Bennett, C., Neuberger, H., Pomeau, Y. and Vichniac, G., Private communications.
33. Martin, O., Odlyzko, A. and Wolfram, S., Comm. Math. Phys. **93**, 219 (1984).
34. Grassberger, P., Krause, F. and von der Twer, T., "A new type of kinetic critical phenomena", University of Wuppertal preprint WU B 83-22 (October 1983).
35. Domany, E. and Kinzel, W., Phys. Rev. Lett. **53**, 311 (1984).
36. Ingerson, T. E. and Buvel, R. L., Physica **10D**, 59 (1984).
37. Kauffman, S., Physica **10D**, 145 (1984).
38. Hopcroft, J., "An *n* log *n* algorithm for minimizing states in a finite automaton", in Proc. Int. Symp. on the Theory of Machines and Computations, Academic Press (1971).
39. Harary, F., Graph Theory, Chapter 15, Addison-Wesley (1972).
40. Hurd, L., "Formal language characterizations of cellular automaton limit sets", to be published.
41. Hasslacher, B., Private communication.
42. Yaku, T., J. Comput. System Sci. 7, 481 (1973); Golze, U., "Differences between 1- and 2-dimensional cell spaces", in A. Lindenmayer and G. Rozenberg (eds.), Automata, Languages, Development, North-Holland (1976).
43. Berger, R., Mem. Amer. Math. Soc., no. 66 (1966); Robinson, R., Inventiones Math. **12**, 177 (1971).
44. Walters, P., An Introduction to Ergodic Theory, Springer (1982).
45. Knuth, D., Seminumerical Algorithms, 2nd. ed., Addison-Wesley (1981).
46. Furstenberg, H. and Lind, D., Private communication.
47. Berlekamp, E. R., Conway, J. H. and Guy, R. K., Winning Ways for Your Mathematical Plays, vol. 2, chap. 25, Academic Press (1982); Gardner, M., Wheels, Life and Other Mathematical Amusements, Freeman (1983).
48. Smith, A. R., J. ACM **18**, 339 (1971).
49. Banks, E. R., "Information processing and transmission in cellular automata", MIT project MAC report no. TR-81 (1971).
50. Smith, A. R., J. Comput. Sys. Sci. **6**, 233 (1972); Sommerhalder, R. and van Westrhenen, S. C., Acta Inform. **19**, 397 (1983).
51. Steiglitz, K., Private communication.
52. Bennett, C. H., "On the logical "depth" of sequences and their reducibilities to random sequences", Info. & Control, to be published.
53. Garey, M. R. and Johnson, D. S., Computers and Intractability: a Guide to the Theory of *NP*-Completeness, Freeman (1979).
54. Sewelson, V., Private communication.
55. Cole, S. N., IEEE Trans. Comput. **C-18**, 349 (1969).
56. Preston, K. *et al.*, Proc. IEEE **67**, 826 (1979).
57. Hillis, D. and Wolfram, S., Work in progress.

2. Computation theoretical approaches

2.1: S. Wolfram, "Computation theory of cellular automata", Commun. Math. Phys. 96 (1984) 15.

2.2: N. Margolus, "Physics-like models of computation", Physica 10D (1984) 81.

2.3: S. Wolfram, "Random sequence generation by cellular automata", Adv. Applied Math. 7 (1986) 123.

2.4: S. Wolfram, "Undecidability and intractability in theoretical physics", Phys. Rev. Lett. 54 (1985) 735.

2.5: S. Wolfram, "Origins of randomness in physical systems", Phys. Rev. Lett. 55 (1985) 449.

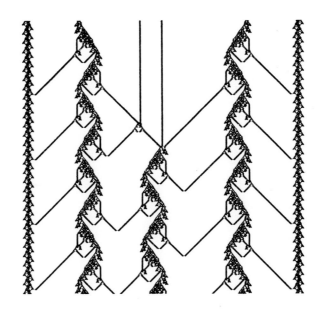

Information processing structures in a one-dimensional cellular automaton. The pattern was produced by evolution according to the $k=2$, $r=3$ totalistic cellular automaton rule with code 88. This class 4 rule supports complex localized structures whose evolution and interactions can be used to implement certain logical functions. One suspects that the cellular automaton is capable of universal computation. (Picture by James Park.)

Commun. Math. Phys. 96, 15–57 (1984)

Communications in
**Mathematical
Physics**
© Springer-Verlag 1984

Computation Theory of Cellular Automata

Stephen Wolfram*

The Institute for Advanced Study, Princeton, NJ 08540, USA

Abstract. Self-organizing behaviour in cellular automata is discussed as a computational process. Formal language theory is used to extend dynamical systems theory descriptions of cellular automata. The sets of configurations generated after a finite number of time steps of cellular automaton evolution are shown to form regular languages. Many examples are given. The sizes of the minimal grammars for these languages provide measures of the complexities of the sets. This complexity is usually found to be non-decreasing with time. The limit sets generated by some classes of cellular automata correspond to regular languages. For other classes of cellular automata they appear to correspond to more complicated languages. Many properties of these sets are then formally non-computable. It is suggested that such undecidability is common in these and other dynamical systems.

1. Introduction

Systems that follow the second law of thermodynamics evolve with time to maximal entropy and complete disorder, destroying any order initially present. Cellular automata are examples of mathematical systems which may instead exhibit "self-organizing" behaviour[1]. Even starting from complete disorder, their irreversible evolution can spontaneously generate ordered structure. One coarse indication of such self-organization is a decrease of entropy with time. This paper discusses an approach to a more complete mathematical characterization of self-organizing processes in cellular automata, and possible quantitative measures of the "complexity" generated by them. The evolution of cellular automata is viewed as a computation which processes information specified as the initial state. The structure of the output from such information processing is then described using

---

* Work supported in part by the U.S. Office of Naval Research under contract number N 00014-80-C-0657
1   An introduction to cellular automata in this context, together with many references is given in [1]. Further results are given in [2, 3], and are surveyed in [4, 5]

the mathematical theory of formal languages (e.g. [6–8]). Detailed results and examples for simpler cases are presented, and some general conjectures are outlined. Computation and formal language theory may in general be expected to play a role in the theory of non-equilibrium and self-organizing systems analogous to the role of information theory in conventional statistical mechanics.

A one dimensional cellular automaton consists of a line of sites, with each site taking on a finite set of possible values, updated in discrete time steps according to a deterministic rule involving a local neighbourhood of sites around it. The value of site i at time step t is denoted $a_i^{(t)}$ and is a symbol chosen from the alphabet

$$S = \{0, 1, ..., k-1\}. \tag{1.1}$$

The possible sequences of these symbols form the set Σ of cellular automaton configurations $A^{(t)}$. Most of this paper concerns the evolution of infinite sequences $\Sigma = S^Z$; finite sequences $\Sigma = S^N$ flanked by quiescent sites (with say value 0) may also be considered. At each time step each site value is updated according to the values of a neighbourhood of $2r+1$ sites around it by a local rule

$$\phi : S^{2r+1} \to S \tag{1.2}$$

of the form[2]

$$a_i^{(t)} = \phi[a_{i-r}^{(t-1)}, a_{i-r+1}^{(t-1)}, ..., a_{i+r}^{(t-1)}]. \tag{1.3}$$

This local rule leads to a global mapping

$$\Phi : \Sigma \to \Sigma \tag{1.4}$$

on complete cellular automaton configurations. Then in general

$$\Omega^{(t+1)} = \Phi\Omega^{(t)} \subseteq \Omega^{(t)}, \tag{1.5}$$

where

$$\Omega^{(t)} = \Phi^t \Sigma \tag{1.6}$$

is the set (ensemble) of configurations generated after t iterated applications of Φ (t time steps).

Formal languages consist of sets of words formed from strings of symbols in a finite alphabet S according to definite grammatical rules. Sets of cellular automaton configurations may thus be considered as formal languages, with each word in the language representing a cellular automaton configuration. Such infinite sets of configurations are then completely specified by finite sets of grammatical rules. (This descriptive use of formal grammars may be contrasted with the use of their transformation rules to define the dynamical evolution of developmental or L systems (e.g. [9]).)

Figure 1.1 gives typical examples of the evolution of cellular automata from disordered initial states according to various rules ϕ. Structure of varying complexity is seen to be formed. Four basic classes of behaviour are found in

---

2   The notation used here differs slightly from that of [2]. In particular, F in [2] is denoted here as ϕ

Cellular Automata   17

Fig. 1.1. Evolution of cellular automata with various typical local rules ϕ. The initial state is disordered; successive lines show configurations obtained at successive time steps. Four qualitative classes of behaviour are seen. (The first five rules shown have $k=2$ and $r=1$, and rule numbers 18, 22, 76, 90 and 128, respectively [1]. The last rule has $k=2$, $r=2$, and totalistic code number 20 [2])

these and other cellular automata [2]. In order of increasing apparent complexity, qualitative characterizations of these classes are as follows:

1. Tends to a spatially homogeneous state.
2. Yields a sequence of simple stable or periodic structures.
3. Exhibits chaotic aperiodic behaviour.
4. Yields complicated localized structures, some propagating.

Approaches based on dynamical systems theory (e.g. [10, 11]) suggest some quantitative characterizations of these classes: the first three are analogous to the limit points, limits cycles and chaotic ("strange") attractors found in continuous dynamical systems. The fourth class exhibits more complex behaviour, and, as discussed below, is conjectured [2] to be capable of universal computation (e.g. [6,

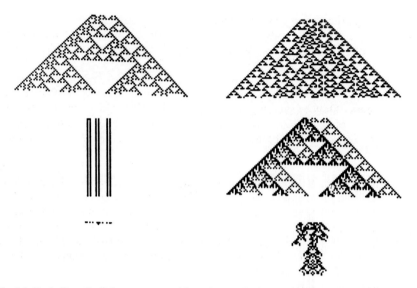

Fig. 1.2. Evolution of cellular automata with various typical local rules from finite initial states. (The rules shown are the same as in Fig. 1)

7, 8]). The formal language theory approach discussed in this paper provides more precise and complete characterizations of the classes and their complexity.

The four classes of cellular automata generate distinctive patterns by evolution from finite initial configurations, as illustrated in Fig. 1.2:

1. Pattern disappears with time.
2. Pattern evolves to a fixed finite size.
3. Pattern grows indefinitely at a fixed rate.
4. Pattern grows and contracts with time.

The classes are also distinguished by the effects of small changes in initial configurations:

1. No change in final state.
2. Changes only in a region of finite size.
3. Changes over a region of ever-increasing size.
4. Irregular changes.

"Information" associated with the initial state thus propagates only a finite distance in classes 1 and 2, but may propagate an infinite distance in classes 3 and 4. In class 3, it typically propagates at a fixed positive speed.

The grammar of a formal language gives rules for generating or recognizing the words in the language. An idealized computer (such as a Turing machine) may be constructed to implement these rules. Such a computer may be taken to consist of a "central processing unit" with a fixed finite number of internal states, together with a "memory" or "tape." Four types of formal language are conventionally identified, roughly characterized by the size of the memory in computers that implement them (e.g. [7]):

0. Unrestricted languages[3]: indefinitely large memory.

1. Context-sensitive languages: memory proportional to input word length.

2. Context-free languages: memory arranged in a stack, with a fixed number of elements available at a given time.

3. Regular languages: no memory.

These four types of languages (essentially) form a hierarchy, with type 0 the most general. Only type 0 languages require full universal computers; the other three types of language are associated with progressively simpler types of computer (linear-bounded automata, pushdown automata, and finite automata, respectively).

The grammatical rules for a formal language may be specified as "productions" which define transformations or rewriting rules for strings of symbols. In addition to the set S of "terminal" symbols s_i which appear directly in the words of the language, one introduces a set U of intermediate "non-terminal" symbols u_i. To generate words in the language, one begins with a particular non-terminal "start" symbol, then uses applicable productions in turn eventually to obtain strings containing only terminal symbols. The different types of languages involve productions of different kinds:

0. Arbitrary productions.

1. Productions $\alpha_1 \to \alpha_2$ for which $|\alpha_2| \geq |\alpha_1|$, where α_i is an arbitrary string of terminal and non-terminal symbols, and $|\alpha_i|$ is its length.

2. Productions of the form $u_i \to \alpha_j$ only (with a fixed bound on $|\alpha_j|$).

3. Productions of the form $u_i \to s_j u_k$ or $u_i \to s_j$ only.

Words in languages are recognized (or "parsed") by finding sequences of inverse productions that transform the words back to the start symbol.

The grammars for regular (type 3) languages may be specified by the finite state transition graphs for finite automata that recognize them. Each arc in such a graph carries a symbol s_i from the alphabet S. The nodes in the graph are labelled by non-terminal symbols, and connected according to the production rules of the grammar. Words in the language correspond to paths through the state transition graph. The (set) entropy of the language, defined as the exponential rate of increase in the number of words with length (see Sect. 3), is then given by the logarithm of the largest eigenvalue of the adjacency matrix for the state transition graph. This eigenvalue is always an algebraic integer.

The set of all possible sequences of zeroes and ones forms a trivial regular language, corresponding to a finite automaton with the state transition graph of Fig. 1.3a. Exclusion of all sequences with pairs of adjacent ones (so that any 1 must be followed by a 0) yields the regular language of Fig. 1.3b. The set of sequences in which, say, an even number of isolated ones appear between every 0110 block, again forms a regular language, now specified by the graph of Fig. 1.3c.

Regular expressions provide a convenient notation for regular languages. For example, $((0*)(1*))*$ represents all possible sequences of zeroes and ones, corresponding to Fig. 1.3a. Here $\alpha*$ denotes an arbitrary number of repetitions of the

---

3   Also known as general, phrase-structure, and semi-Thue languages

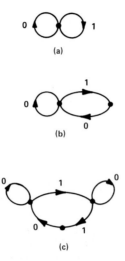

Fig. 1.3a–c. State transition graphs for deterministic finite automata (DFA) corresponding to some regular languages: **a** the set of all possible sequences of zeroes and ones; **b** sequences in which 11 never occurs; **c** sequences in which an even number of isolated 1's appear between each 0110 block. Words in the languages correspond to sequences of symbols on arcs in paths through the DFA state transition graphs. The three DFA shown have successively larger numbers of states Ξ, and the sets of symbol sequences they represent may be considered to have successively larger "regular language complexities"

string α. With this notation, $(0*(10)*)*$ represents Fig. 1.3b, and $(0(0*)1(0*)1)*$ represents Fig. 1.3c.

Many regular grammars may in general yield the same regular language. However, it is always possible to find the simplest grammar for a given regular language (Myhill-Nerode theorem (e.g. [7])), whose corresponding finite automaton has the minimal number of states (nodes). This minimal number of states provides a measure of the "complexity" Ξ of the regular language. The regular languages of Fig. 1.3a–c are thus deemed to have progressively greater regular language complexities.

Section 2 shows that the sets of configurations $\Omega^{(t)}$ generated by any finite number of steps in the evolution of a cellular automaton form a regular language. For some cellular automata, the complexities of the regular languages obtained tend to a fixed limit after a few time steps, yielding a large time limiting set of configurations corresponding to a regular language. In general, it appears that the limit sets for all cellular automata that exhibit only class 1 or 2 behaviour are given by regular languages. For most class 3 and 4 cellular automata, however, the regular language complexities $\Xi^{(t)}$ of the sets $\Omega^{(t)}$ increase rapidly with time, presumably leading to non-regular language limit sets.

Set of symbol sequences analogous to sets of cellular automaton configurations are obtained from the "symbolic dynamics" of continuous dynamical systems, in which the values of real number parameters are divided into discrete bins, labelled by symbols (e.g. [10, 11]). The simplest symbol sequences obtained in

Cellular Automata   21

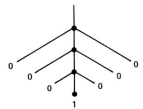

Fig. 1.4. The derivation tree for a word in the context-free language consisting of sequences of the form $0^n 10^n$

this way are "full shifts," corresponding to trivial regular languages Σ containing all possible sequences of symbols. More complicated systems yield finite complement languages, or "subshifts of finite type," in which a finite set of fixed blocks of symbols is excluded. "Sofic" systems, equivalent to general regular languages, have also been studied [12]. There is nevertheless evidence that, just as in cellular automata, regular languages are inadequate to describe the complete symbolic dynamics of even quite simple continuous dynamical systems.

Context-free (type 2) languages are generalizations of regular languages. Words in context-free languages may be viewed as sequences of terminal nodes (leaves) in trees constructed according to context-free grammatical rules. Each non-terminal symbol in the context-free grammar is taken to correspond to a type of tree node. The production rules for the non-terminal symbol then specify its possible descendents in the tree. For each word in the language, there corresponds such a "derivation" tree, rooted at the start symbol. (In most context-free languages, there are "ambiguous" words, obtained from multiple distinct derivation trees.) The syntax for most practical computer languages is supposed to be context-free. Each grammatical production rule corresponds to a subexpression with a particular structure (such as $u \bigcirc v$); the subexpressions may be arbitrarily nested [as in $((a \bigcirc (b \bigcirc c)) \bigcirc d) \bigcirc e)$], corresponding to arbitrary derivation trees.

Regular languages correspond to context-free languages whose derivation trees consist only of a "trunk" sprouting a sequence of leaves, one at a time. An example of a context-free language not represented by any regular grammar is the sequence of strings of the form $0^n 10^n$ for any n. (Here, as elsewhere, α^n represents n-fold repetition of the string α.) A derivation tree for a word in this language is shown in Fig. 1.4. In general, the productions of any context-free language may in fact be arranged so that all derivation trees are binary (Chomsky normal form)[4].

At each point in the generation of a word in a regular language, the next symbol depends only on the current finite automaton state, and not on any previous history. (Regular language words may thus be considered as Markov chains.) To generate words in a context-free language, however, one must maintain a "stack" (last-in first-out memory), which at each point represents the part of the derivation tree above the symbol (tree leaf) just generated. In this way, words in context-free languages may exhibit certain long-range correlations, as illustrated in Fig. 1.4. (In

47 Compare many implementations of the LISP programming language. Also, compare with models of multiparticle production cascade processes (e.g. [13])

practical computer languages, these long-range correlations are typically manifest in the pairing of parentheses separated by many subexpressions.)

The production rules of a context-free grammar specify transformations for individual non-terminal symbols, independent of the "context" in which they appear. Context-sensitive grammars represent a generalization in which transformations for a particular symbol may depend on the strings of symbols that precede or follow it (its "context"). However, the transformation or production rule for any string α_1 is required to yield a longer (or equal length) string α_2. The set of all strings of the form $0^n 1^n 0^n$ for any n forms a context-sensitive language, not represented by any context-free or simpler language. The words in a context-sensitive language may be viewed as formed from sequences of terminal nodes in a directed graph. The graph is a derivation tree rooted at the start symbol, but with connections representing context sensitivities added. The requirement $|\alpha_2| \geq |\alpha_1|$ implies that there are progressively more nodes at each stage: the length of a word in context-sensitive language thus gives an upper bound on the number of nodes that occur at any stage in its derivation. A machine that recognizes words in a context-sensitive language by enumerating all applicable derivation graphs need therefore only have a memory as large as the words to be recognized.

Unrestricted (type 0) languages are associated with universal computers. A system is considered capable of "universal computation" if, with some particular input, it can simulate the behaviour of any other computational system[5]. A universal computer may thus be "programmed" to implement any finite algorithm. A universal Turing machine has an infinite memory, and a central processing unit with a particular "instruction set." (The "simplest" known universal Turing machine has seven internal states, and a memory arranged as a line of sites, each having four possible values, and with one site accessible to the central processing unit at each time step (e.g. [8]).) Several quite different systems capable of universal computation have also been found. Among these are string manipulation systems which directly apply the production rules of type 0 languages; machines with one, infinite precision, arithmetic register; logic circuits analogous to those of practical digital electronic computers; and mathematical systems such as λ-calculus (general recursive functions). Some cellular automata have also been proved capable of universal computation. For example, a one-dimensional cellular automaton with $k=18$ and $r=1$ is equivalent to the simplest known universal Turing machine (e.g. [14]). (A two-dimensional cellular automata, the "Game of Life", with $k=2$ and a nine site neighbourhood, has also been proved computationally universal (e.g. [15]).) It is conjectured that all cellular automata in the fourth class indicated above are in fact capable of universal computation [2].

There are many problems which can be stated in finite terms, but which are "undecidable" in a finite time, even for a universal computer[6]. An example is the "halting problem": to determine whether a particular computer will "halt" in a finite time, given particular input. The only way to predict the behaviour of some

5   Although there are some mathematically-defined operations which they cannot perform (as discussed below), it seems likely that the usual class of "universal computers" can simulate the behaviour of any physically-realizable system

6   This is a form of Godel's theorem, in which the processes of mathematical proof are formalized in the operation of a computer

system **S** is to execute some procedure in a universal computer; but if, for example, **S** is itself a universal computer, then the procedure must reduce to a direct simulation, and can run no more than a finite amount faster than the evolution of **S** itself. The infinite time behaviour of **S** cannot therefore be determined in general in a finite time. For a cellular automaton, an analogue of the halting problem is to determine whether a particular finite initial configuration will ultimately evolve to the null configuration.

Any problem which depends on the results of infinite information processing may potentially be undecidable. However, when the information processing is sufficiently simple, there may be a finite "short-cut" procedure to determine the solution. For example, the information processing corresponding to the evolution of cellular automata with only class 1 or 2 behaviour appears to be sufficiently simple that their infinite time behaviour may be found by finite computation. Many problems concerning the infinite time behaviour of class 3 and 4 cellular automata may, however, be undecidable. For example, the entropies of the invariant sets for class 3 and 4 cellular automata may in general be non-computable numbers. This would be the case if the languages corresponding to these limit sets were of type 0 or 1.

It seems likely, in fact, that the consequences of infinite evolution in many dynamical systems may not described in finite mathematical terms, so that many questions concerning their limiting behaviour are formally undecidable. Many features of the behaviour of such systems may be determined effectively only by explicit simulation: no general predictions are possible.

Even for results that can in principle be obtained by finite computation there is a wide variation in the magnitude of time (or memory resources) required. Several classes of finite computations may be distinguished (e.g. [7]).

The first class (denoted P) consists of problems that can be solved by a deterministic procedure in a time given by some polynomial function of the size of their input. For example, finding the successor of a length n sequence in a cellular automaton takes (at most) a time linear in n, and is therefore a problem in the class P. Since most universal computers can simulate any other computer in a polynomial time, the times required on different computers usually differ at most by a polynomial transformation, and the set of problems in class P is defined almost independent of computer.

Nondeterministic polynomial time problems (NP) form a second class. Solutions to such problems may not necessarily be obtained in a polynomial time by a systematic procedure, but the correctness of a candidate solution, once guessed, can be tested in a polynomial time. Clearly $P \subseteq NP$, and there is considerable circumstantial evidence that $P \neq NP$. The problem of finding a pre-image for a length n sequence under cellular automaton evolution is in the class NP.

The problem classes P and NP are characterized by the times required for computations. One may also consider the class of problems PSPACE that require memory space given by a polynomial function of the size of the input, but may take an arbitrary time. There is again circumstantial evidence that $P \subset PSPACE$.

Just as there exist universal computers which, when given particular input, can simulate any other computer, so, analogously, there exist "NP-complete" (or

"PSPACE-complete") problems which, with particular input, correspond to any NP (or PSPACE) problem of a particular size (e.g. [6, 7]). Many NP and PSPACE complete problems are known. An example of an NP-complete problem is "satisfiability": finding truth values for n variables which make a particular Boolean expression true. If $P \neq NP$ then there is essentially no faster method to solve this problem than to try the 2^n possible sets of values. (It appears that any method must at least require a time larger than any polynomial in n.) As discussed in Sect. 6, it is likely that the problem of finding pre-images for sequences in certain cellular automata, or of determining whether particular sequences are ever generated, is NP-complete. This would imply that no simple description exists even for some finite time properties of cellular automata: results may be found essentially only by explicit simulation of all possibilities.

2. Construction of Finite Time Sets

This section describes the construction of the set of configurations $\Omega^{(t)}$ generated after a finite number of time steps t of cellular automaton evolution, starting from the set $\Omega^{(0)} = \Sigma$ of all possible configurations. It is shown that $\Omega^{(t)}$ may be represented as a regular language (cf. [2, 16]), and an explicit construction of the minimal grammar for this language is given. Section 3 describes some properties of such grammars, and Sect. 4 discusses their form for a variety of cellular automata.

To describe the construction we begin with a simple example. The procedure followed may be generalized directly.

Consider the construction of the set $\Omega^{(1)}$ generated by one time step in the evolution of the $k = 2$, $r = 1$ cellular automaton with a local rule ϕ given by ("rule number 76" [1])

$$111 \to 0, \quad 110 \to 1, \quad 101 \to 0, \quad 100 \to 0, \quad 011 \to 1, \quad 010 \to 1, \quad 001 \to 0, \quad 000 \to 0.$$

$$(2.1)$$

The value $a_i^{(1)}$ of a site at position i in a configuration $A^{(1)} = \Phi A^{(0)} \in \Omega^{(1)}$ depends on a neighbourhood of three sites $\{a_{i-1}^{(0)}, a_i^{(0)}, a_{i+1}^{(0)}\}$ in the preceding configuration $A^{(0)} = \Omega^{(0)}$. The adjacent site $a_{i+1}^{(1)}$ depends on the overlapping neighbourhood $\{a_i^{(0)}, a_{i+1}^{(0)}, a_{i+2}^{(0)}\}$. The dependence of $a_{i+1}^{(1)}$ on $a_i^{(0)}$ associated with this two-site overlap in neighbourhoods may be represented by the graph g of Fig. 2.1 (analogous to a de Bruijn graph [17]). The nodes in the graph represent the overlaps $\{a_i^{(0)}, a_{i+1}^{(0)}\}$. These nodes are joined by directed arcs corresponding to three-site neighbourhoods. The local cellular automaton rule ϕ of Eq. (2.1) defines a transformation for each three-site neighbourhood, and thus associates a symbol with each arc of g. Each possible path through g corresponds to a particular initial configuration $A^{(0)}$. The successor $A^{(1)}$ of each initial configuration is given by the sequence of symbols associated with the arcs on the path. The sequences of symbols obtained by following all possible paths through g thus correspond to all possible configurations $A^{(1)}$ obtained after one time step in the evolution of the cellular automaton (2.1). The complete set $\Omega^{(1)}$ may thus be represented by the graph g. It is clear that not all possible sequences of 0's and 1's can appear in the configurations of $\Omega^{(1)}$. For example, no path in g can include the sequence 111, and thus no configuration in $\Omega^{(1)}$ can contain a block of sites 111.

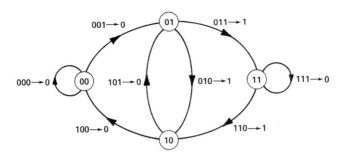

Fig. 2.1. The state transition graph g for a non-deterministic finite automaton (NDFA) that generates configurations obtained after one time step in the evolution of the $k=2$, $r=1$ cellular automaton with rule number 76 [Eq. (2.1)]. Possible sequences of site values are represented by possible paths through the graph. The nodes in the graph are labelled by pairs of initial site values; the arcs then correspond to triples of initial site values. Each such triple is mapped under rule number 76 to a particular site value. The graph with arcs labelled by these site values corresponds to all possible configurations obtained after one time step. Note that the basic graph is the same for all $k=2$, $r=1$ cellular automata; only the images of the initial site value triples change from one rule to another

The graph g of Fig. 2.1 may be considered as the state transition graph for a finite automaton which generates the formal language $\Omega^{(1)}$. Each node of g corresponds to a state of the finite automaton, and each arc to a transition in the finite automaton, or equivalently to a production rule in the grammar represented by the finite automaton. The set $\Omega^{(1)}$ thus forms a regular language. Labelling the states in g as u_0, u_1, u_2, u_3, the productions in the grammar for this language are:

$$u_0 \to 0u_0, \quad u_0 \to 0u_1, \quad u_1 \to 1u_2, \quad u_1 \to 1u_3, \quad u_2 \to 0u_0,$$
$$u_2 \to 0u_1, \quad u_3 \to 0u_3, \quad u_3 \to 1u_2. \tag{2.2}$$

This finite set of rules provides a complete specification of the infinite set $\Omega^{(1)}$.

Each path through g corresponds uniquely to a particular initial configuration $A^{(0)}$. But several different paths may yield the same successor configuration $A^{(1)}$. Each such path corresponds to a distinct inverse image of $A^{(1)}$ under Φ. Enumeration of paths in g shows, for example, that there are 5 distinct inverse images for the sequence 00 under the cellular automaton mapping (2.1), 5 also for 01 and 10, and 1 for 11.

The finite automaton g of Fig. 2.1 is not the only possible one that generates the language $\Omega^{(1)}$. An alternative finite automaton $\bar{g}$ is shown in Fig. 2.2, and may be considered "simpler" than g since it has fewer states. $\bar{g}$ is obtained from g by combining the 00 and 10 nodes, which are equivalent in that only paths carrying the same symbol sequences pass through these nodes. The complete set of symbol sequences generated by the possible paths through $\bar{g}$ is identical to that generated by possible paths through g.

The finite automata g and $\bar{g}$ are non-deterministic in the sense that multiple arcs carrying the same symbol emanate from some nodes, so that several distinct paths may generate the same word in the formal language. It is convenient for many purposes to find deterministic finite automata (DFA) equivalent to the non-

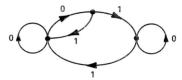

Fig. 2.2. The state transition graph $\bar{g}$ for an alternative NDFA that generates the language $\Omega^{(1)}$ obtained after one time step of evolution according to rule 76. This NDFA is obtained by combining two equivalent states in the NDFA g of Fig. 2.1

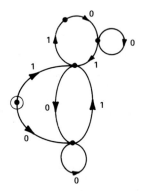

Fig. 2.3. The state transition graph G for a deterministic finite automaton (DFA) obtained from the non-deterministic finite automaton of Fig. 2.1 by the subset construction [and represented by the productions of Eq. (2.3)]. Here and in other DFA graphs, the start node ψ_S is shown encircled. Words in the regular language $\Omega^{(1)}$ correspond to paths through G, starting at ψ_S

deterministic finite automata (NDFA) g and $\bar{g}$. Such DFA may always be found by the standard "subset construction" (e.g. [6, 7]).

Consider for example the construction of a DFA G equivalent to the NDFA g of Fig. 2.1. Let ψ be the set of all possible subsets of the set of nodes $\{u_i\}$ (the power set of $\{u_i\}$). There are $2^4 = 16$ elements ψ_i of ψ; each potentially corresponds to a state in G. The construction of G begins from the "start node" $\psi_S = \{u_0, u_1, u_2, u_3\}$. This node is joined by a 0 arc to the node $\{u_0, u_1, u_3\}$ corresponding to the set of NDFA states reached by a 0 arc according to (2.2) from any of the u_i in ψ_S. An analogous procedure is applied for each arc at each node in G. The resulting graph is shown in Fig. 2.3, and may be represented by the productions

$$\psi_S = \{u_0, u_1, u_2, u_3\} \to 0\{u_0, u_1, u_3\}, \quad \{u_0, u_1, u_2, u_3\} \to 1\{u_2, u_3\},$$
$$\{u_0, u_1, u_3\} \to 0\{u_0, u_1, u_3\}, \quad \{u_0, u_1, u_3\} \to 1\{u_2, u_3\},$$
$$\{u_2, u_3\} \to 0\{u_0, u_1, u_3\}, \quad \{u_2, u_3\} \to 1\{u_2\}, \tag{2.3}$$
$$\{u_2\} \to 0\{u_0, u_1\}, \quad \{u_2\} \to 1\{\},$$
$$\{u_0, u_1\} \to 0\{u_0, u_1\}, \quad \{u_0, u_1\} \to 1\{u_2, u_3\}.$$

Notice that only 5 of the 16 possible ψ_i are reached by transitions from ψ_S. The production in Eq. (2.3) yielding the null set $\{\}$ (often denoted ε) signifies the absence of an arc carrying the symbol 1 emanating from the $\{u_2\}$ node.

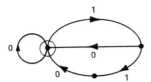

Fig. 2.4. The state transition graph $\bar{G}$ for the minimal DFA that generates the regular language $\Omega^{(1)}$ obtained after one time step of evolution according to cellular automaton rule 76. The graph is obtained by combining equivalent nodes in the DFA G of Fig. 2.3. It has the smallest possible number of nodes

The DFA G of Fig. 2.3 provides an alternative complete description of the language $\Omega^{(1)}$ represented by the NDFA g and $\bar{g}$ of Figs. 2.1 and 2.2. Possible sequences of symbols in words of $\Omega^{(1)}$ correspond to possible paths through G, starting at ψ_S. Consider the procedure for recognizing whether a sequence α can occur in $\Omega^{(1)}$. If α can occur, then it must correspond to a path through the NDFA g, starting at some node. The set of possible paths through g is represented by a single path through the DFA G. The start state ψ_S in G corresponds to the set of all possible states in g. As each symbol in the sequence α is scanned, the DFA G makes a transition to a state representing the set of states that g could reach at that point. The sequence α can thus occur in a word of $\Omega^{(1)}$ if and only if it corresponds to a path in G. The deterministic nature of G ensures that this path is unique.

Complete cellular automaton configurations consist of infinite sequences of symbols, and correspond to infinite paths in the DFA graph G. The possible words in $\Omega^{(1)}$ may thus be generated by following all possible paths through G.

Just as for the NDFA g, some of the states in the DFA G are equivalent, and may be combined. Two states are equivalent if and only if transitions from them with all possible symbols (here 0 or 1) lead to equivalent states. An equivalent DFA $\bar{G}$ shown in Fig. 2.4 may thus be obtained by representing each equivalence class of states in G by a single state. It may be shown that this DFA is the minimal one that recognizes the language $\Omega^{(1)}$ [18, 6, 7]. It is unique (up to state relabellings), and has fewer states than any equivalent DFA. Such a procedure yields the minimal form for any DFA; the analogous procedure for NDFA does not, however, necessarily yield a minimal form.

In most cases, the minimal DFA that generates all (two-way) infinite words of a regular language is the same as the minimal DFA constructed above that recognizes all finite (or one-way infinite) sequences of symbols in words of the language. In some cases, such as that of Fig. 2.5 (the set $\Omega^{(1)}$ for rule number 18), however, the latter DFA may contain additional "transient" subgraphs rooted at ψ_S, feeding into the main graph. The set of infinite paths through these transient subgraphs is typically a subset of the set of infinite paths in the main graph.

The minimal DFA $\bar{G}$ of Fig. 2.4 provides a simple description of the regular language $\Omega^{(1)}$. Regular expressions, mentioned in Sect. 1, provide a convenient notation for this and other regular languages. In terms of regular expressions,

$$\Omega^{(1)} = ((0^*)1(0 \vee 10)), \tag{2.10}$$

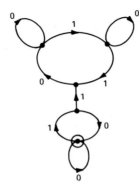

Fig. 2.5. The state transition graph $\bar{G}$ for the minimal DFA corresponding to the regular language $\Omega^{(1)}$ obtained after one time step in the evolution of cellular automaton rule 18. This graph contains a "transient" subgraph rooted at the start state, feeding into the main graph. All symbol sequences occurring at any point in a word $\Omega^{(1)}$ may be recognized as corresponding to paths through $\bar{G}$ beginning at the start state. Complete words in $\Omega^{(1)}$ may nevertheless be generated as possible infinite paths in $\bar{G}$, with the transient subgraph removed

where infinite repetition to form each infinite word is understood. Here α^* represents an arbitrary number (possibly zero) of repetitions of the string α, and $\alpha_1 \vee \alpha_2$ stands for α_1 or α_2.

The example discussed so far generalizes immediately to show that the set $\Omega^{(t)}$ of configurations generated by t time steps of evolution according to any cellular automaton rule forms a regular language. Constructions analogous to those described above give grammars for these languages. The number of states in the initial NDFA g is in general k^{2rt}. (Two examples are shown in Fig. 2.6; graphs for successively larger values of rt may be obtained by a recursive construction [17].) The size of the DFA G obtained from g by the subset construction may be as large as $2^{k^{2rt}} - 1$, but is usually much smaller. (Note that the "reject" state $\{\}$ is not counted in the size of the grammar.)

As an example, consider the language $\Omega^{(2)}$ generated by two time steps in the evolution of the cellular automaton (2.1). The original NDFA g which corresponds to this language has 16 states, and the DFA G obtained from it by the subset construction has nine states. Nevertheless, the resulting minimal DFA $\bar{G}$ has just three states, and is in fact identical to that found for $\Omega^{(1)}$ as shown in Fig. 2.4. Since $\bar{G}$ gives a complete (finite) specification of the languages $\Omega^{(t)}$, this implies that

$$\Omega^{(1)} = \Omega^{(2)} = \Phi\Omega^{(1)} \tag{2.11}$$

in this case. $\Omega^{(1)}$ is thus the limit set for the evolution of the cellular automaton of Eq. (2.1).

3. Properties of Finite Time Sets

This section discusses some properties of the regular language sets $\Omega^{(t)}$ generated by a finite number of steps of cellular automaton evolution, and constructed by the procedure of Sect. 2.

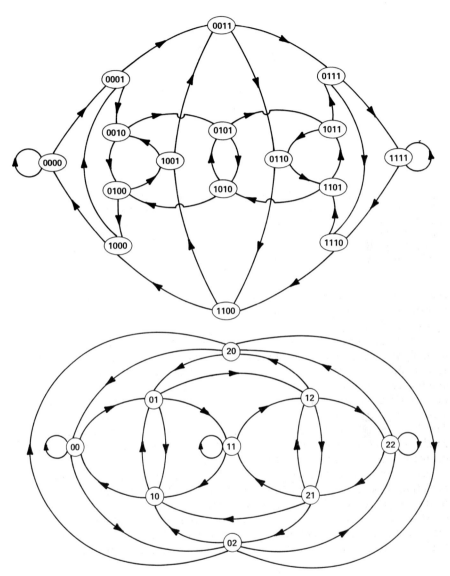

Fig. 2.6. Non-deterministic finite automaton graphs (de Bruijn graphs) analogous to Fig. 2.1 for the cases $k=2$, $r=2$, and $k=3$, $r=1$

We consider as a sample set the 32 "legal" cellular automaton rules with $k=2$ and $r=1$. A rule ϕ is considered legal if it is symmetric, and maps the null configuration (with all site values 0) to itself. Each of the 256 possible $k=2$, $r=1$ cellular automaton rules is conveniently labelled by a "rule number," defined as the decimal equivalent of the sequence of binary digits $\phi[1, 1, 1]$, $\phi[1, 1, 0]$, ..., $\phi[0, 0, 0]$ (analogous to Eq. (2.1)) [1].

Table 1. Numbers of nodes $\Xi^{(t)}$ (and arcs) in minimal deterministic finite automata (DFA) representing regular languages corresponding to sets of configurations $\Omega^{(t)}$ generated after t time steps in the evolution of legal $k = 2$, $r = 1$ cellular automata. Each configuration corresponds to a path through the DFA state transition graph. The construction of Sect. 2 yields the DFA with the minimal number of nodes (states) $\Xi^{(t)}$ that generates a given regular language $\Omega^{(t)}$. This DFA may be considered to give the shortest specification of $\Omega^{(t)}$ viewed as a regular language. Its size $\Xi^{(t)}$ measures the "complexity" of $\Omega^{(t)}$. The initial ($t = 0$) set of configurations include all possible sequences of zeroes and ones, and correspond to a trivial regular language. Cellular automata with only class 1 or 2 behaviour yield regular languages whose complexities become constant, or increase as polynomials in t. Cellular automata capable of class 3 or 4 behaviour usually lead to rapidly-increasing complexities. Bounds on these complexities are given when their exact calculation exceeded available computational resources. Some of the results in this table were obtained using the methods of [52] and [53]

Rule	$\Xi^{(0)}$	$\Xi^{(1)}$	$\Xi^{(2)}$	$\Xi^{(3)}$	$\Xi^{(4)}$
0	1 (2)	1 (1)	1 (1)	1 (1)	1 (1)
4	1 (2)	2 (3)	2 (3)	2 (3)	2 (3)
18	1 (2)	5 (9)	47 (91)	143 (270)	$\gtrsim 20000$
22	1 (2)	15 (29)	280 (551)	4506 (8963)	$\gtrsim 20000$
32	1 (2)	2 (3)	5 (7)	7 (9)	9 (11)
36	1 (2)	3 (5)	3 (4)	3 (4)	3 (4)
50	1 (2)	3 (5)	8 (14)	10 (17)	12 (20)
54	1 (2)	9 (16)	17 (32)	94 (179)	675 (1316)
72	1 (2)	5 (9)	5 (8)	5 (8)	5 (8)
76	1 (2)	3 (5)	3 (5)	3 (5)	3 (5)
90	1 (2)	1 (2)	1 (2)	1 (2)	1 (2)
94	1 (2)	15 (29)	230 (455)	3904 (7760)	$\gtrsim 20000$
104	1 (2)	15 (29)	265 (525)	2340 (4647)	1394 (2675)
108	1 (2)	9 (16)	11 (19)	11 (19)	11 (19)
122	1 (2)	15 (29)	179 (347)	5088 (9933)	$\gtrsim 20000$
126	1 (2)	3 (5)	13 (23)	107 (198)	2867 (5476)
128	1 (2)	4 (6)	6 (8)	8 (10)	10 (12)
132	1 (2)	5 (9)	7 (12)	9 (15)	11 (18)
146	1 (2)	15 (29)	92 (177)	1587 (3126)	$\gtrsim 20000$
150	1 (2)	1 (2)	1 (2)	1 (2)	1 (2)
160	1 (2)	9 (15)	16 (24)	25 (35)	36 (48)
164	1 (2)	15 (29)	116 (227)	667 (1310)	1214 (2363)
178	1 (2)	11 (20)	15 (26)	19 (32)	23 (38)
182	1 (2)	15 (29)	92 (177)	1587 (3126)	$\gtrsim 20000$
200	1 (2)	3 (5)	3 (5)	3 (5)	3 (5)
204	1 (2)	1 (2)	1 (2)	1 (2)	1 (2)
218	1 (2)	15 (29)	116 (227)	667 (1310)	1214 (2363)
222	1 (2)	5 (9)	7 (12)	9 (15)	11 (18)
232	1 (2)	11 (20)	15 (26)	19 (32)	23 (38)
236	1 (2)	3 (5)	3 (5)	3 (5)	3 (5)
250	1 (2)	9 (15)	16 (24)	25 (35)	36 (48)
254	1 (2)	4 (6)	6 (8)	8 (10)	10 (12)

Cellular Automata   31

Table 2. Characteristic polynomials $\chi^{(1)}(\lambda)$ for the adjacency matrices of state transition graphs for minimal DFA representing regular languages generated after one time step in the evolution of legal $k=2$, $r=1$ cellular automata. The nonzero roots of these polynomials determine the number of distinct symbol sequences that can appear in configurations generated by the cellular automaton evolution. The maximal root λ_{max} determines the limiting entropy of the sequences

Rule	$\chi^{(1)}(\lambda)$	λ_{max}
0	$1-\lambda$	1.000
4	$-1-\lambda+\lambda^2$	1.618
18	$(1-\lambda-\lambda^2)(-1+\lambda-2\lambda^2+\lambda^3)$	1.755
22	$\lambda(1-\lambda)(2-2\lambda^2+6\lambda^3-3\lambda^4-5\lambda^5+10\lambda^6-5\lambda^7-3\lambda^8+6\lambda^9-2\lambda^{10}+2\lambda^{11}-3\lambda^{12}+\lambda^{13})$	1.917
32	$-1-\lambda+\lambda^2$	1.618
36	$1-\lambda+2\lambda^2-\lambda^3$	1.755
50	$1+\lambda+\lambda^2-\lambda^3$	1.839
54	$\lambda^3(1+\lambda^2)(1-\lambda+2\lambda^3-\lambda^4)$	1.867
72	$(1+\lambda-\lambda^2)(-1+\lambda-2\lambda^2+\lambda^3)$	1.755
76	$1+\lambda+\lambda^2-\lambda^3$	1.839
90	$2-\lambda$	2.000
94	$-\lambda(2-2\lambda+2\lambda^2-\lambda^3+2\lambda^4-5\lambda^5+13\lambda^6-16\lambda^7+10\lambda^8-3\lambda^{10}-\lambda^{11}+5\lambda^{12}-4\lambda^{13}+\lambda^{14})$	1.883
104	$\lambda(1-\lambda)(2-2\lambda^2+6\lambda^3-3\lambda^4-5\lambda^5+10\lambda^6-5\lambda^7-3\lambda^8+6\lambda^9-2\lambda^{10}+2\lambda^{11}-3\lambda^{12}+\lambda^{13})$	1.917
108	$\lambda^3(1+\lambda^2)(1-\lambda+2\lambda^3-\lambda^4)$	1.867
122	$-\lambda(2-2\lambda+2\lambda^2-\lambda^3+2\lambda^4-5\lambda^5+13\lambda^6-16\lambda^7+10\lambda^8-3\lambda^{10}-\lambda^{11}+5\lambda^{12}-4\lambda^{13}+\lambda^{14})$	1.883
126	$1-\lambda+2\lambda^2-\lambda^3$	1.755
128	$(-1-\lambda+\lambda^2)(1-\lambda+\lambda^2)$	1.618
132	$1-\lambda^2+2\lambda^4-\lambda^5$	1.785
146	$-\lambda(-2+4\lambda-6\lambda^2+4\lambda^3+\lambda^4-7\lambda^5+12\lambda^6-13\lambda^7+9\lambda^8-4\lambda^9+\lambda^{10}-2\lambda^{11}+5\lambda^{12}-4\lambda^{13}+\lambda^{14})$	1.887
150	$2-\lambda$	2.000
160	$(1-\lambda^2-\lambda^3)(1-\lambda^2+\lambda^3)(-1+\lambda-2\lambda^2+\lambda^3)$	1.755
164	$-\lambda(2-\lambda^2-2\lambda^4+5\lambda^5-9\lambda^6+14\lambda^7-9\lambda^8+2\lambda^9-6\lambda^{10}+5\lambda^{11}+3\lambda^{12}-4\lambda^{13}+\lambda^{14})$	1.915
178	$\lambda(1-\lambda^2+\lambda^5)(1-\lambda^2+2\lambda^4-\lambda^5)$	1.785
182	$-\lambda(-2+4\lambda-6\lambda^2+4\lambda^3+\lambda^4-7\lambda^5+12\lambda^6-13\lambda^7+9\lambda^8-4\lambda^9+\lambda^{10}-2\lambda^{11}+5\lambda^{12}-4\lambda^{13}+\lambda^{14})$	1.887
200	$1-\lambda+2\lambda^2-\lambda^3$	1.755
204	$2-\lambda$	2.000
218	$-\lambda(2-\lambda^2-2\lambda^4+5\lambda^5-9\lambda^6+14\lambda^7-9\lambda^8+2\lambda^9-6\lambda^{10}+5\lambda^{11}+3\lambda^{12}-4\lambda^{13}+\lambda^{14})$	1.915
222	$1-\lambda^2+2\lambda^4-\lambda^5$	1.785
232	$\lambda(1-\lambda^2-\lambda^5)(-1+\lambda^2-2\lambda^4+\lambda^5)$	1.785
236	$1-\lambda+2\lambda^2-\lambda^3$	1.755
250	$(1-\lambda^2-\lambda^3)(1-\lambda^2+\lambda^3)(-1+\lambda-2\lambda^2+\lambda^3)$	1.755
254	$(-1-\lambda+\lambda^2)(1-\lambda+\lambda^2)$	1.618

Tables 1, 2 and 3 give some properties of the sets $\Omega^{(t)}$ generated by a few time steps in the evolution of the 32 legal $k=2$, $r=1$ cellular automata[7]. These properties are deduced from the minimal DFA which describe the $\Omega^{(t)}$, obtained according to the construction of Sect. 2.

The minimal DFA corresponding to the trivial language $\Omega^{(0)}=\Sigma$ illustrated in Fig. 1.1(a) has just one state. The minimal DFA corresponding to the minimal regular grammars for more complicated languages have progressively more states.

---

7   Requests for copies of the *C* language computer program used to obtain these and other results in this paper should be directed to the author

Table 3. The length $L^{(t)}$ ov the shortest distinct blocks of site values newly-excluded after exactly t time steps in the evolution of legal $k=2$, $r=1$ cellular automata. The notation * indicates that the set of cellular automaton configurations $\Omega^{(t)}$ forms a finite complement language (finite number of distinct excluded blocks). The notation – signifies no new excluded blocks

Rule	$L^{(1)}$	$L^{(2)}$	$L^{(3)}$	$L^{(4)}$
0	1*	–	–	–
4	2*	–	–	–
18	3	11	12	13
22	8	7	11	9
32	2*	4*	6*	8*
36	3*	2*	–	–
50	3*	5*	9*	11*
54	5	9	9	7
72	3	3*	–	–
76	3*	–	–	–
90	–	–	–	–
94	5	7	11	11
104	8	8	8	7
108	5	4*	–	–
122	5	7	8	10
126	3*	12	13	14
128	3*	5*	7*	9*
132	4*	5*	6*	7*
146	6	6	8	8
150	–	–	–	–
160	5*	7*	9*	11*
164	9	9	8	9
178	5*	6*	7*	8*
182	6	6	8	8
200	3*	–	–	–
204	–	–	–	–
218	9	9	8	9
222	4*	5*	6*	7*
232	5*	6*	7*	8*
236	3*	–	–	–
250	5*	7*	9*	11*
254	3*	5*	7*	9*

The total number of states $\Xi^{(t)}$ in the minimal DFA that generates a set $\Omega^{(t)}$ provides a measure of the "complexity" of the set $\Omega^{(t)}$, considered as a regular language. $\Xi^{(t)}$ gives the size of the shortest specification of the set $\Omega^{(t)}$ in terms of regular languages: this shortest specification becomes longer as the complexity of the set increases.

Table 1 gives the "regular language complexities" $\Xi^{(t)}$ for the sets $\Omega^{(t)}$ generated at the first few time steps in the evolution of the legal $k=2$, $r=1$ cellular automata.

In all the cases given, $\Xi^{(t)}$ is seen to be non-decreasing with time. Cellular automata with only class 1 or 2 appear to give $\Xi^{(t)}$ which tend to constants after one or two time steps, or increase linearly or quadratically with time. Class 3 and 4 cellular automata usually give $\Xi^{(t)}$ which increase rapidly with time. In general,

$$1 \leq \Xi^{(t)} \leq 2^{k^{2rt}} - 1. \tag{3.1}$$

The upper bound is found to be attained in several cases for $t = 1$; for larger t, $\Xi^{(t)}$ appears to grow at most exponentially with t.

All possible sequences of symbols occur in the trivial language Σ. In more complicated regular languages, only some number $N(X)$ of the k^X possible sequences of X symbols may occur. Each sequence which occurs corresponds to a distinct path in the minimal DFA graph for the language. (Note that all distinct paths in a DFA correspond to different symbol sequences; this need not be the case in a NDFA graph.) The number of such paths is conveniently computed using a matrix representation for the DFA.

Consider as an example the set $\Omega^{(1)}$ obtained by one time step in the evolution of the cellular automaton (2.1). The minimal DFA graph $\bar{G}$ for this set is given in Fig. 2.4, and may be represented by the adjacency matrix

$$M = \begin{pmatrix} 1 & 1 & 0 \\ 1 & 0 & 1 \\ 1 & 0 & 0 \end{pmatrix}. \tag{3.2}$$

The elements of M^X give the numbers $N(X)$ of possible length X paths in $\bar{G}$. For lengths from 1 to 10 these numbers are 2, 4, 7, 13, 24, 44, 81, 149, 274, and 504. In general, at least for large X,

$$N(X) \simeq \mathrm{Tr}[M^X] = \sum \lambda_i^X \sim \lambda_{\max}^X, \tag{3.3}$$

where the λ_i are the eigenvalues of M, and $\lambda_{\max}$ is the largest of them. These eigenvalues are determined from the characteristic polynomial $\chi(\lambda)$ for the minimal DFA adjacency matrix, given in the case of Eq. (3.2) by [8]

$$\chi(\lambda) = 1 + \lambda + \lambda^2 - \lambda^3. \tag{3.4}$$

The largest (real) root of this characteristic polynomial (known as the "index" of the graph [19]) is given by the cubic algebraic integer

$$\lambda_{\max} = [1 + \kappa + 4/\kappa] \simeq 1.83929, \tag{3.5}$$
$$\kappa = [(38 + \sqrt{1188})/2]^{1/3}$$

The set of infinite configurations $\Omega^{(t)}$ generated by cellular automaton evolution may be considered to form a Cantor set. The dimension of this Cantor set is given by

$$s = \lim_{X \to \infty} \frac{1}{X} \log_k N(X), \tag{3.6}$$

---

8    $1/\chi(\lambda)$ is related to the generating function for the sequence $N(X)$ (e.g. [19, Sect. 1.8])

and is equal to the topological entropy of the shift mapping restricted to this set (e.g. [20]). For any regular language, this entropy is given according to Eqs. (3.2) by [21]

$$s = \log_k \lambda_{max}. \tag{3.7}$$

For the case of Eq. (3.2), the entropy is thus

$$s \simeq \log_2 1.83929 \simeq 0.87915. \tag{3.8}$$

Table 2 gives the characteristic polynomials $\chi^{(1)}(\lambda)$ for the regular languages $\Omega^{(1)}$ obtained after one time step in the evolution of the 32 legal $k = 2, r = 1$ cellular automata, together with their largest real roots λ_{max}. All the nonzero roots of the $\chi(\lambda)$ appear in the expression (3.3) for $N(X)$, and are therefore the same for all possible DFA corresponding to a particular regular language. (They may thus be considered "topological invariants.") Additional powers of λ may appear in the characteristic polynomials obtained from non-minimal DFA.

The characteristic polynomials $\chi(\lambda)$ such as those in Table 2 obtained from regular languages are always monic (the term with the highest power of λ that appears in them always has unit coefficient). The largest roots λ_{max} of the $\chi(\lambda)$ for regular languages are thus always algebraic integers (e.g. [22])[9], so that the entropies for regular languages are always the logarithms of algebraic integers. The minimal polynomial with λ_{max} as a root has a degree not greater than the size $\Xi^{(t)}$ of the minimal DFA for a regular language $\Omega^{(t)}$. This bound is usually not reached, since the characteristic polynomial $\chi(\lambda)$ is usually reducible, as seen in Table 2. Notice that in many cases, $\chi(\lambda)$ has several factors with equal degrees. (The factorizations of the $\chi(\lambda)$ are related to the colouring properties of the corresponding graphs [19]. Note that graphs corresponding to minimal DFA always have trivial automorphism groups.) Factors (other than λ^n) with smaller degrees appear to be associated with transient subgraphs in the minimal DFA graph.

The entropy (3.6) characterizes the number of distinct symbol sequences generated by cellular automaton evolution, without regard to the probabilities with which they occur. One may also define a measure entropy (e.g. [20])

$$s_\mu = - \lim_{X \to \infty} \sum_{i=1}^{k^X} p_i \log_k p_i \tag{3.9}$$

in terms of the probabilities p_i for length X sequences. Starting from an initial ensemble in which all symbol sequences of a given length occur with equal probabilities, the probability for a sequence i after t time steps is given by

$$p_i = \xi_i / k^{X + 2rt}, \tag{3.10}$$

where ξ_i is the number of (length $X + 2rt$) t-step preimages of the sequence i under the cellular automaton mapping Φ. This number is equal to the number of distinct paths through the NDFA graph analogous to g in Fig. 2.1 that yield the sequence i. It may also be computed from reduced NDFA graphs analogous to $\bar{g}$ of Fig. 2.2 by

---

9   The λ_{max} are always Perron numbers [23]. Any Perron number may be obtained from some regular language, and in fact also from some finite complement language [23]

including a weight for each path, equal to the product of weights giving the number of unreduced nodes combined into each node on the path.

The set of configurations generated by cellular automaton evolution always contracts or remains unchanged with time, as implied by Eq. (1.5). The entropies associated with the sets $\Omega^{(t)}$ are therefore non-increasing with time. Class 1 cellular automata are characterized by (spatial) entropies that tend to zero with time [2]. Class 2, 3 and 4 cellular automata generate sets of configurations with nonzero limiting spatial entropy. (Class 2 cellular automata nevertheless yield patterns essentially periodic in time, with zero temporal entropy.)

Some cellular automata have the special property that

$$\Phi \Sigma = \Sigma, \tag{3.11}$$

so that all possible configurations can occur at any time in their evolution, and the entropies of the $\Omega^{(t)}$ are always equal to one. Such surjective cellular automaton rules may be recognized by the presence of all k possible outgoing arcs at each node in a DFA representing the grammar of the set $\Omega^{(1)}$ obtained after one time step in their evolution. The finite maximum size $2^{k^{2r}}$ for such a DFA, constructed as in Sect. 2, ensures that this procedure (cf. [24–27, 2]) for determining the surjectiveness of any cellular automaton rule is a finite one[10].

Since there are k outgoing arcs at each node in the original NDFA analogous to Fig. 2.1 for any cellular automaton rule, the rule is surjective if in all cases these arcs carry distinct symbols (so that the NDFA is in fact a DFA). This occurs whenever the local cellular automaton mapping ϕ is injective with respect to its first or last argument (as for additive rules [29, 16] such as 90, 150 or 204 in Tables 1–3). However, at least when $k > 2$ or $r > 1$, there exist surjective cellular automata for which this does not occur [25, 30]. Since all surjective cellular automata must yield the same trivial minimal DFA, it is possible that a reversal of the minimization and subset algorithms discussed in Sect. 2 could be used to generate all NDFA analogous to Fig. 2.1 that correspond to surjective rules.

Surjective cellular automata yield trivial regular languages, in which all possible blocks of symbols may appear. Some cellular automata generate the slightly more complicated "finite complement" regular languages, in which a finite set of distinct blocks are excluded. (Such languages are equivalent to "subshifts of finite type" (e.g. [10, 11]).) An example of a finite complement language, illustrated in Fig. 1.1b, consists of all sequences from which the block of sites 11 is absent. To construct the grammar for a finite complement language in which blocks of length b are excluded, first form a graph analogous to Fig. 2.1, but with sequences of length $b-1$ at each node. Each arc then corresponds to a length b sequence, and may be labelled by the last symbol in the sequence. With this labelling, one arc carrying each of the k possible symbols emanates from each of the k^{b-1} nodes, so that the graph represents a DFA. Removing arcs corresponding to the excluded length b blocks then yields the graph for a DFA that recognizes the finite

---

10   The algorithm essentially involves testing whether a NDFA with k^{2r} states is equivalent to a NDFA that generates the trivial language Σ. This problem is known to be PSPACE-complete [28], and therefore presumably cannot be solved in a time polynomial in k^{2r}

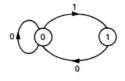

a

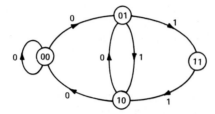

b

Fig. 3.1a and b. Non-deterministic finite automata (NDFA) corresponding to finite complement regular languages consisting of sequences of zeroes and ones in which **a** the block 11 is excluded, and **b** the block 111 is excluded. The graphs are constructed from analogues of Fig. 2.1 by dropping arcs corresponding to excluded blocks

complement language with these blocks absent. Examples of the resulting graphs for two simple cases are shown in Fig. 3.1.

The minimal DFA for a finite complement language with a maximal distinct excluded block of length b has at most k^{b-1} states, and at least b states. An excluded block is considered "distinct" if it contains no excluded sub-blocks. (Hence, for example, in the language of Figs. 1.1a and 3.1a, the excluded block 11 is considered distinct, but 110, 111 and so on, are not.)

Any path through the minimal DFA graph for a regular language of length greater than $\varXi^{(t)}$ must contain a cycle, which retraverses some arcs. If no symbol sequence of length less than $\varXi^{(t)}$ is excluded, then no sequence of any length can therefore be excluded, and the corresponding language must be trivial. If some symbol sequences of length less than $\varXi^{(t)}$ are excluded, but no distinct sequences with lengths between $\varXi^{(t)}$ and $2\varXi^{(t)}$ are excluded, then no longer distinct sequences can be excluded, and the corresponding language must be a finite complement one. If further distinct excluded blocks with lengths between $\varXi^{(t)}$ and $2\varXi^{(t)}$ are found, then an infinite series of longer distinct excluded blocks must exist, and the language cannot be a finite complement one.

The language of Fig. 2.4, generated by the evolution of the $k=2$, $r=1$ cellular automaton with rule number 76, is a finite complement one, in which 111 is the only distinct excluded block. The language of Fig. 2.5, obtained after one time step in the evolution of rule number 18, is not a finite complement one. The block 111 is the shortest excluded in this case. But the distinct length 7 block 1101011 is also excluded, as are the two distinct length 8 blocks 11001011 and 11010011, three distinct length 9 blocks (11010100011, 110001011, 110010011), four distinct length 10 blocks, and so on.

The length $L^{(t)}$ of the shortest excluded block in a language $\varOmega^{(t)}$ generated by cellular automaton evolution (denoted X_c in [2]) is in general given by the shortest distance from the start node in the corresponding DFA graph to an "incomplete" node, with less than k outgoing arcs. If the cellular automaton rule is not surjective, then

$$0 < L^{(t)} \leqq \Xi^{(t)}. \tag{3.12}$$

Whenever cellular automaton evolution is irreversible, the set of configurations $\Omega^{(t)}$ generated contracts with time, and progressively more distinct blocks are excluded. One may define $L^{(t)}$ to be the length of the shortest newly-excluded block at time step t in the evolution of a cellular automaton. The values of $L^{(t)}$ obtained in the first few time steps of evolution according to the 32 legal $k = 2, r = 1$ cellular automaton rules are given in Table 3. In most cases, $L^{(t)}$ is seen to increase with time, indicating that progressively finer subsets of Σ are excluded, and qualitatively reflecting the increase of $\Xi^{(t)}$. In general, however, $L^{(t)}$ need not increase monotonically with time. A length l block is excluded after t time steps if there is no initial length $l + 2rt$ block that evolves into it. A length l block is newly excluded at time step t if and only if no length $l + 2r$ blocks allowed at time step $t - 1$ evolve to it, but at least one length $l + 2r$ block newly excluded at time step $t - 1$ would evolve to it. The length $L^{(t)}$ of the shortest newly excluded block at time t is thus bounded by

$$L^{(t)} \geqq L^{(t-1)} - 2r. \tag{3.13}$$

Table 3 includes several cases for which the lower bound is realized.

The sets of infinite symbol sequences $\Omega^{(t)}$ generated by cellular automaton evolution are characterized in part by the numbers and lengths of allowed and excluded finite blocks which appear in them. A further characterization may be given in terms of the number $\Pi(p)$ of infinite sequences with (spatial) period p that appear. This number is related to the number of distinct cycles in the minimal DFA graph for $\Omega^{(t)}$. Cycles are considered distinct if the sequences of symbols that appear in them are distinct. The enumeration of cycles thus requires knowledge of the arc labelling as well as connectivity of the DFA graph.

Just as the number of finite blocks $N(X)$ for all X may be summarized in the characteristic polynomial $\chi(\lambda)$, so also the number of periodic configurations $\Pi(p)$ may be summarized in the zeta function (e.g. [10, 11])

$$\zeta(\lambda) = \exp\left(\sum_{p=1}^{\infty} \Pi(p)\lambda^p/p \right). \tag{3.14}$$

For all regular languages $\zeta(\lambda)$ is a rational function of λ [31]. For the special case of finite complement languages,

$$\zeta(\lambda) = 1/\chi(\lambda). \tag{3.15}$$

A finite procedure may be given [32] to compute $\zeta(\lambda)$ for any regular language.

4. Evolution of Finite Time Sets

Tables 1–3 gave several properties of the sets of configurations generated by a finite number of steps in the evolution of legal $k = 2, r = 1$ cellular automata. This section discusses these results, identifies several types of behaviour, and considers analogies with classes of cellular automaton behaviour defined by dynamical systems theory means [2].

In the simplest cases, the set $\Omega^{(t)}$ generated by a cellular automaton evolves to a fixed form after a small number of time steps T (the case of surjective cellular automata, with $\Omega^{(t)} = \Sigma$ for all t, is considered separately). The minimal DFA corresponding to $\Omega^{(t)}$ for all $t \geq T$ are then identical, and the values of $\Xi^{(t)}$ and $\chi^{(t)}(\lambda)$ are thus constant. (Notice that $\Xi^{(t)} = \Xi^{(t+1)}$ does not necessarily imply $\Omega^{(t)} = \Omega^{(t+1)}$, as seen for rule 36 in Table 1.) In addition, for $t \geq T$, no more distinct blocks of sites are excluded. Such behaviour occurs in the trivial case of rule 0, under which all initial configurations are mapped to the null configuration after one time step. It also occurs for many other rules: one example is rule 76, discussed in Sects. 2 and 3. All the examples of this behaviour in Tables 1–3 have $T = 1$ (e.g. rule 76) or $T = 2$ (e.g. rule 108). In the trivial case of rule 0, only a single configuration (the null configuration) can appear when $t \geq T$. More complicated single configurations are sometimes generated, represented by minimal DFA consisting of a single cycle. In most cases (such as rule 76), however, $\Omega^{(T)}$ contains an infinite number of configurations. However, it appears that even in these cases, all configurations occur on finite cycles: each configuration is invariant under the cellular automaton mapping, or some finite iteration of it. (A result given in Sect. 5 then shows that the $\Omega^{(T)}$ must form finite complement languages in these cases.) This implies that changes in the initial state for such cellular automata propagate a distance of at most rT sites. A small initial change can thus ultimately affect a region no larger than $2rT$ sites. Such cellular automata must therefore exhibit class 1 or 2 behaviour [2].

For a second set of cellular automata, the form of the minimal DFA does not become fixed after a few time steps, but exhibits a simple growth with time, maintaining a fixed overall structure. The $L^{(t)}$ for such cellular automata typically increases linearly with time, and $\Xi^{(t)}$ increases as some polynomial function of t (linear or quadratic for legal $k = 2$, $r = 1$ rules). Rule 128 gives an example of this behaviour. Under this rule $111 \rightarrow 1$, but all other neighbourhoods map to 0. Any initial sequence of ones thus decreases steadily in length by one site on each side at each time step. After t time steps, any pair of ones must be separated by at least $2t + 1$ zeroes; all blocks of the form 10^j1 for $1 \leq j \leq 2t$ are thus excluded. The first few languages $\Omega^{(t)}$ in the sequence generated by successive time steps in the evolution of rule 128 are shown in Fig. 4.1. The minimal DFA are seen to maintain the same overall structure, but include a linearly increasing number of nodes at each time step. The characteristic polynomials corresponding to these DFA are given by

$$\chi^{(t)}(\lambda) = (1 - \lambda^t + \lambda^{t+1})(-1 - \lambda^t + \lambda^{t+1}), \qquad (4.1)$$

yielding a set entropy which tends to zero at large times, roughly as $1/t$. Rule 160 provides another example in which the minimal DFA maintains the same overall structure, but increases in size with time. In this case, sequences of the form $1[(0 \vee 1)0]^j(0 \vee 1)1$ for all $j \leq t$, are excluded after t time steps, and the size $\Xi^{(t)}$ of the corresponding minimal DFA grows quadratically with time.

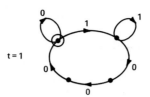

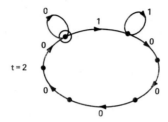

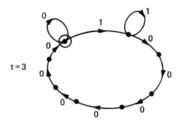

Fig. 4.1. Minimal deterministic finite automata (DFA) corresponding to the regular languages $\Omega^{(t)}$ generated in the first few time steps of evolution according to cellular automaton rule 128. The DFA maintain the same structure, but increase in size with time. They correspond to finite complement languages, with all blocks of the form $10^{j}1$ excluded for $1 \leq j < 2t$

Many cellular automata generate sets $\Omega^{(t)}$ whose corresponding minimal DFA become much more complicated at each successive time step, and appear to exhibit no simple overall structure.

Figure 4.2 shows the minimal DFA obtained after one and two time steps in the evolution of rule 126. No simple progression in the form of these minimal DFA is seen. $\Omega^{(1)}$ is a finite complement language, with only the block 010 excluded, yielding a characteristic polynomial

$$\chi^{(1)}(\lambda) = 1 - \lambda + 2\lambda^2 - \lambda^3 , \qquad (4.2)$$

giving $\lambda_{\max} \simeq 1.7549$. After two time steps, an infinite sequence of distinct blocks is excluded, starting with the length 12 block 011101101110. The corresponding

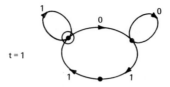

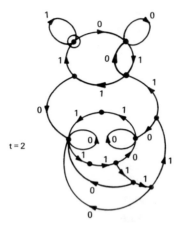

Fig. 4.2. Minimal deterministic finite automata corresponding to the regular languages generated in the first two time steps of evolution according to the class 3 cellular automaton rule 126. A considerable increase in complexity with time is evident, characteristic of cellular automata which can exhibit class 3 behaviour

characteristic polynomial is

$$\chi^{(2)}(\lambda) = -1 + \lambda - \lambda^2 + 2\lambda^3 - 4\lambda^4 + \lambda^5 + 3\lambda^6 - 5\lambda^7$$

$$+ 3\lambda^8 - 3\lambda^9 + 5\lambda^{10} - 6\lambda^{11} + 4\lambda^{12} - \lambda^{13}, \tag{4.3}$$

with $\lambda_{max} \simeq 1.7321$. The minimal DFA for $\Omega^{(3)}$ has 107 states, and the shortest newly-excluded block is 1011100011101 (length 13). $\Xi^{(t)}$ increases rapidly with time. After four time steps, the shortest newly-excluded blocks are 10111000011101, 10111000001110 and its reversal (length 14), and $\Xi^{(4)} = 2876$.

Figures 2.5 and 4.3 give the minimal DFA obtained after one and two time steps in the evolution of rule 18. A considerable increase in complication with time is again evident. After one time step, the shortest of an infinite number of distinct excluded blocks is 1101011 (length 7); after two time steps, the shortest newly-excluded block is 10011011001 (length 11); after three time steps, it is 110010010011 (length 12), and after four time steps it is 1001000010011 (length 13). In this case, as

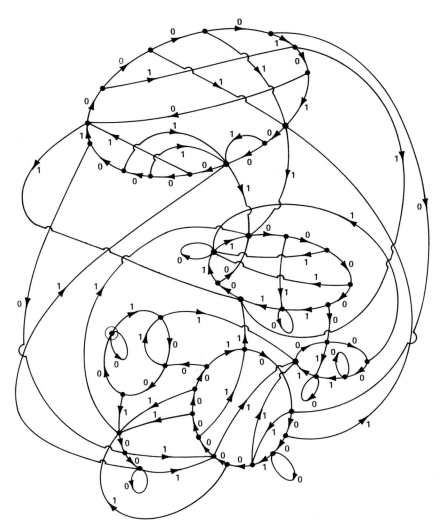

Fig. 4.3. Minimal deterministic finite automaton (DFA) corresponding to the regular language generated after two time steps of evolution according to the class 3 cellular automaton rule 18. The minimal DFA for $t = 1$ is shown in Fig. 2.5. Rapidly-increasing complexity is again evident. The DFA illustrated here has 47 states

for rule 126, $L^{(t)}$ is found to increase monotonically over the range of times investigated. Progressively larger neighbourhoods of the start state are therefore left unchanged in the corresponding minimal DFA. However, as discussed in Sect 3, $L^{(t)}$ need not increase with time, but must in general only satisfy the inequality (3.13). Rule 22 provides an example in which $L^{(t)}$ decreases with time. The minimal DFA for $\Omega^{(1)}$ in this case is shown in Fig. 4.4; the shortest excluded

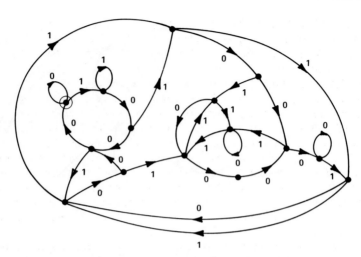

Fig. 4.4. Minimal deterministic finite automaton (DFA) corresponding to the regular language $\Omega^{(1)}$ obtained after one time step in the evolution of the class 3 cellular automaton rule 22. The DFA has all 15 possible states. The shortest excluded block in $\Omega^{(1)}$ has length 8, and corresponds to the shortest path from the encircled start state to the one "incomplete" node in the DFA graph

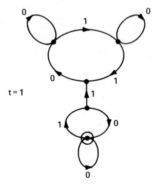

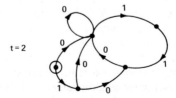

Fig. 4.5. Minimal deterministic finite automata corresponding to the regular languages $\Omega^{(t)}$ generated in the first two time steps of evolution under rule 72. $\Omega^{(1)}$ is an infinite complement regular language, with the infinite sequence of distinct blocks 111, 1101011, 11001011, ... excluded. $\Omega^{(2)}$ is a finite complement language, with only the blocks 010 and 111 excluded

blocks are 10101001 and 10010101 (length 8). After two time steps, the blocks 1110101 and 1010111 (length 7) are also excluded. The shortest newly-excluded blocks after three time steps are 01000010101, 01000110101, 10000010101 and their reversals (length 11). After four time steps, the shortest newly-excluded blocks are 010110011 and 110011010 (length 9), realizing the equality in (3.13).

Rule 126 provides an example in which the set generated after one time step is a finite complement language, but the sets generated at subsequent times are not. Rule 72 exhibits the opposite behaviour[11], as shown in Fig. 4.5. After one time step, it yields a set in which the infinite sequence of distinct blocks 111, 1101011, 11001011, ... are excluded (as in $\Omega^{(1)}$ for rule 18). After two times steps, however, the block 010 is also excluded. The exclusion of this single block implies exclusion of the infinite set of blocks excluded from $\Omega^{(1)}$. The resulting set thus corresponds to a finite complement language. In general, it can be shown that if a cellular automaton evolves to a finite complement language limit set, then it must do so in a finite number of time steps [34].

The sets $\Omega^{(t)}$ generated by most cellular automata never appear to become simpler with time. One exception is rule 72, in which the number of arcs in the minimal DFA for $\Omega^{(2)}$ is less than in that for $\Omega^{(1)}$. In most cases, the regular language complexity $\Xi^{(t)}$ appears to be non-decreasing with time. In fact, whenever the set of configurations generated continues to contract with time, a different regular language must be obtained at each time step. Since there are a limited number of regular languages with complexities below any given value (certainly less than $2^{k\Xi^2}$), the complexity must on average increase at least slowly with time in this case.

Table 1 suggests that a definite set of cellular automata (including rules 18, 22 and 126) yield regular language complexities $\Xi^{(t)}$ that grow on average more rapidly than any polynomial in time (perhaps exponentially with time). Many of the cellular automata in this set generically exhibit class 3, chaotic, behaviour, suggesting that rapidly-increasing $\Xi^{(t)}$ are a signal for class 3 behaviour in cellular automata.

In a few cases, such as rule 94, $\Xi^{(t)}$ increases rapidly with time, but almost all initial configurations are found to give ultimately periodic behaviour. Nevertheless, special initial conditions (in this case, those in which successive pairs of sites have equal values) can yield chaotic behaviour. Since the set $\Omega^{(t)}$ includes all configurations that ever occur, it includes those that give chaotic behaviour, even though they occur with vanishingly small probability. Presumably these configurations would not affect a probabilistic grammar for the set $\Omega^{(t)}$ that included only nonzero probability configurations. But the $\Xi^{(t)}$ for the grammars discussed here appear to increase rapidly with time whenever any set of configurations in the cellular automaton yield class 3 behaviour.

Some exceptional cases are surjective class 3 cellular automata, such as the additive rules 90 and 150, in which every possible configuration can be generated at any time. The complexity of these and other cellular automata could perhaps be measured by constructing a grammar for the set of possible space-time patterns generated in their evolution. Such a grammar could presumably be characterized

---

11   A more complicated example of this behaviour was given in [33]

in terms of computers with memories arranged in a two-dimensional lattice (cf. [35])[12].

The local rules ϕ for the 32 legal $k = 2$, $r = 1$ cellular automata of Tables 1–3 are all distinct. Yet in many cases sets of configurations with the same structure or properties are found to be generated. In some cases, there may exist bijective mappings which transform configurations evolving according to one cellular automaton rule into configurations evolving according to another rule. Several properties of the sets $\Omega^{(t)}$ are invariant under such mappings. One example is the set of non-zero roots of the characteristic polynomials $\chi^{(t)}(\lambda)$. While after one time step several of the cellular automata in Tables 1–3 yield the same sets of configurations $\Omega^{(1)}$, there are few examples of complete equivalence between pairs of cellular automaton rules. One simple example is rules 146 and 182, which are related by interchange of the roles of 0 and 1.

5. Some Invariant Sets

Section 2 showed that the set of configurations generated after a finite number of steps in the evolution of any cellular automaton forms a regular language. Sections 3 and 4 discussed some properties of such sets. This section and the next one consider the limiting sets of configurations generated after many time steps of cellular automaton evolution.

For all configurations A that appear in the limit set for a cellular automaton, there must exist some configuration A' such that $A = \Phi^t A'$ for any t. Any set of configurations invariant under the cellular automaton rule therefore appear in its limit set. This section considers some simple examples of invariant sets; Sect. 6 gives some comments on the complete structure of limit sets for cellular automata.

Periodic Sets

A simple class of invariant sets consist of configurations periodic with time under cellular automaton evolution. Such sets are found to form finite complement languages.

Consider the set of configurations that are stable (have temporal period 1) under a cellular automaton rule with $k = 2$ and $r = 1$. The set of such configurations is exactly those which contain only neighbourhoods $\{a_{i-1}, a_i, a_{i+1}\}$ for which

$$\phi[a_{i-1}, a_i, a_{i+1}] = a_i. \tag{5.1}$$

Only the finite set of distinct three-site blocks that violate (5.1) are forbidden, so that the complete set forms a finite-complement language, with a maximum distinct excluded block of length 3. A NDFA that generates the set of stable configurations is represented by a graph analogous to Fig. 3.1 in which only those

---

12   This paper concentrates on one-dimensional cellular automata. Such cellular automata potentially correspond most directly with conventional formal languages. Two and higher dimensional cellular automata show some differences. For example the set of configurations obtained after a finite number of time steps in their evolution need not form a regular language and may in fact be nonrecursive [36, 51]

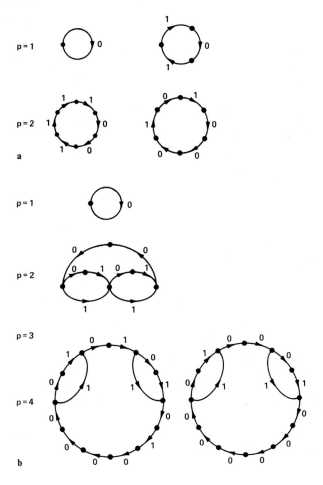

arcs satisfying (5.1) are retained. The minimal grammar for this set is obtained by constructing the minimal equivalent DFA, as described in Sect. 2.

The procedure generalizes immediately to arbitrary cellular automaton rules, and to sets of configurations with any finite period (cf. [37]). The distinct excluded blocks in the finite complement languages corresponding to sets of configurations with period p have maximum length $2pr + 1$.

Figure 5.1 shows the minimal grammars for sets of configurations with various periods under the $k = 2$, $r = 1$ cellular automata with rule numbers 90, 18 and 22. The grammars are represented by graphs containing several disconnected pieces, each corresponding to a disjoint set of configurations.

Figure 5.1a suggests that only a finite number of configurations, all spatially periodic, are found with each temporal period in the surjective cellular automaton rule 90. For this and other surjective cellular automata whose local mappings ϕ are injective in their first and last arguments, the number of distinct configurations

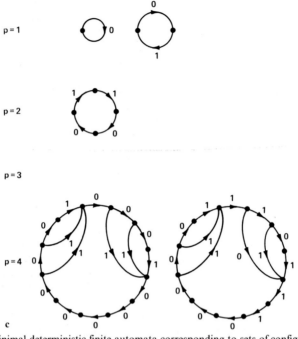

Fig. 5.1a–c. Minimal deterministic finite automata corresponding to sets of configurations with (temporal) periods exactly p under cellular automaton rules **a** 90, **b** 18 and **c** 22

with any period p is always finite, and is exactly k^{hp}, where **h** is the invariant entropy of the cellular automaton mapping (**h** = 2 for rule 90)[13]. This result follows from the fact that the complete space-time pattern generated by the evolution of such a cellular automaton is completely determined by any patch of site values with infinite temporal extent, but spatial width **h** (typically equal to $2r$). Moreover, any possible set of site values may occur in this patch. If the complete space-time pattern is to have period p, then so must the patch; but there are exactly k^{hp} possible patches with period p. (For large p, this result is as expected for any expansive homeomorphism (e.g. [10, 11]).)

   In general, the sets of configurations with a particular periodicity under a cellular automaton rule are infinite, as illustrated for rules 18 and 22 in Figs. 5.1b and 5.1c. Presumably there are sets of this kind with arbitrarily large periods. These infinite sets are nevertheless finite complement languages. For example, for the set of configurations with period two under rule 18, only the distinct blocks 111, 1011, 1101 and 10101 are excluded. It is common in class 3 cellular automata to find configurations with almost every possible period; for class 4 cellular automata, only some periods are typically found.

---

13   The actual configurations with particular periods may be found by methods analogous to those used in [29] for the complementary problem of determining the temporal periods of configurations with given spatial period

Periodic configurations form a small subset of all the configurations in the limit sets for cellular automata. Their entropy nevertheless provides a lower bound on the entropy of the complete limit sets. For rule 90, the set of periodic configurations has zero entropy, yet the complete limit contains all possible configurations, and thus has entropy 1. For rule 18, the period 2 set has entropy $\simeq 0.4057$ (given as the logarithm of the largest root of $\lambda^3 - \lambda - 1$), while the period 4 set has entropy $\simeq 0.1824$ ($\lambda^6 - \lambda - 1$). For rule number 22, the period 4 set has entropy $\simeq 0.3219$ ($\lambda^5 - \lambda^4 + \lambda^3 - \lambda^2 - 1$). Since irreversible cellular automaton mappings are contractive, the entropy of the set obtained after a finite number of time steps gives an upper bound on the entropy of the complete limit set. Using results from Table 3 one then finds

$$0.4057 \lesssim s_{[18]}^{(\infty)} \lesssim 0.8114,$$
$$0.1824 \lesssim s_{[22]}^{(\infty)} \lesssim 0.9390.$$
(5.2)

Simulation Sets

The complete invariant sets for many cellular automata Φ are very complicated. Parts of these invariant sets may however have a simpler structure, and may consist of configurations for which Φ "simulates" a simpler cellular automaton rule. Thus for example stable configurations under Φ may be considered as those for which Φ "simulates" the identity mapping.

One class of configurations for which a cellular automaton rule Φ_1 may simulate a rule Φ_2 are those obtained by "blocking transformations." Each symbol in the possible configurations of Φ_2 is replaced by a length b_X block of symbols in Φ_1, and each time step in the evolution of Φ_2 is simulated by b_T time steps of evolution under Φ_1. Thus, for example, rule 18 simulates rule 90 under the ($b_X = 2, b_T = 2$) blocking transformation $00 \to 0$, $01 \to 1$ [1, 38, 5]. The evolution of an arbitrary configuration under rule 90 is thus simulated by the evolution under rule 18 of a configuration consisting of the digrams 00 and 01. But since rule 90 is surjective, all possible configurations correspond to an invariant set. Thus configurations containing only 00 and 01 digrams form an invariant set for rule 18. The entropy of these configurations is $1/2$, so that

$$0.5 \leq s_{[18]}^{(\infty)} \lesssim 0.8114.$$
(5.3a)

Rule 22 simulates rule 90 under the (4,4) blocking transformation $0000 \to 0$, $0001 \to 1$, implying that

$$0.25 \leq s_{[22]}^{(\infty)} \lesssim 0.9390.$$
(5.3b)

A cellular automaton rule may simulate other rules with the same values of k and r under different blocking transformations (cf. the simulation network given in [5]). Some rules, apparently only surjective ones such as rule 90, simulate themselves, and thus correspond to fixed points of the blocking transformation. In other cases, one rule may simulate another under several distinct blocking transformations. For example, rule 18 simulates rule 90 under both $00 \to 0$, $01 \to 1$, and $00 \to 0$, $10 \to 1$, while rule 22 simulates rule 90 under any permutation of $0000 \to 0$, $0001 \to 1$. One may consider the sets of blocks appearing in these blocking transformations to represent different "phases." An initial configuration then consists of several "domains," each of which contains blocks of one phase. The domains are separated by "walls." For rule 18, these walls appear to execute random walks, and annihilate in pairs, yielding progressively larger domains of a single phase [38]. The simulation of rule 90 by rule 18 may thus be considered "attractive" [3]. For rule 22, no such simple behaviour is observed.

Blocking transformations yield a particular class of configurations, corresponding to simple finite complement languages. Other classes of configurations, specified by more general grammars, may also yield simulations. (An example occurs for rule number 73, in which configurations containing only odd-length sequences of 0 and 1 sites simulate rule 90.) In addition, a set of configurations evolving under one rule may simulate an invariant set of configurations evolving under another rule.

6. Comments on Limiting Behaviour

Section 2 showed that after any finite number of time steps, the set of configurations $\Omega^{(t)}$ generated by any cellular automaton forms a regular language. Some cellular automata yield regular languages even in the infinite time limit; others appear to generate limit sets corresponding to more complicated formal languages. Cellular automata which exhibit different classes of overall behaviour appear to yield characteristically different limiting languages.

As discussed in Sect. 4, some cellular automata in Tables 1–3 yield regular languages which attain a fixed form after a few time steps. The limit sets for such cellular automata are thus regular languages. In fact, except for surjective rules, the limit sets found appear to contain only temporally periodic configurations, and are therefore finite complement languages. These cellular automata exhibit simple large time behaviour, characteristic of classes 1 and 2.

Rule 128 provides a more complicated example, discussed in Sect. 4. After t time steps, any pair of ones in configurations generated by this rule must be separated by at least $2t$ sites. The complete set of possible configurations forms a finite complement regular language, with a minimal DFA illustrated in Fig. 4.1 whose size $\Xi^{(t)}$ increases linearly with time. After many time steps, almost all initial configurations evolve to the null configuration. However, even after an arbitrarily long time, configurations containing just a single block of ones may still appear. A block of n ones, flanked by infinite sequences of zeroes, is generated after any number of time steps t from a block of $n + 2t$ ones. Such configurations therefore have exactly one predecessor under any number of time steps of the cellular automaton evolution. They thus appear in the limit set for rule 128, although if all initial configurations are given equal weight, they are generated with zero probability. Once generated, their evolution is never periodic. An increasing number of distinct blocks are excluded from the successive $\Omega^{(t)}$ obtained by evolution under rule 128. The set of configurations generated in the infinite time limit does not, therefore, correspond to a finite complement language. Nevertheless, the set does form a regular language, shown in Fig. 6.1. While the set contains an infinite number of configurations, its entropy vanishes, as given by the limit of Eq. (4.1).

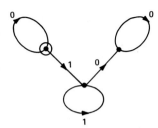

Fig. 6.1. The deterministic finite automaton representing the regular language corresponding to the limit set for cellular automaton rule 128. This infinite complement regular language is obtained as the infinite time limit of the series of finite complement regular languages illustrated in Fig. 4.1. It contains an infinite number of configurations, but has zero limiting entropy

Several rules given in Tables 1–3 exhibit behaviour similar to rule 128: they generate (finite complement) regular languages whose minimal grammars increase in size linearly or quadratically with time, but in the infinite time limit, yield regular language limit sets. These limit sets contain one or a few periodic configurations, together with an infinite number of aperiodic configurations, generated from a set of initial configurations of measure zero. The sets have zero entropy, and do not correspond to finite complement languages. (Only trivial finite complement languages can have zero entropy.) All the class 1 cellular automata (except for the trivial rule 0) in Tables 1–3 exhibit such limiting behaviour. The generation of limit sets corresponding to regular languages that are not finite complement languages appears to be a general feature of class 1 cellular automata.

Tables 1–3 suggest the result, discussed in Sect. 4, that cellular automata capable of class 3 or 4 behaviour give rise to sets of configurations represented by regular languages whose complexity increases rapidly with time. The limit sets for such cellular automata are therefore presumably not usually regular languages. If a finite description of them can be given, it must be in terms of more complicated formal languages.

Any language that can be described by a regular grammar must obey the regular language "pumping lemma" (e.g. [7]). This requires that it be possible to write all sufficiently long symbol sequences α appearing in the language in the form $\alpha_1\alpha_2\alpha_3$ so that for any n the symbol sequence $\alpha_1\alpha_2^n\alpha_3$ also appears in the language. (This result follows from the fact that any sufficiently long sequence must correspond to a path containing a cycle in the DFA. This cycles may then be traversed any number of times, yielding arbitrarily repeated symbol sequences.) The sets generated after a finite number of time steps in cellular automaton evolution always obey this condition: arbitrary repetitions of the string α_2 are obtained by evolution from initial configurations containing arbitrarily-repeated sequences evolving to α_2.

It is possible to construct cellular automata for which the regular language pumping lemma fails in the large time limit, and which therefore yield non-regular language limit sets. In one class of examples [39, 34], there are pairs of localized structures which propagate with opposite velocities from point sources. After t time steps, such cellular automata generate configurations consisting roughly of repetitions of sequences

$$(10^j 20^j 1) \qquad j \leqq t. \qquad\qquad (6.1)$$

In the infinite time limit, arbitrarily long identical pairs of symbol sequences thus appear. The limit sets for such cellular automata are therefore not regular languages. Instead it appears that they correspond to context-free languages.

The pumping lemma for regular languages may be generalized to context-free languages. Any sufficiently long sequence in a context-free language must be of the form $\alpha_1\alpha_2\alpha_3\alpha_4\alpha_5$ such that $\alpha_1\alpha_2^n\alpha_3\alpha_4^n\alpha_5$ is also in the language for any n. The possibility for separated equal length identical substrings is a reflection of the non-local nature of context-free languages, manifest for example in the indefinitely large memory stacks required in machines to recognize them.

Limiting sets of configurations of the form (6.1) that violate the regular language pumping lemma nevertheless obey the context-free language pumping lemma, and thus correspond to context-free languages.

The correspondence between sets of infinite cellular automaton configurations and context-free languages is slightly more complicated than for regular languages. In all cases, the cellular automaton configurations correspond to infinite symbol sequences generated according to a formal grammar. For regular languages, it it also possible to construct finite automata which recognize words in the language, starting at any point. The necessity for a stack memory in the generation of context-free languages makes their recognition starting at any point in general impossible. Infinite configurations generated by context-free grammars must thus be viewed as concatenations of finite context-free language words. Only at the boundaries between these words is the stack memory for the machine generating the configuration empty, so that sequences of symbols may be recognized. Configurations generated by context-sensitive and more complicated grammars must be considered in an analogous way.

If the limit set for a cellular automaton is a context-free language, whose generation requires a computer with an indefinitely large stack memory, then one expects that the regular language sets obtained at successive finite time steps in the evolution of the cellular automaton would require progressively larger finite size stack memories. If the limiting context-free grammar contains say Q (non-terminal) productions, then there are $O(Q^t)$ possible stack configurations after t time steps, and the set of configurations obtained may be recognized by a finite automaton with about Q^t states. In addition, the context-free pumping lemma is satisfied for repetitions of substrings of length up to about t. Regular languages that approximate context-free languages for t time steps should have comparatively simple repetitive forms. The regular languages of Fig. 4.1 generated at finite times by rule 128 have roughly the expected form, but their limit is in fact a regular language. The absence of obvious patterns in the regular grammars such as Figs. 4.2–4.4 generated by typical class 3 cellular automata after even a few time steps suggests that the limiting languages in these cases are not context-free. They are presumably context-sensitive or unrestricted languages.

The entropies of regular languages are always logarithms of algebraic integers [as in Eq. (3.5)]. Context-free languages may, however, have entropies given by logarithms of general algebraic numbers (whose minimal polynomials are not necessarily monic). The enumeration of words in a formal language may be cast in algebraic terms by considering the sequence of words in the language as a formal power series satisfying equations corresponding to the production rules for the language (e.g. [40]). For the simple regular language $((0^*)10)$ (repetition understood) of Fig. 1.3b, with production rules

$$u_0 \to s_0 u_0 , \qquad u_0 \to s_1 u_1 , \qquad u_1 \to s_0 u_0 , \qquad (6.2)$$

(where the terminal symbols s_0 and s_1 represent 0 and 1 respectively), the corresponding equations are

$$u_0 = s_0 u_0 + s_1 u_1 + 1 , \qquad u_1 = s_0 u_0 + 1 . \qquad (6.3)$$

Solving for u_0 as the start symbol one obtains

$$u_0 = (s_1 + 1)/(1 - s_0 - s_1 s_0) . \qquad (6.4)$$

The expansion of this generating function (accounting for the non-commutative nature of symbol string concatenation) yields the sequence of possible words in the language. Replacing all terminal symbols in the generating function by a dummy variable x, the coefficient of x^n in its expansion gives the number of distinct symbol sequences of length n in the language. The asymptotic growth rate of this number, and thus the entropy of the language, are then determined by the smallest real root of the (monic) denominator polynomial. The generating function for any regular language is always a rational function of x. For a context-free language, however, the equations analogous to (3.8) are in general non-linear in the u_i. At least for unambiguous languages, the positions of the leading poles in the resulting generating functions obtained by solving these simultaneous polynomial equations are nevertheless algebraic numbers [41].

There is a finite procedure to find the minimal regular grammar that generates a given regular language, as described in Sect. 2. No finite procedure exists in general, however, to find the minimal context-free or other grammar corresponding to a more complicated language. The analogue of the regular language complexity is thus formally non-computable for context-free and more complicated languages. This is an example of the result that no finite procedure can in general determine the shortest input program that generates a particular output when run for an arbitrarily long time on some computer (e.g. [42]). Explicit testing of successively longer programs is inadequate, since the insolubility of the halting problem implies that no upper bound can in general be given for the time before the required sequence is generated. Particular simple cases of this problem are nevertheless soluble, so that, for example, the minimal grammars for regular languages are computable.

The entropies for regular and context-free languages may be computed by the finite procedures described above. The entropies for context-sensitive (type 1) and unrestricted (type 0) languages are, however, in general non-computable numbers [43]. Bounds on them may be given. But no finite procedure exists to calculate them to arbitrary precision. (They are in many respects analogous to the non-computable probabilities for universal computers to halt given random input.) If many class 3 and 4 cellular automata do indeed yield limit sets corresponding to context-sensitive or unrestricted languages, then the entropies of these sets are in general non-computable.

The discussion so far has concerned the generation of infinite configurations by cellular automaton evolution. One may also consider the evolution of configurations in which nonzero sites exist only in a finite region. Then for class 3 cellular automata with almost all initial states, the region of nonzero sites expands linearly with time. (Such expansion is guaranteed if, for example, $\phi[1, 0, ..., 0] = 1$ and so on.) For class 4 cellular automata, the region may expand and contract with time. One may characterize the structures generated by considering the set of finite sequences generated at any time by evolution from a set of finite initial configurations. For class 3 cellular automata, this set appears to be no more complicated than a context sensitive language, while for class 4 cellular automata, it may be an unrestricted language. Notice that the set generated after a fixed finite number of time steps always corresponds to a regular language, just as for infinite configurations. (The regular grammar for these finite configurations consists of all paths with the relevant length that begin and end at the 00...0 node of the NDFA analogous to Fig. 2.1.)

Consider the language formed by the set of sequences of length n generated after any number of time steps in the evolution of a class 3 cellular automaton from all possible initial configurations with size n_0^{14}. This language appears to be at most context-sensitive, since a word of length n in it can presumably be recognized in a finite time by a computer with a memory of size at most n. In its simplest form, the computer operates by testing configurations generated by evolution from all k^{n_0} possible initial states. Since the configurations expand steadily with time, the evolution of each configuration need be traced only until it is of size n; the required configuration of length n is either reached at that time, or will never be reached.

In a class 4 cellular automaton, evolution from an initial configuration of size n_0 may yield arbitrarily large configurations, but then ultimately contract to give a size n configuration. No upper bound on the time or memory space required to generate the size n configuration may therefore be given. The problem of determining whether a particular finite configuration is ever generated in the evolution of a class 4 cellular automaton from one of a finite set of initial configurations may therefore in general be formally undecidable. No finite computation can give all the structures of a particular size ultimately generated in the evolution of a class 4 cellular automaton.

The procedure for recognizing finite configurations generated by class 3 cellular automata, while finite in principle, may require large computational resources. Whenever the context-sensitive language corresponding to the set of finite configurations cannot be described by a context-free or simpler grammar, the problem of recognizing words in the language is PSPACE-complete with respect to the lengths of the words (e.g. [28]). It can thus presumably be performed

---

14   This is analogous to but distinct from the problem of finding all initial configurations which ultimately evolve to a particular complete final configuration, such as the null configuration (cf. [2, 44, 45])

essentially no more efficiently than by testing the structures generated by the evolution of each of the k^{n_0} possible finite initial configurations.

As well as considering the evolution of finite complete configurations, one may also consider the generation of finite sequences of symbols in the evolution of infinite configurations. Enumeration of sets of length n sequences that can and cannot occur provide partial characterizations of sets of infinite configurations. However, even for configurations generated at a finite time t, such enumeration in general requires large computational resources. A symbol sequence of length n appears only if at least one length $n_0 = n + 2rt$ initial block evolves to it after t time steps. A computation time polynomial in n and t suffices to determine whether a particular candidate initial block evolves to a required sequence. The problem of determining whether any such initial block exists is therefore in the class NP. One may expect that for many cellular automata, this problem is in fact NP-complete. (The procedure of Sect. 2 provides no short cut, since the construction of the required DFA is an exponential computational process.) It may therefore effectively be solved essentially only by explicit simulation of the evolution of all exponentially-many possible initial sequences.

In the limit of infinite time, the problem of determining whether a particular finite sequence is generated in the evolution of a cellular automata becomes in general undecidable. For a cellular automaton with only class 1 or 2 behaviour, the limit set always appears to correspond to a regular language, for which the problem is decidable. But for class 3 and 4 cellular automata, whose limit sets presumably correspond to more complicated formal languages, the problem may be undeciable. (The problem is in general in the undecidability class Π_1 [46]; the set of finite sequences that occur is thus recursively enumerable, but not necessarily recursive.) Even when the general problem is undecidable, the appearance of particular finite sequences in the limit set for a cellular automaton may be decidable. The fraction of particular sequences whose appearance in the limit set is undecidable provides a measure of the degree of unpredictability or "computational achievement" of the cellular automaton evolution (presumably related to "logical depth" [47]).

7. Discussion

This paper has taken some preliminary steps in the application of computation theory to the global analysis of cellular automata. Cellular automata are viewed as computers, whose time evolution processes the information specified by their initial configurations. Many aspects of this information processing may be described in terms of computation theory. The intrinsic discreteness of cellular automata allows for immediate identifications with conventional computational systems; but the basic approach and many of the results obtained should be applicable to many other dynamical systems.

Self-organization in cellular automata involves the generation of distinguished sets of configurations with time. These sets are described as formal languages in computation theory terms. Each configuration corresponds to a word in a language, and is formed from a sequence of symbols according to definite grammatical rules. These grammatical rules provide a complete and succinct specification of the sets generated by the cellular automaton evolution.

Section 2 showed that, starting with all possible initial configurations, the sets generated by a finite number of time steps of cellular automaton evolution always correspond to regular formal languages. Such languages are recognized by finite automata. These finite automata are specified by finite state transition graphs; words in the languages correspond to all possible paths through these graphs. The (limiting) set entropies of such regular languages are then given as logarithms of the algebraic integers corresponding to the largest eigenvalues of the incidence matrices for their state transition graphs.

In genral, several different finite automata or regular grammars may yield the same regular language. However, it is always possible to find a simplest finite automaton, or set of grammatical rules, which correspond to any particular regular language. This simplest finite automaton provides a canonical representation for sets generated by cellular automaton evolution, and its size (number of states) gives a measure of their "complexity." The larger the "regular language complexity" for a set of configurations, the more complicated is the minimal set of grammatical rules necessary to describe it as a regular language.

Section 4 suggests the general result that the regular language complexity is non-decreasing with time for all cellular automata. This result gives a quantitative characterization of progressive self-organization in cellular automata. It may give a first indication of a generalization of the second law of thermodynamics to irreversible systems.

Entropy may be estimated from experimental data by fitting parameters in simple models which reproduce the data. Extraction of regular language complexities from experimental data requires the identification of maximal (regular language) patterns in the data, or the construction of a minimal (finite automaton) model that generates the data. Given perfect data (and an upper bound on the regular language complexity), a direct method may be used (e.g. [48]). In practice, it will probably be convenient to construct stochastic finite automata which provide probabilistic reproductions of the available data (cf. estimates for the structure of Markovian sources (e.g. [49])).

Dynamical systems theory methods were used in [2] to identify four general classes of cellular automaton behaviour. Sections 4 and 6 suggested computation theory characterizations of these classes. The limit sets for cellular automata with only class 1 or 2 behaviour are regular languages. For most class 3 and 4 cellular automata, the regular language complexity increases steadily with time, so that the set of configurations obtained in the large time limit does not usually form a regular language. Instead (at least for appropriate finite size configurations) the limit sets for class 3 cellular automata appear to correspond to context-sensitive languages, while those for class 4 cellular automata correspond to general languages.

Regular languages are sufficiently simple that their properties may be determined by finite computational procedures. Properties of context-free and more complicated languages are, however, often not computable by finite means. Thus, for example, the minimal grammars for such languages (whose sizes would provide analogues of the regular language complexity) cannot in general be found by finite computations. Moreover, for context-sensitive and general languages, even quantities such as entropy are formally non-computable.

When cellular automaton evolution is viewed as computation, one may consider that the limiting properties of a cellular automaton are determined by an infinite computational process. One should not expect in general that the results of this infinite process can be summarized in finite mathematical terms. For sufficiently simple cellular automata, apparently those of classes 1 and 2, however, it is nevertheless possible to "short cut" the infinite processes of cellular automaton evolution, and to give a finite specification of their limiting properties. For most class 3 and 4 cellular automata, no such short cut appears possible: their behaviour may in general be determined by no procedure significantly faster than explicit simulation, and many of their limiting properties cannot be determined by any finite computational process. (Such non-computable limiting behaviour would be an immediate consequence of the universal computation capability conjectured for class 4 cellular automata, but does not depend on it.)

Non-computability and undecidability are common phenomena in the systems investigated in pure mathematics, logic and computation. But they have not been identified in the systems considered in theoretical physics. In many physical theories one can in fact imagine constructing complicated systems which behave, for example, as universal computers, and for which undecidable propositions may be formulated. Cellular automata (and other dynamical systems) may be considered as simple physical theories. This paper has suggested that in fact even simple, natural, questions concerning the limiting behaviour of cellular automata are often undecidable (except for very simple systems such as those corresponding to class 1 and 2 cellular automata). One may speculate that undecidability is common in all but the most trivial physical theories. Even simply-formulated problems in theoretical physics may be found to be provably insoluble.

Undecidability and non-computability are features of problems which attempt to summarize the consequences of infinite processes. Finite processes may always be carried out explicitly. For some particularly simple processes, the consequences of a large, but finite, number of steps may be deduced by a procedure involving only a small number of steps. But at least for many computational processes (e.g. [28]), it is believed that no such short cut exists: each step (or each possibility) must in fact be carried out explicitly. It was suggested that this phenomenon is common in cellular automata. One may speculate that it is widespread in physical systems. No simple theory or formula could ever be given for the overall behaviour of such systems: the consequences of their evolution could not be predicted, but could effectively be found only by direct simulation or observation.

Acknowledgements. I am grateful to A. Aho, C. Bennett, J. Conway, D. Hillis, L. Hurd, D. Lind, O. Martin, M. Mendes France, J. Milnor, A. Odlyzko, N. Packard, J. Reeds, and many others for discussions. A preliminary version of this paper was presented at a workshop on "Coding and Isomorphisms in Ergodic Theory," held at the Mathematical Sciences Research Institute, Berkeley (December 8–13, 1983). I thank M. Boyle, E. Coven, J. Franks, and many of the other participants for their comments. Some of the results given above were obtained using the computer mathematics system SMP [50].

References

1. Wolfram, S.: Statistical mechanics of cellular automata. Rev. Mod. Phys. **55**, 601 (1983)
2. Wolfram, S.: Universality and complexity in cellular automata. Physica **10D**, 1 (1984)
3. Packard, N.H.: Complexity of growing patterns in cellular automata, Institute for Advanced Study preprint (October 1983), and to be published in Dynamical behaviour of automata. Demongeot, J., Goles, E., Tchuente, M., (eds.). Academic Press (proceedings of a workshop held in Marseilles, September 1983)
4. Wolfram, S.: Cellular automata as models for complexity. Nature (to be published)
5. Woram, S.: Cellular automata. Los Alamos Science, Fall 1983 issue
6. Beckman, F.S.: Mathematical foundations of programming. Reading, MA: Addison-Wesley 1980
7. Hopcroft, J.E., Ullman, J.D.: Introduction to automata theory, languages, and computation. Reading, MA: Addison-Wesley 1979
8. Minsky, M.: Computation: finite and infinite machines. Englewood Cliffs, NJ: Prentice-Hall 1967
9. Rozenberg, G., Salomaa, A. (eds.): L systems. In: Lecture Notes in Computer Science, Vol. 15 Rozenberg, G., Salomaa, A.: The mathematical theory of L systems. New York: Academic Press 1980
10. Guckenheimer, J., Holmes, P.: Nonlinear oscillations, dynamical systems, and bifurcations of vector fields. Berlin, Heidelberg, New York: Springer 1983
11. Walters, P.: An introduction to ergodic theory. Berlin, Heidelberg, New York: Springer 1982
12. Weiss, B.: Subshifts of finite type and sofic systems. Monat. Math. **17**, 462 (1973); Coven, E.M., Paul, M.E.: Sofic systems. Israel J. Math. **20** 165 (1975)
13. Field, R.D., Wolfram, S.: A QCD model for e^+e^- annihilation. Nucl. Phys. B **213**, 65 (1983)
14. Smith, A.R.: Simple computation-universal cellular spaces. J. ACM **18**, 331 (1971)
15. Berlekamp, E.R., Conway, J.H., Guy, R.K.: Winning ways for your mathematical plays. New York: Academic Press, Vol. 2, Chap. 25
16. Lind, D.: Applications of ergodic theory and sofic systems to cellular automata. Physica **10**D, 36 (1984)
17. de Bruijn, N.G.: A combinatorial problem. Ned. Akad. Weten. Proc. **49**, 758 (1946); Good, I.J.; Normal recurring decimals. J. Lond. Math. Soc. **21**, 167 (1946)
18. Nerode, A.: Linear automaton transformations. Proc. Am. Math. Soc. **9**, 541 (1958)
19. Cvetkovic, D., Doob, M., Sachs, H.: Spectra of graphs. New York: Academic Press 1980
20. Billingsley, P.: Ergodic theory and information. New York: Wiley 1965
21. Chomsky, N., Miller, G.A.: Finite state languages. Inform. Control **1**, 91 (1958)
22. Stewart, I.N., Tall, D.O.: Algebraic number theory. London: Chapman & Hall 1979
23. Lind, D.A.: The entropies of topological Markow shifts and a related class of algebraic integers. Ergodic Theory and Dynamical Systems (to be published)
24. Milnor, J.: Unpublished notes (cited in [2])
25. Hedlund, G.A.: Endomorphisms and automorphisms of the shift dynamical system. Math. Syst. Theor. **3**, 320 (1969); Hedlund, G.A.: Transformations commuting with the shift. In: Topological dynamics. Auslander, J., Gottschalk, W.H. (eds.). New York: Benjamin 1968
26. Amoroso, S., Patt, Y.N.: Decision procedures for surjectivity and injectivity of parallel maps for tessellation structures. J. Comp. Syst. Sci. **6**, 448 (1972)
27. Nasu, M.: Local maps inducing surjective global maps of one-dimensional tessellation automata. Math. Syst. Theor. **11**, 327 (1978)
28. Garey, M.R., Johnson, D.S.: Computers and intractability: a guide to the theory of NP-completeness. San Francisco: Freeman 1979, Sect. A10
29. Martin, O., Odlyzko, A.M., Wolfram, S.: Algebraic properties of cellular automata. Commun. Math. Phys. **93**, 219 (1984)
30. Hedlund, G.: Private communication

Cellular Automata

31. Manning, A.: Axiom A diffeomorphisms have rational zeta functions. Bull. Lond. Math. Soc. **3**, 215 (1971);
    Coven, E., Paul, M.: Finite procedures for sofic systems. Monat. Math. **83**, 265 (1977)
32. Franks, J.: Private communication
33. Coven, E.: Private communication
34. Hurd, L.: Formal language characterizations of cellular automata limit sets (to be published)
35. Rosenfeld, A.: Picture languages. New York: Academic Press (1979)
36. Golze, U.: Differences between 1- and 2-dimensional cell spaces. In: Automata, Languages and Development, Lindenmayer, A., Rozenberg, G. (eds.). Amsterdam: North-Holland 1976
    Yaku, T.: The constructibility of a configuration in a cellular automaton. J. Comput. System Sci. **7**, 481 (1983)
37. Grassberger, P.: Private communication
38. Grassberger, P.: A new mechanism for deterministic diffusion. Phys. Rev. A (to be published)
    Chaos and diffusion in deterministic cellular automata. Physica **10**D, 52 (1984)
39. Hillis, D., Hurd, L.: Private communications
40. Salomaa, A., Soittola, M.: Automata-theoretic aspects of formal power series. Berlin, Heidelberg, New York: Springer 1978
41. Kuich, W.: On the entropy of context-free languages. Inform. Cont. **16**, 173 (1970)
42. Chaitin, G.: Algorithmic information theory. IBM J. Res. Dev. **21**, 350 (1977)
43. Kaminger, F.P.: The non-computability of the channel capacity of context-sensitive languages. Inform. Cont. **17**, 175 (1970)
44. Smith, A.R.: Real-time language recognition by one-dimensional cellular automata. J. Comput. Syst. Sci. **6**, 233 (1972)
45. Sommerhalder, R., van Westrhenen, S.C.: Parallel language recognition in constant time by cellular automata. Acta Inform. **19**, 397 (1983)
46. Rogers, H.: Theory of recursive functions and effective computability. New York: McGraw-Hill 1967
47. Bennett, C.H.: On the logical "depth" of sequences and their reducibilities to random sequences. Inform. Control (to be published)
48. Conway, J.H.: Regular algebra and finite machines. London: Chapman & Hall 1971
49. Shannon, C.E.: Prediction and entropy of printed English. Bell Syst. Tech. J. **30**, 50 (1951)
50. Wolfram, S.: SMP reference manual. Computer Mathematics Group. Los Angeles: Inference Corporation 1983
51. Packard, N.H., Wolfram, S.: Two dimensional cellular automata. Institute for Advanced Study preprint, May 1984
52. Hopcroft, H.: An $n \log n$ algorithm for minimizing states in a finite automaton. In: Proceedings of the International Symposium on the Theory of Machines and Computations. New York: Academic Press 1971
53. Hurd, L.: Private communication

Communicated by O. E. Lanford

Received December 12, 1983; in revised form April 17, 1984

Physica 10D (1984) 81–95
North-Holland, Amsterdam

PHYSICS-LIKE MODELS OF COMPUTATION*

Norman MARGOLUS

MIT Laboratory for Computer Science, Cambridge Massachusetts 02139, USA

Reversible Cellular Automata are computer-models that embody discrete analogues of the classical-physics notions of space, time, locality, and microscopic reversibility. They are offered as a step towards models of computation that are closer to fundamental physics.

1. Introduction

Reversible Cellular Automata (RCA) are computer-models that embody discrete analogues of the classical-physics notions of space, time, locality, and microscopic reversibility.

In this paper, I will describe some RCA, explain how they can be used as computer models, and discuss RCA analogues of energy and entropy – concepts that are fundamental in physics, but have not played a fundamental role in computer theory.

2. Cellular automata

In CA, 'space' is a regular lattice of 'cells', each of which contains one of a small allowed set of integers. Only cells that are close together interact in one 'time-step' – the time evolution is given by a rule that looks at the contents of a few neigh-

bouring cells, and decides what should change. At each step, this local rule is applied everywhere simultaneously[10].

The best-known example of such a 'digital-world' is Conway's[5] "Game of Life". On a sheet of graph-paper, fill each cell with a '1' or a '0'. In each three-by-three neighbourhood there is a center cell and eight adjacent cells. The new state of each cell is determined by counting the number of adjacent 1's – if exactly two adjacent cells contain a one, the center is left unchanged. If three are ones, the center becomes a one. In all other cases, the center becomes a zero.

Such a rule gives rise to a set of characteristic patterns that 'move' (reappear in a slightly displaced position after some number of steps) patterns that are stable (unchanging with time) patterns that oscillate (pass through some cycle of configurations) and many very complicated interactions and behaviours. The evolution of a given initial configuration is often very hard to anticipate (see colour plate in [9]).

One way to show that a given rule can exhibit complicated behaviour is to show (as has been done for "Life"[4]) that in the corresponding 'world' it is possible to have computers. If you start the automaton with an appropriate initial state, you will see digits acting as signals moving about and interacting with each other to perform all of the logical operations of a digital computer. Such a computer-automaton is said to be *universal*.**

* This research was supported in part by the Defense Advanced Research Projects Agency and was monitored by the Office of Naval Research under Contracts Nos. N00014–75–C–0661 and N00014–83–K–0125, and in part by NSF Grant No. 8214312-IST.

** Von Neumann[10] was interested in the problem of evolution – could life emerge from simple rules? He exhibited a CA rule that permitted computers, and in which these computers could reproduce and mutate. In this paper, I refer only to the existence of computers when I use the term universal.

3. Reversible cellular automata

Any CA rule can be described by an equation of the form*

$$S_{i,t+1} = f(S_{\{i\},t}), \tag{1}$$

where $S_{i,t+1}$ is the state of the cell at position 'i' and at time '$t + 1$', and $f(S_{\{i\},t})$ is a function of the states of cells in a neighbourhood of i, at time t.

In general, (1) gives rise to a non-invertible dynamics. If f is the 'Life' rule, this evolution is not reversible – if an area now contains only zeros, did it contain zeros one step ago, or were there perhaps some isolated ones that just changed? Its impossible to tell.

It turns out to be very easy to write down CA laws that give an invertible dynamics – just as easy as constructing irreversible ones, in fact. Consider first the following finite difference equation, with x_t a real variable:

$$x_{t+1} = f(x_t) - x_{t-1}. \tag{2}$$

If you want to compute x_{t+1}, you must know x_t and x_{t-1} – these two constitute the complete 'state' of the system. For what functions f will the time evolution be invertible?

$$x_{t-1} = f(x_t) - x_{t+1}, \tag{3}$$

therefore any f at all will do**! Knowing x for two consecutive times allows you to calculate any preceding or any succeeding value of x (To my knowledge Fredkin[2] was the first to study reversibility in finite-difference-equations of this sort.)

The generalization to CA is straightforward – let x in (2) be replaced by c_i, the contents of the cell at position 'i' in our automaton,

$$c_{i,t+1} = f(c_{\{i\},t}) - c_{i,t-1}, \tag{4}$$

where $f(c_{\{i\},t})$ is any function involving the contents of cells near position 'i', at time 't', and the difference is taken mod the number of allowed cell values***. If we let the state of a cell correspond to its contents in two successive steps, then (4) can be reexpressed in the form (1), but its reversibility is not manifest†.

Such rules can be universal (I give an example in the appendix). Reversible computation is a relatively new idea [1, 3, 8] that has been used to show that a fundamental lower bound on dissipation in computers associated with the irreversibility of conventional logic elements[6] can be avoided.

4. Entropy in RCA

If we fill the cells of our automaton with randomly chosen binary values and then evolve it according to the Life rule, we see a complex ebb and flow of structures and activity, with so-called 'gliders' arising here and there, moving across clumps of zeros, and then being drawn back into a complex boiling 'soup' of activity, or perhaps rekindling complicated interactions in an area which had settled down into uncoupled, short period oscillating structures.

If, instead of the Life rule, we follow some invertible time evolution, we invariably find that, at each step, the state of the automaton looks just as random as when we started‡. This is expected

---

*This serves to clarify what sorts of systems we're dealing with, but is often not the simplest or most illuminating way to express the rule.

**Assuming integer addition and subtraction is done without error, if such an equation is iterated on a digital computer, its time evolution remains *exactly* reversible, despite roundoff and truncation errors in computing f.

***Differences mod-k and logical functions can always be re-expressed as ordinary polynomial functions. For example, if A and B are binary variables, then $(A - B)^2$ is the same as $A + B \pmod 2$, $1 - A$ is the same as not(A), $A * B$ is the same as and(A, B), etc. Thus (4) is equivalent to an ordinary real-variable finite difference equation with integer initial conditions.

†The global time evolution generated by (4) is not guaranteed to be invertible unless suitable boundary conditions are chosen, such as no boundary (i.e. an infinite or periodic space) or 'fixed' boundaries (cell values on the boundary are not allowed to change with time).

‡Spatial correlations will not arise if they are initially absent, but time correlations are often very evident, and are characteristic of the particular rule being employed – see the next section.

from a simple counting argument, since most configurations look random (only a very few random-looking initial configurations can be mapped by a given number of invertible steps into the few simple-looking configurations, since the overall mapping is bijective).

This is not meant to imply that RCA are less interesting than irreversible CA. Starting an RCA from a random state is like starting a thermodynamic system in a maximum entropy state – its not allowed to get any simpler since its randomness can't decrease, and it can't get more complicated, since its already as random as it can be, and so nothing much happens.

If we start an RCA from a very non-random state (e.g. some small pattern on a background of zeros) then we can have an interesting time evolution. If we choose a rule and an initial state that allow information to propagate, then what tends to happen is that the state of the RCA becomes more and more complicated. More precisely, if each state of the automaton is viewed as a 'message', with the contents of the cells being the characters of the message, and if only local measures of correlation are applied, then the amount of information* in successive messages is increasing. Of course the automaton is really only repeatedly encrypting its state, and so if all correlations are taken into account the amount of information really never changes. What happens is that the automaton will introduce some redundancy into the message, and use more cells to encode the same information. Information that was initially localized becomes spread out as correlations between the states of many cells, and it

---

*For a discussion of the information content of a message, cf.[7].

**(4) generates a locally invertible time evolution. If we know the values of cells near position i at two successive times, we can tell what the preceding value of the center cell was.

***In mechanics, this corresponds to degrees of freedom that, for certain initial conditions, are decoupled from the rest.

†For rules with 2 states per cell, only two rules, "count the parity of the neighbourhood" and its complement, have no configuration of part of the neighbourhood that makes the remaining neighbours irrelevant.

becomes very difficult for a locally invertible evolution to put the redundant pieces back together**. To use an analogy, an invertible mapping could change two copies of this document into one copy, and several sheets of blank paper. Two separate invertible mappings, each acting only on one of the copies, could not accomplish this end.

From the point of view of creatures 'living' inside an RCA, their inability to make use of complicated correlations between large numbers of cells means that for all practical purposes, the entropy of the automaton increases. To use a thermodynamic analogy, if I want to compress a gas, it doesn't help me to know that the gas was all in one corner of the room just a few minutes ago. I have no ability to make use of the complicated correlations that this statement implies, and so I say that entropy increased when the gas expanded to its current volume.

5. Conservation laws in second-order RCA

In general, an RCA has as many conserved quantities as there are cells – it 'remembers' the initial state of each cell, since you can recover this information by running the system backwards. Do these give rise to any invariants which can be computed in a local manner from the current state of the system? Can we find an invariant that is analogous to a classical mechanical energy?

In RCA, the simplest locally-computable invariants are of course cells whose values never change***. Such situations can arise because many rules ignore the remainder of the neighbours when part of the neighbourhood has some particular configuration†. For example, consider any rule that, in all cases where the center cell of the neighbourhood is 1, ignores the rest of the neighbours and returns a 2. Such a rule, when used with (4), results in a very simple conservation law. If we look at the case in one dimension where the automaton at two consecutive time-steps looks like this:

$t-1$ ...1... '.' indicates a cell whose value is
t ...1... irrelevant to the discussion   (5)

then the center cell here will always be a 1.

For a more interesting 1-dimensional example, consider a 2 state per cell CA with a rule f that returns a 1 iff each of the two cells adjacent to the center is the same as the center:

$$f(c_{\{x\},t}) = \begin{cases} 1, & \text{if } c_{x-1,t} = c_{x,t} = c_{x+1,t} \\ 0, & \text{otherwise.} \end{cases} \quad (6)$$

---

* In irreversible CA, a guarantee that a cell will always be part of such a pair does not guarantee that it always has been.
** An extreme instance of 'decoupling' of entire regions occurs with any rule that doesn't depend on the center cell, but depends on its nearest neighbours. For example, in 1D we might have a region that looks like this:

$t-1$ ...1.1.0.1...
t ...1.0.0.1.1...
$t+1$ ...?.?.?.?...

From (4) it is clear that we have enough information to compute the states of the cells marked with '?' – the system decouples into two entirely independent (but interleaved) sublattices, each evolving without reference to the other.

With this rule, 'a' and 'b' standing for any binary values, and '$\bar{a}$', '$\bar{b}$' their binary complements, the second-order time evolution given by (4) says that

$$\begin{Bmatrix} t-1 & ...a\bar{a}... \\ t & ...b\bar{b}... \end{Bmatrix} \rightarrow \begin{Bmatrix} t & ...b\bar{b}... \\ t+1 & ...a\bar{a}... \end{Bmatrix} \quad (7)$$

which is again of the same form, so these two cells are decoupled from the rest of the automaton. Any cell which is *not* initially part of such a pair will never be (and never was)*; counting all such cells gives us an (invariant) estimate of how many cells are available to represent dynamically changing information (but only an estimate – whole regions may be decoupled from the rest of the automaton because they are surrounded by a wall of decoupled cells**; a local counting wouldn't reveal this).

If we concentrate on the active (as opposed to the decoupled) cells, we can distinguish various kinds of activity, and try to associate conserved quantities with each. As a simple 1-dimensional example, consider 'dislocations' propagating in a regular background pattern of cells. Using the rule (6) again, consider the sequence of steps shown in fig. 1 (light and dark squares stand for 0's and 1's respectively; dislocations are triplets in a background of pairs and are outlined for emphasis). In this evolution, the number of such 'signals' is the

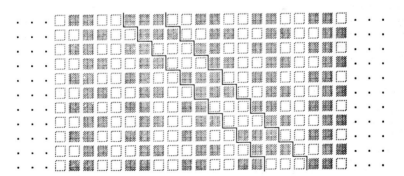

Fig. 1.

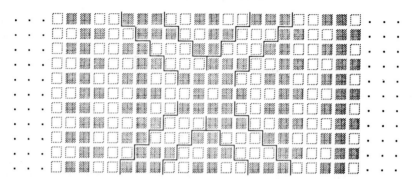

Fig. 2.

same as the number of blocks of cells we start off with the form $\cdots\overset{aa}{\underset{bb}{}}\cdots$, and is conserved.

If two such dislocations collide, we know that they can't just completely stop moving. Proof: if we tried to invert the evolution, we wouldn't know when to start the signals moving again. In fig. 2, the following quantity is the same for every pair of consecutive time-steps:

$$\left(\text{\# of cells in blocks of form } \cdots\overset{\cdots aa\cdots}{\underset{\cdots bb\cdots}{}}\right)$$

$$+\left(\text{\# of cells in blocks of form } \overset{\cdots c.c\cdots}{\underset{..d.d\cdots}{}}\right) \tag{8}$$

The first term counts the number of moving signals, and so could be thought of as a 'kinetic-energy' analogue. The second term accounts for the disappearance of this 'K.E.' during a collision, and so could be considered a potential-energy analogue.

---

* If f_s is the global rule that applies to the solid blocking, and f_d to the dotted blocking, then $S_{i+1} = f_s(f_d(S_i))$ describes the evolution using a time independent rule. By including a small amount of positional information in the state of each cell, this rule can be written in the form (1).

6. First-order RCA

Rather than continue to analyze RCA in order to discover conservation laws, we will now proceed to construct a class of automata that all obey a very simple local conservation law: the total number of 1's never changes, and neither does the number of 0's.

The trick we will use is quite general, but it will be illustrated in 2 dimensions with 2 states per cell. Fig. 3 shows a Cartesian lattice of cells, divided into 2 × 2 blocks of cells. We treat each 2 × 2 block as a conservative-logic[3] gate, with 4 inputs (its current state) and 4 outputs (its next state). These 'gates' are interconnected in an entirely uniform and predictable manner – in applying the rule to the 2 × 2 blocks, we alternate between using the solid blocking in this diagram for one step, and then using the dotted blocking for the next*. Fig.

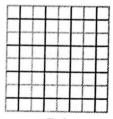

Fig. 3.

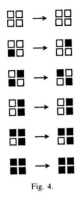

Fig. 4.

4 shows an example of a conservative rule (one that conserves 1's and 0's) that is reversible. In the case of all 0's or all 1's, there is no choice, they remain unchanged. Any rotation of one of the blocks on the left is mapped onto the corresponding rotation of the result to its right – this rule is rotationally symmetrical, and these are all of the possible cases. Since each distinct initial state of a block is mapped onto a distinct final state, this rule is reversible. As will be shown later, the automaton corresponding to this rule is universal.

One could easily have written down an example of a rule that conserved 1's and 0's, but that didn't always map a distinct initial state of a block into a distinct final state – such a rule would be conservative, but not reversible. As the corresponding automaton evolved, it would forget all sorts of details about the initial state, but it would always remember the numbers of 1's and 0's.* Thus the existence of an interesting local conservation law does not depend on the rule being reversible!

In the language of digital logic, a gate from which it is possible to construct any boolean function of any number of input variables is a *universal*

---

* For each gate (block), we can tell after each step how many possible predecessors the result-block has. Thus we can count exactly how much information is lost at each step.
**The BBMCA models space as being uniformly filled with gates, and so a connection is already apparent.

gate. If a logic gate is not universal, then no interconnection of such gates can be a computer. Thus the only candidates for universal CA's in the scheme described above are those whose rule corresponds to a universal logic gate.

In order to promote the CA rule of fig. 4 as a link between classical mechanics and computation, I will first discuss Fredkin's Billiard Ball Model (BBM) of computation[3] – a classical mechanical system that can be used to do digital computation. It will then be easier to discuss this rule, which I call the BBMCA – a purely digital model of computation which is closely related to the BBM.

7. The billiard ball model of computation

The BBM is a classical mechanical system, and obeys a continuous dynamics – positions and velocities, masses and times are all real variables. In order to make it perform a digital computation, we make use of the fact that integers are also real numbers. By suitably restricting the initial conditions we allow the system to have, and by only looking at the system at regularly spaced time intervals, we can make a continuous dynamics perform a digital process. In this case, we begin with a 2-dimensional gas of identical hard spheres. If the center of a sphere is present at a given point in space at a given point in time, we will say that there is a '1' there, otherwise there is a '0' there. The 1's can move from place to place, but their number never changes.

The key insight behind the BBM is this: *every place where a collision of finite-diameter hard spheres might occur can be viewed as a boolean logic gate***. What path a ball follows depends upon whether or not it hits anything – it makes a decision.

To see how to use this decision to do boolean logic, consider fig. 5. At points A and B and at time t_i, we either put balls at A, B, or both, or we put none. Any balls present are moving as indicated with a speed 's'. If balls are present at *both* A and B, then they will collide and follow the outer

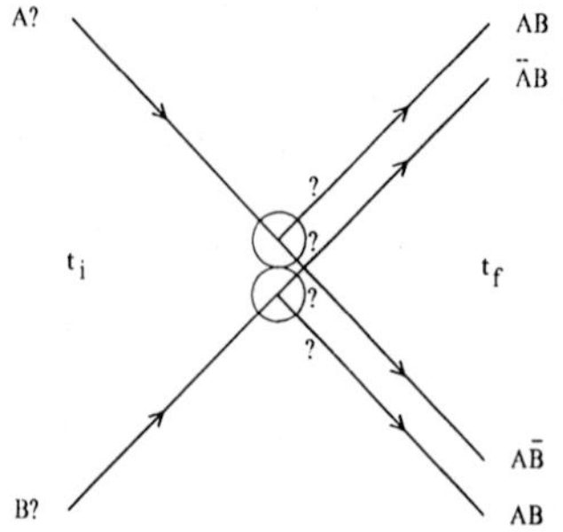

Fig. 5.

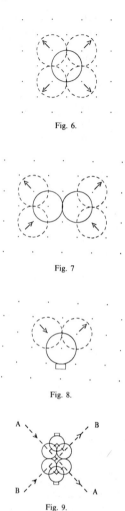

Fig. 6.

Fig. 7

Fig. 8.

Fig. 9.

outgoing paths. Otherwise, only the inner outgoing paths will be used. At time $t = t_i$, position A is a 1 if a ball is there, and 0 otherwise (similarly for position B). At $t = t_f$, the four labeled spots have a ball or no ball – which they have is given by the logical function labeling the spot. For example, if $A = 1$ and $B = 0$, then the ball coming from A encounters no ball coming from B, and ends up at the point labeled "A and not B". A place where a collision might occur acts as a reversible, universal [3] 1-conserving logic-gate, with two inputs and four outputs. A path that may or may not contain balls acts as a signal-carrying wire. Mirrors (reflectors) allows bends in the paths. In order to be able to use the outputs from such a collision-gate as inputs to other such gates, we need to very precisely control the angle and timing of the collisions, as well as the relative speeds of the balls. We make this simple to do by severely restricting the allowed initial conditions. Each ball must start at a grid point of a Cartesian lattice, moving 'along' the grid in one of 4 allowed directions. See fig. 6. All balls move at the same speed. The time it takes a ball to move from one grid point to another we call our unit of time. The grid spacing is chosen so that balls collide while at grid-points. See fig. 7. All collisions are right angle collisions, so that one time-step after a collision, balls are still on the grid. Fixed mirrors are positioned so that balls hit them while at a grid point, and so stay on the grid. See fig. 8. By using mirrors, signals can be

routed and delayed as required to perform digital logic. The configuration of mirrors in fig. 9 solves the problem of making two signals cross without affecting each other. (Notice that if two balls come in together, the signals cross but the balls don't!).

Mirrors and collisions determine the possible paths that signals may follow ('wires'). In order to ensure that all collisions will be right-angle collisions (and not head-on, for example, which would take us off our grid) we can label all 'wires' with arrows, and restrict initial conditions and interconnections so that a ball found on a given 'wire' always moves in the labeled direction.

Thus our universal gates can be connected as required to 'build' a computer. Computations can be pipelined – an efficient 'assembly-line' way of doing things, where questions flow in one end and finished products (answers) flow out the other, while all the stages in between are kept busy. Reversibility turns out not to be a great hindrance – unwanted intermediate results can be mostly 'erased' by copying the answer once you have it, and then running the computation backwards to get rid of everything but a copy of the inputs.

This then, in brief, is the BBM. Kinetic energy is conserved, since all collisions are elastic. Momentum is not conserved, since the mirrors are assumed to be fixed (infinitely massive).

8. The BBM cellular automaton

When viewed only at integer time-steps, the BBM consists of a Cartesian lattice of points, each of which may 'contain' a 0 or a 1, evolving according to a local rule. It would therefore seem to be a straightforward matter to find a CA rule that duplicates this digital time evolution.

Unfortunately, the most direct translation of the BBM into a CA has several problems. First of all, to have separate states of a cell to represent 4 kinds of balls (4 directions) an empty cell and a mirror, and to have the balls absolutely conserved (as they are in the original BBM) would require a standard "change the center cell" rule with 6 states per cell, and a 17 cell neighbourhood. Such a rule has a very large number of possible configurations for its neighbourhood, which makes it unwieldy. Moreover, many of these configurations involve such events as head-on collisions, which were disallowed in the BBM – a CA rule, however, should be defined for all configurations. It is not at all clear how to extend the BBM rule to these extra cases, and still have it remain reversible and 'energy' conserving.

At the expense of making collisions cause a slight delay, we can get away with the very simple rule of fig. 4, which involves only 2 states per cell in a 4 cell neighbourhood, is reversible, and conserves the number of ones (and zeros) in all cases.

The ⊞ → ⊞ (and rotations) case in fig. 4 is the one that causes an isolated '1' to propagate in a straight line, in one of four directions (depending on which of the four corners of its starting block you put it in). See fig. 10. The legend "solid" or "dotted" below each of these automaton configurations tells you whether the grouping of cells into blocks for the next application of the rule

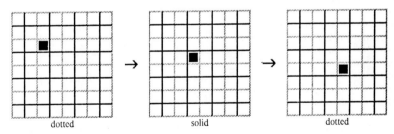

dotted   solid   dotted

Fig. 10.

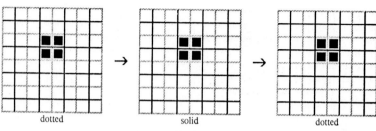

Fig. 11.

is indicated by the solid or the dotted lines. In the diagrams, a 0 is shown as an empty (blank) cell, a one is shown as a filled-in cell. Since ⊞ → ⊞ (and rotations), a square of four ones straddling the boundary of two adjacent blocks will be stable – we will use such squares to construct mirrors. See fig 11. The four 1's straddle two dotted blocks horizontally, then two solid blocks vertically, and then two dotted again. Since ⊞ → ⊞ (and rotations), pairs of travelling ones perform a billiard-ball type collision. See fig. 12. In all of these figures, the paths the ones were originally following have been lightly drawn in, to show that the 'and' case shown results in an outward displacement, just as in the BBM. (Unlike the BBM, there is a delay in such a collision, which we'll have to worry about in synchronizing signals). Finally, ⊞ → ⊞ (and rotations) permits the reflection of double signals by a mirror. See fig. 13. The 'mirror' consists of two adjacent stable squares (notice that a square is stable no matter what you put next to it – its

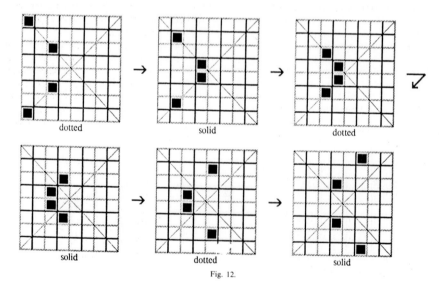

Fig. 12.

90   *N. Margolus / Physics-like models of computation*

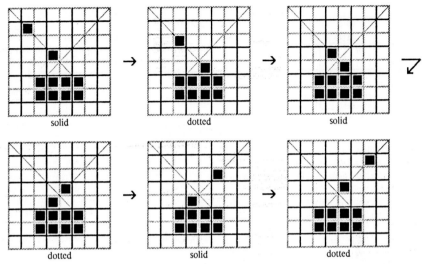

Fig. 13.

'decoupled'). Again, the signal path has been lightly drawn in. After each reflection such as that shown above, the signal has been delayed by a distance of one block along the plane of the mirror (in this picture, the signal winds up one block-column behind where it would have been had it not hit the mirror).

---

* We can tell how many steps a signal will take to traverse a given path (from one position where the signal is moving freely to another) by simply drawing the path joining the two points (including all points that may be visited by at least one '1') and counting how many cells are on the path.

In the BBM, such a reflection would cause no horizontal delay. We can compensate for such extra delays, as well as add any desired horizontal delay of 2 or more blocks. See fig. 14.

Suppose we want to arrange for two signals to collide, with the plane of the collision being horizontal. If we get the two signals aligned vertically and they are approaching each other as they move forward, they will collide properly. We can adjust the time it takes one or both signals to reach a given vertical column by using delays such as those in fig. 14*.

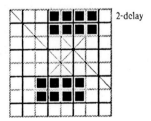

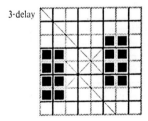

Fig. 14.

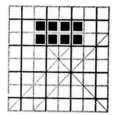

Fig. 15.

In order to allow signal-paths to cross without interacting, we use signal timing. By leaving a gap long enough for one signal (2 blocks) between all signals, we need only delay one of the paths by 2 blocks along the plane of the collision we're avoiding, in order to allow the signals to pass each other harmlessly. This gap is also enough to allow us to separate parallel output paths from a collision. See fig. 15. After the collision (fig. 12) the upper path already has a 1-block horizontal delay relative to the lower path. The mirror introduces a further 1-block delay, and so the upper signal passes through the timing-gap left in the lower signal path. With the addition of some extra synchronization and crossover delays, any BBM circuit can now be translated into a BBMCA circuit. Since the BBM has been shown to be a universal computer, the BBMCA is also.

There are many rules similar to the BBMCA that are also universal – for example, if we take the BBMCA rule of fig. 4 and modify it so that for each case shown, the result (right-hand side) is rotated 90-degrees clockwise on the 'dotted' steps, and counterclockwise on the 'solid' steps (i.e.

*The idea for this BBMCA variation arose out of a discussion with Tommaso Toffoli.

$\begin{smallmatrix}\blacksquare\square\\\square\square\end{smallmatrix} \rightarrow \begin{smallmatrix}\square\square\\\blacksquare\square\end{smallmatrix}$ on dotted steps, and $\begin{smallmatrix}\blacksquare\square\\\square\square\end{smallmatrix} \rightarrow \begin{smallmatrix}\square\blacksquare\\\square\square\end{smallmatrix}$ on the solid steps, etc.) then we get another rule that is also computation universal. Its universality can be shown in a direct manner by using this rule to simulate the BBMCA (this rule can simulate a given BBMCA computation isomorphically using eight times as much space, and four times as much time)*.

9. Relationship of BBMCA to conservative logic

The collision-gate of fig. 5 has two inputs and four outputs. If we wish to consider it to be a conservative-logic gate (one that conserves both 0's and 1's) then we must regard it as a gate with four inputs and four outputs, two of the inputs being constrained to always be zeros.

The gate upon which the BBMCA is based also has four inputs and four outputs. Is there some connection here? Let us redraw the BBMCA rule in a different form (see fig. 16). The mapping of input variables onto output variables of the BBMCA has been redrawn as if the inputs all arrive and leave in a vertical column. If we use this correspondence to draw the four possible cases with $a = d = 0$, drawing $\square$ for 0, $\blacksquare$ for 1, and showing each input/output case, we get an evolution (see fig. 17) which is logically the same as the collision gate. Thus the BBMCA rule of fig. 4 can be regarded as a completion of the collision-gate to a (reversible) conservative-logic gate!

$$\begin{matrix}ab\\cd\end{matrix} \rightarrow \begin{matrix}AB\\CD\end{matrix} \quad \text{becomes} \quad \begin{matrix}a\text{-}| & |\text{-}A\\b\text{-}| & |\text{-}B\\c\text{-}| & |\text{-}C\\d\text{-}| & |\text{-}D\end{matrix}$$

Fig. 16.

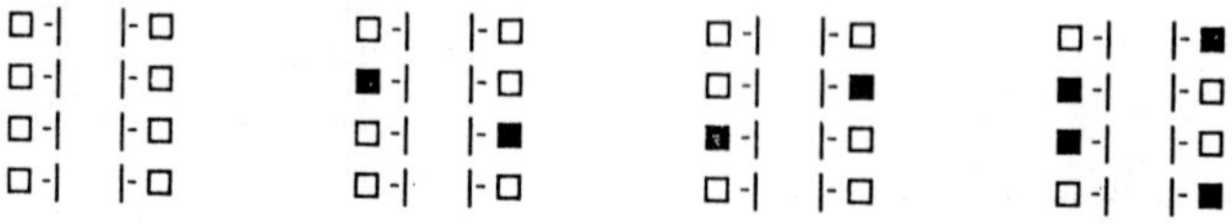

Fig. 17.

10. Energy in the BBMCA

In the BBM, the kinetic energy is proportional to the number of moving 1's. In the BBMCA, if we let $p_{x,y,t\text{-}1/2} = c_{x,y,t} - c_{x,y,t-1}$, then $\Sigma_{xy}(p^2_{x,y,t\text{-}1/2}/2)$ counts the number of moving ones (each moving one disappears from one cell, and appears in another, so $\Sigma_{xy}p^2$ – which counts how many places change – would count each moving one twice).

The ones that aren't moving are at those places that were a one at $t-1$, and still are one at time t. Thus the number of stationary ones is $\Sigma_{xy}c_{x,y,t}c_{x,y,t-1}$. A complicated way of writing the (constant) total number of ones is

$$E_{t\text{-}1/2} = \sum_{xy}\frac{p^2_{x,y,t\text{-}1/2}}{2} + \sum_{xy} c_t c_{t-1}. \qquad (9)$$

During a collision, some of the 'kinetic-energy' changes into 'potential-energy', and then it changes back again.*

Since (9) is a constant for *any* rule for which $\Sigma_{xy}c^2_{x,y,t}$ is constant, it is not possible to derive the particular rule from this expression. We might (for example) introduce the rule into (9) by using it to eliminate c_t (thus writing E as a function of $p_{t\text{-}1/2}$ and c_{t-1}) and see if we can push the mechanics analogy further.

Using number-of-ones to play the role of energy in BBMCA circuits and considering circuits for which we have only a statistical knowledge of what the different inputs will be, elaborate thermodynamic analogues can be established, but this will be discussed elsewhere. Although the overall system has a single deterministically evolving state, from the point of view of small pieces of the system, their inputs may appear random.

11. Conclusion

The laws of nature are the ultimate computing resource — the most efficient computation imag-

---

* One can think of mechanical models of the BBMCA for which the two terms of (9) are proportional to the physical kinetic and potential energy of the system midway between two steps.

inable would make the most direct possible use of the physical interactions and degrees of freedom available. Physical quantities and concepts would have a direct computational interpretation. Computer scientists cannot hope to find the right quantities to use to talk about efficient computation until they have models of computation that are much closer to fundamental physics. Reversible Cellular Automata are offered as a step towards this end.

Appendix A

A second-order, reversible, universal automaton

This appendix describes another BBM-type automaton. As before, we begin with a 2-dimensional cartesian lattice, this time with 3 states per cell, which we can designate as $-1, 0, +1$, and which we will draw as '\', blank, and '/' respectively in diagrams.

The time evolution will be given by $c_{x,y,t+1} = f(c_{\{x,y\},t}) - c_{x,y,t-1}$, where $f(c_{\{x,y\},t})$ is a function that 'looks' at the 3×3 neighbourhood with $c_{x,y}$ as its center cell, and '$-$' is taken mod3.

For each possible configuration of the neighbourhood, f will return a value of $-1, 0,$ or $+1$. Just as head-on collisions never arise in BBM computations, many configurations of this RCA need not arise in order to 'build' a universal computer. We will leave these cases undefined – each choice for these undefined cases defines a distinct universal RCA.

An isolated '/' or '\' will correspond to a travelling billiard ball – if only the cases defined here arise, the number of such 'balls' will be conserved. An isolated '/' will propagate along a positively sloped diagonal – its evolution will be governed by the following cases:

000	000	/00	000	0/0	000	000	000
000	0/0	000	000	000	/00	00/	000
000	000	000	00/	000	000	000	0/0

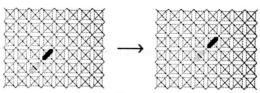

Fig. 18.

all return a '0' as the value for f;

```
00/  000
000  000
000  /00
```

both yield a value of '/' (i.e. $+ 1$). A sample time evolution (using halftones to show a cell's contents at time $t - 1$ and solid lines for time t, with diagonals lightly drawn through all cells) is shown in fig. 18. Intuitively, this rule at time t tries to make the '/' travel both forwards and backwards along its diagonal–subtracting away a '/' where it was at time $t - 1$ just leaves a '/' in the forwards direction.

We define this rule to be rotationally symmetric. It will be helpful to adopt the following convention: the 90-degree clockwise rotation of

$$
\begin{matrix} 000 \\ 000 \\ /00 \end{matrix} \rightarrow / \quad \text{is} \quad \begin{matrix} \backslash00 \\ 000 \\ 000 \end{matrix} \rightarrow \backslash.
$$

Inversions are defined analogously. Thus an isolated '\' will follow a negatively sloped diagonal path if the propagation of signals is governed by the cases:

$$
\begin{matrix} 000 & 000 & /00 & 0/0 \\ 000 & 0/0 & 000 & 000 \\ 000 & 000 & 000 & 000 \end{matrix} \rightarrow 0,
$$

$$
\begin{matrix} 000 \\ 000 \\ /00 \end{matrix} \rightarrow /
$$

(and rotations and inversions).

For compactness in writing the complete rule, we adopt the convention that inversions as well as rotations of the cases given are mapped onto the corresponding inversions or rotations of the result given.

These cases become zero:

```
\\\  \\0  \\0  \\0  \\0  \\/  \\/  \0\  \0\  \0\  \00  \00  \00
000  000  /00  /00  //0  000  /0/  000  /00  /00  000  /\0  /\/ ,
///  //0  000  00/  00/  //\  /\\  /0/  000  00/  /00  000  000
```

```
\00  \00  \00  \00  \00  \00  \00  \00  \00  \00  \00  \00  \00
/0\  /0\  /00  /00  /00  /00  /00  /00  /0/  /0/  //\  //0  //0 '
000  00/  \00  \0/  000  00/  0/0  0//  000  00/  000  000  00/
```

```
\0/  \0/  \0/  00\\  0\\  0\0  0\0  0\0  0\0  00\  00\  00\  00\
000  /0\  /0\  000  000  000  000  000  /0/  0\0  00\  000  000 '
/0\  \0/  000  000  0//  000  00/  0/0  0\0  000  00/  000  00/
```

```
00\  000  000  000  /\\  /\\  /0\
00/  0\0  000  /0\  000  /0/  000
000  000  000  \0/  \//  \\/  \0/
```

These cases become one:

```
\00  \00  \00  \00  \0/  \0/  \0/  \/\  0\0  0\/  0\/  0\/
/\0  /00  /00  /0\  /00  /00  /00  /\0  \00  00\  000 '
\/0  /\0  /00  /0/  00/  000  00/  000  \/0  /00  000  /00
```

```
00\  000  000  000  000  000
/\0  000  000  /\\  /\0  /\0  /\/
\/0  /00  /0/  \//  \/0  \//  \/0
```

(plus rotations and inversions). There are 2617 undefined cases.

Using this rule, a mirror is shown in fig. 19. We needed to define certain cases just to allow a mirror to remain unchanged when no signals are nearby.

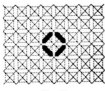

Fig. 19.

A signal bouncing on a mirror is shown in fig. 20. (Notice that there is no horizontal delay, as there was in the BBMCA). If this signal had been shifted one column to the right, it would have passed the mirror unaffected. We put some mirrors near places where signals might collide, so that (with its small neighbourhood) this rule can simulate an attractive collision – the signal paths will be displaced inward in a collision, rather than outward as in the BBM. See fig. 21. (If a signal arrives on

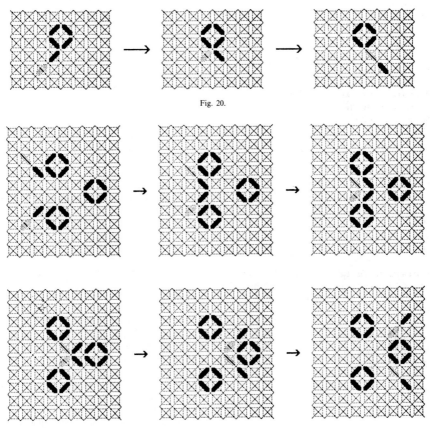

Fig. 20.

Fig. 21.

just one path, it goes through without any displacement). Two such gates, back to back, can be used to make signals cross over without affecting each other. See figs. 22 and 23.

Since all collisions occur without any delay along the plane of the collision, considerations of synchronization are very similar to those in the

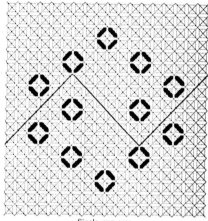

Single one case
Fig. 22.

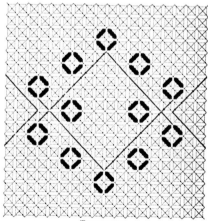

Two ones case
Fig. 23.

BBM. The proof of this automaton's universality is essentially the same as for the BBM.

To give another example of a universal second-order RCA, we can begin with the BBMCA rule. If f_s is the global rule that applies to the solid blocking and changes an entire configuration into the next configuration, and similarly f_d applies to the dotted blocking, then we can describe the BBMCA evolution by

$$S_{t+1} + S_{t-1} = f(S_t), \tag{10}$$

where $S_{t+1} + S_{t-1}$ is taken to be the configuration obtained by taking the cell-by-cell sum (mod 2) of S_{t+1} and S_{t-1}, and $f(S_t) = f_s(S_t) + f_d(S_t)$ is also such a sum.

(10) can be rewritten in the form (4) with a 3×3 neighbourhood and a dependence on the parity of the center cell's position (parity of $x + y$). In a similar manner, any invertible CA rule can be written in the form (10), and in the form (4) if it is locally invertible.

References

[1] C.H. Bennet, "Logical reversibility of computation," IBM Journal of Research and Development 6 (1973) 85–91; "The Thermodynamics of Computation," Int. J. of Theo. Phys. 21 (1982) 905–940.
[2] E. Fredkin, private communication.
[3] E. Fredkin and T. Toffoli, "Conservative Logic," Int. J. of Theo. Phys. 21 (1982) 219–253.
[4] E. Berlekamp, J. Conway and R. Guy, "Winning Ways for your Mathematical Plays", vol. 2 (Academic Press, New York, 1982).
[5] M. Gardner, "The Fantastic Combinations of John Conway's New Solitaire Game 'Life'," Scientific American 223:4 (1970) 120–123.
[6] R. Landauer, "Irreversibility and heat generation in the computing process," IBM Journal of Research and Development 5 (1961) 183–191.
[7] C. Shannon and W. Weaver, The Mathematical Theory of Communication (Univ. of Illinois Press, Illinois 1949).
[8] T. Toffoli, "Computation and construction universality of reversible cellular automata," Journal of computer systems science 15 (1977) 213–231.
[9] T. Toffoli, "CAM: A High-Performance Cellular-Automaton Machine," Physica 10D (1984) 195–204 (these proceedings).
[10] J. Von Neumann, Theory of Self-Reproducing Automata, (ed. and compiled by A.W. Burks) (Univ. Illinois Press, Illinois, 1966).

ADVANCES IN APPLIED MATHEMATICS **7**, 123–169 (1986)

Random Sequence Generation by Cellular Automata

Stephen Wolfram

Thinking Machines Corporation, 245 First Street, Cambridge, Massachusetts 02142 and
The Institute for Advanced Study, Princeton, New Jersey 08540

A 1-dimensional cellular automaton which generates random sequences is discussed. Each site in the cellular automaton has value 0 or 1, and is updated in parallel according to the rule $a_i' = a_{i-1}$ XOR $(a_i$ OR $a_{i+1})$ $(a_i' = (a_{i-1} + a_i + a_{i+1} + a_i a_{i+1})$ mod 2). Despite the simplicity of this rule, the time sequences of site values that it yields seem to be completely random. These sequences are analysed by a variety of empirical, combinatorial, statistical, dynamical systems theory and computation theory methods. An efficient random sequence generator based on them is suggested. © 1986 Academic Press, Inc.

1. Random Sequence Generation

Sequences that seem random are needed for a wide variety of purposes. They are used for unbiased sampling in the Monte Carlo method, and to imitate stochastic natural processes. They are used in implementing randomized algorithms which require arbitrary choices. And their unpredictability is used in games of chance, and potentially in data encryption.

To generate a random sequence on a digital computer, one starts with a fixed length seed, then iteratively applies some transformation to it, progressively extracting as long as possible a random sequence (e.g., [1]). In general one considers a sequence "random" if no patterns can be recognized in it, no predictions can be made about it, and no simple description of it can be found (e.g., [2]). But if in fact the sequence can be generated by iteration of a definite transformation, then a simple description of it certainly does exist.[1] The sequence can nevertheless seem random if no computations done on it reveal this simple description. The original seed must be transformed in such a complicated way that the computations cannot recover it.

---

[1] A stricter definition of randomness can be based on the non-existence of simple descriptions [3], rather than merely the difficulty in finding them. None of the sequences discussed here, nor many generally considered random, would qualify according to this definition.

123

124   STEPHEN WOLFRAM

The degree of randomness of a sequence can be defined in terms of the classes of computations which cannot discern patterns in it. A sequence is "random enough" for application in a particular system if the computations that the system effectively performs are not sophisticated enough to be able to find patterns in the sequence. So, for example, a sequence might be random enough for Monte Carlo integration if the values it yields are distributed sufficiently uniformly. The existence say of particular correlations in the sequence might not be discerned in this calculation. Whenever a computation that uses a random sequence takes a bounded time, there is a limit to the degree of randomness that the sequence need have. Statistical tests of randomness emulate various simple computations encountered in practice, and check that statistical properties of the sequence agree with those predicted if every element occurred purely according to probabilities. It would be better if one could show in general that patterns could not be recognized in certain sequences by any computation whatsoever that, for example, takes less than a certain time. No such results can yet be proved, so one must for now rely on more circumstantial evidence for adequate degrees of randomness.

The fact that acceptably random sequences can indeed be generated efficiently by digital computers is a consequence of the fact that quite simple transformations, when iterated, can yield extremely complicated behaviour. Simple computations are able to produce sequences whose origins can apparently be deduced only by much more complex computations.

Most current practical random sequence generation computer programs are based on linear congruence relations (of the form $x' = ax + b \bmod n$) (e.g., [1]), or linear feedback shift registers [4] (analogous to the linear cellular automata discussed below). The linearity and simplicity of these systems has made complete algebraic analyses possible and has allowed certain randomness properties to be proved [1, 4]. But it also leads to efficient algebraic algorithms for predicting the sequences (or deducing their seeds), and limits their degree of randomness.

An efficient random sequence generator should produce a sequence of length L in a time at most polynomial in L (and linear on most kinds of computers). It is always possible to deduce the seed (say of length s) for such a sequence by an exhaustive search which takes a time at most $O(2^s)$. But if in fact such an exponentially long computation were needed to find any pattern in the sequence, then the sequence would be random enough for almost any practical application (so long as it involved less than exponential time computations).

No such lower bounds on computational complexity are yet known. It is however often possible to show that one problem is computationally equivalent to a large class of others. So, for example, one could potentially show that the problem of deducing the seed for certain sequences was NP-complete [5]: special instances of the problem would then correspond to arbitrary problems in the class NP, and the problem would in general be as difficult as any in NP. (One should also show some form of uniform reducibility to ensure that the problem is difficult almost always, as well as in the worst case.) The class NP (nondeterministic polynomial time) includes many well-studied problems (such as integer factorization), which involve finding objects (such as prime factors) that satisfy polynomial-time-testable conditions, but for which no systematic polynomial time (P) algorithms have ever been discovered.

Random sequence generators have been constructed with the property that recognizing patterns in the sequences they produce is in principle equivalent to solving certain difficult number theoretical problems [2] (which are in the class NP, but are not NP-complete). An example is the sequence of least significant bits obtained by iterating the transformation $x' = x^2$ mod (pq), where p and q are large primes (congruent to 3 modulo 4) [6]. Making predictions from this sequence is in principle equivalent to factoring the integer pq [6, 7].

There are in fact many standard mathematical processes which are simple to perform, yet produce sequences so complicated that they seem random. An example is taking square roots of integers. Despite the simplicity of its computation, no practical statistical procedures have revealed any regularity in say the digit sequence of $\sqrt{2}$ (e.g., [8]). (Not even its normality or equidistribution has however actually been proved.) An even simpler example is multiplication by $\frac{3}{2}$, say in base 6.[2] Starting with 1, one obtains the pattern shown in Fig. 1.1. The center vertical column of values, corresponding to the leading digit in the fractional part of $(\frac{3}{2})^n$, seems random [10]. (Though again not even its normality has actually been proved.) Given the complete number obtained at a particular stage, multiplication by $(\frac{2}{3})^n$ suffices to reproduce the original seed. But given only the center column, it seems difficult to deduce the seed.

Many physical processes also yield seemingly random behaviour. In some cases, the randomness can be attributed to the effects of external random input. Thus, for example, "analog" random sequence generators such as noise diodes work by sampling thermal fluctuations associated with a heat bath containing many components. Coin tossings and Roulette wheels

---

[2] This operation can be performed locally on a base 6 digit sequence, and so can be implemented as a cellular automaton. Given particular finite boundary conditions, it acts like a linear congruential sequence generator (e.g. [1]). But in an infinite region, its behaviour is more complicated, and is related to the so-called $3N + 1$ problem [9].

126   STEPHEN WOLFRAM

```
      1.
      1.3
      2.13
      3.213
      5.0213
     11.33213
     15.220213
     25.0303213
     41.34350213
    102.235433213
    133.3553520213
    222..25525003213
    333.425113050213
    522.3414514133213
   1203.53241532220213
   2005.521025203303213
   3012.5013420051350213
   4321.13223301152433213
  10501.520351315510520213
  14132.5005451554442003213
  23221.13124155541030050213
  35031.51510255531343113321 3
  54345.454434255152344522021 3
 123542.4240534245505412030321 3
 205534.040122341244232004350213
 312523.10020353211035001054332 13
 451204.430305520143543014235202 13
1115011.043442500235534323355003213
1454314.40540413035552350525430502 13
2423454.012310213555505442123441332 13
4035423.020443322555442403205402220213
10055334.331105204255404005012303330321 3
13125223.514442010425310011320435213502 13
21512035.45410301404214301520105502243321 3
32450055.423134323103234325001424334052021 3
51113125.3345235044350535113023405231200321 3
114451512.224205441054422445133531204500050213
154115450.3403124012420341115225150111301133213
253155413.530451002340053145504154314513152202 13
421555322.5141143035301215424402534541515503032 13
```

FIG. 1.1. Successive powers of $3/2$ in base 6. The leading digits in the fractional parts of
these numbers form a sequence that seems random. The process of multiplication by $3/2$ in
base 6 corresponds to a $k = 6$, $r = 1$ cellular automaton rule.

produce outcomes that depend sensitively on initial velocities determined by
complex systems with many components. It seems however that in all such
cases, sequences extracted sufficiently quickly can depend on only a few
components of the environment, and must eventually show definite correlations.

One suspects in fact that randomness in many physical systems (probably
including turbulent fluids) arises not from external random input, but rather
through intrinsic mathematical processes [11]. This paper discusses the
generation of random sequences by simple procedures which seem to
capture many features of this phenomenon. The investigations described
may not only suggest practical methods for random sequence generation,
but also provide further understanding of the nature and origins of randomness in physical processes.

CELLULAR AUTOMATA   127

2. CELLULAR AUTOMATA

A 1-dimensional cellular automaton [12, 13] consists of a line of sites with values a_i between 0 and $k - 1$. These values are updated in parallel (synchronously) in discrete time steps according to a fixed rule of the form

$$a'_i = \phi(a_{i-r}, a_{i-r+1}, \cdots, a_{i+r}). \tag{2.1}$$

Much of this paper is concerned with the study of a particular $k = 2$, $r = 1$ cellular automaton, described in Section 3.

For mathematical purposes, it is often convenient to consider cellular automata with an infinite number of sites. But practical implementations must contain a finite number of sites N. These are typically arranged in a circular register, so as to have periodic boundary conditions, given in the $r = 1$ case by

$$a'_1 = \phi(a_N, a_1, a_2)$$
$$a'_N = \phi(a_{N-1}, a_N, a_1). \tag{2.2}$$

It is also possible to arrange the sites in a feedback shift register (cf. [4]), with boundary conditions

$$a'_1 = \phi(\phi(a_2, a_3, a_4), \phi(a_3, a_4, a_5), a_1),$$
$$a'_2 = \phi(\phi(a_3, a_4, a_5), a_1, a_2). \tag{2.3}$$

Cellular automata can be considered as discrete approximations to partial differential equations, and used as direct models for a wide variety of natural systems (e.g. [14]). They can also be considered as discrete dynamical systems corresponding to continuous mappings on the Cantor set (e.g. [15]). Finally they can be viewed as computational systems, whose evolution processes information contained in their initial configurations (e.g. [16]).

Despite the simplicity of their construction, cellular automata are found to be capable of diverse and complex behaviour. Figure 2.1 shows some patterns generated by evolution according to various cellular automaton rules, starting from typical disordered initial conditions. Four basic outcomes are seen [15]: (1) the pattern becomes homogeneous (fixed point), (2) the pattern degenerates into simple periodic structures (limit cycles), (3) the pattern is aperiodic, and appears chaotic, and (4) complicated localized structures are produced. The first two classes of cellular automata yield readily predictable behaviour, and show no seemingly random elements. But

128   STEPHEN WOLFRAM

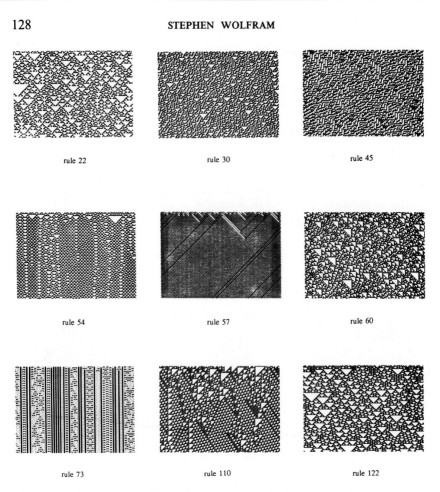

FIG. 2.1.   Patterns generated by evolution of various $k = 2$, $r = 1$ cellular automata from disordered initial states. Successive lines give configurations obtained on successive time steps, with white and black squares representing sites with values 0 and 1 respectively. The coefficient of 2^i in the binary decomposition of each rule number gives the value of the function ϕ in Eq. (2.1) for the neighbourhood whose site values form the integer i (cf. [17]).

the third class gives rise to behaviour that is more complex. They can produce patterns whose features cannot readily be predicted in detail, and in fact often seem completely random. Such cellular automata can be used as models of randomness in nature. They can also be considered as abstract mathematical systems, and used for practical random sequence generation.

Figure 2.1 showed patterns produced by evolution according to various cellular automaton rules, starting from typical disordered initial conditions, in which the value of each site is randomly chosen to be zero or one. Figure 2.2 shows some patterns obtained instead by evolution from a very simple

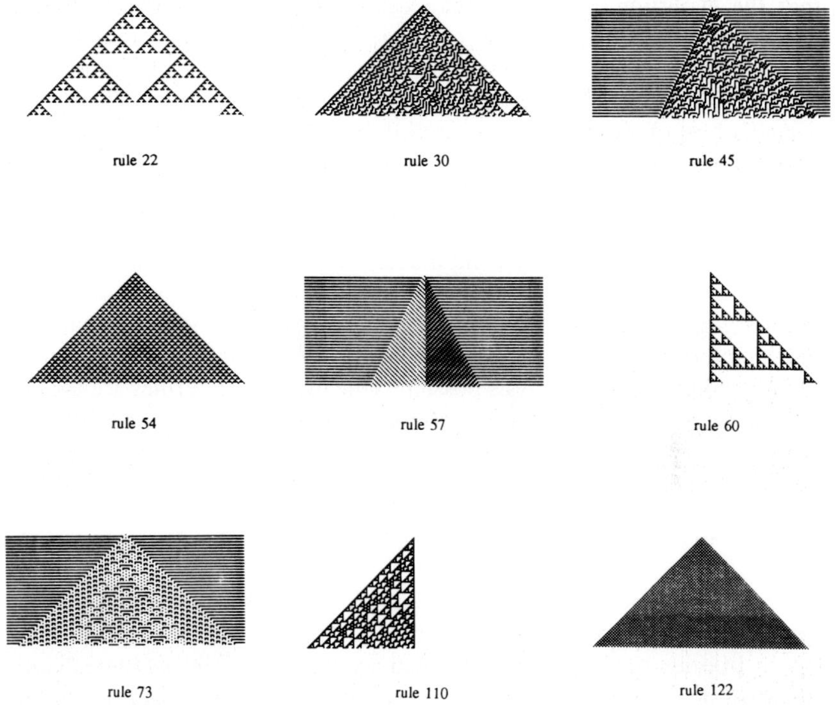

CELLULAR AUTOMATA   129

FIG. 2.2. Patterns generated by evolution of various $k = 2$, $r = 1$ cellular automata from an initial state containing a single nonzero site. Complex patterns are seen to be produced even with such simple initial conditions.

initial condition containing a single nonzero site. With such simple initial conditions, some class 3 cellular automata yield rather simple patterns, which are typically periodic or at least self similar (almost periodic). There are nevertheless class 3 cellular automata which yield complex patterns, even from simple initial states. Their evolution can intrinsically produce apparent randomness, without external input of random initial conditions. It is such "autoplectic" systems [11] which seem most promising for explaining randomness in nature, or for use as practical random sequence generation procedures.

Many class 3 cellular automata seem to perform very complicated transformations on their initial conditions. Their evolution thus corresponds to a complicated computation. But any predictions of the cellular automaton behaviour must also be obtained through computations. Effective predictions require computations that are more sophisticated than those corresponding to the cellular automaton evolution itself. One suspects however that the evolution of many class 3 cellular automata is in fact computationally as sophisticated as that of any (physically realizable) system can be [18, 19]. It is thus "computationally irreducible," and its outcome can effectively be found only by direct simulation or observation. There are no general computational shortcuts or finite mathematical formulae for it. As a consequence, many questions concerning infinite time or infinite size limits cannot be answered by bounded computations, and must be considered formally undecidable. In addition, questions about finite time or finite size behaviour, while ultimately computable, may be computationally intractable, and could require, for example, exponential time computations.

Most class 3 cellular automata are expected to be computationally irreducible. A few rules however have special simplifying features which make predictions and analysis possible. One class of such rules are those for which the function ϕ is linear (modulo k) in the a_{i+j}. Such cellular automata are analogous to linear feedback shift registers [4]. An example with $k = 2$ is

$$a_i' = (a_{i-1} + a_i) \bmod 2 = (a_{i-1} \text{ XOR } a_i), \qquad (2.4)$$

where XOR stands for exclusive disjunction (this is rule number 60 in the scheme of [17]). Linear cellular automata satisfy a superposition principle, which implies that patterns generated with arbitrary initial states can be obtained as appropriate superpositions of the self-similar pattern produced with a single non-zero initial site (as illustrated in Fig. 2.2). As a result, it is possible to give a complete algebraic description of the behaviour of the system [20], and to deduce the outcome of its evolution by a much reduced computation.

Most class 3 cellular automata are however nonlinear. No general methods to predict their behaviour have been found, and from their likely computational irreducibility one expects that no such methods even in principle exist. In studying such systems one must therefore to a large extent forsake conventional mathematical techniques and instead rely on empirical and experimental mathematical results.

3. A RANDOM SEQUENCE GENERATOR

There are a total of $2^{2^3} = 256$ cellular automaton rules that depend on three sites, each with two possible values ($k = 2, r = 1$). Among these are several linear rules similar to that of Eq. (2.4). But the two rules that seem best as random sequence generators are nonlinear, and are given by

$$a_i' = a_{i-1} \text{ XOR } (a_i \text{ OR } a_{i+1}) \qquad (3.1a)$$

or, equivalently,

$$a_i' = (a_{i-1} + a_i + a_{i+1} + a_i a_{i+1}) \bmod 2 \qquad (3.1b)$$

(rule number 30 [17]; equivalent to rule 86 under reflection), and

$$a_i' = a_{i-1} \text{ XOR } (a_i \text{ OR } (\text{NOT } a_{i+1})) \qquad (3.2a)$$

or

$$a_i' = (1 + a_{i-1} + a_{i+1} + a_i a_{i+1}) \bmod 2 \qquad (3.2b)$$

(rule 45; reflection equivalent to rule 75). Here XOR stands for exclusive disjunction (addition modulo two); OR for inclusive disjunction (Boolean addition), and NOT for negation. The patterns obtained by evolution from a single nonzero site with each of these rules were shown in Fig. 2.2. It is indeed remarkable that such complexity can arise in systems of such simple construction. A first indication of their potential for random sequence generation is the apparent randomness of the center vertical column of values in the patterns of Fig. 2.2.

This paper concentrates on the cellular automaton of Eq. (3.1). The methods used carry over directly to the cellular automaton of Eq. (3.2), but some of the results obtained in this case are slightly less favourable for random sequence generation.

The cellular automaton rule (3.1) is essentially nonlinear. Nevertheless, its dependence on a_{i-1} is in fact linear. This feature (termed "left permutivity" in [21], and also studied in [22]) is the basis for many of its properties. In the form (3.1), the rule gives the new value a_i' of a site in terms of the old values a_{i-1}, a_i and a_{i+1}. But the linear dependence on a_{i-1} allows the rule to be rewritten as

$$a_{i-1} = a_i' \text{ XOR } (a_i \text{ OR } a_{i+1}), \qquad (3.3)$$

giving a_{i-1} in terms of a_i', a_i and a_{i+1}. This relation implies that the spacetime patterns shown, for example, in Figs. 2.1 and 2.2 can be found not only by direct time evolution according to (3.1) from a given initial configuration, but also by extending spatially according to (3.3), starting with the temporal sequence of values of two adjacent sites.

Random sequences are obtained from (3.1) by sampling the values that a particular site attains as a function of time. In practical implementations, a finite number of sites are considered, and are typically arranged in a circular register. Given almost any initial "seed" configuration for the sites in the

132   STEPHEN WOLFRAM

register, a long and seemingly random sequence can apparently be obtained. This paper discusses several approaches to the analysis of the cellular automaton (3.1) and the sequences it produces. While little can rigourously be proved, the overwhelming weight of evidence is that the sequences indeed have a high degree of randomness.

4. GLOBAL PROPERTIES

This section considers the behaviour of the cellular automaton (3.1) starting from all possible initial states. The basic approach is to count the possible sequences and patterns that can occur, and to characterize them using methods from dynamical systems theory (e.g. [23]). The next section discusses the behaviour obtained by evolution from particular initial configurations. For purposes of simplicity, this section concentrates on the infinite size limit; Section 9 considers finite size effects.

Figure 4.1 shows a spacetime pattern produced by evolution according to (3.1) starting from a typical disordered initial state. While definite structure

FIG. 4.1. Pattern produced by evolution according to the cellular automaton rule (3.1) from a typical disordered initial state.

is evident, one may suspect that a single line of sites at any angle in the pattern can have an arbitrary sequence of values. Below we shall show that this is in fact the case: given an appropriate initial condition, any sequence can be generated in an infinite cellular automaton with the rule (3.1).

The rule (3.1) can be considered as a mapping from one (say infinite) cellular automaton configuration to another. An important property of this mapping is that it is surjective or onto. Any configuration A can thus always be obtained as the image of some configuration A^-, according to $A = \phi A^-$. A possible configuration A^- (not necessarily unique) can be found by starting with a candidate pair of site values, then extending to the left using Eq. (3.3). So if all possible initial configurations are considered, then any configuration can be generated at any time step. Thus with appropriate initial conditions, any spatial sequence of site values can be produced.

Every length X spatial sequence of site values that occurs is determined by a length $X + 2$ sequence on the previous time step. The surjectivity of the rule (3.1) implies that such a predecessor exists for any length X sequence. But Eq. (3.3) also implies that there are exactly four predecessors for any sequence. Given values a_i, a_{i-1}, and so on, in one sequence, the values a_{i+1}^- and a_i^- in its predecessor can be chosen in all the four possible ways; in each case the remaining a_{i-j}^- are then uniquely determined by Eq. (3.3). Thus starting from an ensemble that contains all possible (infinite) cellular automaton configurations with equal probabilities, each configuration will be generated with equal probability throughout the evolution of the cellular automaton, and so every possible spatial sequence of a particular length will occur with equal frequency.

One may also consider sequences of values attained by a single site as a function of time. Starting from an initial ensemble which contains all configurations with equal probabilities, all such sequences again occur with equal frequencies. For, given any temporal sequence, iteration of Eq. (3.3) yields an equal number of initial configurations which evolve to it. The same is true for sequences of site values on lines at any angle in the spacetime pattern.

Entropies provide characterizations of the number of possible sequences that occur. First, let the number of distinct length n blocks in these sequences be $N(n)$, and let the ith such sequence appear with probability p_i. Then the topological entropy of the sequence is given by (e.g. [15])

$$s = \lim_{n \to \infty} \frac{1}{n} \log_2 N(n), \tag{4.1}$$

and the measure entropy by

$$s_\mu = \lim_{n \to \infty} \frac{-1}{n} \sum_i^{2^n} p_i \log_2 p_i. \tag{4.2}$$

If the cellular automaton configurations are considered as elements of a Cantor set, then these entropies give respectively the Hausdorff (strictly Kolmogorov) and measure dimensions of this set. If the sequences are considered as "messages," then the entropies give respectively their capacity and Shannon information content.

For the cellular automaton of Eq. (3.1), all possible sequences occur with equal probabilities (given an equal probability initial ensemble) so both entropies are maximal:

$$s_\mu = s = 1. \tag{4.3}$$

Any reduction in entropy would reveal redundancy in the sequences, and would imply a lack of randomness. Equation (4.3) is thus a necessary (though not sufficient) condition for randomness. (It is related to statistical test A of Sect. 10 and Appendix A.)

Although Eq. (4.3) implies that all possible sequences of values for single sites can occur along any spacetime direction, the deterministic nature of the cellular automaton rule (3.1) implies that only certain spacetime patches of values can occur. In fact, all the site values in a particular patch are completely determined by the values that appear on its upper, left and right boundaries. Once these boundaries are specified, the values of remaining sites in the patch are redundant, and can be found simply by applying (3.1) and (3.3).

In general the degree of redundancy in such spacetime patterns can be characterized by the invariant topological and measure entropies for the cellular automaton mapping, given by (e.g. [15, 24])

$$\mathbf{h} = \lim_{X \to \infty} \lim_{T \to \infty} \frac{1}{T} \log_2 N(X, T) \tag{4.4}$$

and

$$\mathbf{h}_\mu = \lim_{X \to \infty} \lim_{T \to \infty} \frac{-1}{T} \sum_{i=1}^{2^{XT}} p_i \log_2 p_i, \tag{4.5}$$

where $N(X, T)$ gives the total number of distinct $X \times T$ spacetime patches of site values that occur, and the p_i give their probabilities.

It is clear from the locality of the rule (3.1) that

$$\mathbf{h}_\mu \leq \mathbf{h} \leq 2. \tag{4.6}$$

A calculation based on the method of [25] in fact shows that*

$$\mathbf{h}_\mu \leq 1.20. \tag{4.7}$$

*Recent results [45] suggest in fact that $\mathbf{h}_\mu \simeq 1 + T^{-(0.6 \pm 0.1)}$, yielding a final value of 1.

CELLULAR AUTOMATA   135

Hence a knowledge of the time sequences of values of about 1.2 sites suffice in principle to determine the values of all other sites. In practice however the function which gives the initial configuration in terms of these temporal sequences seems rapidly to become intractably complicated, as discussed in Section 7.

5. STABILITY PROPERTIES

Section 4 considered properties of possible patterns generated by evolution with the cellular automaton rule of Eq. (3.1), starting from all possible initial configurations. This section considers the change in the patterns produced by small perturbations in the initial state. Figure 5.1 shows the differences resulting from reversal of a single site value in a typical disordered initial configuration. The region affected increases in size with time, reflecting the instability of the patterns generated.

This instability implies that information on localized changes eventually propagates throughout the cellular automaton. The rates of information transmission to the left and right are determined by the slopes of the difference pattern in Fig. 5.1. These in turn give left and right Lyapunov exponents λ_L and λ_R for the cellular automaton evolution [15, 26]. (The sequence of site values in a configuration, starting from a particular point, can be represented as a real number. Linear growth of the difference pattern

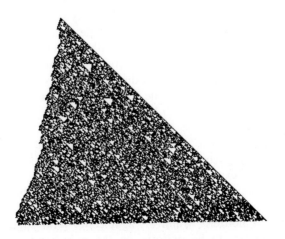

FIG. 5.1. Differences in patterns produced by evolution according to the cellular automaton rule of Eq. (3.1) from two typical disordered states which differ by reversal of the centre site value. the growth of the region of differences reflects the instability of the cellular automaton evolution.

136   STEPHEN WOLFRAM

in Fig. 5.1 then implies exponential divergence of the numbers representing nearby configurations.)

The form of the cellular automaton rule (3.1) immediately implies that

$$\lambda_R = 1. \tag{5.1}$$

For consider a configuration in which the difference pattern has reached site -1. Whatever the current values of sites 0 and 1, the XOR in (3.1) leads to a change in the new value of site 0. The value (5.1) is the maximum allowed by the locality of the rule (3.1).

Empirical measurements suggest that the left-hand side of the difference pattern expands at an asymptotically linear rate, with a slope [45]

$$\lambda_L = (0.2428 \pm 0.0003). \tag{5.2}$$

A simple statistical estimate for λ_L can be given. Consider a pair of configuations for which the front of the difference pattern has reached site 0. As a first approximation, one may assume that the motion of this front depends only on the neighbouring values a_{-1} and a_{+1}, where, by construction, a_{-1} is the same for the two configurations. When $a_{-1} = 0$, the front advances (left) by one site, independent of the values of the a_1. When $a_{-1} = 1$, the front remains stationary if the a_{+1} for the two configurations are equal, and retreats by one site if they are unequal. If possible sets of site values occured with equal probabilities, the front should thus follow a biased random walk, advancing at average speed 1/4. In practice, however, Fig. 5.1 shows that the front can retreat by many sites in a single time step. This occurs when the cellular automaton rule yields the same image for multiple site value sequences, as for say 10100 and 11001. Such phenomena make the probabilities for different difference patterns unequal, and invalidate this purely statistical approach discussed. (The values of λ_L obtained in this approach by considering the effects of between 1 and 5 sites on the right are 0.25, 0.1875, 0.15625, 0.140625 and 0.134766.)

The result (5.2) gives the average speed of the left-hand side of the difference pattern. As the random walk interpretation suggests, however, one can choose initial configurations for which a single site change leads to differences which expand at speed 1 on the left. In general, one can construct the analog of a Green's function, giving the probability that a site at a particular position and time will be affected by an initial perturbation. This function is nonzero within a "light cone" with edges expanding at speed 1. It appears to be uniform on the right-hand side. But on the left-hand side, it appears to be determined by a diffusion equation which gives the average behaviour of the biased random walk. The difference pattern can thus extend beyond the line given by Eq. (5.2), but with an exponentially damped probability.

Lyapunov exponents measure the rate of information transmission in cellular automata, and provide upper bounds on entropies, which measure the information content of patterns generated by cellular automaton evolution. For surjective cellular automata it can be shown, for example, that [15]

$$\mathbf{h}_{\mu} \le (\lambda_L + \lambda_R), \tag{5.3}$$

consistent with Eqs. (4.6) and (5.2). The existence of positive Lyapunov exponents is a characteristic feature of class 3 cellular automata.

The difference pattern of Fig. 5.1, and the related Green's function, measure the effect of initial perturbations on the values of individual sites. In studying random sequence generation, one must also consider the effect of such perturbations on time sequences of site values, say of length T. These sequences are always completely determined from the initial values of $2T + 1$ sites. But not all these initial values necessarily affect the time sequences. A change in any of the $T + 1$ left-hand initial sites necessarily leads to a change in at least one element of the time sequence. But some changes in the T right-hand initial sites have no effect on any element of the time sequence. It seems that the probability for a particular initial site to affect the time sequence decreases exponentially with distance to the right. The average number of sites on the right which affect the time sequence is found to be approximately 0.26 + 0.19T. Thus the total number of initial sites on which a length T time sequence depends is on average approximately 1.91 + 1.19T. This result is presumably related to the entropy (4.6).

6. PARTICULAR INITIAL STATES

Sections 4 and 5 have discussed some properties of the patterns produced by evolution according to Eq. (3.1) from generic initial conditions. This section considers evolution from particular special initial configurations.

Figure 6.1 shows on two scales the pattern produced by evolution from a configuration containing a single nonzero site. (This could be considered a difference pattern for the special time-invariant state in which all sites have value zero.) Remarkable complexity is evident.

There are however some definite regularities. For example, diagonal sequences of sites on the left-hand side of the pattern are periodic, with small periods. In general, the value of a site at a depth N from the edge of the pattern depends only on sites at depths N or less; all the other sites on which it could depend always have value 0 because of the initial conditions given. As a consequence, the sites down to depth N are independent of those deeper in the pattern, and in fact follow a shifted version of the cellular automaton rule (3.1), with boundary conditions that constrain two sites at one end to have value zero. Since such a finite cellular automaton has a total of 2^N possible states, any time sequence of values in it must have a period of at most 2^N. The corresponding diagonal sequences in the pattern of Fig. 6.1 must therefore also have periods not greater than 2^N.

Table 6.1 gives the actual periods of diagonal sequences found at various depths on the left- and right-hand sides of the pattern in Fig. 6.1. These are compared with those for the self-similar pattern shown in Fig. 2.2 generated by evolution according to the linear cellular automaton rule (2.4).

The short periods on the left-hand side of the pattern in Fig. 6.1 are related to the high degree of irreversibility in the effective cellular automaton rule for diagonal sequences in this case [27]. Starting with any possible initial configuration, this cellular automaton always yields cycles with period 2^j. The maximum value of j increases very slowly with N, yielding maximum cycle lengths which increase in jumps, on average slower than linearly with N. (Between the N values at which the maximum cycle length increases, a single additional cycle of maximal length seems to be added each time N increases by one. The total number of cycle states thus increases at most quadratically with N, implying an increasing degree of irreversibility.) The actual sequences that occur near the left-hand boundary of the pattern in Fig. 6.1 correspond to a particular set of those possible in this effective cellular automaton. In a first approximation, they can be considered uniformly distributed among possible N-site configurations, and their periods increase very slowly with N.

The effective rule for the right-hand side diagonal pattern in Fig. 6.1 is a shifted version of Eq. (3.1)

$$a_i' = a_i \text{ XOR } (a_{i+1} \text{ OR } a_{i+2}), \tag{6.1a}$$

with boundary conditions

$$a_{N-1}' = a_{N-1} \text{ XOR } a_N,$$
$$a_N' = a_N. \tag{6.1b}$$

This system is exactly reversible: all of its 2^N possible configurations have unique predecessors. All the configurations thus lie on cycles, and again the cycles have periods of the form 2^j. Figure 6.2 shows the lengths of longest cycles as a function of N. These lengths increase roughly exponentially with

CELLULAR AUTOMATA 139

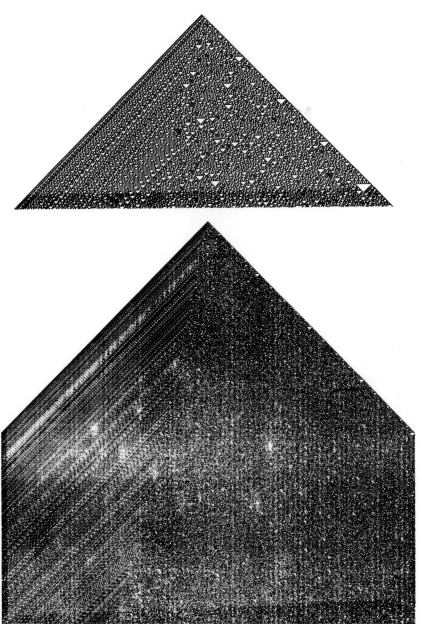

FIG. 6.1. Patterns generated by evolution for 250 and 2000 generation, respectively, according to the cellular automaton rule (3.1) from an initial state containing a single nonzero site. (The second pattern was obtained by Jim Salem using a prototype Connection Machine computer.)

STEPHEN WOLFRAM

TABLE 6.1

Depth	CA30 π_R	π_L	CA60 π_R
0	1	1	1
1	2	1	2
2	2	1	4
3	4	2	4
4	8	1	8
5	8	2	8
6	16	2	8
7	32	1	8
8	32	4	16
9	64	1	16
10	64	4	16
11	64	4	16
12	64	4	16
13	64	4	16
14	64	4	16
15	128	4	16
16	256	4	32
32		8	64
64		4	128
128		8	256
256		8	512
512		16	1024
1024		16	2048

Period lengths for diagonal sequences in patterns generated by evolution from a single nonzero site according to the cellular automaton rules of Eqs. (3.1) and (2.4). π_R and π_L signify respectively periods for diagonal sequences on the right and left of the patterns, at the specified depth. (The entries left blank were not found.)

N; a least squares fit to the data of Fig. 6.2 yields

$$\log_2 \Pi_N \cong 0.5(N + 1). \tag{6.2}$$

This length is small compared to the total number of states 2^N; few states in fact lie on such longest cycles. Nevertheless, the periods of the right-hand diagonal sequences in Fig. 6.1 do seem to increase roughly exponentially with depth, as suggested by Table 6.1.

CELLULAR AUTOMATA   **141**

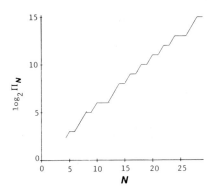

FIG. 6.2. Maximal period lengths Π_N for the effective cellular automaton which gives the right-hand diagonal sequences in Fig. 6.1 down to depth N. Points plotted at integer N are joined for pictorial purposes.

The boundary in Fig. 6.1 between regular behaviour on the left and irregular behaviour on the right seems to be asymptotically linear, and to move to the left with speed 0.25. A statistical argument for this result can be given in analogy with that for Eq. (5.2). Each site at depth d on the left-hand side of the pattern could in principle be affected by sites down to depth d arbitrarily far up in the pattern. In practice, however, it is unaffected by changes in sites outside a cone whose boundary propagates at speed $\lambda_L \cong 0.25$. Thus the irregularity on the right spreads to the left only at this speed.

While diagonal sequences at angles ± 1 in Fig. 6.1 must ultimately become periodic, sequences closer to the vertical need not. In fact, no periodicity has been found in any such sequences. The center vertical (i.e., temporal) sequence has, for example, been tested up to length $2^{19} \cong 5 \times 10^5$, and no periodicity is seen. One can prove in fact that only one such vertical sequence (obtained from any initial state containing a finite number of nonzero sites) can possibly be periodic [22]. For if two sequences were both periodic, then it would follow that all sequences to their right must also be, which would lead to a contradiction at the edge of the pattern.

Not only has no periodicity been detected in the center vertical sequence of Fig. 6.1.; the sequence has also passed all other statistical tests of randomness applied to it, as discussed in Section 10.

While individual sequences seem random, there are local regularities in the overall pattern of Fig. 6.1. Examples are the triangular regions of zero sites. Such regularities are associated with invariants of the cellular automaton rule.

142   STEPHEN WOLFRAM

TABLE 6.2

Period	Element
1	0
	01
3	000011111001
4	0000001
	0000111
	0010011
	0111111

Configurations periodic under the cellular automaton mapping (3.1) consist of infinite repetitions of the elements given. Notice that the four elements given for period four correspond simply to different phases in a cycle. The patterns generated by these periodic configurations are shown in Fig. 6.3.

The particular configuration in which all sites have value 0 is invariant under the cellular automaton rule of Eq. (3.1). As a consequence, any string of zeroes that appears can be corrupted only by effects that propagate in from its ends. Thus each string of zeroes that is produced leads to a uniform triangular region.

Table 6.2 and Fig. 6.3 give other configurations which are periodic under the rule (3.1). (They can be considered as invariant under iterations of the rule.) Again, any string that contains just the sequences in these configurations can be corrupted only through end effects, and leads to a regular region in spacetime patterns generated by Eq. (3.1).

In general, there is a finite set of configurations with any particular period p under a permutive cellular automaton rule such as (3.1). The configura-

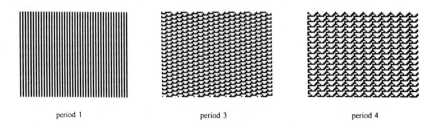

period 1   period 3   period 4

FIG. 6.3. Periodic patterns for the cellular automaton rule of Eq. (3.1). The form of these patterns is given in Table 6.2.

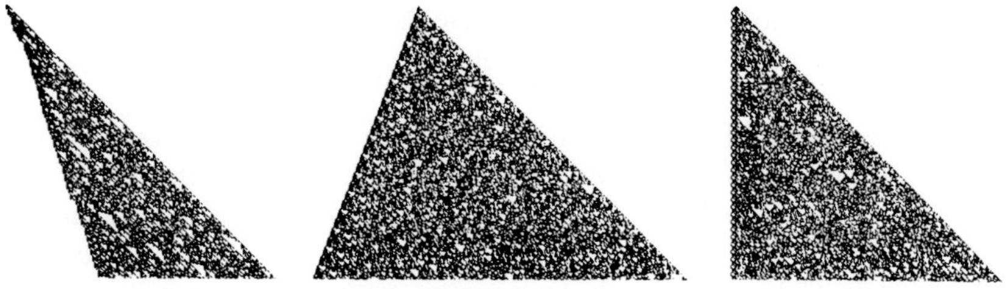

CELLULAR AUTOMATA **143**

period 1   period 3   period 4

FIG. 6.4.   Patterns produced by evolution according to the cellular automaton rule (3.1) by single site initial defects in the periodic patterns of Fig. 6.2 and Table 6.2.

tions may be found by starting with a candidate length $2p$ string, then testing whether this and the string it yields through Eq. (3.3) on the left are in fact invariant under ϕ^p. The string to be tested need never be longer than 2^{2p}, since such a string can contain all possible length $2p$ strings. Thus the periodic configurations consist of repetitions of blocks containing 2^{2p} or less site values. (For an arbitrary cellular automaton rule, the set of invariant configurations forms a finite complement language which contains in general an infinite number of sequences with the constraint that certain blocks are excluded [16].)

The pattern in Fig. 6.1 can be considered the effect of a single site "defect" in the periodic pattern resulting from a configuration with all sites 0. Figure 6.4 shows difference patterns produced by single site defects in the other periodic configurations of Table 6.2 and Fig. 6.3

The periodic configurations of Table 6.2 and Fig. 6.3 can be viewed as special states in which the cellular automaton of Eq. (3.1) behaves just like the identity rule. Concatenations of other blocks could simulate other cellular automata: one block might correspond to a value 0 site, and another to a value 1 site in the effective cellular automaton. Some cellular automata (such as that of Eq. (2.4)) simulate themselves under such "blocking transformations," and thus evolve to self-similar patterns. The cellular automata of Eqs. (3.1) and (3.2) are unique among $k = 2$, $r = 1$ rules in simulating no other rules, at least with blocks of length up to eight [14].

7. FUNCTIONAL PROPERTIES

Cellular automaton rules such as (3.1) can be considered as functions ϕ which map three Boolean values to one. Iterations of these rules for say t

steps correspond to functions of $2t + 1$ Boolean values. The complexity of these functions reflects the intrinsic complexity of the cellular automaton evolution.

The complexity of a Boolean function can be characterized by the number of logic gates that would be needed to evaluate it with a particular kind of circuit, or the number of terms that it would have in a particular symbolic representation. Explicit evolution according to the cellular automaton rule (3.1) corresponds to a circuit with $O(t^2)$ components and depth t. But for purposes of comparison, it is convenient to consider fixed depth representations. One such representation is disjunctive normal form (DNF), in which the function is written as a disjunction of conjunctions. A two-level circuit can be constructed in direct correspondence with this form (as programmable logic arrays often are).

For the function of Eq. (3.1), the DNF is

$$\phi(a_{-1}, a_0, a_1) = (\overline{a_{-1}a_0}) + (a_{-1}\overline{a_0}\,\overline{a_1}) + (\overline{a_{-1}a_1}), \qquad (7.1)$$

where $+$ stands for OR, concatenation for AND, and bar for NOT. Notice that by using in addition an XOR operation, Eq. (3.1) itself gives a shorter form for this function.

The general problem of finding the absolute shortest representation for an arbitrary Boolean function, even in DNF, is NP-complete (e.g. [5]), and so presumably requires an exponential time computation. But a definite approximation can be found in terms of "prime implicants" (e.g. [28]). A Boolean function of n variables can be considered as a colouring of the Boolean n-cube. Prime implicants give the hyperplanes (with different dimensions) in the n-cube which must be superimposed to obtain the region with value 1. Each prime implicant can thus be used as a term in a DNF for the function. The number of prime implicants required gives a measure of the total number of "holes" in the colouring of the n-cube, and thus of the complexity of the function.

The minimal DNF obtained with prime implicants for the function corresponding to two iterations of the cellular automaton mapping (3.1) is

$$
\begin{aligned}
\phi^2 &(a_{-2}, a_{-1}, a_0, a_1, a_2) \\
&= (\overline{a_{-2}\,a_{-1}\,a_0}\,a_1\,\overline{a_2}) + (\overline{a_{-2}}\,a_{-1}a_0a_1\,\overline{a_2}) \\
&\quad + (a_{-2}\overline{a_{-1}}\,a_0a_1\,\overline{a_2}) + (a_{-2}a_{-1}a_0\,\overline{a_1}\,\overline{a_2}) \\
&\quad + (a_{-2}\overline{a_{-1}}\,\overline{a_1}\,\overline{a_2}) + (\overline{a_{-2}\,a_{-1}\,a_0}a_2) \\
&\quad + (a_{-2}\overline{a_{-1}}\,a_0a_2) + (\overline{a_{-2}}\,a_{-1}a_0a_2) + (a_{-2}a_{-1}\,\overline{a_0}). \quad (7.2)
\end{aligned}
$$

TABLE 7.1

t	CA30		CA60
	P.I.	Min.	P.I./Min.
1	3	3	2
2	9	7	2
3	23	17	8
4	76	41	2
5	185	105	8
6	666	272	8

Number of terms in disjunctive normal form Boolean expressions corresponding to iterations of the mappings (3.1) (CA30) and (2.4) (CA60). P.I. gives the number of prime implicants; min. the number of terms obtained by [29]. (The two numbers are the equal in the case of Eq. (2.4).)

Table 7.1 gives the number of prime implicants for successive iterations of the mapping (3.1). These results are plotted in Fig. 7.1. For arbitrary Boolean functions of $2t + 1$ variables, the number of prime implicants could increase like 4^t. In practice, however, a least squares fit to the data of Table 7.1 suggests growth like $4^{0.77t}$.

Various efficient methods are known to find DNF that are somewhat simpler than those obtained using prime implicants. With one such method

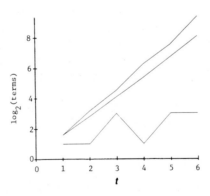

FIG. 7.1. Number of terms in disjunctive normal form Boolean expressions for t step iterations of the mappings (3.1) and (2.4). The upper curve gives the number of prime implicants for iterations of Eq. (3.1). The next curve gives the minimal number of terms obtained in this case using [29]. The lowest curve gives the minimal number of terms for the linear cellular automaton mapping (2.4).

[28, 29], the DNF of Eq. (7.2) can be reduced to

$$
\begin{aligned}
\phi^2 & (a_{-2}, a_{-1}, a_0, a_1, a_2) \\
& = (\overline{a_{-2}}\,\overline{a_{-1}}\,\overline{a_0}\,a_1) + (\overline{a_{-2}}\,a_{-1}a_0a_1) \\
& \quad + (\overline{a_{-2}}\,\overline{a_{-1}}\,\overline{a_0}\,a_2) + (\overline{a_{-2}}\,a_{-1}a_0a_2) \\
& \quad + (a_{-2}\,\overline{a_1}\,\overline{a_2}) + (a_{-2}\,\overline{a_{-1}}\,a_0) + (a_{-2}a_{-1}\,\overline{a_0}). \quad (7.3)
\end{aligned}
$$

The sizes of the minimal DNF obtained by this method for iterations of Eq. (3.1) are shown in Table 7.1 and Fig. 7.1. They are seen to grow more slowly than those obtained with prime implicants; the data given are however again fit by exponential growth like $4^{0.65t}$.

Table 7.1 and Fig. 7.1 also give the size of the minimal DNF for iterations of the linear cellular automaton mapping (2.4). This number remains much smaller, apparently increasing like $2^{2\#_1(t)-1} < t^2$, where $\#_1(t)$ gives the number of ones in the binary representation for the integer t (cf. [30]).

The rapid increase in the size of the minimal DNF found for iterations of Eq. (3.1) indicates the increasing computational complexity of determining the result of evolution according to (3.1), and supports the conjecture of its computational irreducibility. (Note however that even the parity function cannot be computed by any DNF, or in general fixed-depth, circuit of polynomial size [3.1].)

Equation (7.3) gives the function which determines the value of a single site after two iterations of the cellular automaton rule (3.1). One can also construct a function which gives the length t sequence of values of a particular site attained through time by evolution from a given length $2t + 1$ initial sequence. The minimal DNF representation for this function is found (using [29]) to grow in size approximately as $2^{1.36t}$.

The results of Table 7.1 and Fig. 7.1 concern the difficulty of finding the outcome of cellular automaton evolution according to Eq. (3.1) from a given initial state. One may also consider the problem of deducing the initial state from time sequences of site values produced in the evolution. Given say t steps in the time sequence of values for two adjacent sites, the initial configuration up to t sites to the left can be deduced directly by iteration of Eq. (3.3). The combinatorial results of Section 4 indicate in fact that only about 1.2 such temporal sequences should on average be required. And in principle from a single sufficiently long temporal sequence, it should be possible to deduce a complete initial configuration for a finite cellular automaton. In practice, however, the necessary computation seems to become increasingly intractable as the size of the system increases.

Given a particular temporal sequence, say at position 0, Eq. (3.3) uniquely determines the values of all sites in a triangle to the left as a function of

TABLE 7.2

n	$\langle$Var.$\rangle$	$\langle$P.I.$\rangle$	Max. P.I.
2	0.5	0.75	1
3	1	1.125	2
4	1.375	1.375	3
5	1.125	1.219	3
6	2.281	2.719	12
7	2.828	3.539	17
8	3.164	4.105	26
9	3.699		
10	4.254		

Properties of Boolean expressions for leftmost initial site values deduced from length n time sequences, obtained by evolution according to Eq. (3.1). The average number of variables appearing in the Boolean expressions is given, together with the number of prime implicants in the disjunctive normal form for the expression. The maximum number of variables which can appear is always $n - 1$. (Results for $n \geq 9$ were obtained by Carl Feynman using a Symbolics 3600 LISP machine. The entries left blank were not found.)

values in the temporal sequence at position 1. The number of values in the position 1 temporal sequence on which a given site depends varies with the form of the position 0 sequence [32]. For example, if the position 0 sequence consists solely of ones, then the whole triangle of sites is completely determined, entirely independent of the position 1 sequence. Table 7.2 gives some results from considering the dependence of the site value a_{-t} at position $-t$ (the apex of the triangle) on the position 1 sequence, for all 2^t possible position 0 sequences. The number of values in the position 1

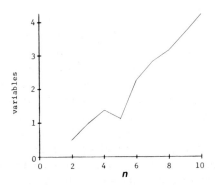

FIG. 7.2. Average number of additional site values necessary to "back-track" and determine uniquely the initial site value a_{-n} given the sequence of values a_0 for n subsequent time steps.

sequence on which a_{-t} depends seems to be roughly Poisson distributed, with a mean that grows like $0.4t$, as shown in Fig. 7.2. This is consistent with the combinatorial result (4.6).

Table 7.2 also gives some properties of the prime implicant forms for a_{-t}. It is clear that the complexity of the function that determines a_{-t} from temporal sequences grows with t, probably at an increasingly rapid rate. Again this suggests that the problem of deducing the initial sequence for evolution according to Eq. (3.1), while combinatorially possible, is computational complex.

By comparison, the corresponding problem for evolution according to the linear rule (2.4) is quite straightforward. For each possible position 0 sequence, there are only two possible forms for the dependence of a_{-t} on the position 1 sequence, and each of them involves exactly $2^{\#_1(t-1)}$ prime implicants. This simplicity can be viewed as a consequence of the algebraic structure associated with this system.

8. Computation Theoretical Properties

The discussion of the previous section can be considered as giving a characterization of the computational complexity of iterations of the cellular automaton mapping (3.1) in a particular simple model of computation. The results obtained suggest that at least in this model, there is no shortcut method for finding the outcome of the evolution: the computations required are no less than for an explicit simulation of each time step. As discussed above, one suspects in fact that the evolution is in general computationally irreducible, so that no possible computation could find its outcome more efficiently than by direct simulation.

This would be the case if the cellular automaton of Eq. (3.1) could act as an efficient universal computer (e.g. [33]), so that with an appropriate initial state, its evolution could mimic any possible computation. In particular, it could be that the problem of finding the value of a particular site after t steps (given say a simply-specified initial state, as in Fig. 6.1) must take a time polynomial in t on any computer. (Direct simulation takes $O(t^2)$ time on a serial-processing computer, and $O(t)$ time with $O(t)$ parallel processors.) For a linear cellular automaton such as that of Eq. (2.4), this problem can be solved in a time polynomial in $\log(t)$; but for the cellular automaton of Eq. (3.1) it quite probably cannot [18].

In addition to studying cellular automaton evolution from given initial configurations, one may consider the problem of deducing configurations of the cellular automaton from partial information such as temporal sequences.

In particular, one may study the computational complexity of finding the seed for a cellular automaton in a finite region from the temporal sequences it generates.

There are 2^N possible seeds for a size N cellular automaton, and one can always find which ones produce a particular sequence by trying each of them in turn. Such a procedure would however rapidly become impractical. The results in Section 7 suggest a slightly more efficient method. If it were possible to find two adjacent temporal sequences, then the seed could be found easily using Eq. (3.3). Given only one temporal sequence, however, some elements of the seed are initially undetermined. Nevertheless, in a finite size system, say with periodic boundary conditions, one can derive many distinct equations for a single site value. The site value can then be deduced by solving the resulting system of simultaneous Boolean equations. The equations will however typically involve many variables. As discussed in Section 7, the number of variables seems to be Poisson-distributed with a mean around $0.4N$.

The general problem of solving a Boolean equation in n variables is NP-complete (e.g. [5]), and so presumably cannot be solved in a time polynomial in n. In addition, it seems likely that the average time to solve an arbitrary Boolean equation is correspondingly long. To relate the problem of deducing the seed discussed above to this would however require a demonstration that the Boolean equations generated were in a sense uniformly distributed over all possibilities. Out of all 2^{2^n} n-variable equations, the problem here typically involves $O(2^n)$, but these seem to have no special simplifying features. At least with the method discussed above, it is thus conceivable that the problem of deducing the seed is equivalent to the general problem of solving Boolean equations, which is NP-complete.

9. FINITE SIZE BEHAVIOUR

Much of the discussion above has concerned the behaviour of the cellular automaton (3.1) in the idealized limit of an infinite lattice of sites. But practical implementations must use finite size registers, and certain global properties can depend on the size and boundary conditions chosen.

The total number of possible states in a size N cellular automaton is 2^N. Evolution between these states can be represented by a finite state transition diagram. Figure 9.1 gives some examples of such diagrams for the cellular automaton of Eq. (3.1) with periodic boundary conditions, as in Eq. (2.2). Table 9.1 summarizes some of their properties. The results are seen to depend not only the magnitude of N, but also presumably on its number theoretical properties.

150   STEPHEN WOLFRAM

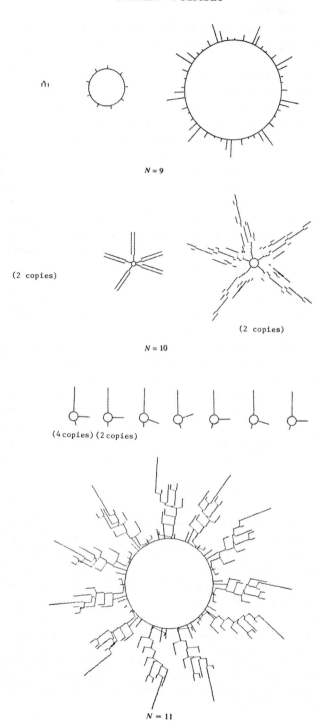

Each state transition diagram contains a set of cycles, fed by trees representing transients. The cycles may be considered as "attractors" to which states in their "basins of attraction" irreversibly evolve.

There are many regularities in the structure of the state transition diagrams obtained from Eq. (3.1). The evolution is thus not well-approximated by a random mapping between 2^N states.

A first observation is that most configurations have unique predecessors under the mapping (3.1) (as mentioned for infinite lattices in Sect. 4), so there is little branching in the state transition diagram. In fact, it can be shown [32] that a configuration has a unique predecessor unless it contains a pair of value zero sites separated by a sequence of $3n + 1$ value one sites (with $n \geq 0$), or unless N is divisible by 3, and all sites have value one. In the former case, the configuration has exactly zero or two predecessors; in the latter case, it has three. The numbers of configurations with zero and two predecessors are equal when N is not divisible by 3; there are two more with zero predecessors when $3|N$. For large N, the number of configurations with zero or two predecessors behaves as [32] κ^N, where $\kappa \cong 1.696$ is the real root of $4\kappa^3 - 2\kappa^2 - 1 = 0$. Since the total number of configurations grows like 2^N, the fraction of nodes in the state transition diagram that are branch points thus tends exponentially to zero.

A second observation is that there are often many identical parts in the state transition diagrams of Table 9.1 and Fig. 9.1. This is largely a consequence of shift invariance. States in a cellular automaton with periodic boundary conditions that are related by shifts (translations) evolve equivalently. Thus, for example, there are often several identical cycles, related by shifts in their configurations. In addition, the periods of the cycles are often divisible by N or its factors, since they contain several sequences of configurations related by shifts. The transient trees that feed each of these sequences are then identical.

The evolution of a finite cellular automaton with periodic boundary conditions is equivalent to the evolution of an infinite cellular automaton with a periodic initial configuration. Thus the results on cycle length distributions in Table 9.1 can be considered as inverse to those in Table 6.2 on configurations with given temporal periods. Cycles of lengths corresponding to these temporal periods occur whenever N is divisible by the spatial periods of these configurations. Such short cycles are absent if N has none of these factors.

FIG. 9.1.   State transition diagrams for configurations of cellular automata evolving according to Eq. (3.1) in circular registers of size N. Each node represents one of the 2^N possible length N configurations, and is joined by an arc to its successor under the cellular automaton mapping. Transients corresponding to trees in the graph are seen ultimately to evolve to periodic cycles. Some properties of these state transition diagrams are given in Table 9.1. (Graphics by Steve Strassmann.)

152   STEPHEN WOLFRAM

TABLE 9.1

N	Cycles	Frac. longest	Cyc. frac.	$\langle$Transient$\rangle$
4	$1 \times 8, 3 \times 1$	0.75	0.69	0.5
5	$1 \times 5, 1 \times 1$	0.94	0.19	4.3
6	3×1	1.00	0.05	3.3
7	$1 \times 63, 7 \times 4, 1 \times 1$	0.60	0.72	0.4
8	$1 \times 40, 1 \times 8, 3 \times 1$	0.88	0.20	3.1
9	$1 \times 171, 1 \times 72, 1 \times 1$	0.81	0.48	1.1
10	$2 \times 15, 1 \times 5, 3 \times 1$	0.82	0.04	14.8
11	$1 \times 154, 11 \times 17, 1 \times 1$	0.76	0.17	3.3
12	$4 \times 102, 1 \times 8, 4 \times 3, 3 \times 1$	0.93	0.11	4.4
13	$1 \times 832, 1 \times 260, 1 \times 247, 1 \times 91, 1 \times 1$	0.32	0.17	2.2
14	$1 \times 1428, 2 \times 133, 1 \times 112, 2 \times 84, 1 \times 63, 1 \times 14, 3 \times 1$	0.84	0.13	2.7
15	$1 \times 1455, 5 \times 30, 5 \times 9, 15 \times 7, 4 \times 5, 1 \times 1$	0.93	0.05	5.7
16	$1 \times 6016, 1 \times 4144, 3 \times 40, 1 \times 8, 3 \times 1$	0.50	0.16	
17	$1 \times 10846, 1 \times 1632, 1 \times 867, 1 \times 306, 1 \times 136, 1 \times 17, 1 \times 1$	0.96	0.11	

Properties of state transition diagrams for the cellular automaton rule of Eq. (3.1) in a circular register of size N. The multiplicity and length of each cycle is given, followed by the fraction of initial states which evolve to a longest cycle (size of attractor basin), the total fraction of all 2^N states which lie on cycles, and the average length of transient before a cycle is reached in evolution from an arbitrary initial state. (Results for $N \geq 16$ were obtained by Holly Peck.)

For large N, the state transition diagrams for Eq. (3.1) appear to be increasingly dominated by a single cycle. This cycle is longer than the others, and its basin of attraction is large enough that most arbitrarily chosen initial states evolve to it. The low degree of branching in the transient trees implies that the points reached from arbitrary initial states should be roughly uniformly distributed around the cycle.

The shorter cycles in Table 9.1 can be considered as related to subsets of states invariant under the cellular automaton rule. With N even, for example, configurations which consist of two identical length $N/2$ subsequences can evolve only to configurations of the same type. Once such a configuration has been reached, the evolution is "trapped" within this subset of configurations, and must yield shorter cycles. (This phenomenon also occurs for cellular automata with essentially trivial rules, such as the shift mapping $a_i' = a_i$. All states are on cycles in this case. The different cycles correspond to the possible "necklaces" with N beads of two kinds, which are inequivalent under shifts or rotations. These necklaces in turn correspond to cyclotomic polynomials; there are $\sum_{d|N} \phi(d) 2^{N/d}$ of them, where ϕ the Euler totient function (e.g. [4]).) In general, there may exist subsets of states with certain special symmetry properties that are preserved by the cellular automaton rule. Initial states with particular, symmetrical,

CELLULAR AUTOMATA   153

forms can be expected to have these properties, and thus to be trapped in subsets of state space, and to yield short cycles. For example, with $N = 36$, a configuration containing a single nonzero site evolves to a length 2844 cycle, while most initial configurations evolve to the longest cycle, with 2237472 states.

In the infinite size limit, patterns such as that of Fig. 6.1 generated by the cellular automaton of Eq. (3.1) never become periodic. But with a total of N sites, a cycle must occur after 2^N or less steps. Table 9.2 and Fig. 9.2 give the actual maximal cycle lengths Π_N found. A roughly exponential increase of Π_N with N is seen, and a least squares fit to the data of Table 9.2 yields

$$\log_2 \Pi_N \cong 0.61(N + 1). \tag{9.1}$$

Note that if the state transition diagram corresponded to an entirely random mapping between the 2^N cellular automaton states, then cycles of average length $2^{N/2}$ would be expected [34]. The cycles actually obtained are significantly longer. The exponent in Eq. (9.1) may be related to the entropy (4.6) as a result of the expansivity or instability of the mapping discussed in Section 5.

If there were very short cycles, then the sequences produced by the cellular automaton would readily be predictable. So if in fact no such prediction can be made by any polynomial time computation, the length of the cycles that occur should in general increase asymptotically faster than polynomial in N (cf. [2]). This behaviour is supported by Eq. (9.1).

If indeed the evolution of cellular automata such as (3.1) is computationally irreducible, then a complex computation may always be required to determine for example the lengths of cycles that appear. For in this case, there can effectively be no better way to find the succession of states that occur, except by explicit application of the rule (3.1). One expects in fact that the problem of finding say whether two configurations lie on the same cycle is PSPACE-complete, and so presumably cannot be solved in a time polynomial in N, but rather essentially requires a direct simulation of the cellular automaton evolution. (Note that if the lengths of the cycles studied are $O(2^M)$, where both 2^{N-M} and 2^M are large, then parallel processing is essentially of no avail in this problem.)

While the determination of cycle lengths and structures may be computationally intractable for cellular automata such as (3.1), it should be much easier for linear cases such as (2.4). From the algebraic theory of these systems it is possible to show for example that the maximal cycle length Π_N satisfies [20]

$$\Pi_N \mid 2^{\mathrm{ord}_N(2)} - 1, \tag{9.2}$$

where $n \mid m$ states that the integer n exactly divides m. Here $\mathrm{ord}_N(k)$ is the

TABLE 9.2

N	CA30 Π_N	$\log_2\Pi_N$	CA60 Π_N	$\log_2\Pi_N$
4	8	3.0	1	0.0
5	5	2.3	15	3.9
6	1	0.0	6	2.6
7	63	6.0	7	2.8
8	40	5.3	1	0.0
9	171	7.4	63	6.0
10	15	3.9	30	4.9
11	154	7.3	341	8.4
12	102	6.7	12	3.6
13	832	9.7	819	9.7
14	1428	10.5	14	3.8
15	1455	10.5	15	3.9
16	6016	12.6	1	0.0
17	10845	13.4	255	8.0
18	2844	11.5	126	7.0
19	3705	11.9	9709	13.2
20	6150	12.6	60	5.9
21	2793	11.4	63	6.0
22	3256	11.7	682	9.4
23	38249	15.2	2047	11.0
24	185040	17.5	24	4.6
25	588425	19.2	25575	14.6
26	312156	18.3	1638	10.7
27	67554	16.0	13797	13.7
28	249165	17.9	28	4.8
29	1466066	20.5	475107	18.9
30	306120	18.2	30	4.9
31	2841150	21.4	31	5.0
32	2002272	20.9	1	0.0
33	2038476	21.0	1023	10.0
34	5656002	22.4	510	9.0
35	18480630	24.1	4095	12.0
36	2237472	21.1	252	8.0
37	49276415	25.6	3233097	21.6
38	9329228	23.2	19418	14.2
39	961272	19.9	4095	12.0
40	19211080	24.2	120	6.9
41	51151354	25.6	41943	15.4
42	109603410	26.7	126	7.0
43	93537212	26.5	5461	12.4
44	192218312	27.5	1364	10.4
45	75864495	26.2	4095	12.0
46	261598274	28.0	4094	12.0
47	811284813	29.6	8388607	23.0
48	3035918676	31.5	48	5.6
49	9937383652	33.2	2097151	21.0
50	593487780	29.1	51150	15.6
51	3625711023	31.8	255	8.0
52	20653434880	34.3	3276	11.7
53	40114679273	35.2	3556769739	31.7
54	7551779562	32.8	27594	14.8

Maximum cycle lengths Π_N found for the cellular automata of Eqs. (3.1) (CA30) and (2.4) (CA60) in circular registers of size N. In the former case, a selection of seeds, including single nonzero sites, were used. In the latter case, maximal length cycles are always obtained with single nonzero site seeds. The results are plotted in Fig. 9.2. (Results for $N \geq 32$ were obtained by Holly Peck and Tsutomu Shimomura with an assembly-language program on a Celerity C-1200 computer.)

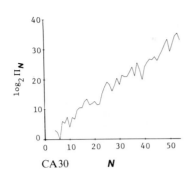
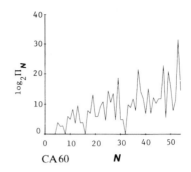

FIG. 9.2. Maximal cycle lengths Π_N for the cellular automaton of Eqs. (3.1) (CA30) and (2.4) (CA60) in circular registers of size N.

multiplicative order function, equal to the minimum integer j such that $k^j = 1 \bmod N$. This function divides the totient function $\phi(N)$ (equal to the number of integers less than N which are relatively prime to N), which is maximal for prime N. Table 9.2 and Fig. 9.2 give the actual maximal periods found in this case. Equation (9.2) rarely holds as an equality, and the Π_N found are usually much shorter than the corresponding ones for the nonlinear rule (3.1).

The cycle structures of finite cellular automata depend in detail on the boundary conditions chosen. Table 9.3 gives the maximal cycle lengths found for rules (3.1) and (2.4) with shift register boundary conditions. The results differ substantially from those with periodic boundary conditions given in Table 9.2. One notable feature is the presence of length $2^N - 1$ cycles in the linear cellular automaton (2.4) for certain N. These correspond to maximal length linear feedback shift registers, and can be identified by a direct algebraic procedure [4].

Other boundary conditions may also be considered. Among them are twisted ones, in which the sites a_1 and a_N are negated in Eq. (2.2). The maximum cycle lengths found with such boundary conditions seem typically shorter than in the purely periodic case.

One may in addition consider boundary conditions in which the boundary site values are fixed, rather than being periodically identified. Section 6 (particularly Fig. 6.2) gave some examples of results with such boundary conditions. Different cycles are obtained in different cases; all those investigated nevertheless give maximal cycle lengths shorter than those of Table 9.2 found with periodic boundary conditions.

What has been discussed so far are cycles in complete finite cellular automaton configurations. But in obtaining random sequences one samples single sites. The sequences found could potentially have periods which were

　　　　　　　　　　　STEPHEN WOLFRAM

TABLE 9.3

	CA30		CA60	
N	Π_N	$\log_2 \Pi_N$	Π_N	$\log_2 \Pi_N$
4	5	2.3	15	3.9
5	2	1.0	21	4.4
6	7	2.8	21	4.4
7	4	2.0	127	7.0
8	17	4.1	63	6.0
9	65	6.0	73	6.2
10	6	2.6	889	9.8
11	57	5.8	1533	10.6
12	50	5.6	1085	10.1
13	118	6.9	7905	12.9
14	185	7.5	11811	13.5
15	257	8.0	32767	15.0
16	481	8.9	255	8.0
17	907	9.8	273	8.1
18	1681	10.7	253921	18.0
19	707	9.5	413385	18.7
20	2679	11.4	761763	19.5
21	5630	12.5	5461	12.4
22	1368	10.4	4194303	22.0
23	31241	14.9	2088705	21.0
24	3567	11.8	2097151	21.0
25	60503	15.9	2192337	21.1
26	4752	12.2	22995	14.5
27	46519	15.5	41943035	25.3
28	35569	15.1	17895697	24.1
29	207197	17.7		
30	149899	17.2		
31	482717	18.9		

Maximum cycle lengths Π_N found for the cellular automata of Eqs. (3.1) (CA30) and (2.4) (CA60) in shift registers of size N (with boundary conditions given by Eq. (2.3)).

sub-multiples of the periods for the complete configuration. For permutive rules such as (3.1) (or (2.4)) this cannot, however, occur.

The state transition diagrams summarized in Table 9.1 give the number of complete N-site configurations that can occur at various stages in the evolution of the cellular automaton (3.1). One may also consider the number of single site temporal sequences that can occur. Table 9.4 gives the fraction of the 2^L possible length L temporal sequences that are actually generated from any of the 2^N possible initial states in a size N cellular automaton

CELLULAR AUTOMATA   157

TABLE 9.4

L	3	4	5	6	7	8	9	10	11	12	13	14	15
3	0.500	1.000	1.000	1.000	1.000	1.000	1.000	1.000	1.000	1.000	1.000	1.000	1.000
4	0.250	0.625	0.875	0.938	1.000	1.000	1.000	1.000	1.000	1.000	1.000	1.000	1.000
5	0.125	0.313	0.656	0.844	1.000	1.000	1.000	1.000	1.000	1.000	1.000	1.000	1.000
6	0.063	0.156	0.344	0.594	0.906	1.000	1.000	1.000	1.000	1.000	1.000	1.000	1.000
7	0.031	0.078	0.180	0.352	0.609	0.891	1.000	1.000	1.000	1.000	1.000	1.000	1.000
8	0.016	0.039	0.094	0.188	0.328	0.633	0.949	0.992	1.000	1.000	1.000	1.000	1.000
9	0.008	0.020	0.047	0.094	0.168	0.361	0.668	0.895	0.996	1.000	1.000	1.000	1.000
10	0.004	0.010	0.023	0.047	0.085	0.195	0.386	0.644	0.917	0.989	1.000	1.000	1.000
11	0.002	0.005	0.012	0.023	0.042	0.102	0.204	0.377	0.666	0.897	0.995	1.000	1.000
12	0.001	0.002	0.006	0.012	0.021	0.052	0.105	0.204	0.387	0.651	0.911	0.995	1.000
13	0.000	0.001	0.003	0.006	0.011	0.026	0.054	0.105	0.209	0.385	0.669	0.913	0.995
14	0.000	0.001	0.001	0.003	0.005	0.013	0.027	0.053	0.109	0.209	0.397	0.671	0.906
15	0.000	0.000	0.001	0.001	0.003	0.007	0.013	0.027	0.055	0.109	0.215	0.399	0.668

Fraction of length L temporal sequences generated from all possible seeds by evolution according to Eq. (3.1) in a length N circular register. Results for successive values of N are given in successive columns. The results are plotted in Fig. 9.3.

evolving according to Eq. (3.1) (with periodic boundary conditions). The results are plotted in Fig. 9.3. Whenever $N \geq L + 2$, all possible sequences seem to be generated. They appear with roughly equal frequencies.

10. STATISTICAL PROPERTIES

The sequences generated by the cellular automaton of Eq. (3.1) may be considered effectively random if no feasible procedure can identify a pattern

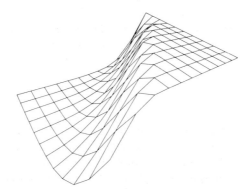

FIG. 9.3. Fraction of length L sequences obtained by evolution from all possible seeds according to Eq. (3.1) in a size N circular register. The three-dimensional view is from the point $N = L = 20$, with elevation 2.

in them, or allow their behaviour to be predicted. Even though it may not be possible to prove that no such procedure can exist, circumstantial evidence can be accumulated by trying various statistical procedures and finding that they reveal no regularities. The basic approach is to compare statistical results on sequences generated by (3.1) with those calculated for sequences whose elements occur purely according to probabilities.

To establish the validity of (3.1) as a general-purpose random sequence generator, one should apply a variety of statistical procedures, related to various different kinds of calculations. The choice of tests is necessarily as ad hoc as the choice of calculations done. Appendix A lists those used here. (But see also [35].) Some can be considered related to Monte Carlo simulations of physical and other systems. Others to statistical analyses that would be done on data from various kinds of measurements. While quite ad hoc, the tests seem to be sensitive, and reasonably independent.

As an example, consider the "equidistribution" or "frequency" test. If a sequence of zeroes and ones is to be random, the digits zero and one must occur in it with equal frequency. In general, in fact, all 2^n possible length n blocks of digits must also occur with equal frequency. (The measure entropy of (4.2) is maximal exactly when such equidistribution occurs.) However, in a finite sample of length m, there are expected to be statistical fluctuations, which lead to slightly different numbers of zeroes and ones. (The value of entropy deduced from a finite sample is thus almost always not maximal, even if it would be maximal were the sequence to be continued forever.) As a consequence, one can never definitively conclude by studying a finite sample that the complete sequence is not random. One can however calculate the probability that a truly random sequence would have the properties seen in the finite sample.

To do this, (e.g. [36]), one evaluates χ^2, defined in terms of the observed and expected frequencies p_0 and p_e as

$$\chi^2 = \sum_1^\nu (p_0 - p_e)^2/p_e. \qquad (10.1)$$

Here ν gives the number of degrees of freedom, or number of distinct objects whose frequencies are included in the sum. If blocks of length n are studied then $\nu = 2^n$. Now one must find the probability that a value of χ^2 larger than that observed would occur for a random sequence. This "confidence interval" is obtained immediately from the integral of the χ^2 distribution (e.g. [36]).

If the confidence interval is very close to zero or one, then the observed χ^2 is unlikely to be produced from a random sequence, and one may infer that the observed sequence is not random. Of course, if say a total of k tests

CELLULAR AUTOMATA   159

TABLE 10.1

	CA30 $N = 17$ $L = 8k$	CA30 $N = 17$ $L = 64k$	CA30 $N = 23$ $L = 64k$	CA30 $N = 29$ $L = 64k$	CA30 $N = 37$ $L = 64k$	CA30 $N = 49$ $L = 64k$
A	**0.0039**	**1.0000**	**0.0456**	0.7375	0.3852	0.8003
B	**0.0171**	**0.9944**	0.3391	0.4888	0.1010	0.1494
C	0.4164	0.4783	0.7256	0.4847	0.4083	0.9407
D	0.3227	**0.9998**	0.1506	0.1434	0.1678	0.6074
E	0.4576	0.4484	0.6790	0.8492	0.5414	0.7991
F	0.4306	0.8644	0.8751	0.5590	0.6681	0.6606
G	0.2942	**0.9944**	0.1232	0.7359	0.4448	0.6961

Results of the statistical tests described in Appendix A for sequences of length L ($k = 1024$) generated by the cellular automaton of Eq. (3.1) (rule number 30) in circular registers of length N. In each case, the seed used consists of a single nonzero site. The numbers given are the probabilities (confidence intervals) for statistical averages of truly random sequences to exceed those of the sequences analysed. The numbers should be uniformly distributed between 0 and 1 if the sequences analysed are indeed truly random. Results below 0.05 and above 0.95 are shown in bold type. Accumulations close to 0 or 1 suggest deviations from randomness. Such accumulations are seen in this case only when the period of the cellular automaton is comparable to the length of the sequence sampled. (The statistical test programs used here were written in C by Don Mitchell.)

are done, it is to be expected that the confidence interval for at least one of them will be less than $1/k$. Evidence for nonrandomness in a sequence must come from an excess of confidence interval values close to zero or one, over and above the number expected for a uniform distribution.

Table 10.1 gives results from the statistical tests described in Appendix A for sequences generated by the cellular automaton (3.1) in a finite circular register. Except when the sample sequence is comparable in length to the period of the system, as given by Table 9.2, no significant deviations from randomness are found.

Table 10.2 gives statistical results for sequences generated by other procedures. Those obtained from linear feedback shift registers, while provably random in some respects (e.g. [4]), are revealed as significantly nonrandom by several of the tests used here. Many sequences obtained from linear congruential generators are also found to be significantly nonrandom with respect to these tests. No regularities are detected in the digit sequence of $\sqrt{2}$ (and other surds tried) (cf. [37]). There is, however, some possible evidence for nonrandomness in the digit sequences of e and π (cf. [38]). (This will be explored elsewhere.)

160   STEPHEN WOLFRAM

TABLE 10.2

	CA60 $N = 29$ $L = 64k$	LFSR $N = 17$ $L = 64k$	LFSR $N = 29$ $L = 64k$	LCG $N = 32$ $L = 64k$	$\sqrt{2}$ $L = 51906k$	e $L = 9501k$	π $L = 26755k$
A	**1.0000**	**0.0390**	**0.9998**	**0.0167**	0.6255	0.5505	0.1441
B	**1.0000**	**0.9773**	0.4378	0.0841	0.0801	0.4556	**0.9525**
C	**1.0000**	0.2654	**1.0000**	0.1676	0.0582	0.8615	0.2799
D	**1.0000**	0.8797	0.8400	0.8322	0.8553	0.7605	**0.9986**
E	0.9256	**1.0000**	0.9435	0.5850	0.6363	0.6890	**0.0049**
F	**0.9998**	**1.0000**	**0.9674**	0.9248	0.8499	0.7031	0.1297
G	**1.0000**	**0.9790**	0.3476	0.3137	0.8465	0.4086	0.5473

Results of statistical tests for sequences generated by various procedures. CA60 is the linear cellular automaton rule of Eq. (2.4), in a size N circular register. LFSR is a linear feedback shift register of length N with period $2^N - 1$. For $N = 17$ the shift register taps are at positions 14 and 17; for $N = 29$ they are at positions 27 and 29. For CA60 and LFSR seeds consisting of a single nonzero site were used. LCG is the linear congruential generator $x' = (1103515245x + 12345) \bmod 2^{31}$ (used, for example, in many implementations of the UNIX operating system). The seed $x = 1$ was used. The behaviour of CA60, LFSR and LCG are illustrated in Fig. 11.1. $\sqrt{2}$, e, and π are the binary digit sequences of the square root of two, the exponential constant, and pi, respectively. (These digit sequences were obtained by R. W. Gosper using a Symbolics 3600 LISP machine.)

TABLE 10.3

	$i = 0$ $L = 8k$	$i = 0$ $L = 64k$	$i = 0$ $L = 512k$	$i = 1$ $L = 512k$	$i = -1$ $L = 512k$	$i = 32$ $L = 512k$	$i = -32$ $L = 512k$
A	0.1536	0.2234	0.6453	0.8629	0.8630	0.8733	0.2677
B	0.5996	0.0637	0.4891	0.7639	0.8343	0.2525	0.1751
C	0.6448	0.6538	0.5443	0.5887	0.4000	0.8271	0.8815
D	0.5921	0.2643	**0.0051**	**0.0105**	0.7030	0.4550	0.7832
E	0.1358	0.1348	0.6631	0.8430	0.7498	0.1264	0.8353
F	0.2622	0.1957	0.9385	0.4324	0.9009	0.4736	0.8022
G	0.4542	0.8773	0.6658	0.1080	0.7169	0.7744	0.2364

Results of statistical tests for vertical sequences at position i in the pattern of Fig. 6.1 generated by evolution according to Eq. (3.1) from a single nonzero initial site on an infinite lattice. Leading zeroes in each sequence were truncated. (The sequences were obtained by Jim Salem using a prototype Connection Machine computer.)

Table 10.3 gives statistical results for temporal sequences in the pattern of Fig. 6.1 obtained by evolution according to Eq. (3.1) from a single nonzero initial site on an infinite lattice. Once again, no significant deviations from randomness are seen.

If deviations from randomness were detected by some statistical procedure, then this procedure could be used to make statistical predictions about the sequence. In addition, it could be used to obtain a compressed representation for the sequence, and would thus demonstrate that the sequence did not have maximal information content. The fact that deviations from randomness have not been found by any of the statistical procedures considered lends strong support to the belief that sequences produced by Eq. (3.1) with large N are indeed random for practical purposes.

11. PRACTICAL IMPLEMENTATION

The simplicity and intrinsic parallelism of the cellular automaton rule (3.1) makes possible efficient implementation on many kinds of computers.

On a serial-processing computer, each site could be updated in turn according to (3.1). But in practice, site values can be represented by single bits in say a 32-bit word, and updated in parallel using standard word-wise Boolean operations. (Additional bit-wise operations are often needed for boundary conditions.)

On a synchronous parallel-processing computer, different sites or groups of sites in the cellular automaton can be assigned to different processors. They can then be updated independently (though synchronously), using the same instructions, and with only local communications.

Very efficient hardware implementations of (3.1) should also be possible. For short registers, explicit circuitry can be included for each site. And for long registers, a pipelined approach analogous to a feedback shift register can be used (cf. [39]).

The evidence presented above suggests that the cellular automaton of Eq. (3.1) can serve as a practical random sequence generator. The most appropriate detailed choices of parameters depend on the application intended. The most obvious constraint is one of cycle length. To obtain a cycle length larger than $2^{32} \cong 4 \times 10^9$, Table 9.2 shows that a circular register of length $N = 49$ can be used. Cycle lengths tend to increase with N, but Table 9.2 shows some irregularities. Thus it is not clear, for example, how large N need be to obtain a cycle length larger than $2^{64} \cong 10^{19}$. But based on Eq. (9.1), a value $N = 127$ should certainly suffice.

Random sequences can be obtained by sampling the sequence of values of a particular site in a register updated according to Eq. (3.1). The theoretical and statistical studies described above support the contention that such sequences show no regularities. For some critical applications, it may be best however, to sample site values only say on alternate time steps. While this method generates a sequence more slowly, it should foil prediction procedures along the lines discussed in Section 7.

Sequences could potentially be obtained more quickly by extracting the values of several sites in the register at each time step. But Eq. (4.6) implies that some statistical correlations must exist between these values. The correlations are probably minimized if the sites sampled are equally spaced around the register. Nevertheless, in some applications where only a low degree of randomness is needed, it may even be satisfactory to use all site values in the register. (An example appears to be approximation of partial differential equations, where randomness can be used to emulate additional low-order digits.)

The random sequences obtained from Eq. (3.1) have an equal fraction of 0 and 1. Many applications, however, involve random binary choices with unequal probabilities. There is nevertheless a simple algorithm [40] to obtain digits with arbitrary probabilities. First write the probability p for outcome 1 as a binary number. Then generate a random binary sequence s with a length equal to this number. The output is obtained by an iterative proce-dure. Begin with a "current result" of 1. Then, starting from the least significant digit in p, successively find a new result by combining the old result with the corresponding digit of s, using a function AND or OR, depending on whether the digit in p is 0 or 1, respectively. The final result thus obtained is equal to 1 with probability exactly p.

Configurations in two length N registers with slightly different seeds should become progressively less correlated under the action (3.1) as a result of the instability discussed in Section 5. The characteristic time for this process is governed by Eqs. (5.1) and (5.2), and should be $\cong 0.8\ N$. Thus, if several sequences are to be generated with seeds that differ only slightly (obtained for example from addresses of computer elements), then (3.1) should applied at least $O(N)$ times to the seeds before beginning to extract random sequences.

One may compare the scheme for random sequence generation described here with the linear methods now in common use (e.g. [1]). Figure 11.1 shows patterns produced by these various schemes. The primary feature of linear schemes is that they can be analysed by algebraic methods. As a consequence, certain randomness properties can be proved for the sequences they generate, and cases that give long cycles can be identified. But the simplicity in structure which underlies this analysis also limits the degree of randomness that such schemes can produce. The nonlinear scheme de-

CELLULAR AUTOMATA   163

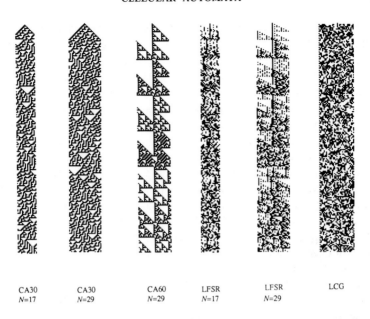

<table>
| CA30
$N=17$ | CA30
$N=29$ | CA60
$N=29$ | LFSR
$N=17$ | LFSR
$N=29$ | LCG |
</table>

FIG. 11.1.   Patterns obtained by various procedures in registers of size N. CA30 stands for the cellular automaton of Eq. (3.1), with periodic boundary conditions. CA60 is the linear cellular automaton of Eq. (2.4), again with periodic boundary conditions. LFSR is a linear feedback shift register with size N and period $2^N - 1$. For $N = 17$ the taps are at positions 14 and 17; for $N = 29$, they are at positions 27 and 29. LCG is a linear congruential sequence generator, operating on the 32-bit integers whose binary digit sequences are given. The seed in all cases consists of a single nonzero bit in the centre of the register. Statistical properties of the sequences produced are given in Tables 10.1 and 10.2.

scribed here is not readily amenable to complete analysis, and no significant limits on the degree of randomness it yields are known. But on the other hand, no conventional mathematical proofs for particular randomness properties can be given, and it must be investigated by largely empirical methods.

12. ALTERNATIVE SCHEMES

The cellular automaton of Eq. (3.1) is one of the simplest that seems good for random sequence generation. But other cellular automata may also be considered, and some potentially have certain advantages.

Among $k = 2$, $r = 1$ cellular automata, Eq. (3.2) is the only other serious contender. No direct equivalence between this rule and that of Eq. (3.1) is

known, but their properties are very similar. Equation (3.2) gives however
[45]

$$\lambda_L = (0.1724 \pm 0.0004), \tag{12.1}$$

slightly smaller than the corresponding result (5.2) for Eq. (3.1). In addition,
it gives a slightly smaller invariant entropy $\mathbf{h}_\mu$. It seems to have no
advantages over (3.1).

Cellular automata with $k > 2$ or $R > 3$ may also be studied. (Here R is
defined as the total number of sites in the neighbourhood for the rule.) Any
class 3 (chaotic) cellular automaton rule can be considered a candidate
random sequence generator. Autoplectic rules which produce complex patterns even from simple initial conditions are probably best. Some of these
rules have larger Lyapunov exponents and invariant entropies than Eq.
(3.1), but they are also more difficult to compute. In addition, many rules
that seem to produce chaotic overall patterns nevertheless yield sequences
that show definite regularities, resulting, for example, in non-maximal
temporal entropies. Permutive chaotic rules avoid such problems, but are
very similar in character to the rule of Eq. (3.1), and so potentially share any
of its possible deficiencies.

One possibility is to consider bijective cellular automaton rules, which are
invertible, so that each configuration has both a unique successor in time,
and a unique predecessor. The state transition diagrams for such cellular
automata in finite regions with periodic boundary conditions can contain
only cycles, and no transients. But only a very small fraction of all cellular
automaton rules are bijective, and very few of those that are exhibit chaotic
behaviour. Table 12.1 gives some non-trivial bijective cellular automaton
rules with $k = 2$ and $R \le 5$ (cf. [41]). None of those with $R \le 4$ are
chaotic.

With larger effective k, it is nevertheless possible to construct chaotic
bijective rules explicitly. One method [42] yields cellular automaton rules
that are most easily stated in terms of dependence on second-to-last as well
as immediately preceding site values:

$$a_i^{(t)} = \phi\left(a_{i-r}^{(t-1)}, \cdots, a_{i+r}^{(t-1)}\right) \text{XOR } a_i^{(t-2)}. \tag{12.2}$$

Such rules may be stated in the standard form (2.1) by considering sites with
k^2 possible values. Some examples of patterns generated by rules of the
form (12.2) are shown in Fig. 12.1. The rules are bijective, so that all states
lie on cycles. However, there are often many distinct cycles, each quite
short, making the system unsuitable for random sequence generation.

CELLULAR AUTOMATA   165

TABLE 12.1

ϕ	ϕ^{-1}
$k = 2, R = 4$	
1kng	1kng
1s5k	1s5k
1hmc	1hmc
1j4s	1j4s
$k = 2, R = 5$	
3nh1vo0	3nh1vo0
3ug5vo0	3ug5vo0
39gtvo0	f20nv1jogtvo0

Bijective cellular automata rules with k possible values for each site and depending on strictly R previous site values. The rules given are "totally quiescent," so that $\phi(a, a, \ldots, a) = a$ for all a. The rules are specified by giving the values of ϕ as digits in a binary number indexed by a number formed from the arguments of ϕ. The binary number is then stated in base 32, with letters of the alphabet representing successive digits greater than 9. Leading zeroes are not truncated. Long specifications correspond to rules with larger values of R.

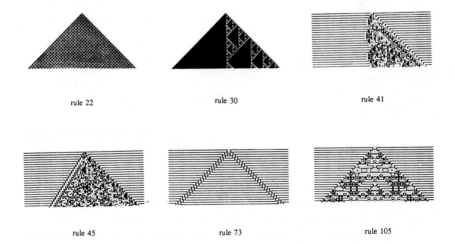

rule 22   rule 30   rule 41

rule 45   rule 73   rule 105

FIG. 12.1. Patterns generated by various bijective (reversible) $k = 2$, $r = 1$ cellular automata with rules of the form (12.1).

13. DISCUSSION

This paper has used methods from several disciplines to study the behaviour of the nonlinear cellular automaton of Eq. (3.1). Despite the simplicity of its construction, all the approaches taken support the conjecture that its behaviour is so complicated as to seem random for practical purposes. It is remarkable that such a simple system can give rise to such complexity. But it is in keeping with the observation that mathematical systems with few axioms, or computers with few intrinsic instructions, can lead to essentially arbitrary complexity. And it seems likely that the mathematical mechanisms at work are also responsible for much of the randomness and chaos seen in nature.

The simplicity of Eq. (3.1) makes it amenable to highly efficient practical implementation. And the analyses carried out here suggest that the sequences it produces have a high degree of randomness. In fact, if any regularity could be found in these sequences, it would probably have substantial consequences for studies of many complex and seemingly random phenomena.

APPENDIX A: STATISTICAL PROCEDURES

This Appendix describes the statistical randomness testing procedures used in Section 10. The procedures are mostly taken from [1], although their numbering has been changed slightly. The basic method in each case is to compare an observed distribution with that calculated for a purely probabilistic sequence.

The sequences studied consist of strings of binary bits. In many of the tests, these bits are grouped into blocks: either length 8 (non-overlapping) bytes, or length 4 (non-overlapping) nybbles. The possible bit sequences in these blocks can be represented by integer "values" between 0 and 255 or 16, respectively.

A. *Block Frequency Distribution.* Each of the 2^n possible n-blocks should occur with equal frequency. ($n = 8$ is used.)

B. *Gap Length Distribution.* The lengths of runs of n-blocks whose values are all greater than i_2 or less than i_1 should follow a binomial distribution. ($n = 8$, $i_1 = 100$, $i_2 = 200$ are used; runs longer than 16 blocks are lumped together.)

C. *Distinct Blocks Distribution.* The frequencies with which p out of q successive m-blocks are distinct should follow a definite distribution. ($m = 4$, $q = 4$ are used.)

D. *Block Accumulation Distribution.* The number of successive n-blocks necessary for all possible m-blocks to appear in order as their first m elements should follow a definite distribution. ($n = 8$, $m = 3$ are used; numbers greater than 40 are lumped together.)

E. *Permutation Frequency Distribution.* The values of q successive n-blocks should occur in all $q!$ possible orderings with equal frequency. ($n = 8$, $q = 5$ are used.)

F. *Monotone Sequence Length Distribution.* The lengths of sequences in which successive n-blocks have monotonically increasing values should follow a definite distribution. ($n = 8$ is used; lengths greater than 6 are lumped together; elements immediately following each run are discarded to make successive runs statistically independent.)

G. *Maxima Distribution.* The maximum values of n-blocks in sequences of q n-blocks should follow a power law distribution. ($n = 8$, $q = 8$ are used.)

ACKNOWLEDGMENTS

Many people have contributed in various ways to the material presented here. For specific suggestions I thank: Persi Diaconis, Carl Feynman, Richard Feynman, Shafi Goldwasser, Peter Grassberger, Erica Jen, and John Milnor.

For discussions I thank: Lenore Blum, Manuel Blum, Whit Diffie, Rolf Fiebrich, Danny Hillis, Doug Lind, Silvio Micali, Marvin Minsky, Andrew Odlyzko, Steve Omohundro, Norman Packard, and Jim Reeds.

For help with computational matters I thank: Keira Bromberg, Bill Gosper, Don Mitchell, Bruce Nemnich, Holly Peck, Jim Salem, Tsutomu Shimomura, Steve Strassmann, and Don Webber.

The computer mathematics system SMP [43] was used for some of the calculations. I thank the Science Office of Sun Microsystems for the loan of a SUN workstation on which most of the graphics and many of the calculations were done. And finally I thank Thinking Machines Corporation for the use of a prototype Connection Machine computer [44], without which much more about the cellular automaton of Eq. (3.1) would still be unknown.

Note added in proof. Eq. (3.1) can also be used to generate efficiently a key sequence for stream encryption [46].

REFERENCES

1. D. KNUTH, "Seminumerical Algorithms," Addison–Wesley, Reading, Mass., 1981.
2. A. SHAMIR, "On the generation of cryptographically strong pseudorandom sequences," Lecture Notes in Computer Science Vol. 62, p. 544, Springer-Verlag, New York/Berlin, 1981; S. GOLDWASSER AND S. MICALI, Probabilistic encryption, *J. Comput. System Sci.* **28**, (1984) 270; M. BLUM AND S. MICALI, How to generate crytographically strong sequences of pseudorandom bits, *SIAM J. Comput.* **13** (1984) 850; A. YAO, Theory and applications of trapdoor functions, *in* "Proc. 23rd IEEE Symp. on Foundations of Computer Science," 1982.

168   STEPHEN WOLFRAM

3. G. CHAITIN, On the length of programs for computing finite binary sequences, I, II, *J. Assoc. Comput. Mach.* **13** (1966) 547; **16**, (1969) 145, Randomness and mathematical proof, *Sci. Amer.* **232**, No. 5 (1975) 47; A. N. KOLMOGOROV, Three approaches to the concept of "the amount of information," *Problems Inform. Transmission* **1** (1965) 1; R. SOLOMONOFF, A formal theory of inductive inference, *Inform. Control* **7** (1964) 1; P. MARTIN-LOF, The definition of random sequences, *Inform. Control* **9** (1966) 602; L. LEVIN, On the notion of a random sequence, *Soviet Math. Dokl.* **14** (1973), 1413.

4. S. W. GOLOMB, "Shift Register Sequences," Holden–Day, San Francisco, (1967).

5. M. GAREY AND D. JOHNSON, "Computers and Intractability: A Guide to the Theory of NP-Completeness," W. H. Freeman, San Francisco, 1979.

6. L. BLUM, M. BLUM, AND M. SHUB, Comparison of two pseudorandom number generators, *in* "Advances in Cryptology: Proc. of CRYPTO-82" (D. Chaum, R. Rivest, and A. T. Sherman, Eds.), Plenum, New York, 1983.

7. W. ALEXI, B. CHOR, O. GOLDREICH, AND C. SCHNORR, RSA/Rabin bits are $\frac{1}{2} + 1/\text{poly}(\log N)$ secure, *in* "Proc. Found. Comput. Sci.," (1984); U. VAZIRANI AND V. VAZIRANI, Efficient and secure pseudorandom number generation, *in* "Proc. Found. Comput. Sci.," 1984.

8. L. KUIPERS AND H. NIEDERREITER, "Uniform Distribution of Sequences," Wiley, New York, 1974.

9. J. LAGARIAS, The $3x + 1$ problem and its generalizations, *Amer. Math. Monthly* **92** (1985), 3.

10. K. MAHLER, An unsolved problem on the powers of $\frac{3}{2}$, *Proc. Austral. Math. Soc.* **8** (1968), 313; G. CHOQUET, Repartition des nombres $k(\frac{3}{2})^n$; mesures et ensembles associes, *C. R. Acad. Sci. Paris A* **290** (1980), 575.

11. S. WOLFRAM, Origins of randomness in physical systems, *Phys. Rev. Lett.* **55** (1985), 449.

12. S. WOLFRAM, Cellular automata as models of complexity, *Nature* **311** (1984), 419.

13. D. FARMER, T. TOFFOLI, AND S. WOLFRAM, (Eds.), Cellular automata, *Physica D* **10** Nos. 1, 2, (1984).

14. S. WOLFRAM, Cellular automata and condensed matter physics, *in* Proc. NATO Advanced Study Institute on Scaling phenomena in disordered systems, April 1985.

15. S. WOLFRAM, Universality and complexity in cellular automata, *Physica D* **10** (1984), 1.

16. S. WOLFRAM, Computation theory of cellular automata, *Comm. Math. Phys.* **96** (1984), 15.

17. S. WOLFRAM, Statistical mechanics of cellular automata, *Rev. Modern Phys.* **55** (1983), 601.

18. S. WOLFRAM, Undecidability and intractability in theoretical physics, *Phys. Rev. Lett.* **54** (1985) 735.

19. S. WOLFRAM, Computer software in science and mathematics, *Sci. Amer.*, September 1984.

20. O. MARTIN, A. ODLYZKO, AND S. WOLFRAM, Algebraic properties of cellular automata, *Comm. Math. Phys.* **93** (1984), 219.

21. J. MILNOR, Notes on surjective cellular automaton-maps, Institute for Advanced Study preprint, June 1984.

22. E. JEN, "Global Properties of Cellular Automata," Los Alamos report LA-UR-85-1218, 1985; *J. Statist. Phys.*, in press.

23. J. GUCKENHEIMER AND P. HOLMES, "Nonlinear Oscillations, Dynamical Systems, and Bifurcations of Vector Fields", Springer-Verlag, New York/Berlin, 1983.

24. J. MILNOR, Entropy of cellular automaton-maps, Institute for Advanced Study preprint, May 1984; Directional entropies of cellular automaton maps, Institute for Advanced Study preprint, October 1984.

25. YA. SINAI, An answer to a question by J. Milnor, Comment Math. Helv. **60** (1985), 173.

26. N. PACKARD, Complexity of growing patterns in cellular automata, *in* "Dynamical systems and cellular automata," (J. Demongeot, E. Goles, and M. Tchuente, Eds.), Academic Press, 1985.

CELLULAR AUTOMATA 169

27. R. FEYNMAN, private communication.
28. R. BRAYTON, G. HACHTEL, C. MCMULLEN, AND A. SANGIOVANNI-VINCENTELLI, "Logic Minimization Algorithms for VLSI Synthesis," Kluwer, 1984.
29. R. RUDELL, "*Espresso* software program," Computer Science Dept., University of California, Berkeley, 1985.
30. S. WOLFRAM, Geometry of binomial coefficients, *Amer. Math. Monthly* **91** (1984), 566.
31. M. FURST, J. SAXE, AND M. SIPSER, Parity, circuits, and the polynomial-time hierarchy, *Math Systems Theory* **17** (1984), 13.
32. C. FEYNMAN AND R. FEYNMAN, private communication.
33. M. MINSKY, "Computation: Finite and Infinite Machines," Prentice–Hall, Englewood Cliffs, N.J., 1967.
34. B. HARRIS, Probability distributions related to random mappings, *Ann. Math. Statist.* **31** (1960), 1045.
35. G. MARSAGLIA, A current view of random number generators, *in* "Proc. Comput. Sci. and Statistics, 16th Sympos. on the Interface," Atlanta, March 1984.
36. G. W. SNEDECOR AND W. G. COCHRAN, "Statistical Methods," Iowa State Univ. Press, Ames, 1967.
37. W. BEYER, N. METROPOLIS AND J. R. NEERGAARD, Statistical study of digits of some square roots of integers in various bases, Math. Comp. **24** (1970), 455.
38. S. WAGON, Is π normal?, *Math. Intelligencer*, **7** (1985), 65.
39. T. TOFFOLI, CAM: A high-performance cellular-automaton machine, *Physica D* **10** (1984), 195; K. STEIGLITZ AND R. MORITA, A multi-processor cellular automaton chip, *in* "Proc. 1985 IEEE International Conf. on Acoustics, Speech, and Signal Processing," March 1985.
40. J. SALEM, Thinking Machines Corporation report, to be published.
41. G. HEDLUND, Endomorphisms and automorphisms of the shift dynamical system, *Math. Systems Theory* **3** (1969), 320; G. HEDLUND, private communication.
42. N. MARGOLUS, Physics-like models of computation, *Physica D* **10** (1984), 81.
43. S. WOLFRAM, "SMP Reference Manual," Computer Mathematics Group, Inference Corporation, Los Angeles, 1983.
44. D. HILLIS, "The Connection Machine," MIT Press, Cambridge, Mass., 1985.
45. P. GRASSBERGER, "Towards a quantitative theory of self-generated complexity," Wuppertal preprint (1986).
46. S. WOLFRAM, Cryptography with cellular automata, *in* "Proc. CRYPTO 85," August 1985.

VOLUME 54, NUMBER 8   PHYSICAL REVIEW LETTERS   25 FEBRUARY 1985

Undecidability and Intractability in Theoretical Physics

Stephen Wolfram

The Institute for Advanced Study, Princeton, New Jersey 08540
(Received 26 October 1984)

Physical processes are viewed as computations, and the difficulty of answering questions about them is characterized in terms of the difficulty of performing the corresponding computations. Cellular automata are used to provide explicit examples of various formally undecidable and computationally intractable problems. It is suggested that such problems are common in physical models, and some other potential examples are discussed.

PACS numbers: 02.90.+p, 01.70.+w, 05.90.+m

There is a close correspondence between physical processes and computations. On one hand, theoretical models describe physical processes by computations that transform initial data according to algorithms representing physical laws. And on the other hand, computers themselves are physical systems, obeying physical laws. This paper explores some fundamental consequences of this correspondence.[1]

The behavior of a physical system may always be calculated by simulating explicitly each step in its evolution. Much of theoretical physics has, however, been concerned with devising shorter methods of calculation that reproduce the outcome without tracing each step. Such shortcuts can be made if the computations used in the calculation are more sophisticated than those that the physical system can itself perform. Any computations must, however, be carried out on a computer. But the computer is itself an example of a physical system. And it can determine the outcome of its own evolution only by explicitly following it through: No shortcut is possible. Such computational irreducibility occurs whenever a physical system can act as a computer. The behavior of the system can be found only by direct simulation or observation: No general predictive procedure is possible. Computational irreducibility is common among the systems investigated in mathematics and computation theory.[2] This paper suggests that it is also common in theoretical physics. Computational reducibility may well be the exception rather than the rule: Most physical questions may be answerable only through irreducible amounts of computation. Those that concern idealized limits of infinite time, volume, or numerical precision can require arbitrarily long computations, and so be formally undecidable.

A diverse set of systems are known to be equivalent in their computational capabilities, in that particular forms of one system can emulate any of the others. Standard digital computers are one example of such "universal computers": With fixed intrinsic instructions, different initial states or programs can be devised to simulate different systems. Some other examples are Turing machines, string transformation systems, recursively defined functions, and Diophan-

tine equations.[2] One expects in fact that universal computers are as powerful in their computational capabilities as any physically realizable system can be, so that they can simulate any physical system.[3] This is the case if in all physical systems there is a finite density of information, which can be transmitted only at a finite rate in a finite-dimensional space.[4] No physically implementable procedure could then short cut a computationally irreducible process.

Different physically realizable universal computers appear to require the same order of magnitude times and information storage capacities to solve particular classes of finite problems.[5] One computer may be constructed so that in a single step it carries out the equivalent of two steps on another computer. However, when the amount of information n specifying an instance of a problem becomes large, different computers use resources that differ only by polynomials in n. One may then distinguish several classes of problems.[6] The first, denoted P, are those such as arithmetical ones taking a time polynomial in n. The second, denoted $PSPACE$, are those that can be solved with polynomial storage capacity, but may require exponential time, and so are in practice effectively intractable. Certain problems are "complete" with respect to $PSPACE$, so that particular instances of them correspond to arbitrary $PSPACE$ problems. Solutions to these problems mimic the operation of a universal computer with bounded storage capacity: A computer that solves $PSPACE$-complete problems for any n must be universal. Many mathematical problems are $PSPACE$-complete.[6] (An example is whether one can always win from a given position in chess.) And since there is no evidence to the contrary, it is widely conjectured that $PSPACE \neq P$, so that $PSPACE$-complete problems cannot be solved in polynomial time. A final class of problems, denoted NP, consist in identifying, among an exponentially large collection of objects, those with some particular, easily testable property. An example would be to find an n-digit integer that divides a given $2n$-digit number exactly. A particular candidate divisor, guessed nondeterministically, can be tested in polynomial time, but a systematic solution may require almost all $O(2^n)$ possible candidates to be

735

VOLUME 54, NUMBER 8   PHYSICAL REVIEW LETTERS   25 FEBRUARY 1985

tested. A computer that could follow arbitrarily many computational paths in parallel could solve such problems in polynomial time. For actual computers that allow only boundedly many paths, it is suspected that no general polynomial time solution is possible.[5] Nevertheless, in the infinite time limit, parallel paths are irrelevant, and a computer that solves NP-complete problems is equivalent to other universal computers.[6]

The structure of a system need not be complicated for its behavior to be highly complex, corresponding to a complicated computation. Computational irreducibility may thus be widespread even among systems with simple construction. Cellular automata (CA)[7] provide an example. A CA consists of a lattice of sites, each with k possible values, and each updated in time steps by a deterministic rule depending on a neighborhood of R sites. CA serve as discrete approximations to partial differential equations, and provide models for a wide variety of natural systems. Figure 1 shows typical examples of their behavior. Some rules give periodic patterns, and the outcome after many steps can be predicted without following each intermediate step. Many rules, however, give complex patterns for which no predictive procedure is evident. Some CA are in fact known to be capable of universal computation, so that their evolution must be computationally irreducible. The simplest cases proved have $k=18$ and $R=3$ in one dimension,[8] or $k=2$ and $R=5$ in two dimensions.[9] It is strongly suspected that "class-4" CA are generically capable of universal computation: There are such CA with $k=3$, $R=3$ and $k=2$, $R=5$ in one dimension.[10]

Computationally, irreducibility may occur in systems that are not full universal computers. For inability to perform, specific computations need not allow all computations to be short cut. Though class-3 CA and other chaotic systems may not be universal computers, most of them are expected to be computationally irreducible, so that the solution of problems concerning their behavior requires irreducible amounts of computation.

As a first example consider finding the value of a site in a CA after t steps of evolution from a finite initial seed, as illustrated in Fig. 1. The problem is specified by giving the seed and the CA rule, together with the $\log t$ digits of t. In simple cases such as the first two shown in Fig. 1, it can be solved in the time $O(\log t)$

necessary to input this specification. However, the evolution of a universal computer CA for a polynomial in t steps can implement any computation of length t. As a consequence, its evolution is computationally irreducible, and its outcome found only by an explicit simulation with length $O(t)$: exponentially longer than for the first two in Fig. 1.

One may ask whether the pattern generated by evolution with a CA rule from a particular seed will grow forever, or will eventually die out.[11] If the evolution is computationally irreducible, then an arbitrarily long computation may be needed to answer this question. One may determine by explicit simulation whether the pattern dies out after any specified number of steps, but there is no upper bound on the time needed to find out its ultimate fate.[12] Simple criteria may be given for particular cases, but computational irreducibility implies that no shortcut is possible in general. The infinite-time limiting behavior is formally undecidable: No finite mathematical or computational process can reproduce the infinite CA evolution.

The fate of a pattern in a CA with a finite total number of sites N can always be determined in at most k^N steps. However, if the CA is a universal computer, then the problem is $PSPACE$-complete, and so presumably cannot be solved in a time polynomial in N.[13]

One may consider CA evolution not only from finite seeds, but also from initial states with all infinitely many sites chosen arbitrarily. The value $a^{(t)}$ of a site after many time steps t then in general depends on $2\lambda t \lesssim Rt$ initial site values, where λ is the rate of information transmission (essentially Lyapunov exponent) in the CA.[9] In class-1 and -2 CA, information remains localized, so that $\lambda=0$, and $a^{(t)}$ can be found by a length $O(\log t)$ computation. For class-3 and -4 CA, however, $\lambda > 0$, and $a^{(t)}$ requires an $O(t)$ computation.[14]

The global dynamics of CA are determined by the possible states reached in their evolution. To characterize such states one may ask whether a particular string of n site values can be generated after evolution for t steps from any (length $n+2\lambda t$) initial string. Since candidate initial strings can be tested in $O(t)$ time, this problem is in the class NP. When the CA is a universal computer, the problem is in general NP-complete, and can presumably be answered essentially only by testing all $O(k^{n+2\lambda t})$ candidate initial

FIG. 1. Seven examples of patterns generated by repeated application of various simple cellular automaton rules. The last four are probably computationally irreducible, and can be found only by direct simulation.

VOLUME 54, NUMBER 8   PHYSICAL REVIEW LETTERS   25 FEBRUARY 1985

strings.[15] In the limit $t \to \infty$, it is in general undecidable whether particular strings can appear.[16] As a consequence, the entropy or dimension of the limiting set of CA configurations is in general not finitely computable.

Formal languages describe sets of states generated by CA.[17] The set that appears after t steps in the evolution of a one-dimensional CA forms a regular formal language: each possible state corresponds to a path through a graph with $\Xi^{(t)} < 2^{k^{Rt}}$ nodes. If, indeed, the length of computation to determine whether a string can occur increases exponentially with t for computationally irreducible CA, then the "regular language complexity" $\Xi^{(t)}$ should also increase exponentially, in agreement with empirical data on certain class-3 CA,[17] and reflecting the "irreducible computational work" achieved by their evolution.

Irreducible computations may be required not only to determine the outcome of evolution through time, but also to find possible arrangements of a system in space. For example, whether an $x \times x$ patch of site values occurs after just one step in a two-dimensional CA is in general NP-complete.[18] To determine whether there is any complete infinite configuration that satisfies a particular predicate (such as being invariant under the CA rule) is in general undecidable[18]: It is equivalent to finding the infinite-time behavior of a universal computer that lays down each row on the lattice in turn.

There are many physical systems in which it is known to be possible to construct universal computers. Apart from those modeled by CA, some examples are electric circuits, hard-sphere gases with obstructions, and networks of chemical reactions.[19] The evolution of these systems is in general computationally irreducible, and so suffers from undecidable and intractable problems. Nevertheless, the constructions used to find universal computers in these systems are arcane, and if computationally complex problems occurred only there, they would be rare. It is the thesis of this paper that such problems are in fact common.[20] Certainly there are many systems whose properties are in practice studied only by explicit simulation or exhaustive search: Few computational shortcuts (often stated in terms of invariant quantities) are known.

Many complex or chaotic dynamical systems are expected to be computationally irreducible, and their behavior effectively found only by explicit simulation. Just as it is undecidable whether a particular initial state in a CA leads to unbounded growth, to self-replication, or has some other outcome, so it may be undecidable whether a particular solution to a differential equation (studied say with symbolic dynamics) even enters a certain region of phase space, and whether, say, a certain n-body system is ultimately stable. Similarly, the existence of an attractor, say,

with a dimension above some value, may be undecidable.

Computationally complex problems can arise in finding eigenvalues or extremal states in physical systems. The minimum energy conformation for a polymer is in general NP-complete with respect to its length.[21] Finding a configuration below a specified energy in a spin-glass with particular couplings is similarly NP-complete.[22] Whenever the stationary state of a physical system such as this can be found only by lengthy computation, the dynamic physical processes that lead to it must take a correspondingly long time.[5]

Global properties of some models for physical systems may be undecidable in the infinite-size limit (like those for two-dimensional CA). An example is whether a particular generalized Ising model (or stochastic multidimensional CA[23]) exhibits a phase transition.

Quantum and statistical mechanics involve sums over possibly infinite sets of configurations in systems. To derive finite formulas one must use finite specifications for these sets. But it may be undecidable whether two finite specifications yield equivalent configurations. So, for example, it is undecidable whether two finitely specified four-manifolds or solutions to the Einstein equations are equivalent (under coordinate reparametrization).[24] A theoretical model may be considered as a finite specification of the possible behavior of a system. One may ask for example whether the consequences of two models are identical in all circumstances, so that the models are equivalent. If the models involve computations more complicated than those that can be carried out by a computer with a fixed finite number of states (regular language), this question is in general undecidable. Similarly, it is undecidable what is the simplest such model that describes a given set of empirical data.[25]

This paper has suggested that many physical systems are computationally irreducible, so that their own evolution is effectively the most efficient procedure for determining their future. As a consequence, many questions about these systems can be answered only by very lengthy or potentially infinite computations. But some questions answerable by simpler computations may still be formulated.

This work was supported in part by the U. S. Office of Naval Research under Contract No. N00014-80-C-0657. I am grateful for discussions with many people, particularly C. Bennett, G. Chaitin, R. Feynman, E. Fredkin, D. Hillis, L. Hurd, J. Milnor, N. Packard, M. Perry, R. Shaw, K. Steiglitz, W. Thurston, and L. Yaffe.

---

[1]For a more informal exposition see: S. Wolfram, Sci. Am. **251**, 188 (1984). A fuller treatment will be given else-

where.

[2]E.g., *The Undecidable: Basic Papers on Undecidable Propositions, Unsolvable Problems, and Computable Functions,* edited by M. Davis (Raven, New York, 1965), or J. Hopcroft and J. Ullman, *Introduction to Automata Theory, Languages, and Computations* (Addison-Wesley, Reading, Mass., 1979).

[3]This is a physical form of the Church-Turing hypothesis. Mathematically conceivable systems of greater power can be obtained by including tables of answers to questions insoluble for these universal computers.

[4]Real-number parameters in classical physics allow infinite information density. Nevertheless, even in classical physics, the finiteness of experimental arrangements and measurements, implemented as coarse graining in statistical mechanics, implies finite information input and output. In relativistic quantum field theory, finite density of information (or quantum states) is evident for free fields bounded in phase space [e.g., J. Bekenstein, Phys. Rev. D **30**, 1669 (1984)]. It is less clear for interacting fields, except if space-time is ultimately discrete [but cf. B. Simon, *Functional Integration and Quantum Physics* (Academic, New York, 1979), Sec. III.9]. A finite information transmission rate is implied by relativistic causality and the manifold structure of space-time.

[5]It is just possible, however, that the parallelism of the path integral may allow quantum mechanical systems to solve any NP problem in polynomial time.

[6]M. Garey and D. Johnson, *Computers and Intractability: A Guide to the Theory of NP-Completeness* (Freeman, San Francisco, 1979).

[7]See S. Wolfram, Nature **311**, 419 (1984); *Cellular Automata,* edited by D. Farmer, T. Toffoli, and S. Wolfram, Physica **10D**, Nos. 1 and 2 (1984), and references therein.

[8]A. R. Smith, J. Assoc. Comput. Mach. **18**, 331 (1971).

[9]E. R. Banks, Massachusetts Institute of Technology Report No. TR-81, 1971 (unpublished). The "Game of Life," discussed in E. R. Berlekamp, J. H. Conway, and R. K. Guy, *Winning Ways for Your Mathematical Plays* (Academic, New York, 1982), is an example with $k = 2$, $R = 9$. N. Margolus, Physica (Utrecht) **10D**, 81 (1984), gives a reversible example.

[10]S. Wolfram, Physica (Utrecht) **10D**, 1 (1984), and to be published.

[11]This is analogous to the problem of whether a computer run with particular input will ever reach a "halt" state.

[12]The number of steps to check ("busy-beaver function") in general grows with the seed size faster than any finite formula can describe (Ref. 2).

[13]Cf. C. Bennett, to be published.

[14]Cf. B. Eckhardt, J. Ford, and F. Vivaldi, Physica (Utrecht) **13D**, 339 (1984).

[15]The question is a generalization of whether there exists an assignment of values to sites such that the logical expression corresponding to the t-step CA mapping is true (cf. V. Sewelson, private communication).

[16]L. Hurd, to be published.

[17]S. Wolfram, Commun. Math. Phys. **96**, 15 (1984).

[18]N. Packard and S. Wolfram, to be published. The equivalent problem of covering a plane with a given set of tiles is considered in R. Robinson, Invent. Math. **12**, 177 (1971).

[19]E.g., C. Bennett, Int. J. Theor. Phys. **21**, 905 (1982); E. Fredkin and T. Toffoli, Int. J. Theor. Phys. **21**, 219 (1982); A. Vergis, K. Steiglitz, and B. Dickinson, "The Complexity of Analog Computation" (unpublished).

[20]Conventional computation theory primarily concerns possibilities, not probabilities. There are nevertheless some problems for which almost all instances are known to be of equivalent difficulty. But other problems are known to be much easier on average then in the worst case. In addition, for some NP-complete problems the density of candidate solutions close to the actual one is very large, so approximate solutions can easily be found [S. Kirkpatrick, C. Gelatt, and M. Vecchi, Science **220**, 671 (1983)].

[21]Compare *Time Warps, String Edites, and Macromolecules,* edited by D. Sankoff and J. Kruskal (Addison-Wesley, Reading, Mass., 1983).

[22]F. Barahona, J. Phys. A **13**, 3241 (1982).

[23]E. Domany and W. Kinzel, Phys. Rev. Lett. **53**, 311 (1984).

[24]See W. Haken, in *Word Problems,* edited by W. W. Boone, F. B. Cannonito, and R. C. Lyndon (North-Holland, Amsterdam, 1973).

[25]G. Chaitin, Sci. Am. **232**, 47 (1975), and IBM J. Res. Dev. **21**, 350 (1977); R. Shaw, to be published.

VOLUME 55, NUMBER 5   PHYSICAL REVIEW LETTERS   29 JULY 1985

Origins of Randomness in Physical Systems

Stephen Wolfram

The Institute for Advanced Study, Princeton, New Jersey 08540

(Received 4 February 1985)

Randomness and chaos in physical systems are ususally ultimately attributed to external noise. But it is argued here that even without such random input, the intrinsic behavior of many nonlinear systems can be computationally so complicated as to seem random in all practical experiments. This effect is suggested as the basic origin of such phenomena as fluid turbulence.

PACS numbers: 05.45.+b, 02.90.+p, 03.40.Gc

There are many physical processes that seem random or chaotic. They appear to follow no definite rules, and to be governed merely by probabilities. But all fundamental physical laws, at least outside of quantum mechanics, are thought to be deterministic. So how, then, is apparent randomness produced?

One possibility is that its ultimate source is external noise, often from a heat bath. When the evolution of a system is unstable, so that perturbations grow, any randomness introduced through initial and boundary conditions is transmitted and amplified with time, and eventually affects many components of the system.[1] A simple example of this "homoplectic" behavior occurs in the shift mapping $x_t = 2x_{t-1} \bmod 1$. The time sequence of bins, say, above and below $\frac{1}{2}$ visited by x_t is a direct transcription of the binary-digit sequence of the initial real number x_0.[2] So if this digit sequence is random (as for most x_0 uniformly sampled in the unit interval) then so will the time sequence be; unpredictable behavior arises from a sensitive dependence on unknown features of initial conditions.[3] But if the initial condition is "simple," say a rational number with a periodic digit sequence, then no randomness appears.

There are, however, systems which can also generate apparent randomness internally, without external random input. Figure 1 shows an example, in which a cellular automaton evolving from a simple initial state produces a pattern so complicated that many features of it seem random. Like the shift map, this cellular automaton is homoplectic, and would yield random behavior given random input. But unlike the shift map, it can still produce random behavior even with simple input. Systems which generate randomness in this way will be called "autoplectic."

In developing a mathematical definition of autoplectic behavior, one must first discuss in what sense it is "random." Sequences are commonly considered random if no patterns can be discerned in them. But whether a pattern is found depends on how it is looked for. Different degrees of randomness can be defined in terms of the computational complexity of the procedures used.

The methods usually embodied in practical physics experiments are computationally quite simple.[4,5] They correspond to standard statistical tests for random-

ness,[6] such as relative frequencies of blocks of elements (dimensions and entropies), correlations, and power spectra. (The mathematical properties of ergodicity and mixing are related to tests of this kind.) One characteristic of these tests is that the computation time they require increases asymptotically at most like polynomial in the sequence length.[7] So if in fact no polynomial-time procedure can detect patterns in a sequence, then the sequence can be considered "effectively random" for practical purposes.

Any patterns that are identified in a sequence can be used to give a compressed specification for it. (Thus, for example, Morse coding compresses English text by exploiting the unequal frequencies of letters of the alphabet.) The length of the shortest specification measures the "information content" of a sequence with respect to a particular class of computations. (Standard Shannon information content for a stationary process[8] is associated with simple statistical computations of block frequencies.) Sequences are predictable only to the extent that they are longer than their shortest specification, and so contain information that can be recognized as "redundant" or "overdetermined."

Sequences generated by chaotic physical systems often show some redundancy or determinism under simple statistical procedures. (This happens whenever measurements extract information faster than it can be transferred from other parts of the system.[1]) But, typically, there remain compressed sequences in which no patterns are seen.

A sequence can, in general, be specified by giving an algorithm or computer program for constructing it. The length of the smallest possible program measures the "absolute" information content of the sequence.[9] For an "absolutely random" sequence the program must essentially give each element explicitly, and so be close in length to the sequence itself. But since no computation can increase the absolute information content of a closed system [except for $O(\log t)$ from input of "clock pulses"], physical processes presumably cannot generate absolute randomness.[10] However, the numbers of possible sequences and programs both increase exponentially with length, so that all but an exponentially small fraction of arbitrarily chosen sequences must be absolutely random. Nevertheless, it is usually undecidable what the smallest program for any particular sequence is, and thus whether the sequence is absolutely random. In general, each program of progressively greater length must be tried, and any one of them may run for an arbitrarily long time, so that the question of whether it ever generates the sequence may be formally undecidable.

Even if a sequence can ultimately be obtained from a small specification or program, and so is not absolutely random, it may nevertheless be effectively random if no feasible computation can recover the program.[11] The program can always be found by explicitly trying each possible one in turn.[12] But the total number of possible programs increases exponentially with length, and so such an exhaustive search would soon become infeasible. And if there is no better method the sequence must be effectively random.

In general, one may define the "effective information content" Θ of a sequence to be the length of the shortest specification for it that can be found by a feasible (say polynomial time) computation. A sequence can be considered "simple" if it has small Θ. Θ (often normalized by sequence length) provides a measurue of "complexity," "effective randomness," or "computational unpredictability."

Increasing Θ can be considered the defining characteristic of autoplectic behavior. Examples such as Fig. 1 suggest that Θ can increase through polynomial-time processes. The rule and initial seed have a short specification, with small Θ. But one suspects that no polynomial time computation can recover this specification from the center vertical sequence produced, or can in fact detect any pattern in it.[13] The polynomial-time process of cellular automaton evolution thus increases Θ, and generates effective randomness. It is phenomena of this kind that are the basis for cryptogra-

phy, in which one strives to produce effectively random sequences whose short "keys" cannot be found by any practical cryptanalysis.[14]

The simplest mathematical and physical systems (such as the shift mapping) can be decomposed into essentially uncoupled components, and cannot increase Θ. Such systems are nevertheless often homoplectic, so that they transfer information, and with random input show random behavior. But when their input is simple (low Θ), their behavior is correspondingly simple, and is typically periodic. Of course, any system with a fixed finite total number of degrees of freedom (such as a finite cellular automaton) must eventually become periodic. But the phenomena considered here occur on time scales much shorter than such exponentially long recurrences.

Another class of systems widely investigated consists of those with linear couplings between components [such as a cellular automaton in which $a_i^{(t+1)} = (a_{i-1}^{(t)} + a_{i+1}^{(t)}) \bmod 2$]. Given random input, such systems can again yield random output, and are thus homoplectic. But even with simple input, they can produce sequences which pass some statistical tests of randomness. Examples are the standard linear congruence and linear-feedback shift-register (or finite additive cellular automaton[15]) systems used for pseudorandom number generation in practical computer programs.[6,16]

Characteristic of such systems is the generation of self-similar patterns, containing sequences that are invariant under blocking or scaling transformations. These sequences are almost periodic, but may contain all possible blocks of elements with equal frequencies. They can be considered as the outputs of finite-state machines (generalized Markov processes) given the digits of the numerical positions of each element as input.[17] And although the sequences have certain statistical properties of randomness, their seeds can be found by comparatively simply polynomial-time procedures.[18] Such systems are thus not autoplectic (with respect to polynomial-time computations).

Many nonlinear mathematical systems seem, however, to be autoplectic, since they generate sequences in which no patterns have ever been found. An example is the sequence of leading digits in the fractional part of successive powers of $\frac{3}{2}$ [19] (which corresponds to a vertical column in a particular $k = 6$, $r = 1$ cellular automaton with a single site seed).

Despite extensive empirical evidence, almost nothing has, however, been proved about the randomness of such sequences. It is nevertheless possible to construct sequences that are strongly expected to be effectively random.[20] An example is the lowest-order bits of $x_t = x_{t-1}^2 \bmod(pq)$, where p and q are large primes.[20] The problem of deducing the initial seed x_0, or of substantially compressing this sequence, is

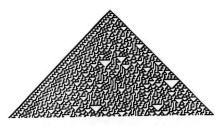

FIG. 1. Pattern generated by cellular automaton evolution from a simple initial state. Site values 0 or 1 (represented by white or black, respectively) are updated at each step according to the rule $a_i' = a_{i-1} \oplus (a_i \vee a_{i+1})$ ($\oplus$ denotes addition modulo 2, and $\vee$ Boolean disjunction). Despite the simplicity of its specification, many features of the pattern (such as the sequence of site values down the center column) appear random.

equivalent to the problem of factoring large integers, which is widely conjectured to require more than polynomial time.[21]

Standard statistical tests have also revealed no patterns in the digit sequences of transcendental numbers such as[22] $\sqrt{2}$, e, and π [22] (or continued-fraction expansions of π or of most cubic irrational numbers). But the polynomial-time procedure of squaring and comparing with an integer does reveal the digits of, say, $\sqrt{2}$ as nonrandom.[23] Without knowing how the sequence was generated, however, such a very special "statistical test" (or program) can probably only be found by explicit enumeration of all exponentially many possible ones. And if a sequence passes all but perhaps exponentially few polynomial-time batteries of statistical tests, it should probably be considered effectively random in practice.

Within a set of homoplectic dynamical systems (such as class 3 or 4 cellular automata) capable of transmitting information, all but the simplest seem to support sophisticated information processing, and are thus expected to be autoplectic. In some cases (quite probably including Fig. 1 [24]) the evolution of the system represents a "complete" or "universal" computation, which, with appropriate initial conditions, can mimic any other (polynomial-time) computation.[21] If short specifications for sequences generated by any one such computation could in general be found in polynomial time, it would imply that all could, which is widely conjectured to be impossible. (Such problems are called *NP*-complete.[21])

Many systems are expected to be computationally irreducible, so that the outcome of their evolution can be found essentially only by direct simulation, and no computational short cuts are possible.[25] To predict the future of these systems requires an almost complete knowledge of their current state. And it seems likely that this can be deduced from partial measurements only by essentially testing all exponentially many possibilities. The evolution of computationally irreducible systems should thus generically be autoplectic.

Autoplectic behavior is most clearly identified in discrete systems such as cellular automata. Continuous dynamical systems involve the idealization of real numbers on which infinite-precision arithmetic operations are performed. For systems such as iterated mappings of the interval there seems to be no robust notion of "simple" initial conditions. (The number of binary digits in images of, say, a dyadic rational grows like p^t, where p is the highest power of x in the map.) But in systems with many degrees of freedom, described for example by partial differential equations, autoplectism may be identified through discrete approximations.

Autoplectism is expected to be responsible for apparent randomness in many physical systems. Some features of turbulent fluid flow,[26] say in a jet ejected from a nozzle, are undoubtedly determined by details of initial or boundary conditions. But when the flow continues to appear random far from the nozzle, one suspects that other sources of effective information are present. One possibility might be thermal fluctuations or external noise, amplified by homoplectic processes.[1] But viscous damping probably allows only sufficiently large-scale perturbations to affect large-scale features of the flow. (Apparently random behavior is found to be almost exactly repeatable in some carefully controlled experiments.[27]) Thus, it seems more likely that the true origin of turbulence is an internal autoplectic process, somewhat like Fig. 1, operating on large-scale features of the flow. Numerical experiments certainly suggest that the Navier-Stokes equations can yield complicated behavior even with simple initial conditions.[28] Autoplectic processes may also be responsible for the widespread applicability of the second law of thermodynamics.

Many discussions have contributed to the material presented here; particularly those with C. Bennett, L. Blum, M. Blum, J. Crutchfield, P. Diaconis, D. Farmer, R. Feynman, U. Frisch, S. Goldwasser, D. Hillis, P. Hohenberg, E. Jen, R. Kraichnan, L. Levin, D. Lind, A. Meyer, S. Micali, J. Milnor, D. Mitchell, A. Odlyzko, N. Packard, I. Procaccia, H. Rose, and R. Shaw. This work was supported in part by the U. S. Office of Naval Research under Contract No. N00014-80-C-0657.

---

[1]For example, R. Shaw, Z. Naturforsch. **36A**, 80 (1981), and in *Chaos and Order in Nature*, edited by H. Haken (Springer, New York, 1981).

[2]An analogous cellular automaton [S. Wolfram, Nature (London) **311**, 419 (1984), and references therein] has evolution rule $a_i^{(t+1)} = a_{i+1}^{(t)}$, so that with time the value of a particular site is determined by the value of progressively more distant initial sites.

[3]For example, *Order in Chaos*, edited by D. Campbell and H. Rose (North-Holland, Amsterdam, 1982). Many processes analyzed in dynamical systems theory admit "Markov partitions" under which they are directly equivalent to the shift mapping. But in some measurements (say of x_t with four bins) their deterministic nature may introduce simple regularities, and "deterministic chaos" may be said to occur. (This term would in fact probably be better reserved for the autoplectic processes to be described below.)

[4]This is probably also true of at least the lower levels of human sensory processing [for example, D. Marr, *Vision* (Freeman, San Francisco, 1982); B. Julesz, Nature (London) **290**, 91 (1981)].

[5]The validity of Monte Carlo simulations tests the random sequences that they use. But most stochastic physical processes are in fact insensitive to all but the simplest equidistribution and statistical independence properties.

(Partial exceptions occur when long-range order is present.) And in general no polynomial-time simulation can reveal patterns in effectively random sequences.

[6]For example, D. Knuth, *Seminumerical Algorithms* (Addison-Wesley, Reading, Mass., 1981).

[7]Some sophisticated statistical procedures, typically involving the partitioning of high-dimensional spaces, seem to take exponential time. But most take close to linear time. It is possible that those used in practice can be characterized as needing $O(\log^p n)$ time on computers with $O(n^q)$ processors (and so be in the computational complexity class NC) [cf. N. Pippenger, in *Proceedings of the Twentieth IEEE Symposium on Foundations of Computer Science* (IEEE, New York, 1979); J. Hoover and L. Ruzzo, unpublished].

[8]For example, R. Hamming, *Coding and Information Theory* (Prentice-Hall, Englewood Cliffs, 1980).

[9]G. Chaitin, J. Assoc. Comput. Mach. **13**, 547 (1966), and **16**, 145 (1969), and Sci. Am. **232**, No. 5, 47 (1975); A. N. Kolmogorov, Problems Inform. Transmission **1**, 1 (1965); R. Solomonoff, Inform. and Control **7**, 1 (1964); L. Levin, Soviet Math. Dokl. **14**, 1413 (1973). Compare J. Ford, Phys. Today **33**, No. 4, 40 (1983). Note that the lengths of programs needed on different universal computers differ only by a constant, since each computer can simulate any other by means of a fixed "interpreter" program.

[10]Quantum mechanics suggests that processes such as radioactive decay occur purely according to probabilities, and so could perhaps give absolutely random sequences. But complete quantum mechanical measurements are an idealization, in which information on a microscopic quantum event is spread through an infinite system. In finite systems, unmeasured quantum states are like unknown classical parameters, and can presumably produce no additional randomness. Suggestions of absolute randomness probably come only when classical and quantum models are mixed, as in the claim that quantum processes near black holes may lose information to space-time regions that are causally disconnected in the classical approximation.

[11]In the cases now known, recognition of any pattern seems to involve essentially complete reconstruction of the original program, but this may not always be so (L. Levin, private communication).

[12]In some cases, such as optimization or eigenvalue problems in the complexity class NP [e.g., M. Garey and D. Johnson, *Computers and Interactability: A Guide to the Theory of NP-Completeness* (Freeman, San Francisco, 1979)], even each individual test may take exponential time.

[13]The sequence certainly passes the standard statistical tests of Ref. 6, and contains all possible subsequences up to length at least 12. It has also been proved that only at most one vertical sequence in the pattern of Fig. 1 can have a finite period [E. Jen, Los Alamos Report No. LA-UR-85-1218 (to be published)].

[14]For example, D. E. R. Denning, *Cryptography and Data Security* (Addison-Wesley, Reading, Mass., 1982). Systems like Fig. 1 can, for example, be used for "stream ciphers" by adding each bit in the sequences produced with a particular seed to a bit in a plain-text message.

[15]For example, O. Martin, A. Odlyzko, and S. Wolfram,

Commun. Math. Phys. **93**, 219 (1984).

[16]B. Jansson, *Random Number Generators* (Almqvist & Wiksells, Stockholm, 1966).

[17]They are one-symbol-deletion tag sequences [A. Cobham, Math. Systems Theory **6**, 164 (1972)], and can be represented by generating functions algebraic over $GF(k)$ [G. Christol, T. Kamae, M. Mendes France, and G. Rauzy, Bull. Soc. Math. France **108**, 401 (1980); J.-M. Deshouillers, Seminar de Theorie des Nombres, Université de Bordeaux Exposé No. 5, 1979 (unpublished); M. Dekking, M. Mendes France, and A. van der Poorten, Math. Intelligencer, **4**, 130, 173, 190 (1983)]. Their self-similarity is related to the pumping lemma for regular languages [e.g., J. Hopcroft and J. Ullman, *Introduction to Automata Theory, Languages and Computation* (Addison-Wesley, Reading, Mass., 1979)]. More complicated sequences associated with context-free formal languages can also be recognized in polynomial time, but the recognition problem for context-sensitive ones is P-space complete.

[18]For example, A. M. Frieze, R. Kannan, and J. C. Lagarias, in *Twenty-Fifth IEEE Symposium on Foundations of Computer Science* (IEEE, New York, 1984). The sequences also typically fail certain statistical randomness tests, such as multidimensional spectral tests (Ref. 6). They are nevertheless probably random with respect to all NC computations [J. Reif and J. Tygar, Harvard University Computation Laboratory Report No. TR-07-84 (to be published)].

[19]For example, G. Choquet, C. R. Acad. Sci. (Paris), Ser. A **290**, 575 (1980); cf. J. Lagarias, Amer. Math. Monthly **92**, 3 (1985). (Note that with appropriate boundary conditions a finite-size version of this system is equivalent to a linear congruential pseudorandom number generator.)

[20]A. Shamir, Lecture Notes in Computer Science, **62**, 544 (1981); S. Goldwasser and S. Micali, J. Comput. Sys. Sci. **28**, 270 (1984); M. Blum and S. Micali, SIAM J. Comput. **13**, 850 (1984); A. Yao, in *Twenty-Third IEEE Symposium on Foundations of Computer Science* (IEEE, New York, 1982); L. Blum, M. Blum, and M. Shub, in *Advances in Cryptology: Proceedings of CRYPTO-82*, edited by D. Chaum, R. Rivest, and A. T. Sherman (Plenum, New York, 1983); O. Goldreich, S. Goldwasser, and S. Micali, in *Twenty-Fifth IEEE Synmposium on Foundations of Computer Science* (IEEE, New York, 1984).

[21]For example, M. Garey and D. Johnson, Ref. 12.

[22]For example, L. Kuipers and H. Niederreiter, *Uniform Distribution of Sequences* (Wiley, New York, 1974).

[23]A polynomial-time procedure is also known for recognizing solutions to more complicated algebraic or trigonometric equations (R. Kannan, A. K. Lenstra, and L. Lovasz, Carnegie-Mellon University Technical Report No. CMU-CS-84-111).

[24]Many localized structures have been found (D. Lind, private communication).

[25]S. Wolfram, Phys. Rev. Lett. **54**, 735 (1985).

[26]For example, U. Frisch, Phys. Scr. **T9**, 137 (1985).

[27]G. Ahlers and R. W. Walden, Phys. Rev. Lett. **44**, 445 (1980).

[28]For example, M. Brachet *et al.*, J. Fluid Mech. **130**, 411 (1983).

3. Some applications

3.1: N. Packard, "Lattice models for solidification and aggregation", to appear in Proc. First International Symposium for Science on Form, (Tsukuba, Japan, 1985).

3.2: B. Madore and W. Freedman, "Computer simulations of the Belousov-Zhabotinsky reaction", Science 222 (1983) 615.

3.3: A. Winfree, E. Winfree and H. Seifert, "Organizing centers in a cellular excitable medium", Physica 17D (1985) 109.

3.4: D. Young, "A local activator-inhibitor model of vertebrate skin patterns", Math. Biosciences 72 (1984) 51.

3.5: Y. Oono and M. Kohmoto, "Discrete model of chemical turbulence", Phys. Rev. Lett. 55 (1985) 2927.

3.6: J. Park, K. Steiglitz and W. Thurston, "Soliton-like behaviour in automata", Physica 19D (1986) 423.

3.7: Y. Pomeau, "Invariant in cellular automata", J. Phys. A17 (1984) L415.

3.8: M. Creutz, "Deterministic Ising dynamics", Ann. Phys. 167 (1986) 62.

3.9: U. Frisch, B. Hasslacher and Y. Pomeau, "Lattice gas automata for the Navier-Stokes equation", Phys. Rev. Lett. 56 (1986) 1505.

3.10: J. Salem and S. Wolfram, "Thermodynamics and hydrodynamics of cellular automata".

3.11: K. Kaneko, "Attractors, basin structures and information processing in cellular automata".

3.12: S. Wolfram, "Approaches to complexity engineering", to be published in Physica D.

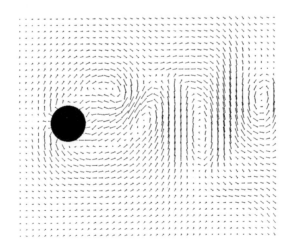

Vortex street formed behind a cylinder moving in a cellular automaton fluid. On a microscopic scale the cellular automaton fluid consists of discrete particles on the links of a two-dimensional lattice, whose configuration is updated according to simple rules which idealize particle motion and collisions. The picture was generated using a 4096×4096 lattice; the macroscopic fluid velocity vectors shown were obtained by averaging particle velocities in 96×96 site regions. The computations were performed on a 65536 processor Connection Machine computer. The hydrodynamic flow corresponds to a Reynolds number around 100.

Lattice Models
for Solidification and Aggregation[*]

Norman H. Packard

The Institute for Advanced Study, Princeton NJ 08540

Keywords: Cellular Automata, Solidification, Aggregation

Simple models with both discrete and continuous variables at every site of a lattice are used to investigate solidification and aggregation in two dimensions. These models display a rich variety of macroscopic forms growing from small seeds. Transitions between macroscopic forms are seen as parameters in the model are varied, and these transitions may be compared with those seen in experiments.

Crystal growth is an excellent example of a physical process that is microscopically very simple (attachment of molecules onto a solid), but that displays a beautiful variety of macroscopic forms. Many local features are predicted from continuum theory,[1] but global features may be analytically inaccessible. For this reason, computer simulation of idealized models for growth processes has become an indispensable tool in studying solidification.[2, 3] Here we present a new class of models that represent solidification by sites on a lattice changing from zero to one according to a local deterministic rule. The strategy is to begin with very simple models that contain few physical elements, and then to add physical elements gradually, with the goal of finding those aspects responsible for particular features of growth. The models display both local and global features that may be compared to the results from solidification experiments.

The simplest deterministic lattice model for solidification is a two dimensional cellular automaton with two states per site to denote presence or absence of solid, and a nearest neighbor transition rule. Further, we will consider only rules which have the property that a site value of one remains one (solidification only; no melting or sublimation). An additional constraint for the rules considered here is that they depend on neighboring site values only through their sum:

$$a_i^{t+1} = f(\sigma_i^t) \qquad \text{with} \qquad \sigma_i^t = \sum_{\delta \ \epsilon \ Nbrhd.} a_{i+\delta}^t \tag{1}$$

The domain of f ranges from zero to the number of neighbors; f takes on values of one or zero.

These rules display four types of behavior for growth from small seeds:^{**} *(i)* No growth at all; This certainly happens for the rule that maps all values of σ to zero. *(ii)* growth into a plate structure with the shape of the plate reflecting the

First published in the Proceedings of the First International Symposium for Science on Form (The University of Tsukuba, 26-30 November 1985), edited by Y Katoh, R Takaki, J Toriwaki and S Ishizaka, KTK Scientific Publishers (1986).

lattice structure; an example is $f(\sigma) = 1$ when $\sigma > 0$. *(iii)* growth of dendritic structure, with sidebranches growing along lattice directions at unit velocity; this type of rule is obtained by adding growth inhibition to the previous rule, *e.g.* with $f(\sigma) = 1$ when $\sigma = 1$. Physically, growth inhibition occurs because of the combined effects of surface tension and radiation of heat of solidification. *(iv)* Growth of an amorphous, asymptotically circular form at less than unit velocity. This form is obtained by adding even more growth inhibition, *e.g.* with $f(\sigma) = 1$ when $\sigma = 2$.

The dendritic forms produced by rules in class *(iii)* exhibit a striking self-similarity: every 2^n time steps, the growing seed forms a plate, then dendritic arms grow from the corners of the plate, sidebranches form, and finally all sidebranches grow into a plate and the process is repeated (Figure 1(a)). This self-similarity may be quantified with a growth dimension that can take on fractional values.[5,6,7] The growth dimension is measurable in experiments, but requires data consisting of the length of the boundary as a function of time, in contrast to the dimension that is often used to characterize other two dimensional patterns like diffusion limited aggregates.[8] It is possible, nevertheless, to see remnants of a snowflake's history embedded in its internal structure, and these sometimes indicate dendritic-plate alternation, with plate boundary length growing exponentially (Figure 1(b)).

The most crucial physical ingredient missing from the cellular automaton model is the flow of heat. This may be modeled with the addition of a continuous variable at each lattice site to represent temperature. The time evolution of the temperature field is given by a discrete approximation to the heat equation, $\dot{T} = c\nabla^2 T$. This amounts to changing a particular site's temperature T_i by taking the average over nearby sites σ_i, and moving T_i toward σ_i by an amount determined by the diffusion constant c. This relaxation method is numerically stable, so that an appropriate choice of diffusion constant assures an accurate simulation of a continuous temperature field in continuous space and time.

The addition of solid again uses a local rule that depends on σ_i, the sum of solid in neighboring sites. Now, however, the local rule yields a continuum temperature threshold value, $T_{thresh} = f(\sigma_i)$. If $T_i < T_{thresh}$ the site is filled with solid; otherwise the site remains empty. The amount of neighboring solid σ_i may be considered as a coarse approximation to the curvature: a boundary site near a convexly curved interface will have fewer neighbors filled with solid than a site near a flat interface. The Gibbs-Thompson effect implies that solidification is more difficult at a convex interface because extra energy must be invested against the force of surface tension. This may be modeled by choosing a function f that takes low values for small σ_i, and high values for larger σ_i. The form of f to be used henceforth will be a quadratic maximum:

$$f(\sigma_i) = \lambda \sigma_i (1 - \sigma_i). \tag{2}$$

**Note that these four classes are quite distinct from the four classes observed in a broader context by Wolfram.[4] Wolfram's classification is on the basis of asymptotic behavior of a cellular automaton rule acting on a random initial condition. Under such circumstances, all the rules discussed here would lead to fixed points, and so would be in Wolfram's class two.

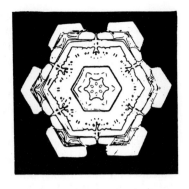

a   b

Figure 1. (a) Growth from a single site initial condition under the action of a solidification cellular automaton rule on a hexagonal lattice. The grey level changes with time, and repeats a cycle of light to dark every 2^n time steps to display the self-similarity of the growth process, as the growing seed alternates between dendrite and plate forms. (b) A picture of a snowflake, showing internal structure reminiscent of the dendrite-plate alternation.

The final ingredient in the dynamical rule is the effect of solidification on the temperature field. When solid is added to a growing seed, latent heat of solidification must be radiated away. This is modeled by causing an increment in the temperature field. The amount of increase corresponds crudely to the increase in the temperature gradient at the interface. In the following simulations, we simply set the temperature to a constant (high) value when new solid is added. This means that heat flows to nearby interface sites and inhibits their solidification.

This model is a hybrid of discrete and continuum elements. The addition of solid happens in a discrete way, which can be only a very coarse approximation to solid deposition on molecular length scales. There is about ten orders of magnitude between the capillary length and the size of a macroscopic crystal; there is only about 2.4 orders of magnitude between the lattice spacing and the size of the macroscopic crystal in the model. Thus, the micro-scales are brought comparatively quite close to the macro-scales, with the hope that many of the generic macroscopic features of the dynamics will remain. In this respect, this model is similar in spirit to recent molecular dynamics simulations of fluid flow using a cellular automaton rule.[9, 10]

The use of a continuum variable at each lattice site to represent temperature gives the model unique features lacking in a purely discrete cellular automaton model. The dynamics may now become parameterized. One parameter is the diffusion rate. Another parameter is the amount of latent heat added upon solidification. Other parameters may characterize the local temperature threshold function. These parameters may be varied to obtain transitions between macroscopic forms that may be compared with experiment. Figure 2 (a-c) shows a sequence of pictures as the parameter λ is varied from low to high values.

When λ is small, more heat must diffuse away before a boundary site will solidify, so the diffusion length is quite long compared to the lattice spacing. Figure 2(a) illustrates such a case. In the limit of infinite diffusion length, solidification has become known as *diffusion limited aggregation*.

The usual simulations of diffusion-limited aggregation model diffusion by random walking particles which can stick to a growing seed.[8, 11] These simulations show the resulting macroscopic form to be a fractal with dimension of ≈ 1.7. Objects with other fractal dimension were also observed. Within the framework of the present model, diffusion limited growth is obtained by having the temperature threshold be small for all values of σ; *i.e.* by setting λ in Eq. (2) to be small. The resulting macroscopic form is displayed in Figure 2(a). The mechanisms for pattern formation in the Witten-Sander model and the present model are slightly different. In the former, voids in the growing structure form because long arms shield regions from subsequent particles. In the latter, voids form because heat is trapped between arms. Nevertheless, the fractal dimension of the two agree to the present accuracy, indicating that they may be in the same universality class. Similar agreement has been indicated in a deterministic simulation of diffusion limited growth using a continuum model.[12]

Figure 2(b) illustrates the effects of raising λ in Eq. (2). There are no longer arbitrarily large voids, but rather a chaotic network of tendrils that appears to have dimension two asymptotically. Though the tendrils are seen to grow in every direction, they show some tendency to grow along lattice directions. The tip splitting instability is apparent,[13] preventing the formation of long dendrites with regular sidebranching.

When λ is raised even further, the tip begins to stabilize, and the tip splitting instability gives way to the sidebranch instability.[1] Anisotropic macroscopic forms have also been seen in stochastic models for diffusion limited aggregation (using an integrated version of random walking particles), if the rule governing the sticking of the particles is made anisotropic.[14] As an anisotropy parameter is varied in the sticking rules, transitions similar to those seen in figure 2 are observed.

Depending on the value of λ and the diffusion constant, the sidebranches can show a variety of structure. In addition to simple, regular sidebranching, the sidebranches can "period double" to display long and short sidebranches alternately. This is evidence for pattern selection mechanism that involves a simple causal relationship between sidebranches rather than filtering of noise at the tip.[15]

The macroscopic forms yielded by this model show remarkable similarity to experiments in pattern formation. The most recent experiments are in two rather different systems: the Hele-Shaw cell, evolution of an interface between two liquids of different viscosity trapped between two plates;[16] and electro-deposition of zinc from an electrolyte solution confined between two plates.[17, 18] Both these experiments show transitions between anisotropic forms such as diffusion limited aggregates and forms that show strong anisotropy. In the case of electro-

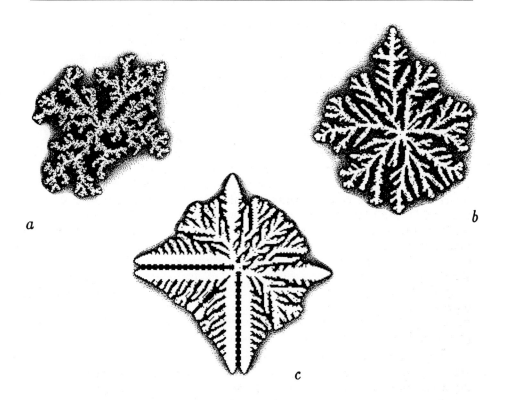

a   *b*

c

Figure 2. As parameters are varied in the deterministic growth rule (*e.g.* λ in Eq. 2) transitions occur between different macroscopic forms: (a) Amorphous, isotropic fractal growth. The form displays fractal scaling over 2.4 orders of magnitude in length, with a fractal dimension of 1.7±.1. (b) Tendril growth, dominated by tip splitting, but no apparent fractal structure. Some anisotropy is evident. (c) A macroscopic form showing strong anisotropy, stable parabolic tip with side branching. The temperature field is denoted by a grey scale.

deposition, the anisotropy comes from the underlying crystal structure; in the Hele-Shaw experiment, the anisotropy is imposed by the boundary conditions (scratches on the two dimensional surfaces containing the fluids). A careful comparison of this model with experiment will require new data analysis techniques based on the processing of spatial images.

I am grateful to many people for discussions and comments; most recently these include J. Crutchfield, N. Goldenfeld, Y. Kuramoto, H. Levine, C. Reiter, L. Sander, Y. Sawada, and R. Shaw. I am especially grateful to D. Farmer and S. Wolfram for discussions and extensive comments on earlier versions of the manuscript. This work was supported in part by an RCA fellowship, and in part by the U.S. Office of Naval Research under contract number N00014-85-K-0045.

References

1. Langer, "Instabilities and Pattern Formation in Crystal Growth," *Rev. Mod. Phys.* **52** p. 1 (1980).

2. D. Kessler, H. Levine, and J. Koplik, "Geometrical Models of Interface Evolution. II. Numerical Simulation," *Phys. Rev.* **A30** p. 3161 (1984).

3. E. Ben-Jacob, N. Goldenfeld, J. S. Langer, and G. Schon, "Boundary-Layer Model of Pattern Formation in Solidification," *Phys. Rev. Lett.* **51** p. 1930 (1985).

4. S. Wolfram, "Universality and Complexity in Cellular Automata," *Physica* **10D** pp. 1-35 (1985).

5. S. Willson, "Growth Rates and Fractional Dimension in Cellular Automata," *Physica* **10D** p. 69 (1984).

6. N. H. Packard and S. Wolfram, "Two Dimensional Cellular Automata," *J. Stat. Phys.* **38**(1985).

7. Y. Sawada, M. Matsushita, M. Yamazaki, and H. Kondo, *Physica Scripta* **T9** pp. 130-132 (1985).

8. T. A. Witten and L. M. Sander, *Phys. Rev. Lett.* **47** p. 1400 (1981).

9. U. Frisch, B. Hasslacher, and Y. Pomeau, "A Lattice Gas Automaton for the Navier-Stokes Equation," *Los Alamos preprint* **LA-UR-85-3503**(1985).

10. J. B. Salem and S. Wolfram, "Thermodynamics and Hydrodynamics with Cellular Automata," *IAS Preprint*, (1986).

11. T. A. Witten and L. M. Sander, *Phys. Rev.* **B27** p. 5686 (1983).

12. L. M. Sander, P. Ramanlal, and E. Ben-Jacob, "Diffusion-Limited Aggregation as a Deterministic Process," *Phys. Rev.* **A32** p. 3160 (1985).

13. D. Kessler, H. Levine, and J. Koplik, "Geometrical Models of Interface Evolution. III. The Theory of Dendritic Growth," *Phys. Rev.* **A31** p. 1712 (1985).

14. J. Nittmann and E. Stanley, "Connection between Tip-splitting Phenomena and Dendritic Growth," *Preprint, Boston University*, (1986).

15. R. Pieters and J. S. Langer, "Noise-Driven Sidebranching in the Boundary-Layer Model of Dendritic Solidification," *ITP preprint*, (1986).

16. E. Ben-Jacob, R. Godbey, N. Goldenfeld, J. Koplik, H. Levine, T. Mueller, and L. Sander, *Phys. Rev. Lett.* **55** p. 1315 (1985).

17. Y. Sawada, A. Dougherty, and J. P. Gollub, "Dendritic and Fractal Patterns in Electrolytic Metal Deposits," *Haverford College preprint*, (1985).

18. D. Grier, E. Ben-Jacob, R. Clarke, and L. M. Sander, "Morphology and Microstructure in Electrochemical Deposition of Zinc," *University of Michigan Preprint*, (1985).

Computer Simulations of the Belousov-Zhabotinsky Reaction

Abstract. *Morphological features of the two-dimensional Belousov-Zhabotinsky reaction were modeled with an algorithm involving only two simple parameters, one describing the productivity of the reaction on a local scale length and the other characterizing the delay or quiescent time after the localized reaction. Self-organizing wavelike structures, including single- and multiarmed spirals, were most easily generated.*

While investigating the range of spatial structures produced by varying the parameters of a self-propagating star formation algorithm (*1*), we found that some of the resulting patterns resemble those found in autocatalytic chemical reactions described in 1958 by Belousov and in 1964 by Zhabotinsky (*2*) and more recently by Winfree (*3*) and others (*4, 5*). The most striking feature of the chemical reaction is the spontaneous appearance of globally coherent structures developing from so-called local oscillators. Although these patterns have been ascribed to symmetry-breaking instabilities caused by diffusion in systems involving more than two variables (*6*), little more than asymptotic analysis of the rate equations has been performed (*7*), and the role of the putative local oscillator or echo waves (*8*) is in dispute. An introduction to the general analytic theory of such autocatalytic reactions can be found in Cohen *et al*. (*9*), Tyson (*8*), and Nicolis and Prigogine (*4*), while a more popular discussion of the spiral-producing reactions is given by Winfree (*10*). The simplicity of our simulations suggests that, while the underlying physics may be very complex, the macroscopic manifestations are easy to understand and model.

The chemical experiments (*3*) show that the following structures can arise. If externally activated (by, say, a hot probe), a single point in the otherwise homogeneous chemical mixture gives rise to an expanding ring of activity. If internally activated (as by a contaminant), the point gives rise to periodic structures of concentric, wavelike rings. However, when a single shell is sheared across its diameter, the end points produce oppositely winding one-armed spirals. Two-, three-, and four-armed spirals can also be produced, depending on the initial conditions (*11*). All these structures are remarkably stable, except for colliding wave fronts of independent origin, which neither penetrate through nor reflect off each other but rather result in mutual annihilation on contact. Our computer model easily generates all these features.

Our algorithm operates on an initially homogeneous hexagonal grid. Initial conditions are set up such that one or more cells are "activated" while others are either "quiescent" or "receptive." In the next time step the activated cells may then stimulate (catalyze) the immediately adjacent cells into activity. The probability that this reaction is completed in one time step is governed by the first parameter, designated productivity. Propagation of activity is restricted by the second parameter, which specifies a period during which a cell must remain quiescent after stimulation. Cells cannot be propagated into during this quiescent phase, nor while they themselves are active. The active phase lasts one time step.

With the probability of propagation set to unity and the quiescent time set to two time steps, the following initial conditions give rise to the structures shown in Fig. 1.

1) An isolated active cell produces a single, circularly symmetric, expanding wavelike front of activity.

2) A single active cell buffered by an adjacent quiescent cell gives rise to a recurring series of concentric wavelike rings.

3) The end points of a line of activated cells buffered on one side by a line of quiescent cells give rise to growing one-armed spiral patterns.

4) The contact point between two lines of activated cells placed end to end but buffered on opposite sides by quiescent cells produces two-armed spiral patterns.

5) Multiarmed spirals (three or more arms) are generated by lines of buffered activators meeting at a common central cell.

These basic structures are extremely stable. When wave fronts from various centers meet, the fronts do indeed annihilate each other, but the structures internal to these surfaces of contact persist throughout the simulation. The only notable variation is illustrated by the time sequence, which shows that the spirals are intermittently connected and disconnected in the core region, as is also observed in active, excitable chemical media (*11*).

Given that the reaction is likely to

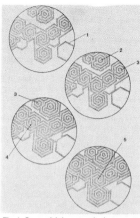

Fig. 1. Sequential time steps in the computer modeling of the Belousov-Zhabotinsky reaction. The structures numbered 1 to 5 result from the initial conditions described in the text.

occur, there is in fact only one free parameter in our model, the quiescent time. It should be emphasized that this parameter does not in any way affect the variety of forms that result but controls only the scaling of the features, such as the interarm separation in fully matured spiral patterns. Thus, the periodicity in these wavelike structures is intrinsic to the diffusion-reaction time scale of the chemical system. Specifically, it is not dependent on the properties of an unspecified local oscillator, nor are the patterns dependent on boundary conditions, of which there are none in this particular simulation. However, any inert boundary in this simulation, as in the chemical reactions, limits only the extent of the propagation.

We hope that these simulations will lead researchers to regard such self-organizing structures as the expected consequences of a wide class of propagating and autocatalytic reactions that can be easily modeled. This broad class, we believe, includes not only the chemical reactions described above but also morphologically equivalent growth patterns in slime molds, certain stages of embryonic development, and the shock-driven models of spiral arm development in galactic systems.

BARRY F. MADORE
WENDY L. FREEDMAN
David Dunlap Observatory,
Department of Astronomy,
University of Toronto,
Toronto, Ontario, Canada M5S 1A5

References and Notes

1. W. L. Freedman and B. F. Madore, *Astrophys. J.* **265**, 140 (1983).
2. B. P. Belousov, *Sb. Ref. Radiats. Med. 1958*, 145 (1959); A. M. Zhabotinsky, *Dokl. Akad. Nauk SSSR* **157**, 392 (1964).
3. A. T. Winfree, *Science* **175**, 634 (1972).
4. G. Nicolis and I. Prigogine, *Self-Organization in Non-Equilibrium Systems* (Wiley, New York, 1977).
5. A. N. Zaikin and A. M. Zhabotinsky, *Nature (London)* **225**, 535 (1970).
6. P. Hanusse, *C. R. Acad. Sci. Ser. C* **277**, 263 (1973).
7. J. Stanshine, thesis, Massachusetts Institute of Technology (1975).
8. J. F. Tyson, *Ann. N.Y. Acad. Sci.* **316**, 279 (1979).
9. D. S. Cohen, J. C. Neu, R. R. Rosales, *SIAM (Soc. Ind. Appl. Math.) J. Appl. Math.* **39**, 8 (1978).
10. A. T. Winfree, *Sci. Am.* **230**, 82 (June 1974).
11. K. I. Agladze and V. I. Krinsky, *Nature (London)* **296**, 424 (1982).
12. This work was supported in part by the Natural Sciences and Engineering Research Council of Canada, Zonta International, and the University of Toronto. We thank S. Shore for helpful comments and for access to his encyclopedic store of references.

6 April 1983; revised 6 April 1983

Physica 17D (1985) 109–115
North-Holland, Amsterdam

ORGANIZING CENTERS IN A CELLULAR EXCITABLE MEDIUM†

A.T. WINFREE
Department of Biological Sciences, Purdue University, West Lafayette, IN 47907, USA

E.M. WINFREE
Evanston Township Highschool, Evanston, IL 60201, USA

and

H. SEIFERT
Mathematisches Institut, Universitat Heidelberg, 6900 Heidelberg, Fed. Rep. Germany

Received 4 January 1985

Excitable media provide much of the subject-matter of physiology, especially of electrophysiology. We simulate excitability in a cubical three-dimensional grid of discrete cells. Topologically distinct organizing centers for self-sustaining rhythmic activity (at period 4) arise from suitable initial conditions. Two are shown: the scroll ring and the linked pair of twisted scroll rings. The first has already been observed in a chemically excitable reagent and possibly in heart muscle; the second, and others of a predicted "periodic table of organizing centers", remain to be observed outside computers.

1. Introduction

The essence of excitability is responsiveness to a threshold-transgressing stimulus from an adjacent excited cell. The response consists of becoming excited in the same way, then, during an interval of exhaustion, recovering the former excitability. In such a medium excitation is contagious and propagates as a pulse. In sheets of excitable medium (cortex of the brain, retina, smooth muscle, atrial muscle, to name only a few) the pulse is a wavefront. In solid blocks of such media (left ventricular muscle, possibly certain parts of the brain) the wavefronts can presumably be two-dimensional surfaces. Is their geometry essentially the same as in wave-propagating media familiar to

†This paper was presented at a Symposium on Nonlinear Oscillations in Physiology, Oxford University, 14 September 1984, and printed in the corresponding book of extended abstracts, ed. Derek Linkens.

physicists, viz. concentric sphere-like "bags" without any edge unless only on the boundary of the medium? It would be of interest to conduct simulations of three-dimensional excitable media to learn what geometrically distinctive varieties of wave might inhabit them, given appropriate initial conditions and boundary conditions. The prospect of encountering qualitatively new kinds of wave should entice anyone intrigued by the qualitatively distinctive peculiarities of excitable media, such as their susceptibility to fibrillation and a host of simpler arrhythmias.

Three-dimensional integration of the pertinent stiff partial differential equations remains prohibitively expensive, even after decades of fast exponential decline in the unit cost of computation. However, the first such have already been reported [1–3]. As such undertakings come within reach, it is important to be ready with potentially interesting initial conditions. Toward this end we simulate an array of *A* by *B* by *C* cubical cells. Each cell may be in one of three states, Q, E, or T. It is normally "quiescent" (Q) and "excitable": if one of its six neighbors is excited, then in the next moment, it too becomes excited (E). In the moment following it is "tired" (T) and therefore cannot be re-excited. In the next moment it is again quiescent and therefore excitable again. An exception to these rules occurs along the boundaries of the array, in that a cell on a face, edge, or corner has only five, four, or three neighbors. This is equivalent to saying in the language of differential equations that we use Dirichlet boundary conditions: the cell just beyond each face is imagined to be held quiescent. The usual alternative to Dirichlet (fixed-state) boundaries is Neumann (no-flux) boundaries: the array is imagined to abut an appropriately mirror-imaged array along each face. In our discrete-state caricature of excitability either rule has the same effect.

Initial conditions consist of an arrangement of E's and T's in a sea of Q's, emplaced by an automatic algorithm or by keyboard entry via an editor subprogram. Then the array is repeatedly scanned, updating each cell's state according to the foregoing rules. The array settles into a period-four repeat after a number of updates no greater than the longest array dimension. This "repeat" consists only of quiescence in one-dimensional simulations ($A = B = 1$) with these absorbing boundary conditions. In two dimensions ($A = 1$, like quadrille paper or window-screen), uniform quiescence may also result, but another frequent result is some arrangement of period-four rotating spirals with one or more arms. The period, in this case, reflects the perimeter of the smallest closed ring in this discrete medium. In a medium with more states intervening between excitation and restored excitability, the shortest closed path of circulation would have at least that perimeter (and period).

Even without digital assistance one can understand these phenomena immediately by simply following the rules with pencil and eraser on quadrille paper. It proves interesting to start from

quiescent initial conditions surrounding a bilayer of active cells, excited on the front side, tired on the rear side, reaching from an edge into the interior of the rectangle array. The dangling endpoint of this plane wave will quickly evolve into the pivot and source of a spiral wave. An even simpler initial condition, which develops a mirror image pair of adjacent spirals, is uniform quiescence (E) punctuated by a single pair of non-quiescent cells: one excited, and one tired neighbor.

Such media and their spiral waves have been thoroughly explored [4–8]. Though obviously quantized in an extreme way, they behave very much like waves in analogous continuous excitable media, such as the Belousov–Zhabotinsky chemical reagent [9–11], various idealizations formulated as partial differential equations [12 and refs. in 13], slime mold [14, 15], thin layers of heart muscle [16–19], cerebral cortex [20, 21], and the retina [22].

Our objective was to determine how far this discrete-state, discrete-space simulation mimics the anticipated behavior of continuous excitable media in three dimensions, as a preliminary to undertaking calculations from continuous differential equations. Anticipations for the three-dimensional case include a great variety of topologically distinct sources, all of the same period except in the regions of extreme curvature. All are generalizations of the two-dimensional spiral. The simplest one (see below) has been observed in vitro [23, 24], in vivo [25], and in numero [1]. The next simplest source has only been described mathematically and animated by computer graphics on videotape.

In this report we demonstrate it dynamically in three dimensions for the first time.

First, as a "control experiment", we demonstrate the simplest scroll ring. In a continuum this is a surface of revolution which can be visualized as a spiral swung about an axle. The source of the spiral in each plane radial to the axle is a point; in the surface of revolution that point becomes a ring, the edge of a wave rolled up around it. This

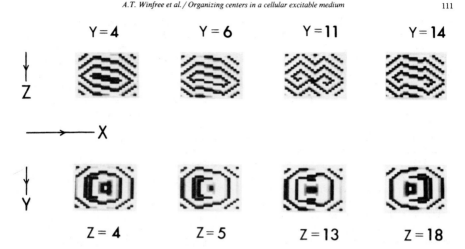

Fig. 1. A $20 \times 20 \times 20$ cube of idealized excitable medium harbors a scroll ring, shown here in 4 planar cross-sections at successive values of Y, and four perpendicular cross-sections at successive values of Z as indicated. The scroll's singularity is a 10×9 rectangular ring slightly tilted out of the XY plane at $Z = 12 + -1$. Excited (E) cells are black, tired (T) cells are grey, and quiescent cells (Q) are blank. Notice in cross-section $Y = 11$ the pair of mirror-image spirals radiating outward to left and right, and inward to center; in perpendicular section $Z = 13$ these waves are seen as concentric rings propagating outward and inward. The repeat period is 4.

source ring (or, in more elaborate cases, an arrangement of mutually linked knotted rings) is called the organizing center for the spatially and temporally periodic activity that radiates from it.

Fig. 1 shows planar sections of the simplest organizing center, a solitary ring in a 20 by 20 by 20 cube. This scroll ring was initiated with a slight tilt relative to the grain of the array in order to expose its cross-sections in a more generic way. This is particularly important for the serial sections at fixed Y. These sections expose concentric inward and outward ring-shaped waves. They are separated by expanding crescents where the section plane grazes one edge of the scroll, i.e. a vertically-travelling broadside has just penetrated the section plane. In fig. 2 a line-drawing outlines waves photographed [24] on fixed and stained serial sections of Belousov–Zhabotinsky reagent. Fig. 3 similarly outlines (as a time sequence rather than serial sections at fixed time) the activation front emerging through the surface in a piece of

heart muscle, recorded by an array of microelectrodes [25]. All six pictures are about 1 cm in diameter.

In all three cases the initial conditions were similar: a wavefront abruptly terminating along a circular edge. In the discrete simulation it was a five by six disk of E cells backed up by a layer of T cells; its perimeter became the organizing center. In the chemical reagent, stimulation at a point produced a hemispherical wave whose circular edge was then abutted against another block of quiescent medium to create the ring source. In the heart muscle, the hemisphere was ruptured by encountering a block of heart muscle artificially made inexcitable; when excitability returned, activity began to issue from the block at a period comparable with that of two-dimensional rotating waves in heart muscle.

Several years ago it was predicted that scroll rings might exist in greater variety, distinguishable by topological indices [26]. In particular, it was

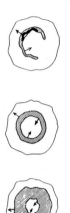

Fig. 2. A piece of rabbit heart muscle appears to harbor a slightly tilted scroll ring, in the interpretation of Medvinsky et al. [25]: waves radiate as concentric inward and outward rings from a circular locus beneath the surface of the muscle. The outer circular wave is about 1 cm diameter. As the inner wave erupts as a broadside into the exposed surface, it is caught at 103, 113, and 143 ms after first stimulation. Their repeat period (106 ms) is close to that of spiral waves in this medium (82 ms).

Fig. 3. Waves from a slightly tilted scroll ring are caught in serial sections across a "pancake" of the Belousov-Zhabotinsky excitable medium. They radiate inward and outward as in figs. 1 and 2. The outer ring is 0.8 mm diameter (from ref. 24).

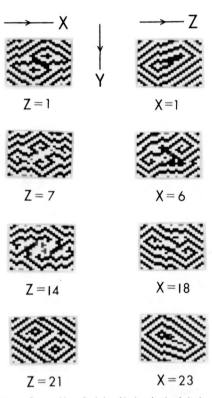

$Z=1$ $X=1$

$Z=7$ $X=6$

$Z=14$ $X=18$

$Z=21$ $X=23$

Fig. 4. Comparable to fig. 1, but this time the singularity is a pair of linked rings, each radiating a twisted scroll ring at period 4. At large X or Z the section planes miss the rings altogether. Sections $Z=7$ and $Z=14$ cut both rings (each twice) to expose mirror-image spirals. Section $X=6$ cuts only one ring; $X=18$ cuts the other.

argued that the least complicated of these more elaborate organizing centers in excitable media would be a linked pair of scroll rings. These conjectures were given more substance in a sequence of papers proceeding from geometry and laboratory arrangements for chemical implementation [13] to topology [27–30] and to computer graphics (but not dynamical simulations) [31, 32]. It was discovered that scroll rings could link if they were also twisted in a topological sense. But no one had

Boundary = ▓

Excited = ▨

Tired = ░

Quiescent = ☐

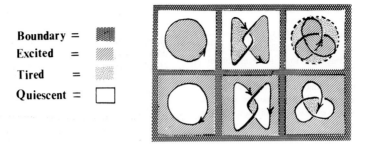

Fig. 5. Top left: initial conditions for a scroll ring consist of a disk of excited (E) cells overlaying (not visible here) a disk of tired (T) cells. This is a piece of wavefront, moving toward the observer where E is exposed, or away where T is exposed. Top Middle: a pair of twisted scroll rings is started from a similar bilayer, but shaped as a twisted band so it has two edges, mutually linked. Top right: initial conditions for a knotted scroll ring are fashioned from two disks of bilayer, joined together by three half-twisted bands. These initial conditions each consist a single surface bounded only by the singular ring(s). Bottom: as in top, but the single bilayer surface is bounded by the singular ring(s) and by the boundary surface of the three-dimensional medium. Arrows orient the wave edges: walking along the edge in the indicated direction you see the wavefront rotating clockwise.

yet computed linked scroll rings to compare with anticipations by topology and computer graphics.

Fig. 4 presents such a simulation in the format of fig. 1, shown at a moment long after initial conditions in a 23 by 23 by 23 cube. Initial conditions consisted, again, of a sheet of wavefront with edges exposed to the surrounding quiescence; those edges became the rotation axes of scroll rings. In this case there were two circular edges and they were linked: the sheet of wavefront was a cylindrical band containing one full twist as shown in fig. 5 (middle column). These edges curled up into counter-rotating scrolls. If a section plane cuts a source ring, it cuts twice, showing spirals of opposite hand as in fig. 1. Some sections in fig. 4 cut no source rings and therefore intercept sheets of wavefront along closed rings; others cut one ring, and so include a pair of opposite spirals; and still others cut both rings, and so show four spirals. This congestion could be alleviated by spacing the initial wavefront edges further apart in a bigger array, ideally of cells packed hexagonally like stacked cannon-balls. But they are far enough apart here to function as independent spiral sources with period four. The integer (unit) twist imposed on the initial band of wavefront persists as the

unit twist of each scroll ring and links them together. This was verified by reconstructing the three dimensional wave from transparencies of its three orthogonal sets of serial sections.

In this discrete medium, initial conditions for all the foreseen organizing centers can be contrived quite simply as a sheet of wavefront (a bilayer of T and E cells) containing appropriate half-twisted bands. Fig. 5 (top) shows in this format suitable initial conditions for the plain scroll ring, the linked twisted pair of scroll rings (of which there is a mirror-image isomer not shown), and for a trefoil-knotted solitary scroll ring (which also has a mirror-image isomer). Although these wave sources function in complete independence of boundaries, like free particles, it is sometimes convenient to start them from a sheet of wavefront that initially touches the boundaries as in fig. 5 (bottom).

These sheets connected by half-twisted bands correspond in the continuum case to the Seifert surface [27–30, 33] bounded by any set of rings. Any Seifert surface satisfies the "exclusion principle" [27–29, 33] which specifies the unique twist associated with the wavefront near any one of its bounding rings. (We thank Professor Herbert Seifert for confirming our derivation of this fact and for the tidier proof contained in the appendix.) The sheet of E cells corresponds to a local maximum of excitation (e.g. $HBrO_2$ in the Belousov–Zhabotinsky reagent), and the sheet of T cells behind it corresponds to a local maximum of inexcitability (e.g. Br^- ions in that chemical analog). By implementing such initial conditions for numerical solution of the equations of cardiac electrophysiology or reaction–diffusion equations, it should be possible to determine the stability of diverse organizing centers, or observe their modes of decay into simpler objects. In the case of fig. 1, for example, it appears that in continuous media the ring typically contracts, ultimately to nothing [1, 11, 24, 34]. What becomes of linked or knotted rings? Possible transmutation pathways have been outlined theoretically [29], but up to now the only such transmutation observed is fission of a single ring into two [34]. Much remains to be discovered.

Such discoveries may illuminate modes of arrhythmia in heart muscle that lead to sudden cardiac death [35]. They may provide hints as to the patterns of stimulation which initiate such lethal waves, and suggest means less damaging than high-current electroconvulsion of the ventricle for terminating them.

Acknowledgements

ATW thanks the National Science Foundation for grants CHE 810322 and PCM 8410752. The contents of this paper were first presented at the German Mathematical Society School on Biological Rhythms and Population Dynamics at Bayreuth University in July 1984, courtesy of Volkswagen Foundation. Acknowledgement is made to the Donors of the Petroleum Research Fund administered by the American Chemical Society for partial support of this work.

Appendix A

A property of Seifert surfaces

by Professor Herbert SEIFERT
with comments by A.T. WINFREE

Notation.

Given a set of R closed oriented (and possibly knotted) rings $\{K_1, K_2, \ldots, K_R\}$ embedded in three-dimensional Euclidean space E^3, denote by L_{ij} the (integer) mutual linkage of ring K_i through ring K_j' traced arbitrarily near to ring K_j along a compact (nonsingular) oriented surface F whose boundary consists of the R oriented closed rings.[†] In [27–29, 33] this F was constructed by the Seifert algorithm to span the collection of rings. It is intended to represent a chemical wavefront, half of a closed surface of uniform chemical concentration. By D_i denote a (singular) disk with boundary K_i. This construct has no direct chemical interpretation. So F together with the set of disks $\{D_i\}$ is a closed singular surface, here denoted by G.

We next consider the intersection number of a curve K_i' with F and with G. The intersection number $I(C, S)$ of an oriented curve C with an oriented surface S is the integer number of times the curve penetrates the surface with same orientation, minus the number of penetrations with opposite orientation. Thus the intersection number of any closed curve with any closed surface is 0 in E^3.

Argument

For each i, $I(G, K_i') = 0$. But G is the sum of F and all the D_j, so:

$$0 = I(G, K_i') = I(F, K_i') + \sum_{j=1}^{R} I(D_j, K_i').$$

The first term on the right is 0, since each K' lies on F and can be approximated by a curve that does not meet (non-singular) F at all. The second term (the sum) is the sum of all L_{ij}: $I(D_j, K_i') = I(D_j, K_i)$ is the linkage of K_i with K_j, the boundary of D_j.

In short, $\sum_{j=1}^{R} L_{ij} = 0$ for each i.

[†] T. Poston suggested this simplification of our usual definition [27–29, 33], which distinguished the case $i = j$.

This is the property remarked on in the text, deduced from physical considerations as an "exclusion principle" delimiting the diversity of chemically realizable "organizing centers" [27–29, 33]. This appendix shows that the exclusion principle, while classifying geometrically consistent organizing centers and assigning quantum numbers L_{ij} to each, has little physical content. Physical principles may further delimit the possibilities by identifying long-term instabilities and modes of decay in some (or all) of these solutions.

References

[1] A.V. Panfilov and A.M. Pertsov, Dokl. Akad. Nauk. USSR 274 (1984) 1500 (in Russian).

[2] A.V. Panfilov, A.N. Rudenko and A.T. Winfree, submitted to Biophysica (in Russian).

[3] A.V. Panfilov and A.T. Winfree, in preparation for Physica D.

[4] S. Hastings, J. Mathe. Biol. 11 (1981) 105.

[5] B.F. Madore and W.L. Freedman, Science 222 (1983) 615.

[6] G.K. Moe, W.C. Rheinboldt and J.A. Abildskov, Amer. Heart J. 67 (1964) 200.

[7] L. Reshodko, J. Gen. Biol. 1 (1973) 80 (in Russian).

[8] L. Reshodko and L.J. Bures, Biol. Cybern. 13 (1974) 181.

[9] J.J. Tyson, The Belousov–Zhabotinsky Reaction, Springer Lecture Notes in Biomathematics, vol. 10, S. Levin, ed. (Springer, New York, 1976).

[10] A.T. Winfree, Science (1973) 937.

[11] A.T. Winfree, Sci. Amer. 230 (1974) 82.

[12] A.T. Winfree, SIAM AMS Proc. 8 (1974) 13.

[13] A.T. Winfree, in Oscillations and Travelling Waves in Chemical Systems, R. Field and M. Burger eds. (Wiley, New York, 1984).

[14] K.N. Tomchik and P.N. Devreotes, Science 212 (1981) 433.

[15] A.C. Newell, in: Fungal Differentiation: a Contemporary Synthesis, J. Smith, ed. (Dekker, New York, 1983), p. 43.

[16] M.A. Allessie, F.I.M. Bonke and F.J.G. Schopman, Circ. Res. 333 (1973) 54.

[17] M.J. Janse, F.J. van Capelle, H. Morsink and A.G. Kleber, Circ. Res. 47 (1980) 151.

[18] V.I. Krinsky, Pharm. Ther. 3 (1978) 539.

[19] F.J. van Capelle and D. Durrer, Circ. Res. 47 (1980) 454.

[20] J. Bures, O. Buresova and V.I. Koroleva, in: Neurophysiological Mechanisms of Epilepsy, V.M. Okujava, ed. (Metsnierba, Tblisi, 1980) pp. 120–130.

[21] V.I. Koroleva and J. Bures, Brain Res. 173 (1979) 209.

[22] N.A. Gorelova and J. Bures, J. Neurobiol. 14 (1983) 353.

[23] B. Welsh, J. Gomatam and A. Burgess, Nature 304 (1983) 611.

[24] A.T. Winfree, Far. Symp. Chem. Soc. 9 (1974) 38.

[25] A.B. Medvinsky, A.M. Pertsov, G.A. Polishuk and V.G. Fast, in: Electrical Field of the Heart, O. Baum, M. Roschevsky and L. Titomir, eds. (Nauka, Moscow, 1983, pp. 38–51 (in Russian).

[26] A.T. Winfree, The Geometry of Biological Time (Springer, New York, 1980).

[27] A.T. Winfree and S.H. Strogatz, Physica 9D (1983) 65.

[28] A.T. Winfree and S.H. Strogatz, Physica 9D (1983) 333.

[29] A.T. Winfree and S.H. Strogatz, Physica 13D (1984) 221.

[30] A.T. Winfree, Physica 12D (1984) 321–332.

[31] S.H. Strogatz, M.L. Prueitt and A.T. Winfree, IEEE Comp. Graphics & Applic. 4 (1984) 66.

[32] A.T. Winfree and S.H. Strogatz, Physica 8D (1983) 35.

[33] A.T. Winfree and S.H. Strogatz, Nature 311 (1984) 611–615.

[34] B. Welsh, Thesis, Glasgow College of Technology (1984).

[35] A.T. Winfree, Sci. Amer. 248 (1983) 144.

A Local Activator-Inhibitor Model of Vertebrate Skin Patterns

DAVID A. YOUNG

University of California, Lawrence Livermore National Laboratory, Livermore, California 94550

Received 12 December 1983; revised 12 May 1984

ABSTRACT

A model for vertebrate skin patterns is presented in which the differentiated (colored) pigment cells produce two diffusible morphogens, an activator and an inhibitor. The concentrations of these two substances at any point on the skin determine whether a pigment cell at that point will be colored or not. Computer simulations with this model show many realistic features of spot and stripe patterns found in vertebrates.

INTRODUCTION

The color patterns on vertebrate skin are of great importance to the survival of the organism because they are involved in camouflage, species identification, and warning patterns. The theoretical problem of the morphogenesis of these patterns is therefore of interest from an evolutionary point of view as well as being a fascinating mathematical problem in its own right.

Typical skin patterns are spots or stripes which are formed by specialized pigment cells (melanocytes). Innumerable variations of spot and stripe patterns are found in fish [2] and mammals [12]. The problem of how skin patterns are formed becomes a description of how the colored pigment cells are distributed on the embryonic skin.

The prevailing theoretical answer to this question is that pattern formation is governed by reaction-diffusion processes of the Turing type [1,8]. In this scheme, the uniformly distributed pigment cells produce two or more species of morphogen molecules which react with each other and diffuse in space to produce a pattern of concentrations with a characteristic wavelength. These morphogen concentration "prepatterns" then induce the differentiation of the pigment cells, producing a permanent pattern similar to the prepattern. The calculated patterns are dependent on initial and boundary conditions, and show many of the characteristic forms observed in vertebrates.

51

52   DAVID A. YOUNG

The Turing model for skin patterns has not yet been tested experimentally, but evidence is accumulating from studies on cold-blooded vertebrates that there are several types of pigment cells which can interact with one another and change each other's properties [5,6,11]. This evidence suggests, contrary to the Turing model, that the intercellular interaction is local, possibly due to short-range diffusion of morphogen molecules or to direct cell contact. In this paper I explore an alternative to the Turing model which involves local cell interactions and which also gives rise to an interesting spectrum of pigment patterns.

THE MODEL

I propose an activator-inhibitor diffusion theory developed originally by Swindale [13] for the study of patterns in the visual cortex of the brain. As a simplified initial condition, I imagine on the early embryonic skin a uniform distribution of pigment cells, containing a mixture of differentiated (colored) cells (DCs) and undifferentiated cells (UCs). A simple mechanism for the production of this mixture might be a slow random process of differentiation in the UC cell population. Each DC produces an inhibitor morphogen which stimulates the dedifferentiation of other nearby DCs, and an activator morphogen which stimulates the differentiation of nearby UCs. The two substances are diffusible, with the inhibitor having the longer range. The UCs are passive and produce no active substances. The fate of each pigment cell, UC or DC, will be determined by the sum of the influences on it from all neighboring DCs.

The processes of production, diffusion, and decay of morphogens can be modeled by a generalized diffusion equation:

$$\frac{\partial M}{\partial t} = \nabla \cdot \mathbf{D} \cdot \nabla M - KM + Q. \tag{1}$$

Here $M = M(\mathbf{r}, t)$ is the morphogen (either inhibitor or activator) concentration, and the terms on the right are diffusion, first-order chemical transformation, and production, respectively.

Each DC produces at constant rate two morphogens, an activator $M^{(1)}$ and an inhibitor $M^{(2)}$, which diffuse away from their source and are uniformly degraded by the neighboring cells, according to Equation (1). The resulting steady-state distributions of the morphogens about a DC are shown schematically in Figure 1(a). Together, the two morphogens constitute a "morphogenetic field" $w(R)$, where R is the distance from the DC, which is "read" by nearby pigment cells in the skin. The "reading" process is modeled by assuming that the net activation effect found close to the DC is represented by a constant positive field value, and the net inhibition effect found further from the DC is represented by a constant negative field value.

VERTEBRATE SKIN PATTERNS   53

This is shown in Figure 1(b). The activation region is a small circular area about the DC with a large constant positive field value. The inhibition region is the outer circular annulus with a small negative field value. The integrated field over the cells in the whole circular area must be close to zero in order to avoid the complete dominance of either activator or inhibitor.

This model is similar to inhibitor theories of patterns formation, for example of hair follicles in mammalian skin [3], or of leaf primordia on the shoot apex of a green plant [7]. However, pure inhibition theories are only capable of producing spatial patterns of pointlike structures. In our case, we need connected regions of differentiated pigment cells, and this requires short-range activation as well as long-range inhibition.

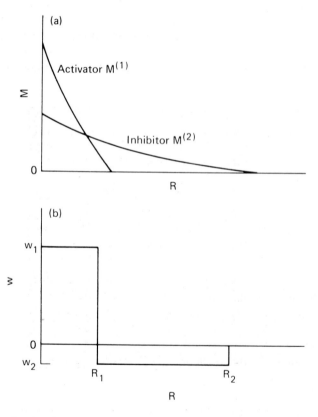

FIG. 1. A schematic illustration of the local activator-inhibitor model. In (a) the steady-state activator and inhibitor concentrations about a differentiated pigment cell are shown. The inhibitor has a longer range than the activator. In (b) the combined (field) effect of activator and inhibitor is modeled with constant positive and negative circular regions.

54   DAVID A. YOUNG

CALCULATIONS

The calculation begins by distributing DCs randomly on a rectangular grid of points representing pigment cells. Then for each grid point at position **R**, the field values due to all nearby DCs at positions $\mathbf{R}_i$ are added up. If $\sum_i w(|\mathbf{R} - \mathbf{R}_i|) > 0$, then the point at **R** becomes (or remains) a DC. If $\sum_i w(|\mathbf{R} - \mathbf{R}_i|) = 0$, the point does not change state, and if $\sum_i w(|\mathbf{R} - \mathbf{R}_i|) < 0$, the point becomes (or remains) a UC. By simplifying the morphogenetic field as shown in Figure 1 and by discretizing the cell positions, I have converted a continuum model [Equation (1)] into a cellular automaton [14]. Cellular automata are very useful for computational purposes because they simplify the problem at hand while retaining the essential features required for exhibiting self-organization phenomena. This is justified by the observation that very nearly the same results are obtained [13] when $w(R)$ is a continuous function, as in Figure 1(a). The process of summing the morphogenetic fields and changing states for each grid point is repeated until the resulting pattern no longer changes. I find that five iterations suffice for convergence to a stable pattern, and that the general form of the final pattern is not sensitive to the initial DC distribution.

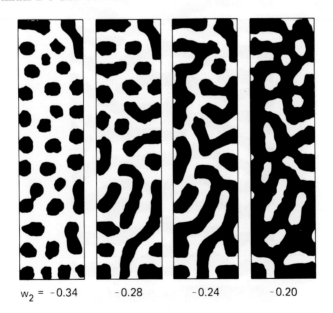

$w_2 =$ -0.34 -0.28 -0.24 -0.20

FIG. 2.   Patterns produced with the activator-inhibitor model. The activation area has a radius of 2.30, and the inhibition area has an outer radius of 6.01. The activation field value w_1 is $+1.0$, and the inhibition field value w_2 is varied as indicated in the four examples. As inhibition is decreased (left to right), the spot pattern connects up into a pattern of stripes. Each panel is 25×100 in the arbitrary grid units.

One useful demonstration with the model is the close relationship of spot and stripe patterns. Many vertebrate species are spotted or striped, and some tropical fish genera show combinations of these patterns on the same individual [2]. Since it is unlikely that stripes and spots arise from wholly different mechanisms, a useful theory must be able to generate both patterns. This is done in Figure 2. Here the outer radius of the inhibitor area is 6.01, the radius of the activator area is 2.30, and the spacing between adjacent grid points is 1.0 in arbitrary units. The activator field value is fixed at 1.0, and the inhibitor field value is varied to produce different patterns. When the inhibitor is strong (Figure 2, $w_2 = -0.34$), the DCs cannot form connected masses, and instead form isolated spots. As the inhibitor weakens, the spots connect up with each other until well-developed stripes appear. With very strong inhibition, the pattern will consist of isolated DCs on an uncolored

FIG. 3. Pattern produced with an anisotropic activator-inhibitor model. The activator area is an ellipse ($x^2/a^2 + y^2/b^2 = 1$) with axes $a_1 = 2.30$, $b_1 = 1.38$, and the inhibitor area is an ellipse with axes $a_2 = 3.61$ and $b_2 = 6.01$. The panel is 60×100 in the arbitrary grid units.

background. With very weak inhibition, the pattern will be a solid mass of color. All of these patterns are found in the vertebrates.

The striping patterns in Figure 2 are isotropic, with no preferred direction, whereas in real skin patterns such as on the zebra, there usually is a preferred direction. Directionality can be introduced by assuming that the diffusion of activator and inhibitor are anisotropic. This was indicated in Equation (1) by the tensor form of the diffusion constant. Anisotropy has been observed in the morphogenesis of pigment patterns in fish [11], and it is also consistent with the concept of polar transport [15], which has been used to explain the effects of tissue polarity in pattern formation. Thus in tissues showing polarity, we expect that the perpendicular components of the diffusion tensor will be unequal, i.e., $D_{xx} \neq D_{yy}$. In the model calculations, this can be introduced by changing the circular areas in the $w(R)$ field to ellipses. An example is shown in Figure 3. The inhibitor is polarized perpendicular to, and the activator parallel to, the stripes. The idea of preferential diffusion of morphogens in perpendicular directions is not new [4,9,16], but it does not have a clear biophysical explanation and must at present be considered only as an interesting hypothesis.

DISCUSSION

Both this model and the Turing reaction-diffusion model (actually consisting of a large class of specific chemical kinetic models) predict basic features of vertebrate skin patterns. In certain details, however, they are different. For example, the activator-inhibitor model pattern in Figure 3 shows both branching and termination of the colored stripes, features which arise from the random initial distribution of DCs, and which are also observed in zebra coat patterns. The Turing models studied to date [1,8] do not show these features, possibly because of the more deterministic nature of these models.

Although the suggested mechanisms of activation and inhibition in this paper are diffusion processes, this assumption is not necessary. The model when reduced to the morphogenetic field concept shown in Figure 1(b) is a simple logical algorithm involving on-off switching of cell differentiation according to a threshold variable. The variable need not be the concentration of a diffusible substance. It might instead have to do with direct cell contacts or with local elastic strains in the underlying tissues. The tissue polarity indicated in Figure 3 might have more to do with unidirectional macromolecular arrays than with anisotropic diffusion. I believe that the diffusion mechanism is the simplest explanation of the pigmentation process, but it is certainly not the only explanation.

Also, the model construct of a static random mixture of differentiated and undifferentiated pigment cells is an oversimplification. A more realistic model might allow the migration of pigment cells toward or away from other

VERTEBRATE SKIN PATTERNS   57

pigment cells according to "attractive" or "repulsive" interactions. Finally, the assumption of only one type of pigment cell is also an oversimplification, since there are known to be several types [5]. Swindale's original model [13] in fact postulated two cell types. The usefulness of the model is not so much in its detailed assumptions, but rather in its logical structure. The fact that a model with a very simple logical structure can reproduce many of the observed features of a pattern strongly suggests that the actual pattern mechanism is also simple, and this is the principal conclusion to be drawn from the present work.

This model could be generalized to generate multicolored patterns, which are common in vertebrates. Also, the topology of striping patterns has been closely studied in connection with the problem of classifying fingerprints [10], and the formal similarity of the patterns on zebra skin and in fingerprints suggests that the morphogenetic mechanisms may be similar.

More detailed and realistic models of vertebrate skin patterns might best arise from experimental work specifically aimed at elucidating the mechanism which controls pigment-cell differentiation. Specifically, one would like to know whether the mechanism involves chemical waves on a uniform cellular substratum, or short-ranged interactions between pigment cells, or something altogether different. Once this is determined, a computer model can then be constructed to investigate how a color pattern emerges from the multicellular system.

REFERENCES

1   J. B. L. Bard, A model for generating aspects of zebra and other mammalian coat patterns, *J. Theoret. Biol.* 93:363–385 (1981).
2   R. H. Carcasson, *A Field Guide to the Coral Reef Fishes of the Indian and West Pacific Oceans*, Collins, London, 1977.
3   J. H. Claxton, the determination of patterns with special reference to that of the central primary skin follicles in sheep, *J. Theoret. Biol.* 7:302–317 (1964).
4   F. W. Cummings and J. W. Prothero, A model of pattern formation in multicellular organisms, *Collect. Phenom.* 3:41–53 (1978).
5   S. K. Frost and G. M. Malacinski, The developmental genetics of pigment mutants in the Mexican axolotl, *Devel. Genet.* 1:271–294 (1980).
6   F. Kirschbaum, Untersuchung über das Farbmuster des Zebrabarbe *Brachydanio rerio* (Cyprinidae, Teleostei), *Wilhelm Roux's Arch.* 177:129–152 (1975).
7   G. J. Mitchison, Phyllotaxis and the Fibonacci series, *Science* 196:270–275 (1977).
8   J. D. Murray, A pre-pattern formation mechanism for animal coat markings, *J. Theoret. Biol.* 88:161–199 (1981).
9   H. G. Othmer and L. E. Scriven, Instability and dynamic pattern in cellular networks, *J. Theoret. Biol.* 32:507–537 (1971).
10 R. Penrose, The topology of ridge systems, *Ann. Human Genetics* 42:435–444 (1979).
11 E. R. Schmidt, Chromatophore development and cell interactions in the skin of Xiphophorine fish, *Wilhelm Roux's Arch.* 184:115–134 (1978).

58   DAVID A. YOUNG

12   A. G. Searle, *Comparative Genetics of Coat Colour in Mammals*, Academic, London, 1968.

13   N. V. Swindale, A model for the formation of ocular dominance stripes, *Proc. Roy. Soc. London Ser. B* 208:243–264 (1980).

14   S. Wolfram, Statistical mechanics of cellular automata, *Rev. Modern Phys.* 55:601–644 (1983).

15   D. A. Young, A polar transport mechanism for biological pattern formation, *Phys. Lett. A* 89A:52–54 (1982).

16   D. A. Young, A polar transport model of planarian regeneration, *Differentiation* 25:1–4 (1983).

PHYSICAL REVIEW

LETTERS

VOLUME 55	30 DECEMBER 1985	NUMBER 27

Discrete Model of Chemical Turbulence

Y. Oono

*Department of Physics and Materials Research Laboratory, University of Illinois at Urbana-Champaign,
Urbana, Illinois 61801, and Schlumberger-Doll Research, Ridgefield, Connecticut 06877*

and

M. Kohmoto[a]

Schlumberger-Doll Research, Ridgefield, Connecticut 06877
(Received 26 September 1985)

Presumably the simplest model of chemical turbulence is proposed, for which the spatial degree of freedom plays an important role. The concentration, as well as space and time, is discretized. As in the real experiment a turbulent phase is found sandwiched between two ordered phases. In addition, the model predicts a solitonlike phase. Thus the model can exhibit a variety of behaviors encountered in partial-differential-equation systems. It is also closely related to cellular automata.

PACS numbers: 05.45.+b, 82.20.Fd

In many physical and chemical systems most interesting phenomena such as pattern formation, turbulence, etc., involve both the spatial and temporal degrees of freedom. The recent reviving interest in the collective behavior of coupled simple subsystems stems from the quest for the role of spatial degrees of freedom. Typical examples being studied are coupled nonlinear oscillators,[1] coupled nonlinear maps,[2] and cellular automata.[3] These examples are closely related to nonlinear partial differential equations (PDE) which are often more realistic models for physical systems. In these examples and PDE systems the coexistence of chaos and ordered coherent structure (order in turbulent states) has been given much attention.[4]

For the coupled limit cycles, Kuramoto, Yamada, and other researchers have been conducting extensive studies for more than ten years.[5,6] One of the most important outcomes is the concept of chemical turbulence. On the basis of the Kuramoto-Yamada-Sivashinsky equation, they predicted the existence of a turbulent phase due to instability induced by diffusion. Notice that this turbulence is conceptually different from the so-called chemical turbulence studied by

Hudson, Roux, and others.[7] In the latter case, reagents are completely stirred, so that there is no spatial degree of freedom.

Following the Kuramoto-Yamada prediction, Yamazaki, Oono, and Hirakawa[8] performed an experimental study of chemical turbulence using the Belousov-Zhabotinsky reaction.[9] In their experiments the regime of the reaction was so chosen that there was purely periodic oscillation when the solution was stirred well. Thus, if a chaotic phenomenon was observed in a vessel without any stirring, it was solely due to the existence of the spatial inhomogeneity. In contrast to experiments in a stirred reactor,[7] the experiments by Yamazaki, Oono, and Hirakawa had difficulties due to the lack of stirring; carbon dioxide bubbles generated by the reaction had to be kept on the wall of the vessel, possible convective flow induced by the exothermic reaction had to be suppressed by a slight temperature gradient, etc. The rough phase diagram constructed experimentally exhibits a disordered regime characterized by rather abrupt changes of frequencies and phases of the concentration oscillation. The main feature of the diagram was that there seemed to be a nonturbulent regime on both sides of the turbulent region along the temperature axis. Since the frequency of the reaction is much more sensitive to the temperature than the diffusion rates, the higher the temperature, the less effective is the diffusion.

The main purpose of the present Letter is to propose presumably the simplest possible model of chemical turbulence. The model is in a class of models whose elements are discrete-state oscillators. We show the existence of a large variety of behaviors in this model: ordered phases, turbulence, solitons, etc. It seems that the model proposed here can exhibit the whole array of behaviors exhibited by (dissipative) PDE sys-

tems in general.

The essence of chemical turbulence is the linear coupling of spatially distributed nonlinear oscillators. The simplest way to represent a cyclic chemical oscillation is to use discrete concentration levels. It has turned out that to have nontrivial behaviors three discrete levels, M, 1, and 0, are sufficient, where M is a positive integer. We also use discrete time and space. Hence, in our simplest model at a given discrete time, one of the discrete concentrations M, 1, and 0 is assigned to each spatial cell. The rule of time evolution must contain the effects of both the spatial coupling and oscillation. Specifically, the rule of our cellular model in one spatial dimension is given by

$$A'(n,t) = \alpha[A(n+1,t) + A(n-1,t)]/2 + (1-\alpha)A(n,t), \qquad (1)$$

and

$$A(n,t+1) = F(A'(n,t)), \qquad (2)$$

where $A(n,t)$ is the concentration at the nth cell at time t, $\alpha \in [0,1]$ is the spatial coupling constant, and the function F is given by

$$F(x) = \begin{cases} 1, & \text{if } 1.5 \leq x, \\ 0, & \text{if } 0.5 \leq x < 1.5, \\ M, & \text{if } x < 0.5. \end{cases} \qquad (3)$$

The diffusion-type spatial coupling is represented by (1), where the parameter α may be regarded as the strength of diffusion. On the other hand, (2) with (3) describes cyclic oscillation; when there is no diffusion, i.e., $\alpha = 0$, each cell makes an intrinsic cycle of 0-M-1-0: This even mimics the time asymmetry of real concentration oscillation in the Belousov-Zhabotinsky reaction. We designate this rule as $0M10$. We can make many variants of this rule, but this is the simplest nontrivial one; the two-state $0M0$ rule gives the simplest oscillating model, but this gives us only ordered phases.

The phase diagram for the $0M10$ model is shown in Fig. 1 and typical behaviors of spatiotemporal patterns are shown in Fig. 2. Both are with stochastic initial conditions. For large values of M one sees four different phases: three-cycle (3 phase), turbulent (T phase), solitonlike (S phase), and second three-cycle (3' phase) in this order as α is increased. The region X is actually divided into many phases, many of which are phases with several different cycles coexisting. There are also some solitonlike phases. Since the maximum concentration M is comparable to the other concentration levels, 0 and 1, the interplay of the spatial coupling and oscillation becomes complex in X. Furthermore, real experiments were never conducted in the regime where the amplitude of the oscillation is small. Hence from here on we focus on large values of M and the four phases mentioned above (3, T, S,

and 3' phases); we expect that phase diagrams with ordered phases separated by turbulent and solitonlike phases would be general features of chemical turbulence.

Our working definition of the turbulent phase is as follows. The nth cell is said to be unpredictable if and only if the Kolmogorov-Sinai entropy[10] of the sequence $\{A(n,t)\}_{t=0}^{\infty}$ of the discrete concentration levels at the cell n is positive. If the majority of the cells are unpredictable, the phase is said to be nontrivial. If the information in the initial concentration distribution is significantly preserved, we say that the nontrivial phase is turbulent.

Although an attempt to estimate the true Kolmogorov-Sinai entropy (or the Shannon entropy

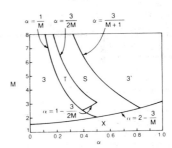

FIG. 1. The phase diagram for the $0M10$ chemical turbulence model defined by Eqs. (1)–(3). M is the peak value of the autonomous oscillation, which is extended to real values by an obvious modification of the rule, and α is the spatial coupling constant. The symbols 3, T, S, and 3' denote, respectively, a periodic phase with period three, the turbulent phase, the solitonlike phase, and a different periodic phase with period three. The letter X collectively denotes various phases which are not discussed in the text.

3   T   S   3'

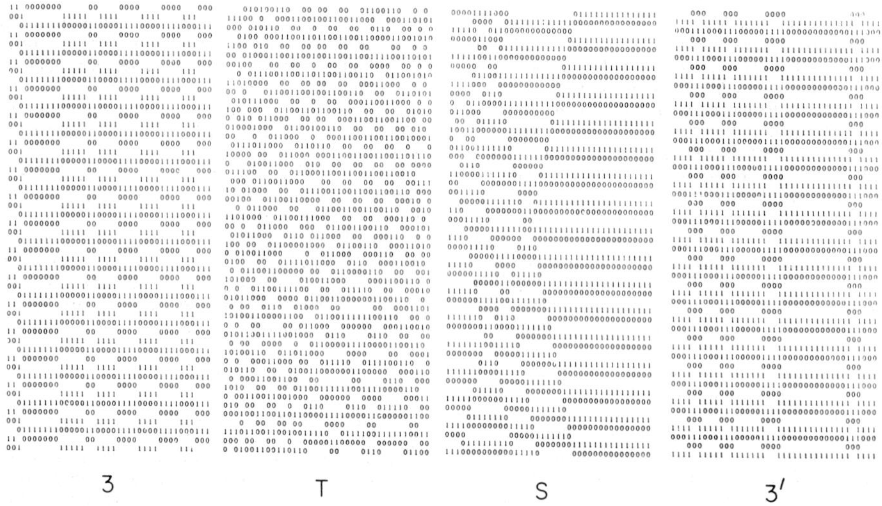

FIG. 2. Typical spatiotemporal patterns of the $0M10$ model. The time flows from top to bottom of the figures. The horizontal direction is the spatial coordinate. An empty cell actually contains M. The symbols 3, T, S, and 3' denote respectively a periodic phase with period three, the turbulent phase, the solitonlike phase, and a different periodic phase with period three.

per letter[11] if we regard the state sequence as a sentence) is being made, here, for a preliminary study, the Markov entropy of the sequence,[12] which is the Kolmogorov-Sinai entropy of the empirical simple Markov process obtained from the statistics of the observed sequences, is used to measure the disorder of the state sequence at a given cell. The spatial average of this entropy is denoted by S_t. The spatial disorder is measured by the Markov entropy of the sequence of states along the spatial axis: This entropy is denoted by S_s. Since our system has only three states, the upper bound of S_t and S_s is $\log_2 3 = 1.58496\ldots$ bits. In addition, as a result of the severe restriction on the time sequence imposed by the structure of F, S_t is likely to be less than 0.7 bit. Thus asymptotic values of $S_t \simeq 0.5$ bit and $S_s \simeq 1.2$ bits specify the turbulent phase; $0 < S_t \le 0.3$ (bit) and $S_s \simeq 0.7$ bit specify the soliton phase; and $S_t = 0$ specifies the periodic ordered phase. Since the number of solitons and their directions of motion strongly depend on the initial condition, so does the value of S_t in the solitonlike phase.

Notice that our system is completely deterministic, so that the temporal disorder is possible only if the (spatial) information in the initial condition is converted to temporal information. This situation is exactly the same as chaos in iterative maps. As an objective of statistical mechanics, such discrete models and cellular automata can be regarded as measure-theoretic dynamical systems. It is well known that any measure-theoretic dynamical system (with finite entropy) is isomorphic to a shift dynamical system with a finite number of symbols.[13]

The boundaries of the four phases correspond to changes of behaviors of fundamental configurations. Consider a wall which divides two ordered spatial domains. The spatial coupling is not strong enough if $\alpha < 1/M$, so that the two domains do not interact, and the wall cannot move. This parameter region corresponds to the ordered 3 phase. If $1/M \le \alpha < 3/(M+1)$, then the wall can move. Suppose we have a domain with M's (M domain) and a domain with 0's (0 domain) at a given time. The 0 domain invades the M domain and the wall between them moves by one spatial unit after three time steps. This parameter region contains the T and the S phases. If $\alpha \ge 3/(M+1)$, the coupling is too strong, so that the wall again cannot move. Notice that although the 0 domain invades the M domain at one time, the 0 domain changes to an M domain and the M domain becomes a 1 domain at the next time step. Now the domain which originally was the 0 domain is invaded by the other domain so that the wall does not move after all. Finally, the boundary of the T and the S phase is related to the existence or nonexistence of self-organization. Consider an impurity in an M domain. If $\alpha < 3/2M$, then the impurity grows. This region, $1/M \leq \alpha < 3/2M$, corresponds to the T phase. On the other hand, if $\alpha \leq 3/2M$, the impurity disappears and the medium becomes completely ordered. Therefore, in the S phase, $3/2M \leq \alpha < 3/(M+1)$, an initial random spatial configuration is self-organized to ordered domains. The domain walls behave like solitons as seen in Fig. 2, S phase.

In the real experiment,[9] only the time sequence at one space point was observed. Therefore, the experiment was not designed to discriminate soliton states from turbulent states. We expect solitonlike propagation of the spatial patterns in real experiments of chemical turbulence. Indeed, we see a suggestive numerical result in Kuramoto's work.[14]

If one more state, 2, is added to the model with the corresponding modification of F in Eq. (2) ($0M210$ model), we clearly observe the intermittent structure in a turbulent phase. We also have an additional phase where chaotic behavior seems to be induced by the existence of very many solitons. Thus there can be a turbulent phase which may be understood in terms of nonlinear modes.

The crucial difference between our model and the model for excitable media,[15] which is also related to the Belousov-Zhabotinsky reaction, is the nonexistence of the quiescent state in our model. In the excitable-medium model the state 0 stays indefinitely at 0 unless at least one M state appears in its neighborhood. There is no turbulent phase in this model (at least in one-space). Although there exist solitary propagating waves, they disappear upon collisions. Hence, these waves are distinctly different from the solitons in our model where they go through each other upon collisions. It is easy to make interpolative models of this excitable-medium model and our $0M10$ model; symbolically we denote them by the $0M10\ldots0$ model. According to our preliminary study, increasing the period diminishes the width of the turbulent region in the phase diagram.

As is mentioned above, our rule does not allow the existence of the quiescent state. Thus as a "cellular automaton," our rule may be said to be illegal.[3] However, for any illegal rule, we can make an equivalent legal rule which allows a quiescent state by suitably redefining the time step and the cell neighborhood. Hence, there is a "legal" cellular automaton which is equivalent to the $0M10$ model. However, our rule, the combination of the autonomous periodic rule and the diffusionlike linear coupling, is a very special one, so that there is an intriguing question on the algorithmic properties of the model, for example, the universality as an automaton.

In conclusion, we have proposed, presumably, the simplest nontrivial model of chemical turbulence (or

coupled limit cycles) which can exhibit many phenomena that we can observe in nonlinear partial differential equations: turbulence, solitons, intermittency, etc. We believe that it is promising to study such a minimal model in order to understand the complex behavior of real partial-differential-equation systems.

One of us (Y.O.) is grateful to Y. Takahashi, S. J. Chang, A. E. Jackson, B. A. Friedman, and S. Puri for useful comments. The authors thank D. Sherrington for carefully reading the manuscript. They are also grateful to Schlumberger-Doll Research for its hospitality. The work is supported, in part, by a National Science Foundation grant through the Illinois Materials Research Laboratory.

(a)Present address: Department of Physics, University of Utah, Salt Lake City, Utah 84112.

[1]Y. Kuramoto and T. Tsuzuki, Prog. Theor. Phys. **52**, 1399 (1974); T. Tsuzuki and Y. Kuramoto, Prog. Theor. Phys. **54**, 687 (1975); T. Yamada and H. Fujisaka, Z. Phys. B **28**, 239 (1977); J. C. Neu, SIAM J. Appl. Math. **36**, 509 (1979), and **38**, 305 (1980).

[2]K. Kaneko, Prog. Theor. Phys. **72**, 480 (1984); R. J. Deissler, Phys. Lett. **100A**, 451 (1984); R. Kapral, Phys. Rev. A **32**, 1076 (1985).

[3]See, e.g., S. Wolfram, Rev. Mod. Phys. **55**, 601 (1983); J. Demongeot, E. Cole, and M. Tchuente, *Dynamical Systems and Cellular Automata* (Academic, New York, 1985).

[4]See, e.g., A. R. Bishop, K. Fesser, P. S. Lomdahl, W. C. Kerr, M. B. Williams, D. F. Dubois, H. A. Rose, and B. Hafizi, Phys. Rev. Lett. **51**, 335 (1983); K. Nozaki and N. Bekki, Phys. Rev. Lett. **51**, 2171 (1983); A. R. Bishop, K. Fesser, and P. S. Lomdahl, Physica (Amsterdam) **7D**, 259 (1983).

[5]For reviews, see Y. Kuramoto, in *STATPHYS 14*, edited by J. Stephenson (North-Holland, Amsterdam, 1981); K. Tomita, Phys. Rep. **86**, 113 (1982).

[6]Y. Kuramoto and T. Yamada, Prog. Theor. Phys. **56**, 679 (1976); T. Yamada and Y. Kuramoto, Prog. Theor. Phys. **56**, 681 (1976); see also Y. Kuramoto, Prog. Theor. Phys. **63**, 1885 (1980).

[7]R. A. Schmits, K. R. Graziani, and J. L. Hudson, J. Chem. Phys. **67**, 3040 (1977); J. L. Hudson, M. Hart, and D. Marinko, J. Chem. Phys. **71**, 1601 (1979); J.-C. Roux, A. Rossi, S. Bachelart, and C. Vidal, Phys. Lett. **77A**, 391 (1980), and Physica (Amsterdam) **2D**, 395 (1981); J.-C. Roux, Physica (Amsterdam) **7D**, 57 (1983); J.-C. Roux, R. H. Simoyi, and H. L. Swinney, Physica (Amsterdam) **8D**, 257 (1983).

[8]H. Yamazaki, Y. Oono, and K. Hirakawa, J. Phys. Soc. Jpn. **44**, 335 (1978), and **46**, 721 (1980).

[9]See, e.g., J. J. Tyson, in *The Belousov-Zhabotinsky Reaction*, edited by S. Levine, Lecture Notes in Biomathematics, Vol. 10 (Springer, Berlin, 1976).

[10]See, e.g., N. F. G. Martin and J. W. England, *Mathematical Theory of Entropy*, Encyclopedia of Mathematics and Its

VOLUME 55, NUMBER 27   PHYSICAL REVIEW LETTERS   30 DECEMBER 1985

Applications, Vol. 12 (Addison-Wesley, Reading, Mass., 1981).

[11]See, e.g., L. Brillouin, *Science and Information Theory* (Academic, New York, 1969).

[12]Y. Oono, T. Kohda, and H. Yamazaki, J. Phys. Soc. Jpn. **48**, 738 (1980).

[13]H. Totoki, *Introduction to Ergodic Theory* (Kyoritsu, Tokyo, 1971); W. Krieger, Trans. Amer. Math. Soc. **149**, 453 (1970). One of us (Y.O.) is grateful to Y. Takahashi for reminding him of the relevant theorem by Krieger. Even if the entropy is not finite, isomorphisms to shift dynamical systems with countably many symbols can still be constructed.

[14]Y. Kuramoto, Prog. Theor. Phys., Suppl. No. 64, 346 (1978).

[15]J. Greenberg, C. Green, and S. Hastings, SIAM J. Algebraic Discrete Methods **1**, 34 (1980); B. F. Madore and W. L. Freedman, Science **222**, 615 (1983).

Physica 19D (1986) 423–432
North-Holland, Amsterdam

SOLITON-LIKE BEHAVIOR IN AUTOMATA†

James K. PARK‡ and Kenneth STEIGLITZ
Dept. of Computer Science, Princeton University, Princeton, NJ 08544, USA

and

William P. THURSTON
Dept. of Mathematics, Princeton University, Princeton, NJ 08544, USA

Received 13 May 1985
Revised 25 October 1985

We propose a new kind of automaton that uses newly computed site values as soon as they are available. We call them *Filter Automata* (FA); they are analogous to Infinite Impulse Response (IIR) digital filters, whereas the usual Cellular Automata (CA) correspond to Finite Impulse Response (FIR) digital filters. It is shown that as a class the FA's are equivalent to CA's, in the sense that the same array of space-generation values can be produced; they must be generated in a different order, however.

A particular class of irreversible, totalistic FA's are described that support a profusion of persistent structures that move at different speeds, and these particle-like patterns collide in nondestructive ways. They often pass through one another with nothing more than a phase jump, much like the solitons that arise in the solution of certain nonlinear differential equations.

Histograms of speed, displacement, and period are given for neighborhood radii from 2 to 6 and particles with generators up to 16 bits wide. We then present statistics, for neighborhood radii 2 to 9, which show that collisions which preserve the identity of particles are very common.

1. Introduction

Cellular automata have attracted attention recently as non-numerical models for nonlinear physical phenomena [1]. Vichniac [2] points out that they "exhibit behaviors and illustrate concepts that are unmistakably physical...", and he goes on to mention "relaxation to chaos through period doublings," "a conspicuous arrow of time in reversible microscopic dynamics," "causality and light-cone," and others. The purpose of this paper is to describe a new kind of automaton that

†This work was supported in part by NSF Grant ECS-8307955, U. S. Army Research–Durham Grant DAAG29-82-K-0095, DARPA Contract N00014-82-K-0549, and ONR Grant N00014-83-K-0275.
‡Now with the Laboratory for Computer Science, Massachusetts Institute of Technology, Cambridge, MA 02139, USA.

supports soliton-like structures in a strikingly clear way.

Scott *et al.* [3] propose as a working definition that "a soliton $\phi_s(x - ut)$ is a solitary wave solution of a wave equation which asymptotically preserves its shape and velocity upon collision with other solitary waves." The simplest examples are provided by solutions to the dispersionless linear wave equation. What is remarkable about solitons is that they can be supported by nonlinear equations with dispersion.

The notion of a solitary wave can be carried over in a natural way to the context of automata (either CA or FA) as follows: The term *solitary wave* or *particle* in an automaton will be taken to mean a periodic pattern of non-zero cell values that propagates with fixed finite velocity. A collision between two particles will be said to be a *soliton collision* if the particles retain their identities after the collision. (For an example see Fig. 3.)

The particular class of filter automata described here, which we call *parity-rule* FA's, support thousands of particles of relatively small size, are irreversible, and totalistic – that is, they belong to the simple class of automata that depend for their next state only on the number of 1's in the argument field of the next-state function (see below). Furthermore, as we will see in what follows, soliton collisions are quite common, occurring for some next-rule radii and ranges of particle widths 99% of the time. In contrast, in the totalistic one-dimensional CA's studied extensively by Wolfram [5] particles are relatively rare and non-destructive collisions extremely rare. There are also two-dimensional CA's that support particles – for example the gliders in the game of Life [7], or the billiard-ball models in the reversible CA's described by Margolus [4]. But the ease with which particles are supported by parity-rule FA's, and their propensity for passing through one another, appear to be unknown in the study of CA's.

2. Filter automata

We will restrict ourselves here to one-dimensional automata with k-valued site values a_i^t, where the subscript i refers to the space variable ($-\infty \le i \le +\infty$), and the superscript t refers to time ($0 \le t \le +\infty$). In the usual CA [5], the evolution of the automation is determined by a fixed rule F of the form

$$a_i^{t+1} = F\big(a_{i-r}^t, a_{i-r+1}^t, \dots, a_i^t, \dots, a_{i+r}^t\big), \qquad (1)$$

with

$$F(0,0,\dots,0) = 0.$$

The next value of site i is a function of the previous values in a neighborhood of size $2r+1$ that extends from $i-r$ to $i+r$. Given initial states at all the sites, which we assume run from

$-\infty$ to $+\infty$, repeated application of the rule F determines the time evolution of the automaton.

In an FA, the next-state rule is of the form

$$a_i^{t+1} = F\big(a_{i-r}^{t+1}, a_{i-r+1}^{t+1}, \dots, a_{i-1}^{t+1}, a_i^t, \dots, a_{i+r}^t\big). \qquad (2)$$

Now the next state is computed using the newly updated values $a_{i-r}^{t+1}, a_{i-r+1}^{t+1}, \dots, a_{i-1}^{t+1}$, instead of $a_{i-r}^t, a_{i-r+1}^t, \dots, a_{i-1}^t$. This is precisely analogous to the operation of an IIR digital filter, whereas a CA corresponds to an FIR digital filter (see, [6], for example).

Although we allow the sites in an FA to extend from $-\infty$ to $+\infty$, we must assume that to the left, anyway, there are only a finite number of sites containing non-zero values. This will then give us an unambiguous way to compute the evolution of the FA, using a left-to-right scan. We always start with an initial configuration that has only a finite number of non-zero site values.

Following Wolfram's terminology [5], when the next-state function F depends only on the sum

$$S(i) = \sum_{j=-r}^{r} a_{i+j} \qquad (3)$$

we say an automation is *totalistic*. This class, although small and easy to specify, appears to exhibit all the interesting kinds of behavior found in general automata, and the particular class of FA's described here will be totalistic.

We will focus attention on the class of filter automata with binary-valued sites ($k=2$) defined by the following next-state rule. If $S(i)$ is the number of 1's in the $i-r$ to $i+r$ window at time t, then the new value of site i is

$$a_i^{t+1} = \begin{cases} 1, & S \text{ even but not } 0, \\ 0, & \text{otherwise.} \end{cases} \qquad (4)$$

These we will call the *parity-rule* filter automata, and we will think of them as parameterized by the single integer $r = 2, 3, \dots,$ the *radius*.

3. Some examples

Fig. 1 shows a typical particle, one that occurs in the $r = 3$ parity-rule FA. The first line corresponds to generation 0 – it indicates the initial values assigned to the array of sites (a black square indicates a site with value 1, and the absence of such a square indicates a site with value 0). Subsequent lines correspond to subsequent generations. It is apparent from this figure that a particle has a well-defined *period* (the number of generations needed for the bit pattern to repeat), and *displacement* (the number of sites moved during a period, with plus measured to the left). In this case the period is 3 and the displacement 1.

We will refer to the succession of states passed through by a particle as its *orbit*. A convenient way to identify a particle uniquely is to view each orbital state as a binary number, and to take the smallest binary number in its orbit as the *canonical code* or *generator* of the particle.

It is not hard to see that the position of the rightmost 1 of a particle can never move right. For this to happen, we must be in the situation where the values of $a_i^t, \ldots, a_{i+r}^t$ are all 0, and there are an even nonzero number of 1's among the values $a_{i-r}^{t+1}, \ldots, a_{i-1}^{t+1}$. As the window slides right, this situation must be repeated, and so an infinite number of 1's would be generated, a contradiction. In fact it has been proved [11] that such situations can never be reached from an initial condition with a finite number of 1's; that is, that the parity-rule FA's are *stable* in this sense.

From the previous observation, we know that particles are either stationary or move left. It is also not hard to see that the maximum speed of a particle is $r - 1$, and this is realized by the particle consisting of $r + 1$ consecutive 1's, which has period 1.

As Hirota and Suzuki [10] describe, one characteristic of solitons is that "A wave packet at any given position dissolves into many solitons each of which travels at its own velocity." Fig. 2 shows a typical evolution from a disordered state for the $r = 3$ parity-rule FA. Exactly the same kind of

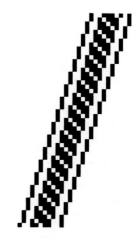

Fig. 1. A typical particle supported by a parity-rule filter automaton, illustrating period and displacement. This example is for $r = 3$, has canonical code 629, displacement 1, and period 3. The code sequence in its orbit is 629, 697, 1241.

dissolution into several particles with different velocities can be observed.

Fig. 3 illustrates pairwise collisions that we call *soliton* – those in which the identity of both particles is preserved, while fig. 4 shows examples of non-soliton collisions. Extensive empirical evidence suggests that in a soliton collision the fast particle cannot be shifted to the right, and the slow particle cannot be shifted to the left. That is, the fast particle may only be pushed forward, and the slow only retarded. The latest collision in fig. 4 results in 2 particles moving in parallel and is particularly interesting because it shows that particle collisions are not always reversible. By *reversible* here, we mean that the picture rotated 180° would be a valid evolution of this or some other automaton. If we turn this picture upside down, it becomes clear that spontaneous splitting would be required for reversibility.

We next want to present some statistics for pairwise particle collisions, but first we need to study the maximum possible number of different

J.K. Park et al. / Soliton-like behavior in automata

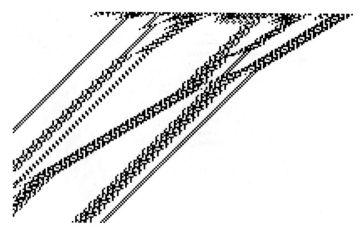

Fig. 2. Typical evolution from a disordered state, for the $r = 3$ parity-rule filter automaton, showing dissolution into several particles with different speeds.

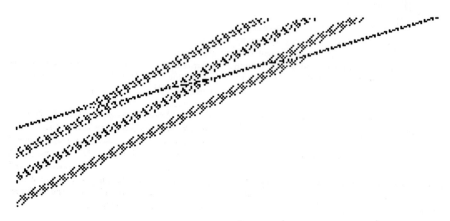

Fig. 3. Some typical soliton collisions, for the $r = 5$ parity-rule filter automaton. The initial particle canonical codes and displacement/periods are, from left to right: 145 (12/6), 201 (12/6), 273 (12/6), and 27 (7/2).

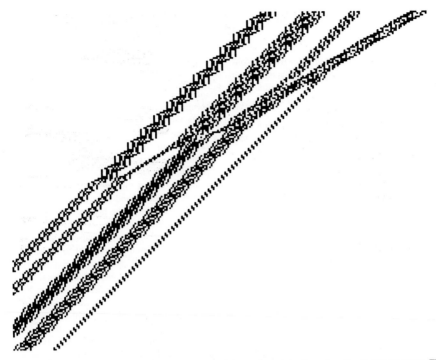

Fig. 4. Some typical collisions in which the identities of both particles are changed, for the $r = 3$ parity-rule filter automaton. The initial particle canonical codes and displacement/periods are, from left to right: 601 (8/8), 9451 (10/10), 43 (6/6), and 967 (20/12).

ways two particles can collide, given their periods and displacements.

4. The determinant of a particle pair

Given any two particles that move at different speeds, it is clear that we can arrange a collision between them by choosing an initial configuration with the faster particle to the right of the slower. If the two particles start close enough together, it may happen that they interact in a way that is impossible when they start far apart. In such cases we say the collision is *improper*; otherwise we say it is *proper*. We will restrict our attention to

proper collisions, because we will always allow an initial spacing adequate for typical interactions.

We will say that two collisions are the *same* if the bit patterns of their history can be put in concordance by shifts in space and time, and *different* otherwise. We can now calculate a strict limit on the number of different proper collisions possible between two particles.

Theorem 1. Let the two particles have displacements d_1, d_2, periods p_1, p_2, and speeds $d_1/p_1 < d_2/p_2$, so that particle 2 hits 1. Let $q = \mathrm{lcm}(p_1, p_2)$, and the difference in speeds be $\Delta s = d_2/p_2 - d_1/p_1$. Then the number of different proper collisions is no larger than

$$DET = p_1 p_2 \cdot \Delta s = p_1 d_2 - p_2 d_1.$$

Proof. After q generations, the same relative configuration of bits in the particles' orbits repeats, so we need only cycle through $q \cdot \Delta s$ initial separations to obtain all possible proper collisions.

Case 1: p_1, p_2 relatively prime. In this case those $q \cdot \Delta s = p_1 p_2 \cdot \Delta s$ situations include all combinations of orbital states.

Case 2: p_1, p_2 not relatively prime. Then in $q \cdot \Delta s$ situations we have covered only $q/(p_1 p_2)$ of the total possible. Therefore there are at most $q p_1 p_2 \cdot \Delta s / q = p_1 p_2 \cdot \Delta s$ possible situations, as before.

We call DET the *determinant* of the collision, because it is precisely the 2×2 determinant with rows $p_1 p_2$ and $d_1 d_2$.

5. Particle and collision statistics

Dictionaries of particles were compiled in the following simple way. All bit patterns with width up to 16 were used as starting configurations of the parity-rule FA's, and the automata played forward in time long enough for particles to be generated and separated. The resulting bit strings were than analyzed to catalog the canonical codes, speeds, displacements, and periods of the particles so generated. Table I shows histograms of the results, sorted by speed for each value of r, for all distinct displacement-period pairs, and for all canonical codes of width ≤ 16.

There is a somewhat hazy distinction between two particles with the same speed traveling in parallel, and one particle. For our purposes we insist that a single particle not have a gap of $2r$ or more consecutive 0's in its canonical code.

The number of particle generators increases sharply with r, growing from only 8 for $r = 2$ to 13109 out of a possible 32768 – the number of odd integers up to 2^{16} – for $r = 6$. Another striking fact

is the tendency for certain displacement/period pairs to be preferred. For example, for $r = 6$ the two pairs $d/p = 44/12$ and $56/14$ account for 65% of the total.

These particle dictionaries were used to study the question of how often a 2-particle collision is *soliton*; that is, how often the two particles pass through one another without change in their identities. Given a set of particles, all possible pairwise collisions were sampled uniformly as follows: a random pair (a, b) was chosen, with b faster than a and all such pairs equally likely. Then the fast particle was played a random number g generations forward in time, where g was chosen uniformly between 0 and the period of the fast particle less 1; that is, the fast particle was put in a random orbital state. The slow particle was then placed in its canonical orbital state with its right end a random number x spaces to the left of the left end of the fast particle, where x was chosen uniformly between k and $k + DET - 1$, k being large enough to ensure a proper collision. Finally, the result of the ensuing collision was weighted by its corresponding DET to make the sample uniform over all possible collisions, rather than over all possible pairs of particles. This was done for 2000 collisions, for various values of radius r and for particle dictionaries with code-widths up to 10, 14, and 16. The results are shown in table II.

The general trend is that the estimated probability of soliton collisions increases with r for fixed code-widths, and decreases with code-width, although there are exceptions. What is perhaps most striking is the high level of the observed frequencies, reaching 99% for code-widths up to 10, and $r = 8$ and 9. It was also the case that mutual annihilation was never observed; every collision resulted in at least one particle.

6. Quasi-equivalence of CA's and FA's

A natural question is whether FA's are essentially different from CA's, or whether any FA can be simulated in some sense by a CA. In this

Table I

Speed, displacement, period, and frequency of occurrence for all particles with canonical code width ≤ 16, radius 2 to 6.

Speed	Disp.	Per.	Freq.	Speed	Disp.	Per.	Freq.	Speed	Disp.	Per.	Freq.
Radius = 2, no. of pars. = 8				*Radius = 4, no. of pars. = 682*				*Radius = 6, no. of pars. = 13109*			
0.000	0	1	1	2.333	35	15	2	1.333	4	3	4
0.500	1	2	1	2.500	5	2	1	2.500	5	2	3
0.500	2	4	1	2.500	25	10	14	2.500	10	4	19
0.500	3	6	1	2.571	18	7	8	2.500	15	6	15
0.500	8	16	1	2.571	36	14	68	2.500	20	8	195
1.000	1	1	3	2.750	11	4	5	2.500	25	10	8
				2.750	22	8	4	3.200	16	5	30
Radius = 3, no. of pars. = 198				2.750	44	16	4	3.200	32	10	1635
0.333	1	3	1	3.000	3	1	1	3.375	27	8	95
0.500	1	2	1					3.667	11	3	5
0.500	4	8	4	*Radius = 5, no. of pars. = 6534*				3.667	22	6	35
1.000	1	1	2	1.000	1	1	1	3.667	44	12	4157
1.000	2	2	1	1.000	3	3	3	3.900	39	10	213
1.000	4	4	4	2.000	2	1	3	4.000	28	7	21
1.000	5	5	1	2.000	4	2	6	4.000	56	14	4353
1.000	6	6	3	2.000	8	4	20	4.250	17	4	5
1.000	8	8	11	2.000	10	5	2	4.250	34	8	7
1.000	10	10	41	2.000	12	6	10	4.250	51	12	231
1.000	16	16	8	2.000	16	8	110	4.250	68	16	1742
1.333	8	6	2	2.000	20	10	4	4.444	40	9	1
1.333	16	12	63	2.600	13	5	41	4.444	80	18	227
1.400	7	5	3	2.600	26	10	926	4.500	9	2	1
1.400	14	10	17	2.750	11	4	3	4.500	63	14	102
1.500	3	2	1	2.750	22	8	46	4.600	23	5	3
1.500	6	4	1	3.000	3	1	2	4.833	29	6	1
1.500	12	8	4	3.000	6	2	1	5.000	5	1	1
1.571	22	14	16	3.000	9	3	51				
1.667	5	3	4	3.000	12	4	4				
1.667	10	6	3	3.000	18	6	69				
1.667	20	12	3	3.000	36	12	2251				
1.750	14	8	1	3.000	45	15	12				
2.000	2	1	3	3.200	32	10	71				
				3.286	23	7	53				
Radius = 4, no. of pars. = 682				3.286	46	14	1929				
0.667	2	3	2	3.500	7	2	2				
1.500	3	2	4	3.500	14	4	35				
1.500	6	4	10	3.500	21	6	3				
1.500	9	6	6	3.500	28	8	32				
1.500	12	8	44	3.500	35	10	1				
1.500	15	10	2	3.500	42	12	37				
2.000	2	1	1	3.500	56	16	693				
2.000	4	2	1	3.500	70	20	4				
2.000	10	5	14	3.667	33	9	7				
2.000	20	10	219	3.667	66	18	96				
2.125	17	8	19	3.800	19	5	2				
2.333	7	3	18	3.800	38	10	1				
2.333	14	6	19	3.800	76	20	2				
2.333	28	12	216	4.000	4	1	1				

Table II
Frequency of collisions that are soliton, in *per cent*. Based on samples of 2000 collisions.

Radius	Width ≤ 10	Width ≤ 14	Width ≤ 16
2	38.74	38.74	38.74
3	24.63	11.02	9.55
4	65.63	30.50	35.38
5	73.97	45.13	45.20
6	80.35	80.76	52.67
7	80.62	84.51	
8	99.78	83.63	
9	99.42	78.16	

section we will show that the latter is true; in particular, we will show that the space-time array generated by any particular FA can also be generated by some CA, and vice-versa.

First consider the constraints imposed on the order in which site values in FA and CA space-time arrays can be generated. Any site value a_i^t in an FA space-time array is determined by $a_{i-r}^t, \ldots, a_{i-1}^t, a_i^{t-1}, \ldots, a_{i+r}^{t-1}$. These site values are in turn determined by the values of other sites, but all sites that could possibly affect a_i^t lie within the region $\{a_{i+j}^{t-k}: k \geq 0, j \leq kr\}$ (this is the shaded region in fig. 5). For the space-time array of a CA, the corresponding region of sites whose values can affect a site value a_i^t is depicted in fig. 6 – it is just the region $\{a_{i+j}^{t-k}: k \geq 0, -kr \leq j \leq kr\}$.

It is clear from figs. 5 and 6 that there exists no 1-to-1 mapping from site values in an FA space-time array to site values in a CA space-time array, with the property that the orientation of the space axis is preserved. Changing the value of some site to the left of a_i^t (but in the same generation) in an FA space-time array can affect this site value, but it cannot in a CA space-time array. Thus, if we wish to simulate an FA with a CA, we must somehow change the orientation of the space axis. This motivates a "tilting" mapping from the FA space-time array to the CA space-time array, a mapping that rotates the region in fig 5 so that it looks more like the region of fig. 6.

Suppose then that we are given an FA, with parameters r and k, and we want to generate the

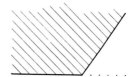

Fig. 5. The region that can affect a point in an FA.

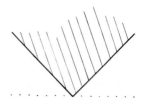

Fig. 6. The region that can affect a point in a CA.

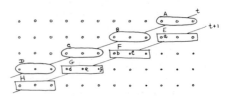

Fig. 7. Illustration of the simulation of an FA by a CA, for the case $r = 2$. The space axis of the CA is tilted, and each cell of the CA is a conglomerate of $(r + 1)$ cells of the FA. To compute the value of CA cell G, we compute in turn the value of FA cell a, b, c, d, e and f.

same array of space-time values using a CA, using parameters r' and k'. The mapping we will use is illustrated in fig. 7 for the case $r = 2$. First, groups of three consecutive cell values in a particular generation of the FA array are coalesced to form one cell in the corresponding CA, so that the CA has $k' = 2^{r+1}$. Next, we take the space axis (constant-time axis) of the CA to be tilted so that cells in the same CA generation correspond to cells along a line rotated counterclockwise from horizontal in the FA space-time array.

Using capital letters to represent the cells in the CA, and lower case for the FA, we next show how the value of the cell G in generation $t + 1$ can be computed from the values in the neighboring cells A, B, C, D in generation t. First, find the value of the FA cell a from the FA components in A and B. Next find the values of cells b and c from the value of a and the components of B and C. Finally, find the values of FA cells d, e, and f from the values of b, c and the components of C and D. This then gives the value of the CA cell G. We can summarize this in the following

Theorem 2. Every space-time array that is generated by an r, k – FA can also be generated by a CA with $r' = r$ and $k' = 2^{r+1}$, provided that we allow the values to be computed in a different order. The FA cell values are also coded in groups of $r + 1$ alphabet symbols.

Next consider the problem of simulating a CA with an FA. A glance at fig. 8 shows that this is easy, provided that we allow a slippage to the left: just compute the value at site $x + r$ at the point x, and choose the FA rule to depend only on the values in the preceding generation. Each generation will therefore be shifted r cells to the left with respect to the CA array, but will otherwise be identical; in this case the time axis is tilted. We summarize this as

Theorem 3. Every space-time array that is generated by a r, k' – CA can also be generated with an

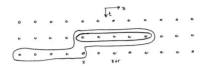

Fig. 8. Illustration of the simulation of a CA by an FA. The value at cell $x + r$ is computed at point x in the FA, so that the time axis of the FA is shifted to the left r cells every generation.

FA with $r = 2r'$ and $k = k'$, provided that we allow each successive row in the FA to be displaced r units to the left with respect to the corresponding row in the CA.

7. Discussion

Many questions about the class of filter automata remain unexplored. Some rules other than the parity rule appear to support particles the way the parity rule does. For example, the following variation of the parity rule seems to support particles and soliton collisions, and the particles are on the average slower:

$$a_i^{t+1} = \begin{cases} 1, & S \text{ even but not 0 or 2,} \\ 0, & \text{otherwise.} \end{cases} \qquad (5)$$

Other rules are unstable in the sense described in section 3. While the question of stability in linear Infinite Impulse Response digital filters is settled by the criterion of characteristic values being inside the unit circle, an analogous general technique for FA's is unknown.

One of the motivations for studying cellular automata in general is to gain insight into the nonlinear phenomena that occur in the solution of differential equations, and in the physical systems they are used to model. We have seen that certain simple one-dimensional automata give rise to solitary waves that very often pass through one another non-destructively. Whether such automata capture the mechanism of soliton generation in differential equations is uncertain, but we feel worthy of further study.

Another reason for studying CA's is the possibility of embedding useful computation within such regular and simple structures. One-dimensional cellular automata can be implemented in VLSI in a highly pipelined and efficient way [8], resulting in what amounts to a cellular automaton machine with an almost unlimited degree of parallelism. Even if the embedding of a useful computation is very inefficient, it may still be more than compensated for in certain applications by the parallelism and efficiency of the VLSI implementation. There exist complex CA's that simulate a universal Turing machine, but we still do not know how to construct a simple one-dimensional CA that does useful computation. It appears, however, that it will help to have particles that can pass through one another, because that will make possible communication between different elements of the automaton.

Carter [9] describes the transmission of information in molecules using physically supported solitons. The general notion of processing information in simple, homogeneous media via solitons opens the way for speculation about such computation at the level of the molecule or the biological cell.

Acknowledgements

We thank Irfan Kamal and Arthur Watson, who helped in many ways, especially with the development and testing of programs. Doug West and Stephen Wolfram took the time to offer useful suggestions.

References

[1] D. Farmer, T. Toffoli and S. Wolfram, Cellular Automata (North-Holland, Amsterdam, 1984).
[2] G.Y. Vichniac, "Simulating Physics with Cellular Automata," pp. 96–116 in [1].
[3] A.C. Scott, F.Y.F. Chu, D.W. McLaughlin, "The Soliton: A New Concept in Applied Science," Proc. IEEE, vol. 61, no. 10, pp. 1443–1483, October 1973.
[4] N. Margolus, "Physics-Like Models of Computation," pp. 81–95 in [1].
[5] S. Wolfram, "Universality and Complexity in Cellular Automata," pp. 1–35 in [1].
[6] A.V. Oppenheim, R.W. Schafer, Digital Signal Processing (Prentice-Hall, Englewood Cliffs, NJ, 1975).
[7] E.R. Berlekamp, J.H. Conway and R.K. Guy, Winning Ways for your Mathematical Plays, Vol. 2: Games in Particular, (Chapter 25, "What is Life?") (Academic Press, New York, NY, 1982).
[8] K. Steiglitz and R.R. Morita, "A Multi-Processor Cellular Automaton Chip," Proc. 1985 IEEE Int. Conf. on Acoustics, Speech, and Signal processing, Tampa, Florida, March 1985.
[9] F.L. Carter, "The Molecular Device Computer: Point of Departure for Large Scale Cellular Automata," pp. 175–194 in [1].
[10] R. Hirota and K. Suzuki, "Theoretical and Experimental Studies of Lattice Solitons on Nonlinear Lumped Networks," Proc. IEEE, vol. 61, no. 10, pp. 1483–1491, October 1973.
[11] C.H. Goldberg, manuscript in preparation.

J. Phys. A: Math. Gen. **17** (1984) L415–L418.

LETTER TO THE EDITOR

Invariant in cellular automata

Y Pomeau*

Schlumberger-Doll Research, PO Box 307, Ridgefield, CT 06877, USA

Received 7 March 1984

Abstract. For a subclass of the set of reversible cellular automata, we give the form of an exact time invariant quantity, which can be seen as a kind of energy. With the ergodic assumption, this is a model of two interacting Ising spin systems.

There is at present a growing interest in cellular automata (Vichniac 1984, Wolfram 1984, Hayes 1984), owing in particular to their physical realisation by inexpensive electronic hardware (Toffoli 1984, see also Margolus 1984). Some of these systems are even able to model dynamical systems such as the flow equations (Hardy *et al* 1976). In this letter we explain the computation of a non-trivial invariant for a class of reversible models invented by Fredkin (see Vichniac 1984). This invariant may be seen as a kind of energy, although its derivation does not use the methods of Hamiltonian mechanics.

Let us consider a lattice of points, regular or not. Indeed finite periodic lattices can be considered as imbedded in infinite periodic lattices. We shall not, therefore, worry about the problem of the thermodynamic limit that does not appear explicitly in our formal computations.

Let us define at each site of this lattice two Boolean variables σ_i and $\hat{\sigma}_i$, where i is the site index. At a given (discrete) time each of these quantities takes either the value 0 or 1. Let ν_i be a neighbourhood of i such that if $j \in \nu_i$ then $i \in \nu_j$. We shall refer to this as the property of symmetry of the neighbourhoods. Now one introduces a deterministic rule for computing the σ's and $\hat{\sigma}$'s at time $(t+1)$ from the σ's and $\hat{\sigma}$'s at time t. Actually, the reversible rules that we want to consider were given originally as two time step rules with a single Boolean variable at each lattice site. However, this may be readily transformed into single time step rules with two Boolean variables at each site. The class of rules that we shall consider has the general form

$$\hat{\sigma}_i^{t+1} = \sigma_i^t \tag{1a}$$

$$\sigma_i^{t+1} = \hat{\sigma}_i^t + A_i^t - 2\hat{\sigma}_i^t A_i^t \tag{1b}$$

or

$$\sigma_i^{t+1} = \hat{\sigma}_i^t + A_i^t(1 - 2\hat{\sigma}_i^t) \tag{1c}$$

where t is the discrete time index and A_i is a function with Boolean values of the σ_j's in the neighbourhood ν_i of site i. The algebra will now be simplified by using the concept of truth value of a statement S, denoted as $(S)_T$. If S is true, then $(S)_T = 1$, otherwise $(S)_T = 0$. Now consider the cases where $A_i = (\Sigma_{j \in \nu_i} \sigma_j = q_i)_T$ where q_i is a natural integer less than or equal to the cardinality of ν_i. Thus the right-hand side of

(1*b*) or (1*c*) is equal to $(A_i' \neq \hat{\sigma}_i')_T$. For an explicit rule, such as the Q2R rule of Vichniac (1984) ($q_i = 2$ in this rule, which explains the 2 in Q2R, the meaning of Q and R will be given below) A_i can be written as a symmetric polynomial with integer coefficients on the σ's in ν_i.

We now want to prove that the quantity

$$\Phi' = \sum_{i,j \in \nu_i} \sigma_i' \hat{\sigma}_j' - \sum_i (\sigma_i' + \hat{\sigma}_j') q_i \tag{2}$$

does not depend on time. In equation (2) the products and sums are to be understood in the usual sense of the operations on integers and not mod 2, although all variables are Boolean. Thus if the cardinality of ν_i is uniformly bounded, Φ' is of the order of the number of sites for large lattices.

To show that Φ is t-independent, let us compute Φ^{t+1}. With rule (1) one has

$$\Phi^{t+1} = \sum_i (\hat{\sigma}_i' + A_i'(1 - 2\hat{\sigma}_i')) \left(\sum_{j \in \nu_i} \sigma_j' - q_i \right) - q_i \sigma_i'. \tag{3}$$

However, the product $A_i'(\sum_{j \in \nu_i} \sigma_i' - q_i)$ is equal to zero because A_i' is the truth value of $(\sum_{j \in \nu_i} \sigma_i' = q_i)$. Thus

$$\Phi^{t+1} = \sum_{i,j \in \nu_i} \hat{\sigma}_i' \sigma_j' - \sum_i q_i (\sigma_i' + \hat{\sigma}_i'). \tag{4}$$

From the symmetry of the neighbourhoods $j \in \nu_i \Leftrightarrow i \in \nu_j$, one may interchange $\hat{\sigma}$ and σ in the quadratic term on the right-hand side of (4) to obtain finally $\Phi' = \Phi^{t+1}$. This derivation raises some questions that we shall now comment on.

(1) Are there other invariants? This is certainly so in a quite trivial sense. As shown by Vichniac (1984) the rule Q2R is consistent with a time dependent behaviour strictly confined in a fixed region of a two-dimensional lattice. Thus by an obvious argument of translational invariance the splitting of the phase space by the invariant is certainly less fine than the one defined by the dynamics. In this case, this points to the existence of local invariants.

(2) Vichniac's set of rules lead us to consider the following type of possible choice for A_i

$$A_i = \left(\sum_{j \in \nu_i} \sigma_j = p_i \right)_T + \left(\sum_{j \in \nu_i} \sigma_j = q_i \right)_T$$

with $p_i \neq q_i$, both p_i and q_i being natural integers less than or equal to the cardinality of ν_j. So the question is: is there an invariant as Φ for this class of rules?

(3) It is quite natural to look for possible connections between this class of automata and other models of statistical mechanics, such as the Ising spin system. Consider for instance the QR rules on a regular square lattice. In this case ν_i is the set of the four (quatre in French, this explains the Q in QR, R being for 'reversible') nearest neighbours of site i. Furthermore, let us put $s = \sigma - \frac{1}{2}$ and $\hat{s} = \hat{\sigma} - \frac{1}{2}$, so that s and $\hat{s}$ are the usual spin-half variables of an Ising model. The invariant becomes

$$\Phi = \sum_{(i,j)} s_i \hat{s}_j + \sum_i (2 - q_i)(s_i + \hat{s}_i) + \sum_i (1 - q_i)$$

where (i, j) means summation over all distinct pairs of nearest neighbours. This is the energy of two distinct Ising models, each model having spins s over a sublattice and spins $\hat{s}$ on the other sublattice. The quantity $(2 - q_i)$ plays the role of an external magnetic field. This could be used to model a deterministic dynamics of the Ising model. Note, however, that as previously mentioned in point (1), such a deterministic dynamics is not necessarily ergodic. It could be ergodic in some weak sense for large systems with 'random' initial conditions, this being enough to make statistical mechanics meaningful in those large systems.

(4) The extension of this to automata on lattices of an arbitrary dimensionality is straightforward. It is also of interest to notice that one may extend the definition of A_i as

$$A_i = \left(\sum_{j \in \nu_i} J_{ij} \sigma_j = q_i \right)_{\mathrm{T}}$$

where J_{ij} are integers such that $J_{ij} = J_{ji}$ (the previous condition of symmetry of the neighbourhoods is a particular formulation of this condition), and where the q_i's are now restricted by

$$|q_i| \leq \sum_{j \in \nu_i} |J_{ij}|.$$

Then the corresponding invariant is

$$\Phi = \sum_{i,j \in \nu_i} J_{ij} \sigma_i \hat{\sigma}_j - \sum_i q_i (\sigma_i + \hat{\sigma}_i).$$

(5) As there is an invariant quantity as time goes on and if the usual assumptions of thermohydrodynamics work, the long wavelength perturbations of the invariant relax according to the Fourier heat equation. The heat conductivity that appears in this equation is given by a Green–Kubo expression, that can be derived as in Hardy *et al* (1974) the shear viscosity of the lattice gas model. Let us give this form of the heat conductivity for the case $q = 2$ and $J_{ij} = 1$ and for a regular lattice. To do this we shall need some more notations. We shall assume that the Boltzmann–Gibbs statistical weight has the usual form: $Z^{-1} \exp(-\Phi/\Theta)$, Z being the partition function Φ the energy as given in equation (4) for $q_i = 2$ and Θ the temperature measured with the same (dimensionless) units as Φ. Furthermore, the formal expression of the α Cartesian component of the microscopic heat flux at time τ and site i is

$$J_{i,\alpha} = \tfrac{1}{2} \sum_{j \in \nu_i} r_{ij,\alpha} (\sigma_i^\tau \hat{\sigma}_j^\tau - \sigma_j^\tau \hat{\sigma}_i^\tau)$$

where $r_{ij,\alpha}$ is the α component of the vector $\mathbf{r}_{ij} = \mathbf{r}_j - \mathbf{r}_i$ whose ends are at neighbouring sites i and j of the lattice. With all these notations the formal expression of the heat conductivity reads

$$\kappa_{\alpha\beta} = \frac{1}{\Theta^2} \sum_{t=0}^{\infty} \left\langle J_{i,\alpha}(0) \sum_k J_{k,\beta}(t) \right\rangle$$

where the average is taken over a Boltzmann–Gibbs distribution of initial conditions for the dynamics in phase space. The heat conductivity tensor is defined in such a way that the heat transport equation reads

$$\frac{\partial \Phi(\mathbf{r}, t)}{\partial t} = \sum_{\alpha,\beta} \kappa_{\alpha\beta} \frac{\partial^2 \Theta}{\partial r_\alpha \, \partial r_\beta}$$

where $\Phi(\mathbf{r}, t)$ is the local value of the energy per site. This is related to the local temperature through the equilibrium equation of state in the near equilibrium situations where the Fourier equation is valid. On the other hand it is well known (Pomeau and Resibois 1975) that transport coefficients often diverge in two dimensions. If one assumes that there is only one global conserved quantity, Φ, in this model it does not seem that those divergences affect the previous expression of the heat conductivity.

(6) A very fascinating result, among many others, was proved by Onsager for the 2D Ising model at the Curie temperature: pair correlation functions between spins on the same line are rational numbers. It has been conjectured since, but as far as we know never proved that any equilibrium correlation is a rational number in the same conditions. If this is true, an immediate consequence is that any time correlation function of the equilibrium fluctuations of our model are also given by rational numbers at the Curie temperature when the cellular automaton is equivalent to two Ising models on a square lattice.

This work was initiated at the 1984 conference on 'Physics and Computation' held at Drake's Anchorage, where I was introduced to the problem of the invariants of the reversible rules by Gerard Vichniac. I have greatly benefited from discussions with him and with Charles Bennett, and have been very much inspired by the work of Tom Toffoli and Norm Margolus with the CAM machine.

References

Hardy J, de Pazzis O and Pomeau Y 1976 *Phys. Rev.* A **13** 1949
Hayes B 1984 *Sci. Am.* to appear
Margolus N 1984 *Physica* D to appear
Pomeau Y and Resibois P 1975 *Phys. Rep.* **19C** 64
Toffoli T 1984 *Physica* D to appear
Vichniac G Y 1984 *Physica* D to appear
Wolfram S 1983 *Los Alamos Science* **9** 2

*Permanent address: SPhT, CEN-Saclay F-91191, Gif-sur-Yvette, Cedex, FRANCE.

Deterministic Ising Dynamics

Michael Creutz

Department of Physics, Brookhaven National Laboratory,
Upton, New York 11973

Received January 17, 1985; revised May 13, 1985

A deterministic cellular automaton rule is presented which simulates the Ising model. On each cell in addition to an Ising spin is a space–time parity bit and a variable playing the role of a momentum conjugate to the spin. The procedure permits study of nonequilibrium phenomena, heat flow, mixing, and time correlations. The algorithm can make full use of multispin coding, thus permitting fast programs involving parallel processing on serial machines. © 1986 Academic Press, Inc.

Introduction

Numerical simulations of statistical systems have become a major tool in the study of phase transitions and critical phenomena. Monte Carlo and molecular dynamics calculations represent two complimentary schemes for such simulations. In the Monte Carlo approach, one generates a Markov chain of configurations using a pseudo-random number generator. The algorithm is constructed, usually using a principle of detailed balance, so that the ultimate probability of encountering any particular configuration is proportional to the Boltzmann weight. The corresponding temperature is a parameter in the program; indeed, the computer is serving as a thermal reservoir at that temperature.

Molecular dynamics calculations, on the other hand, are an attempt to follow the deterministic evolution of a system under an appropriate microscopic Hamiltonian. This approach makes no use of random numbers, the apparent statistical nature of the whole system arising from the complexity of a large phase space. Such algorithms also do not utilize the temperature as an input parameter. Indeed, its value is found after the by fact using the equipartition of energy among the various degrees of freedom. For example, the average kinetic energy of a given molecule should be $\frac{1}{2}kT$ per degree of freedom.

Recently a simulation algorithm interpolating between the Monte Carlo and molecular dynamics techniques was presented [1]. This microcanonical Monte

* The submitted manuscript has been authored under Contract No. DE-AC02-76CH00016 with the U. S. Department of Energy. Accordingly, the U. S. Government retains a nonexclusive, royalty-free license to publish or reproduce the published form on this contribution, or allow others to do so, for U.S. Government purposes.

DETERMINISTIC ISING DYNAMICS   63

Carlo method consists of taking a random walk on a surface of constant energy. To simplify the process of maintaining the constraint that the energy be constant, one or more additional variables, called demons, serve to transfer energy around the system. These variables play a role analogous to the kinetic energy in molecular dynamics in that their average value gives a handle on the temperature of the system. Two advantages of this approach are that it can be programmed to run an order of magnitude faster than conventional Monte Carlo for discrete systems [2], and it does not require high quality random numbers. In addition, generalizing the scheme to several parameters beyond the "temperature," it provides a method for measuring these parameters. This should prove useful in Monte Carlo renormalization group calculations [3].

In this paper we investigate a variation on this microcanonical scheme as applied to the Ising model. Here, however, the "demon" variables do not move around the lattice, but become an integral part of the system. Each site of the lattice is tied to one such variable, which then plays the role of a momentum conjugate to the corresponding spin. Energy is no longer transferred around the lattice by the demons, but can only flow through the bonds via the intrinsic Ising interaction.

In this way we obtain a deterministic Ising dynamics which exactly conserves the total energy. Any localized region can heat or cool only by the transfer of energy from other parts of the lattice. In this respect the algorithm differs from the stochastic Ising dynamics presented by Glauber [4] or represented by conventional Metropolis simulation [5]. Indeed, those schemes contain a parameter representing the temperature. In essence the Glauber system is coupled to a heat reservoir with which energy can be exchanged. In our case, on the other hand, the temperature is a statistical concept which is only defined by averages, which may be over space, time, or both. Because the temperature of the system is internally determined, heat flow and thermal conductivity can be studied numerically. It is not clear that these concepts have any meaning in a conventional Monte Carlo simulation.

A particular advantage of the present scheme is that it is easily implemented by simple bit manipulation. All variables are small integers and no real numbers are used. From a practical point of view, this means that extremely fast programs using multispin coding are possible [2, 6]. This technique uses bit by bit boolean operations to permit parallel processing on a serial machine. The algorithm is also readily amenable to true parallel processing.

From a conceptual point of view, the approach is able to simulate a heat equation via an algorithm in which all bits used by the computer are of comparable importance. This is in sharp contrast to the use of floating point numbers, wherein the first bit of a word is more important that the last. As the heat equation is a rather generic partial differential equation, the computational advantages of this bit manipulation approach may have considerably wider application (a similar point of view has been expressed in [7]).

Our dynamics is set up formally as a collection of cellular automata [8]. Another cellular automaton dynamics discussed in [9], gave several exact results for a variation on the usual three-dimensional Ising model. The approach presented here differs in being a totally deterministic and reversible dynamics. Unfortunately the present dynamics does not appear to be exactly solvable even in the one dimensional case.

THE DYNAMICS

For simplicity we discuss a two-dimensional square lattice. The algorithm is readily generalized to any dimension or other lattice structures. Associated with each site i of the lattice are four binary bits. Time evolution is by discrete steps, with the values for the site variables at time-step $t+1$ being uniquely determined from their values and those of their nearest neighbors at time t. The updating rule, described in detail below, thus defines a system of deterministic cellular automata.

The first of the four bits on each site is the Ising spin. Considered as a bit taking the value 0 or 1, we denote this variable by B_i. When we wish to use the multiplicative representation of the Z_2 group, we write this variable as

$$S_i = 2_i B_i - 1 \in \{\pm 1\}. \tag{1}$$

The energy of the Ising model is

$$H_I = \sum_{\{i,j\}} S_i S_j, \tag{2}$$

where the sum is over all nearest neighbor pairs of lattice sites.

The next two bits on each site represent the demon or momentum variable conjugate to the spin. These bits represent a two bit integer taking values from 0 to 3. Denoting these bits by $D_{1,i}$ and $D_{2,i}$, we associate with them the kinetic energy

$$H_K = 4 \sum_i (D_{1,i} + 2D_{2,i}). \tag{3}$$

The factor of 4 is inserted because flipping any spin in Eq. (2) only changes the Ising energy by a multiple of four, and we wish to keep this property for the kinetic term as well. The updating algorithm presented below exactly conserves the total energy

$$H = H_K + H_I. \tag{4}$$

Actually the number of bits representing the momentum variable is arbitrary. From an analytic point of view, it might be simpler to consider an arbitrary positive integer. At the opposite extreme, one could consider only a single bit, although in this case the following dynamics in more than one dimension cannot change an isolated spin completely surrounded by antiparallel neighbors. We feel that keeping two bits is a reasonable compromise because in equilibrium the kinetic term will be excited with a Boltzmann weight. For temperature near the critical value in the two or three-dimensional models, a two bit demon will be fully excited only a few percent of the time and thus two bits are nearly equivalent to an infinity of them.

The fourth bit associated with each site gives the space–time parity of the site. The sole purpose of this bit is to implement a checkerboard style updating. This is a trivial way of circumventing the result of [10], stating that any cellular automaton rule which updates all spins simultaneously cannot simulate the Ising model. Here at each time step we only consider changing spins on that half of the sites that have a set parity bit. All these parity bits are then inverted for the next time–step. Although we refer to this bit as an extra variable, in practice the computer need not actually be storing its value for each site because of its rather trivial nature.

We now give the dynamical rules for updating the spin and momentum variables. When the parity bit for a step is reset ($=0$) the only change is to invert that bit for the next time–step. On the other hand, if the parity bit is set ($=1$), in addition we use the microcanonical rule of [1]. That is, first, the resulting change in the Ising energy of Eq. (2) upon a flip of the spin S_i is calculated. If this change can be absorbed in the momentum variable associated with the same site in such a manner that the total energy of Eq. (4) is exactly conserved, then both the spin is flipped and the momentum is appropriately changed. If, however, the kinetic term is unable to absorb the energy change, then both the spin and associated momentum remain unchanged.

As discussed in [1], on a large system the values of the kinetic variable should become exponentially distributed with the Boltzmann weight corresponding to the temperature $T = 1/\beta$ of the system. Thus we expect

$$P(E_i) \propto \exp(-4\beta E_i), \tag{5}$$

where we define $E_i = D_{1,i} + 2D_{2,i}$. Thus the expectation value of E_i gives a means of measuring the system temperature

$$\langle E_i \rangle = \sum_{n=0}^{3} n e^{-4\beta} \bigg/ \sum_{n=0}^{3} e^{-4\beta}. \tag{6}$$

This relation is easily inverted to find β. The expectation value in Eq. (6) can be taken either over time or some spatial region or both.

Because of the checkerboard updating implemented with the parity bits, the algorithm requires two time–steps to give every spin of the lattice a chance to change. Thus in comparison to ordinary Monte Carlo simulations, two steps correspond to one full sweep over the system variables.

We close this section by noting that in addition to being deterministic, this dynamics is reversible. A simple inversion of all parity bits between two time–steps will reverse the evolution and send the system exactly backwards through the initial sequence of configurations. This inversion of parity bits amounts to hitting either the red or the black squares of the checkerboard updating procedure twice in a row.

66   MICHAEL CREUTZ

Some Experiments

We now discuss several simple numerical experiments done on the two dimensional model. We use a fully multispin coded program on a CDC 7600 computer. As this uses 60 bit words, we keep our lattice of size 120 in one dimension so that a given row of the lattice occupies two words. In the following we make the other dimension 120 as well. We always work with periodic boundary conditions.

A first question is whether the model actually succeeds in reproducing the Ising model. Figure 1 shows the results of several runs in the vicinity of the critical point of the model. Here we plot the nearest neighbor correlation as a function of the inverse temperature. The points are the results of simulations with the deterministic dynamics and the curve represents the exact solution on an infinite lattice. The simulations represent the average over the last 18,000 of 20,000 time–steps. For initial conditions we took all spins ordered and all momenta as zero except for the second bit of all those lying on sites of even parity. The latter bits were initially set to unity randomly with a given probability for each run. If this probability is a fraction with a denominator which is a power of two, this initial setting of bits can be easily accomplished with logical operations on random words. In Fig. 1 the statistical errors on the measured points are comparable to the size of the dots. Note that the data agree well with the exact solution except near the critical point at $\beta = \frac{1}{2}\log(1 + \sqrt{2}) = 0.44068$. Here finite size effects are presumably coming into play.

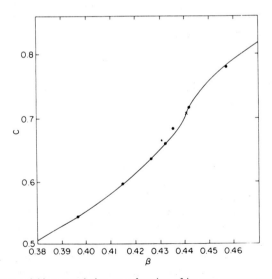

Fig. 1.   The nearest neighbor correlation as a function of inverse temperature for the two–dimensional Ising model. The solid line is the exact solution for an infinite lattice and the points are from simulations using the deterministic dynamcis on a 120 × 120 site lattice. The cross indicates the critical temperature and coupling.

DETERMINISTIC ISING DYNAMICS   67

We now turn to some experiments which conventional Monte Carlo could not do. In Fig. 2 we show the relaxation of a system where the initial energy distribution was not uniform. Here we initialized the lattice as above except on rows 31 to 90 the probability of setting second momenta bits on even sites was $\frac{3}{4}$ and on the remaining rows it was $\frac{1}{2}$. Thus the middle half of the lattice contained more energy per site, corresponding to a higher initial temperature. In Fig. 2 the profile of the temperature is plotted at various times. Each point is obtained from the expectation of the momentum variables averaged over five rows and 500 time–steps. For the average over the first 500 steps we see that the lattice center is substantially hotter than the edges. (Actually because the lattice is periodic, there is no true edge.) As time evolves, this temperature peak diffuses away. By 4000 iterations the initial peak is beginning to dissolve into the fluctuations in the local temperature.

We now consider placing a heat source and a heat sink in the lattice. For this

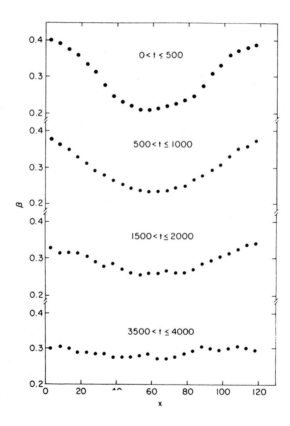

FIG. 2. The evolution of a thermal bump. The initial condition is described in the text. The top points represent the thermal profile averaged over the first 500 time–steps. Descending in the figure, we show $500 < t \leqslant 1000$, $1500 < t \leqslant 2000$, and $3500 < t \leqslant 4000$.

68   MICHAEL CREUTZ

experiment, after each two time–steps we randomize the first bit of all momentum variables on even sites in the first row of the lattice. This effectively couples this row to a high temperature heat source. At the same time we remove heat from row 61 by setting all the momenta on this row to zero. For our initial lattice we procede as before and set all spins and momenta to zero except the second bits of the momenta on even sites, which are set to one with probability $\frac{1}{2}$. In Fig. 3 we show the amount of heat entering and leaving the lattice as a function of time. The quantity plotted here is the change in the energy H per time–step and per spin in the source or the sink row. Each point is an average over 1000 time steps and is divided by two to correct for the two directions heat can flow around our periodic lattice. Note that after 10,000 updates the inflowing and outflowing heats match. Figure 4 shows the final steady state temperature profile as obtained by averaging the momenta in each row over 1000 updates after an initial 19,000.

From the slope in Fig. 4 we can determine a thermal conductivity. We define K by

$$Q = -K \, \Delta T / \Delta x. \qquad (7)$$

Here x is the distance through the lattice and Q is the heat flow entering row 1 per spin and per update. Note from the figure that in the high temperature region β is nearly linear in the distance through the lattice. This slope is approximately 0.0055 units in beta per lattice row. From Fig. 3 we see that the heat is flowing at a rate of about 0.016 units per time–step per site. Thus we obtain

$$K \approx 3\beta^2 \qquad (8)$$

for the thermal conductivity in the high temperature region. The increase in slope in Fig. 4 as the temperature drops indicates a rapid decrease in the conductivity as the

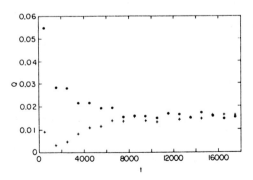

FIG. 3. For the hear flow experiment described in the text, the solid points represent the heat per spin and per time–step entering row 1 and the pluses represent the heat leaving row 61, both plotted as a function of time.

DETERMINISTIC ISING DYNAMICS   69

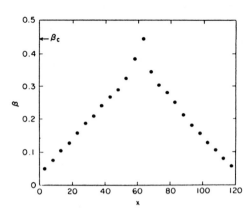

FIG. 4.   The steady state thermal profile of a 120 × 120 lattice heated at row 1 and cooled at row 61.

critical temperature is approached. Indeed, we have found that this conductivity
becomes quite small and difficult to measure below the critical point. It would be
interesting to have some theoretical predictions for the behavior of this quantity at
high and low temperatures.

Given any dynamical rule for evolution, one can ask for correlations between the
dynamical variables at different times. In Fig. 5 we show the correlation between a
spin and itself at a later time as a function of the time difference. Each plotted point
is an average over the lattice and over 5000 time–steps after an initial 1000 to
equilibrate. To obtain this average, we used the trick [11] of using two lattices
where the second is obtained by doing some number of iterations on the first. The
two lattices are then each updated independently and repeatedly compared. This
technique allows one to accumulate high statistics on lattices separated by a large
number of time steps but without storing a large number of intermediate lattices.
Note that the falloff of the correlation with time is initially quite rapid, while eventually a simple exponential behavior sets in. In this figure two values of beta are
shown. As might be naively expected, the run closer to the critical point has the
longer decorrelation time.

A good dynamics for studying statistical phenomena should give a path through
phase–space which is quite sensitive to small disturbances. Indeed, if two trajectories start near one another, they should rapidly diverge from each other if
statistical results are to be independent of initial conditions. This mixing
phenomenon is easily studied with the dynamics considered here. The correlation
between the spins on two lattices gives a simple definition of a distance between two
configurations. In Fig. 6 we show the evolution of the correlation between two lattices which initially differ only by one spin being flipped. After an initial 1000
time–steps to get a single lattice into equilibrium, all of its spins and moment are
copied into a second lattice. Exactly one spin in this second lattice is then flipped.

70   MICHAEL CREUTZ

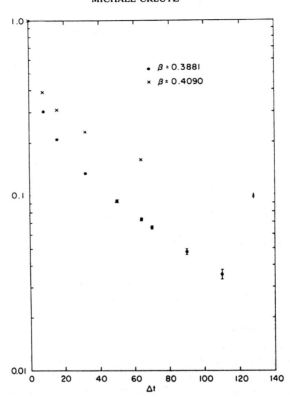

FIG. 5. The temporal correlation of a spin with itself as a function of the time between measurements. The solid points represent a lattice with $\beta = 0.3881$ and the crosses, $\beta = 0.4090$.

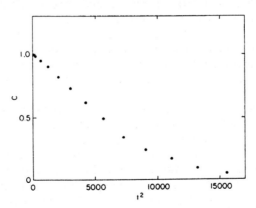

FIG. 6. As a function of time squared, the correlation between two lattices initially differing only in the value of a single spin. The quantity C represents the expectation of a spin on the first lattice times the corresponding spin on the second. The lattices are at $\beta = 0.4090$.

DETERMINISTIC ISING DYNAMICS   71

Finally, both lattices are subjected to the dynamics of this paper and compared. The points represent the correlation between corresponding spins in the two lattices; the measurements are averaged over 20 time–steps.

In Fig. 6 we have plotted the data versus t^2. This gives rise to a linear behavior at short times. To understand this, note that any disturbance can propagate into the lattice by at most one spacing in any unit time interval. Thus there is a "speed of light" or maximum velocity with which the effects of any disturbance can propagate. This means that the maximum dimension which can possibly be affected by our initial spin flip grows linearly with time. As our lattice is two dimensional, the volume included in this dimension grows with t^2. The observed behavior indicates that the disturbance we have introduced indeed grows at a constant velocity, although this speed appears to be somewhat less than the maximum possible of one site per update. This velocity is related to the Lyapunov exponent mentioned in [8]. With the model in d dimensions, this argument suggests that the initial behavior of this correlation will be linear in t^d.

Concluding Remark

We have presented a simple dynamical system which simulates the Ising model. An obvious question is whether this dynamics is ergodic. Indeed, it is easy to show that it is not. The rules for temporal evolution are symmetric under certain translations of the entire lattice and its momentum variables. This includes translations in any single coordinate direction by a multiple of two spacings, or a simultaneous translation in two directions by an odd number of sites in each. In addition the dynamics is symmetric to an inversion of all the lattice spins. Thus, if we start with a lattice configuration which is symmetric under any combination of such symmetries, it will remain so. Of course, conventional molecular dynamics calculations can have similar symmetries. For a generic continuous system, configurations carrying a preserved symmetry will represent a set of measure zero in the entire volume of phase–space. With a finite volume Ising system such configurations necessarily represent a finite part of the discrete phase–space, but this fraction should become insignificant as the volume goes to infinity. It would be interesting to know if this dynamics possesses further hidden symmetries beyond those mentioned above.

References

1. M. Creutz, *Phys. Rev. Letters* **50** (1983), 1411; "Proceedings of the Argonne National Laboratory Workshop on Gauge Theory on a Lattice," pp. 1–11, Argonne National Laboratory, Argonne, Ill., 1984.
2. G. Bhanot, M. Creutz, and H. Neuberger, *Nuc. Phys.* B **235** [FS11], (1984) 417; M. Creutz, P. Mitra, and K. J. M. Moriarty, *Comput. Phys. Comm.* **33** (1984), 361; preprint BNL–35470, 1984.
3. M. Creutz, A. Gocksch, M. Ogilvie, and M. Okawa, *Phys. Rev. Lett.* **53** (1984), 875.
4. R. J. Glauber, *J. Math. Phys.* **4** (1963), 294.

72   MICHAEL CREUTZ

5. N. METROPOLIS, A. ROSENBLUTH, M. ROSENBLUTH, A. TELLER, AND E. TELLER, *J. Chem. Phys.* **21** (1953), 1087; K. Binder, *in* "Phase Transitions and Critical Phenomena," (C. Domb and M. S. Green, Eds.), Vol. 5B, 2, Academic Press, New York, 1976.

6. M. CREUTZ, L. JACOBS, AND C. REBBI, *Phys. Rev. Lett.* **42** (1980), 1390; L. Jacobs and C. Rebbi, *J. Comput. Phys.* **41** (1981), 203.

7. T. TOFFOLI, *Physica D* **10** (1984), 117.

8. N. H. PACKARD AND S. WOLFRAM, *J. Stat. Phys.* **38** (1985), 901.

9. E. DOMANY, *Phys. Rev. Lett.* **52** (1984), 871.

10. G. VICHNIAC, *Physica D* **10** (1984), 96.

11. N. MARGOLIS, T. TOFFOLI, AND G. VICHNIAC, Private communication.

PHYSICAL REVIEW LETTERS

7 April 1986

Lattice-Gas Automata for the Navier-Stokes Equation

U. Frisch

Centre National de la Recherche Scientifique, Observatoire de Nice, 06003 Nice Cedex, France

B. Hasslacher

Theoretical Division and Center for Nonlinear Studies, Los Alamos National Laboratory, Los Alamos, New Mexico 87545

and

Y. Pomeau

*Centre National de la Recherche Scientifique, Ecole Normale Supérieure, 75231 Paris Cedex, France, and
Service de Physique Théorique, Centre d'Etudes Nucléaires de Saclay, 91191 Gif-sur-Yvette, France*

(Received 22 October 1985)

We show that a class of deterministic lattice gases with discrete Boolean elements simulates the Navier-Stokes equation, and can be used to design simple, massively parallel computing machines.

PACS numbers: 89.80.+h

The relatively recent availability of sophisticated interactive digital simulation has led to considerable progress in the unraveling of universal features of complexity generated by nonlinear dynamical systems with few degrees of freedom. In contrast, nonlinear systems with many degrees of freedom, e.g., high-Reynolds-number flow, are understood only on a quite superficial level,[1] and are likely to remain so, unless they can be explored in depth, e.g., by interactive simulation. This is many orders of magnitude beyond the capacity of existing computational resources. There are similar limitations on our ability to simulate many other multidimensional field theories.

Massively parallel architectures and algorithms are needed to avoid the ultimate computation limits of the speed of light and various solid-state constraints. Also, when parameter space must be explored quickly and extreme accuracy is unnecessary, a floating-point representation may not be efficient. For example, to compute the drag due to turbulent flow past an obstacle with a modest accuracy of 5 bits, common experience in computational fluid dynamics shows that intermediate computations require from 32 to 64 bits. Floating-point representations hierarchically favor bits in the most significant places,[2] which is a major cause of numerical instability. In principle, schemes which give bits equal weight would be preferable. Because of roundoff noise, a floating-point calculation can run away to unphysical regimes, in an attempt to treat each bit equally.

A simulation strategy can be devised which both is naturally parallel and treats all bits on an equal footing, for systems which evolve by discrete cellular automaton rules, with only local interactions.[3] This avoids the complex switching networks which limit the computational power of conventional parallel arrays.

There has been speculation that various physically

interesting field equations can be approximated by the large-scale behavior of suitably chosen cellular automata.[4] We shall here construct lattice-gas automata which asymptotically go over to the incompressible 2D and 3D Navier-Stokes equations.

To understand the physics behind lattice gases, we first point out that a fluid can be described on three levels: the molecular level at which motion, usually Hamiltonian, is reversible; the kinetic level, at which irreversible low-density Boltzmann approximation; and the macroscopic level, in the continuum approximation. At the first two levels of description, the fluid is near thermodynamic equilibrium. In the last there are free thermodynamic variables: local density, momentum, temperature, etc. A macroscopic description of the fluid comes about by a patching together of equilibria which are varying slowly in space and time, implying continuum equations for thermodynamic variables as consistency conditions. This was first realized by Maxwell,[5] and put in final form by Chapman and Enskog.[6]

There are many ways of building microscopic models that lead to a given set of continuum equations. It is known that one can build two- and three-dimensional Boltzmann models, with a small number of velocity vectors, which, in the continuum limit, reproduce quite accurately major fluid dynamical features (e.g., shock waves in a dilute gas, etc.[7]). Such Boltzmann models are fundamentally probabilistic, discrete only in velocity, but continuous in space and time. In contrast, we will use lattice-gas models, which have a completely discrete phase space and time and therefore may be viewed as made of "Boolean molecules."

The simplest case is the Hardy, de Pazzis, and Pomeau model[8] (hereafter called HPP) which has an underlying regular, square, two-dimensional lattice with unit link lengths. At each vertex, there are up to four molecules of equal mass, with unit speed, whose velocities point in one of the four link directions. The simultaneous occupation of a vertex by identical molecules is forbidden. Time is also discrete. The update is as follows. First, each molecule moves one link, to the nearest vertex to which its velocity was pointing. Then, any configuration of exactly two molecules moving in opposite directions at a vertex (head-on collisions) is replaced by another one at right angles to the original. All other configurations are left unchanged. The HPP model has a number of important properties.[8] The crucial one is the existence of thermodynamic equilibria. No ergodic theorem is known, but relaxation to equilibrium has been demonstrated numerically.[8] These equilibria have free *continuous* parameters, namely, the average density and momentum. The equilibrium distribution functions are completely factorized over vertices and directions, being independent of vertex position, but dependent on direction, unless the mean momentum vanishes. When density and momentum are varied slowly in space and time, "macrodynamical" equations emerge which differ from the nonlinear Navier-Stokes equations in three respects.

The discrepancies may be classified as (1) lack of Galilean invariance, (2) lack of isotropy, and (3) a crossover dimension problem. Galilean invariance is by definition broken by the lattice; consequently, thermodynamic equilibria with different velocities cannot be related by a simple transformation. This is reflected by the nonlinear term in the momentum equation, containing a momentum flux tensor, which not only has quadratic terms in the hydrodynamic velocity $\mathbf{u}$, as it should be in the Navier-Stokes equation, but also has nonlinear corrections to arbitrarily high order in the velocity. However, these terms are negligible at low Mach number, a condition which also guarantees incompressibility. The HPP automaton is invariant under $\pi/2$ rotations. Such a lattice symmetry is insufficient to insure the isotropy of the fourth degree tensor relating momentum flux to quadratic terms in the velocity. Finally, crossover dimension is a general property of two-dimensional hydrodynamics, when thermal noise is added to the Navier-Stokes equations or to the HPP version of it. Simply put, the viscosity develops a logarithmic scale dependence, which is a dimensional crossover phenomenon, common in phase transitions and field theory.[9] In three dimensions, this difficulty does not exist.

Focusing on the isotropy problem, we note that for the HPP model, the momentum flux tensor has the form

$$P_{\alpha\beta} = p\delta_{\alpha\beta} + T_{\alpha\beta\gamma\epsilon}u_\gamma u_\epsilon + O(u^4). \qquad (1)$$

Here $p = \rho/2$ is the pressure; terms odd in $\mathbf{u}$ vanish by parity. The tensor T is, by construction, pairwise symmetric in both (α,β) and (γ,ϵ). Observe that when the underlying microworld is *two-dimensional* and invariant under the hexagonal rotation group (multiples of $\pi/3$), the tensor T is isotropic and (1) takes the form

$$P_{\alpha\beta} = (p + \mu u^2)\delta_{\alpha\beta} + \lambda u_\alpha u_\beta + O(u^4), \qquad (2)$$

with suitable scalar factors λ and μ. At low Mach number this is the correct form for the Navier-Stokes equation. This observation appears to be new. So, in two dimensions, we will use a triangular instead of a square lattice. Each vertex then has a hexagonal neighborhood (Fig. 1). We will call this model the hexagonal lattice gas (HLG). The setup is the same as in the HPP lattice gas, except for modified collision rules. A suitable set is one given by Harris,[10] in connection with a discrete Boltzmann model, supplemented by a Fermi exclusion condition, of single occupation of each Boolean state. The Fermi-modified Harris rules are as follows: Number the six links out of any vertex counterclockwise, with an index i, defined on the integers (mod6). There are both two- and three-body collisions. For two-body collisions, we have $(i,i+3)$ goes to (a) $(i+1,i-2)$ or (b) $(i-1,i+2)$. Type a and b outcomes have equal *a priori* weights. For three-body collisions we have $(i,i+2,i-2)$ goes to $(i+3,i+1,i-1)$. In these rules, it is assumed that no incident link to a vertex is populated, other than the ones given as initial states. All other configurations remain unaffected by collisions. These rules are designed to conserve particle number and momentum at each vertex, i.e., a total of three scalar conservation relations. Without three-body collisions, there would be four scalar conservation relations, namely mass and

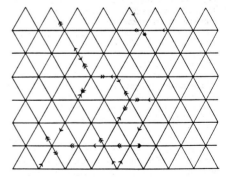

FIG. 1. Triangular lattice with hexagonal symmetry and hexagonal lattice-gas rules. Particles at time t and $t+1$ are marked by single and double arrows, respectively.

VOLUME 56, NUMBER 14   **PHYSICAL REVIEW LETTERS**   7 APRIL 1986

momentum along each of the three lattice directions.

Note that the HPP rules are invariant under duality (interchange of particles and holes), whereas the present rules are not. Duality can be restored by addition of suitable four-particle collision rules, but we will not use them here.

We display a variant of this model where at most one particle is allowed to remain at rest at each vertex. The rest particles are labeled by an asterisk and the previous rules are supplemented with $(i, i+2)$ goes to $(i-2, *)$ and $(i, *)$ goes to $(i+2, i-2)$. Additional variations on the model allow one to define a nontrivial temperature. The remainder of this discussion is concerned only with the basic (HLG) model.

We briefly outline how the hexagonal lattice gas leads to the two-dimensional Navier-Stokes equations. A detailed derivation will be presented elsewhere.[11] Let N_i be the average population at a vertex with velocity in the direction i. The average is over a macroscopic space-time region so that N_i depends slowly on space and time variables. We define a slowly varying density ρ and momentum $\rho\mathbf{u}$ by

$$\rho = \sum_i N_i, \quad \rho\mathbf{u} = \sum_i N_i \mathbf{c}_i, \tag{3}$$

where $\mathbf{c}_i$ is a unit vector in the direction i. Locally, for a given ρ and $\mathbf{u}$, the N_i's can be computed from both these definitions and the detailed-balance equations at thermodynamic equilibrium, which are too involved to present here. This gives a Fermi-Dirac distribution:

$$N_i = \{1 + \exp[\alpha(\rho, u) + \beta(\rho, u)\mathbf{c}_i \cdot \mathbf{u}]\}^{-1}. \tag{4}$$

In general, α and β satisfy equations with no simple solutions. However, when $\mathbf{u} = 0$, it is obvious by symmetry that $N_i = \rho/6$. Therefore, α and β can be expanded in a Taylor series around $\mathbf{u} = 0$. The result can be used to compute mass and momentum flux to first order in the macroscopic gradients. Second-order terms in the gradients (viscous terms) are obtained by Green-Kubo relations or by a Chapman-Enskog expansion.[12] The following set of hydrodynamic equations is thus obtained:

$$\partial\rho/\partial t + \nabla \cdot (\rho\mathbf{u}) = 0, \tag{5}$$

$$\frac{\partial}{\partial t}(\rho u_\alpha) + \sum_\beta \frac{\partial}{\partial x_\beta}[g(\rho)\rho u_\alpha u_\beta + O(u^4)]$$

$$= -\frac{\partial}{\partial x_\alpha}p + \eta_1(\rho)\nabla^2 u_\alpha + \eta_2(\rho)\frac{\partial}{\partial x_\alpha}\nabla \cdot \mathbf{u}, \tag{6}$$

with $g(\rho) = (\rho-3)/(\rho-6)$ and $p = \rho/2$. $\eta_1(\rho)$ and $\eta_2(\rho)$ are the shear and bulk viscosities.[12]

Deletion of the nonlinear and viscous terms gives the wave equation for sound waves propagating isotropically with a speed equal to the "velocity of light" (here set equal to 1) over $\sqrt{2}$, just as for a two-dimensional photon gas. These sound waves have

been observed in simulations on the MIT cellular automaton machine by Margolus, Toffoli, and Vichniac.[13] They used lattice-gas models that yield the same wave equation as above.

The nonlinear system (5) and (6) goes over to the incompressible Navier-Stokes equation by the following limiting procedure: Let the Mach number $M = u\sqrt{2}$ tend to zero, and the hydrodynamic scale L tend to infinity, while keeping their product fixed. As in the usual derivation of the incompressible limit, density fluctuations become irrelevant, except in the pressure term; also, the continuity equation (5) reduces to $\nabla \cdot \mathbf{u} = 0$. Thus, the factor $g(\rho)$ is to leading order a constant and may, for $0 < \rho < 3$, be absorbed in a rescaled time. The resulting Reynolds number is

$$N_R = ML\rho g(\rho)/\sqrt{2}\eta_1(\rho). \tag{7}$$

Note that Galilean invariance, which does not hold at the lattice level, is restored macroscopically.

A straightforward lift of the hexagonal lattice-gas model from two into three dimensions does not work. The reason is that the regular space-filling simplex with the greatest symmetry in three dimensions is the face centered cubic, with twelve equal-speed velocity directions out of each vertex. Unfortunately, the relevant tensors such as $T_{\alpha\beta\gamma\epsilon}$ in Eq. (1) depend now on three constants. This induces a spurious, isotropy-breaking term in the Navier-Stokes equations, proportional to $(\partial/\partial x_\alpha)u_\alpha^2$ (no summation on α).

This obstacle may be removed by a splitting method. The nonlinear term in the three-dimensional Navier-Stokes equation is recast as the sum of two terms, each containing spurious elements and each realizable on a different lattice (for example, a face-centered-cubic lattice and a regular cubic lattice).

In lattice-gas models, as in general cellular automata (CA's), boundary conditions are very easy to implement. Specular reflection of molecules gives so-called "free slip" boundary conditions for the hydrodynamic velocity $\mathbf{u}$. "Rigid" boundary conditions are obtained either by random scattering of particles back into the incoming half plane from a locally planar boundary, or by specular reflection from a microscale roughened version of the macroscopic boundary.

We mention some practical limitations on lattice-gas models. For the hydrodynamic description to hold, there must be a scale separation between the smallest hydrodynamic scale and the lattice link length; as we shall see, this requirement is automatically satisfied. Lattice-gas models must be run at moderate Mach numbers M (say 0.3 to 0.5), to remain incompressible, and to avoid spurious high-order nonlinear terms. For fixed Mach number, the largest Reynolds number associated with a D-dimensional lattice with $O(N)$ sites in each direction is $O(N)$. This is because in our units, the kinematic viscosity of the hexagonal lattice gas is $O(1)$. From standard turbulence theory,[14] it follows that the dissipation scale is $O(N^{1/2})$ in 2D and $O(N^{1/4})$ in 3D. This insures the required scale separation at large Reynolds numbers. It would, however, be desirable to reduce the scale separation, especially in 2D, to avoid excessive storage requirements compared to conventional incompressible floating-point simulations (in the latter, the mesh can be taken comparable to the dissipation scale).

For this, we observe that the viscosity in the lattice gas is decreased by a factor P if we subdivide each cell into a sublattice with links P times smaller. We note also that the sublattice need not be similar to the original lattice. It must have the same collision rules, to preserve local thermodynamic equilibria, but the geometry does not matter since macroscopic quantities may be considered uniform over the cell. Thus, all the sublattice vertices in a given cell may be regarded as indistinguishable and can be coded in $O(\ln P)$ rather than $O(P^D)$ bits; interactions occur between randomly chosen vertex pairs within cells and between neighboring cells; the latter being less frequent by a factor $O(1/P)$.

Simulations of the models discussed here, done on general-purpose computers and exhibiting a variety of known two-dimensional hydrodynamic phenomena, have been made by d'Humières, Lallemand, and Shimomura.[15]

We have given a concrete hydrodynamical example of how CA's can be used to simulate classical nonlinear fields. We expect that further CA implementations will be found for the Navier-Stokes equation and other problems, not necessarily based on thermalized lattice gases and possibly less constrained than ours.

S. Wolfram stimulated our interest in cellular automata as a possible new approach to turbulence phenomena. Acknowledgments are also due to T. Bloch, R. Caflish, D. d'Humières, R. Gatignol, R. Kraichnan, P. Lallemand, N. Margolus, D. Nelson, J. L. Oneto, S. A. Orszag, J. P. Rivet, T. Shimomura, Z. S. She, B. Shraiman, T. Toffoli, and G. Vichniac, as well as the following: Woods Hole Geophysical Fluid Dynamics Summer Program (U.F., Y.P.); Aspen Center for Physics, 1985 Chaos Workshop (B.H.); and

Service de Physique Théorique, Centre d'Etudes Nucleaires de Saclay (B.H.). This work was supported in part by National Science Foundation Grant No. 8442384.

---

[1]U. Frisch, Phys. Scr. **T9**, 131 (1985).

[2]D. E. Knuth, *The Art of Computer Programming: Semi Numerical Algorithms* (Addison-Wesley, Reading, Mass., 1981), Vol. 2, p. 238.

[3]T. Toffoli, Physica (Amsterdam) **10D**, 117 (1984); S. Wolfram, Nature **311**, 419 (1984).

[4]Y. Pomeau, J. Phys. A **17**, L415 (1984); G. Vichniac, Physica (Amsterdam) **10D**, 96 (1984), and references therein; N. Margolus, Physica (Amsterdam) **10D**, 81 (1984).

[5]J. C. Maxwell, *The Scientific Papers, Vol. 2* (Cambridge Univ. Press, Cambridge, England, 1890), p. 681.

[6]G. E. Uhlenbeck and G. W. Ford, *Lectures in Statistical Mechanics,* Lectures in Applied Math Vol. 1 (American Mathematical Society, Providence, R.I., 1963).

[7]J. E. Broadwell, Phys. Fluids **7**, 1243 (1964); R. Gatignol, *Théorie Cinétique des Gaz à Répartition Discrète des Vitesses,* Lecture Notes in Physics Vol. 36 (Springer, Berlin, 1975).

[8]J. Hardy and Y. Pomeau, J. Math. Phys. **13**, 1042 (1972); J. Hardy, Y. Pomeau, and O. de Pazzis, J. Math. Phys. **14**, 1746 (1973); J. Hardy, O. de Pazzis, and Y. Pomeau, Phys. Rev. A **13**, 1949 (1976).

[9]D. Forster, D. R. Nelson, and M. J. Stephen, Phys. Rev. A **16**, 732 (1977).

[10]S. Harris, Phys. Fluids **9**, 1328 (1966).

[11]U. Frisch, B. Hasslacher, and Y. Pomeau, "Hydrodynamics on Lattice Gases," to be published.

[12]J. Rivet and U. Frisch, C.R. Seances Acad. Sci., Ser. 2 **302**, 267 (1986).

[13]N. Margolus, T. Toffoli, and G. Vichniac, private communication, and Massachusetts Institute of Technology Technical Memo No. LCS-TM-296, 1984 (unpublished).

[14]A. N. Kolmogorov, C.R. (Dokl.) Acad. Sci. USSR **30**, 301, 538 (1941); R. Kraichnan, Phys. Fluids **10**, 1417 (1967); G. K. Batchelor, Phys. Fluids **12**, Suppl. 2, 233 (1969).

[15]D. d'Humières, P. Lallemand, and T. Shimomura, "Lattice gas cellular automata, a new experimental tool for hydrodynamics," to be published.

Thermodynamics and Hydrodynamics with Cellular Automata

James B. Salem
Thinking Machines Corporation, 245 First Street, Cambridge, MA 02144

and

Stephen Wolfram
The Institute for Advanced Study, Princeton NJ 08540.

(November 1985)

Simple cellular automata which seem to capture the essential features of thermodynamics and hydrodynamics are discussed. At a microscopic level, the cellular automata are discrete approximations to molecular dynamics, and show relaxation towards
equilibrium. On a large scale, they behave like continuum fluids, and suggest efficient
methods for hydrodynamic simulation.

Thermodynamics and hydrodynamics describe the overall behaviour of many systems, independent of the precise microscopic construction of each system. One can thus study thermodynamics and
hydrodynamics using simple models, which are more amenable to efficient simulation, and potentially
to mathematical analysis.

Cellular automata (CA) are discrete dynamical systems which give simple models for many complex physical processes [1]. This paper considers CA which can be viewed as discrete approximations
to molecular dynamics. In the simplest case, each link in a regular spatial lattice carries at most one
"particle" with unit velocity in each direction. At each time step, each particle moves one link; those
arriving at a particular site then "scatter" according to a fixed set of rules. This discrete system is
well-suited to simulation on digital computers. The state of each site is represented by a few bits, and
follows simple logical rules. The rules are local, so that many sites can be updated in parallel. The
simulations in this paper were performed on a Connection Machine Computer [2] which updates sites
concurrently in each of 65536 Boolean processors [3].

In two dimensions, one can consider square and hexagonal (six links at 60°) lattices. On a square
lattice [4], the only nontrivial local rule which conserves momentum and particle number takes isolated
pairs of particles colliding head on scatter in the orthogonal direction (no interaction in other cases). On
a hexagonal lattice [5], such pairs may scatter in either of the other two directions, and the scattering
may be affected by particles in the third direction. Four particles coming along two directions may also
scatter in different directions. Finally, particles on three links separated by 120° may scatter along the

other three links. At fixed boundaries, particles may either "bounce back" (yielding "no slip" on average), or reflect "specularly" through 120°.

On a microscopic scale, these rules are deterministic, reversible and discrete. But on a sufficiently large scale, a statistical description may apply, and the system may behave like a continuum fluid, with macroscopic quantities, such as hydrodynamic velocity, obtained by kinetic theory averages.

Figure 1 illustrates relaxation to "thermodynamic equilibrium". The system randomizes, and coarse-grained entropy increases. This macroscopic behaviour is robust, but microscopic details depend sensitively on initial conditions. Small perturbations (say of one particle) have microscopic effects over linearly-expanding regions [6]. Thus ensembles of "nearby" initial states usually evolve to contain widely-differing "typical" states. But in addition, individual "simply-specified" initial states can yield behaviour so complex as to seem random [7,8], as in figure 1. The dynamics thus "encrypts" the initial data; given only coarse-grained, partial, information, the initial simplicity cannot be recovered or recognized by computationally feasible procedures [7], and the behaviour is effectively irreversible.

Microscopic instability implies that predictions of detailed behaviour are impossible without ever more extensive knowledge of initial conditions. With complete knowledge (say from a simple specification), the behaviour can always be reproduced by explicit simulation. But if effective predictions are to be made, more efficient computational procedures should be found. The CA considered here can in fact act as universal computers [9]: with appropriate initial conditions, their evolution can implement any computation. Streams of particles corresponding to "wires" can meet in logical gates implemented by fixed obstructions or other streams. As a consequence, the evolution is computationally irreducible [10]; there is no general shortcut to explicit simulation. No simpler computation can reproduce all the possible phenomena.

Some overall statistical predictions can nevertheless be made. In isolation, the CA seem to relax to an equilibrium in which links are populated effectively randomly with a particular average particle density ρ and net velocity (as in figure 1). On length scales large compared to the mean free path λ, the system then behaves like a continuum fluid. The effective fluid pressure is $p=\rho/2$, giving a speed of sound $c=1/\sqrt{2}$. Despite the microscopic anisotropy of the lattice, circular sound wavefronts are obtained from point sources (so long as their wavelength is larger than the mean free path) [11].

Assuming local equilibrium, the large-scale behaviour of the CA can be approximated by average rules for collections of particles, with particular average densities and velocities. The rules are like finite difference approximations to partial differential equations, whose form can be found by a standard Chapman-Enskog expansion [12] of microscopic particle distributions in terms of macroscopic quantities. The results are analogous to those for systems [13] in which particles occur with an arbitrary continuous density at each point in space, but have only a finite set of possible velocities corresponding to the links of the lattice. The hexagonal lattice CA is then found to follow exactly the standard Navier-Stokes equations [5,14]. As usual, the parameters in the Navier-Stokes equations depend on the microscopic structure of the system. Kinetic theory suggests a kinematic viscosity $v \cong \lambda/2$ [15].

Figures 2 and 3 show hydrodynamic phenomena in the large scale behaviour of the hexagonal lattice CA. An overall flow U is obtained by maintaining a difference in the numbers of left- and right-moving particles at the boundaries. Since local equilibrium is rapidly reached from almost any state, the results are insensitive to the precise arrangement used. Random boundary fluxes imitate an infinite region; a regular pattern of incoming particles nevertheless also suffices, and reflecting or cyclic boundary conditions can be used on the top and bottom edges.

The hydrodynamics of the CA is much like a standard physical fluid [16]. For low Mach numbers $Ma=U/c$, the fluid is approximately incompressible, and the flows show dynamical similarity, depending only on Reynolds number $Re=UL/v$ ($L \gg \lambda$). The patterns obtained agree qualitatively with experiment [3]. At low Re, the flows are macroscopically stable; perturbations are dissipated into microscopic "heat".

As Re increases, periodic vortex streets are at first produced, and then vortices are shed in an irregular, turbulent, fashion. Perturbations now affect details of the flow, though not its statistical properties. The macroscopic irregularity does not depend on microscopic randomness; it occurs even if microscopically simple (say spatially and temporally periodic) initial and boundary conditions are used, as illustrated in figure 2. As at the microscopic level, it seems that the evolution corresponds to a sufficiently complex computation that its results seem random [7].

The CA discussed here should serve as a basis for practical hydrodynamic simulations. They are simple to program, readily amenable to parallel processing, able to handle complex geometries easily [17], and presumably show no unphysical instabilities. (Generalization to three dimensions is straight-forward in principle [18].)

Standard finite difference methods [19] consider discrete cells of fluid described by continuous parameters. These parameters are usually represented as digital numbers with say 64 bits of precision. Most of these bits are, however, probably irrelevant in determining observable features of flow. In the CA approach, all bits are of essentially equal importance, and the number of elementary operations per-formed is potentially closer to the irreducible limit.

The difficulty of computation in a particular case depends on the number of cells that must be used. Below a certain dissipation length scale $a \sim \text{Re}^{-d/4}$ (in d dimensions), viscosity makes physical homogeneous turbulent fluids smooth [16]. In finite difference schemes, individual cells can represent fluid regions of this size. But complete calculations with the CA considered here probably require increasing numbers of cells in each region [20]. Approximate "turbulence models" involving fewer cells may however be devised.

Several further extensions of the CA scheme can be considered. First, on some or all of the lat-tice, basic units containing say n particles, rather than single particles, can be used. The properties of these units can be specified by digital numbers with $O(\log n)$ bits, but exact conservation laws can still be maintained. This scheme comes closer to adaptive grid finite difference methods [19], and potentially avoids detailed computation in featureless parts of flows.

A second, related, extension introduces discrete internal degrees of freedom for each particle. These could represent different particle types, directions of discrete vortices [19], or internal energy (giving variable temperature [21]).

This paper has given further evidence that simple cellular automata can reproduce the essential features of thermodynamic and hydrodynamic behaviour. These models make contact with results in dynamical systems theory and computation theory. They should also yield efficient practical simula-tions, particularly on parallel-processing computers.

Cellular automata can potentially reproduce behaviour conventionally described by partial differential equations in many other systems whose intrinsic dynamics involves many degrees of free-dom with no large disparity in scales.

We are grateful to U. Frisch, B. Hasslacher, Y. Pomeau and T. Shimomura for sharing their unpub-lished results with us, and to N. Margolus, S. Omohundro, S. Orszag, N. Packard, R. Shaw, T. Toffoli, G. Vichniac and V. Yakhot for discussions. We thank many people at Thinking Machines Corporation for their help and encouragement. The work of S.W. was supported in part by the U.S. Office of Naval Research under contract number N00014-85-K-0045.

1. See for example S. Wolfram, "Cellular automata as models of complexity", Nature **311**, 419 (1984) where applications to thermodynamics and hydrodynamics were mentioned but not explored.

2. D. Hillis, *The Connection Machine* (MIT press, 1985). This application is discussed in S. Wol-fram, "Scientific computation with the Connection Machine", Thinking Machines Corporation report (March 1985).

3. More detailed results of theory and simulation will be given in a forthcoming series of papers.

4. J. Hardy, Y. Pomeau and O. de Pazzis, "Time evolution of a two-dimensional model system. I. Invariant states and time correlation functions", J. Math. Phys. **14**, 1746 (1973); J. Hardy, O. de Pazzis and Y. Pomeau, "Molecular dynamics of a classical lattice gas: transport properties and time correlation functions", Phys. Rev. **A13**, 1949 (1976).

5.   U. Frisch, B. Hasslacher and Y. Pomeau, "A lattice gas automaton for the Navier-Stokes equation", Los Alamos preprint LA-UR-85-3503.

6.   The expansion rate gives the Lyapunov exponent as defined in N. Packard and S. Wolfram, "Two-dimensional cellular automata", J. Stat. Phys. **38**, 901 (1985). Note that the effect involves many particles, and does not arise from instability in the motion of single particles, as in the case of hard spheres with continuous position variables (e.g. O. Penrose, "Foundations of statistical mechanics", Rep. Prog. Phys. **42**, 129 (1979).)

7.   S. Wolfram, "Origins of randomness in physical systems", Phys. Rev. Lett. **55**, 449 (1985); "Random sequence generation by cellular automata", Adv. Appl. Math. (in press).

8.   Simple patterns are obtained with very simple or symmetrical initial conditions. On a hexagonal lattice, the motion of an isolated particle in a rectangular box is described by a linear congruence relation, and is ergodic when the side lengths are not commensurate.

9.   N. Margolus, "Physics-like models of computation", Physica **10D**, 81 (1984) shows this for some similar CA.

10. S. Wolfram, "Undecidability and intractability in theoretical physics", Phys. Rev. Lett. **54**, 735 (1985).

11. *cf* T. Toffoli, "CAM: A high-performance cellular automaton machine", Physica **10D**, 195 (1984).

12. e.g. A. Sommerfeld, *Thermodynamics and statistical mechanics*, (Academic Press, 1955).

13. J. C. Maxwell, *Scientific Papers II*, (Cambridge University Press, 1890); J. Broadwell, "Shock structure in a simple discrete velocity gas", Phys. Fluids **7**, 1243 (1964); S. Harris, *The Boltzmann Equation*, (Holt, Reinhart and Winston, 1971); J. Hardy and Y. Pomeau, "Thermodynamics and hydrodynamics for a modeled fluid", J. Math. Phys. **13**, 1042 (1972); R. Gatignol, *Theorie cinetique des gaz a repartition discrete de vitesse*, (Springer, 1975).

14. On a square lattice, the total momentum in each row is separately conserved, and so cannot be convected by velocity in the orthogonal direction [4]. Symmetric three particle collisions on a hexagonal lattice remove this spurious conservation law.

15. The symmetric rank four tensor which determines the nonlinear and viscous terms in the Navier-Stokes equations is isotropic for a hexagonal but not a square lattice (*cf* [5]). Higher order coefficients are anisotropic in both cases. In two dimensions, there can be logarithmic corrections to the Newtonian fluid approximation: these can apparently be ignored on the length scales considered, but yield a formal divergence in the viscosity (*cf* [4]).

16. e.g. D. J. Tritton, *Physical fluid dynamics*, (Van Nostrand, 1977).

17. They can also treat microscopic boundary effects beyond the hydrodynamic approximation.

18. Icosahedral symmetry yields isotropic fluid behaviour, and can be achieved with a quasilattice, or approximately by periodic lattices (*cf* D. Levine *et al.*, "Elasticity and dislocations in pentagonal and icosahedral quasicrystals", Phys. Rev. Lett. **54**, 1520 (1985); P. Bak, "Symmetry, stability, and elastic properties of icosahedral incommensurate crystals", Phys. Rev. **B32**, 5764 (1985)).

19. e.g. P. Roache, *Computational fluid dynamics*, (Hermosa, Albuquerque, 1976).

20. S. Orszag and V. Yakhot, "Reynolds number scaling of cellular automaton hydrodynamics", Princeton University Applied and Computational Math. report (November 1985).

21. In simple cases the resulting model is analogous to a deterministic microcanonical spin system (M. Creutz, "Deterministic Ising dynamics", Ann. Phys., to be published.)

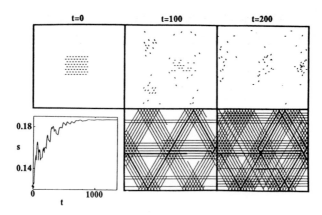

Figure 1. Relaxation to "thermodynamic equilibrium" in the hexagonal lattice cellular automaton (CA) described in the text. Discrete particles are initially in a simple array in the centre of a 32×32 site square box. The upper sequence shows the randomization of this pattern with time; the lower sequence shows the cells visited in the discrete phase space (one particle track is drawn thicker). The graph illustrates the resulting increase of coarse-grained entropy $\sum p_i \log_2 p_i$ calculated from particle densities in 32×32 regions of a 256×256 box.

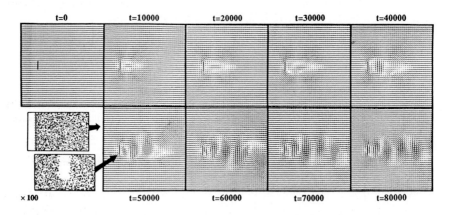

Figure 2. Time evolution of hydrodynamic flow around a plate in the CA of figure 1 on a 4096×4096 site lattice. Hydrodynamic velocities are obtained as indicated by averaging over 96×96 site regions. There is an average density of 0.3 particles per link (giving a total of 3×10^8 particles). An overall velocity U=0.1 is maintained by introducing an excess of particles (here in a regular pattern) on the left hand boundary.

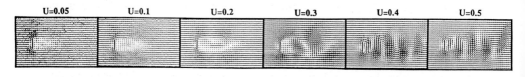

Figure 3. Hydrodynamic flows obtained after 10^5 time steps in the CA of figure 2, for various overall velocities U.

ATTRACTORS, BASIN STRUCTURES AND
INFORMATION PROCESSING IN CELLULAR AUTOMATA

Kunihiko KANEKO

Institute of Physics, College of Arts and Sciences
University of Tokyo, Komaba, Meguro, Tokyo 153, JAPAN

One-dimensional cellular automata (CA) are investigated. An information theory for multi-attractor systems is constructed, and quantities are introduced to characterize the complexity of basin volumes, stability of attractors against noise, information storage in attractors, and connectivity among attractors by noise. These quantities and basin structures are calculated numerically for one-dimensional CA of various classes. The patterns of the main attractors are shown, focusing on the holographic memory of class-3 CA, successive changes of main attractors with size in class-4 CA, and soliton-like attractors in some CA.

1. Introduction

Spatially extended dynamical systems are important tools to understand the complex behavior in nature. The simplest model in such systems is a cellular automaton (CA), where the system is composed of a discrete time n and space i (lattice) and discrete variables $x_n(i)$'s on the lattice. It was introduced for the computer architecture and will also be useful to understand the qualitative nature in turbulence (in a wide sense) or neural networks or some other biological systems[1-3]. In this paper, the complexity in CA with a finite size is investigated.

Computer systems or artificial intelligence has recently been investigated from the viewpoint of a dynamical system theory. When we consider some kind of artificial intelligence on the basis of a dynamical system, informational aspects are important. As was pointed out by Rob Shaw[4], a dynamical system with chaos can be thought of as an information source.

Another aspect of a dynamical system is a storage of information. If a dynamical system has a large number of attractors as is commonly seen in the spatially extended systems, information can be stored in each attractor. Examples can be seen in the neural network models (see e.g., Hopfield[5]). In the following sections, an information theory for a multi-attractor system is constructed, mainly in connection with the CA with a finite lattice size. From the quantities introduced in the following, the capacity for the storage into attractors will be discussed.

Stability of the storage is also important which is related to the problem of self-repair or retrieval[6]. We will consider the stability of each attractor against a noise and define the mutual information between attractors.

In this paper a one-dimensional cellular automaton with a lattice size N is investigated. If a state of each cell can take k values, the total number of states is k^N. Thus, the system finally settles down into a cycle.

From the viewpoint of the creation and storage of information, a cellular automaton can be classified into the following four types. The classification is essentially the same as the one by S. Wolfram[1,7], though the precise definition for the classification is not available at present. In the following, "a large number" means a quantity exponential to the system size, $(O(e^N))$, while "a small number" means a quantity less than some power of the system size $(o(N^a))$. The period of an attractor is said to be long if the period is $O(N)$, while it is short if it is bounded by $O(1)$.

(1) Small number of attractors with short periods: (No creation and small storage of information): class 1

(2) A large number of attractors with short periods: (No creation but large storage of information): class 2

(3) A small number of attractors with long periods: (Positive creation with small storage of information): class 3

(4) A large number of attractors with possible long periods: (Possible positive creation with large storage of information): class 4

In Sections 2 and 3, a framework to study the complexity of multi-attractor systems is introduced, where the complexity in the volume of basins and the dynamical aspects by the jumping process among attractors by a small noise (i.e., single site flip-flop) are investigated.

In Sections 4-7, the structure of attractors and basins are investigated for various classes of CA, mainly focusing on the number of attractors, the distribution of the volume of basins of attraction, the probability distribution at each attractor by the noise, and jumping process among attractors by the noise. The quantities introduced in Sections 2 and 3 are calculated.

2. Information Theory for Multi-attractor Systems

Here we study the number of attractors and the structure of the basin for each attractor in a CA with a finite size[9]. A quantity to characterize the complexity of the basin of attraction is introduced.

(a) Complexity of basins

Let us denote the number of attractors by M. Each attractor is denoted by $\{a_i\}$

$(i = 1, 2, \ldots, M)$. First, we examine how many initial configurations are attracted into each attractor a_i. The number of configurations divided by k^N gives the ratio of the basin volume for the attractor a_i, which is denoted by b_i ($\Sigma b_i = 1$). Let us define the complexity for basins by

$$C_B = -\Sigma b_i \ln b_i \quad ,$$

which characterizes the information for the initial state necessary to predict the final state. If each attractor has an equal volume of basins, $C_B = \ln M$ (maximum). In many cases studied here, the complexity is much less than $\ln M$, since the volume of basin of most attractors is very small. The distribution of the basin volume b_i itself is also important.

(b) Period of attractors and contraction ratio

Let us denote the period of each attractor by T_i. The summation $c = \Sigma T_i$ gives the volume of phase space utilized by CA after the transients have decayed out. If we start from all possible initial configurations, only a limited number of states remain after some iterations. The ratio of contraction for the process is given by $c/2^N$. For a discussion about the contraction of CA from a different point of view, see Hogg and Huberman[8].

3. Dynamical Process Among Attractors

(a) Jumping among attractors by noise[10]

Since most CA have more than one attractor, a unique invariant measure cannot be attained. In a real physical situation, existence of a small noise is expected. By the noise, a unique (or a small number of) measure is selected out. Here we consider the case with a very low noise. The noise takes only an integer value in CA, and the "low" noise here means that the rate of the application of noise is very low. In the low noise case, the dynamics of the system may be decomposed into the following two processes; (i) the state stays at the original attractors (for most of the time) and (ii) the state jumps out to some other attractors by the effect of a noise. The state of a CA stays at the original attractors $\{a_i$'s$\}$ most of the time and the transition among attractors by a noise occurs in a short time interval. Here, we neglect the time for the latter process as small compared with the time for the former. An example of a pattern for a stochastic CA is shown in Fig. 1, where the transitions among the attractors can be seen.

Let us define the transition matrix between attractors. If a low noise is applied on the attractor a_i, a jump from the attractor a_i to a_j occurs with some probability. Here the noise is a process $x_i \to x_i' = x_i + r \pmod{k}$ $(0 < r < k)$. The jumping

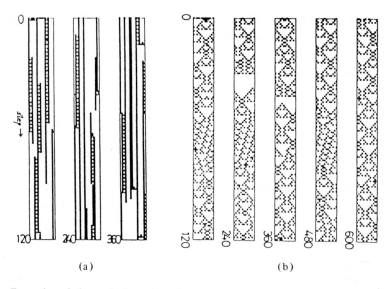

Fig. 1. Examples of the evolution of stochastic cellular automata. The flip-flop $0 \leftrightarrow 1$ or $1 \leftrightarrow 0$ by a noise occurs with the ratio p.

(a) Rule 108:   noise $p = 0.05$:   size $N = 16$
(b) Rule 146:   noise $p = 0.05$:   size $N = 13$

process depends on the state of the CA when the noise is applied (i.e., the phase of the oscillation with the period T_i; T_i possibilities) and on the lattice site at which the noise is applied (N (= system size) possibilities) and on the value of noise $r((k-1)$ possibilities). If the CA has only two states ($k = 2$), there is only one possibility for the noise (i.e., $1 \to 0$ or $0 \to 1$). In the following, we treat the case $k = 2$ (see Fig. 2 schematically).

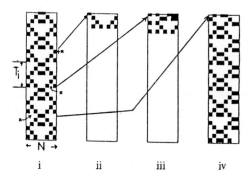

i   ii   iii   iv

Fig. 2. Schematic representation of the transition by the noise on an attractor. Rule 146 and $N = 8$. The attractor (i) has a period 6 and there are 6×8 possibilities on the application of noise. In the figure three examples are shown, where the noise is applied on the position marked x. Two ((ii) and (iii)) go to the attractor all-0, while the other (iv) goes to the same attractor as (i) with 2 shifts.

P_{ij} is defined as the ratio of the transition from a_i to a_j. That is, the number of the events $a_i \rightarrow a_j$ for all the possible flip-flops by the noise, divided by the number of such possibilities $T_i N$.

The probability that a system is in the attractor a_i for a low noise case is given by

q_i = the i-th component of the eigenvector for the matrix P_{ji}

       corresponding to the eigenvalue 1.

       (i.e., $\sum_i q_i P_{ij} = q_j$)

If the eigenvalue 1 is degenerate with the multiplicity m_p, the superposition of each eigenvector can be an invariant measure within the above low-noise limit approximation.

Some attractor a_i can be so weak that P_{ki} is zero for all k, (i.e., there is no flow from other attractors by a low noise) as can be seen in the next section. Such an attractor is repulsive against a noise and may be regarded as an extension of the notion of garden of Eden to attractors. If q_i is zero (which is a weaker condition than the above $P_{ki} = 0$-condition), the residence probability at the attractor a_i is zero, and such an attractor is unimportant in the low-noise limit.

The diagonal part P_{ii} is a measure for the strength of self-repair for the attractor a_i against a one-position flip-flop.

(b) Complexity in the jumping process among attractors by noise

Once the probability measure q_i is attained by a noise, we can define the complexity for the probability distribution for attractors by

$C_M = -\Sigma q_i \ln q_i$,

for a given invariant measure. The meaning of the quantity C_M is as follows: After the transients have decayed out for a low noise system, we make a measurement to determine at which attractor the system stays at that time. The information gain by the above measurement is C_M. The difference between C_B and C_M lies in that the former quantity is concerned with the knowledge about the initial state, while the latter is related to the observation for the aged system with a noise.

Another important quantity is a dynamical information gain by noise. Let us assume that we knew that a system had initially been at the attractor a_i and have observed that the system is now at the attractor a_j after a noise was applied. How much information has been obtained through this observation? We can get some information about the noise, i.e., the phase of the oscillation of CA when the noise is applied and the site where the noise is applied. The amount is given by $\ln(P_{ij}^{-1})$

Thus the dynamical information gain per noise is given by

$$C_D = -\sum_{ij} q_i P_{ij} \ln P_{ij} \quad,$$

since the ratio for the event $a_i \longrightarrow a_j$ is $q_i P_{ij}$.

As is easily seen,

$$C_T = C_M - C_D$$

is non-negative. The quantity C_T corresponds to the mutual information[11] between attractors by noise.

If C_D is large, the information creation by noise is large. That is, the uncertainty about the attractor into which the system settles down after the addition of noise is large. It can also be stated that if the mutual information is large ($C_D \ll C_M$), the structure of the network of the transition among attractors is well organized, while the network of the transition is global and irregular if C_T is small.

(c) Method of the calculation in one-dimensional CA

As a simple example of the theory for multi-attractor systems in this section, one-dimensional cellular automata with two states (0 or 1) are investigated. The periodic boundary condition is used throughout this paper. The models are

(i) legal cellular automata with range 1[1]

(ii) totalistic cellular automata with range 2[7]

(iii) cellular automata with range 2 which have "soliton"-like excitations[11].

As a method for the coding for the rule, the rule number (for the model (i)) or the rule code (for the model (ii)) by Stephen Wolfram[1,7] is used, while the rule code by Aizawa, Nishikawa and the author[12] is used to characterize the rule for the model (iii) (see Appendix). In the following sections, we use the notation Rule *** for (i) and totalistic *** for (ii), where *** is a number which characterizes the rule or code. For (iii), we use $(l_1 l_2 l_3 l_4 l_5 l_6 l_7)(k_1 k_2 k_3 k_4 k_5)$, where k_i and l_i take 0 or 1.

The method of calculations is as follows: (i) Take a one-dimensional cellular automaton with a size N ($7 < N < 23$) and simulate it for all initial configurations (i.e., 2^N possibilities). (ii) Enumerate all possible attractors (find M, a_i ($i = 1, 2, \ldots, M$)), and their periods T_i's and list all the patterns. (iii) Calculate how many initial configurations are attracted into the attractor a_i. The number of such initial configurations divided by 2^N gives b_i, from which the basin complexity C_B is calculated. (iv) Take an attractor a_i and change a value of one lattice site for a_i ($0 \leftrightarrow 1$). There are $N \times T_i$ possibilities for this flip-flop. We simulate the CA starting from the configuration obtained by all these possible flip-flops and check to which attractor a_j the state is attracted. The number of such configurations divided by $N \times T_i$ gives P_{ij}. The left eigenvector for P_{ij} corresponding to the eigenvalue 1 gives q_i. From P_{ij} and q_i, measure complexity C_M and dynamical complexity C_D are calculated.

Here, instead of obtaining all possible eigenvectors, we choose an initial vector $(b_1, b_2, b_3, \ldots, b_M)^T$ and multiply the matrix $\{P_{ij}\}$ many times till the set of vectors is settled down into the fixed point, from which we obtain $\{q_1, q_2, \ldots, q_M\}^T$. If the invariant measure is unique, this procedure gives the correct measure. If the measure is not unique (i.e., nonergodic), this procedure selects out one measure closest to the equipartition distribution for all the possible configurations. For a finite one-dimensional CA, such a nonergodic case seems to be rare except the following case; i.e., the all-0 attractor is sometimes disconnected from other attractors and there are two eigenvectors for the eigenvalue 1; one is $q_i^1 = 1$ for the all-0 attractor and $= 0$ otherwise, and the other is $q_i^2 = 0$ for all-0.

For the classification of attractors, the configurations which coincide by the spatial translation are regarded as the same attractor. For example, the patterns 11000001, 11100000 and 00111000 are regarded as the same.

The results for the various classes of CA are shown in the following three sections. See Ref. 13 for the preliminary report.

4. Class-1 and Class-2 CA

(a) Class-1 CA

As is expected, this case is trivial. The number of attractors M remains small (about $1 \sim 3$) even if N is increased. As N goes larger C_B, C_M, and C_D rapidly go to zero. For example, the possible attractor is all-0, and 0101010101 (which is possible only for $N =$ even) for Rule 32.

(b) Class-2 CA

The following rules were numerically investigated for $7 < N < 19$: elementary rules 132, 50, 108, and totalistic rules 24, and 104. Some examples for the number of attractors and three complexities are shown in Figs. 3 and 4. Let us discuss the common properties for these rules.

As is expected, the number of attractors increase as $\exp(\alpha \times N)$. The basin, measure and dynamical complexities increase as $a_B \times N + $ const., $a_M \times N + $ const., and $a_D \times N + $ const. where a's are some constants. See Table I for α, the values of the constants, and the types of basic oscillators.

In one type in class-2 CA such as the Rules 132, 108, and 50, a_B and a_M take almost the same values, while a_D takes a smaller value. In another type in CA, the "all-0" attractor is stable against a single flip-flop and there is a gate from other

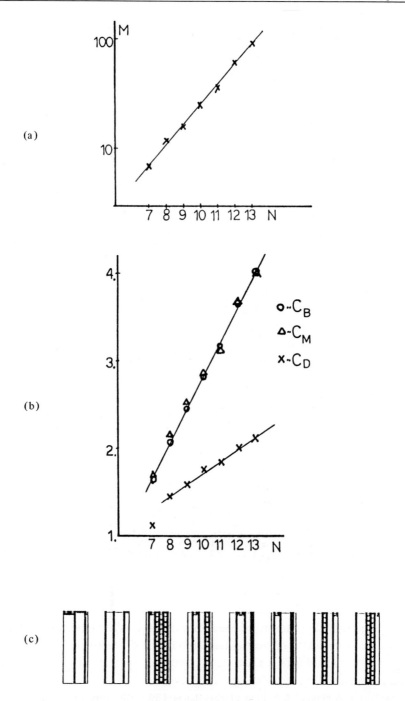

Fig. 3. Number of attractors M (a), C_B, C_M, and C_D (b) as a function of size N for the Rule 108. Examples of attractors with some transients for $N = 12$ (c).

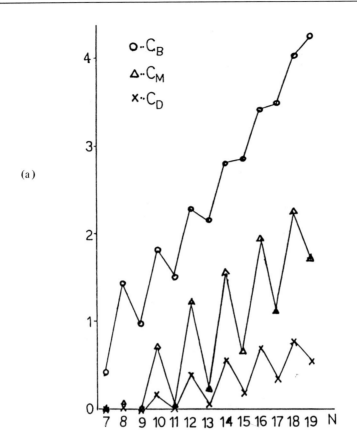

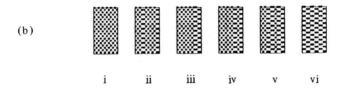

i   ii   iii   iv   v   vi

Fig. 4. C_B, C_M, and C_D (a) as a function of size N for Rule 50. Examples of attractors for $N = 12$ (b) (i-vi). In the paper, O is for C_B, Δ for C_M, and x for C_D.
The volume of the basin b_i is
(i) 15.6%   (ii) 15.2%   (iii) 12.3%   (iv) 16.9% (max.)   (v) 15.8%, and   (vi) 0.1%.
The attractors with the basin volume larger than 3% are restricted to (i)–(v). The basin volume for all-0 is minimum (0.05%) and (vi) is the minimum except the all-0 attractor.

Table I. Properties of attractors for class-2 CA. $\beta = \alpha/\ln 2$, i.e., $M \sim 2^{\beta N}$.

Rule	β	a_B	a_M	a_D	Basic Oscillator and period
132	.5	1.3	1.3	0.8	010 ————1
108	.6	.4	.4	.13	010, 0110, ————1 101 ⟷ 111 ————2
104	.3	.05	0	0	0110 ————1 fall into all-0 by noise
tot24	.3	.08	0	0	01110 ————1 fall into all-0 by noise
50	.5	3.0	3.0	.6	0110 ⟷ 1001 in 010101-structure as kinks; period = 2

attractors to the all-0 state. Thus, the measure and dynamical complexities vanish, because all the other attractors fall into the attractor "all-0" by the effect of a noise. Thus $q_i = 1$ for the all-0 attractor and 0 for other attractors. The Rule 104 and totalistic 24 belong to this type.

The complexity of class-2 CA seems to be classified into the above two cases.

The class 2 behavior is understood by the superposition of local oscillators. If a local oscillator has a period t and spatial range r, the number of attractor is roughly given by $(t + 1)^{N/r}$, since there are $(t + 1)$ possibilities in each r region (put the oscillator or not and put it with which phase of oscillation). This argument is easily extended to the case where there are more than one type of local oscillators. The linear increase of complexity is explained in the same way.

The volume of each attractor changes as size in the following way. The ratio of the volume of the basin for the attractor "all-0" decreases as size and the basin volumes for the attractors with more oscillators increase successively as the size.

In class-2 CA, the increase of dynamical complexity is much smaller, which means that the transition among attractors is organized. An example of the transition matrix P_{ij} is shown in Table II. In the example, the possible change of the number of oscillators by a single flip is only ± 1. In the class-2 CA studied here, the transition is regular. Since an attractor in class-2 CA can be regarded as a superposition of local oscillators, single flip-flop cannot affect the global behavior, and the transition is rather limited.

Some attractors are "repulsive" by the noise, as can be seen in Table II (for such attractor a_i, P_{ki} is zero for all k).

If a local oscillator exists as a kink in a zigzag structure (see Fig. 4b), there appears a difference in the parity of the size N (see Fig. 4a). As the system size goes larger the difference decreases.

Table II. Transition matrix P_{ij} for Rule 132 with $N = 9$ with all attractors $a_1 - a_{12}$. Here the number change by the transition is only ± 1.

	1	2	3	4	5	6	7	8	9	10	11
1	0	1	0	0	0	0	0	0	0	0	0
2	$\frac{3}{9}$	0	$\frac{2}{9}$	$\frac{2}{9}$	$\frac{2}{9}$	0	0	0	0	0	0
3	0	$\frac{5}{9}$	0	0	0	$\frac{2}{9}$	$\frac{1}{9}$	$\frac{1}{9}$	0	0	0
4	0	$\frac{6}{9}$	0	0	0	0	$\frac{1}{9}$	$\frac{1}{9}$	0	0	0
5	0	$\frac{6}{9}$	0	0	0	$\frac{1}{9}$	$\frac{1}{9}$	$\frac{1}{9}$	0	0	0
6	0	0	$\frac{4}{9}$	$\frac{2}{9}$	$\frac{1}{9}$	0	0	0	0	$\frac{2}{9}$	0
7	0	0	$\frac{3}{9}$	$\frac{2}{9}$	$\frac{3}{9}$	0	0	0	0	$\frac{1}{9}$	0
8	0	0	$\frac{3}{9}$	$\frac{2}{9}$	$\frac{3}{9}$	0	0	0	0	$\frac{1}{9}$	0
9	0	0	0	1	0	0	0	0	0	0	0
10	0	0	0	0	0	$\frac{4}{9}$	$\frac{2}{9}$	$\frac{2}{9}$	$\frac{1}{9}$	0	0
11	1	0	0	0	0	0	0	0	0	0	0

attractors: a_1 = 000000000 a_2 = 000000001 a_3 = 000000101

a_4 = 000000001 a_5 = 000010001 a_6 = 000010101

a_7 = 000100101 a_8 = 000101001 a_9 = 001001001

a_{10} = 001010101 a_{11} = 111111111

(all attractors are fixed points)

5. Class-3 CA

The class 3 behavior of CA is characterized by triangles with various sizes. Here we have investigated Rules 22, 18, 54, 146, and totalistic 12 and 22. The number of attractors and complexities as a function of system size are shown in Figs. 5-7, with some patterns of typical attractors (see also Fig. 8). See Ref. 13 for the Rule 54. Though the behavior is very complicated, the following points are common in class-3 CA.

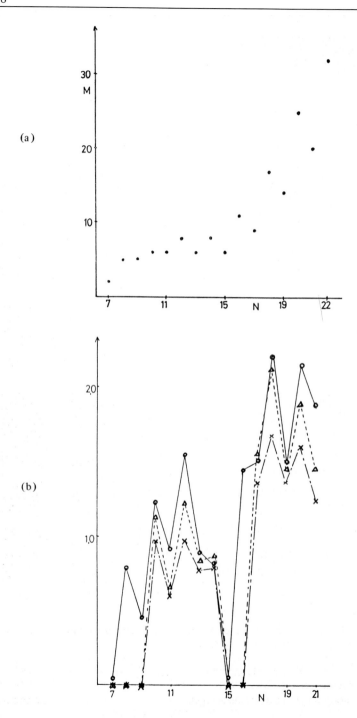

Fig. 5. Number of attractors M (a), C_B, C_M, and C_D (b) as a function of size N for the Rule 146.

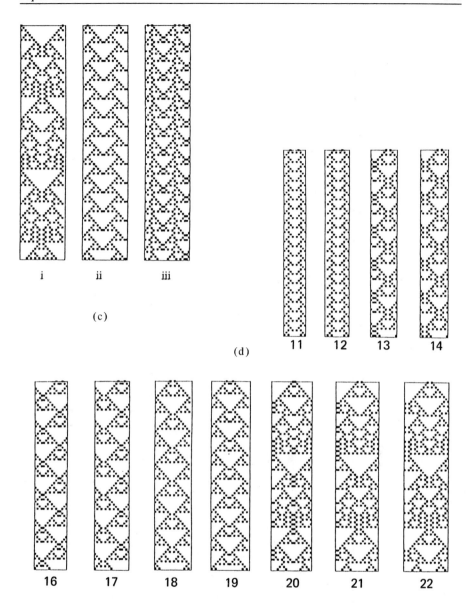

i   ii   iii

(c)

(d)   11   12   13   14

16   17   18   19   20   21   22

Fig. 5. (c) Examples of typical attractors for $N = 21$. The volume of the basin b_i is
    all-0:35%   (i) 17.3%   (ii) 11.0%   (iii) 10.5%
The attractors with the basin volume larger than 10% are restricted to (i) – (iii).
In (d), the attractor with the largest basin volume (or second largest if the one with the largest
volume is all-0) is shown for $10 < N < 23$.
    The basin volume b_i for the depicted attractor is
70% ($N = 11$),   35% ($N = 12$),   70% ($N = 13$),   30% ($N = 14$),
5%   ($N = 16$),   28% ($N = 17$),   12% ($N = 18$),   27% ($N = 19$),
18% ($N = 20$),   40% ($N = 21$),   17% ($N = 22$).
    The attractor for $N = 15$ is omitted since the all-0 has the basin volume more than 99%.

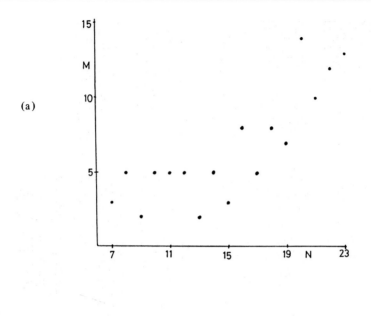

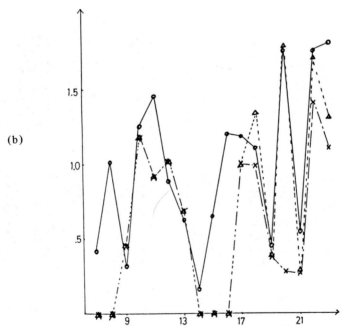

Fig. 6. Numbers of attractors M (a), C_B, C_M, and C_D (b) as a function of size N for the Rule 22.

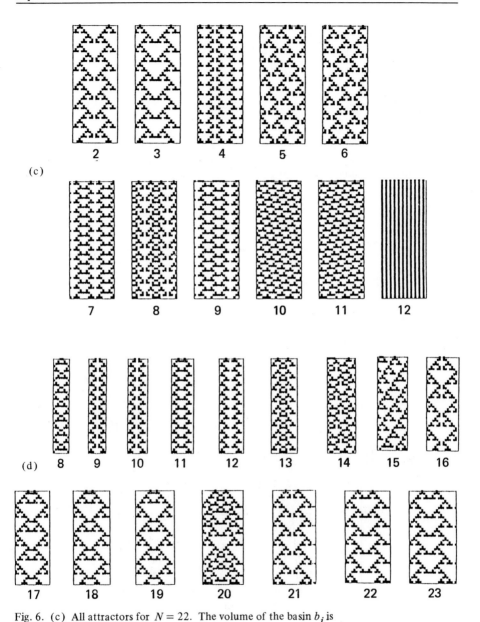

Fig. 6. (c) All attractors for $N = 22$. The volume of the basin b_i is
(1) all-0: 0.4%   (2) 30.3%   (3) 39.4%   (4) 6.5%   (5) 9.6%   (6) 9.6%   (7) 0.9%
(8) 0.8%   (9) 2.3%   (10) 0.01%   (11) 0.01%   (12) 5×10^{-5}%.
In (d), the attractor with the largest basin volume (or second largest if the one with the largest volume is all-0) is shown for $7 < N < 24$.
The basin volume b_i for the depicted attractor is
21.8% ($N = 8$),   89.6% ($N = 9$),   38.1% ($N = 10$),   35.9% ($N = 11$),
65.6% ($N = 12$),   32.4% ($N = 13$),   3.7% ($N = 14$),   10.4% ($N = 15$),
38.7% ($N = 16$),   40.0% ($N = 17$),   51.6% ($N = 18$),   91.3% ($N = 19$),
35.3% ($N = 20$),   88.8% ($N = 21$),   39.4% ($N = 22$),   31.2% ($N = 23$).

(e)

Fig. 6 (e) shows the transition loop by P_{ij} among attractors. Only the attractors with non-vanishing q_i are shown. The arrows indicate the possible transition between attractors by a single flip-flop noise. See Table III for P_{ij}.

(a)

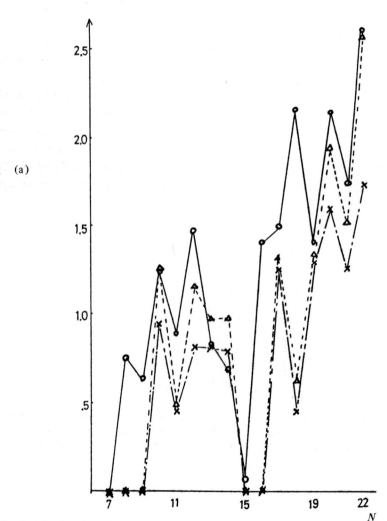

Fig. 7. (a) C_M, C_B, and C_D as a function of size N for the Rule 18. The number of attractors take $10 \sim 20$ for $N < 20$.

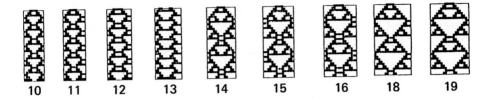

10 11 12 13 14 15 16 18 19

Fig. 8. The attractor with the largest basin volume (or second largest if one with the largest volume is all-0) is shown for $9 < N < 20$ for the totalistic rule 12.

The basin volume b_i for the depicted attractor is

19% ($N = 10$), 63% ($N = 11$), 39% ($N = 12$), 67% ($N = 13$),
38% ($N = 14$), 76% ($N = 15$), 36% ($N = 16$), 23% ($N = 18$),
47% ($N = 19$).

The attractor for $N = 17$ is omitted since it has only 1% volume.

(i) Number of attractors changes irregularly as the system size N. The increase in the size is at most bounded by some power of the system size N.

(ii) The attractors which have a large region of basins are the ones with triangle structures and the all-0 attractor. Among the attractors with triangle structures, the attractor with a larger size of triangles has a larger size of basin of attractions (see Figs. 5(c) and 6(c)). In Fig. 9, the basin volume is shown as a function of the size of the largest triangle in the attractor for the totalistic rule 22. In the example the basin volume is roughly proportional to the (size of the largest triangle

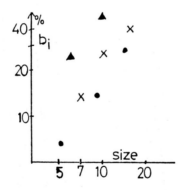

Fig. 9. Basin volume as a size of the largest triangle in the attractor for the totalistic rule 22. The size denotes the longest sequence of 0's in the attractor. Only the typical attractors with class-3-like triangle structures are chosen. ● is for $N = 16$, ▲ for $N = 17$, and x for $N = 19$.

in the attractor).[1.5] For other rules, the similar behavior is observed, but the power fit is not so good (at least for small size N).

(iii) The attractors with the largest basin volume except the "all-0" are shown in Figs. 5(d), 6(d), and 8. The characteristic feature of such attractors is that they start from a simple seed. For example the attractors are formed from the seeds 101, 100001, or 11 for Rule 146, 11, 1111, or 11111 for totalistic rule 12, 1001, 10001, 100001, and 1000001 for Rule 22, and so on. The simplest seed which has a recurrence gives the attractor with the largest basin volume. This kind of choice is quite analogous to the choice of eigenfunction in the Schrodinger equation where the interference of phase is important. In our problem, the possible configuration of triangles is determined by a kind of interference effect.

The analogy with the wave mechanics also implies that the most probable attractor is related to the ground state of wavefunction in the sense that the number of nodes is smallest (in other words, it has only a small number of "seeds", i.e., the structure of 00000***0000). The attractor with smaller basin volume may be related to the excited state in the sense that it has more nodes.

(iv) The irregular behavior as a size change seems to depend on some number theoretic properties of the size and rules. For example, there occurs singular behavior around at $N = 2^k - i$ ($i = 0$, or 1, or -1 which depends on the rule). The number of attractors decreases at some values near $N = 2^k$ and all-0 attractor has a large measure. For Rule 146, the ratio for the attractor with all-0 has 99% measure at $N = 15$ ($= 2^4 - 1$). The complexities C_M and C_D vanish at $N = 15$ and 16 (see Fig. 5(b)). For Rule 22, C_M and C_D vanish at $N = 14$, 15, and 16 (see Fig. 6(b)), since all the attractors fall down into "all-0" attractor by the noise.

At these sizes, the measure for all-0 is close to 1, since the patterns from simple seeds fall into all-0 and the attractor with the large triangles cannot exist (see $N = 14$ and 15 of Fig. 6(d)).

(v) The complexities also change irregularly as the system size. They seem to increase slowly as the system size. Generally speaking, C_D is not so small compared with C_M. That is, the mutual information $C_M - C_D$ is small compared with the cases for the CA in other classes. Thus, the connectivity among attractors by a low noise is random.

An example of P_{ij} is shown in Table III, for Rule 22 with $N = 22$. After the transients, the measure for the attractor a_8, a_{10}, a_{11}, and a_{12} goes to zero and the transitions by small noise occur among attractors $a_1, a_2, \ldots, a_7$ and a_9. The transition diagram is shown in Table III. Here the most probable loop of the transition is $a_2 \to a_3 \to a_2$, while the second most probable loop is $a_1 \to a_2 \to a_3 \to a_1$, and the next is $a_3 \to a_7 \to a_4 \to a_3$, and so on. We note that various transition loops are formed by the addition of a noise (Fig. 6(e)).

Table III. Transition matrix P_{ij} for Rule 22 with $N = 22$. The attractors are shown in Fig. 6.

	1	2	3	4	5	6	7	8	9	10	11	12
1	0	1	0	0	0	0	0	0	0	0	0	0
2	0	$\frac{5}{11}$	$\frac{6}{11}$	0	0	0	0	0	0	0	0	0
3	$\frac{1}{11}$	$\frac{2}{11}$	$\frac{6}{11}$	0	$\frac{1}{22}$	$\frac{1}{22}$	$\frac{1}{11}$	0	0	0	0	0
4	0	$\frac{1}{11}$	$\frac{4}{11}$	$\frac{3}{11}$	$\frac{1}{11}$	$\frac{1}{11}$	0	0	$\frac{1}{11}$	0	0	0
5	0	$\frac{6}{11}$	$\frac{4}{11}$	$\frac{1}{11}$	0	0	0	0	0	0	0	0
6	0	$\frac{6}{11}$	$\frac{4}{11}$	0	0	0	0	0	0	0	0	0
7	0	$\frac{1}{11}$	$\frac{2}{11}$	$\frac{4}{11}$	0	0	$\frac{2}{11}$	0	$\frac{2}{11}$	0	0	0
8	0	$\frac{4}{11}$	$\frac{4}{11}$	0	$\frac{1}{11}$	$\frac{1}{11}$	0	$\frac{1}{11}$	0	0	0	0
9	0	$\frac{3}{11}$	$\frac{9}{22}$	$\frac{1}{11}$	$\frac{3}{22}$	$\frac{1}{22}$	0	0	$\frac{1}{22}$	0	0	0
10	0	0	$\frac{7}{11}$	0	$\frac{3}{11}$	$\frac{1}{11}$	0	0	0	0	0	0
11	0	0	$\frac{7}{11}$	0	$\frac{1}{11}$	$\frac{3}{11}$	0	0	0	0	0	0
12	$\frac{1}{2}$	$\frac{1}{2}$	0	0	0	0	0	0	0	0	0	0

(vi) Another interesting quantity is the period of an attractor. The period of the main attractor increases slowly with an irregular change. The numerical observation shows: The main attractor starts from a small seed and grows with a constant speed till it comes back to the seed pattern after $O(N)$ steps (the position of the seed may be different from the original position). Thus, the increase of the period seems to be bounded at most by some power of the size, which is consistent with the numerical results. The longest period among all the attractors increases faster. It seems to be bounded by some power, though it shows a rapid jump at some values of N.

6. Class-4 CA

The class 4 behavior for CA characterized by Stephen Wolfram is long-time transients and the existence of local oscillators and local propagating patterns and the sensitive dependence of patterns on the initial configurations. We have investigated here the totalistic-52, totalistic 20 and models S1-S2 (see Appendix). The results for other models with soliton-like excitations will be shown in the next section. The characteristic features for the basin structure of the class 4 systems may be summarized as follows:

(i) The number of attractors increase exponentially, though the increase is rather irregular. The pattern of attractor which has a large region of basins changes as size, though "all-0" or "all-1" has a large basin of attractions in many rules. The pattern of the attractor with the large basin of attraction except the all-0 or all-1 changes as the size (for some size, it is global and for other size it is local), which is a main difference from the other classes. As N is increased, attractors of essentially new type appear successively.

(ii) The basin complexity C_B takes a comparatively large value, which changes irregularly as size. The measure complexity C_M is much smaller than C_B, since the probability measure (by a noise) for "all-0" (or "all-1") is much larger than the ratio for the basin of attractions to such states. The dynamical complexity is much smaller, which means that the mutual information is rather large. In other words, the transition between attractors by noise is regularly structurized.

In the following we show three typical examples for class-4 CA.

(1) totalistic rule 52: (see Fig. 10).

The rule is symmetric about the transformation $0\langle-\rangle1$. The main attractors are all-0 and all-1 which have the same basin volume b_i and probability q_i by the symmetry. The ratio of the basin volume for the all-0 (or all-1) attractor is shown in Fig. 10(d). Both global and local attractors coexist, some of which are shown with the basin volume (Fig. 10(c)). The number of attractors increases exponentially with a large increasing rate.

By the noise, the attractors fall into all-1 or all-0 attractor for almost all N. Since the probability for each attractor is 50%, the measure complexity takes ln 2. The dynamical complexity vanishes for almost all N, since the all-1 and all-0 attractors are disconnected by a single flip-flop (see Fig. 10(e) for the transition loop by P_{ij}.)

(2) totalistic rule 20 (see Fig. 11):

The number of attractors increase rather irregularly around $10 < M < 20$, while it has a rapid jump at $N = 12$ and 16 ($M > 50$). The all-0 attractor has a large basin volume which is shown in Fig. 11(c). For $N > 14$, the state falls down into the all-0 attractor and the measure and dynamical complexities vanish.

In Fig. 11(b), the main attractors for $N = 21$ are shown. A local irregular propagating pattern is remarkable. In Fig. 11(c), the attractor pattern with the largest basin volume (or the second largest in the case that all-0 has the largest

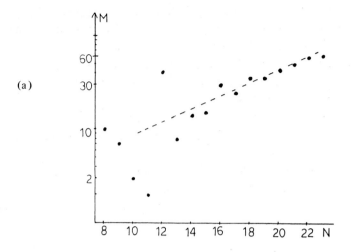

(a)

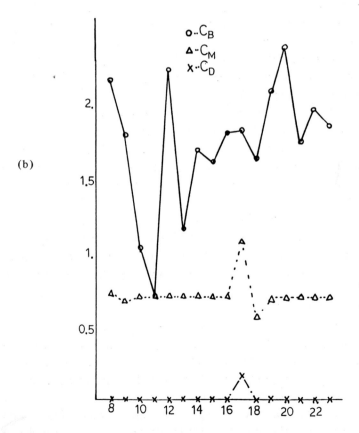

(b)

Fig. 10. Number of attractors M (a), C_B, C_M, and C_D (b) as a function of size N for the totalistic rule 52.

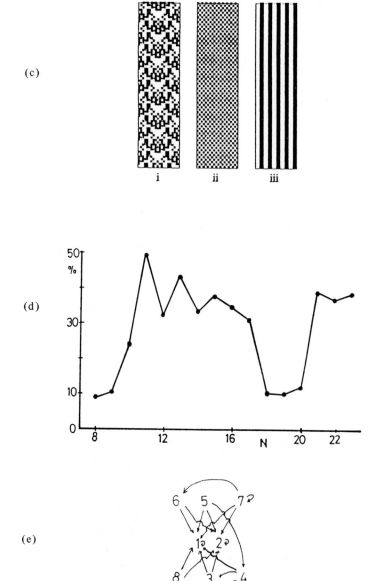

Fig. 10(c) Examples of typical attractors are shown for $N = 20$. The volume of the basin b_i is all-0: 11.9%, all-1 : 11.9%, (i) 23.3% (the attractor obtained by the $0 \langle - \rangle 1$ transformation has the same basin volume), (ii) 4.8% (iii) 7.3%.
The attractors with the basin volume larger than 3% are restricted to (i)–(v).
(d) shows the basin volume all all-0 (or all-1) as a function of N.
(e) shows the transition loop by P_{ij} for totalistic rule 52 with $N = 13$. The arrows indicate the possible transition between attractors by a single flip-flop noise. Only the attractors $a_1 = $ all-0 and $a_2 = $ all-1 have non-zero probability q ($= 50\%$).

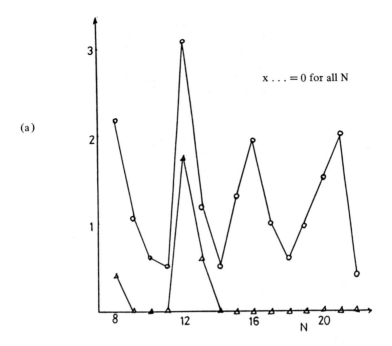

(a)

x . . . = 0 for all N

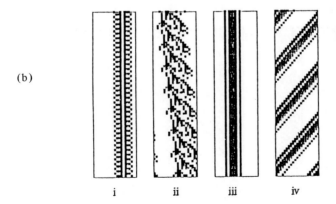

(b)

Fig. 11. C_B, C_M, and C_D (a) as a function of size N for the totalistic rule 20. Examples of typical attractors are shown for $N = 21$ (b).

The volume of the basin b_i is

all-0 : 96.2%.   (i) 1.1%   (ii) 0.3%   (iii) 1.7%   (iv) 0.09%.

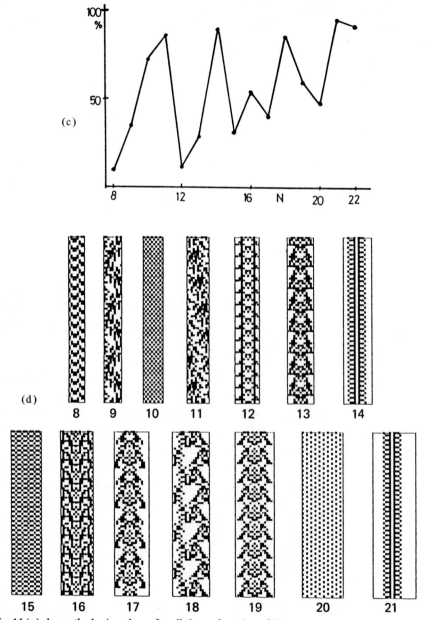

Fig. 11(c) shows the basin volume for all-0 as a function of N.
In (d), the attractor with the largest basin volume (or second largest if the one with the largest volume is all-0) is shown for $7 < N < 22$.
The basin volume b_i for the depicted attractor is
18.8% ($N=8$), 49.2% ($N=9$), 27.5% ($N=10$), 6.9% ($N=11$),
32.5% ($N=12$), 56.9% ($N=13$), 5.5% ($N=14$), 47.8% ($N=15$),
17.7% ($N=16$), 46.0% ($N=17$), 7.1% ($N=18$), 33.1% ($N=19$),
21.3% ($N=20$), 1.1% ($N=21$).

basin volume) is shown. We note that such attractors are global and irregular, and the patterns change their structures as the size in an irregular way, which is remarkably different from the class-3 case.

(3) model S1 $\{0001101\} - \{01011\}$ (see Fig. 12):

The number of attractors increases exponentially. The all-0 and all-1 attractors have large basin volumes. The measure entropy is much smaller than the basin volume, since the measure of all-0 (or all-1) attractor increases by the flip-flop. The dynamical entropy is much smaller and the transition by the flip-flop is regular.

(4) model S2 $\{0011101\} - \{01011\}$ (figures omitted) (for $N < 19$)

The following features are observed: (a) exponential increase of the number of attractors; (b) C_B takes about $1 \sim 1.5$ for $12 < N < 19$; (c) C_M is very small (for most N); (d) $C_D = 0$; (e) all-0 attractor has a large measure.

7. CA with Soliton-like Excitations

As is shown in the Appendix and Ref. 12, a class of CA with solitons shows an interesting behavior such as the integrable-like behavior or soliton turbulence. Here the following two typical examples will be investigated; one is for the integrable-like behavior, and the other for the turbulent-like behavior:

(1) Integrable-like behavior; model S3; $\{0000011\} - \{11000\}$

For this class, the basins for the state of superpositions of "solitons" go larger as the system size is increased. The important difference between this type of behavior and the usual integrable systems studied in the soliton theory is that our system is integrable only after the transients have decayed out. Thus, our system can be regarded as an "integrable system on an attractor".

For small N, however, the basin volume for the superposition of solitons is not necessarily large. Some bound states of solitons have large basin volumes as can be seen in Figs. 13, where the main attractors are shown for $N = 19$, 20, and 21. As the size is increased, however, the ratio for the superposition of soliton-states seems to increase. We have studied some attractors for $N = 30$ or 45 by choosing some samples of initial configurations. More than 80% of such initial configurations are attracted into the ensembles of solitons.

The number of attractors increase exponentially. The dynamical entropy is much smaller than the measure entropy, which shows that the transition among attractors is well-organized.

(2) Soliton turbulence; model S4; $\{0001011\} - \{10011\}$

Some CA show the soliton turbulence[12]. A typical pattern is shown in Fig. 15 (d), where the sensitivity on the phase of collisions of 1101-"solitons" make the turbulent-like phenomena. For the attractors for a CA with a small size, however, it is hard to find such patterns. The main attractors are (A) superposition of

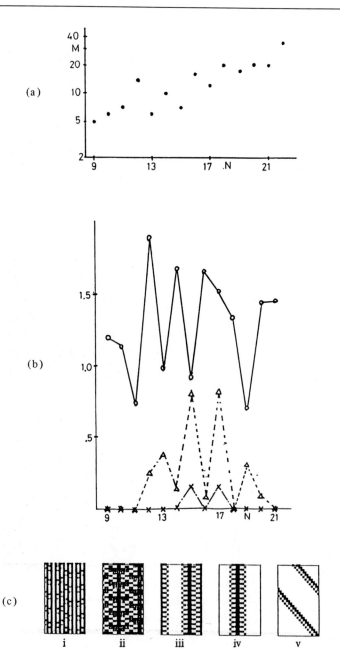

Fig. 12. Number of attractors M (a), C_B, C_M, and C_D (b) as a function of size N for the model S1. Examples of typical attractors for $N = 20$ (c).
The volume of the basin b_i is all-0: 39.5%; all-1: 30.3%; (i) 15.4%   (ii) 10.4%   (iii) 2.0% (iv) 1.1%   (v) 0.3% (and the same volume for the attractor with the converse direction). The attractors with the basin volume larger than 0.3% are shown.

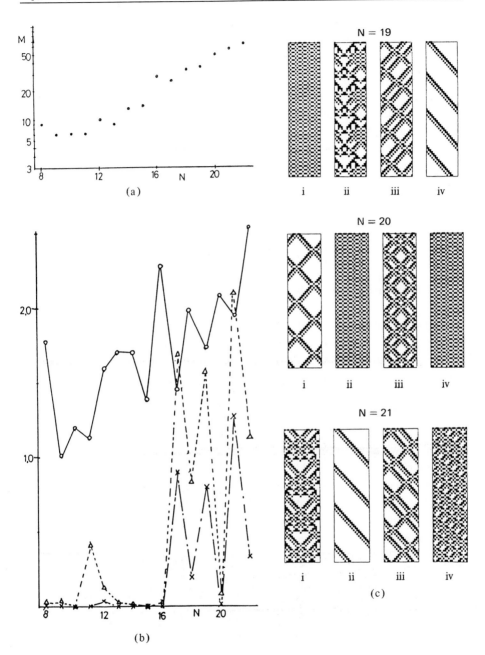

Fig. 13. Number of attractors M (a), C_B, C_M, and C_D (b) as a function of size N for the model S3.

In (c), the main attractors are shown

$N = 19$: (i) 52.2%   (ii) 11.0%   (iii) 5.3%   (iv) 3.4%
$N = 20$: (i) 21.6%   (ii) 19.4%   (iii) 16.6%   (iv) 3.5%
$N = 21$: (i) 52%   (ii) 5.5%   (iii) 4.7%   (iv) 4.1%

of some soliton-states (B) global periodic patterns (see Fig. 14). The latter patterns have the larger region of basins. Some collisions increase the number of solitons till they reach the global patterns (the latter attractors) with small periods and with more than 50% of 1's.

We have performed some simulations for $N = 30$ and 40 by choosing some initial configurations. Still, the attractor with the typical turbulent patterns have small measures (about 10%). The turbulent pattern is seen as the transients before the CA falls into the attractor in the type (B) above (see Fig. 14(d)). The transient time increases rapidly as the size. This observation may imply that the turbulent-like patterns in some CA may be characterized as the transients, the time for which diverges as the system size.

Here, (1) exponential growth of number of attractors (2) large C_B and very small C_D are observed again, which are typical in class-4 CA.

8. Discussion and Future Problems

We have investigated here the storage of information in the attractors of CA and the complexity of networks among attractors connected by a noise.

One important question is "what is the generic behavior for the CA with large N?" In class-1 CA, the attractor is trivial and shows no essential change as the size. In class-2 CA, the attractors are local. Thus the attractors at large N is essentially the superposition of the attractors of small size. In class-3 CA, the attractors are global and the period of the main attractors increases. The generic behavior at large N, however, is characterized by the attractors. The main attractors for a large size can be characterized by those for a small size, since there is self-similarity and the main attractor is generated by a simple seed.

In class-4 CA, however, it may be hard to predict the behavior of the CA with a large size from the result for the attractors, since the transients before the state falls into the attractors increase rapidly as size, as can be seen in the case for soliton-turbulence, where the turbulent behavior (for large N) may be attributed not to the attractors but to the transients.

The memory in the class-3 CA may be used as the holographic memory in the following three points. First, the attractor includes a self-similar triangle structure. Thus, we can construct a pattern similar to the original triangle structure even if the information of some parts is lost. Secondly, the interference effect in the triangle structures discussed in Section 5 is analogous with the interference in the wave mechanics. Thirdly, the attractor's network by the transition matrix P_{ij} is global.

It will be of interest to extend our approach to basin structures in other systems. The direct application is possible for the 2-dimensional case, where the phase

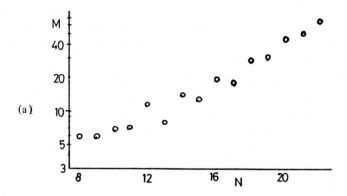

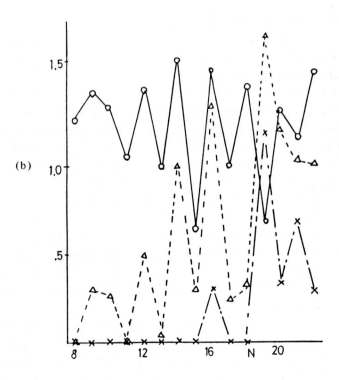

Fig. 14. Number of attractors M (a), C_B, C_M, and C_D (b) as a function of size N for the model S4.

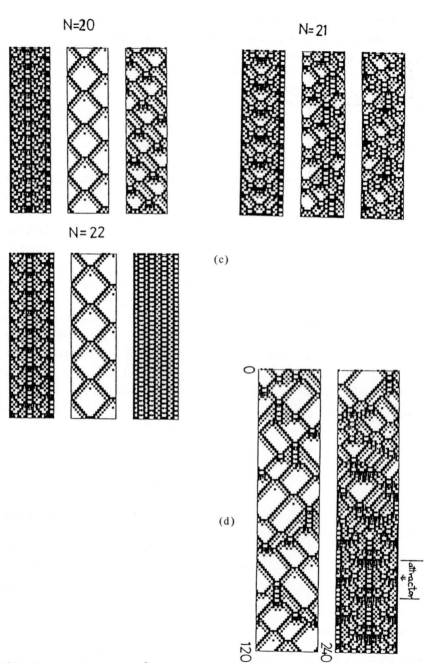

Fig. 14 (c), the main attractors are shown
$N = 20$:   (i) 52.6%   (ii) 31.7%   (iii)   3.6%   (each)
$N = 21$:   (i) 73.8%   (ii)   6.2%   (iii)   5.2%
$N = 22$:   (i) 35%   (ii) 13%   (iii) 17.6%
In (d), an example of an evolution for $N = 30$ is shown. After the time step ~ 200, it falls into
the attractor with small period.

transition by a noise can occur[14]. It is of interest to characterize the low-noise ordered states from the basin structure for the ordered states.

Another important example is a spin glass system, especially Sherington-Kirkpatrick model, where a lot of attractors (fixed points) are hierarchically organized[15].

Also, it is of importance to extend our approach to the dynamical system with continuous variables: Most dissipative systems such as low-dimensional maps[16] with small dissipation or high-dimensional maps such as coupled map lattices[17,18] can have a large number of attractors. The basin structure has recently been investigated intensively. The statistical properties such as the basin volume entropy and the jumping process among attractors by the noise can be investigated in the present paper's line.

Informational aspects in the dynamical system are important in the intelligent network system[19,20] such as the neural or immune network and some artificial intelligence systems.

In addition to the storage of information studied in this paper, the processing of information should be studied in the high-dimensional chaos or CA. In the CA with soliton-like excitations, the information can be transmitted by the "solitons", even if the state is turbulent. The information transmission is calculated by the mutual information flow, in a similar way as the coupled map lattice case[18].

Also, selective propagation of information will be of use in the intelligent system. In CA with soliton-like patterns, some specific patterns can easily propagate to other sites, while others decay out. The quantification of such selectivity will be of importance.

Acknowledgements

The author would like to thank Mr. Yukito Iba and Dr. Shinji Takesue for stimulating discussions and critical comments. He would also like to thank Dr. Yoji Aizawa, Dr. Norman Packard and Dr. Stephen Wolfram for useful discussions and critical comments. He is grateful to the Institute of Plasma Physics at Nagoya for the facility of FACOM M-200 and to the Research Institute of Fundamental Physics at Kyoto University for some financial support.

Appendix: CA with Soliton-like Excitations

In the class-4 cellular automata, the patterns which propagate with some speed is commonly seen. In Ref. 12, a class of CA with specific type of "soliton" (00101100) is investigated in detail. We impose the condition that the pattern

00101100 should move right at the next step (i.e., 00010110). There are 2^{12} possible rules for the legal CA with 2-states and range = 2 which satisfies this type of 1011-soliton conditions.

Simulations for all these rules have been performed. The rule for the two-state CA with range 2 is coded by the 32 numbers $i_k = (0, 1)$; $(00000) \rightarrow i_0$, $(00001) \rightarrow i_1, \ldots, (11111) \rightarrow i_{31}$. The condition that the rule must be legal as stated by Wolfram and the above condition of the existence of 1011-soliton leaves only 12 parameters: $(l_1, l_2, l_3, l_4, l_5, l_6, l_7) - (k_1, k_2, k_3, k_4, k_5) \equiv (i_4, i_{10}, i_{14}, i_{17}, i_{21}, i_{27}, i_{31}) - (i_7, i_9, i_{15}, i_{19}, i_{23})$. Interesting behavior which does not belong to the usual class-4 type is soliton-like behavior. For some rules, the dynamics of system is governed only by the soliton-like excitations (1011) and their collisions. If the "solitons" pass through each other by collisions, the CA can be regarded as a kind of integrable system. In some other rules, the collision of solitons show the sensitive dependence on the phase of the collisions, which induces the turbulence as an ensemble of 1011-solitons. Some of course, show the usual class-4 behavior, while some show class-3 or class-2 behavior for most of the initial conditions. Here, the evolution of the following four rules are shown. See Ref. 12 for more details:

(1) S1:(0001101-01011): class-4 like (see Fig. 15.1)
(2) S2:(0011101-01011): class-4 like (see Fig. 15.2)
(3) S3:(0000011-11000): integrable-like (see Fig. 15.3)
(4) S4:(0001011-10011): soliton-turbulence (see Fig. 15.4)

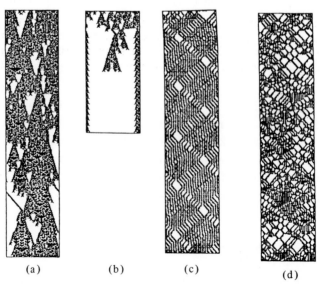

(a)   (b)   (c)   (d)

Fig. 15. Examples of the evolution of CA with "solitons". $N = 100$. Random initial configurations.
(a) model S1   (b) model S2   (c) model S3   (d) model S4.

References

1.   S. Wolfram, *Rev. Mod. Phys.* **55** (1983) 601; *Physica* **10D** (1984) 1.
2.   *Physica* **10D** (1984), ed. D. Farmer, T. Toffoli, and S. Wolfram.
3.   Dynamical Systems and Cellular Automata (1985), Academic Press, ed. J. Demongeot, E. Golès, and M. Tchuente.
4.   R. Shaw, *Zeit. fur Naturforschung* **36a** (1981) 80.
5.   See e.g., J. J. Hopfield, *Proc. Natl. Acad. Sci.* **79** (1982) 2554; **81** (1984) 3088.
6.   T. Hogg and B. A. Huberman, *Proc. Natl. Acad. Sci.* **81** (1984) 6871.
7.   S. Wolfram, in Ref. 2.
8.   T. Hogg and B. A. Huberman, *Phys. Rev.* **A32** (1985) 2338.
9.   See for the structures for Rule 90; O. Martin, A. Odlyzko, and S. Wolfram, *Comm. Math. Phys.* **93** (1984) 219.
10. The jumping among attractors was first investigated for the random Boolean networks, by S. Kauffman, *J. Theor. Biol.* **22** (1969) 437.
11. K. Matsumoto and I. Tsuda, *J. Phys.* **A18** (1985) 3561 and preprint.
12. Y. Aizawa, I. Nishikawa and K. Kaneko, in preparation.
13. K. Kaneko, in "Dynamical Systems and Nonlinear Oscillations" pp. 194-209 (ed. G. Ikegami, World Scientific, 1986).
14. K. Kaneko and Y. Akutsu, *J. Phys.* **A19** (1986) 69.
15. M. Mezard *et al., Phys. Rev. Lett.* **52** (1984) 1156; D. Sherrington and S. Kirkpatrick, *Phys. Rev. Lett.* **35** (1975) 1792.
16. C. Grebogi, E. Ott, and J. Yorke, *Physica* **7D** (1983) 181; S. Takesue and K. Kaneko, *Prog. Theor. Phys.* **71** (1984) 35.
17. K. Kaneko, "Collapse of Tori and Genesis of Chaos in Dissipative Systems" (World Scientific, 1986) Chapter 7; J. P. Crutchfield and K. Kaneko, in preparation.
18. K. Kaneko, "Lyapunov Analysis and Information Processing in Coupled Map Lattices" to appear in *Physica* **D** and references cited therein.
19. J. D. Farmer, N. H. Packard, and A. S. Perelson, *Physica* **D**, to appear.
20. F. J. Varela, "Principles of Biological Autonomy", North-Holland, 1979.

Approaches to Complexity Engineering*

Stephen Wolfram

The Institute for Advanced Study, Princeton NJ 08540.

(December 1985; modified February 1986)

Principles for designing complex systems with specified forms of behaviour are discussed. Multiple scale cellular automata are suggested as dissipative dynamical systems suitable for tasks such as pattern recognition. Fundamental aspects of the engineering of such systems are characterized using computation theory, and some practical procedures are discussed.

The capabilities of the brain and many other biological systems go far beyond those of any artificial systems so far constructed by conventional engineering means. There is however extensive evidence that at a functional level, the basic components of such complex natural systems are quite simple, and could for example be emulated with a variety of technologies. But how a large number of these components can act together to perform complex tasks is not yet known. There are probably some rather general principles which govern such overall behaviour, and allow it to be moulded to achieve particular goals. If these principles could be found and applied, they would make new forms of engineering possible. This paper discusses some approaches to such forms of engineering with complex systems. The emphasis is on general concepts and analogies. But some of the specific systems discussed should nevertheless be amenable to implementation and detailed analysis.

In conventional engineering or computer programming, systems are built to achieve their goals by following strict plans, which specify the detailed behaviour of each of their component parts. Their overall behaviour must always be simple enough that complete prediction and often also analysis is possible. Thus for example motion in conventional mechanical engineering devices is usually constrained simply to be periodic. And in conventional computer programming, each step consists of a single operation on a small number of data elements. In both of these cases, much more complex behaviour could be obtained from the basic components, whether mechanical or logical, but the principles necessary to make use of such behaviour are not yet known.

Nature provides many examples of systems whose basic components are simple, but whose overall behaviour is extremely complex. Mathematical models such as cellular automata (e.g. [1]) seem

---

* Loosely based on an invited talk entitled "Cellular automaton engineering" given at the conference on "Evolution, Games and Learning" held at Los Alamos in May 1985. To be published in *Physica D*. More details will appear in due course.

to capture many essential features of such systems, and provide some understanding of the basic mechanisms by which complexity is produced for example in turbulent fluid flow. But now one must use this understanding to design systems whose complex behaviour can be controlled and directed to particular tasks. From complex systems science, one must now develop complex systems engineering.

Complexity in natural systems typically arises from the collective effect of a very large number of components. It is often essentially impossible to predict the detailed behaviour of any one particular component, or in fact the precise behaviour of the complete system. But the system as a whole may nevertheless show definite overall behaviour, and this behaviour usually has several important features.

Perhaps most important, it is robust, and is typically unaffected by perturbations or failures of individual components. Thus for example a change in the detailed initial conditions for a system usually has little or no effect on the overall outcome of its evolution (although it may have a large effect on the detailed behaviour of some individual elements). The visual system in the brain, for example, can recognize objects even though there are distortions or imperfections in the input image. Its operation is also presumably unaffected by the failure of a few neurons. In sharp contrast, however, typical computer programs require explicit account to be taken of each possible form of input. In addition, failure of any one element usually leads to catastrophic failure of the whole program.

Dissipation, in one of many forms, is a key principle which lies behind much of the robustness seen in natural systems. Through dissipation, only a few features in the behaviour of a system survive with time, and others are damped away. Dissipation is often used to obtain reliable behaviour in mechanical engineering systems. Many different initial motions can for example be dissipated away through viscous damping which brings particular components to rest. Such behaviour is typically represented by a differential equation whose solution tends to a fixed point at large times, independent of its initial conditions. Any information on the particular initial conditions is thus destroyed by the irreversible evolution of the system.

In more complicated systems, there may be several fixed points, reached from different sets of initial conditions. This is the case for an idealized ball rolling on a landscape, with dissipation in the form of friction. Starting at any initial point, the ball is "attracted" towards one of the local height minima in the landscape, and eventually comes to rest there. The set of initial positions from which the ball goes to a particular such fixed point can be considered as the "basin of attraction" for that fixed point. Each basin of attraction is bounded by a "watershed" which typically lies along a ridge in the landscape. Dissipation destroys information on details of initial conditions, but preserves the knowledge of which basin of attraction they were in. The evolution of the system can be viewed as dividing its inputs into various "categories", corresponding to different basins of attraction. This operation is the essence of many forms of pattern recognition: despite small changes, one recognizes that a particular input is in a particular category, or matches a particular pattern. In the example of a ball rolling on a landscape, the categories correspond to different regions of initial positions. Small changes in input correspond to small changes in initial position.

The state of the system just discussed is given by the continuous variables representing the position of the ball. More familiar examples of pattern recognition arise in discrete or digital systems, such as those used for image processing. An image might be represented by a 256×256 array of cells, each black or white. Then a simple image processing (or "image restoration") operation would be to replace any isolated black cell by a white cell. In this way certain single cell errors in the images can be removed (or "damped out"), and classes of images differing just by such errors can be recognized as equivalent (e.g. [3]). The process can be considered to have attractors corresponding to the possible images without such errors. Clearly there are many of these attractors, each with a particular basin of attraction. But in contrast to the example with continuous variables above, there is no obvious measure of "distance" on the space of images, which could be used to determine which basin of attraction a particular image is in. Rather the category of an image is best determined by explicit application of the image processing operation.

Length n sequences of bits can be considered as corners of an n-dimensional unit hypercube. The Hamming distance between two sequences can then be defined as the number of edges of the hypercube that must be traversed to get from one to the other, or, equivalently, the total number of bits that differ between them. It is possible using algebraic methods to devise transformations with basins of attraction corresponding to spheres which enclose all points at a Hamming distance of at most say two bits from a given point [4]. This allows error-correcting codes to be devised in which definite messages can be reconstructed even though they may contain say up to two erroneous bits.

The transformations used in error-correcting codes are specially constructed to have basins of attraction with very simple forms. Most dissipative systems, however, yield much more complicated basins of attraction, which cannot for example be described by simple scalar quantities such as distances. The form of these basins of attraction determines what kinds of perturbations are damped out, and thus what classes of inputs can be recognized as equivalent.

As a first example, consider various idealizations of the system discussed above consisting of a ball rolling with friction on a landscape, now assumed one dimensional. In the approximation of a point ball, this is equivalent to a particle moving with damping in a one-dimensional potential. The attractors for the system are again fixed points corresponding to minima of the potential. But the basins of attraction depend substantially on the exact dynamics assumed. In the case of very large friction, the particle satisfies a differential equation in which velocity is proportional to force, and force is given by the gradient of the potential. With zero initial velocity, the basins of attraction in this case have a simple form, separated by boundaries at the positions of maxima in the potential. In a more realistic model, with finite friction and the inertia of the ball included, the system becomes similar to a Roulette wheel. And in this case it is known that the outcome is a sensitive function of the precise initial conditions. As a consequence, the basins of attraction corresponding for example to different holes around the wheel must have a complicated, interdigitated, form (*cf* [5]).

Complicated basin boundaries can also be obtained with simpler equations of motion. As one example, one can take time to be discrete, and assume that the potential has the form of a polynomial, so that the differential equation of motion is approximated by an iterated polynomial mapping. The sequence of positions found from this mapping may overshoot the minimum, and for some values of parameters may in fact never converge to it. The region of initial conditions which lead to a particular attractor may therefore be complicated. In the case of the complex iterated mapping $z \rightarrow z^2 + c$, the boundary of the basin of attraction (say for the attractor $z = \infty$) is a Julia set, and has a very complicated fractal form (e.g. [6]).

The essentials of the problem of finding basins of attraction already arise in the problem of determining what set of inputs to a function of discrete variables yields a particular output. This problem is known in general to be computationally very difficult. In fact, the satisfiability problem of determining which if any assignments of truth values to n variables in a Boolean expression make the whole expression true is NP-complete, and can presumably be solved in general essentially only by explicitly testing all 2^n possible assignments (e.g. [7]). For some functions with a simple, perhaps algebraic, structure, an efficient inversion procedure to find appropriate inputs may exist. But in general no simple mathematical formula can describe the pattern of inputs: they will simply seem random (*cf* [8]).

Many realistic examples of this problem are found in cellular automata. Cellular automata consist of a lattice of sites with discrete values updated in discrete steps according to a fixed local rule. The image processing operation mentioned above can be considered as a single step in the evolution of a simple two-dimensional cellular automaton (*cf* [9]). Other cellular automata show much more complicated behaviour, and it seems in fact that with appropriate rules they capture the essential features of many complex systems in nature (e.g. [1]). The basic problems of complexity engineering thus presumably already arise in cellular automata.

Most cellular automata are dissipative, or irreversible, so that after many steps, they evolve to attractors which contain only a subset of their states. In some cellular automata (usually identified as classes 1 and 2), these attractors are fixed points (or limit cycles), and small changes in initial conditions are usually damped out [10]. Other cellular automata (classes 3 and 4), however, never settle down to a fixed state with time, but instead continue to show complicated, chaotic, behaviour. Such cellular automata are unstable, so that most initial perturbations grow with time to affect the detailed configuration of an ever-increasing number of sites. The statistical properties of the behaviour produced are nevertheless robust, and are unaffected by such perturbations.

It can be shown that the set of fixed points of a one-dimensional cellular automata consists simply of all those configurations in which particular blocks of site values do not appear [11]. This set forms a (finite complement) regular language, and can be represented by the set of possible paths through a certain labelled directed graph [11]. Even when they are not fixed points, the set of states that can occur after say t time steps in the evolution of a one-dimensional cellular automaton in fact also forms a regular language (though not necessarily a finite complement one). In addition, the basin of attraction, or in general the set of all states which evolve after t steps to a given one, can be represented as a regular language. For class 1 and 2 cellular automata, the size of the minimal graph for this language stays bounded, or at most increases like a polynomial with t. For class 3 and 4 cellular automata, however, the size of the graph often increases apparently exponentially with t, so that it becomes increasingly difficult to describe the basin of attraction. The general problem of determining which states evolve to a particular one after t steps is in fact a generalization of the satisfiability problem for logical functions mentioned above, and is thus NP complete. The basin of attraction in the worst case can thus presumably be found only by explicit testing of essentially all $O(2^t)$ possible initial configurations (cf [12]). Its form will again often be so complicated as to seem random. For two-dimensional cellular automata, it is already an NP-complete problem just to find fixed points (say to determine which $n{\times}n$ blocks of sites with specified boundaries are invariant under the cellular automaton rule) [13].

It is typical of complex systems that to reproduce their behaviour requires extensive computation. This is a consequence of the fact that the evolution of the systems themselves typically corresponds to a sophisticated computation. In fact, the evolution of many complex systems is probably computationally irreducible: it can be found essentially only by direct simulation, and cannot be predicted by any short-cut procedure [12,14]. Such computational irreducibility is a necessary consequence of the efficient use of computational resources in a system. Any computational reducibility is a sign of inefficiency, since it implies that some other system can determine the outcome more efficiently.

Many systems in nature may well be computationally irreducible, so that no general predictions can be made about their behaviour. But if a system is to be used for engineering, it must be possible to determine in advance at least some aspects of its behaviour. Conventional engineering requires detailed specification of the precise behaviour of each component in a system. To make use of complex systems in engineering, one must relax this constraint, and instead require only some general or approximate specification of overall behaviour.

One goal is to design systems which have particular attractors. For the example of an inertialess ball rolling with friction on a landscape, this is quite straightforward (cf [15]). In one dimension, the height of the landscape at position x could be given by the polynomial $\prod_i (x-x_i)^2$, where the x_i are the desired minima, or attractors. This polynomial is explicitly constructed to yield certain attractors in the dynamics. However, it implies a particular structure for the basins of attraction. If the attractors are close to equally spaced, or are sufficiently far apart, then the boundaries of the basins of attraction for successive attractors will be roughly half way between them. Notice, however, that as the parameters of the landscape polynomial are changed, the structure of the attractors and basins of attraction obtained can change discontinuously, as described by catastrophe theory.

For a more complex system, such as a cellular automaton, it is more difficult to obtain a particular set of attractors. One approach is to construct cellular automaton rules which leave particular sequences invariant [16]. If these sequences are say of length L, and are arbitrarily chosen, then it may be necessary to use a cellular automaton rule which involves a neighbourhood of up to $L-1$ sites. The necessary rule is straightforward to construct, but takes up to 2^{L-1} bits to specify.

Many kinds of complex systems can be considered as bases for engineering. Conventional engineering suggests some principles to follow. The most important is the principle of modularity. The components of a system should be arranged in some form of hierarchy. Components higher on the hierarchy should provide overall control for sets of components lower on the hierarchy, which can be treated as single units or modules. This principle is crucial to software engineering, where the modules are typically subroutines. It is also manifest in biology in the existence of organs and definite body parts, apparently mirrored by subroutine-like constructs in the genetic code.

An important aspect of modularity is the abstraction it makes possible. Once the construction of a particular module has been completed, the module can be treated as a single object, and only its overall behaviour need be considered, wherever the module appears. Modularity thus divides the problem of constructing or analysing a system into many levels, potentially making each level manageable.

Modularity is used in essentially all of the systems to be discussed below. In most cases, there are just two levels: controlling (master) and controlled (slave) components. The components on these two levels usually change on different time scales. The controlling components change at most slowly, and are often fixed once a system say with a particular set of attractors has been obtained. The controlled components change rapidly, processing input data according to dynamical rules determined by the controlling components. Such separation of time scales is common in many natural and artificial systems. In biology, for example, phenotypes of organisms grow by fast processes, but are determined by genotypes which seem to change only slowly with time. In software engineering, computer memory is divided into a part for "programs", which are supposed to remain fixed or change only slowly, and another part for intermediate data, which changes rapidly.

Multiple scale cellular automata provide simple but quite general examples of such hierarchical systems. An ordinary cellular automaton consists of a lattice of sites, with each site having say k possible values, updated according to the same definite rule. A two-scale cellular automaton can be considered to consist of two lattices, with site values changing on different characteristic time scales. The values of the sites on the "slow" lattice control the rules used at the corresponding sites on the "fast" lattice. With q possible values for the slow lattice sites, there is an array of q possible rules for each site on the fast lattice. (Such a two-scale cellular automaton could always be emulated by specially chosen configurations in an ordinary cellular automaton with at most qk possible values at each site.)

If the sites on the slow lattice are fixed, then a two-scale cellular automaton acts like a dynamic random field spin system (e.g. [17]), or a spin glass (e.g. [18]) (*cf* [19]). Examples of patterns generated by cellular automata of this kind are shown in figure 1. If instead the "slow" lattice sites change rapidly, and take on essentially random values, perhaps as a result of following a chaotic cellular automaton rule, then the evolution of the fast lattice is like that of a stochastic cellular automaton, or a directed percolation system (e.g. [21]).

With dissipative dynamics, the evolution of the fast lattice in a two-scale cellular automaton yields attractors. The form of these attractors is determined by the control configuration on the slow lattice. By choosing different slow lattice configurations, it is thus possible to engineer particular attractor structures.

In a typical case, a two-scale cellular automaton might be engineered to recognize inputs in different categories. Each category would be represented by a fixed point in the fast lattice dynamics. The system could then be arranged in several ways. Assume that the input is a one-dimensional symbol sequence (such as a text string). Then one possibility would be to consider a one-dimensional cellular automaton whose fixed points correspond to symbol sequences characteristic of each category. But if the required fixed points are arbitrarily chosen, only a few of them can be obtained with a single slow configuration. If the cellular automaton has N sites, then each fixed point of the fast lattice is specified by $N\log_2 k$ bits. A configuration of the slow lattice involves only $N\log_2 q$ bits. As a consequence, the number of arbitrarily-chosen fixed points that can be specified is just $\log q/\log k$, a result independent of N. (More fixed points may potentially be specified if there is redundancy between their symbol sequences.)

It is usually not necessary, however, to give all $N\log_2 k$ bits of a fixed point to specify the form of the attractor for a particular category. The number of bits actually needed presumably increases with the number of categories. It is common to find a small number of possible categories or responses to a wide variety of input data. The responses can then for example be represented by the values of a small number of sites on the fast lattice of a two-scale cellular automaton. The input data can be used to give initial values for a larger number of sites, possibly a different set. (In an analogy with the nervous system, some sites might receive input from afferent nerves while others, typically smaller in number, might generate output for efferent nerves.)

A second possibility is to consider a two-dimensional two-scale cellular automaton, in which the input is specified along a line, and the dynamics of the fast lattice transfers information only in a direction orthogonal to this line [22] (*cf* [23]). This arrangement is functionally equivalent to a one-dimensional two-scale cellular automaton in which the slow lattice configuration changes at each time step. In its two-dimensional form, the arrangement is very similar to a systolic array [24], or in fact to a multistage generalization of standard modular logic circuits. In an $N{\times}M$ system of this kind, a single slow lattice configuration can specify $M\log q/\log k$ length N fixed points in the fast configuration (*cf* [2]).

In the approaches just discussed, input is given as an initial condition for the fast lattice. An alternative possibility is that the input could be given on the slow lattice, and could remain throughout the evolution of the fast lattice. The input might for example then specify boundary conditions for evolution on a two-dimensional fast lattice. Output could be obtained from the final configuration of the fast lattice. However, there will often be several different attractors for the fast lattice dynamics even given boundary conditions from a particular slow lattice configuration. Which attractor is reached will typically depend on the initial conditions for the fast lattice, which are not specified in this approach. With appropriate dynamics, however, it is nevertheless possible to obtain almost unique attractors: one approach is to add probabilistic elements or noise to the fast lattice dynamics so as to make it ergodic, with a unique invariant measure corresponding to a definite "phase" [25].

Cellular automata are arranged to be as simple as possible in their basic microscopic construction. They are discrete in space and time. Their sites are all identical, and are arranged on a regular lattice. The sites have a finite set of possible values, which are updated synchronously according to identical deterministic rules that depend on a few local neighbours. But despite this microscopic simplicity, the overall macroscopic behaviour of cellular automata can be highly complex. On a large scale, cellular automata can for example show continuum features [13,26], randomness [8], and effective long-range interactions [27]. Some cellular automata are even known to be universal computers [28], and so can presumably simulate any possible form of behaviour. Arbitrary complexity can thus arise in cellular automata. But for engineering purposes, it may be better to consider basic models that are more sophisticated than cellular automata, and in which additional complexity is included from the outset (*cf* [2]). Multiple scale cellular automata incorporate modularity, and need not be homogeneous. Further generalizations can also be considered, though one suspects that in the end none of them will turn out to be crucial.

First, cellular automaton dynamics is local: it involves no long-range connections which can transmit information over a large distance in one step. This allows (one or two-dimensional) cellular automata to be implemented directly in the simple planar geometries appropriate, for example, for very large-scale integrated circuits. Long range electronic signals are usually carried by wires which cross in the third dimension to form a complicated network. (Optical communications may also be possible.) Such an arrangement is difficult to implement technologically. When dynamically-changing connections are required, therefore, more homogeneous switching networks are used, as in computerized telephone exchanges, or the Connection Machine computer [29]. Such networks are typically connected like cellular automata, though often in three (and sometimes more) dimensions.

Some natural systems nevertheless seem to incorporate intrinsic long range connections. Chemical reaction networks are one example: reaction pathways can give almost arbitrary connectivity in the abstract space of possible chemical species [30,31,32]. Another example is the brain, where nerves can carry signals over long distances. In many parts of the brain, the pattern of connectivity chosen seems to involve many short-range connections, together with a few long-range ones, like motorways (freeways) or trunk lines [33]. It is always possible to simulate an arbitrary arrangement of long range connections through sequences of short range connections; but the existence of a few intrinsic long range connections may make large classes of such simulations much more efficient [33].

Many computational algorithms seem to involve arbitrary exchange of data. Thus for example, in the fast Fourier transform, elements are combined according to a shuffle-exchange graph (e.g. [29]). Such algorithms can always be implemented by a sequence of local operations. But they seem to be most easily conceived without reference to the dynamics of data transfer. Indeed, computers and programming languages have traditionally been constructed to enforce the idealization that any piece of data is available in a fixed time (notions such as registers and pipelining go slightly beyond this). Conventional computational complexity theory also follows this idealization (e.g. [34]). But in developing systems that come closer to actual physical constraints, one must go beyond this idealization. Several classes of algorithms are emerging that can be implemented efficiently and naturally with local communications (e.g. [35]). A one-dimensional cellular automaton ("iterative array") can be used for integer multiplication [36,37] and sorting [38]. Two dimensional cellular automata ("systolic arrays") can perform a variety of matrix manipulation operations [24].

Although the basic rules for cellular automata are local, they are usually applied in synchrony, as if controlled by a global clock. A generalization would allow asynchrony, so that different sites could be updated at different times (e.g. [39]). Only a few sites might, for example, be updated at each time step. This typically yields more gradual transitions from one cellular automaton configuration to another, and can prevent certain instabilities. Asynchronous updating makes it more difficult for information to propagate through the cellular automaton, and thus tends to prevent initial perturbations from spreading. As a result, the evolution is more irreversible and dissipative. Fixed point and limit cycle (class 1 and 2) behaviour therefore becomes more common.

For implementation and analysis, it is often convenient to maintain a regular updating schedule. One possibility is to alternate between updates of even and odd-numbered sites (e.g. [40]). "New" rather than "old" values for the nearest neighbours of a particular cell are then effectively used. This procedure is analogous to the implicit, rather than explicit, method for updating site values in finite difference approximations to partial differential equations (e.g. [41]), where it is known to lead to better convergence in certain cases. The scheme also yields, for example, systematic relaxation to thermodynamic equilibrium in a cellular automaton version of the microcanonical Ising model [42]: simultaneous updating of all sites would allow undamped oscillations in this case [40].

One can also consider systems in which sites are updated in a random order, perhaps one at a time. Such systems can often be analysed using "mean field theory", by assuming that the behaviour of each individual component is random, with a particular average (cf [2]). Statistical predictions can then often be made from iterations of maps involving single real variables. As a result, monotonic approach to fixed points is more easily established.

Random asynchronous updating nevertheless makes detailed analysis more difficult. Standard computational procedures usually require definite ordering of operations, which can be regained in this case only at some cost (cf [43]).

Rather than introducing randomness into the updating scheme, one can instead include it directly in the basic cellular automaton rule. The evolution of such stochastic cellular automata can be analogous to the steps in a Monte Carlo simulation of a spin system at nonzero temperature [44]. Randomness typically prevents the system from being trapped in metastable states, and can therefore accelerate the approach to equilibrium.

In most practical implementations, however, supposedly random sequences must be obtained from simple algorithms (e.g. [37]). Chaotic cellular automata can produce sequences with a high degree of randomness [8], presumably making explicit insertion of external randomness unnecessary.

Another important simplifying feature of cellular automata is the assumption of discrete states. This feature is convenient for implementation by digital electronic circuits. But many natural systems seem to involve continuously-variable parameters. There are usually components, such as molecules or vesicles of neurotransmitter, that behave as discrete on certain levels. But very large numbers of these components can act in bulk, so that for example only their total concentration is significant, and this can be considered as an essentially continuous variable. In some systems, such bulk quantities have simple behaviour, described say by partial differential equations. But the overall behaviour of many cellular automata and other systems can be sufficiently complex that no such bulk or average description is adequate. Instead the evolution of each individual component must be followed explicitly (cf [12]).

For engineering purposes, it may nevertheless sometimes be convenient to consider systems which involve essentially continuous parameters. Such systems can for example support cumulative small incremental changes. In a cellular automaton, the states of n elements are typically represented by $O(n)$

bits of information. But bulk quantities can be more efficiently encoded as digital numbers, with only $O(\log n)$ bits. There may be some situations in which data is best packaged in this way, and manipulated say with arithmetic operations.

Systems whose evolution can be described in terms of arithmetic operations on numbers can potentially be analysed using a variety of standard mathematical techniques. This is particularly so when the evolution obeys a linear superposition principle, so that the complete behaviour can be built up from a simple superposition of elementary pieces. Such linear systems often admit extensive algebraic analysis (*cf* [45]), so that their behaviour is usually too simple to show the complexity required.

Having selected a basic system, the problem of engineering consists in designing or programming it to perform particular tasks. The conventional approach is systematically to devise a detailed step-by-step plan. But such a direct constructive approach cannot make the most efficient use of a complex system.

Logic circuit design provides an example (e.g. [46]). The task to be performed is the computation of a Boolean function with n inputs specified by a truth table. In a typical case, the basic system is a programmable logic array (PLA): a two-level circuit which implements disjunctive normal form (DNF) Boolean expressions, consisting of disjunctions (ORs) of conjunctions (ANDs) of input variables (possibly negated) [24,47]. The direct approach would be to construct a circuit which explicitly tests for each of the 2^n cases in the truth table. The resulting circuit would contain $O(n2^n)$ gates. Thus for example, the majority function, which yields 1 if two or more of its three inputs a_i are one would be represented by the logical circuit corresponding to $a_1a_2a_3+a_1a_2\bar{a}_3+a_1\bar{a}_2a_3+\bar{a}_1a_2a_3$, where multiplication denotes AND, addition OR, and bar NOT. (This function can be viewed as the $k=2$, $r=1$ cellular automaton rule number 232 [20]).

Much smaller circuits are, however, often sufficient. But direct constructive techniques are not usually appropriate for finding them. Instead one uses methods that manipulate the structure of circuits, without direct regard to the meaning of the Boolean functions they represent. Many methods start by extracting prime implicants [46,47]. Logical functions of n variables can be considered as colourings of the Boolean n-cube. Prime implicants represent this colouring by decomposing it into pieces along hyperplanes with different dimensionalities. Each prime implicant corresponds to a single conjunction of input variables: a circuit for the original Boolean function can be formed from a disjunction of these conjunctions. This circuit is typically much smaller than the one obtained by direct construction. (For the majority function mentioned above, it is $a_1a_2+a_1a_3+a_2a_3$.) And while it performs the same task, it is usually no longer possible to give an explicit step-by-step "explanation" of its operation.

A variety of algebraic and heuristic techniques are used for further simplification of DNF Boolean expressions [47]. But it is in general very difficult to find the absolutely minimal expression for any particular function. In principle, one could just enumerate all possible progressively more complicated expressions or circuits, and find the first one which reproduces the required function. But the number of possible circuits grows exponentially with the number of gates, so such an exhaustive search rapidly becomes entirely infeasible. It can be shown in fact that the problem of finding the absolute minimal expression is NP hard, suggesting that there can never be a general procedure for it that takes only polynomial time [7]. Exhaustive search is thus effectively the only possible exact method of solution.

Circuits with a still smaller number of gates can in principle be constructed by allowing more than two levels of logic. (Some such circuits for the majority function are discussed in ref. [48].) But the difficulty of finding the necessary circuits increases rapidly as more general forms are allowed, and as the absolute minimum circuit is approached. A similar phenomenon is observed with many complex systems: finding optimal designs becomes rapidly more difficult as the efficiency of the designs increases (*cf* [49]).

In most cases, however, it is not necessary to find the absolutely minimal circuit: any sufficiently simple circuit will suffice. As a result, one can consider methods that find only approximately minimal circuits.

Most approximation techniques are basically iterative: they start from one circuit, then successively make changes which preserve the functionality of the circuit, but modify its structure. The purpose is to find minima in the circuit size or "cost" ("fitness") function over the space of possible circuits. The effectiveness of different techniques depends on the form of the circuit size "landscape".

If the landscape was like a smooth bowl, then the global minimum could be found by starting at any point, and systematically descending in the direction of the local gradient vector. But in most cases the landscape is presumably more complicated. It could for example be essentially flat, except for one narrow hole containing the minimum (like a golf course). In such a case, no simple iterative procedure could find the minimum.

Another possibility, probably common in practice, is that the landscape has a form reminiscent of real topographical landscapes, with a complicated pattern of peaks and valleys of many different sizes. Such a landscape might well have a self similar or fractal form: features seen at different magnifications could be related by simple scalings. Straightforward gradient descent would always get stuck in local minima on such a landscape, and cannot be used to find a global minimum (just as water forms localized lakes on a topographical landscape). Instead one should use a procedure which deals first with large-scale features, then progressively treats smaller and smaller scale details.

Simulated annealing is an example of such a technique [50]. It is based on the gradient descent method, but with stochastic noise added. The noise level is initially large, so that all but the largest scale features of the landscape are smeared out. A minimum is found at this level. Then the noise level ("temperature") is reduced, so that smaller scale features become relevant, and the minimum is progressively refined. The optimal temperature variation ("annealing schedule") is probably determined by the fractal dimension of the landscape.

In actual implementations of the simulated annealing technique, the noise will not be truly random, but will instead be generated by some definite, and typically quite simple, procedure. As a consequence, the whole simulated annealing computation can be considered entirely deterministic. And since the landscape is probably quite random, it is possible that simple deterministic perturbations of paths may suffice (*cf* [51]).

In the simulated annealing approach, each individual "move" might consist of a transformation involving say two logic gates. An alternative procedure is first to find the minimal circuit made from "modules" containing many gates, and then to consider rearranging progressively smaller submodules. The hierarchical nature of this deterministic procedure can again mirror the hierarchical form of the landscape.

The two approaches just discussed involve iterative improvement of a single solution. One can also consider approaches in which many candidate solutions are treated in parallel. Biological evolution apparently uses one such approach. It generates a tree of different genotypes, and tests the "fitness" of each branch in parallel. Unfit branches die off. But branches that fare well have many offspring, each with a genotype different by a small random perturbation ("genetic algorithm" [52]). These offspring are then in turn tested, and can themselves produce further offspring. As a result, a search is effectively conducted along many paths at once, with a higher density of paths in regions with greater fitness. (This is analogous to decision tree searching with, say, $\alpha\beta$-pruning [53].) Random perturbations in the paths at each generation may prevent getting stuck in local minima, but on a fractal landscape of the type discussed above, this procedure seems less efficient than one based on consideration of progressively finer details.

In the simplest iterative procedures, the possible changes made to candidate solutions are chosen from a fixed set. But one can also imagine modifying the set of possible changes dynamically [54] (*cf* [55]). To do this, one must parametrize the possible changes, and in turn search the space of possibilities for optimal solutions.

The issues discussed for logic circuit design also arise in engineering complex systems such as two-scale cellular automata. A typical problem in this case is to find a configuration for the slow lattice that yields particular fixed points for evolution on the fast lattice. With simple linear rules, for example, a constructive algebraic solution to this problem can be given. But for arbitrary rules, the problem is in general NP hard. An exact solution can thus presumably be found only by exhaustive search. Approximation procedures must therefore again be used.

The general problem is to find designs or arrangements of complex systems that behave in specified ways. The behaviour sought usually corresponds to a comparatively simple, usually polynomial time, computation. But to find exactly the necessary design may require a computation that effectively tests exponentially many possibilities. Since the correctness of each possibility can be tested in polynomial time, the problem of finding an appropriate design is in the computational complexity class NP (non-deterministic polynomial time). But in many cases, the problem is in fact NP complete (or at least NP hard). Special instances of the problem thus correspond to arbitrary problems in NP; any general solution could thus be applied to all problems in NP.

There are many NP complete problems, all equivalent in the computational difficulty of their exact solution [7]. Examples are satisfiability (finding an assignment of truth values to variables which makes a Boolean expression true), Hamilton circuits (finding a path through a graph that visits each arc exactly once), and spin glass energy minima (finding the minimum energy configuration in a spin glass model). In no case is an algorithm known which takes polynomial time, and systematically yields the exact solution.

Many approximate algorithms are nevertheless known. And while the difficulty of finding exact solutions to the different problems is equivalent, the ease of approximation differs considerably. (A separate consideration is what fraction of the instances of a problem are difficult to solve with a particular algorithm. Some number theoretical problems, for example, have the property that all their instances are of essentially equivalent difficulty [56].) Presumably the "landscapes" for different problems fall into several classes. There is already some evidence that the landscapes for spin glass energy and the "travelling salesman" problem have a hierarchical or ultrametric, and thus fractal, form [57]. This may explain why the simulated annealing method is comparatively effective in these cases.

Even though their explicit forms cannot be found, it could be that particular, say statistical, features of solutions to NP problems could easily be predicted. Certainly any solution must be distinguished by the P operation used to test its validity. But at least for some class of NP hard problems, one suspects that solutions will appear random according to all standard statistical procedures. Despite the "selection" process used to find them, this would imply that their statistical properties would be typical of the ensemble of all possibilities (*cf* [58]).

There are many potential applications for complex systems engineering. The most immediate ones are in pattern recognition. The basic problem is to take a wide variety of inputs, say versions of spoken words, and to recognize to which category or written word they correspond (e.g. [59]).

In general, there could be an arbitrary mapping from input to output, so that each particular case would have to be specified explicitly. But in practice the number of possible inputs is far too large for this to be feasible, and redundancy in the inputs must be used. One must effectively make some form of model for the inputs, which can be used, for example, to delineate categories that yield the same output. The kinds of models that are most appropriate depend on the regularities that exist in the input data. In human language and various other everyday forms of input, there seem for example to be regularities such as power laws for frequencies (e.g. [60]).

In a simple case, one might take inputs within one category to differ only by particular kinds of distortions or errors. Thus in studies of DNA sequences, changes associated with substitution, deletion, insertion, or transposition of elements are usually considered [61]. (These changes have small effects on the spatial structure of the molecule, which determines many of its functions.) A typical problem of pattern recognition is to determine the category of a particular input, regardless of such changes.

Several approaches are conventionally used (e.g. [59]).

One approach is template matching. Each category is defined by a fixed "template". Then inputs are successively compared with each template, and the quality of match is determined, typically by statistical means. The input is assigned to the category with the best match.

A second approach is feature extraction. A fixed set of "features" is defined. The presence or absence of each feature in a particular input is then determined, often by template-matching techniques. The category of the input is found from the set of features it contains, typically according to a fixed table.

In both these approaches, the pattern recognition procedure must be specially designed to deal with each particular set of categories considered. Templates or features that are sufficiently orthogonal must be constructed, and small changes in the behaviour required may necessitate large changes in the arrangement used.

It would be more satisfactory to have generic systems which would take simple specifications of categories, and recognize inputs using "reasonable" boundaries between the categories. Dissipative dynamical systems with this capability can potentially be constructed. Different categories would be specified as fixed points (or other attractors). Then the dynamics of the system would determines the forms of the basins of attraction. Any input within a particular basin would be led to the appropriate fixed point. In general, however, different inputs would take varying numbers of steps to reach a fixed point. Conventional pattern recognition schemes typically take a fixed time, independent of input. But more flexible schemes presumably require variable times.

It should be realized, however, that such schemes implicitly make definite models for the input data. It is by no means clear that the dynamics of such systems yield basin structures appropriate for particular data. The basins are typically complicated and difficult to specify. There will usually be no simple distance measure or metric, analogous to the quality of template matches, which determines the basin for a particular input from the fixed point to which it is "closest". Figure 2 shows a representation of the basins of attraction in a two-scale cellular automaton. No simple metric is evident.

While the detailed behaviour of a system may be difficult to specify, it may be possible to find a high-level phenomenological description of some overall features, perhaps along the lines conventional in psychology, or in the symbolic approach to artificial intelligence (e.g. [62]). One can imagine, for example, proximity relations for attractors analogous to semantic networks (e.g. [62]). This high level description might have the same kind of relation to the underlying dynamics as phenomenological descriptions such as vortex streets have to the basic equations of fluid flow.

To perform a particular pattern recognition task, one must design a system with the appropriate attractor structure. If, for example, categories are to represented by certain specified fixed points, the system must be constructed to have these fixed points. In a two-scale cellular automaton with $k=2$ and $q=2$, a single fixed point on the whole lattice can potentially be produced with by an appropriate choice of the slow configuration. Arbitrary fixed points can be obtained in this way only with particular pairs of rules. (The rules that take all configurations to zero, and all configurations to one, provide a trivial example.) But even in this case, it is common for several different slow configurations to yield the same required fixed point, but to give very different basin structures. Often spurious additional fixed points are also produced. It is not yet clear how best to obtain only the exact fixed points required.

It would be best to devise a scheme for "learning by example" (*cf* [63]). In the simplest case, the fixed points would be configurations corresponding to "typical" members of the required categories. In a more sophisticated case, many input and output pairs would be presented, and an iterative algorithm would be used to design an attractor structure to represent them. In a multiple scale cellular automaton, such an algorithm might typically make "small" incremental changes of a few sites in the slow configuration. Again such a procedure involves inferences about new inputs, and requires a definite model.

Acknowledgements

I am grateful for discussions with many people, including Danny Hillis, David Johnson, Stuart Kauffman, Alan Lapedes, Marvin Minsky, Steve Omohundro, Norman Packard, Terry Sejnowski, Rob Shaw, and Gerry Tesauro.

References

1. S. Wolfram, "Cellular automata as models of complexity", Nature 311 (1984) 419.

2. J. Hopfield, "Neural networks and physical systems with emergent collective computational abilities", Proc. Natl. Acad. Sci. 79 (1982) 2554.

3. W. Green, *Digital image processing*, Van Nostrand (1983).

4. R. Hamming, *Coding and information theory*, Prentice-Hall (1980).

5. V. Vulovic and R. Prange, "Is the toss of a true coin really random?", Maryland preprint (1985).

6. S. McDonald, C. Grebogi, E. Ott and J. Yorke, "Fractal basin boundaries", Physica 17D (1985) 125.

7. M. Garey and D. Johnson, *Computers and intractability: a guide to the theory of NP-completeness*, Freeman (1979).

8. S. Wolfram, "Random sequence generation by cellular automata", Adv. Applied Math. (in press).

9. K. Preston and M. Duff, *Modern cellular automata*, Plenum (1984).

10. S. Wolfram, "Universality and complexity in cellular automata", Physica 10D (1984) 1.

11. S. Wolfram, "Computation theory of cellular automata", Commun. Math. Phys. 96 (1984) 15.

12. S. Wolfram, "Undecidability and intractability in theoretical physics", Phys. Rev. Lett. 54 (1985) 735.

13. N. Packard and S. Wolfram, "Two-dimensional cellular automata", J. Stat. Phys. 38 (1985) 901.

14. S. Wolfram, "Computer software in science and mathematics", Sci. Amer. (September 1984).

15. R. Sverdlove, "Inverse problems for dynamical systems in the plane", in *Dynamical systems*, A. R. Bednarek and L. Cesari (eds.), Academic Press (1977).

16. E. Jen, "Invariant strings and pattern-recognizing properties of one-dimensional cellular automata", J. Stat. Phys., to be published; Los Alamos preprint LA-UR-85-2896.

17. J. Villain, "The random field Ising model", in *Scaling phenomena and disordered systems*, NATO ASI, Geilo, Norway (April 1985).

18. Proc. Heidelberg Colloq. on Spin Glasses, Heidelberg (June 1983); K. H. Fischer, Phys. Status Solidi 116 (1983) 357.

19. G. Vichniac, P. Tamayo and H. Hartman, "Annealed and quenched inhomogeneous cellular automata", J. Stat. Phys., to be published.

20. S. Wolfram, "Statistical mechanics of cellular automata", Rev. Mod. Phys. 55 (1983) 601.

21. W. Kinzel, "Phase transitions of cellular automata", Z. Phys. B58 (1985) 229.

22. T. Hogg and B. Huberman, "Parallel computing structures capable of flexible associations and recognition of fuzzy inputs", J. Stat. Phys. 41 (1985) 115.

23. T. Sejnowski and C. R. Rosenberg, "NETtalk: A parallel network that learns to read aloud", Physica D, to be published (Johns Hopkins Elec. Eng. and Comput. Sci. Tech. Report 86-01).

24. C. Mead and L. Conway, *An introduction to VLSI systems*, Addison-Wesley (1980).

25. D. Ackley, G. Hinton and T. Sejnowski, "A learning algorithm for Boltzmann machines", Cognitve Sci. 9 (1985) 147.

26. U. Frisch, B. Hasslacher and Y. Pomeau, "A lattice gas automaton for the Navier-Stokes equation", Los Alamos preprint LA-UR-85-3503; J. Salem and S. Wolfram, "Thermodynamics and hydrodynamics with cellular automata", IAS preprint (November 1985).

27. S. Wolfram, "Glider gun guidelines", report distributed through Computer Recreations section of Scientific American; J. Park, K. Steiglitz and W. Thurston, "Soliton-like behaviour in cellular automata", Princeton University Computer Science Dept. report (1985).

28. A. Smith, "Simple computation-universal cellular spaces", J. ACM 18 (1971) 331; E. R. Berlekamp, J. H. Conway and R. K. Guy, *Winning ways for your mathematical plays*, Academic

Press (1982).

29. D. Hillis, *The Connection Machine*, MIT press (1985).

30. S. Kauffman, "Metabolic stability and epigenesis in randomly constructed genetic nets", J. Theoret. Biol. 22 (1969) 437; "Autocatalytic sets of proteins", J. Theor. Biol. (in press).

31. A. Gelfand and C. Walker, "Network modelling techniques: from small scale properties to large scale systems", University of Connecticut report (1982).

32. E. Goles Chacc, "Comportement dynamique de reseaux d'automates", Grenoble University report (1985).

33. C. Leiserson, "Fat trees: universal networks for hardware efficient supercomputing", IEEE Trans. Comput. C-36 (1985) 892.

34. J. Hopcroft and J. Ullman, *Introduction to automata theory, languages and computation*, Addison-Wesley (1979).

35. S. Omohundro, "Connection Machine algorithms primer", Thinking Machines Corporation (Cambridge, Mass.) report in preparation.

36. A. J. Atrubin, "A one-dimensional real-time iterative multiplier", IEEE Trans. Comput. EC-14 (1965) 394.

37. D. Knuth, *Seminumerical algorithms*, Addison-Wesley (1981).

38. H. Nishio, "Real time sorting of binary numbers by 1-dimensional cellular automata", Kyoto university report (1981).

39. T. E. Ingerson and R. L. Buvel, "Structure in asynchronous cellular automata", Physica 10D (1984) 59.

40. G. Vichniac, "Simulating physics with cellular automata", Physica 10D (1984) 96.

41. C. Gerald, *Applied numerical analysis*, Addison-Wesley (1978).

42. M. Creutz, "Deterministic Ising dynamics", Ann. Phys. (in press).

43. A. Grasselli, "Synchronization of cellular arrays: the firing squad problem in two dimensions", Info. & Control 28 (1975) 113.

44. E. Domany and W. Kinzel, "Equivalence of cellular automata to Ising models and directed percolation", Phys. Rev. Lett. 53 (1984) 311.

45. M. Minsky and S. Papert, *Perceptrons*, MIT press (1969).

46. Z. Kohavi, *Switching and finite automata theory*, McGraw-Hill (1970).

47. R. Brayton, G. Hachtel, C. McMullen and A. Sangiovanni-Vincentelli, *Logic minimization algorithms for VLSI synthesis*, Kluwer (1984).

48. L. Valiant, "Short monotone formulae for the majority function", Harvard University report TR-01-84 (1983).

49. M. Conrad, "On design principles for a molecular computer", Commun. ACM 28 (1985) 464.

50. S. Kirkpatrick, C. Gelatt and M. Vecchi, "Optimization by simulated annealing", Science 220 (1983) 671.

51. J. Hopfield and D. Tank, "Neural computation of decisions in optimization problems", Biol. Cybern. 52 (1985) 141.

52. J. Holland, "Genetic algorithms and adaptation", Tech. Rep. #34, Univ. Michigan (1981).

53. A. Barr and E. Feigenbaum, *The handbook of artificial intelligence*, HeurisTech Press (1983), vol. 1.

54. J. Holland, "Escaping brittleness: the possibilities of general purpose learning algorithms applied to parallel rule-based systems", University of Michigan report.

55. D. Lenat, "Computer software for intelligent systems", Scientific American (September 1984).

56. M. Blum and S. Micali, "How to generate cryptographically strong sequences of pseudo-random bits", SIAM J. Comput. 13 (1984) 850.

57. S. Kirkpatrick and G. Toulouse, "Configuration space analysis of travelling salesman problems", J. Physique 46 (1985) 1277.

58. S. Kauffman, "Self-organization, selection, adaptation and its limits: a new pattern of inference in evolution and development", in *Evolution at a crossroads*, D. J. Depew and B. H. Weber (eds.), MIT press (1985).

59. C. J. D. M. Verhagen *et al.*, "Progress report on pattern recognition", Rep. Prog. Phys. 43 (1980) 785; B. Batchelor (ed.), *Pattern recognition*, Plenum (1978).

60. B. Mandelbrot, *The fractal geometry of nature*, Freeman (1982).

61. D. Sankoff and J. Kruskal (eds.), *Time warps, string edits, and macromolecules: the theory and practice of sequence comparison*, Addison-Wesley (1983).

62. M. Minsky, *Society of mind*, in press.

63. L. Valiant, "A theory of the learnable", Commun. ACM 27 (1984) 1134.

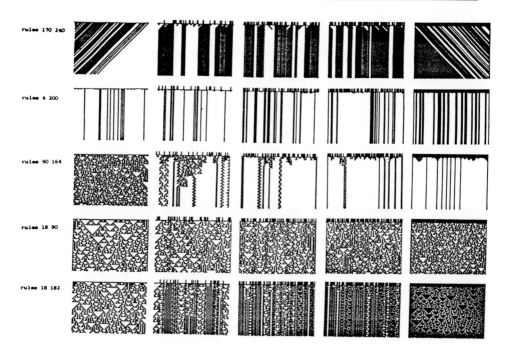

Figure 1. Patterns generated by two-scale cellular automata with $k=2$, $q=2$ and $r=1$. The configuration of the slow lattice is fixed in each case, and is shown at the top. The rule used at a particular site on the fast lattice is chosen from the two rules given according to the value of the corresponding site on the slow lattice. (The rule numbers are as defined in ref. [20].)

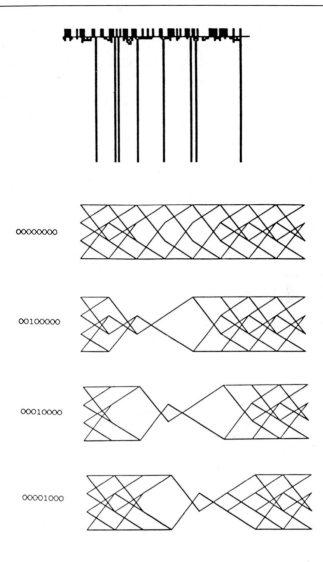

Figure 2. Representation of the basins of attraction for fixed points in a length 8 two-scale cellular automaton with $q=2$ and rules 36 and 72. The configurations in each basin correspond to possible paths traversing each graph from left to right. Descending segments represent value one, ascending segments value zero.

4. Probabilistic cellular automata

4.1: W. Kinzel, "Phase transitions of cellular automata", Z. Phys. B58 (1985) 229.

4.2: P. Grassberger, F. Krause and T. von der Twer, "A new type of kinetic critical phenomenon", J. Phys. A17 (1984) L105.

4.3: K. Kaneko and Y. Akutsu, "Phase transitions in two-dimensional stochastic cellular automata", J. Phys. A19 (1986) L69.

4.4: E. Domany and W. Kinzel, "Equivalence of cellular automata to Ising models and directed percolation", Phys. Rev. Lett. 53 (1984) 311.

4.5: G. Grinstein, C. Jayaprakash and Y. He, "Statistical mechanics of probabilistic cellular automata", Phys. Rev. Lett. 55 (1985) 2527.

4.6: C. Bennett and G. Grinstein, "Role of irreversibility in stabilizing complex and nonergodic behaviour in locally interacting discrete systems", Phys. Rev. Lett. 55 (1985) 657.

Probabilistic cellular automaton rule with non-ergodic behaviour. The picture shows evolution according to a $k=2$, $r=3$ cellular automaton rule, with a probability of 0.01 for complementation of each site value produced. Despite this "noise", the cellular automaton yields increasingly large regions with definite overall behaviour.

Z. Phys. B - Condensed Matter 58, 229-244 (1985)

Phase Transitions of Cellular Automata

W. Kinzel

Institut für Festkörperforschung der Kernforschungsanlage Jülich,
Federal Republic of Germany

Received October 26, 1984

Cellular automata (CA) are simple mathematical models of the dynamics of discrete variables in discrete space and time, with applications in nonequilibrium physics, chemical reactions, population dynamics and parallel computers. Phase transitions of stochastic CA with absorbing states are investigated. Using transfermatrix scaling the phase diagrams, critical properties and the entropy of one-dimensional CA are calculated. The corners of the phase diagrams reduce to deterministic CA discussed by Wolfram (Rev. Mod. Phys. **55**, 601 (1983)). Three-state models are introduced and, for special cases, exactly mapped onto two-state CA. The critical behaviour of other three-state models with one or two absorbing states and with immunization is investigated. Finally CA with competing reactions and/or with disorder are studied.

I. Introduction

Discrete lattice models like Ising, Potts or lattice gas models [1] have been very useful to understand cooperative phenomena in thermal equilibrium. For these models a dynamics may be defined by changing the discrete local variable according to the local Boltzmann weight [2]. Then also the relaxation into thermal equilibrium can be studied. Since such models are well suited for numerical simulations their applications to more complicated systems like disordered materials or incommensurable structures are still extensively investigated [3].

There is a very similar class of discrete models which are called cellular automata (CA) and which find wide and general applications in mathematics, physics, chemistry, biology and computer science [4]. As before these models have a finite set of variables on a lattice. The dynamics is defined in discrete time steps with rules depending on the local neighbourhood only. However now the dynamics is not restricted to the usual Boltzmann weight and detailed balance. Therefore it can model for instance chemical reactions, population changes or other nonlinear processes far from thermal equilibrium.

The local rules of CA may be stochastic or deterministic. In the latter case even one-dimensional CA show a complex interesting behaviour [5, 6].

Starting from a disordered state the system iterates into stationary patterns which appear to fall into four classes [6]:

1) Evolution leads to a homogeneous state

2) Evolution leads to a set of separatet simple stable or periodic structures

3) Evolution leads to a chaotic pattern

4) Evolution leads to complex localized structures.

Thus simple local rules lead to a kind of selforganization. Even simple kinds of selfreproduction of local patterns are observed. Hence the investigation of CA may contribute to our understanding of spontaneous pattern formation [10].

The evolution of the states of CA may be considered as processing of information. If, as in Refs. [5] and [6], all variables are updated at the same time CA may be considered as parallel computers the general theory of which may be useful for constructing new computer generations. Wolfram [6] suggests that local rules of class 4 are capable of universal computation. This means, if the initial state is considered as a program and initial data the CA is capable of evaluating any (computable) function.

If the local rules of CA are stochastic then again

new complex behaviour is observed [7]. In contrast to equilibrium dynamics even one-dimensional CA show continuous phase transitions with universal critical exponents and scaling laws [7, 8]. At least some of these transitions are in the same universality class as those directed percolation and Reggeon field theory [9].

Although D-dimensional CA describe processes far from thermal equilibrium they can be mapped on to $D+1$-dimensional statistical mechanics [11–13]. However, in the interesting case of absorbing states the configuration space is very restricted [12].

In this paper the phase transitions of stochastic one-dimensional CA are investigated in more detail. In Sect. II the models are defined and duality relations derived. Using transfermatrix scaling [14] phase diagrams, critical properties and the entropy are calculated in Sects. III, IV and V. In particular the transitions between deterministic CA of different classes are investigated. The question of universality of CA transitions is addressed in the two following sections. Although increasing the number of local variables should not change the universality class [15], new kinds of transitions were recently reported for reactions with immunization [16] and reactions with two absorbing states [17]. Therefore 3-state CA are introduced in Sect. VI and exact mappings to 2-state CA are derived. Section VII investigates universality. CA with competing and random reaction rates are shortly discussed in Sects. VIII and IX. Finally the summary is presented in the last section. Details of the numerical method may be found in the appendix.

II. Definition of the Model

Consider an one-dimensional chain of N lattice sites. Each site may be in one of the k-states $S_v = 0, 1, 2, \ldots k-1$. Thus the whole chain has k^N states. In the following a given state $(S_1, S_2, \ldots, S_N)$ is labelled by the integer

$$i = \sum_{v=1}^{N} S_v k^{v-1}. \tag{1}$$

Now consider an ensemble of such states where each state i occurs with the probability P_i. This probability $P_i(t)$ evolves in discrete time steps $t = 0, 1, 2, 3 \ldots$ according to some transition probabilities T_{ij}

$$P_i(t) = \sum_{j=0}^{k^N-1} T_{ij} P_j(t-1). \tag{2}$$

Thus the transfermatrix T_{ij} is the probability to get a state i if the system is in state j one time step before.

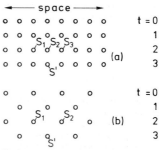

Fig. 1a and b. Geometries of one-dimensional CA studied in this paper. The probability to find variable S' at time $t=3$ depends only on the local neighbourhood S_1, S_2, S_3 (or S_1, S_2) at time $t=2$

In CA T_{ij} is defined by local rules:

$$T_{ij} = \prod_{v=1}^{N} p(S_{v-1}^j, S_v^j, S_{v+1}^j | S_v^i) \tag{3}$$

where S_v^j and S_v^i are the site variables of states j and i, respectively. Thus the transition of the variable S_v^j depends only on its nearest neighbours and on itself (compare with Fig. 1a). The generalizations to more neighbours is straightforward. The whole time evolution is defined by the k^3 by k matrix p. Since the variable S_v^j has to go in one of the k possibilities S_v^i in the next time step, the matrix p has only $k-1$ independent elements per row. If left-right symmetry is used or in addition p depends only on the sum $S_{v-1}^j + S_v^j + S_{v+1}^j$ (totalistic rules [6]) then the number of elements of p is reduced even more. Note that with rule (3) each site is updated simultaneously, while in usual Monte-Carlo-dynamics only one site is updated per time step [3].

In the deterministic limit the elements of p are only one or zero, therefore each state i has a unique path as a function of time. In this case the local rules $p(S_1 S_2 S_3 | S')$ are defined by a single integer number n [5]

$$n = \sum_{\alpha=0}^{k^3-1} S'(\alpha) k^\alpha \tag{4}$$

where $\alpha = S_1 k^2 + S_2 k + S_3$ denotes the neighbourhood of the variable S_2 and $S'(\alpha)$ is the variable S' which the system takes at time $t+1$ if the system is in state (S_1, S_2, S_3) at time t (compare with Fig. 1(a)). For example, if for $k=2$ the deterministic rule table reads

111	110	101	100	011	010	001	000
0	0	0	1	0	1	1	0

then one gets rule $n = (00010110)_2 = 22$, etc.

W. Kinzel: Phase Transitions of Cellular Automata   231

Note that the example above is a totalistic rule; in this case the rule can be denoted by another integer m

$$m = \sum_{\alpha=0}^{3(k-1)} S'(\alpha) k^\alpha \qquad (5)$$

where $\alpha = S_1 + S_2 + S_3$ and $S'(\alpha)$ is again the new variable S' given (S_1, S_2, S_3) in the preceding time step. In the example above one has $m = (0,0,1,0)_2 = 2$. In Refs. [5] and [6] only symmetric CA with absorbing state 0 $(p(000|0) = 1)$ are considered, thus for $k = 2$ there remain 32 general and 8 totalistic rules which fall into the first three classes mentioned in the Introduction. In Sects. III and IV phase transitions between these rules are investigated.

Note that $p(S_1, S_2, S_3 | S')$ may be written as $\exp[-\mathcal{H}(S_1, S_2, S_3, S')/k_B T]$. Therefore T_{ij} may be considered as a row-to-row transfermatrix of a two-dimensional k-state model of equilibrium statistical mechanics [12, 13]. In general $\mathcal{H}$ contains all kinds of interactions between $S_1 S_2 S_3$ and S'. The interesting case of an absorbing state, i.e. $p(000|0) = 1$ transforms into infinitely strong couplings in $\mathcal{H}$ or, equivalently, to a restricted phase space in the partition sum [12].

Consider a $k = 2$ reaction on the lattice of Fig. 1(b), which is equivalent to an alternating odd-even reaction on the process of Fig. 1(a) [12]. In this case the matrix $p(S_1 S_2 | S')$ is a 4 by 2 matrix, only. For special reaction rates this process is equivalent to two-dimensional directed percolation [7, 14]. In this case $S_v = 1$ (0) means that the site v can (cannot) be reached by a directed path from the top of the system (or from a seed point at $t = 0$). If the bond and site probabilities are given by p_b and p_s, respectively, one has

$$p(11|1) = p_s p_b (2 - p_b)$$
$$p(01|1) = p(10|1) = p_s p_b \qquad (5)$$
$$p(00|1) = 0.$$

Also the case of different left-right-probabilities $p(01|1) \neq p(10|1)$ has been studied which for $p(01|1) = p(11|1) = 1$ was solved exactly [18]. There is a simple symmetry relation from permutating the notations of the variables S_v^i. For instance for $k = 2$ this "duality" relation maps a CA with rule $\underline{p}$ to the same one with rule $\underline{p}'$ given by

$$p'(S_1 S_2 S_3 | S') = 1 - p(\bar{S}_1 \bar{S}_2 \bar{S}_3 | S') \qquad (6)$$

where $\bar{S} = (S + 1)$ modulo 2. As shown below for some special cases critical points can be obtained from selfdual points $\underline{p} = \underline{p}'$.

III. Phase Diagrams

In this section phase boundaries of one-dimensional two-state $(k = 2)$ CA are calculated by transfermatrix scaling (see Appendix). In Fig. 2 the phase diagram of the reaction process of Fig. 1(b) is shown for $p_1 = p(01|1) = p(10|1)$, $p_2 = p(11|1)$ and $p_0 = p(00|1) = 0$ [19].

For a finite system of N lattice sites and for any values of $p_1 < 1$ and $p_2 < 1$ any initial state $P_i(t = 0)$ decays exponentially fast to the stationary state $P_0(t = \infty) = 1$ and $P_i(t = \infty) = 0$ for $i \neq 0$. This is shown in the following:

(i) With $p_0 = 0$ one has $T_{00} = 1$ and $T_{0j} = 0 (j \neq 0)$

(ii) From this one obtains the eigenvector $\varphi_0 = 1$ and $\varphi_i = 0 (i \neq 0)$ for the largest eigenvector $\lambda_0 = 1$.

(iii) Then one has $P(t) = \underline{T}^t \cdot \underline{P}(0) \sim \lambda_0^t \cdot \varphi \left[1 + 0 \left(\frac{\lambda_v}{\lambda_0} \right)^t \right]$

where the last equation holds asymptotically for large t. Since all other eigenvalues λ_v are smaller than $\lambda_0 = 1$ (which follows from Frobenius [20]) one obtain the desired result with the relaxation time τ given by the second largest eigenvalue $\lambda_1 : \tau = (-\ln \lambda_1)^{-1}$.

In the thermodynamic limit $N \to \infty$ the situation changes: Then for p_1 and p_2 large enough λ_0 is degenerate with λ_1 and the system has two stationary states $(\varphi_i^0)_i$ as above and $(\varphi_i^1)_i$, where all states $P_i(t = 0)$ (except φ_i^0) decay exponentially fast to the state $P_0(t = \infty) = 0$ and $P_i(t = \infty) = -\varphi_i^1/\varphi_0^1$. Thus in the left part of Fig. 2 (= absorbing phase) the variable 1 always dies out, while in the right part (in the active phase) the stationary state has a nonzero fraction of variable 1. In between there is a sharp phase transition whose properties are discussed in the following section. Note that according to (5) directed bond $(p_s = 1)$, site $(p_b = 1)$ and mixed $(p_s = p_b)$ percolation are single lines shown in Fig. 2.

The corners of the phase diagram, Fig. 2, are deterministic CA. The three corners $(p_1, p_2) = (0,0)$, $(0,1)$ and $(1,1)$ (rules $m = 0$, 4 and 6) belong the class-1 CA. In the first two cases an initial disordered state decays to the state $i = 0 = (000...00)_2$. For the rule $m = 6$ the initial state decays to $i = 2^N - 1 = (111...11)_2$. The CA$(p_1, p_2) = (1,0)$ (rule $m = 2$) belongs to class 3. Since in this case $S' = S_1 + S_2$ (modulo two) it behaves like the rule 90 of [5]:

A single occupied site creates a Pascal's triangle modulo two, as shown in Fig. 3. It is a selfsimilar structure with fractal dimension $\log_2 3 \simeq 1.5850$. If the initial state is disordered then one still obtains triangles on all length scales although the pattern is not selfsimilar.

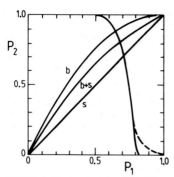

Fig. 2. Phase diagram of a 1-dim. 2-state CA constructed from Fig. 1(b). The full line is the result from scaling systems with 10, 9 and 8 sites. The expected result is indicated by the dashed line. The lines b, $b+s$ and s mark the subspaces of directed bond, mixed bond-site and site percolation, respectively

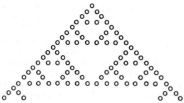

Fig. 3. Pattern created from a 1-dimensional CA (Fig. 1(b)) with deterministic rule $m=2$. The circles show the variable $S=1$ while the empty sites belong to $S=0$. The structure is a Pascal's triangle modulo two, it is selfsimilar with a fractal dimension $\log_2 3 \simeq 1.586$ [5]

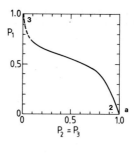

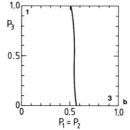

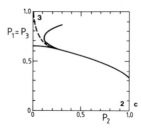

Fig. 4a–c. Phase diagram of a 1-dim. CA (Fig. 1(a)) with totalistic rules and one absorbing state ($p_0=0$). The full lines show the results from transfermatrix-scaling ($N \leq 15$), the error estimate is less than the thickness of the lines. The splitted line of **c** shows the results from $N=9$ and $N=8$, showing that the method fails close to class-3 CA. The expected extrapolations are indicated by dashed lines. The corners of the phase diagrams are deterministic CA which belong to different classes [6], the class number is shown in the figure and in Table 1

For finite chains with N sites the class-3 system evolves into periodic structures which sensitively depend on N. It is presumably for this reason that finite-size scaling does not work close to the deterministic CA with rule $m=2$, (note that one has to compare correlation lengths of strips with $N-2$, $N-1$, and N to get the critical point). For some triple of strips one gets a transition on the $p_2=0$ boundary of Fig. 2 while for other triples the phase boundary bends up again. Presumably the ordered state of Fig. 3 is not dense enough to support a transition away from the deterministic limit, and the phase boundary behaves as shown by the dashed line in Fig. 2. However, this has still to be shown, may be by Monte Carlo simulations on larger systems [8].

The line $p_2=1$ of Fig. 2, is mapped to itself by the duality transformation Eqs. (6), in particular one has

$p_1'=1-p_1$. Thus the phase boundary has to end at the point $p_{1c}=1/2$. In fact this upper boundary has two absorbing states $i=0$ and $i=2^k-1$ $=(111\ldots 11)_2$. An initial state decays into domains of $i=0$ and $i=2^k-1$. The domain walls diffuse with a drift given by $p_1-0.5$. For zero drift ($p_1=1/2$) the

Table 1. Deterministic corners of the (p_2, p_2, p_1)-cube of a 2-state CA with totalistic rules and one absorbing state. The partition into several classes is taken from Ref. 6

p_3	p_2	p_1	p_0	Rule numbers		Class
				m	n	
0	0	0	0	0	0	1
0	0	1	0	2	22	3
0	1	0	0	4	104	2
0	1	1	0	6	126	3
1	0	0	0	8	128	1
1	0	1	0	10	150	3
1	1	0	0	12	232	2
1	1	1	0	14	254	1

walls move like an annihilating random walk. On the average it takes a time $t \sim l^2$ before two walls with a distance l meet and annihilate [12, 21].
In addition this point is an endpoint of a disorder line of an exactly solved two dimensional Ising model on a triangular lattice [12]. Therefore this model can be solved exactly on the line $p_2 = 1$ and on a special line in the (p_0, p_1, p_2)-cube. However, phase transitions occur only for $p_0 = 0$, i.e. if the system has absorbing states [8]. p_0 may be considered as a source of particles $(S=1)$ or as a ghost field in percolation [12, 22] which destroys the phase transition like an external field in ferromagnets.
Now we consider the CA on the geometry of Fig. 1(a). Some cuts of the phase diagrams are shown in Fig. 4. Only totalistic rules are considered, thus one has only 4 independent reaction probabilities $p_0 = p(000 \mid 1)$,

$$p_1 = p(100 \mid 1) = p(010 \mid 1) = p(001 \mid 1),$$

$$p_2 = p(011 \mid 1) = p(101 \mid 1) = p(110 \mid 1)$$

and

$$p_3 = p(111 \mid 1).$$

The eight possible deterministic CA (with absorbing state, $p_0 = 0$) are shown in Table 1 and have extensively been discussed by Wolfram [5]. His numerical investigations suggest that these CA belong to one of the first three classes mentioned in the Introduction (to get class 4 one needs $k \geq 3$ or more than nearest neighbours [6]). These classes are indicated in Table 1 and in Fig. 4. Figure 4 shows that there is always a phase boundary between rule $m=0$ ($p_1 = p_2 = p_3 = 0$) and $m=14$ ($p_1 = p_2 = p_3 = 1$), i.e. between a completely absorbing phase $(0, 0, 0 \ldots)$ and active phase with a nonzero fraction of variable $S_v = 1$. However, even the qualitative behaviour of the phase boundary cannot be related to the class of the deterministic corners. Thus Fig. 4(b) shows a phase

transition to a CA of class 3 while in Figs. 4(a) and 4(b) the phase boundaries seem to end at the class-3 CA [23]. In Fig. 4(a) the transition ends at a class-2 CA while in Fig. 4(c) one has only an absorbing stationary state close to the class-2 corner. Note that in Fig. 4(b) the transition ends exactly at $p_3 = 1$, $p_1 = p_2 = 1/2$ due to duality, Eq. (6).
A simple mean field approximation is obtained by stirring the system at every time step before the next reaction occurs [7]. From this the fraction x_t of sites with variable $S_v = 1$ evolves as a function of time t like a discrete cubic map

$$x_{t+1} = 3 p_1 x_t (1 - x_t)^2 + 3 p_2 x_t^2 (1 - x_t) + p_3 x_t^3 \qquad (7)$$

This maps gives a continuous phase transition for $p_1 = 1/3$ independent of p_2 and p_3. Figure 4 shows that spatial fluctuations change the mean field phase boundary drastically.
The corners of Figs. 2 and 4 which belong to a class-3 deterministic CA show a certain structure [5, 6]. For finite chains this behaviour is reflected in the occurrence of periodic patterns, which shows up in the degeneracy of eigenvalues of the transfermatrix T_{ij}. However, at least for one-dimensional CA such structures are destroyed by stochastic rules, as already noted in Ref. [5]. We have confirmed this by calculating the correlation times of the oscillating parts given by the eigenvalue $\lambda_2, \lambda_3, \ldots$ of the transfermatrix T_{ij}, Eq. (3). While the correlation of the monotonic decay $\tau_1 = -(\ln \lambda_1)^{-1}$ is large and grows exponentially with N in the active phase, all other correlation times τ_i are very small (although all approach infinity at the deterministic limits) and we could not detect any scaling behaviour. It would be interesting to see whether such structures are stable in high spatial dimensions.

IV. Critical Properties

The phase transitions of stochastic CA with one absorbing state are continuous transitions. As in critical phase transitions of equilibrium systems [24] observable quantities have a singular part which obeys power laws with universal critical exponents and scaling relations [7, 8, 15, 25]. Below $D=4$ space dimensions these exponents differ from their mean field values [9, 25].
For $D=1$ the critical exponents and reaction rates can be determined from transfermatrix-scaling (see Appendix) [14]. The correlation time $\tau \left(p - p_c, p_0, \frac{1}{N} \right)$ of a system of N sites scales asymptotically for large N as (for any scale factor b)

$$\tau \left(\varepsilon, p_0, \frac{1}{N} \right) = b^z \tau \left(b^{1/\nu_\perp} \varepsilon, b^{\omega/\nu_\perp} p_0, \frac{b}{N} \right). \qquad (8)$$

Here ε means some deviation from the critical surface in the $p_0 = 0$ space. In addition it is assumed that the source p_0 of $S = 1$ states is the scaling field in the "magnetic" space. This is similar to usual percolation where p_0 is the probability to percolate to a "ghost site" [22]. $z = v_\parallel/v_\perp$ is the dynamic exponent which relates a change of length scale to a change of time scale. $v_\parallel$ and $v_\perp$ describe how the correlation time τ, the correlation length ξ and reaction velocity v behave at the critical surface; one has

$$\tau(\varepsilon) \sim \varepsilon^{-v_\parallel}$$
$$\xi(\varepsilon) \sim \varepsilon^{-v_\perp} \qquad (9)$$
$$v(\varepsilon) \sim \varepsilon^{v_\parallel - v_\perp}.$$

The number $c(\varepsilon)$ of "active" sites $S = 1$ and the probability $P(t)$ of having an active site $S_v = 1$ at time t if one has $S_v = 1$ at $t = 0$ is described by the exponent β:

$$c(\varepsilon) \sim \varepsilon^\beta$$
$$P(t) \sim t^{\beta/v_\parallel} \qquad \text{(at criticality } \varepsilon = 0\text{)}. \qquad (10)$$

Using the usual scaling relations [7, 24] β is related to the "magnetic" scaling power ω by

$$\beta = v_\parallel + v_\perp - \omega. \qquad (11)$$

The exponents z, $v_\parallel$ and $v_\perp$ have been calculated from (8) in [14] for the three cases of directed percolation in Fig. 2. One obtains

$$v_\parallel = 1.734 \pm 0.002,$$
$$v_\perp = 1.100 \pm 0.005, \qquad (12)$$
$$z = 1.582 \pm 0.001.$$

Table 2 shows the results for $z + \omega/v_\perp$ obtained from (8) for $p_1 = p_2 = 0.7058$ in Fig. 2 [14], and the resulting β calculated from Eqs. (11) and (12). The extrapolated result agrees with the value of β collected from other methods [7]:

$$0.273 \pm 0.002 \qquad (13)$$

Due to universality these critical exponents are expected to describe the critical properties along all phase boundaries in Figs. 2 and 4. In fact, for the three percolation points on the transition line of Fig. 2 the numerical results are consistent with the values of (12) [7, 14]. Figure 5 shows the exponent z calculated from chains with $N - 1$, N and $N + 1$ sites for three other examples of the reaction process shown in Fig. 4: *(i)* all diagonals on Fig. 4, $p_1 = p_2 = p_3 = p$ with a critical point $p_c = 0.5385 \pm 0.0005$, *(ii)* the bottom line of Fig. 4(b) $p_1 = p_2 = p$, $p_3 = 0$ with $p_c = 0.573$

Table 2. Estimate for the critical exponent $z + \omega/v_\perp$ obtained from transfermatrix-scaling of the $k = 2$ state CA of Fig. 2 with $p_1 = p_2$. β is given by the scaling relation Eq. (11) and the exponents z and $v_\perp$ taken from Ref. 14

$N - 1/N/N + 1$	$z + \dfrac{\omega}{v_\perp}$	β
3/4/5	3.787	0.415
4/5/6	3.801	0.399
5/6/7	3.830	0.367
6/7/8	3.845	0.351
7/8/9	3.851	0.344
8/9/10	3.865	0.329
9/10/11	3.872	0.321
10/11/12	3.878	0.315
11/12/13	3.883	0.309
12/13/14	3.887	0.304
13/14/15	3.898	0.293

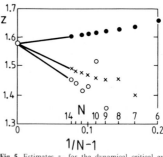

Fig. 5. Estimates z_N for the dynamical critical exponent z as a function of system size N for $p_1 = p_2 = p_3$ (dots); $p_1 = p_2$, $p_3 = 0$ (crosses); $p_1 = p_3$, $p_2 = 1$ (circles) of the model Fig. 4

± 0.005 and *(iii)* the right endpoint of Fig. 4(b), $p_2 = 1$, $p_1 = p_3 = p$ with $p_c = 0.333 \pm 0.002$. Although there is still a strong and sometimes a nonmonotonic N-dependence, the results are consistent with the value of z in (12) and thus consistent with universality.

However, at the selfdual endpoints with two absorbing states there is a different critical behaviour given by an annihilating random walk: In Fig. 2 for $p_2 = 1$, $p_1 = 1/2$ the exact solution gives $v_\parallel = 1$, $v_\perp = 1/2$, $z = 2$ [12]; in Fig. 4(b) for $p_3 = 1$ and $p_1 = p_2 = 1/2$ we have calculated the exponent z from (8), the results are consistent with $z = 2$.

V. Entropy, Coverage and Kink-Density

As discussed in Sect. III, the $(k = 2)$ stochastic CA of Figs. 2 and 4 have a transition from an absorbing phase which has only the absorbing state

$i=(0,0,0,\ldots)$ to an active phase with a nonzero coverage c defined by

$$c = \sum_{i=0}^{2^N-1} \frac{c_i}{N} P_i^0 \tag{14}$$

where $P_i^0 = P_i \,(t=\infty)$ is the stationary probability to find state i and $c_i = \sum_{v=1}^{N} S_v^i$ is the number of active sites $S_v^i = 1$. As a measure for the amount of order in the active phase one may define an entropy S by (measure or metric entropy in Ref. 6)

$$S = - \sum_{i=0}^{2^N-1} P_i^0 \ln P_i^0. \tag{15}$$

Another quantity of interest may be the average number n of 01 or 10 "kinks" on the chain given by

$$n = \sum_{i=0}^{2^N-1} n_i P_i^0 \tag{16}$$

where n_i is the fraction of kinks of state i.
All of these quantities c, S and n may be considered as order parameters since they are zero in the absorbing phase; there only the state $i=0$ is occupied. In particular they are zero in *finite* systems of N sites since an active phase appears in the thermodynamic limit only (compare with Sect. III).
This is similar to a ferromagnet where a spontaneous magnetization can occur in the thermodynamic limit only. Nevertheless there is a possibility to get the order parameter from the eigenvectors of the transfermatrix [26]. Here we use a similar method to get c, S and n from the eigenvector (φ_i^1) of the second largest eigenvector λ_1 of the matrix T_{ij}, Eq. (3).
Any initial distribution $P_i(t=0)$ may be expanded in (right) eigenvectors φ_i^v and eigenvalues λ_v of the transfermatrix T_{ij}

$$P_i(t=0) = \sum_v a_v \varphi_i^v. \tag{17}$$

Then the stationary state is given by

$$P_i^0 = \lim_{t \to \infty} \sum_v \lambda_v^t a_v \varphi_i^v = a_0 \varphi_i^0 + a_1 \varphi_i^1. \tag{18}$$

The last equation holds in the active phase only since there one has $\lambda_0 = \lambda_1 = 1$. Since $\varphi_i^0 = 0$ for $i \neq 0$ (compare with Sect. III) the stationary distribution is proportional to φ_i^1 for $i \neq 0$. Taking $P_0(t=0)=0$ one has $P_0^0 = 0$ since $T_{i0} = 0$ for $i \neq 0$. Thus

$$P_i^0 = \begin{cases} 0 & i=0 \\ -\varphi_i^1/\varphi_0^1 & i \neq 0 \end{cases}. \tag{19}$$

The normalization factor follows from that fact that $\phi_i^0 = 1$ is the left eigenvector of λ_0, therefore $\sum_i \varphi_i^1 = 0$.

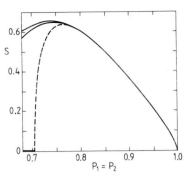

Fig. 6. Entropy S on the line s of Fig. 2 for $N=8$ (upper) and $N=10$ (lower full line). The dashed line sketches the law $S \sim (p-p_c)^\beta$ obtained from finite size scaling ($p_c = 0.7058 \pm 0.0001$ [7])

The situation here is different from the one of usual statistical mechanics like in ferromagnets where a symmetry is broken spontaneously. There φ_i^0 still describes the disordered phase since states with positive and negative magnetization have the same coefficients. But in the ordered phase one needs both φ_i^0 and φ_i^1 in (18) to get the statistical weight for states with only positive magnetization and therewith the magnetization.
We have used Eq. (19) for a *finite* system to calculate c, S and n approximately for the *infinite* system with Eqs. (14), (15) and (16). Figure 6 shows the result for the entropy S as a function of $p=p_1=p_2$ of the CA of Fig. 2. Close to $p=p_c$ S has its maximum which is close to the entropy of the completely disordered case, $S = \ln 2 = 0.6931\ldots$ In the infinite system, $N=\infty$, one has $S=0$ below the critical point $p_c = 0.7058 \pm 0.0001$ [14]. However, for $N=8$ and 10 S is still very large below p_c but slowly decreases with increasing N. We have tried to estimate the critical behaviour from the N-dependence at $p=p_c$. Following the theory of finite size scaling [27] one has for $N \to \infty$, neglecting corrections to scaling and the regular part

$$\begin{aligned} S(p_c, N) &\sim N^{-x/v_\perp} \\ c(p_c, N) &\sim N^{-\beta/v_\perp} \\ n(p_c, N) &\sim N^{-y/v_\perp} \end{aligned} \tag{20}$$

where the exponents, x, β and y describe the critical behaviour in the infinite system for $p \gtrsim p_c$

$$\begin{aligned} S(p, \infty) &\sim (p-p_c)^x \\ c(p_c, \infty) &\sim (p-p_c)^\beta \\ n(p, \infty) &\sim (p-p_c)^y. \end{aligned} \tag{21}$$

Table 3. Entropy S, coverage c and kink density n of the $k=2$ state CA of Fig. 2 at the critical point $p_1 = p_2 = 0.7508$ for different sizes N

N	S	c	n
6	0.6587	0.5597	0.4162
7	0.6503	0.5261	0.3977
8	0.6407	0.5206	0.3822
9	0.6309	0.4981	0.3690
10	0.6231	0.4921	0.3575
11	0.6120	0.4757	0.3475
12	0.6032	0.4700	0.3386
13	0.5948	0.4573	0.3306
14	0.5869	0.4521	0.3235
15	0.5794	0.4419	0.3170

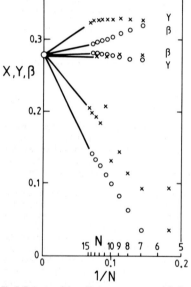

Fig. 7. Estimates of the critical exponents x, y and β of entropy, kink density and coverage, respectively obtained from scaling systems with sizes N and $N-1$ at the critical point. The crosses are from the $k=2$ CA of Fig. 2 with $p_1 = p_2 = 0.7058$ and absorbing boundary condition, the circles are from the $k=2$ CA of Fig. 4 with $p_1 = p_2 = p_3 = 0.5385$ and periodic boundary conditions

If the influence of correlations can be neglected then one has $S \sim c \ln c$ and $n \sim c$, in this case one gets $y = x = \beta$.

Table 3 and Fig. 7 show the results for two different models. In fact all exponents seem to converge for $N \to \infty$ to the value of β of (13). The slow converge of $x(N)$ seems to be due to the logarithmic cor-

rections. Hence the coverage, the entropy and the kink density decay to zero at p_c with the critical exponent β. Note that β is determined from the properties of zero source $p_0 = 0$ while in Table 2 the scaling of p_0 was used to derive β.

VI. Three State Models:
Exact Mappings to Two State Processes

In this section we investigate $k=3$ state stochastic CA for two reasons: *(i)* those models may have applications on multicomponent reactions *(ii)* the problem of universality of the phase transitions of those models is considered in the following section. Here we show that special $k=3$ processes can be mapped onto $k=2$ CA.

To see this consider the transfermatrix T_{ij} of (2) with $S_v^i \varepsilon \{0, 1, 2\}$. In the following the CA is restricted to processes which see the configurations of variable $S_v^i = 0$, only, independent of the distribution of $S_v^i = 1$ and $S_v^i = 2$. For example for the local rule $p(S_1, S_2 | S')$ of the geometry, Fig. 1(b) one has

$$p(11|0) = p(12|0) = p(22|0) = \tilde{p}(11|0)$$
$$p(01|0) = p(02|0) = \tilde{p}(01|0). \tag{22}$$

Now all $k=3$ states $\alpha = (S_1, S_2, \ldots, S_N)$ are mapped to $k=2$ states $\tilde{\alpha}$ by changing the variable $S_v = 2$ to $\tilde{S}_v = 1$, for example if $\alpha = (0, 1, 0, 2)$ then $\tilde{\alpha} = (0, 1, 0, 1)$. Thus a whole class of states α is mapped into a single state $\tilde{\alpha}$. This mapping defines a transfermatrix $\tilde{T}_{\tilde{\alpha}\tilde{\beta}}$ of a $k=2$ CA as follows

$$\tilde{T}_{\tilde{\alpha}\tilde{\beta}} = \sum_{\alpha \varepsilon \tilde{\alpha}} T_{\alpha\beta} \tag{23}$$

where the sum runs over all states α of class $\tilde{\alpha}$ and β is a state of class $\tilde{\beta}$. Equation (23) is a valid definition since the right side does not depend on the state β of class $\tilde{\beta}$. Namely one has with (3)

$$\tilde{T}_{\tilde{\alpha}\tilde{\beta}} = \sum_{\alpha \varepsilon \tilde{\alpha}} \prod_v p(S_v^\beta S_{v+1}^\beta | S_v^\alpha) \tag{24}$$

which may be written as a single product of terms $p(S_v^\beta S_{v+1}^\beta | 0)$ if $S_v^\tilde{\alpha} = 0$ or

$$[p(S_v^\beta S_{v+1}^\beta | 1) + p(S_v^\beta S_{v+1}^\beta | 2)] = [1 - p(S_v^\beta S_{v+1}^\beta | 0)]$$

if $S_v^\tilde{\alpha} \neq 0$. In both cases, by the restriction (22), the expressions depend on the configuration of zero's only, that is they depend on $\tilde{\beta}$, only. Therefore (23) and (24) do not depend on the special state β but only on $\tilde{\beta}$.

Now let λ be an eigenvalue of $T_{\alpha\beta}$ with right eigenvector $(\varphi_\alpha)_\alpha$. Then one has with Eq. (23)

$$\sum_{\beta} T_{\alpha\beta}\,\varphi_\beta = \lambda\varphi_\alpha$$

$$\Rightarrow \sum_{\alpha\in\tilde{\alpha}}\sum_{\beta} T_{\alpha\beta}\,\varphi_\beta = \lambda\sum_{\alpha\in\tilde{\alpha}}\varphi_\alpha$$

$$\Rightarrow \sum_{\beta}\sum_{\beta\in\tilde{\beta}} T_{\tilde{\alpha}\tilde{\beta}}\,\varphi_\beta = \lambda\sum_{\alpha\in\tilde{\alpha}}\varphi_\alpha$$

$$\Rightarrow \sum_{\tilde{\beta}}\tilde{T}_{\tilde{\alpha}\tilde{\beta}}\,\tilde{\varphi}_{\tilde{\beta}} = \lambda\tilde{\varphi}_{\tilde{\alpha}}, \qquad \tilde{\varphi}_{\tilde{\alpha}} = \sum_{\alpha\in\tilde{\alpha}}\varphi_\alpha. \qquad (25)$$

Therefore, if $\tilde{\varphi}_{\tilde{\alpha}} \neq 0$, λ is an eigenvalue of $\tilde{T}$ with eigenvector $\tilde{\varphi}$. From this it follows, that these special $k=3$ CA are equivalent to $k=2$ CA with local rules $\tilde{p}$ defined in Eq. (22).

VII. Universality Classes

Critical phase transitions in thermal equilibrium have universal properties like critical exponents and scaling functions which do not depend on microscopic details of the system [24]. In particular they do not depend on the kind of the lattice nor the range of the interactions, but e.g. for Potts models they depend on the number of variables per site [1].

Field theoretic investigations indicate [15] that for nonequilibrium phase transitions with absorbing states the critical poperties do *not* depend on the number k of local variables, in contrast to equilibrium statistical mechanics. On the other side, for CA with two absorbing states [17] and CA with long time memory [16] new universality classes were reported. Here we study these problems for one-dimensional CA.

a) $k=3$ CA with one Absorbing State

First consider a $k=3$ state CA with one absorbing state $i=0$. For simplicity we use the geometry of Fig. 1(b), thus the reaction rate is given by a 9×3 matrix $p(S_1 S_2 | S')$, Eq. (3). Using the symmetries between left and right and between variables $S_v^i=1$ and $S_v^i=2$ [e.g. $p(12|1)=p(12|2)$] only five independent transition probabilities are left:

$$p(12|1)=p_1, \quad p(01|1)=p_2, \quad p(02|1)=p_3,$$
$$p(11|1)=p_4, \quad p(22|1)=p_5. \qquad (26)$$

In this five dimensional parameter set there is a subspace which can be mapped onto a $k=2$ state CA. Namely, according to Sect. VI, the $k=3$ CA has the same critical behaviour of the CA of Fig. 2 (with $\tilde{p}_1$ and $\tilde{p}_2$) if

$$2p_1 = p_4 + p_5. \qquad (27)$$

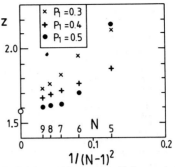

Fig. 8. Estimates of the dynamical critical exponent z from scaling systems with sizes $N, N-1$ and $N-2$ with periodic boundary conditions. The model is the $k=3$ state CA of Fig. 1(b) with one absorbing state and $p_2=p_3=p_4=p_5$ [see (26)]. The results for different values of p_1 are shown. The circle shows the value of z for $k=2$ CA

In this case, according to (22), the rules of the $k=2$ CA are given by

$$\tilde{p}_1 = p_2 + p_3,$$
$$\tilde{p}_2 = 2p_1 + p_4 + p_5. \qquad (28)$$

Hence still a four dimensional parameter subspace can be mapped to the $k=2$ CA of Fig. 2, in particular for $p_1=p_2=p_3=p_4=p_5=p$ this problem has the same critical behaviour as directed site percolation with $p_c=p_c^D/2$ where $p_c^D=0.7058\pm0.0001$ is the percolation threshold.

If the parameters do not obey (28), i.e. if the dynamics of the variable $S_v^i=0$ sees the structure of the dynamics of variables $S_v^i=1$ and $S_v^i=2$ one has to apply numerical methods to check universality. For the case $p=p_2=p_3=p_4=p_5$ we have calculated the exponent z for different probabilities p_1 using transfermatrix scaling (Note that the size of the matrix is $3^{N+1}\times3^{N+1}$, thus we could use only strips with $N\leq9$). Figure 8 shows the results. In all cases the values of $z(N)$ seem to converge to the known result $z=1.581$ [7], confirming the field theoretic predictions [15].

b) CA with Two Absorbing States

The problem with two absorbing states [17] is more difficult to handle by numerical methods. In the absorbing phase an initial state decays to a state which consists of domains of both of the absorbing states. Since the domain walls move like a random walk and annihilate when two of them meet this state decays algebraically to the stationary state which consists of one of the absorbing states. Thus the whole absorbing phase behaves like a critical point with power law decay of correlations. From the random walk behaviour one obtains $\tilde{a}=2$.

In Ref. [17] the kink density n (16) was investigated by Monte Carlo methods for a $k=2$ CA with two absorbing states $(0, 1, 0, 1. \ldots)$ and $(1, 0, 1, 0, 1, \ldots)$. In the absorbing phase n decays with time like $n \sim t^{-1/2}$. At the critical point p_c one finds $n \sim t^{-\alpha}$ with $\alpha = 0.27 \pm 0.08$ and an exponent $\beta = 0.6 \pm 0.2$ which obviously differs much from the value of (13). However, as can be seem from Fig. 4 of Ref. [17], the determination of the critical behaviour is difficult due to fluctuations, finite site effects and the presence of algebraic decay in the absorbing phase, thus an alternative way of deriving critical properties would be useful.

We have investigated the same model with the transfermatrix method. Unfortunately, in the active phase the closeness of a class 3 deterministic CA lead to irregular behaviour of $\tau(N)$ and did not allow any conclusions to be drawn from finite size scaling; similar effects have been discussed in Sect. III. Therefore we have studied this problem for a different model, namely a $k=3$ CA with two absorbing states $i_1 = (1, 1, 1, \ldots, 1)$ and $i_2 = (2, 2, 2, \ldots, 2)$ on the geometry of Fig. 1(b). Again, by using symmetries of left-right and variables $S_v^i = 1$ and $S_v^i = 2$ only four parameters of the 9×3 matrix $p(S_1 S_2 | S')$ are left:

$$p_1 = p(12 | 1),$$
$$p_2 = p(01 | 1),$$
$$p_3 = p(02 | 1),$$
$$p_4 = p(00 | 1). \tag{29}$$

Note that $p(11|1) = p(22|2) = 1$ to get the two absorbing states $i_1 = (3^N - 1)/2$ and $i_2 = 3^N - 1$. As before there is a subspace of probabilities for which the model can be mapped onto the $k=2$ process of Fig. 2. Namely if one has $p_1 = 1/2$ then, with Eq. (22), the dynamics of variable $S_v^v = 0$ does not see the difference between $S_v^i = 1$ and $S_v^i = 2$, and the system is mapped onto the $k=2$ CA of Fig. 2 with one absorbing state $\tilde{p}_0 = 0$ and

$$\tilde{p}_1 = 1 - p_2 - p_3,$$
$$\tilde{p}_2 = 1 - 2p_4. \tag{30}$$

In particular the problem with $p_1 = 1/2$ and $p_2 = p_3 = p_4 = p$ is again equivalent to directed site percolation with $p_c = (1 - p_c^D)/2 \simeq 0.148$.

Hence at least for this subspace of CA with two absorbing states one has the same critical behaviour as for usual $k=2$ CA.

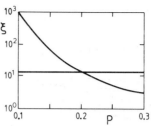

Fig. 9. Correlation times ξ for a $k=3$ state CA with two absorbing state ($N=6$, $p_1 = 1/2$, $p_2 = p_3 = p_4 = p$)

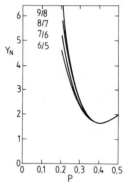

Fig. 10. Logarithmic ratio Y_N of correlation times, Eq. (A.9) as a function of $p = p_1 = p_2 = p_3 = p_4$ for a $k=3$ state CA with two absorbing states. This figure indicates a transition at $p \approx 0.35$

Although this subspace with two absorbing states can be mapped onto a problem with one absorbing state the mechanism of randomly moving and annihilating domain walls is still present. In the absorbing phase $p > p_c$ there is still an algebraic decay of correlations. In a finite system there is an eigenvalue λ_2 and a correlation time $\tau_2 = -(\ln \lambda_2)^{-1}$ corresponding to this mechanism. This eigenvalue λ_2 cannot be mapped onto the $k=2$ CA described in Sect. VI, since $\tilde{\varphi}_{\tilde{a}} = 0$ as easily can be shown. However λ_2 can be obtained from a submatrix of $T_{\alpha\beta}$ which corresponds to a different $k=2$ CA. Therefore one has two independ correlation times as shown in Fig. 9: τ_1 for the decay of variable $S_v^i = 0$ and τ_2 for the domain wall annihilation. In the active phase τ_1 increases exponentially with strip width N while τ_2 increases like N^2.

In the case $p_1 < 1/2$ the domain walls create the state $S_v^i = 0$. Therefore the decoupling mentioned above and the mapping onto $k=2$ CA does not hold any

W. Kinzel: Phase Transitions of Cellular Automata   239

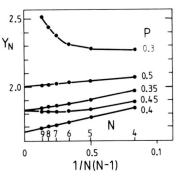

Fig. 11. Y_N as in Fig. 10 as a function of system size N. The extrapolated value $\lim_{N \to \infty} Y_N$ is the dynamical exponent z in the absorbing phase

Table 4. Transition probabilities $p(S_1 S_2 S_3 | S')$ of a $k=3$ state CA which describes an infection with immunization. The values not shown are given by symmetries

$S_1 S_2 S_3	S'$	0	1	2
000	1	0	0
001	$1-p_1$	p_1	0
002	1	0	0
010	0	p_1	$1-p_1$
011	0	p_1	$1-p_1$
012	0	p_1	$1-p_1$
020	0	0	1
021	0	p_2	$1-p_2$
022	0	0	1
100	–	–	–
101	$1-p_1$	p_1	0
102	$1-p_1$	p_1	0
110	–	–	–
111	0	p_1	$1-p_1$
112	0	p_1	$1-p_1$
120	–	–	–
121	0	p_2	$1-p_2$
122	0	p_2	$1-p_2$
200	–	–	–
201	–	–	–
202	1	0	0
210	–	–	–
211	–	–	–
212	0	p_1	$1-p_1$
220	–	–	–
221	–	–	–
222	0	0	1

more. We have studied this problem for $p_1 = p_2 = p_3 = p_4 = p$ numerically with transfermatrix scaling. In Fig. 10 $Y_N = \ln[\tau(p, N)/\tau(p, N-1)]/\ln(N/N-1)$ (compare with Eq. (A.9)) is shown as a function of p. For $p \lesssim 0.3$ the results are consistent with $Y_N \sim N$ indicating an active phase. For $p \lesssim 0.35$ Y_N seems to converge to a constant $z = \lim_{N \to \infty} Y_N < \infty$ indicating a phase with algebraic decay of correlations due to the motion of domain walls of the two absorbing states. However, as shown in Fig. 11, the value of z seems to depend on p. For $p = 1/2$ one gets $z = 2$ since one has an annihilating random walk. But for $p < 1/2$, the variable $S_v^i = 0$ is created and annihilated at the domain walls. Figure 11 indicates that for this case one gets different nonuniversal critical behaviour with $z < 2$. At the transition point p_c, which from Fig. 11 is estimated to have the value $p_c = 0.33 \pm 0.03$, the value of z seems to be much larger than $z = 1.581$ from (12). However this was also the case for the problem which could be mapped onto a $k = 2$ CA, thus this does not necessarily mean new critical behaviour. Therefore we cannot determine the critical properties at the transition from the present method.

c) Infection with Immunization

Cardy recently investigated the following epidemic process [16]: (i) Infected sites infect their neighbours with probability p in the next time step and become immun. (ii) Immun sites have the probability $p - q$ of being reinfected by infected neighbours. A field theory was developed for this process and it was shown that for $q \neq 0$ one gets new critical properties.

Here we study such a process for an one-dimensional CA on the geometry of Fig. 1(a). The system consists of normal $S_v^i = 0$ and of infected, $S_v^i = 1$ states. To describe immunization we introduce a third variable $S_v^i = 2$ of immune states. Thus the memory effect of the $k = 2$ state process is equivalent to a $k = 3$ state CA. The rules $p(S_1, S_2, S_3 | S')$ are defined in Table 4. We have used two probabilities $p_1 = p$ and $p_2 = p - q$ with the meaning described above. Note that one has infinitely many absorbing states, since by definition all states which do not have infected sites do not change with time.

If j is a state without normal sites and i is a state with normal sites, $S_v^i = 0$, then one has $T_{ij} = 0$. That means that there is no transition from states without normal sites to states with normal sites. Therefore the matrix T_{ij} has all the eigenvalues of the corresponding $k = 2$ CA with infected and immune states, only. If the other eigenvalues are not larger then the model has the same critical behaviour as the $k = 2$ CA investigated in the previous sections. In fact we have checked numerically for different values of p and q that the correlation time of the $k = 2$ CA is exactly the one of the $k = 3$ model.

We expect the same behaviour for any such immunization process described above, at least in the active phase: After a transient time all sites will be either infected or immune, therefore the asymptotic behaviour is a usual process with infection probability $p - q$. This argument and the results above for the one dimensional model are at variance with the field theory of Ref. [16].

VIII. Competing Reactions

In this section we are interested in the problem whether stochastic CA may show any structure in time direction. The mean field equation of (7) and Ref. [7] are iterations of simple nonlinear maps which may have oscillations with any period and even may show chaotic motion [28]. However, for all cases of $k = 2$ and $k = 3$ CA we have studied so far we have found time independent stationary states, only.

This is not true for the decay into stationary states. An initial state may relax to the final state with an oscillatory component. In the transfermatrix T_{ij} this shows up as a complex conjugate eigenvalue pair λ_1 and λ_2. The asymptotic decay of $P_i(t=0)$ is given by

$$P_i(t) \sim \varphi_i^0 + (\lambda_1^t + \bar{\lambda}_1^t) \varphi_i^1$$
$$= \varphi_i^0 + \exp(-t/\tau) \cos(qt) \varphi_i^1 \qquad (31)$$

with $\tau = -\ln|\lambda_1|^{-1}$ and $\lambda_1 = |\lambda| \cdot \exp(iq)$. Similar effects occur in equilibrium statistical mechanics at "disorder lines" [29]. For instance in lattice gas models, in the disordered phase of the temperature-coverage-phase diagram there is a line where the second and third largest eigenvalues λ_1 and λ_2 merge to a complex conjugate pair yielding an oscillating and incommensurate asymptotic decay of correlations [30]. Such lines are also found in one dimensional lattice gases [31]. Only systems with competing interactions show such a behaviour.

For CA a merging of eigenvalues λ_1 and λ_2 can occur for nonzero source $p_0 = p(000|1) \neq 0$, only. Namely for $p_0 = 0$ the matrix T_{ij} has elements $T_{00} = 1$ and $T_{j0} = 0$ for $j > 0$. This gives the largest eigenvalue $\lambda_0 = 1$, and the other eigenvalues are eigenvalues of the submatrix T_{ij}' with $i > 0$ and $j > 0$. Since T' has only positive elements and in general is irreducible its largest eigenvalue λ_1 is positive and nondegenerate due to the theorems of Frobenius [20].

But for $p_0 > 0$ transitions form monotonic to oscillating transient decay are possible. We want to demonstrate this for 0-dimensional CA but we have also found such a behaviour for finite chains by investigating T_{ij} numerically. Consider a CA on a single site with two states $S = 0$ and $S = 1$. With $p(0|1) = p_0$ and $p(1|1) = p_1$ the transfermatrix T_{ij} has the form

$$\begin{pmatrix} 1 - p_0 & 1 - p_1 \\ p_0 & p_1 \end{pmatrix} \qquad (32)$$

with eigenvalue $\lambda_0 = 1$ and $\lambda_1 = p_1 - p_0$. The decay of an initial state to the stationary state $P_0^0 = (1 - p_1)/(1 - p_1 + p_0)$, $P_1^0 = p_0/(1 - p_1 + p_0)$ is monotonic for $p_1 > p_0$ and oscillating with period 2 for $p_1 < p_0$, but it is always commensurate with respect to the time step. A "disorder line" does not exist for this case. But for a $k = 3$ state CA competing reactions can be introduced yielding a transition to oscillating decay. For instance consider the reaction of states $S = 0, 1$ and 2 giving by the following transfermatrix T_{ij}

$$\begin{pmatrix} 1 - p_0 & 1 - p_1 - p_3 & 1 - p_1 - p_2 \\ p_0 & p_1 & p_2 \\ 0 & p_3 & p_1 \end{pmatrix}. \qquad (33)$$

For $p_0 p_3 \neq 0$ one obtains nondegenerate eigenvalues $\lambda_1 > \lambda_2$. This means one needs the sequence $0 \rightarrow 1 \rightarrow 2 \rightarrow 0$ competing with $0 \rightarrow 1 \rightarrow 0$ to obtain oscillating decay. The eigenvalues are given by

$$\lambda_0 = 1,$$
$$\lambda_{1/2} = p_1 - p_0/2 \pm ((p_0/2)^2 + p_2 p_3 - p_3 p_0)^{1/2}. \qquad (34)$$

This gives the line where λ_1 and λ_2 merge by

$$p_0^2/4 + p_2 p_3 = p_3 p_0. \qquad (35)$$

The correlation time τ and the wavevector q of the oscillations are

$$\tau^{-1} = -\tfrac{1}{2} \ln(p_1^2 + p_0(p_3 - p_1) - p_2 p_3), \qquad (36)$$
$$q = \text{arctg} \left[(p_1 - p_0/2)/(p_3 p_0 - p_0/4 - p_2 p_3)^{1/2} \right]. \qquad (37)$$

IX. Random Probabilities

So far we have studied stochastic CA with homogeneous local reaction probabilities. In this section CA are introduced which have probabilities $p_v'(S_1 S_2 | S^i)$ which are randomly distributed either in space v in time t or both in space and time. If there is still an absorbing state, i.e. $p'(00|1) = 0$, then we expect that these disordered CA still have a phase transition. However its critical properties may change.

For equilibrium phase transitions a heuristic argument of Harris shows for which cases the critical behaviour is changed [32]. We apply this argument to nonequilibrium critical points of disordered CA.

First consider a D-dimensional CA with reaction probabilities which are distributed in space, only. The distribution may be continuously parameterized by x with $x=0$ being the pure system. The critical point $p_c(x)$ is assumed to be a smooth function of x. The Harris argument checks the consistency of the existence of a single typical correlation length ξ which obeys Eq. (9), i.e. $\xi \sim (p - p_c(x))^{-v_\perp}$. Since the system can only feel the properties within a region of size ξ, it locally sees a fluctuation δx of x with $\delta x \sim \xi^{-\frac{D}{2}}$. This gives

$$\xi^{-\frac{1}{v_\perp}} \sim p - p_c(x + \delta x) = p - p_c(x) - A\xi^{-\frac{D}{2}}$$
$$\sim \xi^{-\frac{1}{v}} \left(1 - A\xi^{\frac{1}{v_\perp} - \frac{D}{2}}\right). \tag{38}$$

Thus for $p \to p_c(x)$, i.e. $\xi \to \infty$, Eq. (38) can only be valid if

$$2 - Dv_\perp < 0. \tag{39}$$

In equilibrium statistical mechanics the left side is just the specific heat exponent α. According to Harris for each random system (39) is valid, in particular if the pure system $x=0$ does not obey (39) then the critical properties are changed. Renormalization group arguments indicate that also the reverse is true: If the pure system has $\alpha < 0$ then randomness is irrelevant and the critical properties are not changed [24]. From this and the values of $v_\perp$ of CA with one absorbing state [7] we conclude that spatial disorder does change the critical behaviour of CA in any dimension (the mean field case is just marginal).

Similar arguments hold for the other two cases of randomness, one just has to replace $Dv_\perp$ by $v_\parallel$ and $Dv_\perp + v_\parallel$ for randomness in time and space plus time, respectively. From this one obtains the following result: Space or time randomness is relevant while randomness in both space and time together give the same critical properties as in the pure system.

So far we have discussed the case of weak disorder. For strong disorder with competing reactions new qualitative effects occur. In this case the system acts as a filter to many different states, similar to pure class-2 CA. If the dynamics is given by a Hamiltonian with random competing interactions such a system is called "spin glass" [33]. For infinite-range interactions the system has a phase transition [34] with the consequence that even such *stochastic* CA stay infinitely long in one of the many stationary states. The number of stationary states grows exponentially with system size [35] and the states have an interesting overlap structure ("ultrametric topology" [36]). Recently these CA found applications as models for memory [37] and prebiological evolution [38].

For short range spin glasses, at least in two dimensions there is no phase transition [39]. Nevertheless

the filter effect is seen in the slow decay of the remanent magnetization at low temperatures [40]: Due to the symmetry of the Hamiltonian this means that almost any initial state decays fast to a state with high overlap to the inital one; then this new state decays very slowly with time t. The average relaxation time τ drastically increases with decreasing temperature T like $\ln \tau \sim T^{-vz}$ [41] with a large exponent vz (whose precise value is still not known [42]). In the context of CA this means, for any finite observation time τ there is a temperature T below which the stochastic CA behaves as a effective filter. Furthermore, for a given set of states one can construct interactions ($\hat{=}$ rules) such that these states are stationary ones for infinite range rules or quasi-stationary ones for short range rules [37].

X. Summary

Cellular automata (CA) are simple mathematical models for the dynamics of cooperative phenomena far from thermal equilibrium. They consist of k states on each site of a lattice reacting in discrete time steps by transition probabilities which depend on nearest neighbour states, only. Such models have been discussed in mathematics, physics, biology, ecology, chemistry and computer science.

In this paper phase transitions of such stochastic CA have been investigated. If the model has absorbing states, i.e. states which cannot be left according to the transition probabilities then CA have a sharp transition between qualitatively different stationary phases: An absorbing phase containing the absorbing states, only and an active phase containing all but the absorbing states with some probability distribution.

The phase diagrams of several one-dimensional $k=2$ CA with one absorbing state have been calculated by transfermatrix scaling. The corners of the phase diagrams are deterministic CA which belong to three different classes [6]. There does not seem to exist any connection between the topology of the phases and the class of the deterministic limits. The structures created by deterministic rules are destroyed within the active phase. Several transition points have exactly been determined by duality relations.

These phase transition of CA with one absorbing state are continuous, i.e. for example approaching the transitions the fraction of active sitesgoes continuously to zero. Universal critical exponents, scaling functions and scaling relations describe the singularities of the observable quantities at the transition. In particular it has been demonstrated that finite size scaling of the correlation time can be applied and yields three independent critical exponents from which all other ones can be calculated. The two scaling fields have been identified as the source p_0 of the active states and the reaction probability for $p_0 = 0$.

The critical exponents have been calculated for several points on the phase boundaries. The numerical results support universality of the critical properties. Only special points corresponding to annihilating random walks have different critical behaviour.

An entropy S has been defined and calculated numerically. It is nonzero only in the active phase, has a maximum close at the transition and goes to zero with the power law $S \sim (p - p_c)^\beta$. The exponent β has been determined from finite size scaling demonstrating that all exponents can be calculated without using the scaling field p_0 conjugate to the order parameter. β has also been obtained from the coverage of active sites and the kink density.

The universality of such phase transitions has been explored further by introducing $k = 3$ state CA. For special subspaces of the set of local reaction probabilities the $k = 3$ models have exactly been mapped onto $k = 2$ ones. There are $k = 3$ CA with one or two absorbing states which have the same critical properties as $k = 2$ CA with one absorbing state only.

For transition rates where such a mapping does not apply the critical properties have been calculated numerically. In the case of $k = 3$ CA with two absorbing states a phase transition was found. In the absorbing phase the correlation decays algebraically with a critical exponent which seems to vary continuously with the model parameters. Due to this "phase of critical points" the present method did not allow a determination of the properties of the transition. In the case of CA with one absorbing state the numerical results were consistent with an universal behaviour of $k = 3$ and $k = 2$ CA. A simple model of an infection process with immunization was shown to have the same critical behaviour as usual $k = 2$ CA. Arguments were given that this holds for all such processes, in contrast to a recent field theoretic investigation [16].

The case of competing reactions was shortly discussed. Similar to the "disorder lines" of equilibrium systems a transition from monotonic to incommensurate oscillating decay to the stationary state was found for $k = 3$ CA.

If the local reaction rates are distributed randomly then CA with absorbing states still show a phase transition. Applying the Harris criterium of equilibrium systems to CA one obtains the following result: If the system is random either in space or in time then randomness changes the critical properties

in any dimension; if the system is random both in space and in time then the critical properties are the same as those of the pure system.

Strong randomness with competing reaction rates leads to spin glass behaviour: Even stochastic CA may have infinitely many stationary (or for short range models very long living) states. Hence those CA filter the initial information which decays to a final state with large overlap to the initial one.

I would like to thank E. Domany, P. Grassberger and H.K. Janssen for stimulating discussions and correspondence. This work was initiated at a summer institute of the Weizmann Institute of Science at Rehovot/Israel; I thank the Einstein Center for Theoretical Physics for support.

Appendix

A.1. Transfermatrix

The transfermatrix T_{ij}, Eq. (3) of a k-state CA on a chain of N sites is a k^N by k^N matrix. The integers i and j label the k^N states as defined in (1). For the numerical evaluation of the largest eigenvalues it is convenient to write T_{ij} as a product of N sparse matrices [43]

$$T = L M^{N-2} R. \tag{A.1}$$

Due to the boundaries the matrices L, M, and R differ slightly. Each matrix has only k^{N+1} elements which are nonzero, and each element is given by the local probabilities $p(S_1 S_2 S_3 | S')$ of (3). Neither T nor L, M, or R have to be stored in the computer, all what is needed is the multiplication of a vector by T which is a few lines computer program as shown below.

T_{ij} is the probability to find state i at time $t + 1$ given state j at time t. Thus T_{ij} adds a whole row (= whole time step) to the total probability. This process is split up into N steps by (A.1), each steps adding a single site only. L adds the left site, R the right one and M one site in the middle. For example, the meaning of the matrix M is illustrated in Fig. 12 for $N = 5$: Given state $j = (S_6, S_5, S_4, S_3, S_2, S_1)_k$ at time $t + 1$, M_{ij} is the probability to find state $i = (S'_6, S'_5, S'_4, S'_3, S'_2, S'_1)_k$ at time t. Note that at least $N + 1$ variables are necessary to transfer the probability from site to site, hence M is a k^{N+1} by k^{N+1} matrix. From Fig. 12 one immediately obtains that M_{ij} is zero unless one has

$$S_v = S'_{v-1} \qquad v = 2, \ldots, N. \tag{A.2}$$

Hence, for a given row i only the variable S_6 is independent and there are only k states $j_1, \ldots, j_k$ for

W. Kinzel: Phase Transitions of Cellular Automata   243

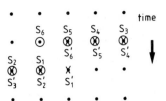

Fig. 12. States $i=(S'_v)$ and $j=(S_v)$ of the sparse transfermatrix M_{ij} as described in the text

which M_{ij} is nonzero. With Eq. (1) one obtains

$$j_\mu = [i/k] + (\mu - 1)\, k^N, \qquad 1 \le \mu \le k \qquad \text{(A.3)}$$

where $[a]$ means the largest integer below a. From Fig. 12 and the meaning of M_{ij} it follows that the nonzero elements of M_{ij} are given by

$$M_{ij} = p(S_{N+1}, S_N, S_{N-1} \,|\, S'_1) \qquad \text{(A.4)}$$

with $p(S_1 S_2 S_3 \,|\, S')$ of Eq. (3). The set S_{N+1}, S_N, S_{N-1} can again be written as a single integer which is given by

$$(S_{N+1}, S_N, S_{N-1})_k = [j/k^{N-2}] \qquad \text{(A.5)}$$

while S'_1 is given by

$$S'_1 = (i/k) \text{ modulo } k. \qquad \text{(A.6)}$$

Therefore each element M_{ij} can be easily determined from the integer representation of states i and j, from simple integer operations (A.3), (A.5) and (A.6) and from the local rules $p(S_1 S_2 S_3 \,|\, S')$. The largest eigenvalue λ_0 of $\underline{T}$ is $\lambda_0 = 1$, since $\sum_i T_{ij} = 1$ by definition. The second largest eigenvalue λ_1 determines the correlation length. In the case of absorbing states, for example for $T_{0j} = 0$ $(j \ne 0)$, λ_1 is easily obtained by a simple iteration:

(i) Start with an (almost) arbitrary vector u_i^0 $(1 \le i \le k^{N-1})$;

(ii) Generate vectors $\underline{u}^n$ by $\underline{u}^n = \underline{T} u^{n-1}$ and replacing u_0^n by $u_0^n = 0$.

(iii) Then λ_1 is obtained from

$$\lambda_1 = \lim_{n \to \infty} \frac{u_i^{n+1}}{u_i^n}. \qquad \text{(A.7)}$$

In practice the iteration is stopped if the relative change of u_i^n is smaller than an given ε; usually we have used $\varepsilon = 10^{-8}$.

If the system has more absorbing states i, their weight u_i^n has to be replaced by $u_i^n = 0$ at each iteration step. If there are no absorbing states, one has to iterate two independent vectors $\underline{u}$ and $\underline{v}$ keeping

them linear independent at each iteration step. In general, to obtain the m largest (in absolute value) eigenvectors one has to iterate m independent vectors $\underline{u}_1^n, \ldots, \underline{u}_m^n$. The eigenvalues are given for $n \to \infty$ by the m by m matrix $u_{\nu\mu} = \langle u_\nu^n \,|\, u_\mu^n \rangle$ [44].

The iteration $\underline{u}^{n+1} = \underline{T} u^n$ is obtained from the matrix product (A.1), each term given by (A.2) to (A.6). For instance (for $k = 2$) the product $\underline{v} = \underline{M}\,\underline{u}$ is given by calling the following FORTRAN subroutine

```
SUBROUTINE MAT(U, V, P, N)
DIMENSION U(1), V(1), P(8, 2)
N1 = 2xx(N + 1)
N2 = 2xxN
N3 = 2xx(N - 2)
DO 1 I = 1, N1
J1 = (I - 1)/2 + 1
J2 = J1 + N2
K1 = (J1 - 1)/N3 + 1
K2 = (J2 - 1)/N3 + 1
K3 = MOD(I - 1, 2) + 1
1 V(I) = P(K1, K3) x U(J1) + P(K2, K3) x U(J2)
RETURN
END
```

Note that in (old) FORTRAN the indices I have to start with I = 1, hence we have added an one the each index. The matrix P(I, J) is the matrix $p(S_1, S_2, S_3 \,|\, S')$ of (3) which is stored once in the beginning. Of course, also the states J1 and J2 and the indices K1, K2 and K3 may be calculated for each I in the beginning and stored as a vector; depending on the architecture of the computer this may be somewhat faster.

A.2. Scaling

Following Nightingale [45] the critical properties of the infinite system are determined from scaling the correlation time τ with the size N of the system, using the asymptotic form Eq. (8). In practice one obtains τ from the eigenvalue λ_1 of the matrix T_{ij}

$$\tau = -(\ln \lambda_1)^{-1}. \qquad \text{(A.8)}$$

Then the quantity

$$Y_N = \ln \left[\tau(p, N) / \tau(p, N-1) \right] / \ln \left[N/N-1 \right] \qquad \text{(A.9)}$$

is calculated for a line in the parameter space parameterized by p. If $p > p_c$ denotes the active phase then one has [44]

$$\begin{aligned} Y_N &\sim N & p > p_c \\ Y_N &= z & p = p_c \\ Y_N &\to 0 & p < p_c \end{aligned} \qquad \text{(A.10)}$$

where z is the dynamical critical exponent. Hence from the intersection of the functions $Y_{N+1}(p)$ and $Y_N(p)$ at the point (z_N, p_N) one obtains an estimate for z and p_c, respectively. From the N-dependence one can estimate the quality of the results, the extrapolation to $N \to \infty$ gives the final estimate. In principle one may also obtain universal irrelevant exponents from z_N and p_N, but in practive this is difficult due to cancellation effects [46].

The other exponents $v_\parallel$ and ω can be obtained from the scaling of the corresponding partial derivates of τ, as can be seen from (8).

References

1. Wu, F.Y.: Rev. Mod. Phys. **54**, 235 (1982)
2. Glauber, R.: J. Math. Phys. **4**, 234 (1963)
3. Binder, K.: Monte Carlo methods in statistical physics. In: Topics in Current Physics. Binder, K. (ed.), Vol. 7. Berlin, Heidelberg, New York: Springer 1979
4. Physica **10** D, 1–247 (1984)
5. Wolfram, S.: Rev. Mod. Phys. **55**, 601 (1983)
6. Wolfram, S.: in Ref. [4], p. 1
7. For a review see Kinzel, W.: In: Percolation structures and processes. Deutsch, G., Zallen, R., Adler, J. (eds.). Bristol: Adam Hilger 1983
8. Grassberger, P., Torre, A. de la: Ann. Phys. **122**, 373 (1979); Grassberger, P.: Z. Phys. B – Condensed Matter **47**, 365 (1982)
9. Grassberger, P.: Nucl. Phys. B **125**, 91 (1977); Cardy, J.L., Sugar, R.: J. Phys. A **13**, L423 (1980)
10. Haken, H.: Synergetics. Berlin, Heidelberg, New York: Springer 1978; Nicolis, G., Prigogine, I.: Self organization in non-equilibrium. New York: Wiley 1977
11. Verhagen, A.M.W.: J. Stat. Phys. **15**, 213 (1976); Enting, I.G.: J. Phys. C **10**, 1379 (1977); A **10**, 1023 (1977); A **11**, 2001 (1978); Rujan, P.: J. Stat. Phys. **29**, 247 (1982); **34**, 615 (1984)
12. Domany, E., Kinzel, W.: Phys. Rev. Lett. **53**, 311 (1984)
13. Choi, M.Y., Huberman, B.A.: J. Phys. A **17**, L765 (1984)
14. Kinzel, W., Yeomans, J.: J. Phys. A **14**, L163 (1981)
15. Janssen, H.K.: A multistate CA belongs to the same universality class as a twostate CA if it has a single absorbing state and if all its components can diffuse (unpublished)
16. Cardy, J.: J. Phys. A **16**, L709 (1983)
17. Grassberger, P., Krause, F., Twer, T. v.d.: J. Phys. A **17**, L105 (1984)
18. Domany, E., Kinzel, W.: Phys. Rev. Lett. **47**, 5 (1981)
19. Some of the following results have already been presented in Ref. [12]
20. Frobenius, S.B.: Preuss. Akad. Wiss. 471 (1908)
21. Grassberger, P.: in Ref. [4], p. 52
22. Griffiths, R.B.: J. Math. Phys. **8**, 484 (1967)
23. As in Fig. 2 the determination of the phase boundaries close to class 3-CA is not possible due to strong finite size effects
24. see e.g. Ma, S.K.: Modern theory of critical phenomena. London: Benjamin 1976
25. Janssen, H.K.: Z. Phys. B – Condensed Matter **42**, 151 (1981)
26. Hamer, C.J.: J. Phys. A **15**, L675 (1982)
27. Barber, M.N.: In: Phase transitions and critical phenomena. Domb, C., Lebowitz, J.L. (eds.), Vol. 8. New York: Academic Press 1983
28. see e.g. Ott, E.: Rev. Mod. Phys. **53**, 655 (1981)
29. Stephenson, J.: J. Math. Phys. **11**, 420 (1970); Phys. Rev. B **1**, 4405 (1970)
30. Kinzel, W., Selke, W., Binder, K.: Surf. Sci. **121**, 13 (1982)
31. Caroll, C.E.: Surf. Sci. **32**, 119 (1972)
32. Harris, A.B.: J. Phys. C **7**, 1671 (1974)
33. Edwards, S.F., Anderson, P.W.: J. Phys. F **5**, 965 (1975) for a recent review see: Fischer, K.H.: Phys. Status Solidi B **116**, 357 (1983)
34. Sherrington, D., Kirkpatrick, S.: Phys. Rev. Lett. **35**, 1792 (1975)
35. Tanaka, E., Edwards, S.F.: J. Phys. F **10**, 2471 (1980); Bray, A.J., Moore, M.A.: J. Phys. C **13**, L469 (1980)
36. Mezard, M., Parisi, G., Sourlas, N., Toulouse, G., Virasoro, M.: Phys. Rev. Lett. **52**, 1156 (1984)
37. Hopfield, J.J.: Proc. Natl. Acad. Sci. USA **79**, 2554 (1982)
38. Anderson, P.W.: Proc. Natl. Acad. Sci. USA **80**, 3386 (1983)
39. Morgenstern, I., Binder, K.: Phys. Rev. Lett. **43**, 1615 (1979); Phys. Rev. B **22**, 288 (1980) see also: Binder, K., Kinzel, W.: Lectures Notes in Physics. Vol. 192, p. 279. Berlin, Heidelberg, New York: Springer 1983
40. Binder, K., Schröder, K.: Phys. Rev. B **14**, 2142 (1976); Kinzel, W.: Phys. Rev. B **19**, 4594 (1979)
41. Kinzel, W., Binder, K.: Phys. Rev. B **29**, 1300 (1984)
42. Young, A.P.: J. Phys. C **18**, L517 (1984)
43. Domb, C.: Proc. R. Soc. London Ser. A **196**, 36 (1949)
44. Wilkinson, J.H.: Algebraic eigenvalue problems. Oxford: Claredon Press 1965
45. Nightingale, P.: J. Appl. Phys. **53**, 7927 (1982)
46. Privman, V., Fisher, M.E.: J. Phys. A **16**, L295 (1983)

W. Kinzel
Institut für Festkörperforschung
Kernforschungsanlage Jülich GmbH
Postfach 1913
D-5170 Jülich 1
Federal Republic of Germany

Note Added in Proof

Randomness in both space and time can be integrated out to give a pure CA with averaged reaction probabilities. I thank the referee and M. Schreckenberg for pointing this out to me.

J. Phys. A: Math. Gen. **17** (1984) L105–L109.

LETTER TO THE EDITOR

A new type of kinetic critical phenomenon

Peter Grassberger, Friedrich Krause and Tassilo von der Twer

Physics Department, University of Wuppertal, Fachbreich 8-Physik, Gaußstrasse 20, D-5600 Wuppertal, Germany

Received 28 November 1983

Abstract. We study a new critical phenomenon in a non-thermal one-dimensional lattice model. It is characterised by the transition from stability to instability of kinks between ordered states. Below the critical point, the kinks are stable and move by annihilating random walks. Above the critical point, they are unstable against creation of kink–antikink pairs. (The *spontaneous* production of pairs is assumed to be absent.) At the critical point $p = p_{cr}$, the density of kinks decreases like $n \sim t^{\alpha}$, with $\alpha = 0.27 \pm 0.08$. Above the critical point, the density of kinks in the stationary state is approximately $(p - p_{cr})^{\beta}$ with $\beta = 0.6 \pm 0.2$. Possible extensions to two or more dimensions and possible applications are discussed.

It is well known that thermal critical phenomena cannot occur in one-dimensional systems. The same is not true for non-thermal systems, provided they have absorbing† (or 'quiescent') states.

The typical example is directed percolation in one space plus one time direction (Durrett 1982, Kinzel and Yeomans 1981), which shows the same critical phenomenon as reggeon field theory (Grassberger and de la Torre 1979, Cardy and Sugar 1980) and the basic contact model (Griffeath 1979). This class of phenomena is characterised by a single absorbing state (all sites not 'wetted'), and thus there is no symmetry breakdown related to them. After realising (Grassberger 1982) that the same critical phenomenon seems to occur also in Schlögl's second model (Schlögl 1972) (contrary to previous investigations which claimed that model to be Ising-like (Nicolis and Malek-Mansour 1980, Brachet and Tirapegui 1981, Borckmans *et al* 1977, 1981)), it was conjectured (Grassberger 1982) that all models with a single absorbing state should show this same critical phenomenon.

In the present letter, we shall study a class of models with two absorbing states. These two states are mutually symmetrical, and thus the transition is accompanied by a spontaneous breakdown of symmetry.

Technically, we shall study two models which are both one-dimensional 'elementary' cellular automata (in the sense of Wolfram (1983)) with very specific added noise. Space and time are discrete in these models. The states are $\{S_i | i \in Z\}$; $S_i = 0, 1$ and the transition rules depend on next neighbours only.

Specifically, the models are chasracterised by the rules

$$
\begin{array}{llllll}
t: & 111 & 101 & 010 & 100 & 001 \quad \overbrace{\begin{array}{cc} 011 & 110 \end{array}} \quad 000 \\
t+1: & 0 & 0 & 1 & 1 & 1 \quad \begin{array}{l} 0 \text{ with prob. } p \\ 1 \text{ with prob. } 1-p \end{array} \quad 0 \quad \text{model A}
\end{array}
$$

† We call a state 'absorbing' if it can be entered but cannot be left. Note that spatially infinite systems can admit several absorbing states in this sense.

L106   *Letter to the Editor*

and

t:	111	101	010	100	001	$\overbrace{011 \quad 110}$	000	
$t+1$:	0	1	0	1	1	1 with prob. p 0 with prob. $1-p$	0	model B.

For $p=0$, these are rule number 94 (in the notation of Wolfram (1983)) (model A) and 50 (model B). Both of these rules are 'simple' in the sense of Wolfram (1983): when starting with a random initial condition, the system very soon settles in a stationary state (rule 94), resp. in a state of period 2 (rule 50). This is shown also in figure 1.

Consider now very small values of p, i.e. a small probability for $011 \rightarrow 0$ in model A and a small probability for $011 \rightarrow 1$ in model B. As seen from figure 2, the system now orders itself spontaneously: there are two symmetric absorbing states in both models. For model A, they correspond to a pattern consisting of vertical stripes:

$$S_i = \begin{cases} 0 & i \text{ even} \\ 1 & i \text{ odd} \end{cases} \quad \text{and} \quad S_i = \begin{cases} 1 & i \text{ even} \\ 0 & i \text{ odd} \end{cases}$$

and for model B, they correspond to a chess-board pattern.

After a random start, there are small ordered domains separated by kinks. For $p=0$, these kinks are stationary, but for small $p \neq 0$ they move by annihilating random

      (a)   (b)

Figure 1. Patterns created, with $p=0$ from (a) model A and (b) model B. The starting configuration was random.

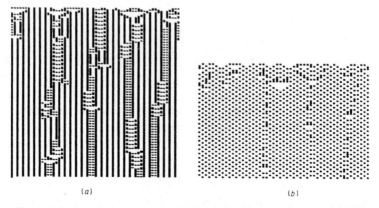

      (a)   (b)

Figure 2. Patterns created, from a random initial configuration, with $p \neq 0$. Again (a) corresponds to model A with $p = 0.05$, and (b) corresponds to model B with $P = 0.2$.

walks. The diffusion coefficient in both cases is $\propto p$ for small p, and the kink density decreases correspondingly (Griffeath 1979, Grassberger 1983) like $(p \cdot t)^{-1/2}$.

In addition to enhancing the random walk, increasing p has another effect: it leads to a splitting of kinks,

$$\text{kink} \rightarrow \text{kink} + (\text{kink} + \text{antikink}).$$

For small p, this creation of new kinks is outpowered by their annihilation. But above a critical value p_{cr}, a single kink in the initial state is sufficient to create a completely disordered state (see figure 3). For $p = 1$, in particular, model A is just rule 22, while model B is rule 122, in the notation of Wolfram (1983). Both these rules are known to be chaotic (Wolfram 1983, Grassberger 1983).

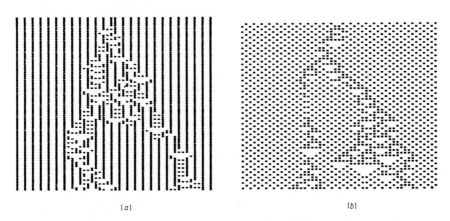

(a)   (b)

Figure 3. Patterns created from initial states containing a single kink, and with $p > p_{cr}$. (a) shows evolution according to model A with $p = 0.25$ and (b) shows model B with $p = 0.65$.

In order to study the behaviour near $p = p_{cr}$ more precisely, we have performed detailed Monte Carlo simulations, the results of which are shown in figure 4. They are based on counting the number of doubly occupied neighbours. In the ordered states, both models have no doubly occupied neighbours, thus this number is proportional to the number of kinks. Lattice sizes were 5,000 (model A) and 20,000 (model B) sites. Periodic boundary conditions were chosen in both cases. The critical probability was found to be

$$p_{cr} = \begin{cases} 0.13 \pm 0.02 & \text{model A} \\ 0.555 \pm 0.01 & \text{model B.} \end{cases}$$

At $p = p_{cr}$, in both models the density of doubly occupied neighbours (i.e. the density of kinks) decreases like

$$n_{\text{kink}} \sim t^{-\alpha}, \qquad \alpha = 0.27 \pm 0.08.$$

Above $p = p_{cr}$, the density of kinks in the stationary state goes to zero for $p \rightarrow p_{cr}$ like

$$n_{\text{kink}} \sim (p - p_{cr})^{\beta}, \qquad \beta = 0.6 \pm 0.2.$$

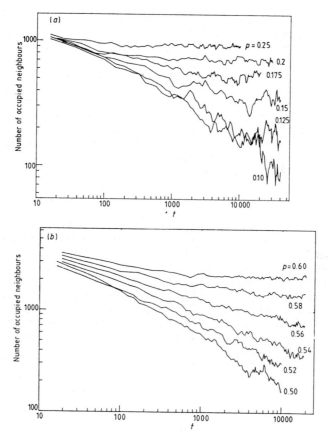

Figure 4. Numbers of occupied neighbouring pairs (measuring the density of kinks), close to the critical probability, as a function of time. For large times, the numbers were averaged over many time steps in order to suppress fluctuations. Lattice sizes were (*a*) 5000 sites (model A) and (*b*) 20 000 sites (model B).

As in other kinetic critical phenomena, there is a third critical exponent measuring the increase of the correlation length when either $t \to \infty$ (for $p = p_{cr}$) or $p \to p_{cr}$ (for $t = \infty$). Our simulations were not precise enough to give a meaningful estimate of this exponent.

Let us now discuss these results.

First, we should point out that the very existence of the transitions depends on the fact that the ordered states were absolutely stable. Even the smallest amount of 'thermal' noise would destroy them. Adding such noise to cellular automata has previously been studied by Wolfram (1983) and Schulman and Seiden (1978).

Secondly, it is not difficult to envisage models with different symmetries, by considering e.g. more than two states per lattice site or with next-nearest neighbour interactions. As a result, one should obtain whole classes of critical phenomena, each with different critical exponents.

When going from one to two dimensions, one can replace the kinks either by point defects or by domain walls. For models with two states per site and NN interactions, the latter is the more natural, as only a domain wall is topologically stable in the same way as a single kink was stable for $0 < p < p_{cr}$ in the above models. Nevertheless, one can artificially construct lattice models in which a single defect can give rise to a defect pair but cannot vanish.

In contrast to the cellular automata studied above, there are automata whose behaviour is extremely complex. The best known example is Conway's game of life (Gardner 1979), but there exist many more similar examples (Farmer and Wolfram 1983). They share with the above rules 94 and 50 the property that they have infinitely many stationary estates. It seems plausible (although we have not yet found an example for such behaviour) that adding *specific* noise to these automata leads to the selection of much less trivially ordered states than in the above models A, B. In the game of life, e.g., extremely complex patterns exist but are very rarely generated from a random start. Adding some transitions with a suitable rate might enhance their production considerably, making the game of life resemble real life much more closely.

Finally, let us suggest a somewhat less speculative application of these ideas. It concerns drawing of single crystals out of a melt. Assume that the original seed contains some defects which prevent ordered growth in their vicinity, leading to randomly moving defects in the plane of growth. On the other hand assume that the melt is sufficiently pure so that spontaneous generation of defects can be neglected. For small drawing velocity, the defects will move until they either annihilate or reach the boundary—leading in this way to a perfect single crystal. For large drawing velocity, any defect might create further defects in the next deposited layer, and formation of a perfect crystal is impossible.

It is of course known that the velocity of drawing has a strong effect on the perfectness of single crystals, but the existence of a sharp transition with associated universal scaling laws has not yet been observed, to our knowledge.

References

Borckmans P, Dewel G and Walgraef D 1977 *Z. Phys. B* **28** 235
—— 1981 *J. Stat. Phys.* **24** 119
Brachet M E and Tirapegui E 1981 *Phys. Lett.* **81A** 211
Cardy J and Sugar R L 1980 *J. Phys. A: Math. Gen.* **13** L423
Durrett R 1982 *Oriented percolation in two dimensions, Los Angeles preprint*
Farmer D and Wolfram S (eds) 1983 *Proc. Conf. Cellular Automata, Los Alamos, March 1983*
Gardener M 1971 *Sci. Am.* **224** 112
Grassberger P 1982 *Z. Phys. B* **47** 365
—— 1983 *Phys. Rev. A* **28** to be published
Grassberger P and de la Torre A 1979 *Ann. Phys., NY* **122** 373
Griffeath D 1979 *Lecture Notes in Mathematics, vol* 724 (Berlin: Springer)
Kinzel W and Yeomans J M 1981 *J. Phys. A: Math. Gen.* **14** L163
Nicolis G and Malek-Mansour M 1980 *J. Stat. Phys.* **22** 495
Schlögl F 1972 *Z. Phys.* **253** 147
Schulman L and Seiden P E 1978 *J. Stat. Phys.* **19** 293
Wolfram S 1983 *Rev. Mod. Phys.* **55** 601

J. Phys. A: Math. Gen. **19** (1986) L69–L75.

LETTER TO THE EDITOR

Phase transitions in two-dimensional stochastic cellular automata

Kunihiko Kaneko and Yasuhiro Akutsu

Institute of Physics, College of Arts and Sciences, University of Tokyo, Komaba, Meguro-ku, Tokyo 153, Japan

Received 1 August 1985, in final form 15 October 1985

Abstract. Two-dimensional stochastic cellular automata with nearest-neighbour couplings are investigated. Depending on the rules, low-level noise phases are classified into ferro, glassy, roll, glassy-roll, and antiferro, glassy-antiferro, antiferro-roll, glassy-antiferro-roll, 'coexistent phase of ferro and antiferro', 'labyrinth' and 'turbulence'. Also, the oscillating phases with period 2 for the above rules are observed in other rules. Transformations from ferro to antiferro and from fixed point to periodic patterns are shown. The nature of phase transitions due to a change of noise is investigated.

Phase transitions are common phenomena both in equilibrium and non-equilibrium systems. In equilibrium statistical mechanics, Ising models have played important roles. In non-equilibrium systems, the transition phenomena are more abundant (Nicolis and Prigogine 1977, Haken 1978), though a simple general model with discrete states is not available. In this letter, we consider a class of stochastic cellular automata (SCA), in order to study the general aspects of phase transitions both for equilibrium and non-equilibrium systems. The model might be regarded as a generalisation of Ising models and may also be a typical discrete model for non-equilibrium transitions.

Recently, Wolfram (1983, 1984) and Packard and Wolfram (1985) have investigated one- and two-dimensional cellular automata (CA) from the viewpoint of dynamical systems theory. A phase transition is observed in one-dimensional CA if the rules are changed at some lattice points (Grassberger *et al* 1984, Kinzel 1985), while it is found in the coupled map lattices as the coupling parameter is changed (Kaneko 1984, 1985a, Aizawa *et al* 1985). In the present letter a class of two-dimensional stochastic cellular automata with nearest-neighbour couplings is investigated on a square lattice with a periodic boundary condition, mainly focusing on the transition phenomena caused by a change of noise strength.

Inclusion of noise in cellular automata is important for the following reasons: first, noise plays the role of temperature in equilibrium systems. Thus, a phase transition as the noise level changes is expected for a system with a dimension higher than one. Second, deterministic cellular automata can have a huge number of attractors. Inclusion of noise brings about a jump among attractors and leads to the selection of a small number of physical states (Kauffman 1969, Kaneko 1985b). Lastly noise plays an important role for the formation of patterns in non-equilibrium systems.

L70 *Letter to the Editor*

The rule for the evolution for the SCA in the present letter is given by

$$s_{i,j}^{t+1} = \begin{cases} I(n_{i,j}^t, s_{i,j}^t) & \text{with probability } 1-p \\ 1-I & \text{with probability } p \end{cases} \tag{1}$$

where $n_{i,j} = s_{i,j-1} + s_{i,j+1} + s_{i-1,j} + s_{i+1,j}$ with $s_{i,j}^t = 0$ or 1. The suffices i,j denote a lattice site while the superscript t shows the discrete time step. In the present letter the system size is chosen to be 32×32. The 'rule' I is a function which takes the value 0 or 1. Noise is added so that $s_{i,j}$ changes its values with probability p. Since the value of I has two choices for each possible 5×2 states ($n = 0, 1, 2, 3, 4$ and $s = 0, 1$), the number of possible rules is 2^{10}.

We have investigated 2^{10} rules for low-level noises ($p \sim 10^{-4}$). The essential feature, however, can be seen in the following symmetric rules.

The symmetric rule is defined as the one which has symmetry about the transformation of 0 and 1. This gives the condition $I(n, s) = 1 - I(4-n, 1-s)$. Hereafter we restrict ourselves to the symmetric rules, which have 2^5 choices.

Furthermore, the number of independent rules is reduced by the method of sublattices. The first transformation is the ferro–antiferro transformation (FAF trsf). Let us consider the sublattice $s_{i,j}^o$ with $i+j =$ odd and $s_{i,j}^e$ with $i+j =$ even. If we apply the transformation $s_{i,j}'^o = 1 - s_{i,j}^o$ and $s_{i,j}'^e = s_{i,j}^e$ for the system with a rule $\bar{I}(n, s') = I(4-n, s')$, it is shown that the dynamics is equivalent to the system for the rule $I(n, s)$. Thus, the result for the rule $\bar{I}(n, s)$ is automatically obtained from that of the rule $I(n, s)$ by the above transformation. The transformation changes a pattern

$$\begin{matrix} 1 & 1 & 1 & 1 \\ 1 & 1 & 1 & 1 \\ 1 & 1 & 1 & 1 \end{matrix} \quad \text{into} \quad \begin{matrix} 1 & 0 & 1 & 0 \\ 0 & 1 & 0 & 1 \\ 1 & 0 & 1 & 0 \end{matrix}$$

and may be called a FAF trsf (ferro–antiferro transformation).

Another transformation is obtained by the use of a temporal sublattice. If the transformation $s_{i,j}'^{2t} = s_{i,j}^{2t}$ and $s_{i,j}'^{2t+1} = 1 - s_{i,j}^{2t+1}$ is applied to a system with a rule $\bar{I}(n, s') = I(4-n, 1-s')$, it is shown that the dynamics are equivalent to a system with a rule $I(n, s)$. The transformation changes a time series 1 1 1 1 1 1 (fixed point) to 1 0 1 0 1 0 (period-2 oscillation) and may be called a periodic–fixed point transformation (PFP trsf).

Taking the above two transformations into account, the number of independent symmetric rules is reduced to 10, which will be studied in the following, where a rule is represented by a code $(i_0 i_1 i_2 i_3 i_4)$, where $i_j = I(j, 0)$ ($I(j, 1) = 1 - I(4-j, 0)$ from symmetry) or by a rule number defined by $\Sigma_{k=0}^4 2^k i_k$.

Patterns at low-level noises are shown in table 1 for rules from 0 to 15. The pattern 'ferro' means a phase with a long range order with broken symmetry about 0 and 1, i.e. the ferromagnetic phase in usual Ising systems. The magnetisation $m = \Sigma_{i,j}(2s_{i,j}-1)$ appears for $p < p_c$, where p_c is the 'transition noise level'. In the 'glassy' phase there appears a short range order for a weak noise, i.e. the correlation $c_{k,1} = \langle(2s_{i+k,j+1}-1)(2s_{i,j}-1)\rangle$ is not small if neither k nor j is large, though the long range order cannot be attained even if the noise level p goes to zero (see figure 1). No phase transition appears as p goes to zero. Since the relaxation becomes slower and slower as $p \to 0$ and the dynamics include a topological constraint, then here we tentatively

Table 1. Patterns of elementary SCA; rule numbers, their codes and patterns at low-level noise, and the presence or absence of long range order (LRO) at low noise are shown. If a rule is derived from another rule number which gives a simpler pattern (i.e. fixed point or ferro is simpler than periodic or antiferro (AF)) by FAF or PFP transformation, the type of transformation and the primary rule are written as 'transformation number'. If a rule is invariant against the FAF transformation, it is written as self-dual (SD).

Rules 0, 1, 2, 3, 4, 5, 10, 17, 18, 19 are primary in the sense that they have neither 'periodic' nor 'AF' phases. Other rules are obtained by FAF or PFP transformation from the above rules.

Number	Code	Pattern	LRO	Transformation
0	0 0 0 0 0	Trivial	No	
1	0 0 0 0 1	Glassy	No	
2	0 0 0 1 0	Ferro	Yes	
3	0 0 0 1 1	Ferro	Yes	
4	0 0 1 0 0	Coexistence of ferro and AF	?	(SD)
5	0 0 1 0 1	Ferro	Yes	
6	0 0 1 1 0	Periodic-AF-roll	Yes	PFP * FAF(19)
7	0 0 1 1 1	Periodic-AF	Yes	PFP * FAF(3)
8	0 1 0 0 0	AF	Yes	FAF(2)
9	0 1 0 0 1	Glassy-AF-roll	No	FAF(18)
10	0 1 0 1 0	⟨Additive turbulence⟩	No	(SD)
11	0 1 0 1 1	Periodic-AF	Yes	PFP * FAF(5)
12	0 1 1 0 0	Periodic-roll	Yes	PFP(19)
13	0 1 1 0 1	Periodic-glassy-roll	No	PFP(18)
14	0 1 1 1 0	Periodic-⟨labyrinth⟩	No	PFP(17) (SD)
15	0 1 1 1 1	Periodic-glassy-AF	No	PFP * FAF(1)

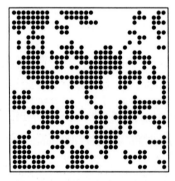

Figure 1. Snapshot of SCA with the rule 00001 (*Glassy*) for $p = 0.005$.
● shows the site (i, j) with $s_{i,j} = 1$, while blank sites mean $s_{i,j} = 0$.

call the phase 'glassy'. The 'roll' phase is shown in figure 2(b), which is the pattern with

$$1 \quad 1 \quad 0 \quad 0 \quad 1 \quad 1 \quad 0 \quad 0$$
$$1 \quad 1 \quad 0 \quad 0 \quad 1 \quad 1 \quad 0 \quad 0.$$
$$1 \quad 1 \quad 0 \quad 0 \quad 1 \quad 1 \quad 0 \quad 0$$

A glassy-roll pattern is a roll phase only with a short range order.

L72   *Letter to the Editor*

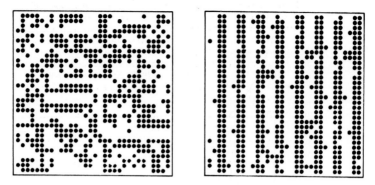

Figure 2. Snapshot of SCA with the rule 1 0 0 1 (roll) for $p = 0.0045$ (*a*) from random initial configurations (after 10^4 steps), (*b*) from ordered initial configurations (after 10^4 steps).

'Periodic' means a pattern which repeats 1 and 0 in time at each site (i.e. the period is 2). All periodic patterns are obtained by PFP trsf. For example, the ordered phase for the rule 1 0 0 1 1 is roll. Applying PFP trsf, the rule is changed to 0 1 1 0 0, which gives the periodic-roll pattern as shown in table 1. Since the rule with a number $n(n > 15)$ is obtained by the PFP trsf from the rule with a number $31 - n$, phases for the rules with $n(n > 15)$ are omitted from the table.

FAF trsf is also useful. If we apply the transformation to a rule for which the low-level noise phase is a pattern such as ferro, glassy, roll or glassy-roll, the low-level noise phase for the transformed rule is the corresponding pattern AF

$$\text{antiferro;} \begin{pmatrix} 1 & 0 & 1 & 0 \\ 0 & 1 & 0 & 1 \\ 1 & 0 & 1 & 0 \end{pmatrix}, \text{glassy-AF, AF-roll} \begin{pmatrix} 1 & 1 & 0 & 0 \\ 0 & 0 & 1 & 1 \\ 1 & 1 & 0 & 0 \end{pmatrix},$$

or glassy-AF-roll respectively. For example, if we apply FAF trsf and PFP trsf successively to the rule 0 1 1 0 1 (periodic-glassy-roll), rule 0 1 0 0 1 is obtained, which yields glassy-AF-roll for low-level noise.

The rules 0 0 0 0 0 (0), 0 0 1 0 0 (4), 0 1 0 1 0 (10), 0 1 1 1 0 (14) (and those obtained by PFP trsf from these rules) are invariant under FAF trsf. Except for the trivial rule 0 (which preserves the initial condition), these rules show interesting behaviour for low-level noise, which will be discussed next.

Here some typical patterns in table 1 are discussed in a little more detail, though the complete accounts will be reportted elsewhere (Akutsu *et al* 1985).

(a) Glassy (rule 0 0 0 0 1). As the noise level p is lowered, the relaxation becomes slower and slower. We calculated the overlap function

$$C(T) = (1/N^2) \sum_{ij} \langle (2s_{i,j}^{t_0+T} - 1)(2s_{i,j}^{t_0} - 1) \rangle \tag{2}$$

where $\langle \ \rangle$ indicates long time average and $N = 32$, after transients have decayed out $(t_0 \sim 10^3)$. $C(T)$ cannot be fitted by a single exponential function and it seems to be represented by $a_1 e^{-T/\tau_1} + a_2 e^{-T/\tau_2} + \cdots$ (see figure 3). The fastest relaxation τ_1 is estimated as follows: the probability that the neighbouring sites take the same values (0 0 or 1 1) is defined as q. For small noise p, the self-consistent approximation for q

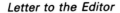

Letter to the Editor   **L73**

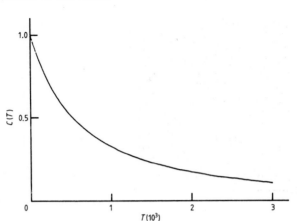

Figure 3. Overlap function $C(T)$ for the rule 0 0 0 0 1 for $p = 0.001$. The data can be fitted well by $a_1 e^{-T/\tau_1} + a_2 e^{-T/\tau_2}$ with $a_1 = 0.37$, $\tau_1 = 3.2 \times 10^2$, $a_2 = 0.63$ and $\tau_2 = 1.4 \times 10^3$.

gives the equation $(1-q)^4 = 8p$, from which the relaxation time τ_1 is given by $(3p)^{-1}$. The numerical results confirm the above estimation. The detailed explanation for the form of $C(T)$, however, remains a problem for the future.

(b) Roll (rule 1 0 0 1 1). The order–disorder transition occurs as can be seen in figure 3. The transition is first order in the sense that it has a hysteresis. If the noise level is lowered, the ordered state appears at $p = p_- \sim 0.0044$, while it remains up to $p = p_+ \sim 0.0049$ when the noise level is increased on this ordered phase. The roll magnetisation is defined in a similar way to the staggered magnetisation for the antiferromagnetic order, which shows a large jump at $p \sim p_+$ or p_-.

Here we note that this ordered state breaks the spatial symmetry (rotational symmetry with angle $\pi/2$), since the rule itself is isotropic.

(c) Ferro–antiferro coexistence (rule 0 0 1 0 0). Both the ferro and AF patterns are stable for low-level noise for the rule. If we start from the random initial configurations, some systems fall into ferro and some into AF, while the others show the coexistence of ferro and antiferro clusters even after 4×10^4 steps, and fixed patterns are not attained (see figure 4).

(d) 'Labyrinth' (rule 1 0 0 0 1). The rule 1 0 0 0 1 (0 1 1 1 0 for the periodic case) gives a pattern in figure 5 for low-level noise. Both the roll

```
1  1  0  0  1  1  0  0                1  0  1  0
1  1  0  0  1  1  0  0 and a single roll 1  0  1  0
1  1  0  0  1  1  0  0                1  0  1  0
```

are not destroyed by a change of $s_{i,j}$ at a single site. These two types of rolls form a labyrinth structure only with a short range order. The relaxation of overlap function (2) is quite slow for low-level noise, which is fit by a single exponential decay

(e) 'Additive turbulence' (rule 0 1 0 1 0 or 1 0 1 0 1). The rule 0 1 0 1 0 shows a chaotic behaviour. No local order with periodic or fixed patterns has been observed (see figure

L74   *Letter to the Editor*

Figure 4. Snapshot of SCA with the rule 0 0 1 0 0 (coexistence of ferro and AF) for $p = 0.001$.

Figure 5. Snapshot of SCA with the rule 1 0 0 0 1 (labyrinth) for $p = 0.001$.

6). Simple patterns such as ferro, AF, periodic-roll and single-roll can exist as a special solution for $p = 0$, but they are easily destroyed by a change of $s_{i,j}$ at a single site.

The above rules are additive in the sense of Wolfram (1983), since they are written as

$$s_{i,j}^{n+1} = s_{i+1,j}^n + s_{i-1,j}^n + s_{i,j+1}^n + s_{i,j-1}^n + s_{i,j}^n \qquad (\text{mod } 2) \qquad (\text{rule } 0\ 1\ 0\ 1\ 0)$$

or

$$s_{i+1,j}^n + s_{i-1,j}^n + s_{i,j+1}^n + s_{i,j-1}^n + s_{i,j}^n - 1 \qquad (\text{mod } 2) \qquad (\text{rule } 1\ 0\ 1\ 0\ 1).$$

Thus, the apparently turbulent behaviour for the above rules can be understood by the superposition of the behaviour obtained by a single site excitation, which is analogous to that by rule 90 of one-dimensional elementary CA (Wolfram 1983).

We have clarified various ordered phases for elementary stochastic cellular automata. Though some of the rules show behaviour common to the usual Ising models (such as ferro or AF), other remarkable phases have been observed. For the 'glassy' phase, it is not yet certain whether the phase is similar to the low temperature phase for the spin glass model with short range interactions. Recently, Fredrickson and Andersen (1984) have considered Ising spin systems with kinetics with some constraint

Letter to the Editor   **L75**

Figure 6. Snapshot of SCA with the rule 0 1 0 1 0 (turbulence) for $p = 0.001$.

and expected a glass transition. The rule 0 0 0 0 1 imposes a topological constraint on the dynamics. Monte Carlo simulation or SCA with unusual kinetics may be of relevance for the understanding of some aspects of glassy behaviour.

The periodic patterns may be related to some non-equilibrium phase transitions in an ensemble of oscillator systems (Kuramoto 1981). Roll patterns and their first order transitions may be related to the dislocation in roll patterns in Benard systems (Ahlers and Behringer 1978, Gollub and Steinman 1981, Fauve *et al* 1984).

Stochastic cellular automata are simple and show a variety of new phenomena, which may open a curtain on a new era of phase transition study.

The authors would like to thank Mr Yukito Iba for stimulating discussions and critical comments.

References

Ahlers G and Behringer R P 1978 *Prog. Theor. Phys. Suppl.* **64** 186
Aizawa Y, Nishikawa I and Kaneko K 1985 in preparation
Akutsu Y, Kaneko K and Iba Y 1985 in preparation
Fauve S, Laroche C, Libchaber A and Perrin B 1984 *Phys. Rev. Lett.* **52** 1774
Fredrickson G H and Andersen H C 1984 *Phys. Rev. Lett.* **53** 1244
Gollub J P and Steinman J F 1981 *Phys. Rev. Lett.* **47** 505
Grassberger P, Krause F and Twer T v d 1984 *J. Phys. A: Math. Gen.* **17** L105
Haken H 1978 *Synergetics* (Berlin: Springer)
Kaneko K 1984 *Prog. Theor. Phys.* **72** 480
—— 1985a *Collapse of Tori and Genesis of Chaos in Dissipative Systems* (Singapore: World Scientific)
—— 1985b *J. Stat. Phys.* to be submitted
Kauffman S 1969 *J. Theor. Biol.* **22** 437
Kinzel W 1985 *Z. Phys.* B **58** 229
Kuramoto Y 1981 *Physica* A **106** 128
Nicolis G and Prigogine I 1977 *Self-organisation in Non-equilibrium Systems* (New York: Wiley)
Packard N and Wolfram S 1985 *J. Stat. Phys.* **38** 901
Wolfram S 1983 *Rev. Mod. Phys.* **55** 601
—— 1984 *Physica* D **10** 1

VOLUME 53, NUMBER 4   PHYSICAL REVIEW LETTERS   23 JULY 1984

Equivalence of Cellular Automata to Ising Models and Directed Percolation

Eytan Domany[a]

Department of Applied Physics, Stanford University, Stanford, California 94305

and

Wolfgang Kinzel

Institut für Festkörperforschung der Kernforschungsanlage Jülich GmbH, D-5170 Jülich, West Germany

(Received 23 December 1983)

Time development of cellular automata in d dimensions is mapped onto equilibrium statistical mechanics of Ising models in $d+1$ dimensions. Directed percolation is equivalent to a cellular automaton, and thus to an Ising model. For a particular case of directed percolation we find $\nu_{\parallel} = 2$, $\nu_{\perp} = 1$, $\eta_{\perp} = 0$.

PACS numbers: 05.50.+q, 05.70.Jk

In a most interesting recent article, Wolfram[1] has presented a large body of phenomenological observations and analytic results on the time development of cellular automata[2] (CA). CA provide extremely simple models for a variety of problems in physics,[3] e.g., nonequilibrium stochastic processes, self-organization, crystal-growth models,[4] chemistry (e.g., reaction models[5]), and biology.[6] CA may also become of some use in computer science.[1,7]

The class of CA studied by Wolfram constitute the deterministic limit of a more general class of stochastic CA. d-dimensional stochastic CA were studied under the name of crystal-growth models.[4] Their time development is equivalent to the equilibrium statistical mechanics of $(d+1)$-dimensional Ising models.[8] This exact mapping is presented here for the simple case of one-dimensional peripheral CA (PCA). We then show that the widely studied problem of directed percolation[9] (DP) in $d=2$ can be easily recast[10] as a particular stochastic PCA. Thus the DP problem is equivalent to an equilibrium Ising model. The standard problems of site and bond DP constitute particular cases of a more general class of DP problems. A special case of this more general class is solved exactly by mapping it onto a random-walk process.

Cellular automata: Definitions.—Consider a linear chain or ring of sites i,j; with each site associate a binary variable $v_i = 0, 1$. The "state" V_t of the system at time t is specified by the set of $v_{i,t}$. At odd (even) times, odd- (even-) indexed sites may change their state, according to preassigned probabilistic rules, and even- (odd-) indexed sites stay in the same state. Thus the full space-time history of our CA can be presented on a two-dimensional lattice. One may get $v_{i,t+1} = 0$ or 1, according to the conditional probabilities $P(v_{i,t+1} | v_{i-1,t}, v_{i+1,t})$ defined as follows:

$$P(1|0,0) = x, \quad P(1|1,1) = z, \quad P(1|0,1) = y. \quad (1)$$

Together with $P(0|v,v') = 1 - P(1|v,v')$, these rules define PCA. In general, $0 \leq x,y,z \leq 1$; when only $x,y,z = 0, 1$ are allowed, we obtain deterministic CA (DCA).

Mapping of CA onto Ising models.—Any space-time development of a CA, characterized by a set of values $v_{i,t} = V$, occurs with probability $P(V)$. Viewing the set of space-time indices (i,t) as a regular $d=2$ lattice, we show that for any CA rules (i.e., x,y,z) there exists an Ising Hamiltonian $H(V)$ such that

$$P(V) = \exp[-H(V)] \quad (2)$$

for all V. If so, all space-time correlation functions and statistical averages of our CA can be expressed as an equilibrium average property of an appropriate Ising model.[8] We have

$$P(V) = P(V_1|V_0) P(V_2|V_1) \cdots P(V_{t+1}|V_t), \quad (3)$$

where $P(V_{t+1}|V_t)$ is the conditional probability to find the CA in state V_{t+1} at time $t+1$, given it was in state V_t at time t. However, because of the local nature of the CA rules,

$$P(V_{t+1}|V_t) = \prod{}' P(v_{i,t+1}|v_{i-1,t}, v_{i+1,t}), \quad (4)$$

where the prime indicates that for even (odd) times i runs over even- (odd-) indexed sites. For the local $P(u|v,v')$ of Eq. (1), we identify $H(V)$ as an Ising Hamiltonian on a triangular lattice, with

$$
\begin{aligned}
-H = {} & B\sum_k v_k + J\sum_{\langle kl \rangle}^{(-)} v_k v_l \\
& + D\sum_{\langle kl \rangle}^{(/)} v_k v_l + E\sum_{\langle klm \rangle}^{(\nabla)} v_k v_l v_m,
\end{aligned}
\quad (5)
$$

where $\sum^{(-)}$ denotes sum over all horizontal bonds, $\sum^{(/)}$ all diagonal bonds, and $\sum^{(\nabla)}$ sums over all *down-pointing* triangles (see Fig. 1), with the cou-

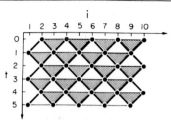

FIG. 1. Triangular Ising lattice, whose equilibrium-statistical mechanics describes the time evolution of a peripheral cellular automaton. Horizontal bonds J (thin lines), diagonal bonds D (heavy lines), three-site interactions E (shaded down-pointing triangles), and field B parametrize the Ising Hamiltonian.

plings given by $A = 1 - x$,

$$e^B = \frac{x(1-y)^2}{(1-x)^3}, \quad e^J = \frac{(1-x)(1-z)}{(1-y)^2},$$

$$e^D = \frac{y(1-x)}{x(1-y)}, \quad e^E = \frac{zx(1-y)^2}{y^2(1-x)(1-z)}. \quad (6)$$

Alternatively, by expressing $v_k = (1 + S_k)/2$, with $S_k = \pm 1$, we obtain a Hamiltonian similar to (5), with couplings $\tilde{B}$, $\tilde{J}$, $\tilde{D}$, and $\tilde{E}$. Thus, time development according to all possible PCA rules is described by equilibrium statistical mechanics of an Ising lattice gas on a triangular lattice.[8]

Time development of DCA.—The eight DCA rules correspond to various manners in which the zero-temperature limit of our Ising model can be taken. Time developments of the DCA correspond to the ground-state configurations of our Ising model, consistent with some preassigned configuration at one of the boundaries. When the ground state of H is nondegenerate (or has finite degeneracy), it will be reached eventually and the DCA time development will correspond to the ground state. However, if the ground state is highly degenerate, complicated time development of the DCA is expected; the system picks one of a multitude of ground states that is consistent with the boundary condition. Thus various properties of DCA can be expressed in terms of correlations at $T = 0$ in degenerate Ising systems.[11]

Phase transitions: Special subspaces.—The point $x = y = z = \frac{1}{2}$ corresponds to $T = \infty$. We did not find phase transitions on any line that connects this point to any DCA point. Moreover, on the surface defined by vanishing three-spin coupling, i.e., $E = 0$, given by $zx/(1-z)(1-x) = y^2/(1-y)^2$, the problem has been treated by Verhagen,[8] and no

transition is found at any finite temperature. If we further impose $\tilde{B} = 0$, the subspace with $S \to -S$ Ising symmetry, given by $x = 1 - z$, $y = 1 - y = \frac{1}{2}$, is obtained. This line is precisely the disorder line[8,11] of the more general triangular Ising model with nearest-neighbor couplings $\tilde{J}$ and $\tilde{D}$; for this model Stephenson has calculated correlation functions in various directions and found diverging correlation length only as $T \to 0$ (i.e., $x \to 0$).[11] We did find phase transitions on the $x = 0$ (and by symmetry, the $z = 1$) surface of our cube.

Directed percolation.—Consider the square lattice of Fig. 1; any site may be present (probability p) or absent $(1-p)$; any bond may be present (probability q) or absent $(1-q)$. Assume that a single site is "wet" at $t = 0$; present bonds conduct "water" only in the downward (increasing t) direction, and only when both sites, connected by the bond, are present. For $q = 1$ this is the site-DP problem, and for $p = 1$ the bond-DP problem.[9] One asks, for given p,q what is the probability of finding wet sites at level (time) t? To reduce this problem to CA rules,[10] note that whether any site (i,t) is wet $(u_{i,t} = 1)$ or dry $(u_{i,t} = 0)$ depends only on the state of the down-pointing triangle whose bottom corner is (i,t), with the CA rules $x = 0$, $y = pq$, $z = pq(2-q)$. Thus the $x = 0$ plane of our CA is a generalized DP problem, in which the standard cases are embedded. Thus we have mapped the DP problem onto an Ising problem on the triangular lattice. The constraint $x = 0$ excludes those Ising configurations in which any down-pointing triangle has $u = 0$ (or $S = -1$) on both of its upper sites, and $u = 1$ (or $S = +1$) on its bottom. Within the space-allowed configurations we have (for bond DP) the Hamiltonian parametrized by $\exp(8\tilde{H}) = q^3(2-q)$, $\exp(8\tilde{J}) = (2-q)/q$, $\exp(8\tilde{D}) = q(2-q)/(1-q)^2$, and $\exp(8\tilde{E}) = (2-q)/q$. The transition line was determined numerically.

The correlation length $\xi(y,z,N)$ is calculated exactly[12] for an infinite strip of width N. The transition point (y_c,z_c) and the critical exponents $\nu_\parallel$ and $\nu_\perp$ in "time" and "space" directions are found using the scaling relation

$$\xi(t,h,1/N) = b^{\nu_\parallel/\nu_\perp} \xi(b^{1/\nu_\perp} t, b^{\omega/\nu_\perp} h, b/N),$$

where t is a scaling field in the y-z plane and b is the change of length scale [usually $b = N/(N-1)$]. We added a symmetry-breaking scaling field h which scales with a new exponent ω. From the usual (anisotropic) scaling relations[9] ω, $\nu_\parallel$, and $\nu_\perp$ give the other critical exponents, $\beta = \nu_\parallel + \nu_\perp - \omega$, $\gamma = 2\omega - \nu_\parallel - \nu_\perp$. The full line of Fig. 2 gives $z_c(y)$ from $N = 10$, 11, and 12. For the upper part of the

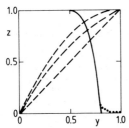

FIG. 2. The $x = 0$ plane of cellular automata rules corresponds to generalized directed percolation. Dashed lines correspond to bond, mixed bond-site ($q = p$), and site percolation. The system percolates to the right of the (solid) transition line. On the $z = 1$ line the model is mapped onto a random-walk problem and solved.

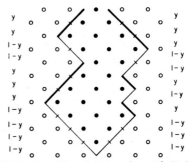

FIG. 3. A typical configuration of a wet (solid dots) domain, for the directed percolation model parametrized by $x = 0$, $z = 1$. The domain boundaries execute a random walk; the weight of each step is indicated on the right (for the right-hand boundary) and on the left (for the left-hand boundary).

phase boundary our results show only little N dependence and can reliably be extrapolated to $N \rightarrow \infty$. However, near the ($z = 0, y = 1$) point irregular variation of $z_c(y)$ with N was observed, reflecting the fact that strips with different widths show very different reaction patterns[1] at the ($z = 0, y = 1$) point, which results in an irregular function $\xi(N)$. The exponents $\nu_{\parallel} = 1.734 \pm 0.002$ and $\nu_{\perp} = 1.100 \pm 0.005$ were obtained for all three percolation problems shown in Fig. 2, indicating that the result is universal along the whole phase boundary (except the end points). To calculate the "magnetic" exponent ω, we have identified h with the probability x, since (i) x destroys the transition as we have seen in our calculations, and (ii) x is a source of $S = 1$ states and thus acts in the percolation language as a ghost field[13] which in usual percolation gives the critical symmetry-breaking scaling field h. At the critical point on the line $y = z$ we obtain $\omega = 2.57 \pm 0.03$, from which we get $\beta = 0.26 \pm 0.03$ and $\gamma = 2.21 \pm 0.05$ in agreement with previous results.[9]

On the line $x = 0, z = 1$ the model can be solved exactly, and we do find a DP transition as y is varied, which occurs at the point with Ising symmetry, i.e., $y = \frac{1}{2}$. Say at $t = 0$ a single site has $u_{l,0} = 1$ while for all others $u_{j,0} = 0$. The probability of finding $u_{k,t} = 1$ for long times t is given by $P_w(t) = \sum_U P(U)\rho_w(U,t)$, where U is any configuration $u_{j,t}$ that appears with probability $P(U)$; $\rho_w(U,t) = 1$ if U contains any site (k,t) with $u_{k,t} = 1$, and $\rho_w(U,t) = 0$ otherwise. On the $x = 0, z = 1$ line only those configurations U are allowed in which *no* down-pointing triangle can have

either (i) $u = 0$ on both the upper corners and $u = 1$ on lowest corner, or (ii) $u = 1$ on both upper corners and $u = 0$ on the bottom. Configurations consistent with these constraints contain a single domain of "wet" sites in a sea of dry ones; furthermore, *no branching* of the wet domain is allowed (see Fig. 3).[14] The weight $P(U)$ of such a graph is determined by the right- and left-hand boundaries of the domain, each of which goes either right or left with increasing t, until they meet and "annihilate." The right-hand boundary carries a weight y for each step to the right, and weight $1 - y$ for each step to the left. The reverse holds for the left-hand boundary. $P_w(t)$ is the sum of all graphs that have not terminated after t steps; i.e., the probability that two random walkers with right/left transition probabilities defined above, starting at a separation of two steps, have not met before t time steps. Instead of dealing with two random walkers, we can consider only the single random walk performed by their relative positions, or the difference walk. This walk is characterized by the transition probabilities $P(d \rightarrow d+1) = p_+ = y^2$, $P(d \rightarrow d) = p_0 = 2y(1-y)$, $P(d \rightarrow d-1) = p_- = (1-y)^2$, where d is half the distance between the walkers. Thus $P_w(t)$ is given by the probability that a gamber with initial capital of 1, with probability p_+ (p_-) of winning (losing) one unit and p_0 of maintaining his holdings, will *not* reach his ruin (zero capital) before t games.[15] We first calculate the probability of ruin after exactly t steps, r_t, summing over all paths starting at 1 and reaching 0 *for the first time* at t. If

VOLUME 53, NUMBER 4   PHYSICAL REVIEW LETTERS   23 JULY 1984

such a path has σ steps with no change we have

$$r_t = \sum_{\sigma=0}^{t-1} {}' \binom{t-1}{\sigma} p_0^\sigma \bar{r}_{t-\sigma},$$

where the prime indicates that σ must have the same parity as $t-1$, and $\bar{r}_k$ is given by[15]

$$\bar{r}_k = \frac{1}{k} p_+^{(k-1)/2} p_-^{(k+1)/2} \binom{k}{(k+1)/2}.$$

Replacing the sum by an integral, and evaluating it by saddle-point integration, we find for $\epsilon = \frac{1}{2} - y \ll 1$ (note that $p_+ - p_- = -2\epsilon$) that $r_t \sim e^{-4t\epsilon^2}/t^{3/2}$ for $t \gg 1$. Therefore the correlation length in the time direction is $\xi_\parallel \sim \epsilon^{-2}$, and $\nu_\parallel = 2$. Also, since

$$P_w(t+1) = 1 - \sum_{t'=1}^{t} r_{t'} = P_w(t) - r_t,$$

we must have at criticality $(\epsilon = 0)$ $P_w(t) \sim t^{-1/2}$. To calculate correlations in the $\perp$ direction, note that the $x=1, z=0$ line is in the subspace of Verhagen's solution.[8] In his notation, on this line $b=0$ and $a=y(1-y) \simeq 1 - 4\epsilon + O(\epsilon^2)$. He found that $P(u_{l,t}=1|u_{0,t}=1) = a^l \cong e^{-4\epsilon l}$ and therefore $\xi_\perp \sim 1/\epsilon$ and $\nu_\perp = 1$, while $\eta_\perp = 0$.

By combining the mapping of one-dimensional peripheral cellular automata onto Ising problems in two dimensions with representation of directed percolation as a particular case of PCA, we have recast the DP problem as an Ising model. A line of DP transitions was found numerically, and a specific DP model was solved exactly.

This work was supported by the National Science Foundation under Grant No. DMR 83-05723-A1 and by U.S. Office of Naval Research Contract No. N00014-82-0699. We thank M. Y. Choi, S. Doniach, B. Huberman, M. Kerszberg, and I. Peschel for discussions, and P. Rujan for bringing Refs. 5 and 8 to our attention.

(a)Permanent address: Department of Electronics, Weizmann Institute of Science, Rehovot, Israel.

[1]S. Wolfram, Rev. Mod. Phys. **55**, 601 (1983). See also Physica (Utrecht) **10D**, 1 (1984).

[2]J. von Neumann, *Theory of Self Reproducing Automata* (University of Illinois, Urbana, 1966).

[3]P. Grassberger, Physica (Utrecht) **10D**, 52 (1984); E. Ott, Rev. Mod. Phys. **53**, 655 (1981); H. Haken, *Chaos and Order in Nature* (Springer, Berlin, 1981).

[4]T. R. Welberry and R. Galbraith, J. Appl. Crystallogr. **6**, 87 (1973); T. R. Welberry and G. H. Miller, Acta Crystallogr., Sec. A **34**, 120 (1978).

[5]G. Nicolis and I. Prigogine, *Self Organization in Nonequilibrium Systems* (Wiley, New York, 1977).

[6]S. Ulam, Annu. Rev. Biochem. **255**, (1974).

[7]B. A. Huberman and T. Hogg, Phys. Rev. Lett. **52**, 1048 (1984).

[8]A. M. W. Verhagen, J. Stat. Phys. **15**, 213 (1976); I. G. Enting, J. Phys. C **10**, 1379 (1977), and J. Phys. A **11**, 555, 2001 (1978).

[9]W. Kinzel, in *Percolation Structures and Processes, Annals of the Israel Physical Society, Vol. 5*, edited by G. Deutscher, R. Zallen, and J. Adler (Bar-Ilan University, Ramat-Gan, Israel, 1983), p. 425.

[10]D. Dhar, *Stochastic Processes, Formalism and Applications*, edited by G. S. Agarwal and S. Dattagupta (Springer, Berlin, 1983).

[11]J. Stephenson, J. Math. Phys. **5**, 1009 (1964), and **11**, 413, 420 (1970), and Phys. Rev. B **1**, 4405 (1970).

[12]W. Kinzel and J. Yeomans, J. Phys. A **14**, L163 (1981).

[13]R. B. Griffiths, J. Math. Phys. **8**, 484 (1967).

[14]Precisely these graphs contribute to the Mauldon process B_i; however, the weights of each graph are different in our model from those assigned by J. G. Mauldon, in *Proceedings of the Fourth Berkeley Symposium on Mathmatics, Statistics and Probability*, edited by Jerzy Neyman (Univ. of California, Berkeley, 1961), Vol. 2, p. 337.

[15]W. Feller, *An Introduction to Probability Theory and its Applications* (Wiley, New York, 1968).

VOLUME 55, NUMBER 23   PHYSICAL REVIEW LETTERS   2 DECEMBER 1985

Statistical Mechanics of Probabilistic Cellular Automata

G. Grinstein

IBM T.J. Watson Research Center, Yorktown Heights, New York 10598

and

C. Jayaprakash and Yu He

Department of Physics, The Ohio State University, Columbus, Ohio 43210
(Received 19 April 1985)

The necessary and sufficient conditions under which fully probabilistic cellular-automata (PCA) rules possess an underlying Hamiltonian (i.e., are "reversible") are established. It is argued that, even for irreversible rules, continuous ferromagnetic transitions in PCA with "up-down" symmetry belong in the universality class of kinetic Ising models. The nonstationary (e.g., periodic) states achieved for asymptotically large times by certain PCA rules in the (mean field) limit of infinite dimension are argued to persist in two and three dimensions, where fluctuations are strong.

PACS numbers: 05.50.+q, 02.50.+s, 64.60.Ht

Cellular automata (CA) are regular arrays of variables, each of which can assume two or more discrete values and evolves in discrete time steps according to a set of local rules which may be either deterministic or probabilistic.[1] They are used to model problems in physics, chemistry, biology, and computer science. The goal is to determine, for any given rule, the nature of the state of the system for asymptotically large time $(t \to \infty)$, and to identify the universality classes of the phase transitions that occur as the rules are varied. In this paper we study probabilistic cellular automata (PCA) with two states per site. Our main results follow:

(1) We establish the necessary and sufficient condition under which a PCA rule is "microscopically reversible," i.e., the transition probabilities obey detailed balance for some underlying Hamiltonian; as $t \to \infty$ the system is therefore described by the stationary (equilibrium) Boltzmann distribution corresponding to that Hamiltonian. The condition depends on whether the rule is applied by updating the spins simultaneously or sequentially.[2]

(2) Continuous transitions into stationary ferromagnetic states of PCA which do not have associated Hamiltonians (i.e., are "irreversible") but which do have the "up-down" symmetry familiar from Ising models, are argued to fall, for both statics and dynamics, in the same universality classes as kinetic Ising models.[3] Thus, at ferromagnetic critical points, fully PCA coarse grained to sufficiently large length scales possess underlying Hamiltonians, even if they do not on microscopic scales.

(3) As in[4] equilibrium statistical mechanics, there exists a systematic expansion for PCA in inverse powers of d, the dimension. In the $d = \infty$ limit one obtains mean-field theory (MFT): The evolution of the CA is described by an iterative map[5] with one variable—the average magnetization—which, for appropriate rules, exhibits time-dependent asymptotic behavior, including limit cycles, chaos, etc.[6] We argue that such time dependence is not an artifact of MFT, but survives the strong fluctuations present in few dimensions. Guided by the analytic results available at $d = \infty$, we construct fully probabilistic, local rules in $d = 2$ which produce, under numerical simulation, nonstationary states, viz., two-, three-, and four-cycles, as $t \to \infty$. We have also found rules which lead to what we believe are chaotic states.

We consider CA on d-dimensional hypercubic lattices with N sites, labeled i, each occupied by an Ising spin $S_i = \pm 1$. With $P(\{S_i\}, t)$ the probability that the system is in the state $\{S_i\}$ at time t, the discrete master equation[7]

$$P(\{\tilde{S}_i\}, t+1) = \sum_{\{S_i\}} P(\{S_i\}, t) \prod_i Q(\tilde{S}_i | S_i, \{S_{i'}\}) \quad (1)$$

describes the PCA's time evolution. Here $Q(\tilde{S}_i | S_i, \{S_{i'}\})$ is the probability that the ith spin assumes the value $\tilde{S}_i$ at time $t+1$, given that this spin and its "neighborhood"—a set of z spins located on nearby sites $\{i'\}$—have the values S_i and $\{S_{i'}\}$, respectively, at time t. The rule is defined by specifying the neighborhood and the 2^{z+1} nonzero independent probabilities Q. We consider only "fully probabilistic" rules, i.e., *all* transition probabilities Q strictly greater than zero. Equation (1) clearly describes simultaneous (synchronous) updating of the spins; one can, alternatively, update single spins in a random sequence, i.e., sequentially. We reduce the number of independent Q's to $z+1$ by considering "totalistic" CA,[1] i.e., $Q = Q(\tilde{S}_i | S_i, \sum_{i'} S_{i'})$, and imposing up-down symmetry: $Q(\tilde{S}_i | S_i, \sum_{i'} S_{i'}) = Q(-\tilde{S}_i | -S_i, -\sum_{i'} S_{i'})$. The rule is then completely specified by the $z+1$ values of the function $f(M_i) = Q(-1|1, zM_i)$, where $M_i = (1/z)\sum_{i'} S_{i'}$ is the average magnetization of the neighborhood of i.

VOLUME 55, NUMBER 23   PHYSICAL REVIEW LETTERS   2 DECEMBER 1985

Kinetic Ising models [i.e., those (reversible) PCA which satisfy detailed balance for some associated Hamiltonian and so approach the Boltzmann distribution as $t \to \infty$] are well understood. Hence it is useful to derive conditions under which CA rules are reversible. First consider simultaneously applied rules. Detailed balance requires the existence of a Hamiltonian H such that for any two states $\{S_i\}$ and $\{\tilde{S}_i\}$,

$$\prod_i Q(\tilde{S}_i|S_i, \{S_{i'}\})/Q(S_i|\tilde{S}_i, \{\tilde{S}_{i'}\})$$
$$= \exp(H\{S_i\} - H\{\tilde{S}_i\}). \quad (2)$$

Consider for simplicity a one-dimensional, totalistic, up-down-symmetric PCA, the neighborhood of S_i consisting of the two near neighbors S_{i+1} and S_{i-1}, so that $z = 2$. Q can then be written in the form

$$Q = A \exp[\tilde{S}_i(aS_i + bM_i + cS_{i+1}S_{i-1})],$$

where A is an arbitrary even function of the S's but is independent of $\tilde{S}_i$, and a, b, and c are arbitrary constants. It follows from (2) that the associated H can consist of at most nearest- and next-nearest-neighbor two-spin interactions. It is then easy to check by considering two specific updatings of the system, viz., (1) only the ith spin flips (i.e., $\tilde{S}_i = -S_i$), and (2) only the ith and $(i+1)$st spins flip, that unless $c = 0$ detailed balance cannot be satisfied for all pairs of states. Similarly, consideration of the updatings in which (1) only the ith spin flips, and (2) only the $(i+1)$st spin flips, shows that detailed balance can be satisfied for all pairs of states for *any* values of a and b. Invoking the fact that Q is a probability, i.e.,

$$\sum_{\tilde{S}_i = \pm 1} Q(\tilde{S}_i|S_i, 2M_i) = 1,$$

then immediately establishes the most general form of Q [or, equivalently, of $f(M)$] for reversible rules as

$$f(M) = [1 - \tanh(a + bM)]/2. \quad (3)$$

Extension of this reasoning to arbitrary dimension and

z is straightforward but tedious; the result (3) continues to hold. Equation (3) implies that no simultaneously applied, up-down symmetric rule with more than two independent parameters can be reversible.

For *sequentially* applied rules the criterion for reversibility is simpler to derive since only single spin flips need be considered. The criterion depends on z. For nearest-neighbor rules ($z = 2d$) the necessary and sufficient condition for the existence of an underlying Hamiltonian is $f(M) = f_e(M)[1 - \tanh(\lambda M)]$, where λ is an arbitrary parameter and f_e an arbitrary even function of M. As the range of the rule increases, the restriction on $f(M)$ required to ensure reversibility becomes progressively less stringent. For infinite-ranged CA ($z = N$) it can be shown that *every* f corresponds to an underlying Hamiltonian, H, which typically involves infinite-ranged, multispin interactions.

These results imply that there are, rather surprisingly, rules [viz., those of the form (3) with $a \ne 0$] which do not obey detailed balance under sequential updating, but which do when simultaneously applied. The converse is, of course, also true, and is not surprising. While we have considered only totalistic rules, our methods can be simply generalized to establish conditions for the existence of underlying Hamiltonians for arbitrary PCA.

We now discuss the universality classes of the continuous ferromagnetic transitions which can occur between stationary states of irreversible PCA with up-down symmetry when the probabilities f are varied. We first discuss MFT for PCA: From (1) one can readily construct an infinite hierarchy of coupled equations involving equal-time correlation functions of progressively higher order. For large z, this hierarchy can, just as in equilibrium statistical mechanics,[4] be systematically decoupled in an expansion in powers of $1/z$. The MF limit, $z = \infty$, is simple, even for *irreversible* CA: The time evolution is completely characterized by the average magnetization $M(t) = \langle S_i \rangle_t$. It is straightforward to verify from (1) that in this limit $M(t)$ obeys the one-variable recursion relation

$$M(t+1) = g(M(t)) = M(t) - 2[f_o(M(t)) + M(t)f_e(M(t))], \quad (4)$$

where f_o and f_e are the odd and even parts of f, respectively. This iterative map g has a fixed point at $M = 0$ (corresponding to the "paramagnetic" state). Other fixed points with $M \ne 0$ ("ferromagnetic" states) may occur, depending on the details of $g(M)$. The stability criterion for a fixed point is[5] $|g'(M^*)| < 1$. Rules for which, at some critical value of the parameters, a stable ferromagnetic and an unstable paramagnetic fixed point coalesce, producing a stable paramagnetic fixed point, lead to continuous ferromagnetic phase transitions. From (4) one obtains the respective values $\frac{1}{2}$ and 1 for the associated critical

exponents β and γ. These are the conventional MF results of equilibrium critical phenomena. Alternative formulations[8] of MFT for PCA in finite d yield the conventional values $\nu = \frac{1}{2}$, $\eta = 0$, and the mean-field dynamical exponent, $z = 2$. In equilibrium critical phenomena, the upper critical dimension, d_c, can be identified as that d at which the hyperscaling relation $\beta = (\nu/2)(d - 2 + \eta)$ is satisfied by the MF exponents.[9] Postulating the validity of this identification for CA we find, as in the static equilibrium case, $d_c = 4$. For $d > 4$, then, all continuous transitions into

VOLUME 55, NUMBER 23   PHYSICAL REVIEW LETTERS   2 DECEMBER 1985

ferromagnetic states in PCA are characterized by the standard MF exponents.

To study the effect of fluctuations for $d < d_c = 4$, consider the Langevin equation,[7]

$$\partial \psi_i / \partial t = Q_i(\{\psi_j\}) + \eta_i(t), \qquad (5)$$

where ψ_i, a classical field at site i, assumes any value between $-\infty$ and $+\infty$, Q_i is an analytic function of the $\{\psi_j\}$, and η_i is a Gaussian random noise variable of zero mean. The critical behavior of kinetic Ising models is described[3] by (time-dependent Ginzburg-Landau) equations of the form (5) with $Q_i(\{\psi_j\})$ $= -\Gamma \partial H(\{\psi_j\})/\partial \psi_i$ and $\overline{\eta_i \eta_j} = 2\Gamma \delta_{ij} \delta(t-t')$. Here $H(\{\psi_j\})$ is the Ginzburg-Landau representation of the underlying Hamiltonian, Γ is the dissipation constant, and this specific choice for $\overline{\eta_i \eta_j}$ ensures[10] that in the $t \to \infty$ limit the system is described by the Boltzmann distribution, $e^{-H(\{\psi_j\})}$. Critical phenomena in kinetic Ising models have been intensively studied[3] through application of the ϵ expansion to this special case of (5).

It is natural to hypothesize that the critical behavior of PCA which do *not* admit underlying Hamiltonians can likewise be described[11] by Eq. (5), with Q analytic in $\{\psi_j\}$ but *not* expressible as $-\Gamma \partial H/\partial \psi_i$ for any H (i.e., with $\partial Q_i/\partial \psi_j \neq \partial Q_j/\partial \psi_i$). For short-ranged PCA with up-down symmetry, Q_i must then be an odd function of ψ_i and of some appropriate neighborhood, $\{\psi_{j'}\}$, of i. For example,

$$Q_i = A \sum_{(i')} \psi_{i'} + B \sum_{(i'_1, i'_2, i'_3)} \psi_{i'_1} \psi_{i'_2} \psi_{i'_3} + O(\psi_i^5) \qquad (6)$$

(the various i' being summed over all nearest neighbors of i) represents, for arbitrary coefficients A and B, a nearest-neighbor rule. The absence of an underlying Hamiltonian for $B \neq 0$ eliminates the fluctuation-dissipation theorem[3]; $\overline{\eta_i \eta_j}$ can thus be taken to be an arbitrary even function of the $\{\psi_j\}$, e.g.,[12] $\overline{\eta_i(t)\eta_j(t')} = \delta_{ij} \delta(t-t')[\Gamma_0 + \Gamma_1 \psi_i^2 + \ldots]$, for constants Γ_0 and Γ_1.

Note that the difference between the $\{Q_i\}$ which result from CA with and without underlying Hamiltonians is quite subtle. For example, the standard nearest-neighbor ψ^4 Ginzburg-Landau Hamiltonian gives rise[3] to a Q_i which consists only of terms linear in ψ_i and $\psi_{i'}$ and a cubic term, ψ_i^3. It is therefore very similar to the Q_i of Eq. (6). The sole difference lies in the wave-vector dependence of the cubic terms in (6). Such wave-vector dependence is *irrelevant* under the renormalization group in $d = 4 - \epsilon$.[3,13] Indeed, it is easily shown that in $d = 4 - \epsilon$, the dynamical fixed point of the standard kinetic Ising model with no conserved variables is stable with respect to *all* additional analytic terms introduced by elimination of the underlying Hamiltonian without breaking of either the lat-

tice or the up-down symmetry. One concludes (subject to the usual caveats concerning one's inability to establish more than *local* stability of fixed points and the dangers of extrapolating from $d = 4 - \epsilon$ to physical dimensions) that fully PCA with up-down symmetry and a nonconserved order parameter fall, for both statics and dynamics, in the universality class of the ordinary Ising model with no conservation laws. Similar arguments show that rules which conserve the order parameter (and so are not fully probabilistic) give rise to ferromagnetic transitions in the universality class of the Ising model with conserved order parameter. Thus, near second-order phase transitions, irreversibility on microscopic length scales renormalizes away, producing, on large scales, reversible systems.

We now discuss nonstationary asymptotic behavior of PCA. It is known from the theory[5] of one-variable iterative maps that by suitable variation of $g(M)$ [viz., $g'(M^*) < -1$] the ferromagnetic fixed point of the (*simultaneously* updated) MF model (4) can be rendered unstable. At its stability limit, this fixed point can bifurcate to a two-cycle, wherein the average magnetization alternates in time between two distinct nonzero values. Indeed, for appropriate choices of g, it is possible to find all the diverse features of single-variable maps,[5] notably bifurcation sequences accumulating to states wherein M is a chaotic function of time. The occurrence of such nonstationary states in MF approximations of PCA has been previously pointed out,[6] but little is known about their stability with respect to fluctuations. [Note that the time-dependent asymptotic behavior in Refs. 6a and 6b occurs for *reversible* rules, either simultaneously (Ref. 6a) or sequentially (Ref. 6b) applied, treated in uncontrolled MF-like approximations. We believe that this is an artifact of the particular approximations employed, and that only *irreversible* PCA can exhibit time dependence. For simultaneously applied PCA it is easy to verify from (4) and the monotonicity of f in (3) that in the $d = \infty$ limit no reversible rule can produce time dependence.]

Since, for totalistic nearest-neighbor (i.e., $z = 2d$) rules, MFT is exact in the $d = \infty$ limit, fluctuation corrections to it are conveniently studied in the systematic $1/d$ (i.e., $1/z$) expansion mentioned earlier. As in statistical mechanics it is easy to show that[4] to $O(1/d)$ only $M(t)$ and the *nearest-neighbor* correlation function $G(t) = \langle S_i S_j \rangle_t - M^2(t)$ are required for a complete characterization of the time evolution; all other correlations are of $O(1/d^2)$. Similarly, to any finite order in $1/d$ the hierarchy of coupled equations reduces to an n-variable iterative map, for some finite n. Two-variable maps are known to exhibit the bifurcation route to chaos[5]; the qualitative features of MFT are thus preserved to $O(1/d)$ for appropriate rules. While the behavior of maps with more variables can be

Volume 55, Number 23   PHYSICAL REVIEW LETTERS   2 December 1985

extraordinarily complex and few analytic results are known, one expects, as in statistical mechanics, that, for systems with discrete symmetry far from critical points (i.e., when spatial correlations are short) MFT should remain a good guide even for $d = 2$ or 3; time-dependent states ought, therefore, to occur for appropriate rules.

To test this expectation we have performed Monte Carlo simulations on simultaneously updated PCA on square lattices in $d = 2$. We computed $M(t)$ for $t = 1, 2, \ldots, t_{max}$, and studied the power spectrum of M. Peaks in this spectrum which occur at rational frequency and which sharpen and grow as the sample size is increased, and do not broaden or shrink with increasing t_{max}, were taken as the signature of periodic states. The parameter space of possible rules is, of course, vast; selecting simple rules in a relatively arbitrary way never led us to time-dependent states. We systematized the search by considering totalistic CA with neighborhoods consisting of $(2n + 1) \times (2n + 1)$ squares of sites for $n = 1, 2, \ldots, 8$; thus $z = 4(n^2 + n)$. Using the known criterion [i.e., large $|g'(M^*)|$] for the occurrence of nonstationary states in MFT (which provides a progressively better description as z increases, becoming exact at $z = \infty$) as a guide, we succeeded in constructing rules which undergo limit cycles with periodicities 2, 3, and 4 on samples of size up to 150×150 and t_{max} of up to 100 000. The three- and four-cycles occurred, however, only for $n \geq 3$ (i.e., $z \geq 48$) and $n \geq 8$ ($z \geq 288$), respectively. We have also found rules (for $z = 80$) which yield what we believe are chaotic states, as evidenced by a broad power spectrum which persists for the largest samples that we studied (200×200), but have yet to observe a periodic state with period > 4, much less a complete bifurcation sequence, which MFT predicts. We therefore consider the evidence for the stability with respect to fluctuations of the chaotic state less compelling than for the two-, three-, and four-cycles. A noteworthy feature of the simulations is the strong stabilizing effect of fluctuations on stationary states: Rules which produce time-dependent states within MFT typically give stationary behavior, or at most two-cycles, until z gets rather large (~ 50). This does not preclude nonstationary states with large periods for smaller z, but indicates that they occupy progressively smaller regions of the huge available parameter space as z decreases.

We are grateful to C. H. Bennett, M. Buettiker, J. Hertz, R. Landauer, R. Pandit, and P. Seiden for helpful comments, and for the hospitality of NORDITA, where part of this work was carried out. One of us (C.J.) acknowledges the hospitality of the IBM T.J. Watson Research Center where this project was initiated. This work was supported in part by the National Science Foundation under Grant No. NSF DMR 83-15770, and in part by the A.P. Sloan Foundation.

---

[1]See, e.g., S. Wolfram, Rev. Mod. Phys. **55**, 601 (1983); see also Physica (Amsterdam) **10D**, 1–247 (1984). (CA, as conventionally defined, are always updated simultaneously. We broaden the definition here to allow for sequential updating.)

[2]E. Domany and W. Kinzel, Phys. Rev. Lett. **53**, 311 (1984), have established the rather different result that *any* simultaneously updated PCA in d dimensions is equivalent to an equilibrium statistical mechanics problem in $d + 1$ dimensions. See also E. Domany, Phys. Rev. Lett. **52**, 871 (1984); G. Y. Vichniac, Physica (Amsterdam) **10D**, 96 (1984).

[3]See, e.g., R. J. Glauber, J. Math. Phys. **4**, 294 (1963); P. C. Hohenberg and B. I. Halperin, Rev. Mod. Phys. **49**, 435 (1977).

[4]R. Brout, Phys. Rev. **118**, 1009 (1960).

[5]See, e.g., P. Collett and J.-P. Eckmann, *Iterated Maps on the Interval as Dynamical Systems* (Birkhäuser, Boston, 1980).

[6a]M. Y. Choi and B. A. Huberman, Phys. Rev. A **28**, 1204 (1983).

[6b]M. Y. Choi and B. A. Huberman, Phys. Rev. B **28**, 2547 (1983).

[6c]D. R. Smith and C. H. Davidson, J. Assoc. Comput. Mach. **9**, 268 (1962).

[7]See, e.g., N. G. van Kampen, *Stochastic Processes in Physics and Chemistry* (North-Holland, Amsterdam, 1981).

[8]See, e.g., L. S. Schulman and P. E. Seiden, J. Stat. Phys. **19**, 293 (1978); G. Grinstein, C. Jayaprakash, and Y. He, unpublished.

[9]See, e.g., P. Pfeuty and G. Toulouse, *Introduction to the Renormalization Group and Critical Phenomena* (Wiley, New York, 1977).

[10]G. E. Uhlenbeck and L. S. Ornstein, Phys. Rev. **36**, 823 (1930).

[11]While the microscopic dynamics of the continuous-time Langevin equation (4) appear rather different from those of *simultaneously* applied irreversible rules, we hypothesize that irreversible PCA are like reversible ones (Ising models) in that *universality classes* for continuous ferromagnetic transitions are independent of how the rules are applied. This hypothesis certainly holds in MFT. Numerical results [C. H. Bennett and G. Grinstein, Phys. Rev. Lett. **55**, 657 (1985)] on a simultaneously applied irreversible rule in $d = 2$ give a magnetization exponent β fully consistent with the Ising-model value of $\frac{1}{8}$.

[12]In the case of CA which are not fully probabilistic but have one absorbing state, $\Gamma_0 = 0$. This is what places such models in a different unversality class [see, e.g., W. Kinzel, Z. Phys. B **58**, 229 (1985); P. Grassberger, Z. Phys. B **47**, 365 (1982); H. K. Janssen, Z. Phys. B **42**, 151 (1981); and Ref. 2].

[13]K. G. Wilson and J. Kogut, Phys. Rep. **12C**, 75 (1974).

VOLUME 55, NUMBER 7   PHYSICAL REVIEW LETTERS   12 AUGUST 1985

Role of Irreversibility in Stabilizing Complex and Nonergodic Behavior in Locally Interacting Discrete Systems

Charles H. Bennett[a] and G. Grinstein

IBM T. J. Watson Research Laboratory, Yorktown Heights, New York 10598
(Received 18 March 1985)

Irreversibility stabilizes certain locally interacting discrete systems against the nucleation and growth of a most-stable phase, thereby enabling them to behave in a computationally complex and nonergodic manner over a set of positive measure in the parameter space of their local transition probabilities, unlike analogous reversible systems.

PACS numbers: 05.90.+m

The dynamics of a statistical-mechanical system in contact with a larger environment is often modeled as a random walk on the system's state space (e.g., kinetic Ising model). If the environment is at equilibrium, this random walk will be "microscopically reversible" (its matrix of transition probabilities being of the form DSD^{-1}, where D is diagonal and S symmetric), and the system's stationary distribution of states will be an equilibrium one (e.g., canonical ensemble) defined by a Hamiltonian simply related to the transition probabilities. On the other hand, if the environment is not at equilibrium, the system's transition matrix in general will be irreversible, and the resulting nonequilibrium stationary distribution may be very hard to characterize.

Though an irreversible system's distribution of states is not simply related to the transition probabilities, its distribution of histories is. More specifically, the stationary distribution of histories for any stochastic model, whether reversible or not, may be viewed as a canonical distribution under an effective Hamiltonian on the space of histories, in which each configuration interacts with its predecessor in time with an "interaction energy," equal simply to the logarithm of the corresponding transition probability.

In particular, we consider the case in which the underlying stochastic model is a probabilistic cellular automaton (CA), in other words, a d-dimensional lattice with finitely many states per site, in which each site, at each discrete time step, undergoes a transition depending probabilistically on the states of its neighbors. In this case[1,2] the stationary distribution of histories of the CA is equivalent to the equilibrium statistics of a corresponding generalized Ising model (GIM) in $d+1$ dimensions. This appears paradoxical, because CA's are known to be capable of complex, nonergodic behavior even when all local transition probabilities are positive,[3,4] whereas the behavior of GIM is generally simple and ergodic (a stochastic process is "ergodic" if its stationary distribution is unique). For example, a standard kinetic Ising model, at a generic point in its temperature-magnetic field parameter space, undergoes nucleation and growth of a unique most-stable phase, thereby relaxing to a stationary distribution independent of the initial conditions.

Here we note the resolution of the paradox, and illustrate an essential difference between reversible and irreversible systems by characterizing the phase diagram, equation of state, domain growth kinetics, and equivalent $(d+1)$-dimensional GIM of one of the simplest nonergodic irreversible CA, viz., Toom's north-east-center (NEC) voting model, The resolution of the paradox lies in the fact that when a d-dimensional CA is represented as a $(d+1)$-dimensional GIM, the parameters (coupling constants) of the latter system are not all independent, but are constrained in such a way as to cause the free energy of the $(d+1)$-dimensional system to be identically zero, no matter how the parameters (transition probabilities) of the underlying CA are varied. It is therefore possible for irreversible systems such as the NEC model to be nonergodic, and in particular to have two or more stable phases, over a finite region in their phase diagrams, whereas reversible systems can exhibit this behavior only over a subset of zero measure consisting of points in the phase diagram where two or more phases, by symmetry or accident, have exactly equal free energy.

North-east-center model, and reasons for its nonergodicity.—The NEC model is one of a class of voting rules for CA shown by Toom[3] to be nonergodic in the presence of small but arbitrary probabilistic perturbations. The model consists of a square lattice of spins, each of which may be up or down. The spins are updated synchronously, with a spin's future state decided by majority vote of the spins in an unsymmetrical neighborhood, consisting of the spin itself and its northern and eastern neighbors. The rule just described is deterministic; we consider two-parameter noisy perturbations of the rule, in which a spin whose present neighborhood majority is up, instead of going up with certainty at the next time step, goes up with probability $1-p$ and down with probability p; and a spin whose present neighborhood majority is down goes down with probability $1-q$ and up with probability q. Alternatively, the noise may be characterized by its "amplitude" $p+q$, analogous to temperature, and its "bias" $(p-q)/(p+q)$, analogous to magnetic field.

Because probabilistic CA typified by the NEC model

VOLUME 55, NUMBER 7   PHYSICAL REVIEW LETTERS   12 AUGUST 1985

are not microscopically reversible, not all the formalism of equilibrium statistical mechanics can be applied to them. In particular, the stationary probability measure $\mu(X)$ on configuration space cannot in general be represented as the Boltzmann exponential of any locally additive potential. However, a stable phase of a probabilistic CA can still be defined in the thermodynamic limit as a probability measure μ on the space of configurations of the infinite lattice, that is (1) stationary under the transition rule, and (2) extremal in the sense of not being expressible as a linear combination of other stationary measures. This is analogous to the definition of a phase for Hamiltonian systems as a measure μ which is extremal among measures that can be obtained as the thermodynamic limit of the canonical ensemble under various boundary conditions.[5]

When p and q are small and positive, no local transition is entirely forbidden; hence the model is ergodic on any finite lattice (e.g., an N-by-N torus). However, Toom showed[3] that for sufficiently small p and q, the transition rate between the mostly up and mostly down states of the entire system tends to zero with increasing N, rendering the infinite system nonergodic, with two stable phases, like a conventional Ising model below its critical temperature.

When the noise is unbiased $(p = q)$, the NEC system behaves like an Ising model in zero field: There is a critical noise amplitude at which the spontaneous magnetization vanishes continuously. Where the NEC system differs from equilibrium systems is in its response to biased noise, e.g., $0 < q < p < \frac{1}{2}$. In an equilibrium system, such a symmetry-breaking perturbation (analogous to magnetic field) would cause the system to become ergodic, by rendering it susceptible to nucleation and growth of the more stable phase. The NEC system, on the other hand, remains nonergodic, with two stable phases, even in the presence of biased noise, if the noise amplitude is small enough. This difference in response to a symmetry-breaking perturbation can be understood by comparing the mechanisms by which the two systems suppress fluctuations of the minority phase.

In a zero-field Ising system below its critical point, as in any reversible system at a first-order phase transition, a flat interface between the two equally stable phases must have zero mean propagation velocity. Finite islands of the minority phase nevertheless shrink because of surface tension, an island of radius r shrinking with velocity $- dr/dt$ proportional to $1/r$. The NEC system's irreversibility, by contrast, allows a flat interface to drift even when the noise is unbiased $(p = q)$, and this drift velocity depends on interface orientation in such a way as to cause islands of either phase to shrink at a rate $- dr/dt$ roughly independent of their radius. In both systems, a small symmetry-breaking perturbation s adds a constant term $\propto s$ to dr/dt for islands of the favored phase. This is suffi-

cient to render the Ising system metastable, by favoring growth of islands larger than a certain critical radius $\propto 1/s$; but in the NEC system, dr/dt remains negative for all r, and so the system remains stable.

The NEC system's interface motions are easiest to understand in the case of zero noise. Here a 135° diagonal interface between up and down spins drifts southwestward with unit speed, regardless of which phase is on which side of the interface, because sites just southwest of the interface, at each instant of time, have neighborhood majorities dominated by their north and east neighbors on the other side. On the other hand, a vertical or horizontal interface in the same system does not drift.

These motions enable the noiseless NEC system to eliminate islands of either phase with a linear shrinkage velocity $- dr/dt$ independent of their size r. To see this, consider an island of, say, up spins of arbitrary size and shape. Let an isosceles right triangle, with the hypotenuse on the northeast, be circumscribed about the island, and let all other spins in this triangle also be flipped up, so that we are considering a somewhat larger island of up spins of a particular triangular shape. Because the NEC rule is monotonic, the addition of further up spins to an island of up spins cannot hasten its disappearance. Therefore, the lifetime of the circumscribed triangular island is an upper bound on the lifetime of the original arbitrary-shaped island. The fate of the triangular island in the noiseless NEC system is quite simple: Its southern and western borders remain fixed, while its northeastern border closes in with unit velocity, eliminating the island in time proportional to its original size r.

The same argument holds in the presence of noise, whether biased or unbiased, because, if the noise amplitude is small enough, each of the interface velocities will differ only slightly (linearly in p and q) from its value in the noiseless system. Under these conditions, an island of size r of either phase will disappear in time proportional to r by differential motion of its borders, the lifetime being longer for one phase than the other if the noise is biased.

Phase diagram of the NEC system.—Numerical studies were performed on the NEC model with use of CAM, a fast CA simulator.[6] Besides providing quantitative data, CAM's real-time display was most helpful in assessing qualitative features of the model.

The phase diagram shown in Fig. 1 was obtained by finding pairs of noise parameters (p,q) such that a large minority island of up to (25 000 sites out of total of $65\,536 = 256 \times 256$ sites) neither grew nor shrank on average, during runs of about 50 000 time steps. The two-phase region is bounded by a pair of first-order transitions (solid curves terminating at the critical point $p = q = 0.90 \pm 0.003$), on which one phase becomes marginally stable, losing its ability to eliminate large islands of the other. Beyond the two-

VOLUME 55, NUMBER 7   PHYSICAL REVIEW LETTERS   12 AUGUST 1985

phase region is a narrow metastable zone (demarcated on the right by dashed lines), in which large islands of the favored phase grow but small islands shrink. A critical exponent of 3.0 ± 0.4 was found for the vertical width of the two-phase region as a function of noise amplitude $(p+q)$ below the critical point. Other runs on one-phase systems with unbiased noise $p=q$ yielded the value $\beta = 0.122 \pm 0.01$ for the exponent describing magnetization as a function of noise amplitude below the critical point.

Besides being irreversible, the NEC model differs from conventional kinetic Ising models in having synchronous updating. However, preliminary runs in which only a fraction ($\frac{1}{2}$ to $\frac{1}{16}$) of the spins are updated at each time step indicate that even a fully asynchronous NEC model would have a qualitatively similar phase diagram. We also explored an analytically solvable mean-field approximation[7] to the NEC rule, based on the recurrence relation

$$R(m) = -1 + [p(1-m)^3 + 3p(1-m)^2(1+m) + 3(1-q)(1-m)(1+m)^2 + (1-q)(1+m)^3]/4,$$

where $R(m)$ is the magnetization at time $t+1$ as a function of that at time t. Here, too, the phase diagram was similar, except that there was no metastable zone, and the critical exponents were $\frac{3}{2}$ (for the two-phase region width) and $\frac{1}{2}$ (for β).

The equivalent $(d+1)$-dimensional Hamiltonian, and its free energy.—We now review the construction of an equivalent $(d+1)$-dimensional Hamiltonian model for an arbitrary (in general irreversible) d-dimensional CA,[1,2] and show why the former can have multiple stable phases over a set of finite measure in the parameter space of the latter. The possible time histories $X(0), X(1), \ldots, X(t)$ of a d-dimensional CA can be viewed as configurations of a $(d+1)$-dimensional lattice with one boundary fixed at $X(0)$, the initial state of the CA. The probability $P(X(0), X(1), \ldots, X(t))$ of such a history may be expressed as the product

$$P(X(1)/X(0))P(X(2)/X(1)) \cdots P(X(t)/X(t-1)),$$

where $P(X(i+1)/X(i))$ denotes the conditional probability for the CA to be in state $X(i+1)$ at time $i+1$ given that it was in state $X(i)$ at time i.

By defining

$$H(X(i+1), X(i)) = -\ln[P(X(i+1)/X(i))],$$

we cast the history probability in the familiar form of a Boltzmann factor (taking $kT = 1$):

$$P(X(0), X(1), \ldots, X(t)) = \exp\left(-\sum_{i=0}^{t-1} H(X(i+1), X(i))\right),$$

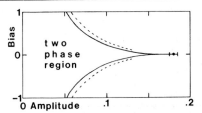

FIG. 1. Phase diagram of the NEC system, for noise parameters p and q, with amplitude $= p+q$ and bias $= (p-q)/(p+q)$.

with H playing the role of an effective Hamiltonian coupling adjacent d-dimensional time slices. All properties of the CA can thus be expressed as canonical-ensemble averages of the $(d+1)$-dimensional system defined by Hamiltonian H. The $(d+1)$-dimensional model has the remarkable feature[8] that its free energy is identically zero regardless of the CA's initial condition or transition probabilities. This follows from the normalization of these probabilities [for each $X(i)$, the sum over $X(i+1)$ of $P(X(i+1)/X(i))$ must be 1], which in turn implies that the $(d+1)$-dimensional partition function is 1. In the thermodynamic limit any stable phase of the d-dimensional system is a stable phase of the $(d+1)$-dimensional system; therefore, if a d-dimensional CA has multiple stable phases, its corresponding $(d+1)$-dimensional Hamiltonian model will also, all with zero free energy.

The preceding argument holds whether $X(0)$, $X(1)$, etc., represent time steps of a synchronous model, or discrete-time snapshots of an asynchronous (master equation) model evolving in continuous time. However, in the synchronous case, because transitions at all sites in the same time slice are independent, the equivalent Hamiltonian is of a generalized Ising form, being a sum of local terms $H(y, \underline{x}) = -\ln P(y/\underline{x})$, where $P(y/\underline{x})$ is the conditional probability that a site will be in state y at time $t+1$, given that its neighborhood was in state $\underline{x}$ at time t. The normalization constraint, viz., that for each $\underline{x}$, the sum over y of $P(y/\underline{x})$ be 1, restricts the GIM to a lower-dimensional surface in the parameter space of its coupling constants, on which the free energy is zero.

In the case of metastable phases the $(d+1)$-dimensional free energy per space-time site is not 0 but $-\ln(1-\Gamma) \approx \Gamma$, where Γ is the nucleation rate for transitions out of the phase. This result, whose derivation will be given elsewhere, may be explained heuristically by saying that the expected number of futures per present is 1 in a stable phase and $1 - \Gamma$ in a metastable phase, if we exclude futures not belonging to the phase. We can now summarize the unusual behavior of the $(d+1)$-dimensional free energy of a nonergodic irreversible system such as the Toom NEC model. Throughout the two-phase region of parameter space, the free energy of both phases is identically zero. Beyond this region, the free energy of the stable phase remains zero, while the other phase's free energy lifts off very smoothly from zero, as $\Gamma \approx \exp(-\text{const}/s^{d-1})$, where s (>0), is the distance in parameter space from the first-order phase boundary.

Irreversibility and complexity.—The relation of irreversibility to complexity has been extensively studied, especially in chemical reaction-diffusion systems.[9] The "dissipative structures" developed by such systems far from equilibrium exhibit macroscopic spacetime ordering which persists over a set of positive measure in parameter space, but because the local interaction lacks the spatial asymmetry of the NEC rule, these systems remain ergodic in the thermodynamic limit. In other words, for a generic choice of parameters, one dissipative structure is stable, and the others are metastable. A closer chemical analog to the NEC system's generic nonergodicity can be seen in the "once for ever" selection exhibited by stirred (i.e., mean-field) nonlinear autocatalytic reaction systems.[10]

Probably the most comprehensive kind of complexity of which cellular automata or other discrete systems are capable is the capacity for universal computation. A computationally universal system[11] is one that can be programmed, through its initial conditions, to simulate any digital computation. Computational universality of course can only occur in a nonergodic system: In a computationally universal system, not only does the indefinite future depend on the initial condition, but it does so in an arbitrarily programmable way. For example, the computational universality of the well-known deterministic CA rule "life" implies that one can find an initial configuration for it that will evolve so as to turn a certain site on if and only if white has a winning strategy at chess. Universal automata can be programmed to mimic arbitrary kinds of nontrivial behavior observed in other systems, e.g., the scale invariance of the Ising model at its critical point.

The property of computational universality was originally demonstrated for systems (e.g., Turing machines, deterministic cellular automata) rather unlike those ordinarily studied in mechanics and statistical mechanics. Later, the property was demonstrated for certain noiseless classical mechanical systems[12] such as hard spheres with appropriate initial and boundary conditions. Very recently,[4] the computa-

tional universality has been shown to hold for certain noisy, locally interacting systems, i.e., irreversible CA in which all local transition probabilities are positive. Notable among these is the three-dimensional CA of Gacs and Reif,[13] which uses the NEC rule in two of its dimensions to correct errors in an arbitrary computation being performed along the third dimension.

Though all universal automata are equivalent in the computations they can perform, there may still be qualitative differences among them in the density of initial states that lead to nontrivial computations. These differences need to be better understood before one can make the tempting assertion that "self-organization" like that observed in nature is a spontaneous tendency of locally interacting irreversible systems, on a set of positive measure in the space of their transition probabilities and initial conditions. Characterizing the generic behavior of homogeneous locally interacting systems capable of universal computation is, we believe, the central problem in what might be called the new field of discrete computational statistical mechanics.

We wish to thank Peter Gacs for helpful discussions during all phases of this work, especially for bringing Toom's models to our attention and emphasizing the robust nature of their nonergodicity, compared to that of the Ising models with which we were familiar.

———————————

[a]Present address: Boston University, 111 Cummington Street, Boston, Mass. 02215.

[1]I. G. Enting, J. Phys. C **10**, 1379 (1977).

[2]E. Domany and W. Kinzel, Phys. Rev. Lett. **53**, 311 (1984).

[3]A. L. Toom, in *Multicomponent Random Systems*, edited by R. L. Dobrushin, in Advances in Probability, Vol. 6 (Dekker, New York, 1980), pp. 549–575.

[4]P. Gacs, University of Rochester Computer Science Department, Technical Report No. 132, 1983 (to be published).

[5]Ya. G. Sinai, *Theory of Phase Transitions: Exact Results* (Pergamon, New York, 1982), pp. 1–10.

[6]T. Toffoli, Physica (Amsterdam) **10D**, 195 (1984).

[7]E. Domany, Phys. Rev. Lett. **52**, 871 (1984).

[8]L. S. Schulman and P. E. Seiden, J. Stat. Phys. **19**, 293 (1978).

[9]G. Nicolis and I. Prigogine, *Self-Organization in Nonequilibrium Systems* (Wiley, New York, 1977); H. Haken, *Synergetics* (Springer-Verlag, New York, 1983), 3rd ed.

[10]M. Eigen and P. Schuster, *The Hypercycle* (Springer, New York, 1979).

[11]S. Wolfram, Rev. Mod. Phys. **55**, 601 (1983), and references therein.

[12]E. Fredkin and T. Toffoli, Int. J. Theor. Phys. **21**, 219 (1982); N. Margolus, Physica (Amsterdam) **10D**, 81–95 (1984).

[13]P. Gacs and J. Reif, in *Proceedings of the Seventeenth ACM Symposium on the Theory of Computing, Providence, Rhode Island, 1985* (Association for Computing Machinery, New York, 1985), pp. 388–395.

An Annotated Bibliography of Cellular Automata

Introduction
1. Some general references
    1.1 Books etc.
    1.2 Recent general readership accounts
2. Pure mathematical approaches
    2.1 Dynamical systems theory
    2.2 Additive cellular automata
    2.3 Combinatorial mathematics approaches
3. Phenomenological and constructional approaches
    3.1 Generic behaviour of cellular automata
    3.2 Measurements of dynamical quantities
    3.3 Geometrical pattern construction
    3.4 The "Game of Life"
    3.5 Structures in other cellular automata
4. Computation theoretical approaches
    4.1 Construction of computationally universal cellular automata
    4.2 Cellular automata as formal models for computation
    4.3 Computability of cellular automaton properties
    4.4 Computation theoretical analyses of cellular automata
5. Probabilistic cellular automata
    5.1 Statistical mechanics approaches
    5.2 Computation theoretical approaches
    5.3 Related studies
6. Physical applications of cellular automata
    6.1 General
    6.2 Growth processes
    6.3 Reaction-diffusion systems
    6.4 Spin systems
    6.5 Hydrodynamics
    6.6 Other
7. Biological applications of cellular automata
    7.1 General
    7.2 Models for self reproduction
    7.3 Specific models for biological structures and processes
8. Practical computation with cellular automata
    8.1 Image processing
    8.2 Algorithms based on cellular automata
    8.3 Computing methodologies based on cellular automata
    8.4 Cellular automaton simulation machines
    8.5 Computer architectures for cellular automata
    8.6 Related studies
9. Further applications of cellular automata
10. Some systems related to cellular automata
    10.1 Lattice dynamical systems
    10.2 Random Boolean networks
    10.3 L systems
Author index

Introduction

This bibliography lists most of the papers on cellular automata that I have seen. The choice of papers is in many cases somewhat arbitrary. For the most part, I have included only papers that consider cellular automata defined in a comparatively strict sense. Boldface references are to papers reprinted in this book. An author index of this bibliography appears at the end.

As is clear from the bibliography, the literature on cellular automata is very diverse. One consequence of this is a problem with nomenclature. Systems that are exactly of the form discussed in this book are often not called cellular automata, while systems which are only vaguely related are sometimes called cellular automata. Alternative names for cellular automata include "tessellation automata", "cell spaces", "block mappings" and "iterative arrays". In the biological literature, cellular automata often refer to assemblies of biological cells viewed as having certain computational capabilities. Cellular automata in the computational literature are sometimes assumed to have components that are much more sophisticated than those discussed in most of this book.

The cellular automaton literature is diverse not only in content but also in origins. Cellular automata have been invented independently many times, and in many cases, independent parts of the literature have developed.

A fair fraction of the literature on two-dimensional cellular automata, particularly in computer science and biology, has its ultimate origins in the work of von Neumann. Much of the mathematical literature on one-dimensional cellular automata (usually called "block mappings" there) probably arose from research on non-linear feedback shift registers. The "Game of Life", invented by J. H. Conway in the early 1970's, and extensively reported by M. Gardner in *Scientific American*, also led to a diverse set of papers.

Cellular automaton research was popular in the 1960's, particularly among computer scientists. But its popularity waned, and by the middle of the 1970's it was considered almost disreputable.

A renaissance in cellular automaton research began a few years ago when interactive graphical computing made it easy to simulate the behaviour of cellular automata. "Experimental mathematics" allowed a wealth of new investigations to be made. This book includes many of the primary papers on these investigations.

Through its recent popularity, cellular automaton research is becoming sufficiently visible that many connections to other lines of research are being recognized. Something like half of the papers listed in this bibliography were for example given to me as a result of articles, talks or reports on my work. (I am particularly grateful to Tom Toffoli for access to his collection of cellular automaton papers.) I would much appreciate receiving copies of other papers that should be included in future bibliographies on cellular automata.

1. Some general references

(Many of the papers contained in this book represent general references.)

1.1 Books etc.

A. Burks (ed.), *Essays on cellular automata*, (Univ. of Illinois Press, 1970).
    Important collection of papers on a variety of aspects of cellular automata, mostly computation theoretical. (Several papers contained in this collection are cited separately below.)

Dynamical systems and cellular automata, (J. Demongeot, E. Goles and M. Tchuente, eds.), (Academic Press, 1985).
    Proceedings of a workshop held in Luminy, France, in September, 1983. Includes some papers on ordinary cellular automata, together with many on various generalizations involving arbitrary networks of finite automata.

Cellular automata, (D. Farmer, T. Toffoli and S. Wolfram, eds.), Physica 10D (1984) nos. 1 and 2, and (North-Holland, 1984).

Important collection of papers on many aspects of the theory, phenomenology and applications of cellular automata (many cited separately below). Based on a multidisciplinary workshop on cellular automata held at Los Alamos in March, 1983. The preface gives a brief outline of cellular automata, together with a short annotated bibliography.

W. Handler, T. Legendi and G. Wolf (eds.), *Parcella '84*: Proc. II Inter. Workshop on Parallel Processing by Cellular Automata and Arrays, (Akademie-Verlag, Berlin, 1985).
Collection of papers on various hardware and software systems based on cellular automata and their generalizations.

H. Nishio, "A classified bibliography on cellular automata theory (with a focus on recent Japanese references)", in Proc. Int. Symp. on Uniform Structures, Automata and Logic, (Tokyo, August, 1975) [available from IEEE, New York, cat. no. 75 CH1052-OC].
A useful but far from complete list of papers, particularly on computational aspects of cellular automata.

T. Toffoli, "Cellular automaton mechanics", PhD Thesis, Logic of Computers Group, University of Michigan (1977).
Survey of mathematical results on cellular automata, and discussion of principles for modelling physical systems by cellular automata. Early results on qualitative generic properties of two-dimensional cellular automata.

T. Toffoli and N. Margolus, *Cellular automata machines: a new environment for modelling*, (MIT press, to be published).
Discussion of the concepts and practicalities of simulating physical and other systems, particularly on the CAM-6 two-dimensional cellular automaton simulation machine.

J. von Neumann, *Theory of self-reproducing automata*, (completed and edited by A. W. Burks), (Univ. of Illinois press, 1966).
Often considered the first discussion of cellular. Discusses some general questions about the use of cellular automata and other computational models, particularly in biology. Also includes detailed construction of a rather complicated two-dimensional cellular automaton shown to be capable of self reproduction.

1.2 Some recent general readership accounts

(see also sect. 3.4)
(Many other articles, mostly shorter, have also appeared.)

K. Dewdney, *Computer recreations*, Sci. Amer., (May, 1985).
Clear discussion of attempts to find computationally universal simple one-dimensional cellular automata, and their significance.

T. Durham, "Explorations in the cellular microworld", Computing Mag. (GB) (January 17, 1985).
Clear discussion of some conceptual bases of cellular automaton models.

B. Hayes, *Computer recreations*, Sci. Amer., (March, 1984).
Sampling of recent research on cellular automata.

S. Levy, "The portable universe", Whole Earth Review (Winter 1985).
Elementary discussion of cellular automata as models of nature.

J. Maddox, "Simulating the replication of life", Nature 305 (1983) 469.
Short discussion of cellular automata as models of self-organizing systems.

C. Reiter, "Life and death on a computer screen", Discover, (August, 1984).
Elementary discussion of research on cellular automata as general computational models.

R. Rucker, "Cellular automata", Isaac Asimov's Science Fiction Magazine (June, 1986).
Discussion of cellular automaton models and their conceptual implications.

J. Tucker, "Cellular automaton machine: the ultimate parallel computer", High Technology (June, 1986).
    Discussion of some computers for efficient cellular automaton simulation, and their uses.

S. Wolfram, "Cellular automata", Los Alamos Science, (Fall 1983).
    General introduction to recent results on the theory and phenomenology of cellular automata.

S. Wolfram, "Cellular automata as models of complexity", Nature 311 (1984) 419.
    Survey of recent results on the theory and phenomenology of cellular automata.

S. Wolfram, "Computer software in science and mathematics", Sci. Amer. (September, 1984).
    Discussion of the practical, methodological and conceptual significance of computational models such as cellular automata.

2. Pure mathematical approaches

2.1 Dynamical systems theory

(see also sects. 3.1 and 4.4)

S. Amoroso and Y. Patt, "Decision procedures for surjectivity and injectivity of parallel maps for tessellation structures", J. Comput. System Sci. 6 (1972) 448.
    Practical procedures for determining surjectivity and injectivity of one-dimensional cellular automaton mappings.

R. Dilao, "Entropies for cellular automata", Bielefeld preprint (1985).
    Calculations of entropies for certain one-dimensional cellular automata.

G. A. Gal'perin, "On an entropy characteristic of homogeneous media with local interaction", Uspeki Math. Nauk. 36 (1981) 207.
    Definition of entropy for one-dimensional cellular automaton mappings.

R. Gilman, "Classes of linear automata", Institute for Advanced Study and Stevens Institute of Technology preprint (1985).
    Attempt at a formal characterization of classes of one-dimensional cellular automata using dynamical systems theory.

M. Harao and S. Noguchi, "Global mapping space of cellular systems", Info. & Control 43 (1979) 241.
    Results on the structures of sets of all possible global cellular automaton mappings with certain properties.

G. A. Hedlund, "Endomorphisms and automorphisms of the shift dynamical system", Math. Systems Theor. 3 (1969) 320.
    Extensive analysis of global mappings corresponding to one-dimensional cellular automata, using symbolic dynamics.

G. A. Hedlund, K. I. Appel and L. R. Welch, "All onto functions of span less than or equal to 5", Communications Research Division (Princeton), Working Paper (July 1963).
    Tables of all one-dimensional cellular automata with $k=2$ and $r\leq2$ which yield surjective global mappings.

D. Lind, "Applications of ergodic theory and sofic systems to cellular automata", Physica 10D (1984) 36.
    Some dynamical systems theory results on $k=2$, $r=1$ cellular automata rules 90 and 18.

A. Maruoka and M. Kimura, "Condition for injectivity of global maps for tessellation automata", Info. & Control 32 (1976) 158.
    Formal conditions for global cellular automaton mappings to be one-to-one.

J. Milnor, "Notes on surjective cellular automaton-maps", Institute for Advanced Study notes, (June 1984).
> Clear description of some classes of surjective cellular automata, with examples.

J. Milnor, "Entropy of cellular automaton-maps", Institute for Advanced Study preprint (1984).
> Definitions and properties of directional entropies for some simple one-dimensional cellular automata.

J. Milnor, "Directional entropies of cellular automaton-maps", in *Disordered systems and biological organization*, (E. Bienenstock, F. Fogelman Soulie and G. Weisbuch, eds.), (Springer, 1986).
> Definitions and some properties of directional entropies for multidimensional cellular automata.

H. Miyazima, "Decision algorithm for k-to-one mapping of cellular automata", Trans. Inst. Electron. and Commun. Eng. Japan, J66D (1983) 812.
> Algorithm for determining whether a one-dimensional cellular automaton rule yields exactly a k-to-one global mapping.

M. Nasu, "Local maps inducing surjective global maps of one-dimensional tessellation automata", Math. Systems Theor. 11 (1978) 327.
> Study of conditions for one-dimensional cellular automaton maps to be surjective.

D. Richardson, "Tesselations with local transformations", J. Comput. System Sci. 6 (1972) 373.
> Discussion of continuity and other properties of global cellular automaton mappings.

Y. G. Sinai, "An answer to a question by J. Milnor", Comm. Math. Helv. 60 (1985) 173.
> Demonstration of the existence of certain directional entropies.

M. Waterman, "Some applications of information theory to cellular automata", Physica 10D (1984) 45.
> Some basic properties of entropies in cellular automata.

S. Willson, "On the ergodic theory of cellular automata", Math. Systems Theor. 9 (1975) 132.
> Some formal conditions for ergodicity in cellular automata are derived.

2.2 Additive cellular automata

(see also sect. 3.3)

H. G. ApSimon, "Periodic forests whose largest clearings are of size 3", Phil. Trans. Roy. Soc. A266 (1970) 113; "Periodic forests whose largest clearings are of size $n \geq 4$", Proc. Roy. Soc. A319 (1970) 399.
> Results on patterns generated by finite additive cellular automata evolving from arbitrary initial conditions.

J. Butler and S. Ntafos, "The vector string descriptor as a tool in the analysis of cellular automata systems", Math. Biosciences 35 (1977) 55.
> Results on the sizes and properties of patterns generated by additive cellular automata.

R. Cordovil, R. Dilao and A. Noronha da Costa, "Periodic orbits for additive cellular automata", CFMC, Lisbon preprint (1985).
> Results on the spatial periods of configurations with particular temporal periods in additive cellular automata.

J.-M. Deshouillers, "La repartition modulo 1 des puissances de rationnels dans l'anneau des series formelles sur un corps fini", Seminaire de Theorie des Nombres, Bordeaux, no. 5 (1980).
> Study of sequences generated by one-dimensional additive cellular automata using methods from number theory and finite automata theory.

P. Guan and Y. He, "Exact results for deterministic cellular automata with additive rules", J. Stat. Phys. (1986).
> Results on global properties of finite additive cellular automata, using circulant matrix methods.

M. Ito, N. Osato and M. Nasu, "Linear cellular automata over $\mathbf{Z}_m$", J. Comput. System Sci. 27 (1983) 125.
>   Conditions for surjectivity and injectivity of additive cellular automaton mappings.

O. Martin, A. Odlyzko and S. Wolfram, "Algebraic properties of cellular automata", Commun. Math. Phys. 93 (1984) 219. [**1.4**]
>   Extensive results on global properties of finite additive cellular automata.

J. C. P. Miller, "Periodic forests of stunted trees", Philos. Trans. Roy. Soc. A266 (1970) 63; A293 (1980) 48.
>   Analysis of patterns produced by finite additive cellular automata with arbitrary initial states, and their relations to number theory.

M. Miyamoto, "An equilibrium state for a one-dimensional life game", J. Math. Kyoto Univ. 19 (1979) 525.
>   Result on generic large-time behaviour of an additive cellular automaton.

Y. Nambu, "Field theory of Galois fields", University of Chicago preprint (1985).
>   Results on global properties of additive cellular automata which give simple approximations for certain physical equations.

W. Pries, A. Thanailakis and H. Card, "Group properties of cellular automata and VLSI applications", EE Dept., University of Manitoba preprint (1985).
>   Algebraic properties of the set of finite additive cellular automaton mappings.

S. Willson, "Cellular automata can generate fractals", Discrete Appl. Math. 8 (1984) 91.
>   Demonstration that simple additive one-dimensional cellular automata can generate fractal patterns.

S. Willson, "The equality of fractional dimensions for certain cellular automata", Physica D (1986).
>   Proofs that several definitions of dimension coincide for patterns generated by additive cellular automata.

S. Willson, "Growth rates and fractional dimensions in cellular automata", Physica 10D (1984) 69.
>   Short mathematical introduction to definitions and properties of dimensions for patterns generated by additive cellular automata.

S. Willson, "Computing fractal dimensions for additive cellular automata", Physica D (1986).
>   Method for computing the fractal dimensions of patterns generated by additive cellular automata.

S. Wolfram, "Geometry of binomial coefficients", Amer. Math. Monthly 91 (1984) 567.
>   Elementary exposition of the generation of fractal patterns by additive cellular automata.

2.3 Combinatorial mathematics approaches

S. Amoroso and I. Epstein, "Indecomposable parallel maps in tessellation structures", J. Comput. System Sci. 13 (1976) 442.
>   Study of conditions under which certain classes of cellular automaton rules cannot be written as compositions of simpler ones.

J. Butler, "A note on cellular automata simulations", Info. & Control 26 (1974) 286.
>   Discussion of equivalences between two-dimensional cellular automata with different neighbourhoods.

J. Butler, "Synthesis of one-dimensional binary cellular automata systems from composite local maps", Info. & Control 43 (1979) 304.
>   Procedure for decomposing certain classes of cellular automaton mappings.

E. Coven, G. Hedlund and F. Rhodes, "The commuting block maps problem", Trans. Amer. Math. Soc. 249 (1979) 113.
>   Study of a class of commuting cellular automaton mappings.

E. Goles and J. Olivos, "Periodic behaviour of binary threshold functions and applications", Discrete Appl. Math. 3 (1981) 93.
   Determines fixed point and cycle behaviour in certain simple cellular automata.

E. Goles, *Comportement dynamique de reseaux d'automates*, University of Grenoble thesis (1985).
   Discussion of a variety of results on cellular automata and related systems, particularly those involving threshold functions.

M. Harao and S. Noguchi, "On some dynamical properties of finite cellular automaton", IEEE Trans. Comput. C-27 (1978) 42.
   Results on the structure of state transition graphs for cellular automata with a finite number of sites.

E. Jen, "Global properties of cellular automata", J. Stat. Phys. (1986).
   Conditions for aperiodic temporal sequences and other properties of $k=2$, $r=1$ cellular automata.

E. Jen, "Invariant strings and pattern-recognizing properties of one-dimensional cellular automata", J. Stat. Phys. (1986).
   Construction of one-dimensional cellular automaton rules that leave particular finite strings invariant.

A. Maruoka and M. Kimura, "Completeness problem of multidimensional tessellation automata", Info. & Control 35 (1977) 52.
   Demonstrations that arbitrary finite blocks of symbols can be generated by appropriate compositions of sufficiently complicated multidimensional cellular automaton mappings.

A. Maruoka, M. Kimura and N. Shoji, "Pattern decomposition for tessellation automata", Theor. Comput. Sci. 14 (1981) 211.
   Results on the construction of cellular automata in which arbitrary initial configurations evolve to configurations with certain simple properties.

M. Nasu and N. Honda, "A completeness property of one-dimensional tessellation automata", J. Comput. System Sci. 12 (1976) 36.
   Demonstrations that arbitrary finite symbol sequences can be generated by appropriate compositions of different cellular automaton mappings.

M. Nasu, "Indecomposable local maps of tessellation automata", Math. Systems Theor. 13 (1979) 81.
   Conditions that one-dimensional cellular automaton rules cannot be written as compositions of simpler rules.

F. Rhodes, "The principal part of a block map", J. Combin. Theor. A33 (1982) 48.
   Conditions under which certain classes of partially-linear cellular automata are irreducible with respect to composition.

F. Rhodes, "The role of the principal part in factorizing block maps", Math. Proc. Camb. Phil. Soc. 96 (1984) 223.
   An approach to finding decompositions of cellular automaton rules.

M. Tchuente, "Dynamics and self organization in one-dimensional arrays", in *Disordered systems and biological organization*, (E. Bienenstock, F. Fogelman Soulie and G. Weisbuch, eds.), (Springer, 1986).
   Results on state transition diagram structures for finite one-dimensional cellular automata, particularly those with the majority function rule.

P. White and J. Butler, "Synthesis of one-dimensional binary scope-2 flexible cellular systems from initial final configuration pairs", Info. & Control 46 (1980) 241.
   Investigation of what fraction of possible finite symbol sequences can be produced by compositions of cellular automaton mappings chosen at each step from a set of possibilities.

H. Yamada and S. Amoroso, "A completeness problem for pattern generation in tessellation automata", J. Comput. System Sci. 4 (1970) 137.
   Study of what finite symbol sequences can be generated by compositions of cellular automaton

mappings.

H. Yamada and S. Amoroso, "Structural and behavioural equivalences of tessellation automata", Info. & Control 18 (1971) 1.
   Study of formal equivalences of cellular automata under various blocking transformations.

3. Phenomenological and constructional approaches

3.1 Generic behaviour of cellular automata

N. Packard and S. Wolfram, "Two-dimensional cellular automata", J. Stat. Phys. 38 (1985) 901 [**1.4**].
   Systematic investigation of simple forms of generic behaviour in two-dimensional cellular automata.

G. Vichniac, "Cellular automata models of disorder and organization", in *Disordered systems and biological organization*, (E. Bienenstock, F. Fogelman Soulie and G. Weisbuch, eds.), (Springer, 1986).
   Survey of various phenomenological properties of two-dimensional cellular automata, particularly related to domain formation.

S. Wolfram, "Statistical mechanics of cellular automata", Rev. Mod. Phys. 55 (1983) 601 [**1.1**].
   Beginning of systematic study of generic properties of one-dimensional cellular automata, together with mathematical characterizations and approaches to modelling natural processes.

S. Wolfram, "Universality and complexity in cellular automata", Physica 10D (1984) 1 [**1.3**].
   Characterization of generic behaviour in one-dimensional cellular automata using qualitative, dynamical systems and computation theoretical methods.

S. Wolfram, "Twenty problems in the theory of cellular automata", Phys. Scripta T9 (1985) 170 [**1.5**].
   Discussion of some outstanding problems concerning the generic behaviour of cellular automata.

3.2 Measurements of dynamical quantities

P. Grassberger, "New mechanism for deterministic diffusion", Phys. Rev. A28 (1983) 3666.
   Discussion of random walks in domain walls, mainly in $k=2$, $r=1$ cellular automaton rule 18.

P. Grassberger, "Chaos and diffusion in deterministic cellular automata", Physica 10D (1984) 52.
   Study of domains and other phenomena, mainly in $k=2$, $r=1$ cellular automaton rule 18.

P. Grassberger, "Long-range effects in an elementary cellular automaton", Wuppertal preprint (1986).
   Study of dynamical systems theory properties of $k=2$, $r=1$ rule 22.

P. Grassberger, "Towards a quantitative theory of self-generated complexity", Wuppertal preprint (1986).
   Definition and measurements of measure-theoretical analogues of regular language complexity (as defined in **2.1**).

N. Packard, "Complexity of growing patterns in cellular automata", in *Dynamical systems and cellular automata*, (J. Demongeot, E. Goles and M. Tchuente, eds.), (Academic Press, 1985).
   Discussion of dynamical systems theory quantities for cellular automata, and computation of some Lyapunov exponents.

S. Wolfram, "Random sequence generation by cellular automata", Adv. Applied Math. 7 (1986) 123 [**2.3**].
   Detailed study and characterization of randomness generated by $k=2$ $r=1$ cellular automaton rule 30.

3.3 Geometrical pattern construction

(see also sects. 2.2 and 3.1)

R. Schrandt and S. Ulam, "On recursively defined geometrical objects and patterns of growth", in A. Burks (ed.), *Essays on cellular automata*, (Univ. of Illinois Press, 1970).
　　Early computer work on patterns generated by two and three-dimensional cellular automata, mostly additive.

S. Ulam, "On some mathematical problems connected with patterns of growth of figures", in A. Burks (ed.), *Essays on cellular automata*, (Univ. of Illinois Press, 1970).
　　Early mathematical work on simple patterns produced by additive two-dimensional cellular automata.

S. Willson, "The growth of configurations", Math. Systems Theor. 10 (1977) 387.
　　Mathematical study of uniformly-growing patterns generated by certain cellular automata.

S. Willson, "Growth patterns of ordered cellular automata", J. Comput. System Sci. 22 (1981) 29.
　　Mathematical results on the size and shape of uniformly-growing patterns in certain cellular automata.

3.4 The "Game of Life"

E. R. Berlekamp, J. H. Conway and R. K. Guy, "Winning ways for your mathematical plays", (Academic Press, 1984), vol. 2.
　　Contains a chapter surveys results on Life, particularly the construction of universal computers.

D. Buckingham, "Some facts of Life", Byte 3 (1978) 54.
　　Summary of properties of many periodic, propagating and other structures in Life.

M. Gardner, *Mathematical games*, Sci. Amer. (February, March, April, 1971; January, 1972).
　　Original announcements of Life and discoveries about it.

M. Gardner, "Wheels, Life and other mathematical amusements", (Freeman, 1983).
　　Contains a survey of discoveries about Life, summarizing *Mathematical Games* columns in *Scientific American* from 1971 and 1972.

R. W. Gosper, "Exploiting regularities in large cellular spaces", Physica 10D (1984) 75.
　　Efficient algorithm for Life simulation, based on storing intermediate partial configurations.

J. Hardouin-Duparc, "Paradis terrestre dans l'automate cellulaire de Conway", R.A.I.R.O. R-3 (1974) 63.
　　Construction of unreachable configurations in Life.

M. Niemiec, "Life algorithms", Byte 4 (1979) 90.
　　Description of some efficient algorithms for simulating Life.

R. T. Wainwright, *Lifeline*, nos. 1-11 (1971-1973).
　　A newsletter for hobbyists working on Life.

R. T. Wainwright, "Life is universal!", Proc. Winter Simulation Conf. (ACM, Washington, 1974).
　　Discussion of the proof of computational universality for Life.

3.5 Structures in other cellular automata

J. Park, "A one-dimensional glider gun", Princeton University report (1985).
　　Discovery of a "glider gun" structure in a simple one-dimensional cellular automaton.

S. Wolfram, "Glider gun guidelines", background to *Computer recreations* column; distributed by *Scientific American* (May, 1985).

Elementary account of results and procedures for finding periodic, propagating and other structures in simple one-dimensional cellular automata.

4. Computation theoretical approaches

4.1 Construction of computationally universal cellular automata

(see also sects. 3.4 and 3.5)

E. R. Banks, "Information processing and transmission in cellular automata", MIT project MAC report no. TR-81 (1971).
> Detailed discussion of a simple two-dimensional cellular automaton which can simulate arbitrary logic circuits.

E. F. Codd, *Cellular automata*, (Academic Press, 1968).
> Detailed discussion of two-dimensional cellular automata capable of universal computation, and of circuits constructed in them.

D. Gordon, "On the computational power of totalistic cellular automata", University of Cincinnati preprint (1985).
> Demonstration that totalistic one-dimensional cellular automata can be computationally universal.

N. Margolus, "Physics-like models of computation", Physica 10D (1984) 81 [**2.2**].
> Discussion of reversible two-dimensional computationally universal cellular automata based on a discrete approximation to a hard-sphere gas.

F. Nourai and R. S. Kashef, "A universal four-state cellular computer", IEEE Trans. Comput. C-24 (1975) 766.
> Explicit construction of a computationally universal two-dimensional cellular automaton.

A. R. Smith III, "Simple computation-universal cellular spaces", J. ACM 18 (1971) 339.
> Construction of one-dimensional cellular automata that are equivalent to universal Turing machines.

J. W. Thatcher, "Universality in the von Neumann cellular model", in A. Burks (ed.), *Essays on cellular automata*, (Univ. of Illinois Press, 1970).
> Discussion of computational universality in the self-reproducing two-dimensional cellular automaton constructed by von Neumann.

T. Toffoli, "Computation and construction universality of reversible cellular automata", J. Comput. System Sci. 15 (1977) 213.
> Demonstration of the formal feasibility of constructing computationally universal reversible cellular automata.

A. Vergis, "Bilateral arrays of combinational cells are universal computers", Princeton University technical report #322 (1984).
> Discussion of computationally universal one-dimensional cellular automata and computational complexity issues related to them.

4.2 Cellular automata as formal models for computation

S. Cole, "Real-time computation by *n*-dimensional iterative arrays of finite state machines", IEEE Trans. Comput. C-18 (1969) 349.
> Computational complexity characterizations of certain simple computations carried out by cellular automata.

J. Pecht, "On the real-time recognition of formal languages in cellular automata", Acta Cybernetica 6 (1983) 33.

Formal constructions of cellular automata that evolve to special states when their input corresponds to strings in particular formal languages.

L. Priese, "A note on asynchronous cellular automata", J. Comput. System Sci. 17 (1978) 237.
Discussion of computational universality in cellular automata where site values can be updated asynchronously.

A. R. Smith III, "Cellular automata complexity trade-offs", Info. & Control 18 (1971) 466.
Investigation of computational efficiency in universal cellular automata.

A. R. Smith III, "Real-time language recognition by one-dimensional cellular automata", J. Comput. System Sci. 6 (1972) 233.
Demonstration that certain formal languages can be recognized more efficiently by cellular automata than by Turing machines.

R. Sommerhalder and S. C. van Westrhenen, "Parallel language recognition in constant time by cellular automata", Acta Informatica 19 (1983) 397.
Characterizations in terms of regular languages of the sets of initial strings which map to particular final strings in one-dimensional cellular automata.

R. Vollmar, "Some remarks on the efficiency of polyautomata", Int. J. Theor. Phys. 21 (1982) 1007.
Definitions and properties of some computational complexity measures for cellular automata.

A. Waksman, "An optimum solution to the firing squad synchronization problem", Info. & Control 9 (1966) 66.
Construction of cellular automata in which global synchronization can be achieved.

4.3 Computability of cellular automaton properties

V. Aladyev, "The behavioural properties of homogeneous structures", Math. Biosciences 29 (1976) 99.
Survey of various computation theoretical aspects of cellular automata, particularly the decidability of reachability problems for configurations.

D. Gajski and H. Yamada, "A busy beaver problem in cellular automata", in Proc. Int. Symp. on Uniformly Structured Automata and Logic (Tokyo, 1975).
Demonstration that certain questions about the growth of configurations in cellular automata are formally undecidable.

H. Takahashi, "Undecidable questions about the maximum invariant set", Info. & Control 33 (1977) 1.
Discussion of the formal undecidability of determining whether the maximum invariant set for a one-dimensional cellular automata can be represented as a regular formal language.

P. Vitanyi, "On a problem in the collective behaviour of automata", Discrete Math. 14 (1976) 99.
Demonstration that questions about the maximum growth of configurations in cellular automata can be formally undecidable.

S. Wolfram, "Undecidability and intractability in theoretical physics", Phys. Rev. Lett. 54 (1985) 735 [2.4].
Discussion of the computational difficulty of predicting the behaviour of cellular automata.

T. Yaku, "The constructability of a configuration in a cellular automaton", J. Comput. System Sci. 6 (1972) 373.
Discussion of the formal undecidability of global surjectivity and injectivity for two-dimensional cellular automata.

T. Yaku, "Inverse and injectivity of parallel relations induced by cellular automata", Proc. Amer. Math. Soc. 58 (1976) 216.
Discussion of certain global properties of two-dimensional cellular automata that are semi-recursive.

4.4 Computation theoretical analyses of cellular automata

(see also sect. 3.1)

V. Aladyev, "Survey of research in the theory of homogeneous structures and their applications", Math. Biosciences 22 (1974) 121.
>   Survey of many computation theoretical aspects of cellular automata.

U. Golze, "Differences between 1- and 2-dimensional cell spaces", in *Automata, Languages, Development* (A. Lindenmayer and G. Rozenberg, eds.), (North-Holland, 1976).
>   Demonstration that in two-dimensional cellular automata there exist finite configurations which can arise only from non-recursive predecessors.

L. Hurd, "Formal language characterizations of cellular automaton limit sets", Princeton University preprint (1986).
>   Demonstration that the limit sets for one-dimensional cellular automata can correspond to various classes of formal languages.

A. R. Smith III, "Introduction to and survey of polyautomata theory", in *Automata, Languages, Development*, (A. Lindenmayer and G. Rozenberg, eds.), (North-Holland, 1976).
>   Survey of mathematical results on cellular automata from a computation theoretical viewpoint.

H. Takahashi, "The maximum invariant set of an automaton system", Info. & Control 32 (1976) 307.
>   Discussion of computation theoretical characterizations of periodic sets of states in finite one-dimensional cellular automata.

S. Wolfram, "Computation theory of cellular automata", Commun. Math. Phys. 96 (1984) 15 [**2.1**].
>   Characterization of sets generated by one-dimensional cellular automaton evolution using the formal theory of regular languages.

S. Wolfram, "Origins of randomness in physical systems", Phys. Rev. Lett. 55 (1985) 449 [**2.5**].
>   Discussion of the nature of randomness produced in cellular automata, based on computation theory.

5. Probabilistic cellular automata

5.1 Statistical mechanics approaches

C. Bennett and G. Grinstein, "Role of irreversibility in stabilizing complex and nonergodic behaviour in locally interacting discrete systems", Phys. Rev. Lett. 55 (1985) 657 [**4.6**].
>   Discussion of the statistical mechanics of non-ergodic behaviour in probabilistic cellular automata.

M. Y. Choi and B. Huberman, "Exact results for multiple state cellular automata", J. Phys. A17 (1984) L765.
>   A generalization of the results of [**4.4**] for cellular automata with more than two states per site.

R. L. Dobrushin, V. I. Kryukov and A. L. Toom (eds.), *Locally interacting systems and their application in biology*, Lecture notes in mathematics no. 653, (Springer-Verlag, 1978).
>   Collection of papers on probabilistic cellular automata.

E. Domany and W. Kinzel, "Equivalence of cellular automata to Ising models and directed percolation", Phys. Rev. Lett. 53 (1984) 311 [**4.4**].
>   Demonstration that the possible spacetime patterns generated by probabilistic cellular automata can form the same ensemble as those produced in equilibrium statistical models.

M. Dresden and D. Wong, "Life games and statistical models", Proc. Natl. Acad. Sci. 72 (1975) 956.
>   Discussion of master equations for simple probabilistic cellular automata.

P. Gach, G. L. Kurdyumov and L. A. Levin, "One-dimensional uniform arrays that wash out finite islands", Probl. Peredachi. Info. 14 (1978) 92.

Construction of a probabilistic cellular automaton with possible non-ergodic behaviour.

P. Grassberger, F. Krause and T. von der Twer, "A new type of kinetic critical phenomenon", J. Phys. A17 (1984) L105 [**4.2**].
Study of domain phenomena in some simple one-dimensional probabilistic cellular automata.

G. Grinstein, C. Jayaprakash and Y. He, "Statistical mechanics of probabilistic cellular automata", Phys. Rev. Lett. 55 (1985) 2527 [**4.5**].
Discussion reversibility conditions and properties for probabilistic cellular automata.

T. Ingerson and R. Buvel, "Structure in asynchronous cellular automata", Physica 10D (1984) 59.
Basic phenomenology of cellular automata with asynchronous updating of different site values.

K. Kaneko and Y. Akutsu, "Phase transitions in two-dimensional stochastic cellular automata", J. Phys. A19 (1986) L69 [**4.3**].
Phenomenology of pattern generation in simple two-dimensional probabilistic cellular automata.

W. Kinzel, "Phase transitions of cellular automata", Z. Phys. B58 (1985) 229 [**4.1**].
Detailed study of statistical mechanical properties of simple probabilistic cellular automata.

G. L. Kurdyumov, "An example of a nonergodic homogeneous random medium with positive transition probabilities", Soviet Math. Dokl. 19 (1978) 211.
Study of a probabilistic cellular automaton which does not visit all possible states even after an arbitrarily long time.

L. Schulman and P. Seiden, "Statistical mechanics of a dynamical system based on Conway's game of Life", J. Stat. Phys. 19 (1978) 293.
Results on statistical properties of a probabilistic version of Life.

G. Vichniac, P. Tamayo and H. Hartman, "Annealed and quenched inhomogeneous cellular automata (INCA)", J. Stat. Phys. (1986).
Study of patterns generated by certain one-dimensional probabilistic cellular automata.

5.2 Computation theoretical approaches

P. Gacs, "Reliable computation with cellular automata", Boston University Computer Science Tech. Rep. 85/009 (1985).
Construction of a rather complicated one-dimensional probabilistic cellular automata claimed to be capable of reliable computation when the noise is small.

P. Gacs and J. Reif, "A simple three-dimensional real-time reliable cellular array", Boston University Computer Science Tech. Rep. 85/002 (1985).
Construction of a comparatively simple three-dimensional probabilistic cellular automaton which is capable of reliable computation in the prescence of noise.

5.3 Related studies

There are several classes of systems which are similar to probabilistic cellular automata, and on which there is extensive literature. Some of these are:

Dynamic spin systems: time-dependent versions of the Ising models and its generalizations. (e.g. K. Kawasaki, "Kinetics of Ising models", in *Phase transitions and critical phenomena: 2* (ed. C. Domb and M. S. Green), (Academic Press, 1972).)

Directed percolation: random filling of lattice elements, with a distinguished direction that can correspond to time. (e.g. W. Kinzel, "Directed percolation", Ann. Israel Phys. Soc. 5 (1983) 425.)

Markov random fields: systems in which probabilities at sites are determined by local rules. (e.g. R. Kindermann and J. L. Snell, *Markov random fields and their applications*, (Amer. Math. Soc.,

1980).)

Particle systems: systems in which discrete particles on a lattice under local interactions according to definite probabilities (e.g. D. Griffeath, *Additive and cancellative interacting particle systems*, (Springer, 1970).)

6. Physical applications of cellular automata

6.1 General

T. Toffoli, "Cellular automata as an alternative to (rather than an approximation of) differential equations in modelling physics", Physica 10D (1984) 117.
  Qualitative discussion of some bases for cellular automaton models of physical systems.

G. Vichniac, "Simulating physics with cellular automata", Physica 10D (1984) 96.
  General discussion of some issues concerning cellular automaton models for physical systems, with a particular emphasis on two-dimensional spin systems.

S. Wolfram, "Cellular automata and condensed matter physics", in *Scaling phenomena in disordered systems* (R. Pynn and A. Skjeltorp, eds.), (Plenum, 1985).
  Survey of some theory and phenomenology of cellular automata relevant to various applications in condensed matter physics.

K. Zuse, *Rechnender Raum*, Schriften fur Datenverarbeitung, vol. 1, (Friedr. Vieweg & Sohn, Brauschweig, 1969) (translated as *Calculating space*, MIT Technical Translation AZT-70-164-GEMIT).
  Qualitative discussion of cellular automata as fundamental models for physical systems.

6.2 Growth processes

G. M. Crisp, "A cellular automaton model of crystal growth: I) Anthracene", Crystallography Unit, University College London preprint (1985).
  Probabilistic cellular automaton model for simple crystal growth and defect production.

W. Good, "Cellular automata formalism: A clue to understand temperature gradient metamorphism in snow?", Swiss Federal Institute for Snow and Avalanche Research, Davos (1985).
  Two-dimensional cellular automaton model for snow grain evolution.

A. Mackay, "Crystal symmetry", Phys. Bull. 27 (1976) 495.
  Qualitative discussion of analogies between cellular automaton and crystal growth processes.

V. Maverick, "Crystalline behaviour in some cellular automata", Organic Chemistry, ETH Zurich, preprint (1985).
  Simple models for regular and irregular dendritic crystal growth in two-dimensional cellular automata.

N. Packard, "Deterministic lattice models for solidification and aggregation", to appear in Proc. First International Symposium for Science on Form, (Tsukuba, Japan, 1985) [3.1].
  Models for various forms of regular, dendritic and random growth, based on two-dimensional cellular automata and related systems.

6.3 Reaction-diffusion systems

J.-P. Allouche and C. Reder, "Spatio-temporal oscillations generated by cellular automata", Discrete Appl. Math. 8 (1984) 215.
  Study of oscillations in simple cellular automaton models for reaction-diffusion systems.

J. Greenberg and S. Hastings, "Spatial patterns for discrete models of diffusion in excitable media", SIAM J. Appl. Math. 34 (1978) 515.
   Simple two-dimensional cellular automaton models of spiral pattern formation in reaction-diffusion systems.

J. Greenberg, B. D. Hassard and S. Hastings, "Pattern formation and periodic structures in systems modeled by reaction-diffusion equations", Bull. Amer. Math. Soc. 84 (1978) 1296.
   Survey of some two-dimensional cellular automaton models for reaction-diffusion systems.

J. Greenberg, C. Greene and S. Hastings, "A combinatorial problem arising in the study of reaction-diffusion equations", SIAM J. Alg. Disc. Math. 1 (1980) 34.
   Characterization of initial conditions leading to particular structures in two-dimensional cellular automaton models of reaction-diffusion systems.

J. Keeler, "An explicit relation between coupled-maps, cellular automata and reaction-diffusion equations", Physics Dept., UC San Diego, preprint (March 1986).
   Relations between statistical averages in cellular automata and reaction-diffusion systems.

B. Madore and W. Freedman, "Computer simulations of the Belousov-Zhabotinsky reaction", Science 222 (1983) 615 [**3.2**].
   Two-dimensional cellular automaton models for pattern formation in reaction-diffusion systems.

Y. Oono, M. Kohmoto, "A discrete model of chemical turbulence", Phys. Rev. Lett. 55 (1985) 2927 [**3.5**].
   One-dimensional cellular automaton models for chemical reaction-diffusion processes exhibiting chaotic behaviour.

A. Winfree, E. Winfree and H. Seifert, "Organizing centers in a cellular excitable medium", Physica 17D (1985) 109 [**3.3**].
   Study of the formation of spiral and other patterns in two-dimensional cellular automaton models of reaction-diffusion processes.

6.4 Spin systems

M. Creutz, "Deterministic Ising dynamics", Ann. Phys. 67 (1986) 62 [**3.8**].
   Two-dimensional cellular automaton version of the dynamic Ising model.

E. Domany, "Exact results for two- and three-dimensional Ising and Potts models", Phys. Rev. Lett. 52 (1984) 871.
   Application of probabilistic cellular automata to obtain new results on various spin system models.

Y. Pomeau, "Invariant in cellular automata", J. Phys. A17 (1984) L415 [**3.7**].
   Definition and discussion of conserved quantities in reversible cellular automata, in analogy with spin systems.

6.5 Hydrodynamics

U. Frisch, B. Hasslacher and Y. Pomeau, "Lattice gas automata for the Navier-Stokes equation", Phys. Rev. Lett. 56 (1986) 1505 [**3.9**].
   Introduction and brief analysis of a two-dimensional cellular automaton model for molecular dynamics which should reproduce the macroscopic Navier-Stokes equations.

J. Hardy, Y. Pomeau and O. de Pazzis, "Time evolution of a two-dimensional model system. I. Invariant states and time correlation functions", J. Math. Phys. 14 (1973) 1746.
   Introduction of a two-dimensional cellular automaton approximation to molecular dynamics, and an analytical study of some its statistical mechanical properties.

J. Hardy, O. de Pazzis and Y. Pomeau, "Molecular dynamics of a classical lattice gas: transport properties and time correlation functions", Phys. Rev. A13 (1976) 1949.

Derivation and computer study of some transport properties of a two-dimensional cellular automaton model for molecular dynamics.

D. d'Humieres, P. Lallemand and T. Shimomura, "An experimental study of lattice gas hydrodynamics", Los Alamos preprint LA-UR-85-4051 (1985).

Results on qualitative flow patterns and viscosity measurements from simulations of two-dimensional cellular automaton fluid models.

D. d'Humieres, Y. Pomeau and P. Lallemand, "Simulation d'allees de Von Karman bidimensionnelles a l'aide d'un gaz sur reseau", C. R. Acad. Sci. Paris II 301 (1985) 1391.

Simulations of vortex streets in cellular automaton hydrodynamic flows.

N. Margolus, T. Toffoli and G. Vichniac, "Cellular automata supercomputers for fluid dynamics modelling", Phys. Rev. Lett. (1986).

Discussion of cellular automaton fluid simulations on the CAM-6 and other cellular automaton machines, together with results on correlation functions.

S. Orszag and V. Yakhot, "Reynolds number scaling of cellular automaton hydrodynamics", Princeton University Applied Math. preprint (1986).

Discussion of asymptotic numbers of sites needed in cellular automaton and conventional hydrodynamic simulations.

J. P. Rivet and U. Frisch, "Automates sur gaz de reseau dans l'approximation de Boltzmann", C. R. Acad. Sci. Paris II 302 (1986) 267.

Derivations of transport coefficients for certain two-dimensional cellular automaton fluid models in the Boltzmann equation approximation.

J. Salem and S. Wolfram, "Thermodynamics and hydrodynamics with cellular automata", Institute for Advanced Study preprint (November 1985) [**3.10**].

Short discussion of the bases for cellular automaton fluid models, together with results of simulations on the Connection Machine massively parallel processing computer.

S. Wolfram, "Cellular automaton fluids 1: Basic theory", Institute for Advanced Study preprint (March, 1986).

A detailed account of the derivation of hydrodynamic behaviour for cellular automaton fluids in two and three dimensions.

V. Yakhot, B. Bayley and S. Orszag, "Analogy between hyperscale transport and cellular automaton fluid dynamics", Princeton University preprint (February 1986).

Discussion of analogies between cellular automaton fluids and fluids with random microscopic forces.

6.6 Other

Y. Aizawa, I. Nishikawa and K. Kaneko, "Soliton cellular automata", Tokyo preprint (1986).

Study of a class of cellular automata found to have propagating structures with certain conservation properties.

M. Y. Choi and B. Huberman, "A dynamical model for the stick-slip behaviour of faults", Xerox PARC preprint (1985).

Probabilistic cellular automaton model for the behaviour of geological faults.

H. Gerola and P. Seiden, "Stochastic star formation and spiral structure of galaxies", Astrophys. J. 223 (1978) 129.

A model for galactic evolution based on a probabilistic cellular automaton.

J. Park, K. Steiglitz and W. Thurston, "Soliton-like behaviour in automata", Physica D19 (1986) [3.6].
Investigation of cellular automata and generalizations which exhibit propagating structures with definite collision properties.

7. Biological applications of cellular automata

7.1 General

R. Baer and H. Martinez, "Automata and biology", Ann. Rev. Biophys. 3 (1974) 255.
Review of cellular and other automata models, and their potential applications in biology.

M. Eigen and R. Winkler, *Laws of the game*, (Harper, 1981).
General readership account of many procedures, including cellular automata, for generating biological structures.

T. Kitagawa, "Cell space approaches in biomathematics", Math. Biosciences 19 (1974) 27.
Survey of finite two-dimensional cellular automata and possible biological applications.

R. Rosen, "Pattern generation in networks", Prog. Theor. Biol. 6 (1981) 161.
Survey of various discrete space models for biological pattern formation, including cellular automata.

7.2 Models for self reproduction

M. Arbib, "Simple self-reproducing universal automata", Info. & Control 9 (1966) 177.
Construction of a self-reproducing cellular automaton slightly simpler than the von Neumann one.

A. Burks, "Von Neumann's self-reproducing automata", in A. Burks (ed.), *Essays on cellular automata*, (Univ. of Illinois Press, 1970).
Summary of von Neumann's elaborate construction of a self-reproducing two-dimensional cellular automaton.

U. Golze, "Destruction and total self-reproduction of universal computers in Codd's cellular space", Prog. Cybernetics and Systems Res. 1 (1975) 58.
Construction of a self-reproducing configuration in a two-dimensional cellular automaton.

C. Langton, "Self-reproduction in cellular automata", Physica 10D (1984) 135.
Demonstration that self-reproduction can occur in rather simple two-dimensional cellular automata, which are not capable of universal computation.

P. Vitanyi, "Sexually reproducing cellular automata", Math. Biosciences 18 (1973) 23.
Schematic construction of cellular automata with sexual self reproduction.

7.3 Specific models for biological structures and processes

A. Burks, "Cellular automata and natural systems", in *Cybernetics and Bionics*, (Oldenbourg, Munich, 1974).
Two-dimensional cellular automaton models for heart fibrillation.

W. Duchting and T. Vogelsaenger, "Aspects of modelling and simulating tumor growth and treatment", J. Cancer Res. Clin. Oncology 105 (1983) 1.
Models for tumor growth based on three-dimensional probabilistic cellular automata.

B. Ermentrout, J. Campbell and G. Oster, "A model for shell patterns based on neural activity", UC Berkeley preprint (September, 1985).
Detailed study of models based on cellular automata and generalizations for mollusc shell pigmentation pattern.

L. V. Reshodko and Z. Drska, "Biological systems of cellular organization and their computer models", J. Theor. Biol. 69 (1977) 563.
    Two-dimensional cellular automaton models for muscle tissue excitation.

S. Smith, R. Watt and R. Hameroff, "Cellular automata in cytoskeletal lattices", Physica 10D (1984) 168.
    Two-dimensional cellular automaton model for complex dynamic processes in arrays of microtubules on cell membranes.

N. Swindale, "A model for the formation of ocular dominance stripes", Proc. Roy. Soc. B208 (1980) 243.
    Discussion of models for visual cortex stripes based on two-dimensional cellular automata.

D. Young, "A local activator-inhibitor model of verterbrate skin patterns", Math. Biosciences 72 (1984) 51 [**3.4**].
    Two-dimensional cellular automaton model for striped and other biological pigmentation patterns.

8. Practical computation with cellular automata

8.1 Image processing

M. Meriaux, "A cellular architecture for image synthesis", Microprocess. and Microprogram. 13 (1984) 179.
    Discussion of cellular automaton methods for basic image processing, and their implementation in VLSI.

K. Preston et al., "Basics of cellular logic with some applications in medical image processing", Proc. IEEE 67 (1979) 826.
    Survey of image processing with cellular automata, mostly for biomedical pattern recognition.

K. Preston and M. Duff, *Modern cellular automata*, (Plenum, 1984).
    Survey of various applications of cellular automata to image processing, and of machines built to implement them.
A. Rosenfeld, "Parallel image processing using cellular arrays", Computer 16 (1983) 14.
    Surveys image processing using various generalizations of cellular automata.

S. Sternberg, "Language and architecture for parallel image processing", in Proc. Conf. in Pattern Recognition in Practice (Amsterdam, May, 1980).
    Survey of image processing by cellular automata, and its relation to computational geometry.

8.2 Algorithms based on cellular automata

(see also sect. 4.1)

A. J. Atrubin, "A one-dimensional real-time iterative multiplier", IEEE Trans. Comput. EC-14 (1965) 394.
    Construction of a one-dimensional cellular automaton for multiplying integers.

R. Balzer, "An 8-state minimal time solution to the firing squad synchronization problem", Info. & Control 10 (1967) 22.
    Discussion of a simple one-dimensional cellular automaton in which all sites attain a particular value a fixed number of steps after input at boundary sites.

P. C. Fischer, "Generation of primes by a one-dimensional real-time iterative array", J. ACM 12 (1965) 388.
    Construction of a one-dimensional cellular automata capable of generating primes.

H. Nishio, "Real time sorting of binary numbers by one-dimensional cellular automata", Proc. Int. Symp. on Uniformly Structured Automata and Logic, (Tokyo, 1975).
Construction of a cellular automaton capable of efficient sorting.

B. Silverman, "Beyond Life", Logo Computer Systems, Montreal (1986).
Discussion of digital circuit simulation by various two-dimensional cellular automata.

H. Umeo and K. Sugata, "Parallel data routing with the firing squad synchronization", Trans. Inf. Process. Soc. Japan 24 (1983) 1.
An algorithm based on cellular automata for synchronization in parallel processing computer systems.

R. Vollmar, *Algorithmen in Zellularautomaten*, (Teubner, 1979).
Survey of cellular automaton algorithms for various kinds of computational problems.

A. Waksman, "An optimum solution to the firing squad synchronization problem", Info. & Control 9 (1966) 66.
Construction of a cellular automaton capable of a form of spatial synchronization.

S. Wolfram, "Cryptography with cellular automata", in Proc. CRYPTO 85 (Santa Barbara, August, 1985).
Short summary of an efficient stream cipher system based on randomness generation by the cellular automaton discussed in [**2.3**].

S. Wolfram, "Cellular automaton supercomputing", in Proc. Scientific Applications and Algorithm Design for High Speed Computing Workshop, (Urbana, Illinois, April, 1986).
Short summary of some principles of cellular automaton algorithms for scientific simulation problems.

8.3 Computing methodologies based on cellular automata

F. C. Hennie, *Iterative arrays of logical circuits*, (MIT press, 1961).
Discussion of arrays of electronic circuits which can be used as cellular automata.

T. Hogg and B. Huberman, "Parallel computing structures capable of flexible associations and recognition of fuzzy inputs", J. Stat. Phys. (1986).
Discussion of an approach to pattern recognition with cellular automata.

F. Hossfeld, "Parallel processes and parallel algorithms", in Proc. Int. Symp. on Synergetics (Schloss Elmau, Germany, May, 1985).
Survey of cellular automata and other approaches to parallel scientific computation.

K. Kaneko, "Complexity in basin structures and information processing by the transition among attractors", in *Dynamical systems and nonlinear oscillations*, (World Scientific, 1986) [**3.11**].
Discussion of fundamental issues related to using attractors in finite cellular automata as bases for computation.

F. Manning, "An approach to highly-integrated, computer-maintained, cellular automata", IEEE Trans. Comput. C-26 (1977) 536.
Approaches to self-repair and related problems in cellular automaton computers.
R. Minnick, "A survey of microcellular research", J. ACM 14 (1967) 203.
Survey of logic circuit construction with two-dimensional cellular automaton arrays.

S. Wolfram, "Approaches to complexity engineering", Physica D (1986) [**3.12**].
Qualitative discussion of applications of multiple-scale cellular automata and other systems relevant to pattern recognition and related problems.

8.4 Cellular automaton simulation machines

N. Margolus and T. Toffoli, "The CAM-7 multiprocessor: a cellular automata machine", MIT Technical Memo. LCS-TM-289 (1985).
>   Discussion of a proposed large-scale pipelined architecture capable of simulating three-dimensional cellular automata.

K. Steiglitz and R. Morita, "A multi-processor cellular automaton chip", in Proc. 1985 IEEE Int. Conf. on Acoustics, Speech and Signal Processing (Tampa, Florida, March 1985).
>   Report on a VLSI chip built for simulating simple one-dimensional cellular automata.

S. Sternberg, "Cytocomputer biomedical image analysis", in Proc. Seventh New England Bioengineering Conf. (Troy, NY, 1979).
>   Survey of biomedical image processing using an image processing computer based on cellular automata.

T. Toffoli, "CAM: A high-performance cellular automaton machine", Physica 10D (1984) 195.
>   Discussion and examples from a pipelined two-dimensional cellular automaton simulator controlled by a personal computer.

K. Porter, "CAM-6 technical manual", distributed by Systems Concepts (San Francisco, 1986).
>   Hardware manual for a two-dimensional cellular automaton simulator for the IBM PC.

8.5 Computer architectures for cellular automata

F. Carter, "The molecular device computer: point of departure for large scale cellular automata", Physica 10D (1984) 175.
>   Discussion of molecular components which could potentially be used as elements in a cellular automaton computer.

W. D. Hillis, "The Connection Machine: a computer architecture based on cellular automata", Physica 10D (1984) 213.
>   Brief discussion of the Connection Machine massively-parallel computer, and constructs for high-level programming on it.

W. D. Hillis, *The Connection Machine*, (MIT press, 1985).
>   Survey of the Connection Machine, its design, programming, and relation to cellular automata and other parallel processing approaches.

S. Omohundro, "Modelling cellular automata with partial differential equations", Physica 10D (1984) 128.
>   Demonstration that cellular automata can be emulated by continuum systems.

8.6 Related studies

There are several related classes of computational systems. Some examples are:

>   Systolic arrays: two-dimensional arrays of locally-connected processors, typically each capable of comparatively sophisticated computations. (e.g. C. Mead and L. Conway, *An introduction to VLSI systems*, (Addison-Wesley, 1980).)

>   Feedback shift registers: related to finite one-dimensional cellular automata; linear feedback shift registers can be close to finite additive cellular automata. (e.g. S. Golomb, *Shift register sequences*, (Holden-Day, 1967).)

9. Further applications of cellular automata

A. Appel and A. J. Stein, "Cellular automata for mixing colors", IBM Tech. Disc. Bull 24 (1981) 2032.
> A cellular automaton algorithm for generating mixed colours on discrete display devices.

A. Appel, C. J. Evangelisti and A. J. Stein, "Animating quantitative maps with cellular automata", IBM Tech. Disc. Bull. 26 (1983) 953.
> A method based on two-dimensional cellular automata for generating cartograms with areas proportional to particular quantities.

W. R. Tobler, "Cellular geography", in *Philosophy in geography* (S. Gale and G. Olsson, eds.), (D. Reidel, 1979).
> Discussion of simple two-dimensional cellular automaton models for geographical organization.

G. Vichniac, "Taking the computer seriously in teaching science (an introduction to cellular automata)", in *Microscience*, Proc. UNESCO Workshop on Microcomputers in Science Education, (G. Marx and P. Szucs, eds.), (Balaton, Hungary, May 1985).
> Discussion of cellular automata as a medium for teaching the concepts and methods of physical science.

10. Some systems related to cellular automata

10.1 Lattice dynamical systems

Lattices of sites whose values are continuous variables, updated according to deterministic rules which depend on their neighbours.

K. Kaneko, "Spatiotemporal intermittency in coupled map lattices", Prog. Theor. Phys. 74 (1985) 1033.
> Study of patterns generated by one-dimensional lattice dynamical systems, and comparison with similar cellular automata.

K. Kaneko, "Lyapunov analysis and information flow in coupled map lattices", Physica D (1986).
> Definitions and computations of Lyapunov exponents for lattices dynamical systems.

R. Kapral, "Pattern formation in two-dimensional arrays of coupled, discrete-time oscillators", Phys. Rev. A31 (1985) 3868.
> Study of patterns generated by two-dimensional lattice dynamical systems.

Y. Kuramoto, "Cooperative dynamics of oscillator community", Prog. Theor. Phys. Suppl. 79 (1984) 223.
> Investigation of one-dimensional lattice dynamical systems using statistical methods.

G.-L. Oppo and R. Kapral, "Discrete models for the formation and evolution of spatial structure in dissipative systems", Chemical Physics Theory preprint, University of Toronto (1986).
> Discussion of domain and other pattern formation in two-dimensional lattice dynamical systems, and relations with cellular automata.

10.2 Random Boolean networks

Systems consisting of logical elements connected in an arbitrary network, updated, usually asynchronously, according to simple logical rules that can differ from one site to another.

J. Hopfield, "Neural networks and physical systems with emergent collective computational abilities", Proc. Natl. Acad. Sci. 79 (1982) 2554.
> Discussion of random Boolean networks used as models for neural networks, and computational devices based on them.

S. Kauffman, ''Emergent properties in random complex automata'', Physica 10D (1984) 145.
Survey of attractor and other properties of random Boolean networks, and their applications to biological systems.

10.3 L systems

Systems which perform transformations on strings of symbols, in which the symbols are modified in parallel according to definite rules which can increase the total number of symbols. Most results are obtained when the rewriting does not depend on neighbouring symbols; but interactions have been considered.

A. R. Smith, ''Plants, fractals and formal languages'', Computer Graphics (July, 1984) (Proc. SIG-GRAPH 84).
Clear discussion of the use of L systems in generating computer graphics models for fractal phenomena including branching patterns in plant growth.

G. Rozenberg and A. Salomaa, ''The mathematical theory of L systems'', (Academic Press, 1980).
Survey of mathematical results on L systems, partcularly those related to formal language theory.

Author index

Aizawa, Y. : 6.6
Akutsu, Y. : 5.1
Aladyev, V. : 4.3, 4.4
Allouche, J.-P. : 6.3
Amoroso, S. : 2.1, 2.3
ApSimon, H. G. : 2.2
Appel, A. : 9
Appel, K. I. : 2.1
Arbib, M. : 7.2
Atrubin, A. J. : 8.2
Baer, R. : 7.1
Balzer, R. : 8.2
Banks, E. R. : 4.1
Bennett, C. : 5.1
Berlekamp, E. R. : 3.4
Buckingham, D. : 3.4
Burks, A. : 1.1, 7.2, 7.3
Butler, J. : 2.2, 2.3
Buvel, R. : 5.1
Carter, F. : 8.5
Choi, M. Y. : 5.1, 6.6
Codd, E. F. : 4.1
Cole, S. : 4.2
Cordovil, R. : 2.2
Coven, E. : 2.3
Creutz, M. : 6.4
Crisp, G. M. : 6.2
Deshouillers, J.-M. : 2.2
Dewdney, K. : 1.2
Dilao, R. : 2.1
Dobrushin, R. L. : 5.1
Domany, E. : 6.4, 5.1
Dresden, M. : 5.1
Drska, Z. : 7.3
Duchting, W. : 7.3
Duff, M. : 8.1
Durham, T. : 1.2
Eigen, M. : 7.1
Epstein, I. : 2.3
Ermentrout, B. : 7.3
Fischer, P. C. : 8.2
Freedman, W. : 6.3
Frisch, U. : 6.5
Gacs, P. : 5.1, 5.2
Gajski, D. : 4.3
Gal'perin, G. A. : 2.1
Gardner, M. : 3.4
Gerola, H. : 6.6
Gilman, R. : 2.1
Goles, E. : 2.3
Golze, U. : 4.4, 7.2
Good, W. : 6.2
Gordon, D. : 4.1
Gosper, R. W. : 3.4
Grassberger, P. : 3.2, 5.1
Greenberg, J. : 6.3
Griffeath, D. : 5.3
Grinstein, G. : 5.1
Guan, P. : 2.2
Handler, W. : 1.1
Harao, M. : 2.1, 2.3
Hardouin-Duparc, J. : 3.4
Hardy, J. : 6.5

Hastings, S. : 6.3
Hayes, B. : 1.2
He, Y. : 2.2
Hedlund, G. A. : 2.1
Hennie, F. C. : 8.3
Hillis, W. D. : 8.5
Hogg, T. : 8.3
Honda, N. : 2.3
Hopfield, J. : 10.2
Hossfeld, F. : 8.3
Huberman, B. : 5.1, 6.6, 8.3
d'Humieres, D. : 6.5
Hurd, L. : 4.4
Ingerson, T. : 5.1
Ito, M. : 2.2
Jen, E. : 2.3
Kaneko, K. : 10.1, 8.3, 5.1
Kapral, R. : 10.1
Kashef, R. S. : 4.1
Kauffman, S. : 10.2
Keeler, J. : 6.3
Kimura, M. : 2.1, 2.3
Kinzel, W. : 5.1
Kitagawa, T. : 7.1
Kuramoto, Y. : 10.1
Kurdyumov, G. L. : 5.1
Langton, C. : 7.2
Levy, S. : 1.2
Lind, D. : 2.1
Mackay, A. : 6.2
Maddox, J. : 1.2
Madore, B. : 6.3
Manning, F. : 8.3
Margolus, N. : 1.1, 4.1, 6.5, 8.4
Martin, O. : 2.2
Martinez, H. : 7.1
Maruoka, A. : 2.1, 2.3
Maverick, V. : 6.2
Meriaux, M. : 8.1
Miller, J. C. P. : 2.2
Milnor, J. : 2.1
Minnick, R. : 8.3
Miyamoto, M. : 2.2
Miyazima, H. : 2.1
Morita, R. : 8.4
Nambu, Y. : 2.2
Nasu, M. : 2.1, 2.3
Niemiec, M. : 3.4
Nishio, H. : 1.1, 8.2
Noguchi, S. : 2.1, 2.3
Nourai, F. : 4.1
Ntafos, S. : 2.2
Olivos, J. : 2.3
Omohundro, S. : 8.5
Oono, Y. : 6.3
Oppo, G.-L. : 10.1
Orszag, S. : 6.5
Packard, N. : 3.1, 3.2, 6.2
Park, J. : 3.5, 6.6
Patt, Y. : 2.1
Pecht, J. : 4.2
Pomeau, Y. : 6.4
Porter, K. : 8.4

Preston, K. : 8.1
Pries, W. : 2.2
Priese, L. : 4.2
Reder, C. : 6.3
Reif, J. : 5.2
Reiter, C. : 1.2
Reshodko, L. V. : 7.3
Rhodes, F. : 2.3
Richardson, D. : 2.1
Rosen, R. : 7.1
Rosenfeld, A. : 8.1
Rozenberg, G. : 10.3
Rucker, R. : 1.2
Salem, J. : 6.5
Salomaa, A. : 10.3
Schrandt, R. : 3.3
Schulman, L. : 5.1
Seiden, P. : 5.1, 6.6
Silverman, B. : 8.2
Sinai, Y. G. : 2.1
Smith, A. R. : 4.1, 4.2, 4.4, 10.3
Smith, S. : 7.3
Sommerhalder, R. : 4.2
Steiglitz, K. : 8.4
Stein, A. J. : 9
Sternberg, S. : 8.1, 8.4
Sugata, K. : 8.2
Swindale, N. : 7.3
Takahashi, H. : 4.3, 4.4
Tchuente, M. : 2.3
Thatcher, J. W. : 4.1
Tobler, W. R. : 9
Toffoli, T. : 1.1, 4.1, 6.1, 8.4
Tucker, J. : 1.2
Ulam, S. : 3.3
Umeo, H. : 8.2
van Westrhenen, S. C. : 4.2
Vergis, A. : 4.1
Vichniac, G. : 3.1, 5.1, 6.1, 9
Vitanyi, P. : 4.3, 7.2
Vogelsaenger, T. : 7.3
Vollmar, R. : 4.2, 8.2
von Neumann, J. : 1.1
Wainwright, R. T. : 3.4
Waksman, A. : 4.2, 8.2
Waterman, M. : 2.1
Welch, L. R. : 2.1
White, P. : 2.3
Willson, S. : 2.1, 2.2, 3.3
Winfree, A. : 6.3
Winkler, R. : 7.1
Wolfram, S. : 1.2, 2.2, 3.1, 3.2, 3.5, 4.3, 4.4,
    6.1, 6.5, 8.2, 8.3
Wong, D. : 5.1
Yakhot, V. : 6.5
Yaku, T. : 4.3
Yamada, H. : 2.3, 4.3
Young, D. : 7.3
Zuse, K. : 6.1

Author Index

Appendix: Properties of the $k=2$, $r=1$ cellular automata

Introduction

1. Rule forms and equivalences
2. Patterns from disordered states
3. Blocked patterns from disordered states
4. Difference patterns
5. Patterns from single site seeds
6. Statistical properties
7. Blocking transformation equivalences
8. Factorizations of rules
9. Lengths of newly-excluded blocks
10. Regular language complexities
11. Measure theoretical complexities
12. Iterated rule expression sizes
13. Finite lattice state transition diagrams
14. Global properties for finite lattices
15. Structures in rule 110
16. Patterns generated by second-order rules

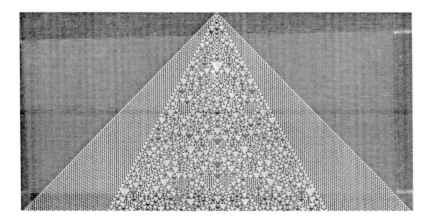

Pattern generated by $k=2$, $r=1$ cellular automaton rule number 73. This complex pattern was produced by evolution from a simple initial state consisting of a single nonzero site. The pattern appears random in many respects.

Introduction

This appendix gives tables of properties of one-dimensional cellular with two possible values at each site ($k=2$), and with rules depending on nearest neighbours ($r=1$). These cellular automata are some of the simplest that can be constructed. Yet they are already capable of a great diversity of highly complex behaviour. The tables in this appendix attempt to capture some of this behaviour, both pictorially and numerically.

There are 256 possible rules for $k=2$, $r=1$ cellular automata. Table 1 gives forms for these rules, together with simple equivalences among them.

Tables 2 and 3 show patterns produced by evolution according to all possible inequivalent rules, starting from "typical" disordered or random initial conditions. Several general classes of qualitative behaviour are seen [1.3]:

1. A fixed, homogeneous, state is eventually reached (e.g. rules 0, 8, 136).

2. A pattern consisting of separated periodic regions is produced (e.g. rules 4, 37, 56, 73).

3. A chaotic, aperiodic, pattern is produced (e.g. rules 18, 45, 146).

4. Complex, localized structures are generated (e.g. rule 110). (This behaviour is clearly visible in the pictures of table 15.)

Much of the data in this appendix can be understood in terms of this classification.

The patterns produced with a particular rule by evolution from different disordered initial states are qualitatively similar. Nevertheless, changes in initial conditions can lead to detailed changes in the configurations produced. Table 4 shows the pattern of differences produced by single-site changes in initial conditions. For class 1 rules, the changes always die out. For class 2 rules, they may persist, but remain localized. Class 3 rules, however, show "instability": small changes in initial conditions can lead to an ever-expanding region of differences. "Information" on the initial state thus propagates, typically at a fixed speed, through the cellular automaton. In class 4 cellular automata, such information transmission occurs irregularly, through motion of specific localized structures.

Table 6 gives the values of some statistical quantities which characterize some of the behaviour seen in tables 2, 3 and 4. The definitions of entropies and Lyapunov exponents for cellular automata [1.3] are closely analogous to those for conventional continuous dynamical systems.

Tables 2, 3, 4 and 6 concern the generic behaviour of cellular automata with "typical" disordered initial conditions. The generation of complexity in cellular automata is however perhaps more clearly illustrated by evolution from particular, simple, initial conditions, as in figure 5. With such initial conditions, some cellular automaton rules yield simple or regular patterns. But other rules yield highly complex patterns, which seem in many respects random.

Tables 2 through 6 suggest that many different $k=2$, $r=1$ cellular automata exhibit similar behaviour. Table 1 gives some simple equivalences between rules. Table 7 gives equivalences arising from more complex transformations. Often different regions in a cellular automaton will form "domains" which show different equivalences.

Table 8 gives further relations between rules, in the form of factorizations which express one rule as compositions of others.

An important feature of cellular automata is their capability for "self organization". Even starting from arbitrary disordered or random initial conditions, their time evolution can pick out particular "ordered" states. Tables 9 through 11 give mathematical characterizations of the sets of configurations that can occur in the evolution of $k=2$, $r=1$ cellular automata. Table 9 concerns blocks of site values which are filtered out by the cellular automaton evolution.

The complete set of configurations produced after any finite number of time steps can be described in terms of regular formal languages [2.1]. Tables 10 and 11 give the values of quantities which characterize the certain aspects of the "complexity" of these languages.

The behaviour of class 3 and 4 cellular automata often seems to be so complex that its outcome cannot be determined except by essentially performing a direct simulation. Tables 10 and 11 may provide some quantitative basis for this supposition. Table 12 gives a more direct measure of the difficulty

Introduction

of computing the outcome of cellular automaton evolution in the context of a simple computational model involving Boolean functions.

The results given for example in table are for cellular automata on lattices with an infinite number of sites. Tables 13 and 14 give some of the more complete results that can be obtained for cellular automata on finite lattices (or with spatially periodic configurations). Table 13 shows fragments of the state transition diagrams which describe the global evolution of finite cellular automata. Table 14 plots some of their overall properties.

Many of the $k=2$, $r=1$ cellular automata show highly complex behaviour. Such behaviour is probably most evident in rule 110. Table 15 gives some properties of the particle-like structures which are found in this rule. One suspects that with appropriate combinations of these structures, it should be possible to perform universal computation.

The final table shows patterns produced by reversible generalizations of the standard $k=2$, $r=1$ cellular automata. Qualitatively similar behaviour is again seen.

It is remarkable that with such simple construction, the $k=2$, $r=1$ cellular automata can show such complex behaviour. The tables in this appendix give some first attempts at characterizing and quantifying this behaviour. Much, however, still remains to be done.

Table 1: Rule forms and equivalences

rule number			boolean expression	dep	equivalent rules			min
dec	binary	hex			conj	refl	c.r.	
0	00000000	00	0	—	255	0	255	0
1	00000001	01	$(\bar{a}_{-1}\bar{a}_0\bar{a}_1)$	•••	127	1	127	1
2	00000010	02	$(\bar{a}_{-1}\bar{a}_0 a_1)$	•••	191	16	247	2
3	00000011	03	$(\bar{a}_{-1}\bar{a}_0)$	••—	63	17	119	3
4	00000100	04	$(\bar{a}_{-1}a_0\bar{a}_1)$	•••	223	4	223	4
5	00000101	05	$(\bar{a}_{-1}\bar{a}_1)$	•—•	95	5	95	5
6	00000110	06	$(\bar{a}_{-1}a_0\bar{a}_1)+(\bar{a}_{-1}\bar{a}_0 a_1)$	•••	159	20	215	6
7	00000111	07	$(\bar{a}_{-1}\bar{a}_1)+(\bar{a}_{-1}\bar{a}_0)$	•••	31	21	87	7
8	00001000	08	$(\bar{a}_{-1}a_0 a_1)$	•••	239	64	253	8
9	00001001	09	$(\bar{a}_{-1}\bar{a}_0\bar{a}_1)+(\bar{a}_{-1}a_0 a_1)$	•••	111	65	125	9
10	00001010	0a	$(\bar{a}_{-1}a_1)$	•—•	175	80	245	10
11	00001011	0b	$(\bar{a}_{-1}\bar{a}_0)+(\bar{a}_{-1}a_1)$	•••	47	81	117	11
12	00001100	0c	$(\bar{a}_{-1}a_0)$	••—	207	68	221	12
13	00001101	0d	$(\bar{a}_{-1}\bar{a}_1)+(\bar{a}_{-1}a_0)$	•••	79	69	93	13
14	00001110	0e	$(\bar{a}_{-1}a_0)+(\bar{a}_{-1}a_1)$	•••	143	84	213	14
15	00001111	0f	$(\bar{a}_{-1})$	o—	15	85	85	15
16	00010000	10	$(a_{-1}\bar{a}_0\bar{a}_1)$	•••	247	2	191	2
17	00010001	11	$(\bar{a}_0\bar{a}_1)$	—••	119	3	63	3
18	00010010	12	$(a_{-1}\bar{a}_0\bar{a}_1)+(\bar{a}_{-1}\bar{a}_0 a_1)$	•••	183	18	183	18
19	00010011	13	$(\bar{a}_0\bar{a}_1)+(\bar{a}_{-1}\bar{a}_0)$	•••	55	19	55	19
20	00010100	14	$(a_{-1}\bar{a}_0\bar{a}_1)+(\bar{a}_{-1}a_0\bar{a}_1)$	•••	215	6	159	6
21	00010101	15	$(\bar{a}_0\bar{a}_1)+(\bar{a}_{-1}\bar{a}_1)$	•••	87	7	31	7
22	00010110	16	$(a_{-1}\bar{a}_0\bar{a}_1)+(\bar{a}_{-1}a_0\bar{a}_1)+(\bar{a}_{-1}\bar{a}_0 a_1)$	•••	151	22	151	22
23	00010111	17	$(\bar{a}_0\bar{a}_1)+(\bar{a}_{-1}\bar{a}_1)+(\bar{a}_{-1}\bar{a}_0)$	•••	23	23	23	23
24	00011000	18	$(a_{-1}\bar{a}_0\bar{a}_1)+(\bar{a}_{-1}a_0 a_1)$	•••	231	66	189	24
25	00011001	19	$(\bar{a}_{-1}a_0 a_1)+(\bar{a}_0\bar{a}_1)$	•••	103	67	61	25
26	00011010	1a	$(a_{-1}\bar{a}_0\bar{a}_1)+(\bar{a}_{-1}a_1)$	•••	167	82	181	26
27	00011011	1b	$(\bar{a}_0\bar{a}_1)+(\bar{a}_{-1}a_1)$	•••	39	83	53	27
28	00011100	1c	$(a_{-1}\bar{a}_0\bar{a}_1)+(\bar{a}_{-1}a_0)$	•••	199	70	157	28
29	00011101	1d	$(\bar{a}_0\bar{a}_1)+(\bar{a}_{-1}a_0)$	•••	71	71	29	29
30	00011110	1e	$(a_{-1}\bar{a}_0\bar{a}_1)+(\bar{a}_{-1}a_0)+(\bar{a}_{-1}a_1)$	o••	135	86	149	30
31	00011111	1f	$(\bar{a}_0\bar{a}_1)+(\bar{a}_{-1})$	•••	7	87	21	7
32	00100000	20	$(a_{-1}\bar{a}_0 a_1)$	•••	251	32	251	32
33	00100001	21	$(\bar{a}_{-1}\bar{a}_0\bar{a}_1)+(a_{-1}\bar{a}_0 a_1)$	•••	123	33	123	33
34	00100010	22	$(\bar{a}_0 a_1)$	—••	187	48	243	34
35	00100011	23	$(\bar{a}_{-1}\bar{a}_0)+(\bar{a}_0 a_1)$	•••	59	49	115	35
36	00100100	24	$(\bar{a}_{-1}a_0\bar{a}_1)+(a_{-1}\bar{a}_0 a_1)$	•••	219	36	219	36
37	00100101	25	$(a_{-1}\bar{a}_0 a_1)+(\bar{a}_{-1}\bar{a}_1)$	•••	91	37	91	37
38	00100110	26	$(\bar{a}_{-1}a_0\bar{a}_1)+(\bar{a}_0 a_1)$	•••	155	52	211	38
39	00100111	27	$(\bar{a}_{-1}\bar{a}_1)+(\bar{a}_0 a_1)$	•••	27	53	83	27
40	00101000	28	$(a_{-1}\bar{a}_0 a_1)+(\bar{a}_{-1}a_0 a_1)$	•••	235	96	249	40
41	00101001	29	$(\bar{a}_{-1}\bar{a}_0\bar{a}_1)+(a_{-1}\bar{a}_0 a_1)+(\bar{a}_{-1}a_0 a_1)$	•••	107	97	121	41
42	00101010	2a	$(\bar{a}_0 a_1)+(\bar{a}_{-1}a_1)$	•••	171	112	241	42
43	00101011	2b	$(\bar{a}_{-1}\bar{a}_0)+(\bar{a}_0 a_1)+(\bar{a}_{-1}a_1)$	•••	43	113	113	43
44	00101100	2c	$(a_{-1}\bar{a}_0 a_1)+(\bar{a}_{-1}a_0)$	•••	203	100	217	44
45	00101101	2d	$(a_{-1}\bar{a}_0 a_1)+(\bar{a}_{-1}\bar{a}_1)+(\bar{a}_{-1}a_0)$	o••	75	101	89	45
46	00101110	2e	$(\bar{a}_{-1}a_0)+(\bar{a}_0 a_1)$	•••	139	116	209	46
47	00101111	2f	$(\bar{a}_0 a_1)+(\bar{a}_{-1})$	•••	11	117	81	11
48	00110000	30	$(a_{-1}\bar{a}_0)$	••—	243	34	187	34
49	00110001	31	$(\bar{a}_0\bar{a}_1)+(a_{-1}\bar{a}_0)$	•••	115	35	59	35

Table 1: Rule forms and equivalences

rule number			boolean expression	dep	equivalent rules			min
dec	binary	hex			conj	refl	c.r.	
50	00110010	32	$(a_{-1}\bar{a}_0)+(\bar{a}_0 a_1)$	•••	179	50	179	50
51	00110011	33	$(\bar{a}_0)$	–○–	51	51	51	51
52	00110100	34	$(\bar{a}_{-1}a_0\bar{a}_1)+(a_{-1}\bar{a}_0)$	•••	211	38	155	38
53	00110101	35	$(\bar{a}_{-1}\bar{a}_1)+(a_{-1}\bar{a}_0)$	•••	83	39	27	27
54	00110110	36	$(\bar{a}_{-1}a_0 a_1)+(a_{-1}\bar{a}_0)+(\bar{a}_0 a_1)$	•○•	147	54	147	54
55	00110111	37	$(\bar{a}_{-1}\bar{a}_1)+(\bar{a}_0)$	•••	19	55	19	19
56	00111000	38	$(\bar{a}_{-1}a_0 a_1)+(a_{-1}\bar{a}_0)$	•••	227	98	185	56
57	00111001	39	$(\bar{a}_{-1}a_0 a_1)+(\bar{a}_0\bar{a}_1)+(a_{-1}\bar{a}_0)$	•○•	99	99	57	57
58	00111010	3a	$(a_{-1}\bar{a}_0)+(\bar{a}_{-1}a_1)$	•••	163	114	177	58
59	00111011	3b	$(\bar{a}_{-1}a_1)+(\bar{a}_0)$	•••	35	115	49	35
60	00111100	3c	$(a_{-1}\bar{a}_0)+(\bar{a}_{-1}a_0)$	○○–	195	102	153	60
61	00111101	3d	$(\bar{a}_{-1}\bar{a}_1)+(a_{-1}\bar{a}_0)+(\bar{a}_{-1}a_0)$	•••	67	103	25	25
62	00111110	3e	$(\bar{a}_{-1}a_1)+(a_{-1}\bar{a}_0)+(\bar{a}_{-1}a_0)$	•••	131	118	145	62
63	00111111	3f	$(\bar{a}_0)+(\bar{a}_{-1})$	••–	3	119	17	3
64	01000000	40	$(a_{-1}a_0\bar{a}_1)$	•••	253	8	239	8
65	01000001	41	$(\bar{a}_{-1}\bar{a}_0\bar{a}_1)+(a_{-1}a_0\bar{a}_1)$	•••	125	9	111	9
66	01000010	42	$(a_{-1}a_0\bar{a}_1)+(\bar{a}_{-1}\bar{a}_0 a_1)$	•••	189	24	231	24
67	01000011	43	$(a_{-1}a_0\bar{a}_1)+(\bar{a}_{-1}\bar{a}_0)$	•••	61	25	103	25
68	01000100	44	$(a_0\bar{a}_1)$	–••	221	12	207	12
69	01000101	45	$(\bar{a}_{-1}\bar{a}_1)+(a_0\bar{a}_1)$	•••	93	13	79	13
70	01000110	46	$(\bar{a}_{-1}\bar{a}_0 a_1)+(a_0\bar{a}_1)$	•••	157	28	199	28
71	01000111	47	$(a_0\bar{a}_1)+(\bar{a}_{-1}\bar{a}_0)$	•••	29	29	71	29
72	01001000	48	$(a_{-1}a_0\bar{a}_1)+(\bar{a}_{-1}a_0 a_1)$	•••	237	72	237	72
73	01001001	49	$(\bar{a}_{-1}\bar{a}_0\bar{a}_1)+(a_{-1}a_0\bar{a}_1)+(\bar{a}_{-1}a_0 a_1)$	•••	109	73	109	73
74	01001010	4a	$(a_{-1}a_0\bar{a}_1)+(\bar{a}_{-1}a_1)$	•••	173	88	229	74
75	01001011	4b	$(a_{-1}a_0\bar{a}_1)+(\bar{a}_{-1}\bar{a}_0)+(\bar{a}_{-1}a_1)$	○••	45	89	101	45
76	01001100	4c	$(a_0\bar{a}_1)+(\bar{a}_{-1}a_0)$	•••	205	76	205	76
77	01001101	4d	$(\bar{a}_{-1}\bar{a}_1)+(a_0\bar{a}_1)+(\bar{a}_{-1}a_0)$	•••	77	77	77	77
78	01001110	4e	$(a_0\bar{a}_1)+(\bar{a}_{-1}a_1)$	•••	141	92	197	78
79	01001111	4f	$(a_0\bar{a}_1)+(\bar{a}_{-1})$	•••	13	93	69	13
80	01010000	50	$(a_{-1}\bar{a}_1)$	•–•	245	10	175	10
81	01010001	51	$(\bar{a}_0\bar{a}_1)+(a_{-1}\bar{a}_1)$	•••	117	11	47	11
82	01010010	52	$(\bar{a}_{-1}\bar{a}_0\bar{a}_1)+(a_{-1}\bar{a}_1)$	•••	181	26	167	26
83	01010011	53	$(a_{-1}\bar{a}_1)+(\bar{a}_{-1}\bar{a}_0)$	•••	53	27	39	27
84	01010100	54	$(a_{-1}\bar{a}_1)+(a_0\bar{a}_1)$	•••	213	14	143	14
85	01010101	55	$(\bar{a}_1)$	–○	85	15	15	15
86	01010110	56	$(\bar{a}_{-1}\bar{a}_0 a_1)+(a_{-1}\bar{a}_1)+(a_0\bar{a}_1)$	••○	149	30	135	30
87	01010111	57	$(\bar{a}_{-1}\bar{a}_0)+(\bar{a}_1)$	•••	21	31	7	7
88	01011000	58	$(\bar{a}_{-1}a_0 a_1)+(a_{-1}\bar{a}_1)$	•••	229	74	173	74
89	01011001	59	$(\bar{a}_{-1}a_0 a_1)+(\bar{a}_0\bar{a}_1)+(a_{-1}\bar{a}_1)$	••○	101	75	45	45
90	01011010	5a	$(a_{-1}\bar{a}_1)+(\bar{a}_{-1}a_1)$	○–○	165	90	165	90
91	01011011	5b	$(\bar{a}_{-1}\bar{a}_0)+(a_{-1}\bar{a}_1)+(\bar{a}_{-1}a_1)$	•••	37	91	37	37
92	01011100	5c	$(a_{-1}\bar{a}_1)+(\bar{a}_{-1}a_0)$	•••	197	78	141	78
93	01011101	5d	$(\bar{a}_{-1}a_0)+(\bar{a}_1)$	•••	69	79	13	13
94	01011110	5e	$(\bar{a}_{-1}a_0)+(a_{-1}\bar{a}_1)+(\bar{a}_{-1}a_1)$	•••	133	94	133	94
95	01011111	5f	$(\bar{a}_1)+(\bar{a}_{-1})$	•–•	5	95	5	5
96	01100000	60	$(a_{-1}a_0\bar{a}_1)+(a_{-1}\bar{a}_0 a_1)$	•••	249	40	235	40
97	01100001	61	$(\bar{a}_{-1}\bar{a}_0\bar{a}_1)+(a_{-1}a_0\bar{a}_1)+(a_{-1}\bar{a}_0 a_1)$	•••	121	41	107	41
98	01100010	62	$(a_{-1}a_0\bar{a}_1)+(\bar{a}_0 a_1)$	•••	185	56	227	56
99	01100011	63	$(a_{-1}a_0\bar{a}_1)+(\bar{a}_{-1}\bar{a}_0)+(\bar{a}_0 a_1)$	•○•	57	57	99	57

Table 1: Rule forms and equivalences

rule number			boolean expression	dep	equivalent rules			min
dec	binary	hex			conj	refl	c.r.	
100	01100100	64	$(a_{-1}\bar{a}_0 a_1)+(a_0\bar{a}_1)$	•••	217	44	203	44
101	01100101	65	$(a_{-1}\bar{a}_0 a_1)+(\bar{a}_{-1}\bar{a}_1)+(a_0\bar{a}_1)$	••○	89	45	75	45
102	01100110	66	$(a_0\bar{a}_1)+(\bar{a}_0 a_1)$	-○○	153	60	195	60
103	01100111	67	$(\bar{a}_{-1}\bar{a}_0)+(a_0\bar{a}_1)+(\bar{a}_0 a_1)$	•••	25	61	67	25
104	01101000	68	$(a_{-1}\bar{a}_0\bar{a}_1)+(a_{-1}\bar{a}_0 a_1)+(\bar{a}_{-1}a_0 a_1)$	•••	233	104	233	104
105	01101001	69	$(\bar{a}_{-1}\bar{a}_0\bar{a}_1)+(a_{-1}a_0\bar{a}_1)+(a_{-1}\bar{a}_0 a_1)+(\bar{a}_{-1}a_0 a_1)$	○○○	105	105	105	105
106	01101010	6a	$(a_{-1}a_0\bar{a}_1)+(\bar{a}_0 a_1)+(\bar{a}_{-1}a_1)$	••○	169	120	225	106
107	01101011	6b	$(a_{-1}\bar{a}_0\bar{a}_1)+(\bar{a}_{-1}\bar{a}_0)+(\bar{a}_0 a_1)+(\bar{a}_{-1}a_1)$	•••	41	121	97	41
108	01101100	6c	$(a_{-1}\bar{a}_0 a_1)+(a_0\bar{a}_1)+(\bar{a}_{-1}a_0)$	•○•	201	108	201	108
109	01101101	6d	$(a_{-1}\bar{a}_0 a_1)+(\bar{a}_{-1}\bar{a}_1)+(a_0\bar{a}_1)+(\bar{a}_{-1}a_0)$	•••	73	109	73	73
110	01101110	6e	$(\bar{a}_{-1}a_0)+(a_0\bar{a}_1)+(\bar{a}_0 a_1)$	•••	137	124	193	110
111	01101111	6f	$(a_0\bar{a}_1)+(\bar{a}_0 a_1)+(\bar{a}_{-1})$	•••	9	125	65	9
112	01110000	70	$(a_{-1}\bar{a}_1)+(a_{-1}\bar{a}_0)$	•••	241	42	171	42
113	01110001	71	$(\bar{a}_0 a_1)+(a_{-1}\bar{a}_1)+(a_{-1}\bar{a}_0)$	•••	113	43	43	43
114	01110010	72	$(a_{-1}\bar{a}_1)+(\bar{a}_0 a_1)$	•••	177	58	163	58
115	01110011	73	$(a_{-1}\bar{a}_1)+(\bar{a}_0)$	•••	49	59	35	35
116	01110100	74	$(a_0\bar{a}_1)+(a_{-1}\bar{a}_0)$	•••	209	46	139	46
117	01110101	75	$(a_{-1}\bar{a}_0)+(\bar{a}_1)$	•••	81	47	11	11
118	01110110	76	$(a_{-1}\bar{a}_0)+(a_0\bar{a}_1)+(\bar{a}_0 a_1)$	•••	145	62	131	62
119	01110111	77	$(\bar{a}_1)+(\bar{a}_0)$	-••	17	63	3	3
120	01111000	78	$(\bar{a}_{-1}a_0 a_1)+(a_{-1}\bar{a}_1)+(a_{-1}\bar{a}_0)$	○••	225	106	169	106
121	01111001	79	$(\bar{a}_{-1}a_0 a_1)+(\bar{a}_0\bar{a}_1)+(a_{-1}\bar{a}_1)+(a_{-1}\bar{a}_0)$	•••	97	107	41	41
122	01111010	7a	$(a_{-1}\bar{a}_0)+(a_{-1}\bar{a}_1)+(\bar{a}_{-1}a_1)$	•••	161	122	161	122
123	01111011	7b	$(a_{-1}\bar{a}_1)+(\bar{a}_{-1}a_1)+(\bar{a}_0)$	•••	33	123	33	33
124	01111100	7c	$(a_{-1}\bar{a}_1)+(a_{-1}\bar{a}_0)+(\bar{a}_{-1}a_0)$	•••	193	110	137	110
125	01111101	7d	$(a_{-1}\bar{a}_0)+(\bar{a}_{-1}a_0)+(\bar{a}_1)$	•••	65	111	9	9
126	01111110	7e	$(a_{-1}\bar{a}_1)+(\bar{a}_0 a_1)+(\bar{a}_{-1}a_0)$	•••	129	126	129	126
127	01111111	7f	$(\bar{a}_1)+(\bar{a}_0)+(\bar{a}_{-1})$	•••	1	127	1	1
128	10000000	80	$(a_{-1}a_0 a_1)$	•••	254	128	254	128
129	10000001	81	$(\bar{a}_{-1}\bar{a}_0\bar{a}_1)+(a_{-1}a_0 a_1)$	•••	126	129	126	126
130	10000010	82	$(\bar{a}_{-1}\bar{a}_0 a_1)+(a_{-1}a_0 a_1)$	•••	190	144	246	130
131	10000011	83	$(a_{-1}a_0 a_1)+(\bar{a}_{-1}\bar{a}_0)$	•••	62	145	118	62
132	10000100	84	$(\bar{a}_{-1}a_0\bar{a}_1)+(a_{-1}a_0 a_1)$	•••	222	132	222	132
133	10000101	85	$(a_{-1}a_0 a_1)+(\bar{a}_{-1}\bar{a}_1)$	•••	94	133	94	94
134	10000110	86	$(\bar{a}_{-1}a_0\bar{a}_1)+(\bar{a}_{-1}\bar{a}_0 a_1)+(a_{-1}a_0 a_1)$	•••	158	148	214	134
135	10000111	87	$(a_{-1}a_0 a_1)+(\bar{a}_{-1}\bar{a}_1)+(\bar{a}_{-1}\bar{a}_0)$	○••	30	149	86	30
136	10001000	88	$(a_0 a_1)$	-••	238	192	252	136
137	10001001	89	$(\bar{a}_{-1}\bar{a}_0\bar{a}_1)+(a_0 a_1)$	•••	110	193	124	110
138	10001010	8a	$(\bar{a}_{-1}a_1)+(a_0 a_1)$	•••	174	208	244	138
139	10001011	8b	$(\bar{a}_{-1}\bar{a}_0)+(a_0 a_1)$	•••	46	209	116	46
140	10001100	8c	$(\bar{a}_{-1}a_0)+(a_0 a_1)$	•••	206	196	220	140
141	10001101	8d	$(\bar{a}_{-1}\bar{a}_1)+(a_0 a_1)$	•••	78	197	92	78
142	10001110	8e	$(\bar{a}_{-1}a_0)+(\bar{a}_{-1}a_1)+(a_0 a_1)$	•••	142	212	212	142
143	10001111	8f	$(a_0 a_1)+(\bar{a}_{-1})$	•••	14	213	84	14
144	10010000	90	$(a_{-1}\bar{a}_0\bar{a}_1)+(a_{-1}a_0 a_1)$	•••	246	130	190	130
145	10010001	91	$(a_{-1}a_0 a_1)+(\bar{a}_0\bar{a}_1)$	•••	118	131	62	62
146	10010010	92	$(a_{-1}\bar{a}_0\bar{a}_1)+(\bar{a}_{-1}\bar{a}_0 a_1)+(a_{-1}a_0 a_1)$	•••	182	146	182	146
147	10010011	93	$(a_{-1}a_0 a_1)+(\bar{a}_0\bar{a}_1)+(\bar{a}_{-1}\bar{a}_0)$	•○•	54	147	54	54
148	10010100	94	$(a_{-1}\bar{a}_0\bar{a}_1)+(\bar{a}_{-1}a_0\bar{a}_1)+(a_{-1}a_0 a_1)$	•••	214	134	158	134
149	10010101	95	$(a_{-1}a_0 a_1)+(\bar{a}_0\bar{a}_1)+(\bar{a}_{-1}\bar{a}_1)$	••○	86	135	30	30

Table 1: Rule forms and equivalences

dec	binary	hex	boolean expression	dep	conj	refl	c.r.	min
150	10010110	96	$(a_{-1}\bar{a}_0\bar{a}_1)+(\bar{a}_{-1}a_0\bar{a}_1)+(\bar{a}_{-1}\bar{a}_0a_1)+(a_{-1}a_0a_1)$	○○○	150	150	150	150
151	10010111	97	$(a_{-1}a_0a_1)+(\bar{a}_0\bar{a}_1)+(\bar{a}_{-1}\bar{a}_1)+(\bar{a}_{-1}\bar{a}_0)$	●●●	22	151	22	22
152	10011000	98	$(a_{-1}\bar{a}_0\bar{a}_1)+(a_0a_1)$	●●●	230	194	188	152
153	10011001	99	$(\bar{a}_0\bar{a}_1)+(a_0a_1)$	—○○	102	195	60	60
154	10011010	9a	$(a_{-1}\bar{a}_0\bar{a}_1)+(\bar{a}_{-1}a_1)+(a_0a_1)$	●●○	166	210	180	154
155	10011011	9b	$(\bar{a}_{-1}\bar{a}_0)+(\bar{a}_0\bar{a}_1)+(a_0a_1)$	●●●	38	211	52	38
156	10011100	9c	$(a_{-1}\bar{a}_0\bar{a}_1)+(\bar{a}_{-1}a_0)+(a_0a_1)$	●○●	198	198	156	156
157	10011101	9d	$(\bar{a}_{-1}a_0)+(\bar{a}_0\bar{a}_1)+(a_0a_1)$	●●●	70	199	28	28
158	10011110	9e	$(a_{-1}\bar{a}_0\bar{a}_1)+(\bar{a}_{-1}a_0)+(\bar{a}_{-1}a_1)+(a_0a_1)$	●●●	134	214	148	134
159	10011111	9f	$(\bar{a}_0\bar{a}_1)+(a_0a_1)+(\bar{a}_{-1})$	●●●	6	215	20	6
160	10100000	a0	$(a_{-1}a_1)$	●—●	250	160	250	160
161	10100001	a1	$(\bar{a}_{-1}\bar{a}_0\bar{a}_1)+(a_{-1}a_1)$	●●●	122	161	122	122
162	10100010	a2	$(\bar{a}_0a_1)+(a_{-1}a_1)$	●●●	186	176	242	162
163	10100011	a3	$(\bar{a}_{-1}\bar{a}_0)+(a_{-1}a_1)$	●●●	58	177	114	58
164	10100100	a4	$(\bar{a}_{-1}a_0\bar{a}_1)+(a_{-1}a_1)$	●●●	218	164	218	164
165	10100101	a5	$(\bar{a}_{-1}\bar{a}_1)+(a_{-1}a_1)$	○—○	90	165	90	90
166	10100110	a6	$(\bar{a}_{-1}a_0\bar{a}_1)+(\bar{a}_0a_1)+(a_{-1}a_1)$	●●○	154	180	210	154
167	10100111	a7	$(\bar{a}_{-1}\bar{a}_0)+(\bar{a}_{-1}\bar{a}_1)+(a_{-1}a_1)$	●●●	26	181	82	26
168	10101000	a8	$(a_{-1}a_1)+(a_0a_1)$	●●●	234	224	248	168
169	10101001	a9	$(\bar{a}_{-1}\bar{a}_0\bar{a}_1)+(a_{-1}a_1)+(a_0a_1)$	●●○	106	225	120	106
170	10101010	aa	(a_1)	—○	170	240	240	170
171	10101011	ab	$(\bar{a}_{-1}\bar{a}_0)+(a_1)$	●●●	42	241	112	42
172	10101100	ac	$(\bar{a}_{-1}a_0)+(a_{-1}a_1)$	●●●	202	228	216	172
173	10101101	ad	$(\bar{a}_{-1}a_0)+(\bar{a}_{-1}\bar{a}_1)+(a_{-1}a_1)$	●●●	74	229	88	74
174	10101110	ae	$(\bar{a}_{-1}a_0)+(a_1)$	●●●	138	244	208	138
175	10101111	af	$(\bar{a}_{-1})+(a_1)$	●—●	10	245	80	10
176	10110000	b0	$(a_{-1}\bar{a}_0)+(a_{-1}a_1)$	●●●	242	162	186	162
177	10110001	b1	$(\bar{a}_0\bar{a}_1)+(a_{-1}a_1)$	●●●	114	163	58	58
178	10110010	b2	$(a_{-1}\bar{a}_0)+(\bar{a}_0a_1)+(a_{-1}a_1)$	●●●	178	178	178	178
179	10110011	b3	$(a_{-1}a_1)+(\bar{a}_0)$	●●●	50	179	50	50
180	10110100	b4	$(\bar{a}_{-1}a_0\bar{a}_1)+(a_{-1}\bar{a}_0)+(a_{-1}a_1)$	○●●	210	166	154	154
181	10110101	b5	$(a_{-1}\bar{a}_0)+(\bar{a}_{-1}\bar{a}_1)+(a_{-1}a_1)$	●●●	82	167	26	26
182	10110110	b6	$(\bar{a}_{-1}a_0\bar{a}_1)+(a_{-1}\bar{a}_0)+(\bar{a}_0a_1)+(a_{-1}a_1)$	●●●	146	182	146	146
183	10110111	b7	$(\bar{a}_{-1}\bar{a}_1)+(a_{-1}a_1)+(\bar{a}_0)$	●●●	18	183	18	18
184	10111000	b8	$(a_{-1}\bar{a}_0)+(a_0a_1)$	●●●	226	226	184	184
185	10111001	b9	$(a_{-1}\bar{a}_0)+(\bar{a}_0a_1)+(a_0a_1)$	●●●	98	227	56	56
186	10111010	ba	$(a_{-1}\bar{a}_0)+(a_1)$	●●●	162	242	176	162
187	10111011	bb	$(\bar{a}_0)+(a_1)$	—●●	34	243	48	34
188	10111100	bc	$(a_{-1}a_1)+(a_{-1}\bar{a}_0)+(\bar{a}_{-1}a_0)$	●●●	194	230	152	152
189	10111101	bd	$(\bar{a}_0\bar{a}_1)+(a_{-1}a_1)+(\bar{a}_{-1}a_0)$	●●●	66	231	24	24
190	10111110	be	$(a_{-1}\bar{a}_0)+(\bar{a}_{-1}a_0)+(a_1)$	●●●	130	246	144	130
191	10111111	bf	$(\bar{a}_0)+(\bar{a}_{-1})+(a_1)$	●●●	2	247	16	2
192	11000000	c0	$(a_{-1}a_0)$	●●—	252	136	238	136
193	11000001	c1	$(\bar{a}_{-1}\bar{a}_0\bar{a}_1)+(a_{-1}a_0)$	●●●	124	137	110	110
194	11000010	c2	$(\bar{a}_{-1}\bar{a}_0a_1)+(a_{-1}a_0)$	●●●	188	152	230	152
195	11000011	c3	$(\bar{a}_{-1}\bar{a}_0)+(a_{-1}a_0)$	○○—	60	153	102	60
196	11000100	c4	$(a_0\bar{a}_1)+(a_{-1}a_0)$	●●●	220	140	206	140
197	11000101	c5	$(\bar{a}_{-1}\bar{a}_1)+(a_{-1}a_0)$	●●●	92	141	78	78
198	11000110	c6	$(\bar{a}_{-1}\bar{a}_0a_1)+(a_0\bar{a}_1)+(a_{-1}a_0)$	●○●	156	156	198	156
199	11000111	c7	$(\bar{a}_{-1}\bar{a}_1)+(\bar{a}_{-1}\bar{a}_0)+(a_{-1}a_0)$	●●●	28	157	70	28

Table 1: Rule forms and equivalences

rule number			boolean expression	dep	equivalent rules			min
dec	binary	hex			conj	refl	c.r.	
200	11001000	c8	$(a_{-1}a_0)+(a_0a_1)$	•••	236	200	236	200
201	11001001	c9	$(\bar{a}_{-1}\bar{a}_0\bar{a}_1)+(a_{-1}a_0)+(a_0a_1)$	•○•	108	201	108	108
202	11001010	ca	$(a_{-1}a_0)+(\bar{a}_{-1}a_1)$	•••	172	216	228	172
203	11001011	cb	$(\bar{a}_{-1}a_1)+(\bar{a}_{-1}\bar{a}_0)+(a_{-1}a_0)$	•••	44	217	100	44
204	11001100	cc	(a_0)	-○-	204	204	204	204
205	11001101	cd	$(\bar{a}_{-1}\bar{a}_1)+(a_0)$	•••	76	205	76	76
206	11001110	ce	$(\bar{a}_{-1}a_1)+(a_0)$	•••	140	220	196	140
207	11001111	cf	$(\bar{a}_{-1})+(a_0)$	••-	12	221	68	12
208	11010000	d0	$(a_{-1}\bar{a}_1)+(a_{-1}a_0)$	•••	244	138	174	138
209	11010001	d1	$(\bar{a}_0\bar{a}_1)+(a_{-1}a_0)$	•••	116	139	46	46
210	11010010	d2	$(\bar{a}_{-1}\bar{a}_0a_1)+(a_{-1}\bar{a}_1)+(a_{-1}a_0)$	○••	180	154	166	154
211	11010011	d3	$(a_{-1}\bar{a}_1)+(\bar{a}_{-1}\bar{a}_0)+(a_{-1}a_0)$	•••	52	155	38	38
212	11010100	d4	$(a_{-1}\bar{a}_1)+(a_0\bar{a}_1)+(a_{-1}a_0)$	•••	212	142	142	142
213	11010101	d5	$(a_{-1}a_0)+(\bar{a}_1)$	•••	84	143	14	14
214	11010110	d6	$(\bar{a}_{-1}\bar{a}_0a_1)+(a_{-1}\bar{a}_1)+(a_0\bar{a}_1)+(a_{-1}a_0)$	•••	148	158	134	134
215	11010111	d7	$(\bar{a}_{-1}\bar{a}_0)+(a_{-1}a_0)+(\bar{a}_1)$	•••	20	159	6	6
216	11011000	d8	$(a_{-1}\bar{a}_1)+(a_0a_1)$	•••	228	202	172	172
217	11011001	d9	$(a_{-1}a_0)+(\bar{a}_0\bar{a}_1)+(a_0a_1)$	•••	100	203	44	44
218	11011010	da	$(a_{-1}a_0)+(a_{-1}\bar{a}_1)+(\bar{a}_{-1}a_1)$	•••	164	218	164	164
219	11011011	db	$(\bar{a}_0\bar{a}_1)+(\bar{a}_{-1}a_1)+(a_{-1}a_0)$	•••	36	219	36	36
220	11011100	dc	$(a_{-1}\bar{a}_1)+(a_0)$	•••	196	206	140	140
221	11011101	dd	$(\bar{a}_1)+(a_0)$	-••	68	207	12	12
222	11011110	de	$(a_{-1}\bar{a}_1)+(\bar{a}_{-1}a_1)+(a_0)$	•••	132	222	132	132
223	11011111	df	$(\bar{a}_1)+(\bar{a}_{-1})+(a_0)$	•••	4	223	4	4
224	11100000	e0	$(a_{-1}a_0)+(a_{-1}a_1)$	•••	248	168	234	168
225	11100001	e1	$(\bar{a}_{-1}\bar{a}_0\bar{a}_1)+(a_{-1}a_0)+(a_{-1}a_1)$	○••	120	169	106	106
226	11100010	e2	$(a_{-1}a_0)+(\bar{a}_0a_1)$	•••	184	184	226	184
227	11100011	e3	$(a_{-1}a_1)+(\bar{a}_{-1}\bar{a}_0)+(a_{-1}a_0)$	•••	56	185	98	56
228	11100100	e4	$(a_0\bar{a}_1)+(a_{-1}a_1)$	•••	216	172	202	172
229	11100101	e5	$(a_{-1}a_0)+(\bar{a}_{-1}\bar{a}_1)+(a_{-1}a_1)$	•••	88	173	74	74
230	11100110	e6	$(a_{-1}a_0)+(a_0\bar{a}_1)+(\bar{a}_0a_1)$	•••	152	188	194	152
231	11100111	e7	$(\bar{a}_{-1}\bar{a}_1)+(\bar{a}_0a_1)+(a_{-1}a_0)$	•••	24	189	66	24
232	11101000	e8	$(a_{-1}a_0)+(a_{-1}a_1)+(a_0a_1)$	•••	232	232	232	232
233	11101001	e9	$(\bar{a}_{-1}\bar{a}_0\bar{a}_1)+(a_{-1}a_0)+(a_{-1}a_1)+(a_0a_1)$	•••	104	233	104	104
234	11101010	ea	$(a_{-1}a_0)+(a_1)$	•••	168	248	224	168
235	11101011	eb	$(\bar{a}_{-1}\bar{a}_0)+(a_{-1}a_0)+(a_1)$	•••	40	249	96	40
236	11101100	ec	$(a_{-1}a_1)+(a_0)$	•••	200	236	200	200
237	11101101	ed	$(\bar{a}_{-1}\bar{a}_1)+(a_{-1}a_1)+(a_0)$	•••	72	237	72	72
238	11101110	ee	$(a_0)+(a_1)$	-••	136	252	192	136
239	11101111	ef	$(\bar{a}_{-1})+(a_0)+(a_1)$	•••	8	253	64	8
240	11110000	f0	(a_{-1})	○-	240	170	170	170
241	11110001	f1	$(\bar{a}_0\bar{a}_1)+(a_{-1})$	•••	112	171	42	42
242	11110010	f2	$(\bar{a}_0a_1)+(a_{-1})$	•••	176	186	162	162
243	11110011	f3	$(\bar{a}_0)+(a_{-1})$	••-	48	187	34	34
244	11110100	f4	$(a_0\bar{a}_1)+(a_{-1})$	•••	208	174	138	138
245	11110101	f5	$(\bar{a}_1)+(a_{-1})$	•-•	80	175	10	10
246	11110110	f6	$(a_0\bar{a}_1)+(\bar{a}_0a_1)+(a_{-1})$	•••	144	190	130	130
247	11110111	f7	$(\bar{a}_1)+(\bar{a}_0)+(a_{-1})$	•••	16	191	2	2
248	11111000	f8	$(a_0a_1)+(a_{-1})$	•••	224	234	168	168
249	11111001	f9	$(\bar{a}_0\bar{a}_1)+(a_0a_1)+(a_{-1})$	•••	96	235	40	40

Table 1: Rule forms and equivalences

rule number			boolean expression	dep	equivalent rules			min
dec	binary	hex			conj	refl	c.r.	
250	11111010	fa	$(a_{-1})+(a_1)$	•—•	160	250	160	160
251	11111011	fb	$(\overline{a}_0)+(a_{-1})+(a_1)$	•••	32	251	32	32
252	11111100	fc	$(a_{-1})+(a_0)$	••—	192	238	136	136
253	11111101	fd	$(\overline{a}_1)+(a_{-1})+(a_0)$	•••	64	239	8	8
254	11111110	fe	$(a_{-1})+(a_0)+(a_1)$	•••	128	254	128	128
255	11111111	ff	1	—	0	255	0	0

Table 1: Forms of rules and equivalences between rules.

The table lists all 256 possible rules for $k=2$, $r=1$ one–dimensional cellular automata. Such cellular automata consist of a line of sites, each with value 0 or 1. At each time step, the value a_i of a site at position i is updated according to the rule

$$a_i' = \phi(a_{i-1},a_i,a_{i+1}) .$$

This table lists the $2^{2^3}=256$ possible choices of ϕ.

Each digit in the binary representation of the rule number gives the value of ϕ for a particular set of (a_{i-1},a_i,a_{i+1}). The digit corresponding to the coefficient of 2^n in the rule number gives the value of $\phi(n_2,n_1,n_0)$, where $n = 4n_2 + 2n_1 + n_0$. Thus the leftmost digit in the binary representation of the rule number gives $\phi(1,1,1)$, the next gives $\phi(1,1,0)$, and so on, down to $\phi(0,0,0)$.

The table also gives the decimal and hexadecimal representations of the rules numbers.

Each ϕ can be considered a Boolean function of three variables, say a_{-1}, a_0 and a_1. The table gives the minimal disjunctive normal form representations for these Boolean functions. Boolean multiplication and addition are used (corresponding to AND and OR operations). Bar denotes complementation. In each case, the expression with the minimal number of components, using only these operations, is given.

The column labelled "dep" gives the dependence of $\phi(a_{-1},a_0,a_1)$ on each of the a_{-1}, a_0 and a_1. The symbol – indicates no change in ϕ when the corresponding a_j is changed. The symbol ○ denotes linear dependence of ϕ on the corresponding a_j: whenever a_j changes, ϕ also changes. The symbol • denotes arbitrary dependence of ϕ. Rules such as 90 in which only ○ and – dependence occurs, are called additive, and can be represented as linear functions modulo two.

For each rule, the table gives rules equivalent under simple transformations. "conj" denotes conjugation: interchange of the roles of 0 and 1. "refl" denotes reflection. Rules invariant under reflection are symmetric. "c.r." denotes the combined operation of conjugation and reflection.

Many of the properties considered in this Appendix are unaffected by these transformations. The rules form equivalence classes under these transformations, and it is usually convenient to consider only the minimal (lowest–numbered) representatives of each class, as given by the last column in the table.

In some cases, further equivalences between rules can be used. Table 7 gives one important set of such further equivalences.

Some special rules are:

51	complement
170	left shift
204	identity
240	right shift

Table by **Lyman P. Hurd** (*Mathematics Department, Princeton University*). (Boolean expressions by S. Wolfram.)

Table 2: Patterns from disordered states

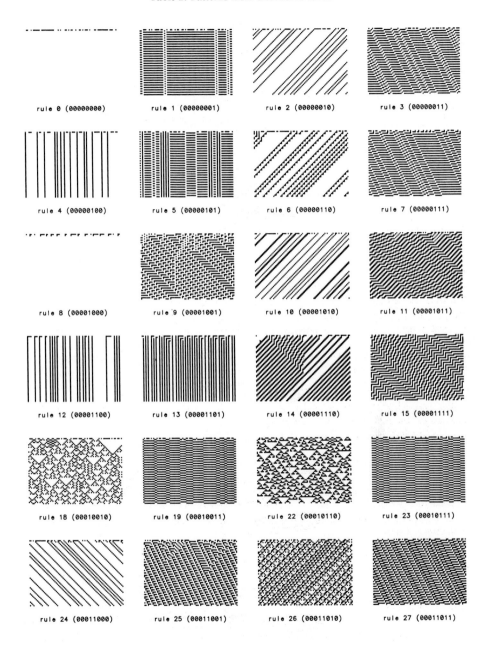

rule 0 (00000000)   rule 1 (00000001)   rule 2 (00000010)   rule 3 (00000011)

rule 4 (00000100)   rule 5 (00000101)   rule 6 (00000110)   rule 7 (00000111)

rule 8 (00001000)   rule 9 (00001001)   rule 10 (00001010)   rule 11 (00001011)

rule 12 (00001100)   rule 13 (00001101)   rule 14 (00001110)   rule 15 (00001111)

rule 18 (00010010)   rule 19 (00010011)   rule 22 (00010110)   rule 23 (00010111)

rule 24 (00011000)   rule 25 (00011001)   rule 26 (00011010)   rule 27 (00011011)

Table 2: Patterns from disordered states

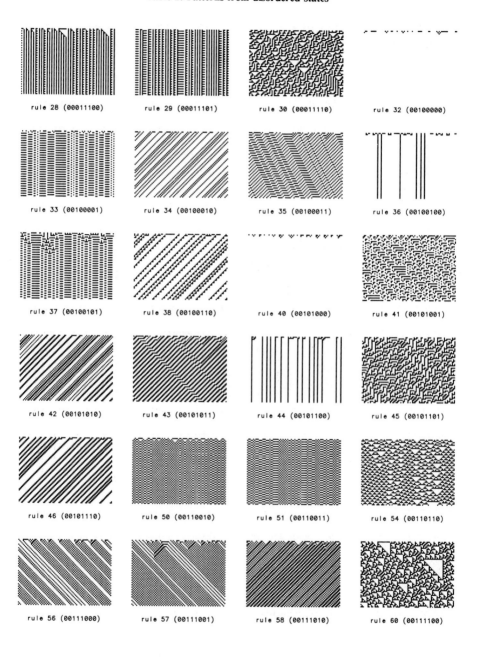

rule 28 (00011100)   rule 29 (00011101)   rule 30 (00011110)   rule 32 (00100000)

rule 33 (00100001)   rule 34 (00100010)   rule 35 (00100011)   rule 36 (00100100)

rule 37 (00100101)   rule 38 (00100110)   rule 40 (00101000)   rule 41 (00101001)

rule 42 (00101010)   rule 43 (00101011)   rule 44 (00101100)   rule 45 (00101101)

rule 46 (00101110)   rule 50 (00110010)   rule 51 (00110011)   rule 54 (00110110)

rule 56 (00111000)   rule 57 (00111001)   rule 58 (00111010)   rule 60 (00111100)

Table 2: Patterns from disordered states

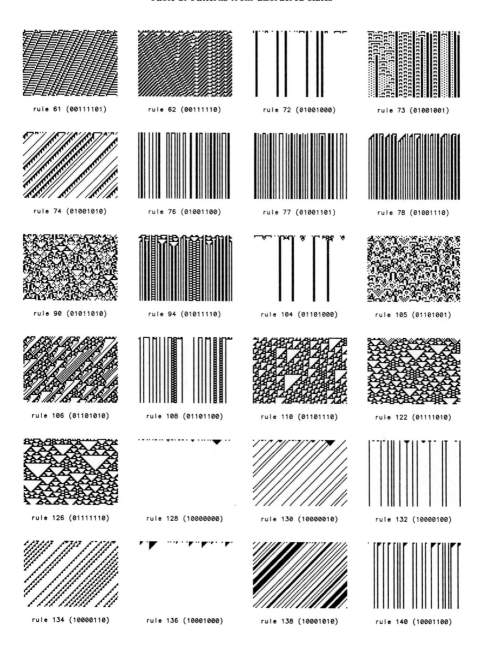

rule 61 (00111101)   rule 62 (00111110)   rule 72 (01001000)   rule 73 (01001001)

rule 74 (01001010)   rule 76 (01001100)   rule 77 (01001101)   rule 78 (01001110)

rule 90 (01011010)   rule 94 (01011110)   rule 104 (01101000)   rule 105 (01101001)

rule 106 (01101010)   rule 108 (01101100)   rule 110 (01101110)   rule 122 (01111010)

rule 126 (01111110)   rule 128 (10000000)   rule 130 (10000010)   rule 132 (10000100)

rule 134 (10000110)   rule 136 (10001000)   rule 138 (10001010)   rule 140 (10001100)

Table 2: Patterns from disordered states

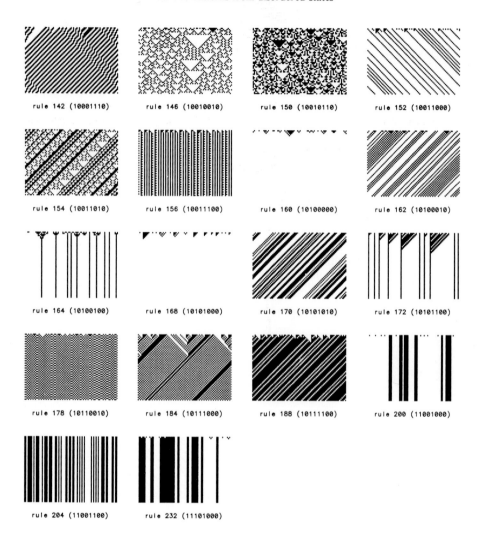

rule 142 (10001110)   rule 146 (10010010)   rule 150 (10010110)   rule 152 (10011000)

rule 154 (10011010)   rule 156 (10011100)   rule 160 (10100000)   rule 162 (10100010)

rule 164 (10100100)   rule 168 (10101000)   rule 170 (10101010)   rule 172 (10101100)

rule 178 (10110010)   rule 184 (10111000)   rule 188 (10111100)   rule 200 (11001000)

rule 204 (11001100)   rule 232 (11101000)

Table 2: Patterns from disordered states

Table 2: Patterns generated by evolution from disordered initial states.

Each picture is for a different rule. All the "minimal representative" rules of table 1 are included. (Other rules have patterns equivalent to those of their minimal representatives.)

Sites with values 1 and 0 are represented respectively by black and white squares. The initial configuration is at the top of each picture. The values of sites in it are chosen randomly to be 0 or 1 with probability 1/2. Successive lines are obtained by applications of the cellular automaton rule.

These pictures show the evolution of cellular automata with 80 sites for 60 time steps. Periodic boundary conditions were imposed on the edges.

Different specific initial configurations for a particular rule almost always yield qualitatively similar patterns. Different rules are however seen to give a wide variety of different kinds of patterns.

Table 3: Blocked patterns from disordered states

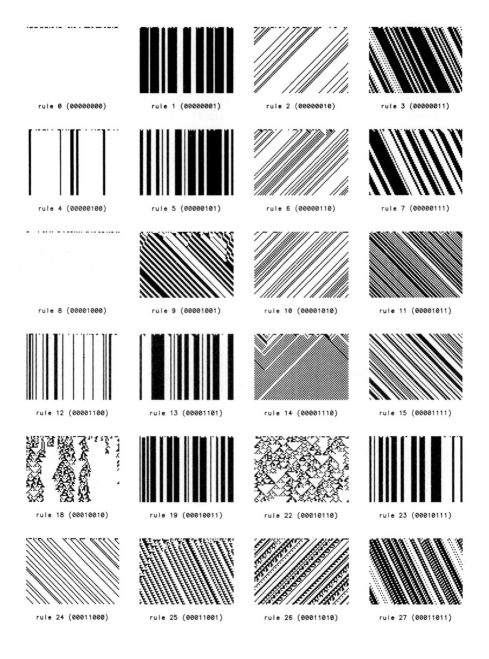

rule 0 (00000000) rule 1 (00000001) rule 2 (00000010) rule 3 (00000011)

rule 4 (00000100) rule 5 (00000101) rule 6 (00000110) rule 7 (00000111)

rule 8 (00001000) rule 9 (00001001) rule 10 (00001010) rule 11 (00001011)

rule 12 (00001100) rule 13 (00001101) rule 14 (00001110) rule 15 (00001111)

rule 18 (00010010) rule 19 (00010011) rule 22 (00010110) rule 23 (00010111)

rule 24 (00011000) rule 25 (00011001) rule 26 (00011010) rule 27 (00011011)

Table 3: Blocked patterns from disordered states

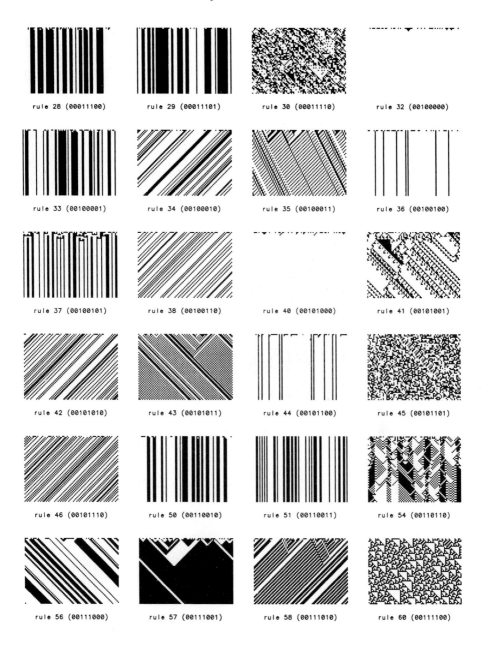

rule 28 (00011100)   rule 29 (00011101)   rule 30 (00011110)   rule 32 (00100000)

rule 33 (00100001)   rule 34 (00100010)   rule 35 (00100011)   rule 36 (00100100)

rule 37 (00100101)   rule 38 (00100110)   rule 40 (00101000)   rule 41 (00101001)

rule 42 (00101010)   rule 43 (00101011)   rule 44 (00101100)   rule 45 (00101101)

rule 46 (00101110)   rule 50 (00110010)   rule 51 (00110011)   rule 54 (00110110)

rule 56 (00111000)   rule 57 (00111001)   rule 58 (00111010)   rule 60 (00111100)

Table 3: Blocked patterns from disordered states

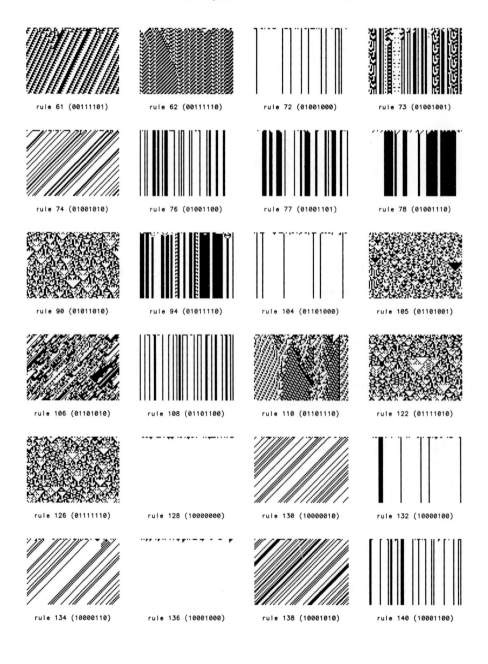

rule 61 (00111101) rule 62 (00111110) rule 72 (01001000) rule 73 (01001001)

rule 74 (01001010) rule 76 (01001100) rule 77 (01001101) rule 78 (01001110)

rule 90 (01011010) rule 94 (01011110) rule 104 (01101000) rule 105 (01101001)

rule 106 (01101010) rule 108 (01101100) rule 110 (01101110) rule 122 (01111010)

rule 126 (01111110) rule 128 (10000000) rule 130 (10000010) rule 132 (10000100)

rule 134 (10000110) rule 136 (10001000) rule 138 (10001010) rule 140 (10001100)

Table 3: Blocked patterns from disordered states

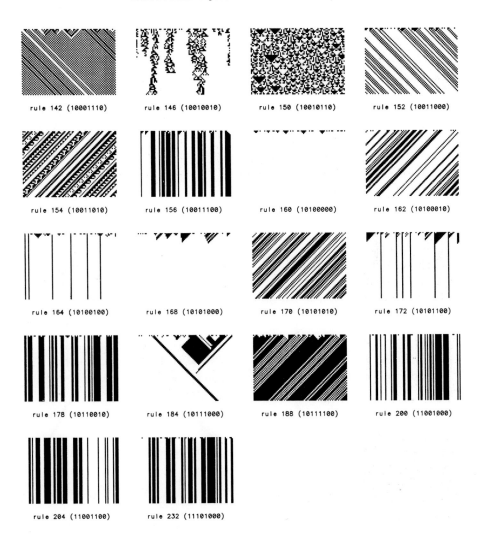

rule 142 (10001110)   rule 146 (10010010)   rule 150 (10010110)   rule 152 (10011000)

rule 154 (10011010)   rule 156 (10011100)   rule 160 (10100000)   rule 162 (10100010)

rule 164 (10100100)   rule 168 (10101000)   rule 170 (10101010)   rule 172 (10101100)

rule 178 (10110010)   rule 184 (10111000)   rule 188 (10111100)   rule 200 (11001000)

rule 204 (11001100)   rule 232 (11101000)

Table 3: Blocks in patterns generated by evolution from disordered initial states.

The pictures in this table are analogous to those in table 2, but show only every other site in both space and time. Certain features become clearer in this "blocked" representation.

It is common for cellular automata to exhibit several "phases". The blocked representation often makes differences between these phases visible.

Table 4: Difference patterns

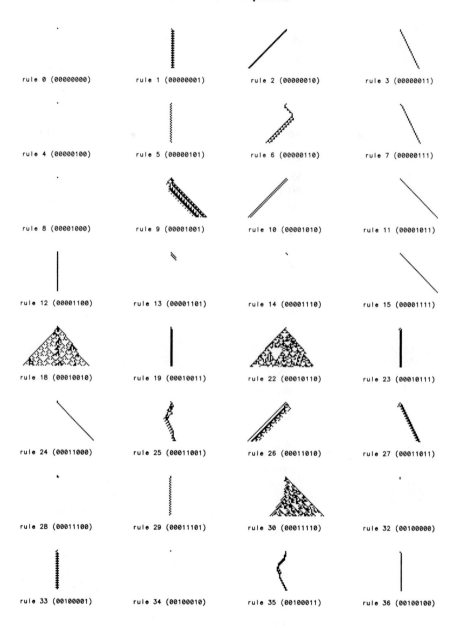

rule 0 (00000000)   rule 1 (00000001)   rule 2 (00000010)   rule 3 (00000011)

rule 4 (00000100)   rule 5 (00000101)   rule 6 (00000110)   rule 7 (00000111)

rule 8 (00001000)   rule 9 (00001001)   rule 10 (00001010)   rule 11 (00001011)

rule 12 (00001100)   rule 13 (00001101)   rule 14 (00001110)   rule 15 (00001111)

rule 18 (00010010)   rule 19 (00010011)   rule 22 (00010110)   rule 23 (00010111)

rule 24 (00011000)   rule 25 (00011001)   rule 26 (00011010)   rule 27 (00011011)

rule 28 (00011100)   rule 29 (00011101)   rule 30 (00011110)   rule 32 (00100000)

rule 33 (00100001)   rule 34 (00100010)   rule 35 (00100011)   rule 36 (00100100)

Table 4: Difference patterns

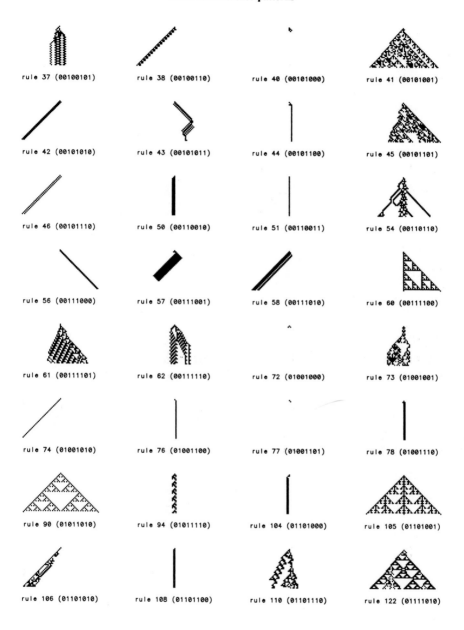

rule 37 (00100101) rule 38 (00100110) rule 40 (00101000) rule 41 (00101001)

rule 42 (00101010) rule 43 (00101011) rule 44 (00101100) rule 45 (00101101)

rule 46 (00101110) rule 50 (00110010) rule 51 (00110011) rule 54 (00110110)

rule 56 (00111000) rule 57 (00111001) rule 58 (00111010) rule 60 (00111100)

rule 61 (00111101) rule 62 (00111110) rule 72 (01001000) rule 73 (01001001)

rule 74 (01001010) rule 76 (01001100) rule 77 (01001101) rule 78 (01001110)

rule 90 (01011010) rule 94 (01011110) rule 104 (01101000) rule 105 (01101001)

rule 106 (01101010) rule 108 (01101100) rule 110 (01101110) rule 122 (01111010)

Table 4: Difference patterns

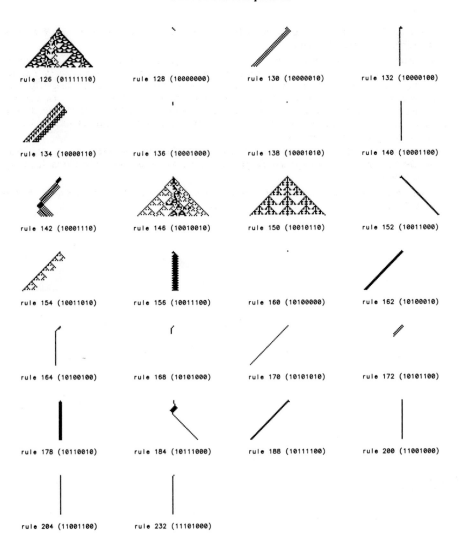

rule 126 (01111110)　　rule 128 (10000000)　　rule 130 (10000010)　　rule 132 (10000100)

rule 134 (10000110)　　rule 136 (10001000)　　rule 138 (10001010)　　rule 140 (10001100)

rule 142 (10001110)　　rule 146 (10010010)　　rule 150 (10010110)　　rule 152 (10011000)

rule 154 (10011010)　　rule 156 (10011100)　　rule 160 (10100000)　　rule 162 (10100010)

rule 164 (10100100)　　rule 168 (10101000)　　rule 170 (10101010)　　rule 172 (10101100)

rule 178 (10110010)　　rule 184 (10111000)　　rule 188 (10111100)　　rule 200 (11001000)

rule 204 (11001100)　　rule 232 (11101000)

Table 4: Difference patterns

Table 4: Differences in patterns produced by evolution from disordered states resulting from changes in single initial site values.

The evolution of small perturbations made in the initial configurations for all the "minimal representative" rules of table 1 are given. In each case, an initial configuration was chosen in which sites had value 0 or 1 with probability 1/2, and the pattern obtained by evolution according to the cellular automaton rule was found. Then the value of the centre site in the initial configuration was complemented, and the resulting pattern obtained by cellular automaton evolution was found. The pictures show as black squares the site values that differed between the patterns found with these initial configurations. Evolution for 40 time steps is shown.

In some cases, the differences die out, or remain localized, with time. In other cases, the differences grow. The left and right growth speeds correspond to the left and right Lyapunov exponents λ_L and λ_R, given in table 6.

For some rules (such as 18), initial perturbations on some configurations may grow, but on others may die out. The pictures show results from a particular trial.

Table 5: Patterns from single site seeds

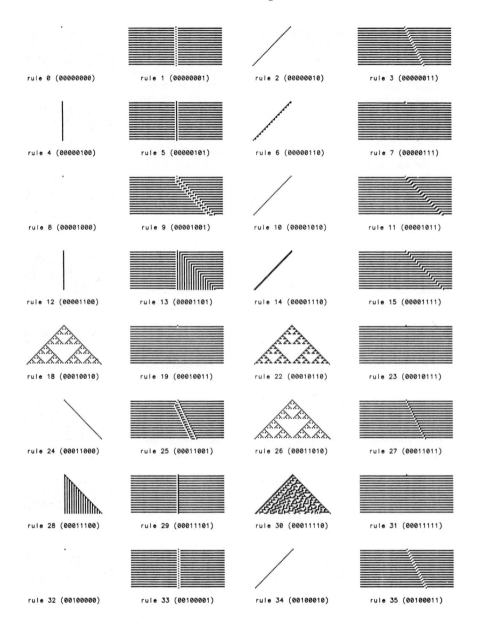

rule 0 (00000000)   rule 1 (00000001)   rule 2 (00000010)   rule 3 (00000011)

rule 4 (00000100)   rule 5 (00000101)   rule 6 (00000110)   rule 7 (00000111)

rule 8 (00001000)   rule 9 (00001001)   rule 10 (00001010)   rule 11 (00001011)

rule 12 (00001100)   rule 13 (00001101)   rule 14 (00001110)   rule 15 (00001111)

rule 18 (00010010)   rule 19 (00010011)   rule 22 (00010110)   rule 23 (00010111)

rule 24 (00011000)   rule 25 (00011001)   rule 26 (00011010)   rule 27 (00011011)

rule 28 (00011100)   rule 29 (00011101)   rule 30 (00011110)   rule 31 (00011111)

rule 32 (00100000)   rule 33 (00100001)   rule 34 (00100010)   rule 35 (00100011)

Table 5: Patterns from single site seeds

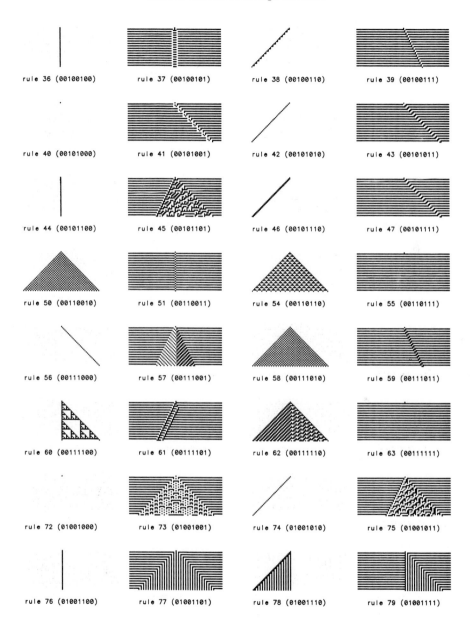

rule 36 (00100100)   rule 37 (00100101)   rule 38 (00100110)   rule 39 (00100111)

rule 40 (00101000)   rule 41 (00101001)   rule 42 (00101010)   rule 43 (00101011)

rule 44 (00101100)   rule 45 (00101101)   rule 46 (00101110)   rule 47 (00101111)

rule 50 (00110010)   rule 51 (00110011)   rule 54 (00110110)   rule 55 (00110111)

rule 56 (00111000)   rule 57 (00111001)   rule 58 (00111010)   rule 59 (00111011)

rule 60 (00111100)   rule 61 (00111101)   rule 62 (00111110)   rule 63 (00111111)

rule 72 (01001000)   rule 73 (01001001)   rule 74 (01001010)   rule 75 (01001011)

rule 76 (01001100)   rule 77 (01001101)   rule 78 (01001110)   rule 79 (01001111)

Table 5: Patterns from single site seeds

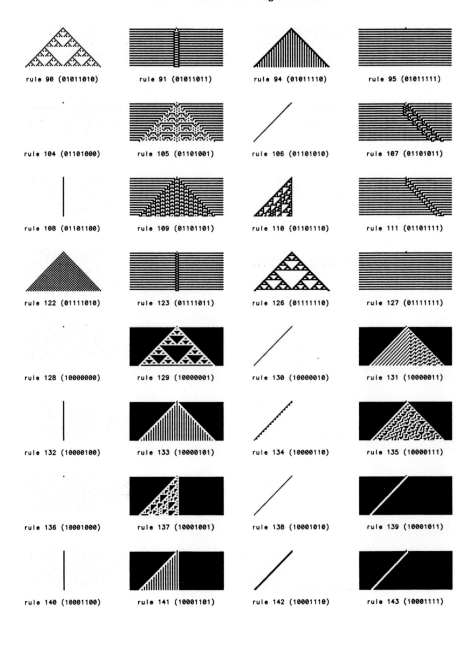

rule 90 (01011010)   rule 91 (01011011)   rule 94 (01011110)   rule 95 (01011111)

rule 104 (01101000)   rule 105 (01101001)   rule 106 (01101010)   rule 107 (01101011)

rule 108 (01101100)   rule 109 (01101101)   rule 110 (01101110)   rule 111 (01101111)

rule 122 (01111010)   rule 123 (01111011)   rule 126 (01111110)   rule 127 (01111111)

rule 128 (10000000)   rule 129 (10000001)   rule 130 (10000010)   rule 131 (10000011)

rule 132 (10000100)   rule 133 (10000101)   rule 134 (10000110)   rule 135 (10000111)

rule 136 (10001000)   rule 137 (10001001)   rule 138 (10001010)   rule 139 (10001011)

rule 140 (10001100)   rule 141 (10001101)   rule 142 (10001110)   rule 143 (10001111)

Table 5: Patterns from single site seeds

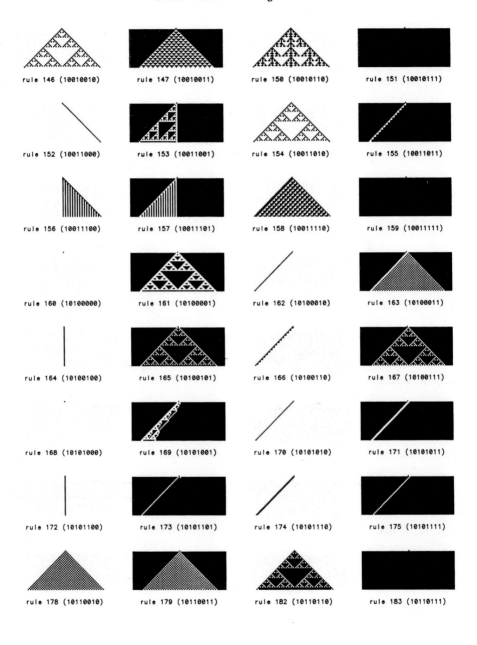

rule 146 (10010010)   rule 147 (10010011)   rule 150 (10010110)   rule 151 (10010111)

rule 152 (10011000)   rule 153 (10011001)   rule 154 (10011010)   rule 155 (10011011)

rule 156 (10011100)   rule 157 (10011101)   rule 158 (10011110)   rule 159 (10011111)

rule 160 (10100000)   rule 161 (10100001)   rule 162 (10100010)   rule 163 (10100011)

rule 164 (10100100)   rule 165 (10100101)   rule 166 (10100110)   rule 167 (10100111)

rule 168 (10101000)   rule 169 (10101001)   rule 170 (10101010)   rule 171 (10101011)

rule 172 (10101100)   rule 173 (10101101)   rule 174 (10101110)   rule 175 (10101111)

rule 178 (10110010)   rule 179 (10110011)   rule 182 (10110110)   rule 183 (10110111)

Table 5: Patterns from single site seeds

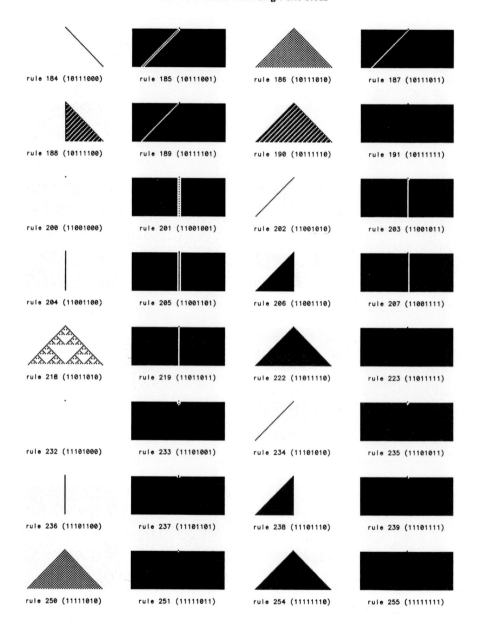

Table 5: Patterns from single site seeds

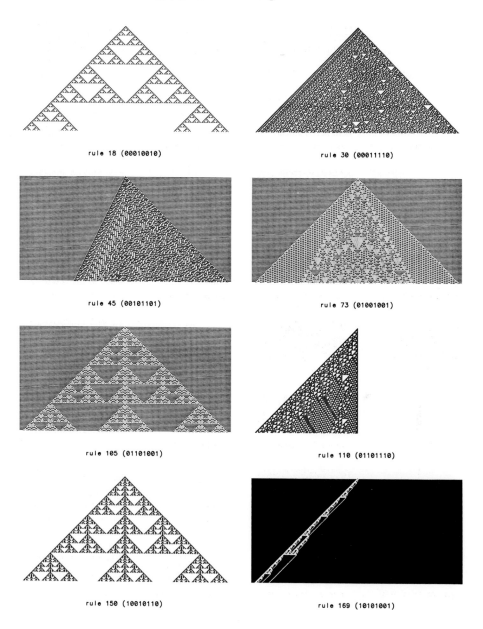

rule 18 (00010010)

rule 30 (00011110)

rule 45 (00101101)

rule 73 (01001001)

rule 105 (01101001)

rule 110 (01101110)

rule 150 (10010110)

rule 169 (10101001)

Table 5: Patterns from single site seeds

Table 5: Patterns generated by evolution from configurations containing a single nonzero site.

The first part of the table shows pictures for all distinct rules. Since the initial configuration is not invariant under complementation, rules which differ by complementation can produce different patterns, and are shown separately. Only the minimal representative is shown for rules related by reflection. In all cases, the patterns correspond to evolution for 38 time steps.

Many rules are seen to yield equivalent patterns. The results of table 7 can often be used to deduce these equivalences.

Some rules (such as 122) yield asymptotically homogeneous patterns. Others (such as 90 and 150) yield asymptotically self similar or fractal patterns. (The fractal dimensions of the patterns obtained from rules 90 and 150 are respectively $\log_2 3 \cong 1.59$ and $\log_2(1+\sqrt{5}) \cong 1.69$.) But some rules (such as 30 and 73) yield irregular patterns which show no periodic or almost periodic behaviour. The second part of the table gives some of the distinct patterns obtained by evolution for 360 time steps. Note that the structure on the right of the pattern generated by rule 110 eventually dies out, leaving an essentially periodic structure.

Table 6: Statistical properties

rule	density	$h_\mu^{(x)}$	λ_L	λ_R	$h_\mu^{(i)}$	h_μ	$h_\mu^{[min]}$
0	0	0	—	—	0	0	0
1	1/8	.43536	0	0	0	0	0
2	1/8	.48752	1	-1	$h_\mu^{(x)}$	$h_\mu^{(x)}$	0
3	1/4	.70121	-1/2	1/2	$h_\mu^{(x)}/2$	$h_\mu^{(x)}/2$	0
4	1/8	.51771	0	0	0	0	0
5	7/16	.702±.001	0	0	0	0	0
6	.241±.001	<.573±.001	1	-1	$h_\mu^{(x)}$	$h_\mu^{(x)}$	0
7	.469±.001	<.502±.001	-1/2	1/2	$h_\mu^{(x)}/2$	$h_\mu^{(x)}/2$	0
8	0	0	—	—	0	0	0
9	.410±.001	<.264±.002	-1	1	$h_\mu^{(x)}$	$h_\mu^{(x)}$	0
10	1/4	.68872	1	-1	$h_\mu^{(x)}$	0	$h_\mu^{(x)}$
11	1/2	<.567±.001	-1	1	$h_\mu^{(x)}$	$h_\mu^{(x)}$	0
12	1/4	.68872	0	0	0	0	0
13	.437±.001	.378±.001	0	0	0	0	0
14	1/2	0	(-1,1)	(1,-1)	0	0	0
15	1/2	1	-1	1	1.0	1.0	0
18	1/4	1/2	1	1	0.5	1.0	1.0
19	1/2	.62351	0	0	0	0	0
22	.35095±.00002	<.795±.001	.7660±.0002	.7660±.0002	<.744±.003	<.9146±.0007	<.9146±.0007
23	1/2	.599±.001	0	0	0	0	0
24	3/16	.55081	-1	1	$h_\mu^{(x)}$	$h_\mu^{(x)}$	0
25	.447±.001	<.180±.001	-1/2	1/2	$h_\mu^{(x)}/2$	$h_\mu^{(x)}/2$	0
26	.386±.001	<.790±.001	1	-1	$h_\mu^{(x)}$	$h_\mu^{(x)}$	0
27	.531±.001	<.800±.001	-1/2	1/2	$h_\mu^{(x)}/2$	$h_\mu^{(x)}/2$	0
28	1/2	.500±.001	0	0	0	0	0
29	1/2	.86742	0	0	0	0	0
30	1/2	1	.2428±.0002	1	1	<1.15436	<.763141
32	0	0	—	—	0	0	0
33	.396±.001	<.637±.001	0	0	0	0	0
34	1/4	.68872	1	-1	$h_\mu^{(x)}$	$h_\mu^{(x)}$	0
35	.375±.001	<.645±.001	-1/2	1/2	$h_\mu^{(x)}/2$	$h_\mu^{(x)}/2$	0
36	1/16	.32483	0	0	0	0	0
37	.384±.001	.506±.001	0	0	0	0	0
38	9/32	.73733	1	-1	$h_\mu^{(x)}$	$h_\mu^{(x)}$	0
40	0	0	—	—	0	0	0
41	.372±.001	<.360±.001	-1	1	$h_\mu^{(x)}$	$h_\mu^{(x)}$	0
42	3/8	.85684	1	-1	$h_\mu^{(x)}$	$h_\mu^{(x)}$	0
43	1/2	0	(-1,1)	(1,-1)	0	0	0
44	.167±.001	<.528±.001	0	0	0	0	0
45	1/2	1	.1724±.0003	1	1	<1.13036	<.673893
46	3/8	.55081	1	-1	$h_\mu^{(x)}$	$h_\mu^{(x)}$	0
50	1/2	.601±.001	0	0	0	0	0
51	1/2	1	0	0	0	0	0
54	.49±.01	<.2720±.0005	.553±.002	.553±.002	<.250±.002	<.250±.002	<.250±.002
56	.376±.001	<.589±.001	-1	1	$h_\mu^{(x)}$	$h_\mu^{(x)}$	0
57	1/2	0	(-1,1)	(1,-1)	0	0	0
58	.625±.001	<.332±.001	1	-1	$h_\mu^{(x)}$	$h_\mu^{(x)}$	0

Table 6: Statistical properties

rule	density	$h_\mu^{(x)}$	λ_L	λ_R	$h_\mu^{(t)}$	h_μ	$h_\mu^{[min]}$
60	1/2	1	0	1	1	2	2
62	.644±.002	<.262±.001	0	0	0	0	0
72	1/8	.32483	0	0	0	0	0
73	.463±.001	<.714±.001	0	0	0	0	0
74	.318±.001	<.629±.001	1	-1	$h_\mu^{(x)}$	$h_\mu^{(x)}$	0
76	3/8	.85060	0	0	0	0	0
77	1/2	.599±.001	0	0	0	0	0
78	.562±.001	.377±.001	0	0	0	0	0
90	1/2	1	1	1	1	2	2
94	.584±.001	<.562±.001	0	0	0	0	0
104	.068±.001	.208±.001	0	0	0	0	0
105	1/2	1	1	1	1	2	2
106	1/2	1	1	-.1335±.0006	1	<1.06985	<.461366
108	5/16	.78025	0	0	0	0	0
110	4/7	0	(.26 —.5)	(-.27 — 0.)	0	0	0
122	1/2	1/2	1	1	0.5	1.0	1.0
126	1/2	1/2	1	1	0.5	1.0	1.0
128	0	0	—	—	0	0	0
130	.167±.001	.525±.001	1	-1	$h_\mu^{(x)}$	$h_\mu^{(x)}$	0
132	1/8	.599±.001	0	0	0	0	0
134	.292±.001	<.533±.001	1	-1	$h_\mu^{(x)}$	$h_\mu^{(x)}$	0
136	0	0	—	—	0	0	0
138	3/8	.806±.001	1	-1	$h_\mu^{(x)}$	$h_\mu^{(x)}$	0
140	1/4	<.678±.001	0	0	0	0	0
142	1/2	0	(-1,1)	(1,-1)	0	0	0
146	1/4	1/2	1	1	0.5	1.0	1.0
150	1/2	1	1	1	1	2	2
152	.185±.001	.515±.001	-1	1	$h_\mu^{(x)}$	$h_\mu^{(x)}$	0
154	1/2	1	1	-1	1	1	0
156	1/2	.502±.001	0	0	0	0	0
160	0	0	—	—	0	0	0
162	.333±.001	.667±.001	1	-1	$h_\mu^{(x)}$	$h_\mu^{(x)}$	0
164	.083±.001	.389±.001	0	0	0	0	0
168	0	0	—	—	0	0	0
170	1/2	-1	1	1	1.0	1.0	0
172	1/8	.485±.001	0	0	0	0	0
178	1/2	.599±.001	0	0	0	0	0
184	1/2	0	(-1,1)	(1,-1)	0	0	0
200	3/8	.70121	0	0	0	0	0
204	1/2	1	0	0	0	0	0
232	1/2	.599±.001	0	0	0	0	0

Table 6: Statistical properties of evolution from disordered states.

Results are given for all the ''minimal representative'' rules of table 1. In all cases, initial configurations were used in which each site has value 0 or 1 with probability 1/2. Some properties of some rules remain unchanged with different kinds of initial configurations.

Rational numbers, or numbers without errors, are quoted whenever analytical arguments yield exact results. In a few cases, the rigour of these arguments may be subject to question.

Table 6: Statistical properties

The column labelled "density" gives the asymptotic density of nonzero sites. For some, but not all, rules this depends on the initial density, here taken to be 1/2. For most rules, the relaxation to the final density appears to be approximately exponentially. For some rules (such as 18), in which particle-like excitations undergo random annihilation, the relaxation may be like $t^{-1/2}$, or slower. Rule 110 shows particularly slow relaxation.

The column labelled $h_\mu^{(x)}$ gives estimates for the asymptotic spatial measure entropy, as defined in [1.3]. This quantity gives a measure of the "information content" of cellular automaton configurations. It is computed by breaking the configuration into blocks of sites, say of length X, then evaluating the quantity $-\frac{1}{X}\sum p_i \log_2 p_i$, where the sum runs over all 2^X possible blocks, which are taken to occur with probabilities p_i. $h_\mu^{(x)}$ is the limit of this quantity as $X \rightarrow \infty$. The values decrease monotonically with X, allowing upper bounds on the $X \rightarrow \infty$ limit to be derived from finite X results. Where errors are quoted, the values or bounds on $h_\mu^{(x)}$ given in the table were obtained after 400 time steps, with blocks up to length $X=11$ considered. (More accurate results were obtained for rules 22 and 54.) Fits to values obtained as a function of X suggest that the exact $h_\mu^{(x)}$ for rules 22 and 54 may in fact be zero.

The definition of $h_\mu^{(x)}$ implies that it achieves its maximal value of 1 only when all possible sequences of site values occur with equal probability, so that each site has value 0 or 1 with independent probability 1/2. $h_\mu^{(x)} = 0$ if only a finite number of complete cellular automaton configurations can occur.

Results for $h_\mu^{(x)}$ given without errors in the table were obtained by explicit construction of probabilistic regular languages which represent the sets of configurations produced by cellular automaton evolution, as in table 11.

The quantities λ_L and λ_R are left and right Lyapunov exponents, which measure the rate of information transmission. They give the slopes of the left and right boundaries of the difference patterns illustrated in figure 4. Thus they measure the rate at which perturbations in cellular automaton configurations spread with to the left and right.

The notation — indicates that almost all changes in initial configurations die out, so that the $\lambda_{L,R}$ are not defined.

The notation (-1,1) indicates that the information propagation direction can alternate, typically as progressively more distant particle-like structures from the initial configuration are encountered. There is probably no definite infinite size limit for the $\lambda_{L,R}$ in such cases.

Rule 110 shows highly complex information transmission properties, associated with the particle-like structures of table 15. The values of $\lambda_{L,R}$ given in the table for this case are possible bounds associated with the fastest and slowest-moving particle-like structures.

The quantity $h_\mu^{(t)}$ is the temporal measure entropy, which measures the information content of time sequences of values of individual sites. It is evaluated by applying the same procedure as for $h_\mu^{(x)}$ but to sequences of values of a single site attained on many successive time steps. It can be shown [1.3] that $h_\mu^{(t)} \leq (\lambda_L + \lambda_R) h_\mu^{(x)}$.

The quantities $h_\mu^{(x)}$ and $h_\mu^{(t)}$ measure respectively the information content of spatial and temporal sequences that are one site wide. The quantity $\mathbf{h}_\mu$ gives the entropy associated with spacetime patches of sites of arbitrary width. (Nevertheless, for many rules, the exact value of $\mathbf{h}_\mu$ is in fact obtained from patches of width 1 or 2.) In general, $\mathbf{h}_\mu \leq 2 h_\mu^{(t)}$, and $h_\mu^{(t)} \leq \mathbf{h}_\mu \leq (\lambda_L + \lambda_R) h_\mu^{(x)}$.

The quantity $\mathbf{h}_\mu$ is evaluated by considering spacetime patches of sites that extend in the time direction. The last column of the table uses a generalization in which the patches can extend in any spacetime direction. It gives the minimum value $\mathbf{h}_\mu$ obtained as a function of direction. (The actual bounds given in the table were obtained from vertical or diagonal patches; other directions may yield stricter bounds.)

Table by **Peter Grassberger** (*Physics Department, University of Wuppertal*).

Table 7: Blocking transformation equivalences

#				#		
0	0:	00 10 (1)		25	0:	1101000 1111000 (7)
1	0:	11 10 (2)			240:	0000000 1101000 (14)
	200:	00 11 (2)		26	90:	00 10 (2)
	204:	000 111 (2)			85:	010 110 (3)
2	34:	00 10 (2)			170:	100100 101100 (6)
	170:	000 100 (3)			0:	10110100 10111100 (8)
	0:	1000 1100 (4)			240:	11100100 10011100 (16)
3	0:	11 10 (2)		27	48:	11 10 (4)
	240:	00 11 (4)			85:	010 110 (3)
4	204:	00 10 (1)			240:	000 100 (6)
	0:	00 11 (1)			0:	0100 1100 (8)
5	200:	00 10 (2)			170:	100100 101100 (6)
	204:	000 100 (2)		28	192:	00 10 (2)
	0:	111 110 (2)			200:	10 01 (2)
	51:	00010 11010 (1)			51:	100 110 (1)
6	184:	00 10 (2)			204:	100 110 (2)
	34:	00 11 (2)			0:	1010 1100 (1)
	170:	0000 1000 (4)		29	204:	00 10 (2)
	128:	0100 1100 (4)			200:	10 01 (2)
	240:	1000 1010 (4)			51:	100 110 (1)
	85:	10000 11000 (5)			0:	1010 1100 (1)
	0:	11000 11100 (10)		30		
7	192:	00 10 (2)		32	0:	00 11 (1)
	0:	000 100 (2)			128:	00 10 (2)
	240:	000 111 (6)		33	132:	00 10 (2)
8	0:	00 10 (1)			200:	00 11 (2)
9	0:	0010 1110 (2)			0:	111 100 (2)
	170:	1000 0010 (6)			204:	000 111 (2)
	34:	1000 0011 (6)			128:	0000 1010 (4)
	204:	0100000 1100000 (5)		34	170:	00 10 (2)
	240:	00000000 11010000 (8)			0:	100 110 (3)
10	34:	00 10 (2)		35	240:	00 11 (4)
	170:	000 100 (3)			0:	100 110 (3)
	0:	010 110 (3)			170:	10100 10010 (5)
11	240:	00 11 (2)		36	0:	00 11 (1)
	0:	010 110 (3)			4:	00 10 (2)
	15:	000 111 (3)			204:	000 100 (1)
	128:	1100 1001 (4)		37	200:	00 11 (2)
	170:	1100100 1001100 (7)			0:	1111 1100 (2)
12	204:	00 10 (1)			204:	0000 1111 (2)
	0:	100 110 (1)			170:	100000 110000 (6)
13	192:	00 10 (2)			240:	010000 110000 (6)
	0:	100 110 (1)			128:	010000 111000 (6)
	204:	10100 10010 (1)		38	34:	00 10 (2)
14	240:	10 01 (2)			85:	100 110 (3)
	34:	00 11 (2)			170:	0000 1000 (4)
	15:	010 101 (3)			0:	1100 1110 (4)
	0:	1100 1000 (4)		40	128:	00 10 (2)
	170:	0000 1100 (4)			0:	00 11 (2)
	128:	1100 0110 (4)			170:	11010 10110 (5)
15	240:	00 10 (2)		41	148:	00 10 (2)
	15:	110 001 (3)			184:	0000 1000 (4)
18	90:	00 10 (2)			176:	0000 1010 (4)
	204:	11000 10100 (2)			170:	11010 10110 (5)
	0:	10100 11100 (2)			240:	00000000 10000000 (8)
19	51:	00 11 (1)			128:	10000000 10100000 (8)
	0:	11 10 (2)			0:	11111000 11001000 (8)
	204:	00 11 (2)			136:	01010000 10101101 (8)
22	146:	00 10 (2)		42	170:	00 10 (2)
	90:	0000 1000 (4)			34:	00 11 (2)
	0:	11011000 11111000 (4)			0:	1100 1110 (4)
23	51:	00 11 (1)		43	170:	10 01 (2)
	128:	00 10 (2)			240:	00 11 (2)
	204:	00 11 (2)			15:	000 111 (3)
	0:	000 100 (2)			0:	1001 1000 (4)
24	48:	00 10 (2)			128:	1100 0110 (4)
	240:	000 100 (3)		44	12:	00 10 (2)
	0:	100 011 (3)			204:	000 100 (1)
					0:	1000 1100 (1)

Table 7: Blocking transformation equivalences

45		
46	34:	00 11 (2)
	0:	110 100 (3)
	170:	000 110 (3)
50	51:	10 01 (1)
	128:	00 10 (2)
	204:	10 01 (2)
	0:	1010 1000 (2)
51	51:	10 01 (1)
	204:	00 10 (2)
54	50:	00 10 (2)
	51:	1000 0010 (2)
	128:	0000 1000 (4)
	204:	1000 0010 (4)
	170:	1000 1110 (4)
	240:	0010 1110 (4)
	0:	000010 111010 (4)
56	240:	00 10 (2)
	128:	10 01 (2)
	184:	010 101 (3)
	0:	1010 0110 (2)
	34:	1010 1101 (4)
	170:	11010 10110 (5)
57	128:	10 01 (2)
	184:	010 101 (3)
	0:	1010 0110 (2)
	48:	1010 0100 (4)
	34:	1010 1101 (4)
	240:	10100 10010 (5)
	170:	11010 10110 (5)
58	128:	00 10 (2)
	0:	101 100 (3)
	240:	1100 1110 (8)
	170:	11010 10110 (5)
60	60:	00 10 (2)
62	240:	1100 1110 (8)
	204:	11000 11010 (3)
	0:	111110 100000 (3)
72	0:	00 10 (1)
	4:	00 11 (2)
	204:	000 110 (1)
73	204:	1100 0110 (2)
	51:	11000 11010 (1)
	0:	10110 10000 (2)
74	34:	00 10 (2)
	170:	000 100 (3)
	0:	10000 10100 (5)
	85:	1110000 1101000 (7)
76	204:	00 10 (1)
	0:	1010 1110 (1)
77	204:	10 01 (1)
	128:	00 10 (2)
	0:	1010 1000 (1)
78	0:	101 100 (1)
	204:	11010 10110 (1)
90	90:	00 10 (2)
94	90:	00 11 (2)
	0:	1010 1110 (1)
	204:	1010 0101 (2)
	51:	10010 11110 (1)
	136:	111100 110110 (6)
	192:	110110 011110 (6)
104	128:	00 10 (2)
	4:	00 11 (2)
	0:	000 100 (1)
	204:	0000 1100 (1)
105	150:	00 10 (2)
106	170:	00 10 (2)

108	76:	00 10 (2)
	204:	000 100 (1)
	51:	10100 11100 (1)
	0:	10010 11110 (2)
110	0:	110100 101100 (9)
	240:	111000 100110 (9)
	170:	10011000 11111000 (16)
122	128:	00 10 (2)
	90:	00 11 (2)
	0:	1010 1000 (2)
	204:	11100 11110 (2)
126	90:	00 11 (2)
	204:	11100 11110 (2)
	0:	01110 10001 (2)
128	0:	00 10 (1)
	128:	00 11 (2)
130	34:	00 10 (2)
	170:	000 100 (3)
	0:	1000 1100 (4)
	128:	1000 1111 (4)
132	204:	00 10 (1)
	128:	00 11 (2)
	0:	000 110 (1)
134	184:	00 10 (2)
	162:	00 11 (2)
	170:	0000 1000 (4)
	128:	0100 1100 (4)
	240:	1000 1010 (4)
	85:	10000 11000 (5)
	0:	100000 101100 (6)
136	0:	00 10 (1)
	136:	00 11 (2)
138	34:	00 10 (2)
	170:	00 11 (2)
	0:	010 110 (3)
140	204:	00 10 (1)
	0:	100 110 (1)
142	240:	10 01 (2)
	170:	00 11 (2)
	15:	010 101 (3)
	0:	1100 1000 (4)
	128:	1100 0110 (4)
146	90:	00 10 (2)
	204:	11000 10100 (2)
	0:	10010 11110 (2)
150	150:	00 10 (2)
152	48:	00 10 (2)
	240:	000 100 (3)
	136:	1000 1111 (4)
	0:	01000 11000 (5)
154	90:	00 10 (2)
	85:	010 110 (3)
	170:	1100 0110 (4)
156	192:	00 10 (2)
	200:	10 01 (2)
	136:	10 11 (2)
	51:	100 110 (1)
	204:	100 110 (2)
	0:	1010 1100 (1)
160	128:	00 10 (2)
	0:	000 100 (1)
162	170:	00 10 (2)
	128:	10 11 (2)
	0:	100 110 (3)
164	90:	11 10 (2)
	128:	00 11 (2)
	204:	000 100 (1)
	0:	0000 1100 (1)

Table 7: Blocking transformation equivalences

168	128:	00 10 (2)
	136:	00 11 (2)
	170:	10 11 (2)
	0:	000 100 (1)
170	170:	00 10 (2)
172	34:	11 10 (2)
	204:	000 100 (1)
	170:	110 111 (3)
	0:	1000 1100 (1)
178	51:	10 01 (1)
	128:	00 10 (2)
	204:	10 01 (2)
	0:	1010 1000 (2)

184	240:	00 10 (2)
	128:	10 01 (2)
	170:	10 11 (2)
	184:	010 101 (3)
	0:	1010 0110 (2)
200	0:	00 10 (1)
	204:	00 11 (1)
204	204:	00 10 (1)
232	204:	00 11 (1)
	128:	00 10 (2)
	0:	000 100 (1)

Table 7: Equivalences between rules under blocking transformations.

When only particular blocks of site values occur, the evolution of one cellular automaton rule (say R) may be equivalent to that of another (say R'). Thus for example, the evolution under rule 1 of configurations consisting of the blocks 000 and 111 is equivalent to evolution under rule 204 in which 000 is replaced by 0, and 111 is replaced by 1. (Two time steps in evolution according to rule 1 are necessary to reproduce one time step of evolution according to rule 204.) Since rule 204 is the identity, this implies that configurations consisting only of the blocks 000 and 111 must be periodic under rule 1 (with period 2).

In general, one may consider replacing site values 0 and 1 in evolution according to rule R by blocks B_0 and B_1. In some cases, the resulting evolution may correspond to T time steps of another rule R'. Evolution according to rule R' can thus be "simulated" by evolution according to rule R, under the blocking transformation $0 \rightarrow B_0$, $1 \rightarrow B_1$. Such blocking transformations can be considered analogous to block spin transformations in the renormalization group approach.

The table gives possible simulations for all the "minimal representative" rules of table 1. The notation R': $B_0 B_1$ (T) indicates simulation of rule R' by replacing 0 with the block B_0, and 1 with B_1; T steps of rule R are needed to reproduce one step of rule R' evolution.

The table includes all simulations for block lengths up to 8. The blocks B_0 and B_1 are always assumed distinct. Only one representative sets of blocks are given for each simulation. (Thus for example, only the blocks 00 and 10 are given for the simulation of rule 90 by rule 18; the blocks 00 and 01 would also suffice.) Simulations with block length 1 are not included; these correspond to transformations given in table 1. No simulations are found for rules 30 and 45 up to block length 8.

Many rules are seen to be equivalent under blocking transformations to simple rules, such as 204 (the identity), 170 (left shift), 240 (right shift), 51 (complementation) and 0. Equivalence is also often found to the additive rules 90 and 150. An important property of all these simple rules is that they simulate themselves under blocking transformations. This has the consequence that patterns generated by these rules are self similar. Fractal patterns are thus produced by evolution according to rules 90 and 150 from single site seeds, as shown in table 5.

The simulations given in the table occur when only particular blocks occur in the configuration of a cellular automata. In disordered configurations, all possible blocks can occur. But since cellular automaton under most rules is irreversible, only a subset of blocks may occur after a sufficiently long time. Often the subset of blocks that occur is, at least approximately, the blocks which correspond to a particular simulation. In this case, the behaviour of one cellular automaton may be considered "attracted" to that of another.

It is common to find "domains" in which only particular blocks occur. Within each such domain, the evolution may correspond to that of a simpler rule. The domains are separated by walls or "defects", whose behaviour is not reproduced by the simpler rule. In some cases, the defects remain stationary; in others, they execute random walks, and, for example, annihilate in pairs. In the latter cases, the sizes of domains grow slowly with time.

Table 7: Blocking transformation equivalences

While a large subset of possible initial configurations for a cellular automaton may be attracted to a particular form of behaviour, there are usually some special initial states (typically occurring among disordered states with probability zero), for which very different behaviour occurs. Such special initial states may for example consist of blocks which yield a simulation to which the rule is not generically attracted.

The blocking transformations considered in the table represent one form of transformation between rules. Many others can also be considered. A general class, which includes the blocking transformations of the table, are those transformations which can be carried out by arbitrary finite state machines.

The blocking transformations used in the table have the property that they reduce the total number of sites. This is a consequence of the fact that the blocks used are always taken not to overlap. An alternative approach is to perform replacements for overlapping blocks, thus obtaining configurations with the same number of sites. An example of such a replacement is $00 \to 0$, $01 \to 1$, $10 \to 1$, $11 \to 0$. For some rules, the resulting transformed configurations show evolution according to other $k=2$, $r=1$ cellular automaton rules. Rules related in this way must have the same global properties, and must yield for example the same entropies. The minimal representative rules from table 1 equivalent under such transformations are:

15, 240	240
23, 232	132
43, 212	184
51, 204	204
77, 178	222
85, 170	170
105, 150	150
113, 142	226

Main table by **John Milnor** (*Institute for Advanced Study*). (Original program by S. Wolfram.) Second table by Peter Grassberger.

Table 8: Factorizations of rules

φ	φ₁	φ₂
0	0	0
	0	12
	0	48
	0	60
	0	192
	0	204
	0	240
	0	252
	17	0
	34	0
	34	192
	51	0
	68	0
	68	192
	85	0
	102	0
	119	0
	136	0
	153	0
	170	0
	187	0
	187	3
	204	0
	221	0
	221	3
	238	0
	255	0
	255	3
	255	12
	255	15
	255	48
	255	51
	255	60
	255	63

φ	φ₁	φ₂
1	17	192
	238	3
2	17	48
	238	12
3	17	240
	51	192
	204	3
	238	15
8	119	48
	136	12
12	34	48
	34	240
	51	48
	204	12
	221	12
	221	15
15	51	240
	204	15
18	17	60
	238	60
19	17	252
	238	63
24	102	48
	153	12
34	34	12
	34	204
	85	48
	170	12
	221	48
	221	51
36	102	192
	153	3

φ	φ₁	φ₂
46	34	60
	34	252
	221	60
	221	63
51	51	204
	85	240
	170	15
	204	51
60	51	60
	102	240
	153	15
	204	60
72	119	60
	136	60
90	102	60
	153	60
126	102	252
	153	63
128	119	3
	136	192
136	85	3
	119	51
	136	204
	170	192
170	85	51
	170	204
200	119	63
	136	252
204	51	51
	85	15
	170	240
	204	204

Table 8: Factorizations into compositions of simpler rules.

The 256 rules in table 1 are stated as functions of three site values $\phi(a_{-1},a_0,a_1)$. Of these, 48 depend only on two of the site values. Some other rules can be formed from compositions of these simpler rules. This table lists rules which can be formed by compositions according to

$$\phi(a_{-1},a_0,a_1) = \phi_2(\phi_1(—,a_{-1},a_0),\phi_1(—,a_0,a_1),—) \ ,$$

where — indicates that the value is irrelevant. Only minimal representative rules from table 1 are included. In each case, all possible compositions are listed. Note that most of the compositions do not commute.

Table by **Erica Jen** (*Los Alamos National Laboratory*).

Table 9: Lengths of newly-excluded blocks

rule	t=1	2	3	4
0	1	—	—	—
1	3	—	—	—
2	2	—	—	—
3	3	—	—	—
4	2	—	—	—
5	5	—	—	—
6	3	6	7	7
7	4	5	5	6
8	2	1	—	—
9	4	7	9	9
10	3	—	—	—
11	3	5	7	9
12	2	—	—	—
13	4	4	6	6
14	3	5	7	9
15	—	—	—	—
18	3	11	12	13
19	3	3	—	—
22	8	7	11	9
23	5	6	7	8
24	2	3	—	—
25	5	6	8	8
26	4	10	8	11
27	4	6	6	9
28	3	6	6	8
29	4	—	—	—
30	—	—	—	—
32	2	4	6	8
33	4	7	6	6
34	2	—	—	—
35	4	6	7	9
36	3	2	—	—
37	9	8	9	8
38	4	3	—	—
40	3	4	5	7
41	5	9	8	9
42	3	—	—	—
43	5	7	9	11
44	4	4	6	6
45	—	—	—	—
46	3	3	—	—
50	3	5	9	11
51	—	—	—	—
54	5	9	9	7
56	3	4	6	8
57	6	5	5	7
58	4	5	5	6

rule	t=1	2	3	4
60	—	—	—	—
62	5	7	8	7
72	3	3	—	—
73	6	6	7	14
74	4	6	6	7
76	3	—	—	—
77	5	6	7	8
78	4	4	6	5
90	—	—	—	—
94	5	7	11	11
104	8	8	8	7
105	—	—	—	—
106	—	—	—	—
108	5	4	—	—
110	5	10	11	11
122	5	7	8	10
126	3	12	13	14
128	3	5	7	9
130	4	6	7	10
132	4	5	6	7
134	5	6	6	8
136	3	4	5	6
138	3	—	—	—
140	4	5	6	7
142	5	7	9	11
146	6	6	8	8
150	—	—	—	—
152	5	5	6	6
154	—	—	—	—
156	6	7	7	9
160	5	7	9	11
162	4	6	8	10
164	9	9	8	9
168	4	5	6	7
170	—	—	—	—
172	4	5	6	7
178	5	6	7	8
184	4	6	8	10
200	3	—	—	—
204	—	—	—	—
232	5	6	7	8

Table 9: Lengths of newly-excluded blocks

Table 9: Lengths of distinct blocks of sites newly excluded at time t.

Most cellular automaton rules are irreversible, so that even starting from all possible initial configurations, only a subset of configurations can occur after t time steps. In this subset of configurations, only certain blocks of site values can occur. The subset can be specified by giving the blocks which are excluded. In some cases (such as rule 128), the number of distinct excluded blocks is finite; in other cases, it is countably infinite. Irreversibility leads to an increase in the size of the set of excluded blocks with time.

The table gives the lengths of the shortest blocks which are newly excluded after exactly t time steps. Such blocks can occur in configurations up to time $t-1$, but cannot occur at time t or after. The lengths $L(t)$ of the shortest blocks newly excluded at time t obey the inequality [2.1] $L(t) \geq L(t-1)-2$.

The notation — in the table indicates that no blocks are newly excluded at a particular time step. This implies that the rule has reached a stable set of configurations, which can occur after any number of steps. It should be noted, however, that this table takes no account of the probabilities with which different configurations may occur.

Table by **Lyman P. Hurd** (*Mathematics Department, Princeton University*). (Original program by S. Wolfram.)

Table 10: Regular language complexities

rule	t=1	t=2	t=3	t=4	t=5	t>5	∞
0	1 [1]	.	.	.	.	1 [1]	1 [1]
1	4 [6]	.	.	.	.	4 [6]	4 [6]
2	3 [4]	.	.	.	.	3 [4]	3 [4]
3	3 [5]	.	.	.	.	3 [5]	3 [5]
4	2 [3]	.	.	.	.	2 [3]	2 [3]
5	9 [15]	.	.	.	.	9 [15]	9 [15]
6	9 [16]	13 [22]	22 [37]	26 [44]	31 [52]		
7	4 [7]	7 [12]	12 [21]	14 [24]	16 [27]		
8	3 [4]	1 [1]	.	.	.	1 [1]	1 [1]
9	9 [16]	22 [40]	44 [80]	106 [198]	266 [500]		
10	4 [6]	.	.	.	.	4 [6]	4 [6]
11	3 [5]	7 [12]	10 [17]	12 [20]	14 [23]		
12	2 [3]	.	.	.	.	2 [3]	2 [3]
13	6 [11]	10 [17]	12 [19]	14 [21]	16 [23]		
14	3 [5]	7 [12]	10 [17]	12 [20]	14 [23]		
15	1 [2]	.	.	.	.	1 [2]	1 [2]
16	3 [4]	.	.	.	.	3 [4]	3 [4]
17	3 [5]	.	.	.	.	3 [5]	3 [5]
18	5 [9]	47 [91]	143 [270]				
19	3 [5]	5 [8]	.	.	.	5 [8]	5 [8]
20	10 [17]	21 [37]	32 [57]	37 [65]	50 [89]		
21	4 [7]	9 [16]	12 [21]	14 [24]	16 [27]		
22	15 [29]	280 [551]	4506 [8963]				
23	11 [20]	15 [26]	19 [32]	23 [38]	27 [44]		
24	2 [3]	3 [4]	.	.	.	3 [4]	3 [4]
25	6 [11]	26 [50]	55 [106]	114 [220]	333 [649]		
26	13 [25]	92 [179]	2238 [4454]				
27	10 [18]	14 [25]	18 [32]	21 [37]	24 [42]		
28	3 [5]	8 [14]	10 [17]	11 [18]	12 [19]		
29	4 [7]	.	.	.	.	4 [7]	4 [7]
30	1 [2]	.	.	.	.	1 [2]	1 [2]
32	2 [3]	5 [7]	7 [9]	9 [11]	11 [13]	$2t+1$ [$2t+3$]	4 [6]
33	5 [9]	11 [20]	26 [47]	40 [68]	41 [68]		
34	2 [3]	.	.	.	.	2 [3]	2 [3]
35	4 [7]	7 [13]	9 [16]	10 [18]	12 [21]		
36	3 [5]	3 [4]	.	.	.	3 [4]	3 [4]
37	15 [29]	194 [376]	870 [1698]	3735 [7290]			
38	5 [9]	5 [8]	.	.	.	5 [8]	5 [8]
40	10 [17]	12 [19]	15 [22]	18 [25]	21 [28]		
41	14 [27]	128 [250]	1049 [2069]				
42	3 [5]	.	.	.	.	3 [5]	3 [5]
43	9 [16]	13 [22]	17 [28]	21 [34]	25 [40]		
44	4 [7]	11 [20]	18 [32]	23 [40]	27 [46]		
45	1 [2]	.	.	.	.	1 [2]	1 [2]
46	3 [5]	5 [8]	.	.	.	5 [8]	5 [8]
48	2 [3]	.	.	.	.	2 [3]	2 [3]
49	4 [7]	6 [10]	7 [11]	9 [14]	10 [15]		
50	3 [5]	8 [14]	10 [17]	12 [20]	14 [23]		
51	1 [2]	.	.	.	.	1 [2]	1 [2]
52	4 [7]	5 [9]	.	.	.	5 [9]	5 [9]

Table 10: Regular language complexities

rule	t=1	t=2	t=3	t=4	t=5	t>5	∞
53	10 [18]	15 [25]	17 [28]	21 [33]	23 [36]		
54	9 [16]	17 [32]	94 [179]	675 [1316]			
56	3 [5]	5 [9]	7 [12]	9 [15]	11 [18]		
57	11 [20]	15 [27]	15 [26]	24 [42]	32 [55]		
58	10 [18]	20 [35]	33 [55]	55 [88]	76 [122]		
60	1 [2]	.	.	.	.	1 [2]	1 [2]
61	5 [9]	16 [30]	40 [76]	94 [177]	185 [350]		
62	5 [9]	21 [39]	61 [114]	81 [150]	129 [240]		
64	3 [4]	1 [1]	.	.	.	1 [1]	1 [1]
65	9 [15]	20 [35]	42 [75]	88 [157]	220 [401]		
66	2 [3]	3 [4]	.	.	.	3 [4]	3 [4]
68	2 [3]	.	.	.	.	2 [3]	2 [3]
69	5 [8]	10 [17]	12 [19]	14 [23]	16 [25]		
70	3 [5]	8 [14]	9 [15]	11 [19]	11 [19]		
72	5 [9]	5 [8]	.	.	.	5 [8]	5 [8]
73	15 [29]	82 [155]	390 [757]	1443 [2796]			
74	13 [25]	45 [85]	66 [123]	69 [125]	75 [135]		
76	3 [5]	.	.	.	.	3 [5]	3 [5]
77	11 [20]	15 [26]	19 [32]	23 [38]	27 [44]		
78	10 [18]	15 [27]	18 [30]	20 [34]	22 [36]		
80	4 [6]	.	.	.	.	4 [6]	4 [6]
81	3 [5]	7 [11]	9 [14]	11 [16]	13 [19]		
82	13 [25]	167 [331]	3134 [6257]				
84	3 [5]	7 [12]	9 [14]	11 [17]	13 [19]		
85	1 [2]	.	.	.	.	1 [2]	1 [2]
86	1 [2]	.	.	.	.	1 [2]	1 [2]
88	13 [25]	63 [117]	114 [210]	117 [213]	1288 [2106]		
89	1 [2]	.	.	.	.	1 [2]	1 [2]
90	1 [2]	.	.	.	.	1 [2]	1 [2]
92	10 [18]	14 [23]	18 [29]	18 [27]	22 [33]		
94	15 [29]	230 [455]	3904 [7760]				
96	9 [16]	11 [17]	14 [20]	17 [23]	20 [26]		
97	14 [27]	99 [195]	626 [1237]				
98	3 [5]	4 [6]	6 [9]	8 [12]	10 [15]		
100	5 [9]	11 [19]	17 [29]	18 [29]	22 [34]		
102	1 [2]	.	.	.	.	1 [2]	1 [2]
104	15 [29]	265 [525]	2340 [4647]	1394 [2675]	1542 [2913]		
105	1 [2]	.	.	.	.	1 [2]	1 [2]
106	1 [2]	.	.	.	.	1 [2]	1 [2]
108	9 [16]	11 [19]	.	.	.	11 [19]	11 [19]
110	5 [9]	20 [38]	160 [312]	1035 [2037]			
112	3 [5]	.	.	.	.	3 [5]	3 [5]
113	9 [16]	13 [22]	17 [28]	21 [34]	25 [40]		
114	10 [18]	20 [35]	33 [56]	50 [82]	72 [115]		
116	3 [5]	5 [8]	.	.	.	5 [8]	5 [8]
118	5 [9]	16 [29]	49 [92]	74 [139]	95 [175]		
120	1 [2]	.	.	.	.	1 [2]	1 [2]
122	15 [29]	179 [347]	5088 [9933]				
124	5 [9]	20 [38]	208 [407]	1356 [2672]			
126	3 [5]	13 [23]	107 [198]	2867 [5476]			

Table 10: Regular language complexities

rule	t=1	t=2	t=3	t=4	t=5	t>5	∞
128	4 [6]	6 [8]	8 [10]	10 [12]	12 [14]	2t+2 [2t+4]	3 [5]
130	9 [15]	14 [21]	18 [25]	22 [29]	26 [33]		
132	5 [9]	7 [12]	9 [15]	11 [18]	13 [21]		
134	14 [27]	44 [82]	99 [182]	125 [224]			
136	3 [5]	4 [6]	5 [7]	6 [8]	7 [9]	t+2 [t+4]	3 [5]
138	3 [5]	.	.	.	.	3 [5]	3 [5]
140	4 [7]	5 [9]	6 [11]	7 [13]	8 [15]		
142	9 [16]	13 [22]	17 [28]	21 [34]	25 [40]		
144	9 [16]	16 [28]	20 [34]	24 [40]	28 [46]		
146	15 [29]	92 [177]	1587 [3126]				
148	14 [27]	68 [127]	113 [209]	188 [347]			
150	1 [2]	.	.	.	.	1 [2]	1 [2]
152	6 [11]	20 [37]	30 [55]	32 [59]	36 [65]		
154	1 [2]	.	.	.	.	1 [2]	1 [2]
156	11 [20]	20 [35]	24 [42]	28 [47]	34 [58]		
160	9 [15]	16 [24]	25 [35]	36 [48]	49 [63]	$(t+2)^2$ [(t+2)(t+4)]	9 [15]
162	5 [8]	7 [10]	9 [12]	11 [14]	13 [16]		
164	15 [29]	116 [227]	667 [1310]	1214 [2363]			
168	4 [7]	5 [8]	6 [9]	7 [10]	8 [11]	t+3 [t+6]	3 [5]
170	1 [2]	.	.	.	.	1 [2]	1 [2]
172	10 [18]	11 [20]	12 [22]	13 [24]	14 [26]		
176	6 [11]	8 [14]	10 [17]	12 [20]	14 [23]		
178	11 [20]	15 [26]	19 [32]	23 [38]	27 [44]		
180	1 [2]	.	.	.	.	1 [2]	1 [2]
184	4 [7]	6 [10]	8 [13]	10 [16]	12 [19]		
188	5 [9]	14 [25]	21 [36]	25 [43]	33 [56]		
192	3 [5]	4 [6]	5 [7]	6 [8]	7 [9]		
196	4 [7]	5 [8]	6 [9]	7 [10]	8 [11]		
200	3 [5]	.	.	.	.	3 [5]	3 [5]
204	1 [2]	.	.	.	.	1 [2]	1 [2]
208	3 [5]	.	.	.	.	3 [5]	3 [5]
212	9 [16]	13 [22]	17 [28]	21 [34]	25 [40]		
216	10 [18]	11 [19]	12 [20]	13 [21]	14 [22]		
224	4 [7]	5 [8]	6 [9]	7 [10]	8 [11]	t+3 [t+6]	3 [5]
232	11 [20]	15 [26]	19 [32]	23 [38]	27 [44]		
240	1 [2]	.	.	.	.	1 [2]	1 [2]

The set of configurations that can appear after t steps in the evolution of a one-dimensional cellular automaton can be shown to form a regular formal language [2.1]. Possible configurations thus correspond to possible paths through a finite graph which represents the grammar for the regular language. The table gives the minimum number of nodes in the graphs for such grammars; the number of arcs is given in brackets in each case. The notation . indicates that the regular language is the same as at the preceding time step.

Entries in the last column of the table give sizes of graphs for regular languages representing limiting sets of states that can be reached after any number of steps.

The size of a regular grammar gives a measure of the "complexity" of the set of configurations it describes. Notice that the grammar specifies merely which configurations can possibly occur; it does not

Table 10: Regular language complexities

account for the probabilities of different configurations.

The graphs used for the table represent possible sequences of site values that occur in configurations read from left to right. Rules related by reflection may in general yield different regular languages. The table thus includes minimal representatives for all rules from table 1 not related by complementation.

Entries in the table for $t{\leq}5$ that have been left blank were not found. They are probably $\geq$20000. The growth of regular language complexities is bounded by $2^{2^{4t}} - 1$.

For some rules, it has been possible to find explicit forms for the regular languages produced after any number of time steps. Formulae for complexities in these cases are listed in the table. In many cases, it is however suspected that the limiting set does not form a regular language, and may in fact be non-recursive.

Table by **Lyman P. Hurd** (*Mathematics Department, Princeton University*). (Original program by S. Wolfram.)

Table 11: Measure theoretical complexities

rule	t=1	t=2	t=3
0	0	0	0
1	0.8223	0.8223	0.8223
2	0.7356	0.7356	0.7356
3	0.9003	0.9003	0.9003
4	0.3768	0.3768	0.3768
5	1.8005	1.8005	1.8005
6	1.783	1.964	1.968
7	1.2707	1.670	1.933
8	0.7356	0	0
9	1.9135	2.598	3.303
10	1.1247	1.1247	1.1247
11	1.0434	1.3442	1.973
12	0.5623	0.5623	0.5623
13	1.4756	1.7879	1.713
14	0.9026	1.3607	1.927
15	0	0	0
16	0.7356	0.7356	0.7356
17	0.9003	0.9003	0.9003
18	0.9026	2.129	3.933
19	0.9026	1.1539	1.1539
20	1.7756	2.759	3.059
21	1.2707	1.8266	1.931
22	2.591	4.601	6.213
23	1.9862	2.153	2.330
24	0.5623	0.9216	0.9216
25	1.643	2.665	3.231
26	2.244	2.659	2.945
27	1.666	2.128	2.392
28	0.8305	1.6009	1.6645
29	1.2652	1.2652	1.2652
30	0	0	0
32	0.3768	0.4957	0.2346
33	1.2930	1.941	2.529
34	0.5623	0.5623	0.5623
35	1.1034	1.748	2.006
36	0.9003	0.4634	0.4634
37	2.518	4.435	5.410
38	1.298	1.256	1.256
40	1.775	1.547	1.127
41	2.332	4.134	5.471
42	0.9003	0.9003	0.9003
43	1.9584	2.269	2.483
44	0.9003	1.574	1.748
45	0	0	0
46	0.5623	0.9216	0.9216
48	0.5623	0.5623	0.5623
49	1.2512	1.451	1.455
50	0.8305	1.589	1.775
51	0	0	0
52	0.9003	0.9216	0.9216
53	1.906	2.057	2.016
54	1.7707	2.609	3.921
56	0.8305	1.3503	1.732
57	2.086	2.190	1.946
58	1.9132	2.132	2.106
60	0	0	0
61	1.430	2.065	3.076
62	1.341	2.406	3.720
64	0.7356	0	0
65	1.5310	2.268	2.970
66	0.5623	0.9216	0.9216

rule	t=1	t=2	t=3
68	0.5623	0.5623	0.5623
69	1.0562	1.790	1.842
70	0.8305	1.7802	1.815
72	0.9003	0.4634	0.4634
73	2.604	3.685	4.473
74	2.461	2.713	2.755
76	0.8305	0.8305	0.8305
77	1.9862	2.153	2.330
78	1.7553	2.111	2.029
80	1.1247	1.1247	1.1247
81	1.0434	1.5837	1.716
82	2.460	3.823	5.375
84	0.8992	1.740	1.984
85	0	0	0
86	0	0	0
88	2.2441	3.081	3.605
89	0	0	0
90	0	0	0
92	1.9132	1.768	1.735
94	2.599	3.682	5.311
96	1.782	1.3689	0.995
97	2.491	3.470	4.846
98	0.8305	0.8932	0.979
100	1.298	1.667	1.632
102	0	0	0
104	2.591	4.379	4.969
105	0	0	0
106	0	0	0
108	1.7707	1.5093	1.5093
110	1.344	2.435	3.407
112	0.9003	0.9003	0.9003
113	1.957	2.266	2.482
114	1.754	2.344	2.825
116	0.5623	0.9215	0.9215
118	1.342	1.945	2.827
120	0	0	0
122	2.600	4.307	5.981
124	1.343	2.321	3.933
126	0.9003	2.049	3.914
128	0.8223	0.457	0.1986
130	1.533	1.263	1.025
132	1.292	1.459	1.637
134	2.496	3.050	3.010
136	0.9003	0.8223	0.641
138	1.0434	1.0434	1.0434
140	1.2512	1.5808	1.7551
142	1.957	2.266	2.482
144	1.913	1.988	2.056
146	2.604	3.742	5.350
148	2.328	3.589	3.815
150	0	0	0
152	1.644	2.626	3.031
154	0	0	0
156	2.083	2.506	2.563
160	1.805	1.633	1.281
162	1.0562	0.8654	0.7355
164	2.520	3.343	3.353
168	1.2707	1.3676	1.369

Table 11: Measure theoretical complexities

rule	$t=1$	$t=2$	$t=3$
170	0	0	0
172	1.909	2.080	2.242
176	1.4757	1.723	1.863
180	0	0	0
184	1.2652	1.575	1.788
188	1.433	1.962	1.733
192	0.9003	0.8113	0.6415
196	1.1034	1.0302	0.9170
200	0.9003	0.9003	0.9003
204	0	0	0
208	1.0434	1.0434	1.0434
212	1.9584	2.269	2.483
216	1.667	2.033	2.065
224	1.2707	1.368	1.369
232	1.9862	2.153	2.330
240	0	0	0

Table 11: Measures of the information content of regular grammars for sets of configurations generated by evolution from disordered initial states.

This table gives values of a probabilistic analogue of the regular language complexity of table 10, in which the nodes of regular language graphs are weighted with the probabilities that they are visited.

Starting from a disordered state in which all possible configurations occur with equal probability, irreversible cellular automaton evolution can lead to ensembles in which different configurations occur with different probabilities. These ensembles can be described by probabilistic analogues of regular languages.

All the configurations that can occur after t steps correspond to possible paths through the standard regular language graphs of table 10. To account for the different probabilities of different configurations, one may weight the nodes of the graph according to the probabilities P_i that they are visited. In terms of these probabilities, one may then compute a measure theoretical complexity $-\sum P_i \log_2 P_i$, where the sum runs over all nodes in the regular language graph. The table gives estimated values for this quantity. The last digit in each estimate is subject to statistical errors.

Table and concept by **Peter Grassberger** (*Physics Department, University of Wuppertal*).

Table 12: Iterated rule expression sizes

rule	t=1	t=2	t=3	t=4	t=5
0	0 (0)	0 (0)	0 (0)	0 (0)	0 (0)
1	1 (1)	6 (4)	1 (1)	6 (4)	1 (1)
2	1 (1)	1 (1)	1 (1)	1 (1)	1 (1)
3	1 (1)	3 (2)	1 (1)	3 (2)	1 (1)
4	1 (1)	1 (1)	1 (1)	1 (1)	1 (1)
5	1 (1)	2 (2)	1 (1)	2 (2)	1 (1)
6	2 (2)	4 (4)	13 (10)	25 (21)	110 (50)
7	2 (2)	7 (4)	6 (6)	18 (8)	10 (10)
8	1 (1)	0 (0)	0 (0)	0 (0)	0 (0)
9	2 (2)	5 (5)	18 (11)	43 (31)	138 (53)
10	1 (1)	1 (1)	1 (1)	1 (1)	1 (1)
11	2 (2)	5 (4)	10 (6)	26 (12)	50 (16)
12	1 (1)	1 (1)	1 (1)	1 (1)	1 (1)
13	2 (2)	6 (4)	9 (5)	13 (7)	17 (8)
14	2 (2)	5 (5)	17 (10)	51 (24)	144 (48)
15	1 (1)	1 (1)	1 (1)	1 (1)	1 (1)
18	2 (2)	4 (4)	18 (18)	35 (26)	140 (108)
19	2 (2)	8 (5)	3 (3)	8 (5)	3 (3)
22	3 (3)	7 (7)	27 (26)	80 (62)	308 (206)
23	3 (3)	8 (5)	7 (7)	33 (9)	11 (11)
24	2 (2)	4 (4)	4 (4)	4 (4)	4 (4)
25	2 (2)	4 (4)	8 (8)	20 (16)	42 (27)
26	2 (2)	6 (6)	21 (17)	56 (43)	192 (100)
27	3 (2)	4 (4)	7 (5)	7 (6)	15 (7)
28	2 (2)	4 (4)	11 (7)	15 (11)	30 (12)
29	3 (2)	4 (4)	3 (2)	4 (4)	3 (2)
30	3 (3)	9 (7)	23 (17)	76 (41)	185 (105)
31	2 (2)	6 (4)	7 (6)	12 (10)	11 (9)
32	1 (1)	1 (1)	1 (1)	1 (1)	1 (1)
33	2 (2)	7 (7)	12 (12)	44 (23)	38 (24)
34	1 (1)	1 (1)	1 (1)	1 (1)	1 (1)
35	2 (2)	4 (3)	8 (4)	13 (8)	30 (11)
36	2 (2)	2 (2)	2 (2)	2 (2)	2 (2)
37	2 (2)	8 (7)	25 (17)	75 (47)	238 (109)
38	2 (2)	4 (3)	4 (4)	4 (3)	4 (4)
39	3 (2)	3 (3)	6 (5)	5 (5)	15 (7)
40	2 (2)	3 (3)	5 (5)	8 (8)	13 (13)
41	3 (3)	8 (8)	26 (24)	92 (69)	283 (218)
42	2 (2)	2 (2)	2 (2)	2 (2)	2 (2)
43	3 (3)	7 (6)	24 (12)	62 (27)	176 (55)
44	2 (2)	3 (3)	4 (4)	8 (5)	10 (6)
45	3 (3)	9 (8)	24 (20)	72 (53)	219 (118)
46	3 (2)	6 (4)	6 (4)	6 (4)	6 (4)
47	2 (2)	5 (4)	13 (6)	28 (8)	64 (16)
50	2 (2)	6 (4)	15 (6)	31 (8)	64 (10)
51	1 (1)	1 (1)	1 (1)	1 (1)	1 (1)
54	3 (3)	7 (6)	18 (15)	59 (38)	165 (85)
55	2 (2)	5 (3)	5 (5)	5 (3)	5 (5)
56	2 (2)	6 (4)	14 (9)	38 (20)	103 (45)
57	3 (3)	7 (6)	17 (12)	41 (23)	130 (50)
58	2 (2)	6 (5)	15 (10)	34 (18)	80 (32)
59	2 (2)	4 (4)	9 (8)	14 (10)	34 (17)
60	2 (2)	2 (2)	8 (8)	2 (2)	8 (8)
61	3 (3)	4 (4)	14 (10)	21 (17)	60 (30)
62	3 (3)	6 (6)	20 (12)	56 (27)	137 (48)
63	2 (2)	2 (2)	2 (2)	2 (2)	2 (2)
72	2 (2)	2 (2)	2 (2)	2 (2)	2 (2)
73	3 (3)	8 (7)	36 (20)	90 (46)	276 (118)
74	2 (2)	5 (5)	13 (11)	30 (22)	77 (45)
75	3 (3)	8 (8)	24 (20)	81 (52)	241 (118)

rule	t=1	t=2	t=3	t=4	t=5
76	2 (2)	2 (2)	2 (2)	2 (2)	2 (2)
77	3 (3)	7 (5)	14 (7)	32 (9)	57 (11)
78	2 (2)	5 (4)	8 (7)	20 (10)	21 (12)
79	2 (2)	7 (4)	9 (5)	13 (6)	20 (7)
90	2 (2)	2 (2)	8 (8)	2 (2)	8 (8)
91	3 (3)	9 (6)	26 (19)	82 (47)	255 (107)
94	3 (3)	8 (8)	26 (19)	106 (46)	276 (106)
95	2 (2)	2 (2)	2 (2)	2 (2)	2 (2)
104	3 (3)	6 (6)	15 (14)	27 (26)	49 (45)
105	4 (4)	4 (4)	16 (16)	4 (4)	256 (256)
106	3 (3)	5 (5)	25 (21)	46 (37)	192 (126)
107	4 (4)	10 (9)	37 (28)	108 (70)	390 (210)
108	3 (3)	5 (5)	9 (8)	5 (5)	9 (8)
109	4 (4)	10 (8)	31 (20)	91 (54)	268 (118)
110	3 (3)	7 (6)	15 (15)	40 (28)	95 (60)
111	3 (3)	7 (6)	21 (14)	57 (25)	139 (56)
122	3 (3)	9 (8)	27 (20)	88 (48)	264 (136)
123	3 (3)	8 (8)	22 (13)	51 (28)	81 (30)
126	3 (3)	8 (8)	22 (19)	103 (67)	221 (116)
127	3 (3)	3 (3)	3 (3)	3 (3)	3 (3)
128	1 (1)	1 (1)	1 (1)	1 (1)	1 (1)
129	2 (2)	9 (8)	26 (20)	93 (78)	250 (120)
130	2 (2)	2 (2)	5 (4)	5 (4)	8 (6)
131	2 (2)	6 (5)	13 (10)	36 (25)	88 (45)
132	2 (2)	3 (3)	4 (4)	5 (5)	6 (6)
133	2 (2)	8 (7)	23 (17)	74 (41)	216 (111)
134	3 (3)	6 (6)	20 (17)	46 (34)	174 (90)
135	3 (3)	9 (7)	22 (17)	66 (41)	202 (107)
136	1 (1)	1 (1)	1 (1)	1 (1)	1 (1)
137	2 (2)	8 (6)	14 (14)	39 (25)	111 (60)
138	2 (2)	2 (2)	2 (2)	2 (2)	2 (2)
139	2 (2)	5 (4)	6 (4)	6 (4)	6 (4)
140	2 (2)	2 (2)	2 (2)	2 (2)	2 (2)
141	2 (2)	7 (4)	10 (6)	22 (8)	28 (9)
142	3 (3)	7 (6)	18 (12)	52 (27)	151 (55)
143	2 (2)	5 (4)	12 (8)	32 (15)	86 (34)
146	3 (3)	6 (6)	29 (29)	61 (48)	224 (193)
147	3 (3)	8 (7)	21 (18)	69 (39)	207 (79)
150	4 (4)	4 (4)	16 (16)	4 (4)	256 (256)
151	4 (4)	10 (7)	33 (29)	104 (63)	372 (201)
152	2 (2)	4 (4)	7 (7)	12 (11)	20 (19)
153	2 (2)	2 (2)	8 (8)	2 (2)	8 (8)
154	3 (3)	4 (4)	28 (15)	6 (6)	42 (19)
155	3 (3)	7 (4)	8 (5)	8 (4)	8 (5)
156	3 (3)	3 (3)	14 (8)	12 (8)	43 (13)
157	3 (3)	7 (4)	11 (7)	24 (8)	23 (10)
158	4 (4)	12 (8)	34 (20)	106 (37)	330 (92)
159	3 (3)	9 (7)	18 (14)	63 (24)	139 (55)
160	1 (1)	1 (1)	1 (1)	1 (1)	1 (1)
161	2 (2)	6 (6)	20 (18)	65 (43)	236 (140)
162	2 (2)	3 (3)	4 (4)	5 (5)	6 (6)
163	2 (2)	5 (5)	15 (9)	32 (15)	73 (25)
164	2 (2)	5 (4)	10 (10)	16 (13)	27 (21)
165	2 (2)	2 (2)	8 (8)	2 (2)	8 (8)
166	3 (3)	4 (4)	24 (15)	6 (6)	32 (19)
167	3 (3)	8 (7)	26 (19)	82 (43)	218 (104)
168	2 (2)	5 (4)	10 (8)	23 (16)	49 (32)
169	3 (3)	6 (5)	28 (21)	76 (37)	244 (124)

Table 12: Iterated rule expression sizes

rule	$t=1$	$t=2$	$t=3$	$t=4$	$t=5$
170	1 (1)	1 (1)	1 (1)	1 (1)	1 (1)
171	3 (2)	3 (2)	3 (2)	3 (2)	3 (2)
172	2 (2)	4 (3)	6 (4)	7 (5)	9 (6)
173	3 (3)	10 (7)	32 (14)	70 (27)	206 (46)
174	2 (2)	2 (2)	2 (2)	2 (2)	2 (2)
175	2 (2)	2 (2)	2 (2)	2 (2)	2 (2)
178	3 (3)	7 (5)	16 (7)	32 (9)	65 (11)
179	2 (2)	4 (4)	8 (6)	15 (8)	31 (10)
182	4 (4)	11 (9)	47 (33)	103 (55)	466 (162)
183	3 (3)	7 (7)	24 (22)	55 (22)	203 (69)
184	2 (2)	5 (5)	15 (13)	48 (37)	161 (111)
185	3 (3)	5 (5)	24 (10)	51 (22)	149 (45)
186	3 (2)	8 (3)	20 (4)	43 (5)	88 (6)
187	2 (2)	2 (2)	2 (2)	2 (2)	2 (2)
188	3 (3)	6 (5)	24 (10)	39 (15)	125 (23)
189	3 (3)	10 (5)	8 (5)	8 (5)	8 (5)
190	4 (3)	5 (4)	23 (6)	21 (7)	91 (9)
191	3 (3)	3 (3)	3 (3)	3 (3)	3 (3)
200	2 (2)	2 (2)	2 (2)	2 (2)	2 (2)
201	3 (3)	10 (5)	23 (10)	10 (5)	23 (10)
202	2 (2)	4 (4)	7 (6)	11 (8)	16 (10)
203	3 (3)	7 (5)	17 (9)	41 (13)	66 (19)

rule	$t=1$	$t=2$	$t=3$	$t=4$	$t=5$
204	1 (1)	1 (1)	1 (1)	1 (1)	1 (1)
205	3 (2)	3 (2)	3 (2)	3 (2)	3 (2)
206	3 (2)	5 (3)	8 (4)	11 (5)	15 (6)
207	2 (2)	2 (2)	2 (2)	2 (2)	2 (2)
218	3 (3)	9 (6)	40 (16)	92 (24)	158 (38)
219	3 (3)	10 (5)	10 (5)	10 (5)	10 (5)
222	4 (3)	10 (5)	19 (7)	31 (9)	46 (11)
223	3 (3)	3 (3)	3 (3)	3 (3)	3 (3)
232	3 (3)	8 (5)	19 (7)	33 (9)	58 (11)
233	4 (4)	10 (7)	39 (20)	112 (34)	307 (63)
234	2 (2)	6 (4)	17 (7)	44 (12)	106 (21)
235	4 (3)	11 (6)	28 (10)	62 (14)	134 (19)
236	3 (2)	3 (2)	3 (2)	3 (2)	3 (2)
237	4 (3)	7 (5)	7 (5)	7 (5)	7 (5)
238	2 (2)	4 (3)	6 (4)	9 (5)	12 (6)
239	3 (3)	1 (1)	1 (1)	1 (1)	1 (1)
250	2 (2)	3 (3)	4 (4)	5 (5)	6 (6)
251	3 (3)	6 (5)	10 (7)	19 (9)	28 (11)
254	4 (3)	11 (5)	24 (7)	45 (9)	76 (11)
255	1 (1)	1 (1)	1 (1)	1 (1)	1 (1)

Table 12: Sizes of Boolean expressions representing functions corresponding to iterations of cellular automaton rules.

Cellular automaton rules with $k=2$ and $r=1$ can be expressed as Boolean functions of three variables, as in table 1. Iterations of these rules for t steps correspond to functions of $2t+1$ variables, which may be expressed as Boolean expressions.

The minimal Boolean expressions obtained after one step were given in table 1. This table gives the numbers of terms in Boolean expressions obtained after t time steps. An increase in these numbers potentially reflects increasing difficulty of computing the outcome of more steps of cellular automaton evolution.

The first number in each case gives the number of prime implicants in the corresponding Boolean expression. The possible values of a set of n Boolean variables correspond to the vertices of a Boolean n-cube. The cases in which a Boolean function has value 1 then correspond to a region of the Boolean n-cube. The number of prime implicants is essentially the number of hyperplanes of various dimensions which must be combined to form this region.

Boolean expressions can conveniently be stated in a disjunctive normal form (DNF), in which they are written as a disjunction (OR) of conjunctions (ANDs). The number of prime implicants gives an upper bound on the number of terms needed in such a form.

Notice that complementation of a function has no simple effect on its DNF expression. As a result, the table includes minimal representatives for all rules from table 1 not related by reflection.

The general problem of finding an absolutely minimal DNF representation for a function appears to be computationally intractable. The table gives in parentheses the numbers of terms in minimal DNF expressions found by the *espresso* computer program (R. Rudell, Computer Science Department, University of California, Berkeley, 1985) which incorporates known algebraic and heuristic techniques. In most cases, the results given are probably absolutely minimal.

Table 13: Finite lattice state transition diagrams

rule 0:
N=9:1x1
N=10:1x1
N=11:1x1

rule 1:
N=9:37x2
N=10:61x2
N=11:100x2

rule 2:
N=9:3x9,
1x3;1X1
N=10;4x10,
1x5,1X1
N=11:6x11,
1x1

rule 3:
N=9:8x18,
1x9;1X3,
1x2
N=10:13x20,
1x10,1x5,
1x2
N=11:21x22,
2x11,1x2

rule 4:
N=9:76x1
N=10:123x1
N=11:199x1

rule 5:
N=9:73x2,
12x1
N=10:136x2,
17x1
N=11:232x2,
22x1

rule 6:
N=9:3x18,
1x1
N=10:11x10,
4x5,3x1
N=11:5x22,
1x1

rule 7:
N=9:2x18,
1x9,1x2
N=10;3x20,
2x1;1x2,
1x11,1x2

rule 8:
N=9:1x1
N=10:1x1
N=11:1x1

rule 9:
N=9:1x1x8,
3x3,1x2
N=10:2x15,
2x10,2x5,
1x2
N=11:2x22,
1x2

rule 10:
N=9:8x9,
1x3;1X1
N=10:11x10,
2x5,1x1
N=11:18x11,
1x1

rule 11:
N=9:4x18,
1x3;1X2
N=10;12x10,
2x5,1x2
N=11:9x22,
1x11,1x2

rule 12:
N=9:76x1
N=10:123x1
N=11:199x1

rule 13:
N=9:1x2,
12x1
N=10:1x2,
17x1
N=11:1x2,
22x1

rule 14:
N=9:4x18,
4x1;1x6,
1x1
N=10;13x10,
3x5,3x1
N=11:9x22,
3x11,1x1

rule 15:
N=9:28x18,
1x6,1x2
N=10:99x10,
5x1,1x2,
2x1
N=11:93x22,
1x2

rule 18:
N=9:9x6,
9x2,1x1
N=10:5x6,
5x4,15x2,
1x1
N=11:2x11,
1x4;11x2,
1x1

rule 19:
N=9:37x2
N=10:61x2
N=11:100x2

rule 22:
N=9:9x4,
1x1
N=10:10x6,
15x4,3x1
N=11:2x11,
11x5,11x4,
1x1

rule 23:
N=9:37x2
N=10:61x2,
2x1
N=11:100x2

Table 13: Finite lattice state transition diagrams

rule 24: N=9:3x9, 1x3,1x1 N=10:4x10, 1x5,1x1 N=11:6x11, 1x1				rule 35: N=9:8x18, 4x3,1x3, 1x2 N=10:13x20, 2x2,2x5, N=11:21x22, 4x11,1x2			
rule 25: N=9:1x18, 1x3,1x3, 1x2 N=10:2x10, 1x2,1x15, N=11:1x33, 1x22,1x11,				rule 36: N=9:31x1 N=10:46x1 N=11:67x1			
rule 26: N=9:2x72, 1x18,1x6, 1x1 N=10:12x20, 1x6,1x5, N=11:4x88, 4x11,1x11,				rule 37: N=9:19x2, N=10:26x2 N=11:2x33, 14x2			
rule 27: N=9:9x18, 1x6,1x2 N=10:13x20, 1x6,1x5, N=11:21x22, 2x11,1x2				rule 38: N=9:6x18, 1x6,1x1 N=10:19x10, N=11:15x22, 1x1			
rule 28: N=9:21x2, 1x1 N=10:30x2, 3x1 N=11:55x2, 1x1				rule 40: N=9:1x9, 1x3,1x1 N=10:1x10, 1x1,1x2, N=11:2x11, 1x1			
rule 29: N=9:121x2 N=10:221x2, 2x1 N=11:408x2				rule 41: N=9:2x36, 1x2,4x3, N=10:2x40, 2x10,1x10, 1x20,1x5, N=11:1x44, 13x11,1x2			
rule 30: N=9:1x171, 1x72,1x1 N=10:2x15, 1x5,3x1 N=11:1x154, 11x7,1x1				rule 42: N=9:26x9, 2x3,1x1 N=10:42x10, 1x1,1x2, N=11:74x11, 1x1			
rule 32: N=9:1x1 N=10:1x2, 1x1 N=11:1x1				rule 43: N=9:4x18, 8x2,2x3, 1x2 N=10:22x10, 4x5,2x2 N=11:9x22, 18x11,1x2			
rule 33: N=9:37x2 N=10:62x2 N=11:100x2				rule 44: N=9:1x3, 31x1 N=10:46x1 N=11:67x1			
rule 34: N=9:6x9, 1x3,1x1 N=10:11x10, 4x5,1x2, 1x1 N=11:18x11, 1x1				rule 45: N=9:1x504, 1x3,1x2, 3x1 N=10:1x430, 1x10,1x20, N=11:1x979, 1x15,1x66, 1x1,11x5, 1x2			

Table 13: Finite lattice state transition diagrams

rule 46:
N=9:3x9,
1x3;1x1,
N=10:4x10,
1x5,1x1
N=11:6x11,
1x1

rule 50:
N=9:39x2,
1x1
N=10:61x2,
1x1
N=11:99x2,
1x1

rule 51:
N=9:256x2
N=10:512x2
N=11:1024x2

rule 54:
N=9:2x27,
9x4,1x1
N=10:2x30,
15x4,1x1
N=11:2x99,
9x11,11x4,
1x1

rule 56:
N=9:9x9,
2x3,1x1
N=10:12x10,
1x5;1x2,
1x1
N=11:20x11,
1x1

rule 57:
N=9:2x9,
2x3,1x2
N=10:2x10,
2x5,2x2
N=11:4x11,
1x2

rule 58:
N=9:1x18,
1x9;1x3,
1x1
N=10:1x20,
1x2,1x1
N=11:2x22,
3x11;1x1

rule 60:
N=9:4x63,
1x3;1x1
N=10:8x30,
1x15,1x1
N=11:3x341,
1x1

rule 62:
N=9:1x18,
22x3,1x1,
N=10:1x20,
1x5,35x3,
1x1
N=11:2x22,
1x1,77x3,
1x1

rule 72:
N=9:31x1
N=10:46x1
N=11:67x1

rule 73:
N=9:18x3,
19x2,3x1
N=10:10x8,
20x3,21x2,
10x1
N=11:11x12,
11x8,11x5,
11x3;34x2,

rule 74:
N=9:2x18,
7x9,1x3,
4x1
N=10:1x30,
1x5,11x10,
1x5,1x1
N=11:1x33,
4x22,6x11,
1x1

rule 76:
N=9:241x1
N=10:443x1
N=11:815x1

rule 77:
N=9:1x2,
78x1
N=10:1x2,
112x1
N=11:1x2,
198x1

rule 78:
N=9:13x1
N=10:18x1
N=11:23x1

rule 90:
N=9:36x7,
4x1
N=10:40x6,
5x3,1x1
N=11:33x31,
1x1

rule 94:
N=9:9x3,
9x2,13x1
N=10:5x6,
18x1;40x2,
18x1
N=11:11x2,
44x2,23x1

rule 104:
N=9:19x1
N=10:1x2,
26x1
N=11:34x1

rule 105:
N=9:9x14,
1x2
N=10:170x6,
2x2
N=11:33x62,
1x2

rule 106:
N=9:3x54,
8x9,4x3,
1x1
N=10:2x205,
1x15,1x10,
1x2,
N=11:1x176,
18x11,1x1

Table 13: Finite lattice state transition diagrams

rule 108:
N=9;54x2,
76x1

N=10;100x2,
121x1

N=11;187x2,
199x1

rule 110:
N=9;9x7,
3x3,1x1

N=10;2x25,
2x15,10x5,
1x1

N=11;1x110,
11x7;1x1

rule 122:
N=9;1x6,
9x2,1x1

N=10;5x6,
9x4,16x2,
1x1

N=11;2x11,
11x4,11x2,
1x1

rule 126:
N=9;1x6,
9x2,1x1

N=10;5x6,
9x4,15x2,
1x1

N=11;2x11,
11x4,11x2,
1x1

rule 128:
N=9:2x1

N=10:2x1

N=11:2x1

rule 130:
N=9;3x9,
1x3,2x1

N=10;4x10,
1x5,2x1

N=11:6x11,
2x1

rule 132:
N=9:77x1

N=10:124x1

N=11:200x1

rule 134:
N=9;3x18,
2x1

N=10;11x10,
4x5,4x1

N=11:5x22,
2x1

rule 136:
N=9:2x1

N=10:2x1

N=11:2x1

rule 138:
N=9;17x9,
1x3,2x1

N=10;26x10,
3x5,2x1

N=11:44x11,
2x1

rule 140:
N=9:77x1

N=10:124x1

N=11:200x1

rule 142:
N=9;4x18,
8x9,1x6,
2x1

N=10;22x10,
4x5,4x1

N=11;9x22,
18x11,2x1

rule 146:
N=9;9x6,
9x2,2x1

N=10;5x6,
9x4,15x2,
2x1

N=11;2x11,
11x4,11x2,
1x1

rule 150:
N=9:18x7,

N=10;160x6,
20x3,4x1

N=11:66x31,
2x1

rule 152:
N=9;3x9,
1x3;2x1

N=10;4x10,
1x5,2x1

N=11:6x11,
2x1

rule 154:
N=9;2x72,
1x36,6x18,
1x6,2x1

N=10;4x40,
10x20,19x10,
2x5,2x5,

N=11;6x88,
27x44,15x22

rule 156:
N=9;21x2,
2x1

N=10;30x2,
4x1

N=11;55x2,
2x1

rule 160:
N=9:2x1

N=10;1x2,
2x1

N=11:2x1

rule 162:
N=9;8x9,
1x3;2x1

N=10;11x10,
2x5,1x2,
4x1

N=11;18x11,
2x1

rule 164:
N=9:32x1

N=10;5x6,
47x1

N=11:68x1

Table 13: Finite lattice state transition diagrams

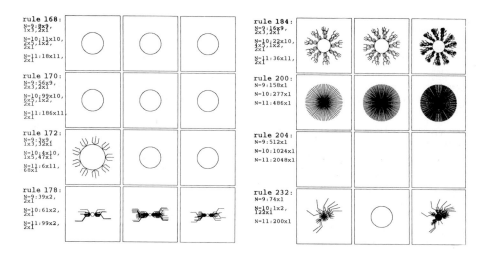

Table 13: State transition diagrams for cellular automata on finite size lattices.

A $k=2$ cellular automaton on a finite lattice with N sites has a total of 2^N possible states. The complete evolution of such a cellular automaton can be represented by a finite diagram which shows the possible transitions between these states. Each node in the diagram corresponds to a complete configuration or state of the finite cellular automaton. A directed arc leads from each such node to its successor under one time step of cellular automaton evolution. The possible time sequences of configurations in the complete evolution of the cellular automaton then correspond to possible paths through the directed graph thus formed.

After a time of at most 2^N steps, a finite cellular automaton must always enter a cycle, periodically visiting a fixed set of states. In general, the complete state transition diagram contains a number of distinct cycles.

The table shows the fragment of the state transition diagram associated with the longest cycle, for all inequivalent $k=2$, $r=1$ rules. Results are given for lattices of sizes $N=9$, $N=10$ and $N=11$. In all cases, the lattices are taken to have periodic boundary conditions, as if their sites were arrranged in a circle.

The table also gives the lengths and multiplicities of all the cycles for each rule. (The notation used is $g{\times}L$, representing g cycles of length L.) Notice that the state transition diagram fragments associated with different cycles of the same length may not be identical. When there are several cycles of maximal length, the fragment shown is the one involving the largest total number of states.

State transition diagram fragments have the general form of cycles fed by trees. The cellular automaton always reaches the cycle after a sufficiently long time. The trees represent transients, and contain states which can occur only after a limited number of time steps. Such transient phenomena are a manifestation of irreversibility in the cellular automaton evolution.

Table 13: Finite lattice state transition diagrams

Some finite cellular automata, such as rule 13, are reversible, so that their state transition diagrams contain no transients, and all states are on cycles.

In some other cases, such as rule 90, highly regular state transition diagrams are obtained, containing for example only balanced trees [2.1]. Many rules, however, yield complicated state transition diagrams.

In the pictures given, individual nodes are not indicated. Nevertheless, the arcs joining nodes are all of equal length in a particular diagram. The overall scale of each diagram can be deduced from the total cycle length given.

The constraint of equal length in some cases forces arcs to intersect in the diagram. In some cases, there are dense areas containing large numbers of arcs. For highly irreversible rules, such as rule 0, large numbers of arcs coverge on a single node, and appear essenitally as a filled black circle.

Notice that the results given here and in table 14 for cellular automata on finite lattices with periodic boundary conditions also apply to infinite cellular automata in which only spatially periodic configurations are considered.

Table by **Holly Peck** (*Los Alamos National Laboratory*).

Table 14: Global properties for finite lattices

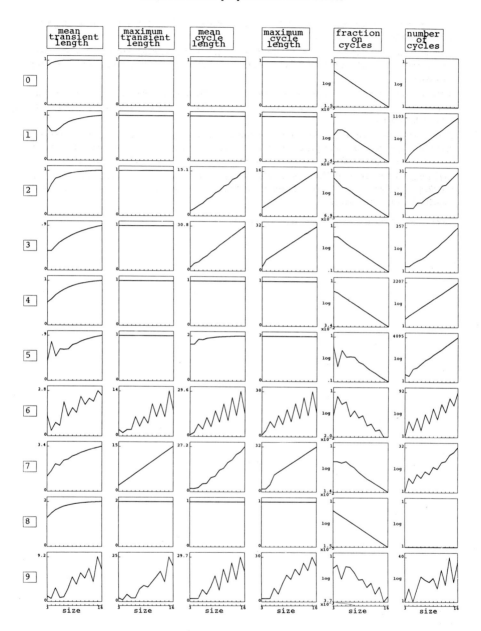

Table 14: Global properties for finite lattices

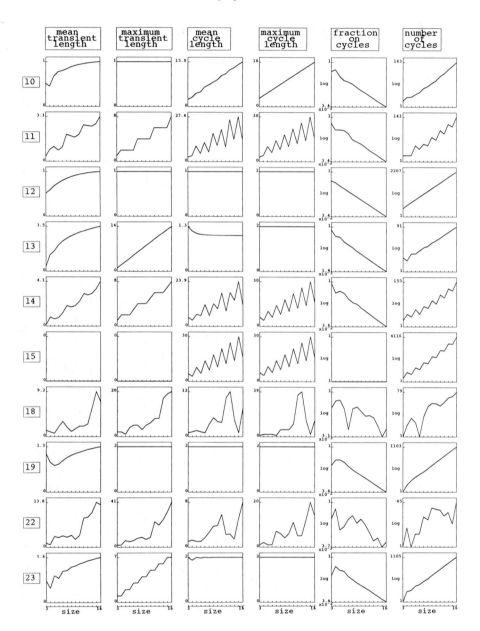

Table 14: Global properties for finite lattices

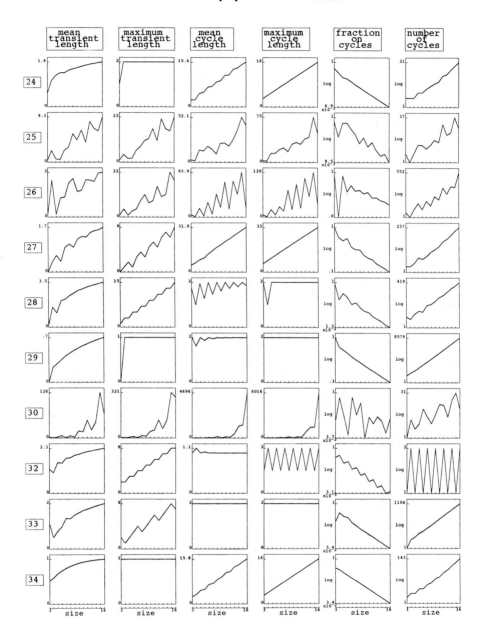

Table 14: Global properties for finite lattices

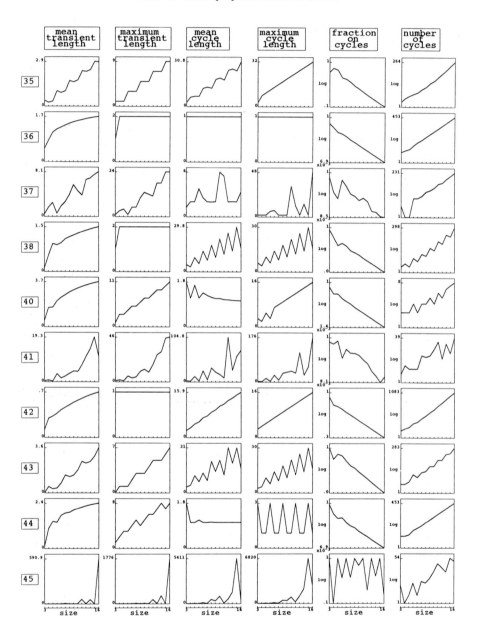

Table 14: Global properties for finite lattices

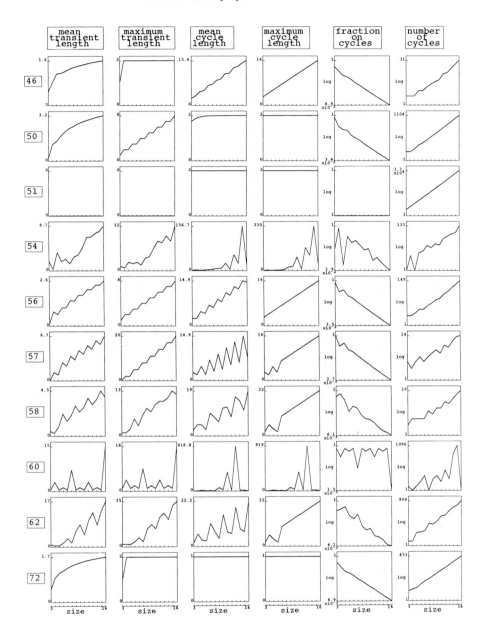

Table 14: Global properties for finite lattices

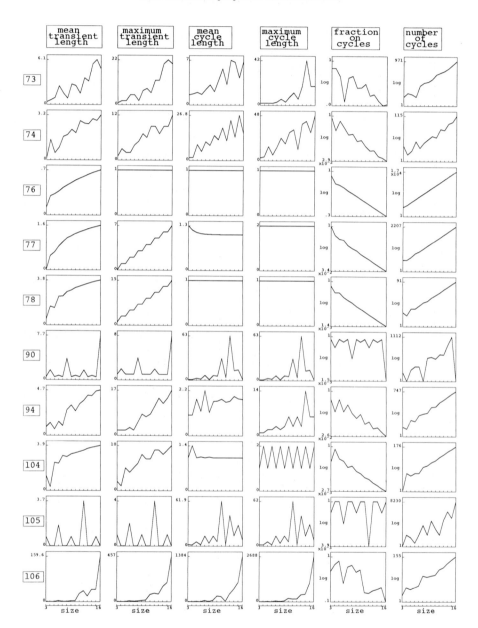

Table 14: Global properties for finite lattices

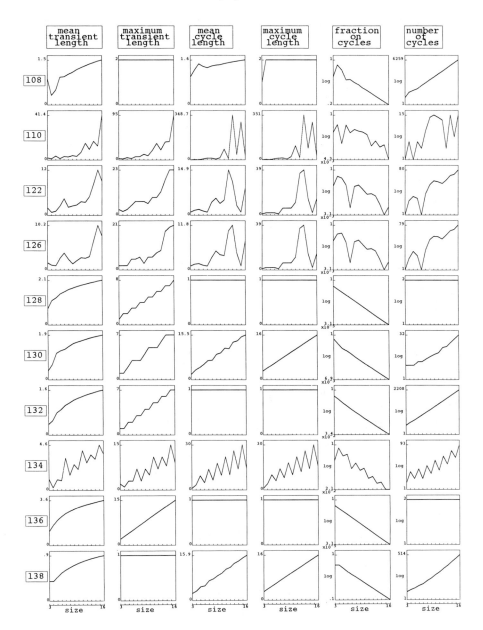

Table 14: Global properties for finite lattices

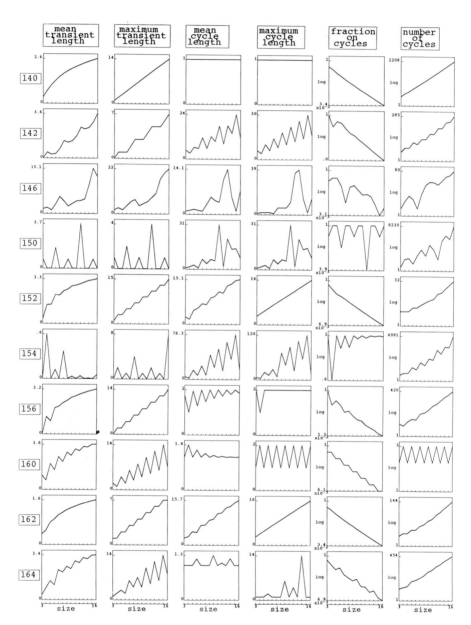

Table 14: Global properties for finite lattices

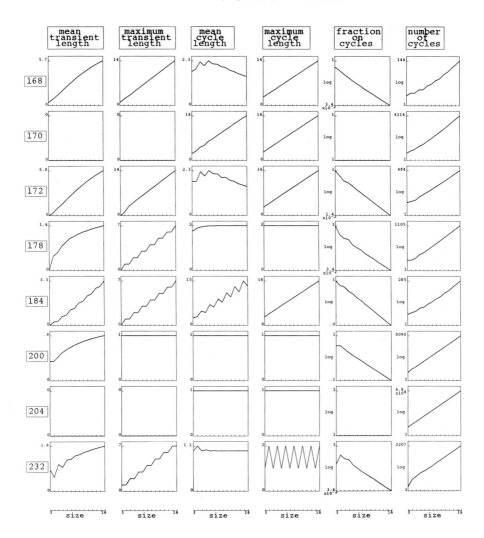

Table 14: Global properties for finite lattices

Table 14: Global properties of cellular automata on finite lattices.

This table gives some properties of state transition diagrams for finite cellular automata. Each picture shows values plotted as a function of lattice size N, varying from 3 to 16. (Periodic boundary conditions are assumed for the cellular automaton evolution.)

The quantities shown are as follows. (In all cases, values for integer sizes N are shown joined by lines.)

"Mean transient length" represents the average number of steps necessary for any particular state to evolve to a cycle. "Maximum transient length" gives the maximum number of steps needed.

"Mean cycle length" gives the average length of the cycle on to which any particular state evolves. (Each cycle is thus weighted in the average with the number of states which evolve to it.) "Maximum cycle length" gives the longest cycle for each value of N. Some such cycles are shown in table 13.

"Fraction on cycles" (given in logarithmic form) represents the fraction of all 2^N possible states which appear on cycles, and thus can occur after a long time. This quantity is related to the set (topological) entropy for invariant set of the evolution.

The last picture for each rule gives the total number of distinct cycles (in logarithmic form). This can be considered as the number of possible distinct attractors for the evolution.

Table by **Holly Peck** (*Los Alamos National Laboratory*). (Original program by S. Wolfram.)

Table 15: Structures in rule 110

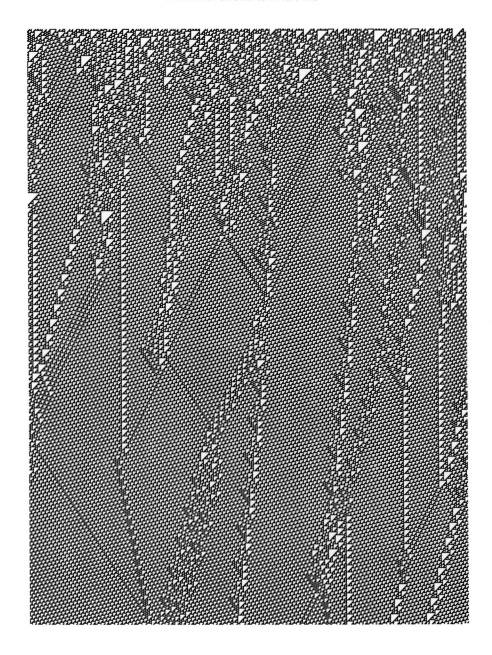

Table 15: Structures in rule 110

Table 15: Structures in rule 110

Table 15: Structures in rule 110

The previous two pages show patterns produced by evolution according rule 110, starting from a disordered initial configuration. The first picture shows all sites on a size 400 lattice. The second picture shows every other site in space and time on a size 800 lattice.

The configurations produced after many steps can be represented in terms of particle-like structures superimposed on a periodic background. The background is found to have spatial period 14 and temporal period 7, and corresponds to repetitions of the block B = 10011011111000. The configurations are then of the form $\cdots$ **BBBBPBBB** $\cdots$, where the particles **P** that have been found so far are:

velocity	P
-6/12	100011001110111111111000
-2/4	11111000
-14/42	11100001110111111111111000
-8/30	100110011000111111000
-4/15	00000
-4/36	111011111111000
-8/20	1111000011000
0/7	11111111000
0/7	100011000
0/7	10011011111111000
2/10	11101011000
2/10	1110100011011111000
2/3	111000

The "velocity" is written as (spatial period)/(temporal period).

One may speculate that the behaviour of rule 110 is sophisticated enough to support universal computation.

Table of particles by **Doug Lind** (*Mathematics Department, University of Washington, Seattle*).

Table 16: Patterns generated by second-order rules

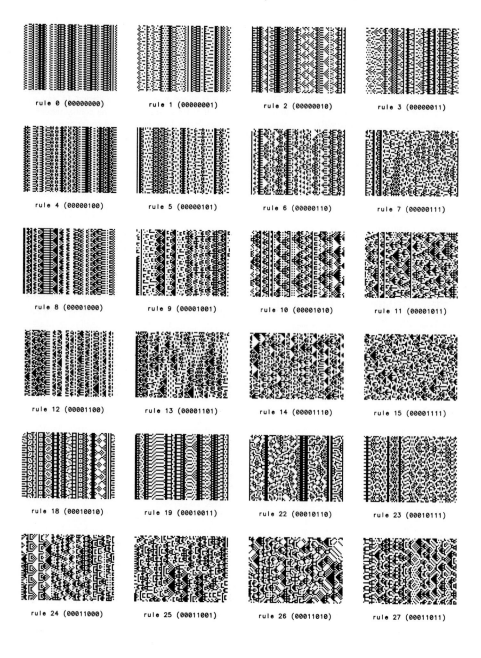

rule 0 (00000000)   rule 1 (00000001)   rule 2 (00000010)   rule 3 (00000011)

rule 4 (00000100)   rule 5 (00000101)   rule 6 (00000110)   rule 7 (00000111)

rule 8 (00001000)   rule 9 (00001001)   rule 10 (00001010)   rule 11 (00001011)

rule 12 (00001100)   rule 13 (00001101)   rule 14 (00001110)   rule 15 (00001111)

rule 18 (00010010)   rule 19 (00010011)   rule 22 (00010110)   rule 23 (00010111)

rule 24 (00011000)   rule 25 (00011001)   rule 26 (00011010)   rule 27 (00011011)

Table 16: Patterns generated by second-order rules

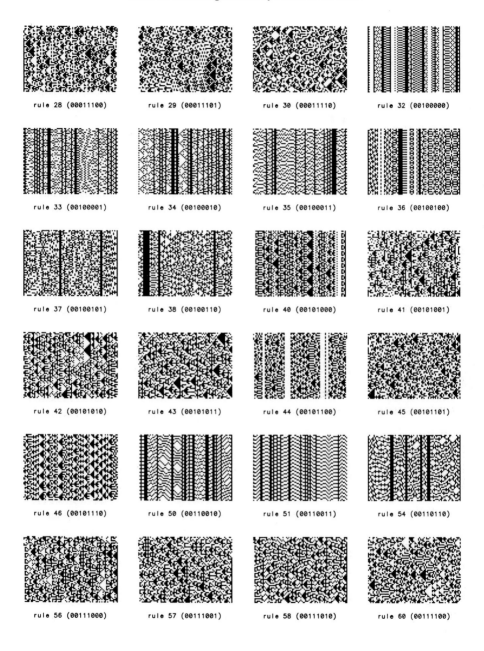

rule 28 (00011100)	rule 29 (00011101)	rule 30 (00011110)	rule 32 (00100000)
rule 33 (00100001)	rule 34 (00100010)	rule 35 (00100011)	rule 36 (00100100)
rule 37 (00100101)	rule 38 (00100110)	rule 40 (00101000)	rule 41 (00101001)
rule 42 (00101010)	rule 43 (00101011)	rule 44 (00101100)	rule 45 (00101101)
rule 46 (00101110)	rule 50 (00110010)	rule 51 (00110011)	rule 54 (00110110)
rule 56 (00111000)	rule 57 (00111001)	rule 58 (00111010)	rule 60 (00111100)

Table 16: Patterns generated by second-order rules

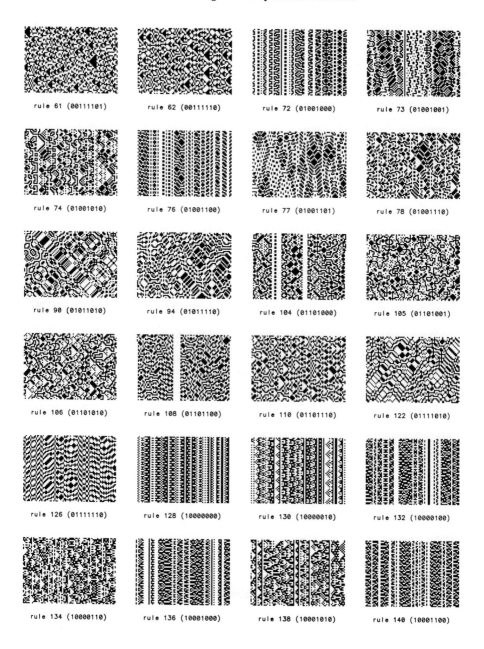

rule 61 (00111101) rule 62 (00111110) rule 72 (01001000) rule 73 (01001001)

rule 74 (01001010) rule 76 (01001100) rule 77 (01001101) rule 78 (01001110)

rule 90 (01011010) rule 94 (01011110) rule 104 (01101000) rule 105 (01101001)

rule 106 (01101010) rule 108 (01101100) rule 110 (01101110) rule 122 (01111010)

rule 126 (01111110) rule 128 (10000000) rule 130 (10000010) rule 132 (10000100)

rule 134 (10000110) rule 136 (10001000) rule 138 (10001010) rule 140 (10001100)

Table 16: Patterns generated by second-order rules

rule 142 (10001110)   rule 146 (10010010)   rule 150 (10010110)   rule 152 (10011000)

rule 154 (10011010)   rule 156 (10011100)   rule 160 (10100000)   rule 162 (10100010)

rule 164 (10100100)   rule 168 (10101000)   rule 170 (10101010)   rule 172 (10101100)

rule 178 (10110010)   rule 184 (10111000)   rule 188 (10111100)   rule 200 (11001000)

rule 204 (11001100)   rule 232 (11101000)

Table 16: Patterns generated by second-order rules

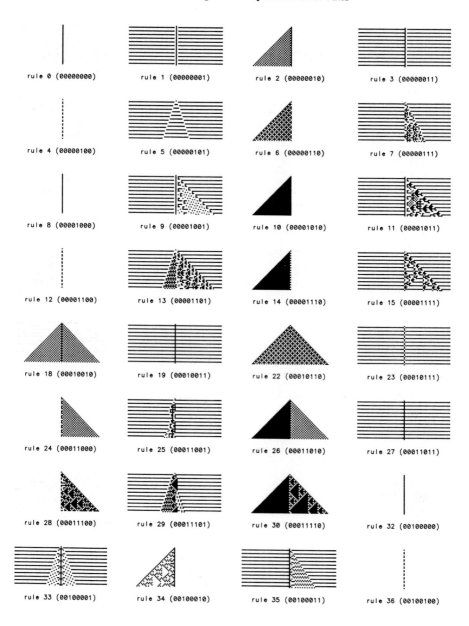

Table 16: Patterns generated by second-order rules

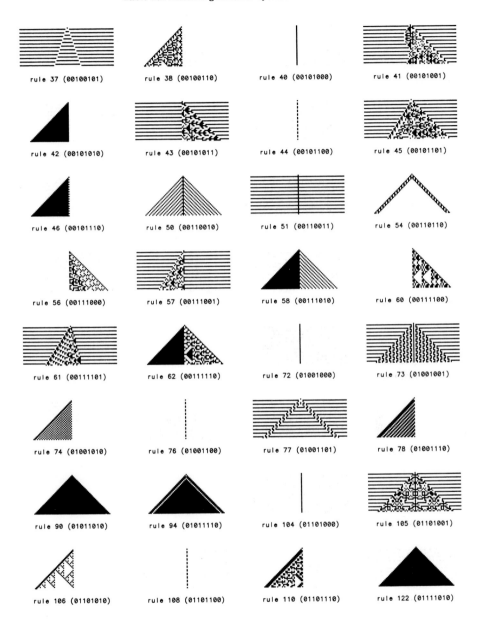

rule 37 (00100101)   rule 38 (00100110)   rule 40 (00101000)   rule 41 (00101001)

rule 42 (00101010)   rule 43 (00101011)   rule 44 (00101100)   rule 45 (00101101)

rule 46 (00101110)   rule 50 (00110010)   rule 51 (00110011)   rule 54 (00110110)

rule 56 (00111000)   rule 57 (00111001)   rule 58 (00111010)   rule 60 (00111100)

rule 61 (00111101)   rule 62 (00111110)   rule 72 (01001000)   rule 73 (01001001)

rule 74 (01001010)   rule 76 (01001100)   rule 77 (01001101)   rule 78 (01001110)

rule 90 (01011010)   rule 94 (01011110)   rule 104 (01101000)   rule 105 (01101001)

rule 106 (01101010)   rule 108 (01101100)   rule 110 (01101110)   rule 122 (01111010)

Table 16: Patterns generated by second-order rules

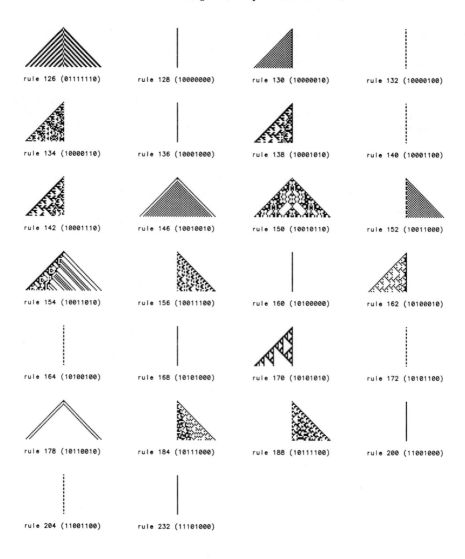

Table 16: Patterns generated by second-order rules

Table 16: Patterns generated by second-order reversible rules.

This table shows patterns produced by second-order generalizations of the $k=2$, $r=1$ cellular automata considered above. The rules are of the form [2.2]

$$a_i^{(t+1)} = \phi(a_{i-1}^{(t)}, a_i^{(t)}, a_{i+1}^{(t)}) + a_i^{(t-1)} \mod 2 \ ,$$

where ϕ is a standard $k=2$, $r=1$ function, as listed in table 1. Such rules determine the configuration at time $t+1$ in terms of the configurations both at time t and time $t-1$. The rules have the special feature that they are reversible: given configurations at times t and $t+1$, the configuration at time $t-1$ can be deduced uniquely according to the rule

$$a_i^{(t-1)} = \phi(a_{i-1}^{(t)}, a_i^{(t)}, a_{i+1}^{(t)}) + a_i^{(t+1)} \mod 2 \ .$$

The first set of patterns were generated with disordered initial configurations at times 0 and -1. In the second set of patterns, the configurations at time 0 and -1 were both taken to be 1.

The forms of behaviour produced by these reversible rules are qualitatively similar to those from standard $k=2$, $r=1$ cellular automata, shown in tables 2 and 5. Evolution to a homogeneous, fixed, pattern is however impossible for reversible systems.

Index

(Bibliography entries not included)

Activator-inhibitor model, 320
Adaptation, 408
Aggregation, 305
Attractors, 91, 208, 367, 401

Billiard ball model, 237
Binomial coefficients, 19
Biology applications, 303
Blocking transformations, 36, 178, 516
Boolean expressions, 487, 529

Cantor set, 29, 94
Cellular automata:
    additive, 51
    classes, 94, 152, 172
    definitions, 1
    finite, 32, 51, 367, 531
    multiple-scale, 404
    one-dimensional, 7, 51, 91, 303, 485
    probabilistic, 26, 35, 417
    reversible, 31, 232, 288, 343, 451, 550
    totalistic, 43, 92
    two-dimensional, 37, 75, 126, 303
Chaos, 113, 160, 328
Complexity engineering, 400
Compositions, 93, 520
Computation theory, 35, 181, 187
Computational irreducibility, 184, 294
Conservation laws, 175, 234
Context-free language, 224
Correlation functions, 21, 355
Critical phenomena, 422, 436
Cryptography, 247
Crystal growth, 131, 305
Cycle lengths, 33, 64, 273, 537

Dendritic growth, 131, 305
Density, 18
Difference patterns, 502
Diffusion, 309
Dimension, 106, 207, 513
Dissipative systems, 7
Domains, 20, 153, 330
Dynamical systems theory, 91

Elementary rules, 9
Entropy, 29, 98, 162, 207, 233, 258, 330, 513
Ergodicity, 455
Excitable media, 311

Factorization, 520
Filter automata, 334

Finite automata, 113, 198
Finite fields, 84
Formal language theory, 181, 192
Fractals, 13, 133

Galois fields, 84
Geometry, 174
Growth dimension, 138
Growth inhibition, 94

Hamiltonian, 349, 452
Hamming distance, 27
Hausdorff dimension, 106
Heat equation, 353, 346
Hydrodynamics, 358, 362

Interfaces, 308, 456
Intractability, 185, 294
Invariants, 175, 343
Inverse problems, 403
Ising models, 344, 347, 447
Iterated mappings, 3

Kinks, 20
Kinks, 435

L systems, 3
Lattice gases, 358
Life game, 43
Limit set, 208
Logic circuits, 407
Logical expressions, 267, 487, 529
Lyapunov exponent, 110, 167, 249, 513

Markov random fields, 3
Master equation, 21
Mean field theory, 21, 453
Modulo two rule, 51
Monte Carlo method, 348

NP completeness, 164, 272, 295
Navier-Stokes equations, 358, 362
Neural networks, 367, 406
Noise, 180, 369, 455
Number theory, 86

Partial differential equations, 3, 179, 320, 328, 358, 362
Particles, 333
Pattern formation, 305, 311, 313, 320, 441
Pattern recognition, 401
Percolation, 3, 447
Period-doubling, 309

Periodic configurations, 218
Phase transitions, 419, 440
Physics applications, 303
Polytopes, 144
Pseudorandom sequences, 247

Random mappings, 34
Randomness, 182, 247, 298
Reaction-diffusion systems, 311, 313, 320, 328
Regular language complexity, 206, 523
Regular languages, 180, 193
Reversibility, 232
Rule 18, 78
Rule 30, 255
Rule 90, 18
Rule 90, 51
Rule 110, 547

Scaling transformations, 177, 516
Self reproduction, 123
Self organization, 7, 458
Shift registers, 3, 53, 252
Solidification, 138, 305
Solitons, 333, 391
Spin systems, 3, 296, 344, 347, 451
Spiral waves, 311, 313
State transition diagrams, 52, 274, 531
Statistical properties, 513
Statistical randomness, 280
Surface tension, 129
Systolic arrays, 3

Temperature, 348
Thermodynamics, 176, 347, 362
Tiling, 165
Time series, 109
Totalistic rules, 43, 92
Transcendental numbers, 249
Turbulence, 328, 360, 363
Turing machines, 3, 35

Undecidability, 164, 182, 294
Universal computation, 35, 120, 182, 232, 295